ADVANCES IN PROTEIN CHEMISTRY

Volume 64

Virus Structure

ADVANCES IN PROTEIN CHEMISTRY

EDITED BY

FREDERIC M. RICHARDS
Department of Molecular Biophysics
and Biochemistry
Yale University
New Haven, Connecticut

DAVID S. EISENBERG
Department of Chemistry
and Biochemistry
University of California, Los Angeles
Los Angeles, California

JOHN KURIYAN
Department of Molecular Biophysics
Howard Hughes Medical Institute
Rockefeller University
New York, New York

VOLUME 64

Virus Structure

EDITED BY

WAH CHIU
National Center for Macromolecular Imaging
Baylor College of Medicine
Houston, Texas

JOHN E. JOHNSON
The Scripps Research Institute
La Jolla, California

ACADEMIC PRESS

An imprint of Elsevier Science

Amsterdam Boston Heidelberg London New York Oxford
Paris San Diego San Francisco Singapore Sydney Tokyo

This book is printed on acid-free paper. ♾

Academic Press
An Elsevier Science Imprint.
525 B Street, Suite 1900, San Diego, California 92101-4495, USA
http://www.academicpress.com

Academic Press
84 Theobald's Road, London WC1X 8RR, UK
http://www.academicpress.com

International Standard Book Number: 0-12-034264-2

PRINTED IN THE UNITED STATES OF AMERICA
03 04 05 06 07 08 9 8 7 6 5 4 3 2 1

CONTENTS

PREFACE xi

Viral Assembly Using Heterologous Expression Systems and Cell Extracts

ANETTE SCHNEEMANN AND MARK J. YOUNG

I. Introduction 1
II. The Driving Force behind the Development of Heterologous Expression Systems for the Study of Viral Assembly and Structure 2
III. Diversity of Heterologous Expression Systems 4
IV. Guidelines for Choosing a Heterologous Expression System 16
V. Representative Examples of Viral Assembly in Heterologous Expression Systems 20
VI. Conclusions 32
References 32

Hybrid Vigor: Hybrid Methods in Viral Structure Determination

ROBERT J. C. GILBERT, JONATHAN M. GRIMES, AND DAVID I. STUART

I. Introduction 37
II. Techniques 38
III. Hybrids 56
IV. Conclusion 83
References 83

Determination of Icosahedral Virus Structures by Electron Cryomicroscopy at Subnanometer Resolution

Z. HONG ZHOU AND WAH CHIU

I. Introduction 93
II. Electron Cryomicroscopy 94
III. Overview of Methods for Subnanometer-resolution Reconstructions 101
IV. Example of Data Collection, Evaluation, and Processing 105
V. Visualization and Structure Interpretation 117
VI. Conclusion 122
References 122

Structural Folds of Viral Proteins

MICHAEL S. CHAPMAN AND LARS LILJAS

I. Introduction 125
II. Terminology 126
III. Virus Families 126
IV. Determination of Structural Fold 126
V. Prototypical Viral Folds 128
VI. Survey Through the Virus Families 132
VII. Common Themes 184
VIII. Phylogenetic Relationships 185
References 187

Virus Particle Dynamics

JOHN E. JOHNSON

I. Introduction 197
II. Particle Fluctuations and Infectivity 199
III. Large-Scale Reversible Quaternary Structure Changes in Viruses 203
IV. Large-Scale Irreversible Quaternary Structure Changes in Double-Stranded DNA Bacteriophage 209
V. Conclusions 216
References 216

Viral Genome Organization

B. V. VENKATARAM PRASAD AND PETER E. PREVELIGE, JR.

I. Introduction 219
II. Single-Stranded RNA Viruses 221
III. Double-Stranded RNA Viruses 229
IV. Single-Stranded DNA Viruses 235
V. Double-Stranded DNA Viruses 240
VI. Conclusions 246
References 248

Mechanism of Scaffolding-Assisted Viral Assembly

BENTLEY A. FANE AND PETER E. PREVELIGE, JR.

I. Introduction 259
II. ϕX174 Morphogenesis 261
III. Prescaffolding Stages: Coat Proteins and Chaperones 261
IV. The ϕX174 Internal Scaffolding Protein 263
V. Genetic Data for Scaffolding Protein Flexibility: ϕX174 and *Herpesviridae* 264
VI. Structural Data for Scaffolding Protein Flexibility: ϕX174, P22, and *Herpesviridae* 266
VII. So What's All This Fuss over These C Termini? 267
VIII. Internal Scaffolding Protein Function in One and Two Scaffolding Protein Systems: ϕX174 versus P22 and Herpesviruses 269
IX. The Assembly Pathway of Bacteriophage P22 270
X. The Role of the P22 Scaffolding Protein 272
XI. Functional Domains of the P22 Scaffolding Protein...... 274
XII. Physical Chemistry of the P22 Scaffolding Protein 280
XIII. The Mechanism of Scaffolding-Assisted Assembly........ 281
XIV. External Scaffolding Proteins 283
XV. The ϕX174 External Scaffolding Protein 284
XVI. P4 Sid Protein 290
XVII. Herpesvirus Triplex Proteins 292
XVIII. Scaffolding-Like Functions 293
References 295

Molecular Mechanisms in Bacteriophage T7 Procapsid Assembly, Maturation, and DNA Containment

Mario E. Cerritelli, James F. Conway, Naiqian Cheng, Benes L. Trus, and Alasdair C. Steven

I. Introduction 301
II. Overexpressed T7 and T3 Connectors have 12- and 13-Fold Symmetry 303
III. The Procapsid Core has 8-Fold Symmetry: Another Symmetry Mismatch 303
IV. Procapsid Structure 305
V. Procapsid Maturation: Expansion is Initiated in the Connector 308
VI. Packaging and Parting of DNA 309
VII. The Mature Capsid Structure: Filled and Empty Shells 310
VIII. Structure of Packaged DNA 315
IX. Summary 319
References 320

Conformational Changes in Enveloped Virus Surface Proteins During Cell Entry

Deborah Fass

I. Introduction: Multiple Stops on the Protein-Folding Landscape 325
II. Influenza Hemagglutinin 326
III. Retroviruses 338
IV. Paramyxoviruses Turn Paradigms Upside Down? 350
V. Oligomerization State Switches in Flaviviruses and Alphaviruses 353
VI. Concluding Remarks 356
References 357

Enveloped Viruses

Richard J. Kuhn and James H. Strauss

I. Introduction 363
II. General Structural Features of Enveloped Viruses 364

III. Alphavirus Structure 365
IV. Flavivirus Structure 367
V. Virus Assembly 369
VI. Virus–Cell Fusion 372
VII. Concluding Remarks 373
References 374

Studying Large Viruses

Frazer J. Rixon and Wah Chiu

I. What is a Large Virus? 379
II. Why Large Viruses? 381
III. Why Study Large Viruses? 384
IV. Methods of Structural Analysis 385
V. Complexity of Organization 386
VI. Structural Folds 393
VII. Assembly Mechanisms 394
VIII. Maturation 399
IX. Accessory Proteins 400
X. Packaging 401
XI. Future Prospects 402
XII. Summary 403
References 404

Structural Studies on Antibody–Virus Complexes

Thomas J. Smith

I. Introduction 409
II. Background 410
III. Structural Studies on Virus–Antibody Complexes 412
IV. Conclusions 439
References 443

Structural Basis of Nonenveloped Virus Cell Entry

PHOEBE L. STEWART, TERENCE S. DERMODY, AND GLEN R. NEMEROW

I. Introduction 455
II. Reovirus Cell Entry, Tissue Tropism, and Pathogenesis 456
III. Reovirus Structure 458
IV. Proteolysis of the σ1 Protein Regulates Viral Growth in the Intestine and Systemic Spread 460
V. The σ1 Tail Binds Cell Surface Sialic Acid 462
VI. The σ1 Head Binds Junctional Adhesion Molecule 463
VII. Reovirus–Receptor Interactions Promote Cell Death by Apoptosis 464
VIII. Picornavirus–Receptor Complexes 465
IX. Poliovirus Cell Entry Mechanisms 466
X. Identification of the Poliovirus Attachment Receptor 468
XI. Poliovirus-Associated Lipid Molecules 469
XII. Receptors for Rhinoviruses 470
XIII. Receptors for Other Picornaviruses 473
XIV. Human Adenoviruses 475
XV. Adenovirus Attachment Receptors 476
XVI. Cell Integrins Promote Adenovirus Internalization 478
XVII. Signaling Events Associated with Adenovirus Internalization 481
XVIII. α_v Integrins Regulate Adenovirus-Mediated Endosome Disruption 482
XIX. Conclusions 482
References 484

AUTHOR INDEX 493
SUBJECT INDEX 531

PREFACE

The study of virus structure and function has led to a fundamental understanding of biology and has been a major force in technological development in biophysics and molecular biology. More recently, viruses have been recognized as a resource for biomaterials and a paradigm for nanotechnology. In this volume we have gathered expert commentaries that cover the full spectrum of modern structural virology. Its goal is to describe the means for defining moderate to high-resolution structures and the biological principles that have emerged from these studies. The articles also define the future of structural virology, as it is clear that the final answers to fundamental questions addressed in most of these presentations have still not been found.

The book is organized so that the first three articles describe the techniques that are leading to discoveries in structural virology. The fourth article summarizes the results of virus x-ray crystallography with detailed descriptions of protein subunit folds and the subsequent articles address fundamental issues of virology with state-of-the-art descriptions of our understanding in each of the areas covered. As editors we were impressed with the diversity of methods used in each of the areas presented in these subsequent articles. Results of different methods, such as electron cryomicroscopy and crystallography, were often explicitly merged in the same investigation to generate pseudoatomic models of exceptionally complex structures. These models in turn led to molecular genetic and mutagenesis studies to validate and extend the biophysical studies. Often data from mass spectrometry or other solution methods would illustrate dynamic aspects of the systems that were not obvious from methods that immobilized specimens in either a crystal lattice or ice before recording data. Data from solution methods often allowed informed interpretations of static data that could not have been achieved by either method in isolation. We were also impressed with advances in both hardware and software associated with electron cryomicroscopy and crystallography that have resulted largely from the needs of structural virology. Specialists in virus structure are often primary users of synchrotron beam lines designed for large unit cell data collection and electron microscopes with field emission guns and cryostages. Experts in software development have been drawn to problems associated with image processing of electron micrographs and, more recently, computational chemists are developing sophisticated models to explain virus particle dynamics. It is clear that much of the innovation, both conceptual and technical, in structural biology is the result of ongoing efforts to understand how viruses "work."

What issues will be addressed in a book on this subject published five years from now? We anticipate that virus particle dynamics will be better understood from both time-resolved experimental studies and theoretical investigations. The features of metastable particles generated in double-stranded DNA phage, herpesvirus, and other complex viruses during their assembly and maturation should be defined, in detail, for at least a few systems. If the "triggers" that drive these particles from their local energy minima can be understood, drugs may be rationally designed to inhibit the transition. It is likely that more individual gene products associated with complex virus capsids, either permanently or transiently, will be studied by X-ray crystallography, probably broadening the spectrum of protein folds associated with viruses. A number of groups are making progress on nonicosahedral components of virus particles with the hope of visualizing portions of the viral genome, or proteins without icosahedral symmetry, in greater detail. It is also clear that we will have "to get our hands dirty" and investigate the virus factories where assembly takes place within the cell. The nave view of viruses assembling from pools of protein, nucleic acid, and their other constituents must be revised as it is becoming clear that RNA replication probably occurs in membrane compartments recruited by polymerase molecules and that genome transcription, replication, and translation are all directly coupled with virus assembly. Only when this process can be visualized at moderate to high resolution can we hope to understand how the beautiful finished products currently investigated are actually created and controlled. Finally, we expect that viruses will play a major role in nanotechnology and biomaterials. Their current use in the development of molecular electronics as scaffolds for self-assembling a molecular circuit board and as platforms for deploying chemically reactive groups in active materials are examples of areas in which they will be harnessed for positive applications.

Wah Chiu
John E. Johnson

VIRAL ASSEMBLY USING HETEROLOGOUS EXPRESSION SYSTEMS AND CELL EXTRACTS

By ANETTE SCHNEEMANN* AND MARK J. YOUNG†

*Department of Molecular Biology, The Scripps Research Institute, La Jolla, California 92037 and
†Departments of Microbiology and Plant Sciences Plant Pathology, Montana State University, Bozeman, Montana 59717

I. Introduction 1
II. The Driving Force behind the Development of Heterologous Expression Systems for the Study of Viral Assembly and Structure 2
III. Diversity of Heterologous Expression Systems 4
A. Prokaryotic Expression Systems 4
B. Eukaryotic Expression Systems 8
C. *In Vitro* Systems 15
IV. Guidelines for Choosing a Heterologous Expression System 16
A. Required End Use 16
B. Required Amount of Single Protein 17
C. Concurrent Synthesis of Multiple Proteins 17
D. Biological Activity 17
E. Expandability 18
F. Turnaround Time 18
G. Complexity of the System 19
V. Representative Examples of Viral Assembly in Heterologous Expression Systems 20
A. Alphaviruses 20
B. Polyomaviruses and Papillomaviruses 21
C. Nodaviruses and Tetraviruses 24
D. Caliciviruses 26
E. Herpes Simplex Virus 1 27
F. Bluetongue Virus 29
G. Respiratory Syncyitial Virus 31
VI. Conclusions 32
References 32

I. INTRODUCTION

Virus structure and assembly have been topics of intense investigation (Chiu *et al.*, 1997; Steven *et al.*, 1997). The increased activity is partly a result of improved technologies that have facilitated analysis of viruses that were refractory to detailed investigations in the past. Greater computational power and improvements in X-ray radiation sources now permit examination of large and complex viruses by single crystal X-ray analysis

0065-3233/03 $35.00

(Grimes *et al.*, 1998; Reinisch *et al.*, 2000; Wikoff *et al.*, 2000). Likewise, advances in electron cryomicroscopy and image reconstruction techniques allow time-resolved investigations of structural transitions associated with capsid assembly and maturation (Conway *et al.*, 2001; Lawton *et al.*, 1997). These developments have been paralleled by refinements in the molecular approaches used for sample preparation, with the result that synthesis of assembly intermediates and end products has become routine for many viruses.

Heterologous expression systems have been of critical importance for the study of viral assembly at the molecular and structural levels. These systems afford enormous flexibility in terms of dissecting the assembly pathway and investigating protein–protein or protein–nucleic acid interactions in the absence of viral transcription and replication. In addition, moderate- to high-resolution structural analyses of assembly precursors, intermediates, and end products, all generated by expression in heterologous systems, have yielded unprecedented molecular details of the structure and function of virus particles. There can be no doubt that the application of heterologous expression systems will continue to provide answers to unresolved questions about viral assembly and structure.

The intent of this article is to provide an overview of the heterologous expression systems that have been used for the study of virus assembly, structure, and function. It provides some guidelines for choosing the appropriate system and highlights the major advantages and disadvantages associated with the various expression strategies. Specific examples illustrate how the use of heterologous expression systems has influenced our understanding of viral assembly and structure. A complete survey of the literature, however, would be impossible within the context of this article and is not intended. Instead, we refer to other review articles that cover specific topics of interest so that the reader may obtain additional information about a particular viral system or experimental strategy. Some heterologous expression systems, particularly the vaccinia–T7 system, were used to develop "reverse genetics systems" for the rescue of infectious viruses from cDNA clones. This aspect is not discussed in this article.

II. The Driving Force behind the Development of Heterologous Expression Systems for the Study of Viral Assembly and Structure

The rationale for developing heterologous expression systems to study viral assembly and structure is rooted in the many limitations and challenges that a homologous system may present. These include poor growth of a virus in tissue culture, low yields of virus particles, inability to trap assembly intermediates, difficulty in separating assembly from other processes such as replication and transcription, and so on. In addition, heterologous expression systems yield viruslike particles (VLPs), which are devoid of the viral genome and therefore cannot cause a productive infection. This feature represents a highly attractive advantage because it permits analysis of a non infectious form of a virus that may normally cause serious disease in humans and other important hosts. This is particularly appreciated when no vaccine or effective antiviral treatment is available for the virus under investigation. For example, assembly and structure of hepatitis C virus (HCV) are currently investigated by heterologous expression of the core protein and glycoproteins in mammalian cells (Mizuno *et al.*, 1995), insect cells (Baumert *et al.*, 1998, 1999; Maillard *et al.*, 2001), *Escherichia coli* (Kunkel *et al.*, 2001; Lorenzo *et al.*, 2001), and yeast (Falcon *et al.*, 1999). Likewise, structure and assembly of human immunodeficiency virus (HIV) cores have been studied extensively in virtually every expression system available (Ehrlich *et al.*, 1992; Ganser *et al.*, 1999; Gheysen *et al.*, 1989; Gross *et al.*, 2000; Karacostas *et al.*, 1989; Li *et al.*, 2000; Spearman and Ratner, 1996). Other examples include Norwalk virus (Jiang *et al.*, 1992), human parvovirus (Brown *et al.*, 1991), Hantaan virus (Betenbaugh *et al.*, 1995), and foot-and-mouth disease virus (Abrams *et al.*, 1995).

Another significant aspect of heterologous expression systems is the ability to generate large quantities of the product of interest. Because many viruses do not grow to high titers in cultured cells or animal models, rigorous analysis of capsid structure and assembly is often not possible. Heterologous expression of the structural proteins, on the other hand, can yield sufficient amounts of fully assembled particles to permit crystallization and X-ray analysis. VLPs of hepatitis B virus (Wynne *et al.*, 1999), Norwalk virus (Prasad *et al.*, 1999), human papillomavirus 16 (Chen *et al.*, 2000), and many others provided the source for crystallization and determination of the high-resolution structure of the capsids of these viruses.

An important advantage of using heterologous expression systems is the ability to separate the assembly process from other events of the viral replication cycle such as transcription and replication, which may complicate the investigation and make interpretation of results difficult. Heterologous expression systems also allow for a more refined analysis of the assembly pathway on the basis of the possibility to express the components required for formation of the particle separately or in various combinations. As described in more detail below, this approach has been particularly useful in analyzing the assembly of large, multicomponent viruses, such as bluetongue virus (BTV) and herpes simplex virus 1 (HSV-1).

Heterologous expression systems also facilitate investigations of the assembly phenotype of mutant coat proteins. For many viruses it is difficult to selectively mutate the genes encoding the structural proteins within the context of a complete virus genome, particularly in large DNA viruses. Alternatively, the nucleotide changes associated with the mutations may interfere with transcription or replication, as is frequently observed in RNA viruses. Such problems are completely circumvented in expression systems, which allow straightforward mutagenesis and independent synthesis of the proteins of interest. The ability to express and study the properties of mutant coat proteins has provided many insights into the requirements of viral assembly, as is described in more detail below. It has permitted isolation and analysis of assembly intermediates, detection of maturation-defective forms of virus particles, and generation of novel types of particles that are not observed in a native infection. Taken together, these results have greatly increased our understanding of the parameters required for viral assembly and the function of the virion.

III. Diversity of Heterologous Expression Systems

A great number of systems are available for the expression of viral structural proteins and analysis of their assembly. These systems include both prokaryotic and eukaryotic environments and the products can be assembled either within the cells or *in vitro*, using cell lysates or purified components. Table I summarizes the systems that have been used for viral assembly and lists their key advantages and disadvantages.

A. *Prokaryotic Expression Systems*

1. *Escherichia coli*

Escherichia coli (*E. coli*) has been the subject of intense investigations for many years. Its genetics, biochemistry, and molecular biology are better understood than those of any other organism. Much of the knowledge accumulated over the past decades has provided the basis for development of a sophisticated protein expression system that encompasses a wide variety of vectors and protocols. *Escherichia coli* can be easily grown and manipulated in the laboratory and the cells have a rapid doubling time, require inexpensive growth conditions, and can be used in large-scale fermentation systems. The diversity of transcriptional and translational control strategies that are available today provides the researcher with a broad range of choices for controlling heterologous protein synthesis. As a result, *E. coli* has been and continues to be the heterologous protein expression system of choice for many applications.

The basic approach to express viral proteins in *E. coli* begins with insertion of the gene of interest into an appropriate expression vector. From the large selection of commercially available vectors, pET vectors have been particularly popular. The next step involves transformation of an appropriate *E. coli* host strain, followed by evaluation of heterologous protein expression and plasmid stability during culture. Once small-scale experiments have identified *E. coli* isolates expressing the foreign protein, the transformed *E. coli* strain can be used in large-scale fermentation systems. Protein production is followed by purification and characterization. On occasion, the nature of the protein to be expressed may require cloning into a specialized vector. For example, many viral coat proteins contain highly basic regions that carry multiple arginine and/or lysine residues. This situation can lead to low levels of protein if the host cell lacks sufficient amounts of the required tRNA. Tellinghuisen *et al.* (1999) circumvented this problem by using the pSBetB vector for expression of the Sindbis virus core protein, which contains a very basic region near its N terminus (Tellinghuisen *et al.*, 1999). The pSBetB vector contains the *E. coli argU* gene for rare arginine codon usage (Schenk *et al.*, 1995) and allowed isolation of several milligrams of protein from a 1-liter culture of bacteria. Standard expression vectors, on the other hand, did not yield appreciable amounts of core protein. A caveat, however, is that such specialized vectors tend to have low copy numbers, thereby reducing the amount of protein produced.

Expression of viral capsid proteins in *E. coli* can result in assembly of particles or intermediates, but this is relatively rare. Usually, the protein remains monomeric or forms small oligomers that must be purified and assembled *in vitro*. Although this adds additional complexity to the system, the approach has been used with great success in many cases. For example, the plant viruses cowpea chlorotic mottle virus (CCMV) (Fox *et al.*, 1998; Zhao *et al.*, 1995) and alfalfa mosaic virus (Yusibov *et al.*, 1996) as well as particles of polyomavirus and papillomavirus (Chen *et al.*, 2000, 2001; Li *et al.*, 1997; Zhang *et al.*, 1998) have been assembled from purified coat protein expressed in *E. coli*. Similarly, cores of the alphaviruses Sindbis virus and Ross River virus (Tellinghuisen *et al.*, 1999) and those of several retroviruses such as HIV (Campbell and Rein, 1999; Gross *et al.*, 2000) and Rous sarcoma virus (Joshi and Vogt, 2000; Yu *et al.*, 2001) were assembled *in vitro*, using *E. coli* as a system for heterologous expression of the capsid protein.

TABLE I

Summary of Major Heterologous Expression Systems for the Assembly of Viruslike Particles

Expression system	Vector	Major advantage(s)	Major disadvantage(s)	Key references
Prokaryotic				
Escherichia coli	Expression plasmid	Inexpensive, simple expression strategy, high yield of protein	Protein frequently insoluble; may not assemble into particles; parallel expression of multiple proteins difficult	Campbell and Rein, 1999; Zhao *et al.*, 1995; Chen *et al.*, 2001
Eukaryotic				
Insect cells	Recombinant baculovirus	Moderate to high yield; assembly of VLPs frequently observed; posttranslational modifications possible; parallel expression of multiple proteins straightforward	Relatively expensive on a large scale, presence of cryptic splice sites in large inserts may lead to inactivation of RNA transcript	Possee, 1997; Roy *et al.*, 1997; Jiang *et al.*, 1992; Dong *et al.*, 1998
Yeast cells	Expression plasmid	Inexpensive, simple expression strategy, posttranslational modifications possible, potential for high yield of protein	Cell lysis difficult; may lead to reduced yield of protein	Sasnauskas *et al.*, 1999; Sasagawa *et al.*, 1995; Vassileva *et al.*, 2000a
Mammalian cells	Recombinant vaccinia virus	Moderate yield; assembly of VLPs frequently observed; posttranslational modifications possible; parallel expression of multiple proteins possible	Relatively expensive on a large scale, vaccination of laboratory workers recommended, little commercial support available	Carroll and Moss, 1997; Hagensee *et al.*, 1993; Zhou *et al.*, 1991; Stauffer *et al.*, 1998

Whole plants	Stable expression via transgenic plants or transient expression via recombinant plant virus vectors	Potential production on an agricultural scale at low cost, expression of foreign protein in many tissues, posttranslational modifications possible; parallel expression of multiple proteins possible	Generation of transgenic plants is time-consuming and requires special expertise, expression levels tend to be low, transgene may be inactivated by gene silencing, size of foreign gene in viral vectors is limited	Porta and Lomonossoff, 1996; Koprowski and Yusibov, 2001; Mason *et al.*, 1996; Modelska *et al.*, 1998
In vitro				
Purified structural components		Presence/absence of assembly components can be easily controlled, ratio of required proteins can be varied	Requires heterologous expression and purification of structural proteins	Newcomb *et al.*, 1999; Tellinghuisen *et al.*, 1999
Reticulocyte lysate	mRNA	Simple and fast	Extremely low yields, assembly rarely observed	Spearman and Ratner, 1996; Iyengar *et al.*, 1996

Abbreviation: VLP, Viruslike particle.

A problem with the *E. coli* system is the tendency of the expressed protein to form inclusion bodies. This problem can sometimes be circumvented by lowering the induction/growth temperature or by reducing the concentration of the agent used for induction. Alternatively, the aggregated protein is purified, fully denatured, and refolded, but this strategy is usually not 100% efficient and leads to loss of a portion of the protein of interest. Another disadvantage of the *E. coli* system is the difficulty of expressing multiple proteins within a single cell, although there are rare cases in which this has been accomplished, for example, in the assembly of mosaic hepatitis B virus (HBV) cores (Preikschat *et al.*, 2000). In most instances, however, expression of multiple proteins in parallel is performed in eukaryotic expression systems.

B. Eukaryotic Expression Systems

1. Baculovirus Expression System

Baculoviruses are a family of insect viruses that have a large double-stranded DNA (dsDNA) genome that can accommodate multiple additional foreign genes. The particular virus that has found widespread application as a vector for expression of heterologous proteins is *Autographa californica* mononuclear polyhedrosis virus (AcMNPV) (for review articles see Possee, 1997; and Kost and Condreay, 1999). This virus grows well in cultured cells such as *Spodoptera frugiperda* (Sf) lines 9 and 21 and *Trichoplusia ni* cells, also known as High Five cells (Invitrogen Life Technologies, Carlsbad CA). The viral replication cycle can be divided into immediate-early, early, and late phases. One of the genes expressed during the late phase is the polyhedrin gene, which has a strong promoter leading to high levels of transcription and massive synthesis of the polyhedrin protein. The baculovirus expression strategy takes advantage of the fact that the polyhedrin protein is dispensable for propagation of the virus in cultured cells and that it can be replaced with a foreign gene of interest. Expression of the foreign gene under control of the polyhedrin promoter results in high levels of protein product, although the yield is usually somewhat lower than that of the polyhedrin protein in wild-type virus.

Because of the large size of the baculovirus genome, unique restriction sites are not available for simple replacement of the polyhedrin gene with the gene of interest. Thus, recombinant viruses are generated by homologous recombination. A number of different strategies are currently employed to this end. In the traditional protocol, Sf cells are cotransfected with purified baculovirus DNA and a transfer vector containing the gene of interest flanked by sequences derived from the polyhedrin locus. Recombination between the two DNA molecules results in replacement of the polyhedrin gene with the foreign gene. Because the recombination event is rare (0.1% recombinants) the resulting progeny virions represent mostly wild-type virus and isolation of the recombinants by screening is extremely tedious. To circumvent this problem, a linearized version of the baculovirus DNA, which lacks a gene essential for replication of the virus, is now commonly used. The essential gene is located on the transfer vector together with the gene of interest and only homologous recombination between the viral DNA and the transfer vector results in viable progeny. This system essentially eliminates the high background of wild-type baculovirus and has accelerated the time required for isolation of recombinants tremendously. Variations of this strategy employ transfer vectors that contain the β-galactosidase gene as a marker, which permits identification of recombinants by formation of blue plaques.

A different protocol relies on recombination in bacteria rather than insect cells (Luckow *et al.*, 1993). In this protocol ("Bac-to-Bac" system; Invitrogen Life Technologies).The gene of interest is cloned into a donor plasmid, which is transfected into bacteria that carry a baculovirus shuttle vector ("bacmid") containing the baculovirus genome. Recombination between the donor plasmid and the bacmid, which is facilitated by the presence of Tn7 attachment sites and transposition proteins encoded on a helper plasmid, transposes the gene of interest into the bacmid. Colonies containing recombinant bacmids can be identified by disruption of the *lacZα* gene. High molecular weight DNA is prepared from the colonies and transfected into insect cells.

Increasingly complex expression strategies have been developed using recombinant baculoviruses. For example, in addition to the polyhedrin promoter, the p10 promoter, another strong very late promoter, has been used by many investigators to overexpress foreign proteins. Transfer vectors that contain both the p10 and polyhedrin promoters followed by separate multiple cloning sites allow construction of recombinant viruses that express two proteins in parallel. Transfer vectors that permit placement of the gene of interest behind a promoter that is active from 4 to 16 h after infection, the so-called immediate-early promoter, or behind a promoter that is active from 4 to 48 h after infection, the gp64 protein promoter, are also available. Even more sophisticated strategies have employed vectors that express three to four proteins simultaneously (Belyaev and Roy, 1993). Because the baculovirus expression system has found such wide application, there is significant commercial support available. Many companies have developed reagents and kits that almost guarantee success of a given project. The kits usually include everything that is required to carry out the project from start to end, including cells, growth medium, baculovirus DNA, transfer vectors, and various control reagents.

In terms of viral assembly and structure the baculovirus system has been used with tremendous success and some representative examples are discussed in more detail below. Generally speaking, the expressed viral protein(s) can be expected to assemble into particles that are structurally similar if not identical to their native counterparts. This has been shown specifically in the case of the nodavirus Flock House virus, where X-ray analysis of native virions and VLPs showed no differences in the structure of the protein capsid (V. Reddy and J. E. Johnson [The Scripps Research Institute, La Jolla, CA], unpublished data). Similarly, structural investigations at lower resolution, using cryoelectron microscopy and three-dimensional image reconstruction, have confirmed the identity of native and "synthetic" virions in many other cases. This feature combined with the large amounts that can be obtained has permitted structural analysis of many viruses for which only limited amounts of native virions were available.

The baculovirus system is generally reliable with regard to the requirements for cellular transport of heterologous proteins and posttranslational modifications such as proteolytic processing, phosphorylation, and glycosylation. Although the glycosylation pattern in insect cells is not identical to that observed in mammalian cells, in most cases this does not appear to affect the function of the protein of interest. A problem occasionally arises on the basis of the fact that replication and transcription of baculovirus genes occur in the nucleus of infected cells. If the gene of interest contains cryptic splice sites, they may result in processing of the RNA transcript and disruption of the open reading frame required for protein synthesis.

2. *Vaccinia Virus System*

Vaccinia virus is a large double-stranded DNA virus that infects many types of mammalian cells and some invertebrate cells. Its replication cycle is confined to the cytoplasm and can be divided into an early phase and a late phase. The late phase is characterized by the synthesis of structural proteins and enzymes that are packaged into progeny virions. Expression of foreign genes from recombinant vaccinia viruses takes advantage of promoters that are active during the late phase in order to maximize yields of the protein of interest. Several features make the vaccinia virus system an attractive system for the study of viral assembly and structure. The large size of the vaccinia virus genome permits insertion of multiple foreign genes under the control of either identical or different promoters. Expression occurs in the cytoplasm, eliminating splicing of transcripts due to the presence of cryptic splice sites. Synthesis of the desired protein(s) occurs in mammalian cells, ensuring appropriate posttranslational modifications and the presence of host factors that may be required for folding and assembly. Cellular transport of glycoproteins to the cell surface and assembly and release of VLPs can be expected to occur normally. A drawback is the fact that the yield of the protein of interest is generally lower when compared with the baculovirus expression system. In addition, although vaccinia virus is considered a safe virus to work with, both the Centers for Disease Control and Prevention (CDC, Atlanta, GA) and the National Institutes of Health (NIH, Bethesda, MD) recommend that laboratory workers be vaccinated before they embark on projects involving exposure to the virus.

Two strategies are used for the expression of foreign proteins employing vaccinia virus vectors. The first involves the generation of recombinant viruses whereas the second relies on expression of proteins from plasmids in which the genes of interest are placed under the control of a vaccinia virus promoter. In the latter case, expression depends on coinfection with a vaccinia helper virus. A detailed description of the system and experimental protocols can be found in several review articles (Carroll and Moss, 1997; Miner and Hruby, 1990; Moss, 1992). Briefly, generation of recombinant vaccinia viruses requires construction of a plasmid in which the gene of interest is placed downstream from a vaccinia virus promoter. Flanking sequences from a vaccinia virus gene reside upstream of the promoter and downstream of the gene of interest. Recombinant viruses are generated by homologous recombination between a wild-type vaccinia virus and the transfer plasmid. To this end, cells are infected with the wild-type virus and cotransfected with the plasmid. Recombinant viruses are subsequently isolated and purified. The recombination event, however, is rare and identification of recombinants requires screening or the presence of selectable markers. A selectable marker gene, such as the β-galactosidase gene, which permits visual selection in the presence of 5-bromo-4-chloro-3-indolyl-β-D-galactopyromiside (X-Gal), or the neomycin phosphotransferase gene, which confers resistance to G418, is therefore often included in the plasmid used for recombination.

In the second strategy cells are transfected with plasmids that carry the gene of interest under the control of a vaccinia virus promoter and superinfected with a vaccinia helper virus. In this case, expression occurs *in trans* by factors provided by the helper virus. The advantage of this strategy is that it is much faster because expression can be achieved as soon as the expression plasmid has been generated whereas generation of vaccinia virus recombinants, isolation, and plaque purification take several weeks. A disadvantage, however, is the fact that the expression levels are lower than those achieved with recombinant viruses. To improve this situation, the T7–vaccinia virus system was developed in the laboratory of B. Moss (NIH). This system employs a recombinant vaccinia virus that contains the gene for the polymerase of bacteriophage T7 and plasmids in which the foreign genes are placed under the control of the T7 promoter. Using this strategy, expression levels 10- to 20-fold higher than those obtained with homologous vaccinia virus promoters have been achieved.

3. Yeast Expression System

Yeast-based expression systems combine many of the advantages of prokaryotic and eukaryotic systems (for reviews see Cregg *et al.*, 2000; Giga-Hama and Kumagai, 1999; Hollenberg and Gellissen, 1997; Trueman, 1995). Similar to *E. coli*, the genetics and biochemistry of yeast are well understood and the cells can be easily manipulated and cultured inexpensively to high density. In addition, yeast cells are capable of secreting proteins and performing secretion-linked protein modifications, such as processing and glycosylation. These features make yeast an attractive system for the expression of heterologous proteins.

Yeast-based heterologous expression systems have been developed for species belonging to *Saccharomyces*, *Pichia*, *Hansenula*, *Candida*, and *Torulopsis*. Of these genera, the *Saccharomyces* and *Pichia* systems have been most widely used to express viral gene products. *Saccharomyces cerevisiae* has been used to express VLPs of polyomaviruses (Sasnauskas *et al.*, 1999), hepatitis B virus (Bitter *et al.*, 1988), papillomaviruses (Hofmann *et al.*, 1995; Sasagawa *et al.*, 1995), and the retrotransposon Ty (Adams *et al.*, 1994). For example, for high-level expression of simian virus 40 (SV40) VLPs the major capsid protein of SV40, VP1, was amplified by polymerase chain reaction (PCR) and cloned into the yeast expression vector pFX7, which contains a *GAL10-PYK1* promoter, a fragment of the 2 μ circle DNA (a naturally occurring yeast plasmid), and a dominant selective marker, the *FDH1* gene, which confers resistance to formaldehyde. This vector also contains an *E. coli* origin of replication and selectable markers for screening recombinants in *E. coli*. Recombinant plasmids are introduced into yeast either by electroporation or by liposome-mediated protocols and expression of the heterologous protein is transcriptionally induced by including galactose in the growth media. The cells are either grown in small-scale flask cultures or to high densities by fermentation methods usually for a period of 20–160 h. Heterologous proteins or VLPs are released from the cells by mechanical and/or enzymatic disruption of the cells (often a low-efficiency step) followed by removal of cell debris. VLPs are then banded on CsCl gradients or sedimented through velocity sucrose gradients.

The methylotrophic yeast *Pichia pastoris* has emerged as a powerful and inexpensive heterologous system for the production of high levels of functionally active recombinant proteins (Cregg *et al.*, 2000). The existence of well-established fermentation methods that can generate high cell densities using defined media and the strong and tightly regulated methanol-inducible alcohol oxidase (*AOX1*) promoter have made *P. pastoris* an excellent heterologous expression system. The *AOX1* transcript and protein are virtually undetectable in cells growing in the presence of nonmethanol carbon sources such as glycerol. Substitution of methanol as the sole carbon source in the growth medium leads to extensive transcription from the *AOX1* promoter, often resulting in protein levels that reach 30% of the total soluble cellular protein. An illustrative example of the power of *P. pastoris* for expression of VLPs is the inexpensive production of hepatitis B virus vaccine in large quantities (Vassileva *et al.*, 2000a,b). It was shown that an increase in the copy number of the HBV surface antigen (HBsAg) gene in *P. pastoris* resulted in a proportional elevation of the level of protein and 22-nm HBsAg particle assembly (Vassileva *et al.*, 2000a). Moreover, it was demonstrated that expression under the control of the newly described glyceraldehyde-3-phosphate dehydrogenase (*GAP*) promoter represents a useful alternative to the *AOX1* promoter (Vassileva *et al.*, 2000b). Use of the *GAP* promoter permits simultaneous biomass generation and antigen production under the control of glycerol or glucose.

4. Plant-Based Expression System

There has been increasing interest in plants as heterologous expression systems, especially for production of biomedically relevant proteins (Koprowski and Yusibov, 2001; Peeters *et al.*, 2001). Plants have several advantages for producing therapeutic proteins, including the lack of contamination with mammalian pathogens, ease of genetic manipulations, eukaryotic protein modification machinery, and economical production systems.

Two basic strategies have been employed for heterologous protein expression in plants. The first is the creation of transgenic plants in which the foreign gene is stably incorporated into the plant genome, transcribed through the nuclear apparatus of the plant, and inherited by the next generation. In contrast, the second strategy uses plant viruses as transient expression systems. Plant viruses are engineered to carry and replicate foreign genes in susceptible host plants. The sequences delivered by the viruses remain part of the viral genome and are replicated by the viral polymerase. The virally expressed foreign genes are not typically inherited. Both transgenic plants and engineered plant viruses have been used successfully to express foreign viral proteins for use as vaccines. However, viral vectors may have certain advantages over transgenic plants including rapid replication time, ease of scale-up, and the possibility of a wide host range for expression in different plants, using the same viral vector.

The most common strategy for producing transgenic plants expressing heterologous gene products involves *Agrobacterium tumefaciens*, a pathogen of dicotyledenous plants, which have evolved a mechanism of DNA transfer from the bacterium to the plant genome. Initially, the viral gene is cloned into a binary plasmid vector (such as pBI121), which places the viral gene under the transcriptional control of the cauliflower mosaic virus (CaMV) 35S promoter and nopaline synthetase terminator. The initial cloning steps are usually accomplished in *E. coli*. Once generated in *E. coli*, the plasmid is mobilized into the *A. tumefaciens* host. *Agrobacterium tumefaciens* harboring the plasmid is subsequently used to infect plant cells, where nonhomologous integration of the viral gene into the plant genome occurs. Subsequent regeneration of the transformed cells into whole plants leads to all the cells containing at least one copy the viral gene under the transcriptional control of the CaMV 35S promoter. Transgenic plant technology has been used to create a heterologous expression system for viral antigens from mammalian pathogens. This includes hepatitis B virus surface antigen (Kapusta *et al.*, 1999; Mason *et al.*, 1992), rabies virus glycoprotein (McGarvey *et al.*, 1995), and Norwalk virus capsid protein (Mason *et al.*, 1996). In general, the disadvantages of the transgenic plant approach are that it is an experimentally time-consuming process (From 1 month to 1 year), expression levels of the foreign gene tend to be relatively low (0.1–1% of the total soluble protein level of the plant), not all plants can be successfully transformed and regenerated into whole plants, and expression of the transgene may be eliminated by the process of gene silencing. The advantages of the transgenic plant approach are that all tissues of the plant may express the foreign gene product, the viral gene is stably inherited, and inexpensive large-scale production systems are often available.

Plant virus-based vectors have a number of advantages as heterologous gene expression tools (for general reviews see Porta and Lomonossoff, 1996; Scholthof *et al.*, 1996). These include the ability to direct rapid and high-level expression of foreign genes in cost-effective and flexible protein production systems. Most approaches for the expression of heterologous genes from plant viral vectors have relied on the expression of the foreign protein as a fusion to a viral structural protein (Belanger *et al.*, 2000; Dalsgaard *et al.*, 1997; Dolja *et al.*, 1992; Fitchen *et al.*, 1995; Gopinath *et al.*, 2000; Hamamoto *et al.*, 1993; McLain *et al.*, 1995; Modelska *et al.*, 1998; Yusibov *et al.*, 1997). Alternatively, the foreign protein is expressed from a duplicated subgenomic mRNA promoter (Donson *et al.*, 1991; Kumagai *et al.*, 1993). Typically, the gene of interest is cloned as a transcriptional and/or translational fusion within the recipient viral genome and the chimeric viral genome is used to directly infect a susceptible host plant. Systemic spread within the plant can lead to high-level expression of the foreign protein.

The coat proteins of plant viruses have been used as carrier molecules for presentation of antigenic epitopes from heterologous proteins (Brennan *et al.*, 2001). In infected plants, the recombinant coat proteins assemble into chimeric particles with the underlying morphology of the native virus and purified particles can elicit specific antibodies to the heterologous epitope when injected into mice (Dalsgaard *et al.*, 1997). Even feeding mice with infected plant material can lead to humoral and mucosal responses to the expressed epitope (Modelska *et al.*, 1998).

The disadvantages of plant–virus-based vectors are that they typically lead to only transient expression of the foreign protein, only a limited number of virus vectors are available, the size of the foreign gene insert may be limited, and recombination within the plant may eliminate expression of the foreign viral protein.

C. *In Vitro Systems*

1. *Purified Structural Components and Cell Lysates*

Expression of viral structural proteins in a heterologous system does not always lead to formation of the desired assembly intermediate or end product. This is particularly the case when the proteins are expressed in *E. coli* or when individual components are expressed separately in different cells. In these cases, the proteins are usually purified and then assembled *in vitro.* Alternatively, assembly is possible using whole cell lysates. For example, assembly of HSV-1 capsids, which requires a minimum of four structural proteins, was observed on mixing of lysates derived from insect cells infected with different baculovirus vectors (Newcomb *et al.*, 1994). Similarly, reovirus cores, obtained from native virions, could be reconstituted into virions by incubation in cell lysates containing the proteins required for formation of the outer shell (Chandran *et al.*, 1999).

2. *Reticulocyte/Wheat Germ Lysate*

Reticulocyte and wheat germ lysates represent relatively expensive and inefficient systems for the study of viral assembly and structure. The level of the synthesized proteins is minute (nanogram range) and frequently insufficient to promote assembly. In some cases, however, formation of particles on *in vitro* synthesis of capsid proteins was observed, for example, for retroviral Gag proteins (Spearman and Ratner, 1996) and the capsid protein of human papillomavirus 16 (Iyengar *et al.*, 1996). The main advantage of using reticulocyte or wheat germ lysates is that they require no subcloning and that they potentially yield results quickly.

IV. Guidelines for Choosing a Heterologous Expression System

A heterologous gene expression system is chosen with regard to both its characteristics and the anticipated application of the expressed protein. Often, both economic and production issues play a major role in choosing a particular expression system. An ideal system provides both low cost and high production capacity.

A. *Required End Use*

One of the most important factors in choosing an expression system is the required end use of the sample to be generated. Are large quantities required for structural analysis? Are posttranslational modifications critical? Is biological activity of the protein or assembly product necessary? Will scale-up become an important issue? Are particles required or are individual subunits or small assembly intermediates sufficient? As described in the previous sections, each expression system has its advantages and disadvantages but the desired properties of the end product may limit the number of systems that are suitable. In general, the highest yields of fully assembled VLPs are obtained with the baculovirus expression system and this system has been the system of choice for generating the large quantities required for crystallization and high-resolution structure determination by X-ray analysis. On the other hand, characterization of assembly pathways and analysis of intermediates can be accomplished with smaller quantities, which are provided by most other systems.

B. *Required Amount of Single Protein*

Single proteins that do not require posttranslational modifications are often overexpressed in *E. coli* during initial studies. The proteins generally remain unassembled or form small oligomers that may represent early assembly intermediates. They are subsequently purified for *in vitro* assembly studies. Because the system is inexpensive, easy to use, and yields large quantities, it has been the system of choice for many investigators. If the desired protein requires posttranslational modifications, it is best expressed in insect or mammalian cells using baculovirus or vaccinia virus vectors. This is particularly important if the posttranslational modification affects the assembly properties of the protein of interest. As indicated above, insect cells have the ability to glycosylate proteins but the glycosylation pattern is not necessarily identical to that observed in mammalian cells. Interestingly, this usually does not interfere with protein function.

C. *Concurrent Synthesis of Multiple Proteins*

In many cases multiple types of proteins must be expressed within a single cell because their interactions are required for assembly of particles or intermediates. Alternatively, the sequence of events that lead to formation of multicomponent capsids can be deduced by expression of subsets of the required proteins. Multiple expression of proteins is difficult in *E. coli* because of the lack of suitable expression vectors. However, it is relatively straightforward in eukaryotic systems using baculovirus or vaccinia virus vectors. These vectors have sufficient capacity to accommodate multiple foreign genes and strategies for generating recombinants carrying multiple inserts are well established. An added advantage of the eukaryotic systems is that transport of proteins to cellular locations, such as the endoplasmic reticulum (ER) or the plasma membrane, is possible. Multiple proteins can also be expressed in parallel by infecting cells with several vectors, each carrying the gene for the protein of interest. However, this becomes increasingly difficult when more than two or three separate recombinant viruses are required.

D. *Biological Activity*

In the context of viral assembly, biological activity refers primarily to the ability of the assembled material to perform some of the functions associated with native virions or their components. This includes the ability of the assembled particle to elicit an immune response similar to native capsids, the ability of the capsid or individual proteins to interact with antibodies in a manner analogous to native material, the binding of assembled particles to the cellular receptor, and so on. Biological activity of the expressed material is usually observed when it is generated in mammalian and insect cells. *Escherichia coli*-expressed protein, on the other hand, can present problems if it forms inclusion bodies in which the protein is aggregated rather than folded correctly. Denaturation and correct refolding of the protein are not trivial. Fortunately, even aggregated protein is often useful for diagnostic purposes such as screening sera for the presence of antibodies by enzyme-linked immunosorbent assay (ELISA) or dot-blot analysis.

E. *Expandability*

Expandability becomes an issue when large quantities of the expressed material are required. Crystallization requires milligram quantities, which can be obtained from as little as a 1-liter insect cell or *E. coli* cell culture. Much larger amounts are needed when the product is to be used commercially, for example, as a vaccine or diagnostic product. In this case, availability of bioreactor technology becomes a critical point of consideration. For most systems—*E. coli*, yeast, and insect and mammalian cells—the possibility of significantly expanding cultures exists. In contrast to industrial settings, however, scale-up within the laboratory is usually limited to several liters. Cost assessment is important, bearing in mind that insect and mammalian cell cultures are significantly more expensive than prokaryotic cell culture or yeast cell culture.

F. *Turnaround Time*

Before embarking on a particular expression strategy it is useful to have some information regarding the hands-on time required before the first results can be obtained. The fastest turnaround times are achieved with cell lysates, such as the reticulocyte and wheat germ lysates, which only need to be primed with *in vitro* synthesized RNA (or DNA in the case of a coupled transcription/translation system). These systems potentially provide answers within a few hours but are rarely used because of the small quantities that they yield. Plasmid-based systems used for expression in *E. coli*, yeast, and in combination with the vaccinia–T7 vector system are relatively fast as well. These strategies require subcloning of the gene(s) of interest into a plasmid vector, which is then transfected into the appropriate cells for expression. Thus, turnaround time is primarily determined by the time it takes to prepare the required vectors. The generation of recombinant baculoviruses and vaccinia viruses, on the other hand, takes significantly longer. In these cases, homologous recombination between a plasmid vector and the viral genome leads to formation of the desired virus, but because the event is rare, screening and several rounds of plaque purification are required to isolate recombinants. It usually takes at least 2–3 weeks until a recombinant virus has been constructed, identified, amplified, and titered. The most time-consuming strategy is the generation of transgenic plants, which can take several months to 1 year.

G. *Complexity of the System*

The various expression systems differ in their experimental complexity and expertise required by the investigator. The most straightforward systems are those that require no more than construction of an expression plasmid and transfection into a suitable host cell such as *E. coli* or yeast. In contrast, the expression systems involving viral vectors are significantly more complex. Not only do they require more sophisticated methods for generating the recombinant virus but they also require familiarity with eukaryotic cell culture, virus propagation, and titration. Moreover, the use of these systems necessitates availability of the appropriate equipment such as tissue culture hoods and incubators. For the baculovirus expression system significant commercial support is available to get the investigator started and ensure success of the project. Many companies offer complete systems that contain all the required components such as cells, growth medium, plasmid vectors, viral stocks, and detailed protocols. For the vaccinia virus system there is little if any commercial support available. Viruses and cells can be obtained from the American Type Culture Collection (ATTC, Manassas, VA), but plasmid vectors usually must be requested from the laboratories in which they were first generated. Because there are many different possibilities for the choice of viral promoters and insertion sites in vaccinia, the investigator needs to thoroughly research which strategy would be likely to yield the best results. The most popular vaccinia virus system in use is the vaccinia–T7 system, which employs a vaccinia virus recombinant that carries the gene for the bacteriophage T7 polymerase and plasmids in which the gene of interest is inserted downstream of the T7 promoter. This virus as well as suitable DNA vectors can be obtained by contacting the laboratory of B. Moss at the NIH.

V. Representative Examples of Viral Assembly in Heterologous Expression Systems

A. *Alphaviruses*

The alphaviruses comprise a genus within the family *Togaviridae* and include Sindbis virus, Semliki Forest virus, and Ross River virus. The alphavirus particle consists of a nucleocapsid core, which encapsidates the single-stranded RNA genome, and a lipid envelope containing virally encoded glycoproteins. The nucleocapsid core is composed of 240 copies of a single type of coat protein arranged on an icosahedral $T = 4$ lattice. Assembly of the nucleocapsid has been studied *in vitro* with coat protein derived from dissociated virions (Wengler *et al.*, 1982, 1984) and, more recently, coat protein overexpressed and purified from *E. coli* (Tellinghuisen *et al.*, 1999). The earlier studies showed that the coat protein can assemble into particles resembling native nucleocapsids under a variety of conditions and in the presence of several types of single-stranded RNAs and other polyanions. Because these studies depended on fully assembled particles as a source of protein, they were limited to investigations of wild-type coat protein. This prevented isolation and analysis of potential assembly intermediates using coat protein mutants. Studies by Tellinghuisen *et al.* (1999) circumvented these limitations by using *E. coli* for expression of wild-type and mutant forms of the nucleocapsid proteins of Sindbis virus (SINV) and Ross River virus (RRV). These studies have provided new and unexpected insights into the mechanism by which this protein assembles to form nucleocapsids.

Using purified SINV proteins, a thorough investigation of the requirements for nucleic acid during assembly revealed that assembly was dependent on their presence and that small single-stranded DNA oligonucleotides served as sufficient substrates for core formation as long as they had a minimum length of at least 14 nucleotides. Interestingly, core formation was not observed in the presence of double-stranded DNA, suggesting that the interactions are not solely based on charge neutralization of the coat protein, which has a basic N-terminal region.

Truncated versions of the nucleocapsid protein that lacked various portions of the N terminus revealed that deletion of the first 32 amino acids interfered with assembly, but this deletion did not eliminate interaction of the coat protein with nucleic acid. When present in small amounts, this mutant coat protein could be incorporated into capsids assembled from the wild-type protein. However, when present in large amounts, the mutant coat protein inhibited assembly of the wild-type protein.

The observation that nucleocapsid assembly in the absence of nucleic acids was inhibited indicated that interaction of the coat protein with the RNA is an essential and early step in the assembly pathway. The availability of mutant coat proteins that retained nucleic acid-binding activity but could not assemble further provided an opportunity to identify a possible coat protein–nucleic acid assembly intermediate. Cross-linking experiments revealed the presence of a coat protein dimer that could be detected only in the presence of nucleic acid and for those types of mutant proteins that had retained nucleic acid-binding activity. The protein dimer itself could not assemble into cores but was incorporated into cores in the presence of wild-type protein. These and other results strongly suggested that the cross-linked dimer represents a genuine intermediate of nucleocapsid core assembly.

Further analysis of the cross-linked intermediate showed that lysine at position 250 of one capsid subunit was covalently linked to the identical amino acid on a second subunit. In a model of the nucleocapsid derived from both cryoelectron microscopy (cryo-EM) analysis and X-ray analysis of the nucleocapsid protein, the covalent bond connects a pentamer of coat proteins with a hexamer of coat proteins, that is, it is an intercapsomer contact rather than an intracapsomer contact. This finding was unexpected because a possible assembly model proposed preassembly of pentameric and hexameric units that would recruit RNA for assembly into the final structure. In light of the new data, however, this scenario is unlikely. Instead, the initial assembly intermediate appears to be a coat protein dimer bound to RNA and the dimer spans the intercapsomere space.

B. *Polyomaviruses and Papillomaviruses*

Polyomaviruses and papillomaviruses are small, nonenveloped viruses that have a double-stranded, circular DNA genome. The viruses were originally grouped into a single family based on similarities in particle size and shape, the double-stranded, circular DNA genome, and nuclear site of assembly. However, molecular cloning of the polyomavirus and papillomavirus genomes showed that the genome organization of the two viruses significantly differs, prompting their separation into two virus families.

Polyoma- and papillomaviruses share a common capsid structure that is assembled from 72 pentameric capsomeres arranged on a $T = 7$ icosahedral lattice. Assembly studies of polyomaviruses were first initiated by expression of the major capsid protein of mouse polyomavirus, VP1, in

E. coli where it forms capsomers composed of five VP1 subunits. These capsomers can be assembled *in vitro* into particles resembling native virions (Garcea *et al.*, 1987; Salunke *et al.*, 1986, 1989). Subsequent studies employed recombinant baculoviruses to investigate assembly in eukaryotic cells. Infection of Sf9 cells with a recombinant baculovirus vector expressing VP1 revealed empty, capsid-like structures in the nucleus whereas VP1 in the cytoplasm remained unassembled (Montross *et al.*, 1991). Nuclear assembly is also observed in native infections and was hypothesized to result from increased availability of calcium in the nucleus relative to the cytoplasm. This hypothesis was supported by the finding that assembly of VLPs could be induced in the cytoplasm on artificially increasing the calcium concentration in the presence of the drug ionomycin (Montross *et al.*, 1991).

An interesting observation was made when VP1 and the minor structural protein VP2 were coexpressed in insect cells. First, immunoprecipitation analyses demonstrated a physical association between VP1 and VP2, which was not surprising given the incorporation of small amounts of this protein into native virions. Second, and more interesting, the phosphorylation pattern of VP1 changed in the presence of VP2. When expressed in insect cells, VP1 is phosphorylated predominantly on serine, whereas in virus-infected mouse cells there is a 2:1 phosphothreonine:phosphoserine ratio. On coexpression of VP2, the nonphysiologic serine phosphorylation was reduced and phosphorylation of Thr-63, the same residue that is phosphorylated in infected mouse cells, was detected (Li *et al.*, 1995). Thus, coexpression of the two proteins in insect cells led to proper substrate phosphorylation.

Papillomaviruses cause warts ("papillomas") at cutaneous and mucosal sites. Some serotypes of the human papillomaviruses (HPVs), in particular HPV-16 and -18, are strongly associated with cervical cancer and other epithelial tumors. Because papillomaviruses cannot be propagated in cultured cells, studies of their life cycle and assembly are difficult. This has prompted many research groups to investigate assembly of particles in heterologous expression systems (for a review see Sapp *et al.*, 1996). It was shown that the major coat protein L1 (the analog of polyomavirus VP1) of various HPV serotypes assembles into viruslike particles in insect cells using recombinant baculoviruses (Kirnbauer *et al.*, 1992, 1993; Rose *et al.*, 1993; Volpers *et al.*, 1994), in mammalian cells using recombinant vaccinia viruses (Hagensee *et al.*, 1993; Zhou *et al.*, 1991), and in yeast cells using plasmid-based expression vectors (Hofmann *et al.*, 1995; Sasagawa *et al.*, 1995). The particles have features similar to those of native virions based on cryo-EM analysis (Hagensee *et al.*, 1994). As in the case of polyomaviruses, papillomavirus VLPs assembled in the nucleus whereas L1 remained unassembled in the cytoplasm. Coexpression of L1 and the minor structural protein L2 (analog of polyomavirus VP2) resulted in formation of VLPs that incorporated L2 into particles. The presence of L2 appeared to increase the yield of VLPs, indicating that it may be important in capsid stabilization (Hagensee *et al.*, 1993). Assembly of papillomavirus VLPs was also demonstrated with L1 expressed in *E. coli.* The L1 protein formed pentamers that could be purified and assembled into VLPs *in vitro* (Li *et al.*, 1997).

VLPs resulting from assembly of L1 were shown to have biological activities similar to those of native virions. For example, they have hemagglutination activity, recognize a specific cell surface molecule (the putative cellular receptor), and are internalized into various types of cells (Muller *et al.*, 1995; Roden *et al.*, 1994; Volpers *et al.*, 1995; Zhou *et al.*, 1993). Moreover, on immunization of animals with VLPs, neutralizing antibodies can be detected and animals are protected from challenge with native virus (Breitburd *et al.*, 1995; Kirnbauer *et al.*, 1992; Suzich *et al.*, 1995). These findings have prompted intense efforts to develop papillomavirus vaccines using VLPs derived from the various expression systems. Several preparations are now in preclinical and early-phase clinical trials with encouraging preliminary results (Schiller and Hidesheim, 2000).

Investigations have focused on generating infectious VLPs, termed pseudovirions. To this end, L1 and L2 were expressed in mammalian cells using vaccinia virus vectors. The cells chosen for VLP formation carried endogenous papillomavirus genomes maintained in episomal form or heterologous plasmids containing the SV40 origin of replication. Assembly of L1 and L2 in the nucleus led to incorporation of the DNA into the particles, albeit at low frequency (Stauffer *et al.*, 1998; Unckell *et al.*, 1997). The resulting particles were able to infect susceptible cells and deliver the packaged DNA to the nucleus.

The structure of a small VLP of human papillomavirus 16 was solved by X-ray crystallography (Chen *et al.*, 2000). The particles were first assembled *in vitro* from pentameric capsomers generated in *E. coli.* To obtain highly diffracting crystals, the N terminus (10 amino acid residues) had to be removed from the L1 protein. While this did not interfere with capsomer formation, the protein assembled into $T = 1$ particles containing 12 capsomers instead of the usual 72 capsomers. Thus the term "small VLP." The crystal structure of the small VLP at 3.5-Å resolution showed that the L1 protein closely resembles the VP1 protein of polyomaviruses and that surface loops contain the sites of sequence variation among different HPV types. These loops also represent dominant, neutralizing epitopes. The crystal structure now serves as a tool for current and future vaccine development.

C. *Nodaviruses and Tetraviruses*

Nodaviruses are small, nonenveloped viruses that have a bipartite, single-stranded RNA genome. Both genomic RNAs are packaged within a single virion. The protein capsid contains 180 copies of a single type of coat protein arranged on a $T = 3$ icosahedral lattice. Following assembly, the particles mature by autocatalytic cleavage of the coat protein subunits. For the insect nodavirus Flock House Virus (FHV) the clevage occurs between residues Asn-363 and Ala-364. This reaction generates the major coat protein β (residues 1–363) and the minor coat protein γ, which contains 44 amino acids (residues 364–407). The maturation cleavage is required for acquisition of infectivity and results in increased particle stability. Expression of the FHV coat protein in the baculovirus system leads to spontaneous assembly of VLPs that are structurally identical to native virions even at high resolution as determined by X-ray crystallography (Schneemann *et al.*, 1993; V. Reddy and J. E. Johnson, unpublished data). This is one of the few examples in which the X-ray structures of both native virions and VLPs have been solved for direct comparison. In contrast to native virions, however, the VLPs contain a heterogeneous mixture of RNAs representing mostly cellular and/or baculoviral RNAs. This observation is in agreement with *in vitro* assembly studies, which indicated that the nodaviral coat protein requires nucleic acid for assembly although a variety of different RNAs can serve as substrates (Schneemann *et al.*, 1994).

The increased particle stability following the maturation cleavage prompted investigations into the structure of the uncleaved precursor particles in order to find a possible structural explanation for this phenomenon. To this end, asparagine at the cleavage site was mutated to aspartate, threonine, and alanine and the mutant coat proteins were expressed in the baculovirus system. The resulting cleavage-defective particles could be obtained in sufficient amounts for crystallization and X-ray analysis (Schneemann *et al.*, 1993). Unexpectedly, the high-resolution structure did not reveal significant differences between the uncleaved and cleaved particles, suggesting that subtle changes may have a significant effect on particle stability (V. Reddy and J. E. Johnson, unpublished data). Alternatively, it is possible that the encapsidated RNA, which is largely invisible in the structure, is reorganized during the maturation process, thereby leading to the increase in stability.

A separate study investigated the function of the N-terminal region of the FHV coat protein in determining the geometry of the particle. It was postulated, on the basis of the position of the N terminus in the high-resolution structure of mature virions, that deletion of the first 31 residues would result in assembly of a particle with $T = 1$ symmetry containing 60 subunits instead of the standard $T = 3$ particle containing 180 subunits. Surprisingly, however, using baculovirus vectors, expression of a coat protein mutant lacking residues 2–31 resulted in assembly of a highly heterogeneous collection of particles that included small bacilliform structures, irregular structures, and wild-type-like $T = 3$ particles (Dong *et al.*, 1998). The anticipated $T = 1$ particles, on the other hand, were not observed. These results showed that the N terminus is not required for formation of $T = 3$ particles, but that it plays as essential role in inhibiting the formation of aberrant particles.

All particles assembled from the N-terminal deletion mutant contained RNA but there was a distinct size distribution, with the smallest particles containing RNAs in a range of 100–300 bases and the largest particles containing RNAs up to 3600 bases. Because packaging of RNA in the baculovirus system is not specific for FHV RNAs, it was postulated that the polymorphism was imposed by the type and size of the RNA that the coat protein selected for packaging. This hypothesis was tested by expressing the N-terminal deletion mutant in *Drosophila* cells in the presence of replicating FHV RNAs. Under these conditions specific recognition of the larger FHV RNA segment, RNA1, prevented formation of the aberrant particles, confirming the notion that selection of nonviral RNAs can significantly alter the assembly process (Marshall and Schneemann, 2001). Interestingly, the N-terminal deletion mutant was unable to recognize the smaller FHV RNA segment, RNA2 (1400 bases), indicating that the N terminus of the FHV coat protein contains important determinants for recognition and packaging of this RNA segment.

Tetraviruses form a family of nonenveloped, $T = 4$ icosahedral viruses with a single-stranded monopartite or bipartite RNA genome. The particles are assembled from 240 copies of a single type of coat protein that undergo assembly-dependent autocatalytic maturation cleavage similar to that observed in nodaviruses. When the coat protein of the tetravirus *Nudaurelia capensis* ω virus (NωV) was expressed in Sf21 cells using a baculovirus vector, VLPs spontaneously formed but, surprisingly, their structure as determined by cryo-EM and image reconstruction was distinctly different from that of native virions (Canady *et al.*, 2000). The diameter was larger and the particles were spherical in contrast to the angular appearance of native virions. In addition, the VLPs, which were purified at pH 7.5, did not show assembly-dependent maturation cleavage and they were more fragile than native virus particles. Interestingly, when the pH was dropped to pH 5, the VLPs underwent a rapid irreversible conformational transition and became structurally indistinguishable from native virions. The diameter of the particles decreased and the internal protein domain became disordered. This transition was followed by autocatalytic cleavage of the coat protein subunits. These results were unexpected because structural rearrangements at this scale had not previously been observed for other small RNA viruses. Because the particles isolated at pH 7.5 appeared to represent a previously unknown precursor particle it was designated a procapsid. This represents one of at least two cases in which a precursor particle was identified with a heterologous expression system. Another example is the procapsid of herpes simplex virus 1, which was first detected by *in vitro* assembly studies, using HSV-1 structural proteins that had been generated with recombinant baculovirus vectors (see Section V.E, below).

D. Caliciviruses

Caliciviruses are small nonenveloped icosahedral viruses that contain a single-stranded positive-sense RNA genome. Norwalk virus (NV), a member of the *Caliciviridae*, is a major cause of epidemic nonbacterial gastroenteritis. There are no cell lines currently available for production of native NV, greatly limiting vaccine development. Both baculovirus (Jiang *et al.*, 1992) and transgenic plant-based heterologous expression systems (Mason *et al.*, 1996) have been utilized to produce NV VLPs. NV expression in plants provides an illustrative example of the general approach of using transgenic plant-based systems to produce VLPs. The NV coat protein was inserted into pBI121, a binary vector for expression of genes in plants using the cauliflower mosaic virus 35S promoter to drive transcription and the nopaline synthetase terminator. The plasmid construct containing the NV coat protein gene was mobilized into *Agrobacterium tumefaciens*, a pathogen of dicotyledерous plants that has evolved a mechanism of DNA transfer from the bacterium to the plant genome. *Agrobacterium tumefaciens* harboring this plasmid was used to infect cells of tobacco and potato with subsequent regeneration of single cells into whole plants. The maximum level of NV coat protein accumulation in tobacco and potato leaves was 0.23% of the total soluble protein (Mason *et al.*, 1996). In potato tubers this corresponded to 10–20 μg of coat protein per gram of tuber weight. The expressed NV coat protein assembled into 28-nm empty VLPs within the cytoplasm of infected cells. These VLPs appeared identical to NV VLPs assembled in insect cells using baculovirus vectors. Plant-derived NV VLPs delivered orally to mice stimulated the production of humoral and mucosal antibody responses.

E. Herpes Simplex Virus 1

Herpes simplex virus 1 (HSV-1) is a large, complex virus that consists of an icosahedral capsid surrounded by a lipid envelope. An additional layer called the tegument is located between the capsid and the envelope. The internal capsid has icosahedral symmetry and can be isolated from mature virions or infected cell nuclei. Capsids are labeled A, B, or C depending on whether they are empty (A capsids), contain scaffolding proteins (B capsids), or contain the viral DNA (C capsids). Assembly of B capsids has been studied extensively using the baculovirus expression system, cell-free lysates, and purified proteins (for a review see Homa and Brown, 1997). Native B capsids contain seven structural proteins: VP5 is the major component and forms 162 capsomers (12 pentons and 150 hexons), which are arranged on a $T = 16$ icosahedral lattice (Newcomb *et al.*, 1993; Trus *et al.*, 1992). VP26 is located at the distal tip of the hexons (Booy *et al.*, 1994), whereas VP19C and VP23 form the triplexes, which hold individual capsomers together (Baker *et al.*, 1990; Newcomb *et al.*, 1993). The three remaining proteins represent the scaffolding proteins (VP21 and VP22a) and a protease (VP24), which is involved in proteolytic processing of the scaffold.

In the initial assembly studies it was shown that Sf9 cells coinfected with six recombinant baculoviruses expressing the seven capsid proteins contained particles that were similar in composition and appearance to native B capsids (Tatman *et al.*, 1994; Thomsen *et al.*, 1994). This result demonstrated that HSV-1 proteins that are not structural components of the capsid are not essential for assembly. In addition, infections in which single recombinant viruses were omitted revealed that only four of the seven proteins are required for capsid formation: VP5, VP19C, VP23, and VP22a. In the absence of the scaffolding protein, VP22a, assembly resulted in the appearance of defective structures such as incomplete capsids or spirals reminiscent of those observed during aberrant assembly of the bacteriophage P22.

Newcomb *et al.* (1994) showed that B capsids also assemble in a cell-free system, using extracts prepared from Sf9 cells infected with recombinant baculoviruses encoding the HSV-1 capsid proteins. Similar to the observations by Tatman *et al.* (1994) and Thomson *et al.* (1994), the capsids formed in the cell-free system resembled native B capsids in morphology, sedimentation rate, protein compositions, and ability to react with HSV-1-specific antibodies. Additional work by Newcomb *et al.* (1996) led to the identification of several assembly intermediates. In the cell-free system the first structures observed were partial capsids that consisted of an arclike segment of the external shell surrounding a region of scaffold. Continued incubation resulted in completion of the partial shells to closed spherical capsids, an intermediate that had not previously been observed in cells infected with native virus. Cryo-EM and image reconstruction of these spherical particles, termed "procapsids," showed that they have $T = 16$ symmetry but are more porous than mature B capsids (Trus *et al.*, 1996). On incubation at 37°C, the procapsids convert to typical polyhedral capsids. Closed spherical capsids and mature polyhedral capsids were found to differ in their stability at 2°C. While spherical capsids disassembled, mature capsids remained intact. Spherical procapsids have also been identified and isolated from HSV-1-infected cells (Newcomb *et al.*, 2000). Cryo-EM analysis confirmed that the native procapsids are structurally indistinguishable from procapsids assembled *in vitro* and that they are unstable at 0°C.

Spencer *et al.* (1998) used the cell-free system to investigate the sequence of protein interactions that leads to formation of the B capsid. They initially showed that the triplexes consist of one copy of VP19C and two copies of VP23 and that the triplexes, but not their individual protein components, bind to a complex of VP5 and scaffolding protein VP22a to form the B capsid. On the basis of this and previous results, they proposed that the major capsid protein VP5 and the scaffolding protein interact in the cytoplasm to form complexes that are transported to the nucleus. Similarly, in the cytoplasm VP19C and VP23 form triplexes that are localized to the nucleus. Assembly occurs by the interaction of the two types of protein complexes, which first form arclike partial capsids that grow continuously to form closed spherical capsid that then mature to polyhedral capsids.

Capsids can also form in the presence of VP5 and VP19C alone (Rixon *et al.*, 1996; Thomsen *et al.*, 1994). These capsids are spherical and much smaller than native B capsids. Cryo-EM reconstructions have shown that they have $T = 7$ symmetry and contain 12 pentons and 60 hexons. A characteristic feature of the hexons is their skewed appearance compared with the hexons of standard B capsids (Saad *et al.*, 1999). These data suggest that the scaffolding protein is important for controlling the size and correct closure of the assembling particle but not the protein interactions per se.

HSV-1 capsid assembly has been accomplished with purified VP5, VP19C, VP23, and a hybrid human cytomegalovirus–HSV scaffold protein (Newcomb *et al.*, 1999). The resulting procapsids were found to be similar in morphology and protein composition to procapsids formed from cell extracts containing the analogous proteins. This newly established system has been exploited to investigate the role of the scaffolding protein during assembly in more detail (Newcomb *et al.*, 2001). Using various amounts of

VP22a in the presence of constant amounts of VP5 and triplex proteins (VP19C and VP23) it was shown that the concentration of scaffolding protein plays an important role in determining capsid morphology. Above a certain threshold concentration typical procapsids having a diameter of 100 nm and $T = 16$ symmetry were observed whereas in the presence of limiting amounts of scaffolding protein smaller particles were observed, having a diameter of 78 nm. It was suggested that these capsids may have $T = 9$ symmetry. Capsids larger than the standard size or smaller than the 78-nm size were not detected. These findings indicate that the concentration of scaffolding protein influences the growing shell and that a minimum amount of scaffolding protein is required for formation of a capsid with the correct size and curvature.

F. *Bluetongue Virus*

Bluetongue virus (BTV) is a large double-stranded RNA virus of the family *Reoviridae*. It belongs to the genus *Orbivirus*, which comprises viruses that are transmitted to vertebrate hosts by certain arthropods including gnats, ticks, and mosquitoes. BTV is endemic in many parts of the world and is of economic importance to the livestock industry because of its ability to cause infection and disease in cattle and sheep. The BTV particle is composed of seven structural proteins, VP1–VP7, which are organized into two shells. VP2 and VP5 form the outer shell of the particle whereas VP3 and VP7 form the inner core, which contains the 10 dsRNA segments and small amounts of the minor proteins VP1, VP4, and VP6. Assembly of BTV particles has been studied extensively with the baculovirus expression system (for reviews see Roy, 1996b; Roy *et al.*, 1997). In fact, much of the pioneering work that established the power of the baculovirus expression system for the study of viral assembly was performed by Roy and co-workers with BTV.

Synthesis of VP3 and VP7 in insect cells using a dual recombinant baculovirus vector that contained the genes for both proteins was found to result in the formation of core-like particles (CLPs) (French and Roy, 1990). These particles were similar in size, general appearance, and stoichiometry of VP3 and VP7 to cores generated from native BTV particles. In contrast to native cores, however, the CLPs were empty and did not contain the minor proteins VP1, VP4, and VP6, which were not part of the expression system. These results showed that VP3 and VP7 are capable of forming particles in the absence of the other structural proteins and the genomic dsRNA segments. In addition, the four nonstructural proteins encoded by the virus appear to have no role in CLP formation.

Cryo-EM studies of the CLPs demonstrated that the organization of VP3 and VP7 is essentially identical to that seen in native cores (Hewat *et al.*, 1992). VP3 forms a thin 120 subunit-containing shell that has $T = 1$ icosahedral symmetry. This shell is surrounded by a layer of 260 VP7 trimers arranged on a $T = 13$ icosahedral lattice. In the CLPs, the second layer composed of VP7 tended to be incomplete, with trimers missing at the fivefold axes, a feature not observed in native cores.

Further experiments in which insect cells were coinfected with the dual recombinant vector expressing VP3 and VP7, and an additional recombinant virus expressing either VP1, VP4, or VP6, showed that VP1 and VP4 were readily incorporated into the CLPs whereas incorporation of VP6 was poor (Le Blois *et al.*, 1991; Loudon and Roy, 1991). Because VP6 is a basic protein it was proposed that it may require the presence of the viral RNAs for efficient packaging.

Additional experiments were performed to investigate the assembly properties of VP2 and VP5. Infection of insect cells with recombinant viruses that carried the gene for either VP2 or VP5 alone, or coinfection of cells with both viruses, did not give rise to particles (French *et al.*, 1990). In addition, attempts to assemble VP2 or VP5 onto CLPs failed. Only when both VP2 and VP5 were coexpressed in the presence of VP3 and VP7 were double-shelled, viruslike particles (VLPs) resembling native virions observed (French *et al.*, 1990). Inability of VP2 or VP5 to interact separately with CLPs suggested that the two proteins interact in order to form the outer shell of the virus.

An interesting technical aspect of the experiments described above was the finding that optimum assembly of VLPs was achieved when a quadruple baculovirus vector carrying the genes for all four proteins, VP2, VP3, VP5, and VP7, was used. This strategy gave rise to a virtually homogeneous population of VLPs whereas coinfection of cells with single or dual vectors, for example, one expressing VP2 and VP5, the other VP3 and VP7, gave rise to mixtures of CLPs and VLPs that contained varying amounts of the capsid proteins (Belyaev and Roy, 1993; Roy, 1996a).

VLPs were similar in structure to native BTV as determined by cryo-EM studies (Hewat *et al.*, 1994). In addition, they exhibited strong hemagglutination activity and antibodies raised against the synthetic particles were highly neutralizing (Roy, 1996a). BTV VLPs also proved to be potent immunogens in sheep and afforded protection against challenge with native virus (Roy, 1996a). On the basis of these and other results, BTV VLPs are currently being developed as a commercial vaccine. CLPs were found to be inferior to VLPs in protecting sheep from infection with native virus, a finding that was not surprising given the observation that antibodies to VP2, which is located in the outer shell, are required for protection. However, CLPs have been used for the synthesis of chimeric particles by insertion of foreign epitopes into surface-exposed regions of VP7. These foreign sequences included epitopes from various pathogens such as bovine leukemia virus, poliovirus, HIV, rabies virus, and foot-and-mouth disease virus (Roy, 1996a). Experiments to test the immunogenicity of such chimeric particles are currently underway.

G. *Respiratory Syncytial Virus*

Respiratory syncytial virus (RSV) is a member of the family *Paramyxoviridae*, which includes Sendai, mumps, and measles viruses. RSV is a major cause of lower respiratory tract illness during infancy and childhood. An effective vaccine is currently not available. The use of the coat protein of alfalfa mosaic virus (AlMV) to present an epitope derived from the glycoprotein of RSV provides an illustrative example of the general approach for using plant–viral vectors to transiently express heterologous antigens. The AlMV genome consists of three genomic RNAs and a subgenomic RNA4. Genomic RNAs 1 and 2 encode viral replicase proteins, whereas RNA3 encodes the cell-to-cell movement protein and the coat protein. The coat protein is translated from the subgenomic RNA4, which is synthesized from the genomic RNA3. AlMV virions are assembled from the 24-kDa coat protein and form particles that can vary in size (30 to 60 nm in length and 18 nm in diameter) and form (spherical, ellipsoid, or bacilliform) depending on the length of the encapsidated RNA. The N terminus of the AlMV coat protein is located on the virion surface and appears to play a critical role in virion assembly. A cDNA encoding a 24-amino acid region of the RSV glycoprotein (amino acid residues 170–191) was ligated into the 5′ end of the AlMV coat protein gene present in a full-length cDNA of RNA3 (Belanger *et al.*, 2000). *In vitro* RNA transcripts generated from the cDNA of the chimeric RNA3 were mechanically applied to the leaves of transgenic tobacco plants already expressing AlMV RNA1 and RNA2. Ten to 14 days postinoculation, recombinant virus systemically spread throughout the plant. On average, the recombinant virus accumulated to 0.75 mg/g of fresh tissue weight and could be purified to homogeneity with yields of 0.5 mg/g of fresh tissue (Belanger *et al.*, 2000). The recombinant particles stimulated production of RSV-specific antibodies in mice, which were protected from subsequent challenge with native RSV.

VI. Conclusions

The use of heterologous expression systems and cell extracts has had an enormous impact on the understanding of viral assembly and structure. The ability to separate assembly and maturation from infection has allowed the analysis of mutants that are not infectious and intermediates in particle maturation that exist only transiently in natural infections. There can be no doubt that these systems will continue to provide significant insights into viral morphogenesis in addition to forming a basis for practical applications such as the development of diagnostics and vaccines.

References

Abrams, C. C., King, A. M., and Belsham, G. J. (1995). *J. Gen. Virol.* **76,** 3089–3098.

Adams, S. E., Richardson, S. M., Kingsman, S. M., and Kingsman, A. J. (1994). *Mol. Biotechnol.* **1,** 125–135.

Baker, T. S., Newcomb, W. W., Booy, F. P., Brown, J. C., and Steven, A. C. (1990). *J. Virol.* **64,** 563–573.

Baumert, T. F., Ito, S., Wong, D. T., and Liang, T. J. (1998). *J. Virol.* **72,** 3827–3836.

Baumert, T. F., Vergalla, J., Satoi, J., Thomson, M., Lechmann, M., Herion, D., Greenberg, H. B., Ito, S., and Liang, T. J. (1999). *Gastroenterology* **117,** 1397–1407.

Belanger, H., Fleysh, N., Cox, S., Bartman, G., Deka, D., Trudel, M., Koprowski, H., and Yusibov, V. (2000). *FASEB J.* **14,** 2323–2328.

Belyaev, A. S., and Roy, P. (1993). *Nucleic Acids Res.* **21,** 1219–1223.

Betenbaugh, M., Yu, M., Kuehl, K., White, J., Pennock, D., Spik, K., and Schmaljohn, C. (1995). *Virus Res.* **38,** 111–124.

Bitter, G. A., Egan, K. M., Burnette, W. N., Samal, B., Fieschko, J. C., Peterson, D. L., Downing, M. R., Wypych, J., and Langley, K. E. (1988). *J. Med. Virol.* **25,** 123–140.

Booy, F. P., Trus, B. L., Newcomb, W. W., Brown, J. C., Conway, J. F., and Steven, A. C. (1994). *Proc. Natl. Acad. Sci. USA* **91,** 5652–5656.

Breitburd, F., Kirnbauer, R., Hubbert, N. L., Nonnenmacher, B., Trin-Dinh-Desmarquet, C., Orth, G., Schiller, J. T., and Lowy, D. R. (1995). *J. Virol.* **69,** 3959–3963.

Brennan, F. R., Jones, T. D., and Hamilton, W. D. (2001). *Mol. Biotechnol.* **17,** 15–26.

Brown, C. S., Van Lent, J. W., Vlak, J. M., and Spaan, W. J. (1991). *J. Virol.* **65,** 2702–2706.

Campbell, S., and Rein, A. (1999). *J. Virol.* **73,** 2270–2279.

Canady, M. A., Tihova, M., Hanzlik, T. N., Johnson, J. E., and Yeager, M. (2000). *J. Mol. Biol.* **299,** 573–584.

Carroll, M. W., and Moss, B. (1997). *Curr. Opin. Biotechnol.* **8,** 573–577.

Chandran, K., Walker, S. B., Chen, Y., Contreras, C. M., Schiff, L. A., Baker, T. S., and Nibert, M. L. (1999). *J. Virol.* **73,** 3941–3950.

Chen, X. S., Garcea, R. L., Goldberg, I., Casini, G., and Harrison, S. C. (2000). *Mol. Cell* **5,** 557–567.

Chen, X. S., Casini, G., Harrison, S. C., and Garcea, R. L. (2001). *J. Mol. Biol.* **307,** 173–182.

Chiu, W., Burnett, R. M., and Garcea, R. L., Eds. (1997). "Structural Biology of Viruses." Oxford University Press, Oxford.

Conway, J. F., Wikoff, W. R., Cheng, N., Duda, R. L., Hendrix, R. W., Johnson, J. E., and Steven, A. C. (2001). *Science* **292,** 744–748.

Cregg, J. M., Cereghino, J. L., Shi, J., and Higgins, D. R. (2000). *Mol. Biotechnol.* **16,** 23–52.

Dalsgaard, K., Uttenthal, A., Jones, T. D., Xu, F., Merryweather, A., Hamilton, W. D., Langeveld, J. P., Boshuizen, R. S., Kamstrup, S., Lomonossoff, G. P., Porta, C., Vela, C., Casal, J. I., Meloen, R. H., and Rodgers, P. B. (1997). *Nat. Biotechnol.* **15,** 248–252.

Dolja, V. V., McBride, H. J., and Carrington, J. C. (1992). *Proc. Natl. Acad. Sci. USA* **89,** 10208–10212.

Dong, X. F., Natarajan, P., Tihova, M., Johnson, J. E., and Schneemann, A. (1998). *J. Virol.* **72,** 6024–6033.

Donson, J., Kearney, C. M., Hilf, M. E., and Dawson, W. O. (1991). *Proc. Natl. Acad. Sci. USA* **15,** 7204–7208.

Ehrlich, L. S., Agresta, B. E., and Carter, C. A. (1992). *J. Virol.* **66,** 4874–4883.

Falcon, V., Garcia, C., de la Rosa, M. C., Menendez, I., Seoane, J., and Grillo, J. M. (1999). *Tissue Cell* **31,** 117–125.

Fitchen, J., Beachy, R. N., and Hein, M. B. (1995). *Vaccine* **13,** 1051–1057.

Fox, J. M., Wang, G., Speir, J. A., Olson, N. H., Johnson, J. E., Baker, T. S., and Young, M. J. (1998). *Virology* **244,** 212–218.

French, T. J., and Roy, P. (1990). *J. Virol.* **64,** 1530–1536.

French, T. J., Marshall, J. J., and Roy, P. (1990). *J. Virol.* **64,** 5695–5700.

Ganser, B. K., Li, S., Klishko, V. Y., Finch, J. T., and Sundquist, W. I. (1999). *Science* **283,** 80–83.

Garcea, R. L., Salunke, D. M., and Caspar, D. L. (1987). *Nature* **329,** 86–87.

Gheysen, D., Jacobs, E., de Foresta, F., Thiriart, C., Francotte, M., Thines, D., and De Wilde, M. (1989). *Cell* **59,** 103–112.

Giga-Hama, Y., and Kumagai, H. (1999). *Biotechnol. Appl. Biochem.* **30,** 235–244.

Gopinath, K., Wellink, J., Porta, C., Taylor, K. M., Lomonossoff, G. P., and van Kammen, A. (2000). *Virology* **267,** 159–173.

Grimes, J. M., Burroughs, J. N., Gouet, P., Diprose, J. M., Malby, R., Zientara, S., Mertens, P. P., and Stuart, D. I. (1998). *Nature* **395,** 470–478.

Gross, I., Hohenberg, H., Wilk, T., Wiegers, K., Grattinger, M., Muller, B., Fuller, S., and Krausslich, H. G. (2000). *EMBO J.* **19,** 103–113.

Hagensee, M. E., Yaegashi, N., and Galloway, D. A. (1993). *J. Virol.* **67,** 315–322.

Hagensee, M. E., Olson, N. H., Baker, T. S., and Galloway, D. A. (1994). *J. Virol.* **68,** 4503–4505.

Hamamoto, H., Sugiyama, Y., Nakagawa, N., Hashida, E., Matsunaga, Y., Takemoto, S., Watanabe, Y., and Okada, Y. (1993). *Biotechnology* **11,** 930–932.

Hewat, E. A., Booth, T. F., Loudon, P. T., and Roy, P. (1992). *Virology* **189,** 10–20.

Hewat, E. A., Booth, T. F., and Roy, P. (1994). *J. Struct. Biol.* **112,** 183–191.

Hofmann, K. J., Cook, J. C., Joyce, J. G., Brown, D. R., Schultz, L. D., George, H. A., Rosolowsky, M., Fife, K. H., and Jansen, K. U. (1995). *Virology* **209,** 506–518.

Hollenberg, C. P., and Gellissen, G. (1997). *Curr. Opin. Biotechnol.* **8,** 554–560.

Homa, F. L., and Brown, J. C. (1997). *Rev. Med. Virol.* **7,** 107–122.

Iyengar, S., Shah, K. V., Kotloff, K. L., Ghim, S. J., and Viscidi, R. P. (1996). *Clin. Diagn. Lab. Immunol.* **3,** 733–739.

Jiang, X., Wang, M., Graham, D. Y., and Estes, M. K. (1992). *J. Virol.* **66,** 6527–6532.

Joshi, S. M., and Vogt, V. M. (2000). *J. Virol.* **74,** 10260–10268.

Kapusta, J., Modelska, A., Figlerowicz, M., Pniewski, T., Letellier, M., Lisowa, O., Yusibov, V., Koprowski, H., Plucienniczak, A., and Legocki, A. B. (1999). *FASEB J.* **13,** 1796–1799.
Karacostas, V., Nagashima, K., Gonda, M. A., and Moss, B. (1989). *Proc. Natl. Acad. Sci. USA* **86,** 8964–8967.
Kirnbauer, R., Booy, F., Cheng, N., Lowy, D. R., and Schiller, J. T. (1992). *Proc. Natl. Acad. Sci. USA* **89,** 12180–12184.
Kirnbauer, R., Taub, J., Greenstone, H., Roden, R., Durst, M., Gissmann, L., Lowy, D. R., and Schiller, J. T. (1993). *J. Virol.* **67,** 6929–6936.
Koprowski, H., and Yusibov, V. (2001). *Vaccine* **19,** 2735–2741.
Kost, T. A., and Condreay, J. P. (1999). *Curr. Opin. Biotechnol.* **10,** 428–433.
Kumagai, M. H., Turpen, T. H., Weinzett, L. N., della-Cioppa, G., Turpen, A. M., Donson, J., Hilf, M. E., Grantham, G. L., Dawson, W. O., and Chow, T. P., et al.(1993). *Proc. Natl. Acad. Sci. USA* **90,** 427–430.
Kunkel, M., Lorinczi, M., Rijnbrand, R., Lemon, S. M., and Watowich, S. J. (2001). *J. Virol.* **75,** 2119–2129.
Lawton, J. A., Estes, M. K., and Prasad, B. V. (1997). *Nat. Struct. Biol.* **4,** 118–121.
Le Blois, H., Fayard, B., Urakawa, T., and Roy, P. (1991). *J. Virol.* **65,** 4821–4831.
Li, M., Delos, S. E., Montross, L., and Garcea, R. L. (1995). *Proc. Natl. Acad. Sci. USA* **92,** 5992–5996.
Li, M., Cripe, T. P., Estes, P. A., Lyon, M. K., Rose, R. C., and Garcea, R. L. (1997). *J. Virol.* **71,** 2988–2995.
Li, S., Hill, C. P., Sundquist, W. I., and Finch, J. T. (2000). *Nature* **407,** 409–413.
Lorenzo, L. J., Duenas-Carrera, S., Falcon, V., Acosta-Rivero, N., Gonzalez, E., de la Rosa, M. C., Menendez, I., and Morales, J. (2001). *Biochem. Biophys. Res. Commun.* **281,** 962–965.
Loudon, P. T., and Roy, P. (1991). *Virology* **180,** 798–802.
Luckow, V. A., Lee, S. C., Barry, G. F., and Olins, P. O. (1993). *J. Virol.* **67,** 4566–4579.
Maillard, P., Krawczynski, K., Nitkiewicz, J., Bronnert, C., Sidorkiewicz, M., Gounon, P., Dubuisson, J., Faure, G., Crainic, R., and Budkowska, A. (2001). *J. Virol.* **75,** 8240–8250.
Marshall, D., and Schneemann, A. (2001). *Virology* **285,** 165–175.
Mason, H. S., Lam, D. M., and Arntzen, C. J. (1992). *Proc. Natl. Acad. Sci. USA* **89,** 11745–11749.
Mason, H. S., Ball, J. M., Shi, J. J., Jiang, X., Estes, M. K., and Arntzen, C. J. (1996). *Proc. Natl. Acad. Sci. USA* **93,** 5335–5340.
McGarvey, P. B., Hammond, J., Dienelt, M. M., Hooper, D. C., Fu, Z. F., Dietzschold, B., Koprowski, H., and Michaels, F. H. (1995). *Biotechnology* **13,** 1484–1487.
McLain, L., Porta, C., Lomonossoff, G. P., Durrani, Z., and Dimmock, N. J. (1995). *AIDS Res. Hum. Retroviruses.* **11,** 327–334.
Miner, J. N., and Hruby, D. E. (1990). *Trends Biotechnol.* **8,** 20–25.
Mizuno, M., Yamada, G., Tanaka, T., Shimotohno, K., Takatani, M., and Tsuji, T. (1995). *Gastroenterology* **109,** 1933–1940.
Modelska, A., Dietzschold, B., Sleysh, N., Fu, Z. F., Steplewski, K., Hooper, D. C., Koprowski, H., and Yusibov, V. (1998). *Proc. Natl. Acad. Sci. USA* **95,** 2481–2485.
Montross, L., Watkins, S., Moreland, R. B., Mamon, H., Caspar, D. L., and Garcea, R. L. (1991). *J. Virol.* **65,** 4991–4998.
Moss, B. (1992). *Curr. Opin. Biotechnol.* **3,** 518–522.
Muller, M., Gissmann, L., Cristiano, R. J., Sun, X. Y., Frazer, I. H., Jenson, A. B., Alonso, A., Zentgraf, H., and Zhou, J. (1995). *J. Virol.* **69,** 948–954.

Newcomb, W. W., Trus, B. L., Booy, F. P., Steven, A. C., Wall, J. S., and Brown, J. C. (1993). *J. Mol. Biol.* **232,** 499–511.
Newcomb, W. W., Homa, F. L., Thomsen, D. R., Ye, Z., and Brown, J. C. (1994). *J. Virol.* **68,** 6059–6063.
Newcomb, W. W., Homa, F. L., Thomsen, D. R., Booy, F. P., Trus, B. L., Steven, A. C., Spencer, J. V., and Brown, J. C. (1996). *J. Mol. Biol.* **263,** 432–446.
Newcomb, W. W., Homa, F. L., Thomsen, D. R., Trus, B. L., Cheng, N., Steven, A., Booy, F., and Brown, J. C. (1999). *J. Virol.* **73,** 4239–4250.
Newcomb, W. W., Trus, B. L., Cheng, N., Steven, A. C., Sheaffer, A. K., Tenney, D. J., Weller, S. K., and Brown, J. C. (2000). *J. Virol.* **74,** 1663–1673.
Newcomb, W. W., Homa, F. L., Thomsen, D. R., and Brown, J. C. (2001). *J. Struct. Biol.* **133,** 23–31.
Peeters, K., De Wilde, C., De Jaeger, G., Angenon, G., and Depicker, A. (2001). *Vaccine* **19,** 2756–2761.
Porta, C., and Lomonossoff, G. P. (1996). *Mol. Biotechnol.* **5,** 209–221.
Possee, R. (1997). *Curr. Opin. Biotechnol.* **8,** 569–572.
Prasad, B. V., Hardy, M. E., Dokland, T., Bella, J., Rossmann, M. G., and Estes, M. K. (1999). *Science* **286,** 287–290.
Preikschat, P., Kazaks, A., Dishlers, A., Pumpens, P., Kruger, D. H., and Meisel, H. (2000). *FEBS Lett.* **478,** 127–132.
Reinisch, K. M., Nibert, M. L., and Harrison, S. C. (2000). *Nature* **404,** 960–967.
Rixon, F. J., Addison, C., McGregor, A., Macnab, S. J., Nicholson, P., Preston, V. G., and Tatman, J. D. (1996). *J. Gen. Virol.* **77,** 2251–2260.
Roden, R. B., Kirnbauer, R., Jenson, A. B., Lowy, D. R., and Schiller, J. T. (1994). *J. Virol.* **68,** 7260–7266.
Rose, R. C., Bonnez, W., Reichman, R. C., and Garcea, R. L. (1993). *J. Virol.* **67,** 1936–1944.
Roy, P. (1996a). *InterVirology* **39,** 62–71.
Roy, P. (1996b). *Virology* **216,** 1–11.
Roy, P., Mikhailov, M., and Bishop, D. H. (1997). *Gene* **190,** 119–129.
Saad, A., Zhou, Z. H., Jakana, J., Chiu, W., and Rixon, F. J. (1999). *J. Virol.* **73,** 6821–6830.
Salunke, D. M., Caspar, D. L. D., and Garcea, R. L. (1986). *Cell* **46,** 895–904.
Salunke, D. M., Caspar, D. L., and Garcea, R. L. (1989). *Biophys. J.* **56,** 887–900.
Sapp, M., Volpers, C., and Streeck, R. E. (1996). *InterVirology* **39,** 49–53.
Sasagawa, T., Pushko, P., Steers, G., Gschmeissner, S. E., Hajibagheri, M. A., Finch, J., Crawford, L., and Tommasino, M. (1995). *Virology* **206,** 126–135.
Sasnauskas, K., Buzaite, O., Vogel, F., Jandrig, B., Razanskas, R., Staniulis, J., Scherneck, S., Kruger, D. H., and Ulrich, R. (1999). *Biol. Chem.* **380,** 381–386.
Schenk, P. M., Baumann, S., Mattes, R., and Steinbiss, H. H. (1995). *Biotechniques* **19,** 196–198.
Schiller, J. T., and Hidesheim, A. (2000). *J. Clin. Virol.* **19,** 67–74.
Schneemann, A., Dasgupta, R., Johnson, J. E., and Rueckert, R. R. (1993). *J. Virol.* **67,** 2756–2763.
Schneemann, A., Gallagher, T. M., and Rueckert, R. R. (1994). *J. Virol.* **68,** 4547–4556.
Scholthof, H. B., Scholthof, K. B., and Jackson, A. O. (1996). *Annu. Rev. Phytopathol.* **34,** 299–323.
Spearman, P., and Ratner, L. (1996). *J. Virol.* **70,** 8187–8194.
Spencer, J. V., Newcomb, W. W., Thomsen, D. R., Homa, F. L., and Brown, J. C. (1998). *J. Virol.* **72,** 3944–3951.

Stauffer, Y., Raj, K., Masternak, K., and Beard, P. (1998). *J. Mol. Biol.* **283,** 529–536.
Steven, A. C., Trus, B. L., Booy, F. P., Cheng, N., Zlotnick, A., Caston, J. R., and Conway, J. F. (1997). *FASEB J.* **11,** 733–742.
Suzich, J. A., Ghim, S. J., Palmer-Hill, F. J., White, W. I., Tamura, J. K., Bell, J. A., Newsome, J. A., Jenson, A. B., and Schlegel, R. (1995). *Proc. Natl. Acad. Sci. USA* **92,** 11553–11557.
Tatman, J. D., Preston, V. G., Nicholson, P., Elliott, R. M., and Rixon, F. J. (1994). *J. Gen. Virol.* **75,** 1101–1113.
Tellinghuisen, T. L., Hamburger, A. E., Fisher, B. R., Ostendorp, R., and Kuhn, R. J. (1999). *J. Virol.* **73,** 5309–5319.
Thomsen, D. R., Roof, L. L., and Homa, F. L. (1994). *J. Virol.* **69,** 2442–2457.
Trueman, L. (1995). *Methods Mol. Biol.* **49,** 341–354.
Trus, B. L., Newcomb, W. W., Booy, F. P., Brown, J. C., and Steven, A. C. (1992). *Proc. Natl. Acad. Sci. USA* **89,** 11508–11512.
Trus, B. L., Booy, F. P., Newcomb, W. W., Brown, J. C., Homa, F. L., Thomsen, D. R., and Steven, A. C. (1996). *J. Mol. Biol.* **263,** 447–462.
Unckell, F., Streeck, R. E., and Sapp, M. (1997). *J. Virol.* **71,** 2934–2939.
Vassileva, A., Chugh, D. A., Swaminathan, S., and Khanna, N. (2000a). *Protein Expr. Purif.* **21,** 71–80.
Vassileva, A., Chugh, D. A., Swaminathan, S., and Khanna, N. (2000b). *J. Biotechnol.* **88,** 21–35.
Volpers, C., Schirmacher, P., Streeck, R. E., and Sapp, M. (1994). *Virology* **200,** 504–512.
Volpers, C., Unckell, F., Schirmacher, P., Streeck, R. E., and Sapp, M. (1995). *J. Virol.* **69,** 3258–3264.
Wengler, G., Boege, U., Wengler, G., Bischoff, H., and Wahn, K. (1982). *Virology* **118,** 401–410.
Wengler, G., Wengler, G., Boege, U., and Wahn, K. (1984). *Virology* **132,** 401–412.
Wikoff, W. R., Liljas, L., Duda, R. L., Tsuruta, H., Hendrix, R. W., and Johnson, J. E. (2000). *Science* **289,** 2129–2133.
Wynne, S. A., Crowther, R. A., and Leslie, A. G. (1999). *Mol. Cell* **3,** 771–780.
Yu, F., Joshi, S. M., Ma, Y. M., Kingston, R. L., Simon, M. N., and Vogt, V. M. (2001). *J. Virol.* **75,** 2753–2764.
Yusibov, V., Kumar, A., North, A., Johnson, J. E., and Loesch-Fries, L. S. (1996). *J. Gen. Virol.* **77,** 567–573.
Yusibov, V., Modelska, A., Steplewski, K., Agadjanyan, M., Weiner, D., Hooper, D. C., and Koprowski, H. (1997). *Proc. Natl. Acad. Sci. USA* **94,** 5784–5788.
Zhang, W., Carmichael, J., Ferguson, J., Inglis, S., Ashrafian, H., and Stanley, M. (1998). *Virology* **243,** 423–431.
Zhao, X., Fox, J. M., Olson, N. H., Baker, T. S., and Young, M. J. (1995). *Virology* **207,** 486–494.
Zhou, J., Sun, X. Y., Stenzel, D. J., and Frazer, I. H. (1991). *Virology* **185,** 251–257.
Zhou, J., Stenzel, D. J., Sun, X. Y., and Frazer, I. H. (1993). *J. Gen. Virol.* **74,** 763–768.

HYBRID VIGOR: HYBRID METHODS IN VIRAL STRUCTURE DETERMINATION

By ROBERT J. C. GILBERT, JONATHAN M. GRIMES, AND DAVID I. STUART

Division of Structural Biology, Nuffield Department of Medicine,
University of Oxford, Oxford OX3 7BN, United Kingdom

I. Introduction 37
II. Techniques 38
A. The Toolkit 38
B. Common Combinations 54
III. Hybrids 56
A. Structural Transformations in the Maturation of a DNA Bacteriophage 56
B. Structure and Dynamics of Bacteriophage ϕ29 58
C. Structures of Membrane-Containing Isometric Viruses: Fusion Machines... 60
D. Difference Imaging and an Internal Membrane 64
E. Structures of Large Viruses: Transcription Machines 67
F. Structural Analyses of Human Immunodeficiency Virus 74
G. Structural Insights from Virus Complexes 78
IV. Conclusion 83
References 83

I. Introduction

Viruses are microcosms of the broader world of cell biology, providing paradigms for the analysis of cellular mechanics arising from protein–protein, protein–nucleic acid, and protein–lipid interaction. In particular, while many cellular processes (e.g., the immunological synapse, nuclear chromatin, and signaling through the cell membrane) remain refractory to structural analysis, some viral models for them are accessible. Viruses are generally formed by the ordered assembly of many copies of one or more proteins. The virus must be robust enough to transmit its genome between hosts, and yet must release nucleic acid in the host to direct the production of progeny virus. This is frequently achieved by the disassembly of the virus particle during the process of host entry. These aspects of virus structure mean that in some cases it is possible to form stable assembly or disassembly intermediates *in vitro*; however, these processes are often profoundly dynamic and hence demand the use of a variety of techniques to tease out structures for representative stages of the life cycle.

We focus mainly on crystallographic and single-particle reconstruction methods, indicate how such methods may be combined to provide routes

0065-3233/03 $35.00

to structure determination, and, finally, provide examples indicating the power of hybrid approaches.

II. Techniques

A. *The Toolkit*

Single-crystal structures of intact viruses have to date been of icosahedrally symmetric particles, which possess 60-fold symmetry. One result of this high symmetry is that there is substantial oversampling of the virion transform so that, despite the weak diffraction from these large assemblies, icosahedral averaging often results in excellent quality electron density maps. Cryoelectron microscopy (cryo-EM) has likewise been used to investigate icosahedral viruses, but has also been successfully applied to pleiomorphic viruses and the analysis of the icosahedral enveloped viruses, which have, for unknown reasons, not yet yielded to crystallographic analysis (Lata *et al.*, 2000; Mancini *et al.*, 2000; Pletnev *et al.*, 2001; Tao *et al.*, 1998). Such cryo-EM analyses tend to be at lower resolution and have frequently been interpreted in terms of the known crystallographic structures of component proteins (Simpson *et al.*, 2000; Tao *et al.*, 1998; Zhang *et al.*, 2001).

The paradigm for the construction of icosahedral viral particles is that they are assembled from 60 chemically and conformationally identical copies of a building block, the icosahedral asymmetric unit (Caspar and Klug, 1962). These 60 copies are related to one another by the set of icosahedral symmetry axes (Crick and Watson, 1956). In practice, most viruses contain more than 60 protein subunits and thus contain more than 1 subunit in the icosahedral asymmetric unit. This is usually a necessity if the virus is to enclose sufficient volume to house its genome. A way in which several copies of the same protein can occupy similar but nonidentical environments in this manner was described by Casper and Klug, who proposed the concept of quasi-equivalence (Caspar and Klug, 1962). In this model the triangular icosahedral asymmetric unit is further divided into *n* subtriangles, *n* being denoted the triangulation number. Although the underlying chemical justification of quasi-equivalence was naive, the theory has had a useful role in interpreting low-resolution structural results. Some viruses have asymmetric units formed from more than one kind of protein—in this case the virus may be denoted pseudo-T*n*, where *n* would be the triangulation number if the proteins were identical (Acharya *et al.*, 1989; Rossmann *et al.*, 1985).

1. X-Ray Crystallography

The first X-ray crystallographic structures of small plant viruses were solved in the late 1970s for tomato bushy stunt virus (TBSV) (Winkler *et al.*, 1977) and southern bean mosaic virus (SBMV) (Suck *et al.*, 1978). Since then structures of tens of different viruses have been solved, progressively achieving results for more complex and larger systems (Johnson, 1999). The successful application of synchrotron radiation for biological structure research began at about the time the first viral crystal structures were being solved. The use of the bright X-ray sources provided by second- and now third-generation synchrotrons has been invaluable in obtaining data of sufficient intensity that viral structures can be solved at resolutions approaching the atomic level (Rossmann, 1999), being first applied to human rhinovirus (Rossmann *et al.*, 1985) and subsequently to an increasing number of viruses. In particular, the largest and/or most difficult structures would have been entirely inaccessible in the absence of synchrotron radiation (Acharya *et al.*, 1989; Grimes *et al.*, 1998; Liddington *et al.*, 1991; Reinisch *et al.*, 2000; Wikoff *et al.*, 2000). The large unit cells that viruses inhabit within crystals mean that there are potentially millions of weak reflections to be recorded from a crystal; bright tunable synchrotron X-rays increase the signal intensity, reduce data collection times, allow use of X-rays with shorter and relatively less damaging wavelengths, and give rise to lower background signals (Acharya *et al.*, 1989; Rossmann, 1999). Although a good range of different viral architectures have been probed crystallographically, a large proportion of the total number of viral crystal structures are of a rather limited group of viruses; for example, many mutants and other modified forms of rhinovirus have been solved, and a good selection of foot-and-mouth disease virus structures also exists. The largest virus structure solved to date is the bluetongue virus core (Grimes *et al.*, 1998), which has been followed by structures of similar-sized related viruses such as the orthoreovirus core (Reinisch *et al.*, 2000). The field of viral crystallography is now set to address the challenges of nonicosahedral viruses and membrane-containing viruses (Cheng *et al.*, 1995; Mancini *et al.*, 2000; Ravantti and Bamford, 1999; Rydman *et al.*, 1999; Simpson *et al.*, 2000; Tao *et al.*, 1998).

The fundamental principles and routes involved in the crystallographic analysis of a virus particle are no different from those of protein crystallography. The atoms in the virus particle will scatter X-rays according to the number of electrons they contain. When millions of identicial viruses are lined up in a three-dimensional crystal they will all scatter X-rays and along certain directions constructive interference occurs, which results in a build-up of X-ray signal as a Bragg reflection or diffraction spot. These diffraction spots lie on the so-called reciprocal lattice, the Fourier transform of the crystal lattice. Each diffracted X-ray is described by an amplitude and phase, which correspond to the Fourier transform of the virus sampled at that point. To reconstruct the electron density cloud that defines the atom positions, the diffracted amplitudes and phases are simply subjected to Fourier transformation. Unfortunately, only the amplitude of the X-ray diffraction spot can be recorded, and all phase information is lost. This is known as the "phase problem" and the reconstitution of phase information is crucial to the analysis. The basic route to solving a crystal structure is (1) growth of single crystals that diffract to suitable resolution, (2) collection of diffraction intensities from single crystals to give a data set of structure factor amplitudes, (3) initial phase determination, and (4) phase refinement in real and reciprocal space to allow the generation of an atomic model of the virus, with minimal errors.

Although there is no fundamental difference between structure determination of a protein and that of a virus, there are clear differences in methodology and experimental approach that require a significant scaling up of resources and effort for virus work. Thus there may be a thousand times more reflections to measure in a virus structure determination than for a small protein, and each reflection will be correspondingly weaker. The weakness of each reflection, and the high density and large number of reflections, place demands not only on the X-ray sources and detectors but on computer hardware and software. We present here only a synopsis of some more recent advances: for a more thorough analysis see, for instance, Fry *et al.* (1999a).

Several technological advances have conspired to render the crystallographic structure determination of viruses, in effect, routine. A key advance has been the availability of synchrotron radiation sources to crystallographers. These facilities provide highly intense, and now with third-generation sources, close-to-parallel X-ray beams, allowing the collection of densely-packed but weak diffraction data. Examples of third-generation sources include the European Synchrotron Radiation Facility (ESRF) in France, the Advanced Photon Source (APS) in the United States, and Spring8 in Japan. An important component of the development of synchrotrons has been the use of so-called undulators, to provide highly brilliant, tuneable, and parallel X-ray beams.

Alongside improvements in synchrotron sources, X-ray detectors have improved enormously. Up to the early 1990s X-ray film was the detector of choice at synchrotrons because of its high spatial resolution, despite its poor efficiency and low dynamic range. Imaging plate and charge coupled device (CCD) detectors swept film aside in the early-to-mid 1990s.

Although CCDs have now superseded imaging plate devices because of their lower point spread function, higher spatial resolution and increased readout speed, both devices have been used successfully in virus structure solution. Both have a large dynamic range and a high-detective quantum efficiency for X-ray wavelengths of about 1 Å, which means that shorter exposure times are required, allowing more data to be gathered before radiation damage becomes obtrusive. Current detectors have good X-ray sensitivity and spatial resolution with duty cycles that match well the exposure times needed for the collection of X-ray data. However, important developments are ongoing, such as the development of solid-state amorphous selenium detectors (Zhao *et al.*, 2001). These detectors promise a reduction in the point-spread function to close to zero, and provide a large active area. Preliminary experiments have demonstrated the enormous potential of this combination for virus crystallography [Marresearch (Norderstedt, Germany), Jules Hendrix, personal communication]; however, there are still problems to be solved before the signal to noise is competitive with CCDs.

Paralleling these technical developments, there have been vast improvements in computing hardware, which enable large amounts of virus data to be analyzed rapidly. With more automated methods for data processing it is in principle possible computationally to keep pace with the rate of production of diffraction data. A number of software packages are now available for data processing, such as MOSFLM (Leslie, 1992) and DENZO (Otwinowski and Minor, 1997), which we have used routinely to process data from unit cells in excess of 1000 Å. The critical step in the processing of X-ray diffraction images is finding the precise orientation of the crystallographic axes with respect to the goniostat. Current auto-indexing algorithms make this routine, with the single caveat that when indexing dense patterns of spots it is vital to know the position of the direct X-ray beam on the detector to within 1 or 2 pixels. The system may be conveniently calibrated with powder rings generated by cytoplasmic polyhedrosis virus (CPV) polyhedra (G. Sutton, J. M. Diprose, J. M. Grimes, and D. I. Stuart, unpublished results).

Whereas protein crystals are routinely cooled to liquid nitrogen temperatures, cryogenic cooling of virus crystals is uncommon. A major problem is that the mosaic spread of crystals tends to increase on cooling, frequently resulting in overlap of the Bragg reflections. An implication of room temperature data collection is that a large number of crystals are often required to collect sufficient data to solve a virus structure. Thus for challenging projects such as bluetongue virus (BTV) (Grimes *et al.*, 1998) more than 1000 crystals may be needed. In addition, for some viruses disease security protocols require that the virus crystals be premounted in capillaries, ruling out the more popular methods for crystal cooling. However, crystals may be successfully mounted in thin tapered quartz capillaries (Luger *et al.*, 1997). Using this technology, together with an annealing protocol, a complete high-resolution data set has been collected from a frozen crystal of foot-and-mouth disease virus, without a large increase in mosaic spread (E. Fry and D. I. Stuart, unpublished results). The cooling of crystals with large unit cells needs more careful examination and a routine solution to this problem would greatly facilitate the process of structure determination.

Once diffraction data have been gathered the next stage in the structure analysis is usually focused on seeking a definition of the orientation of the particle and, often, its position in the crystal cell. This information is essential for both phase refinement and the analysis of isomorphous replacement experiments. Because of the inevitable noncrystallographic redundancy (a minimum of 5-fold) and the fixed relationship between the various icosahedral symmetry axes it is often possible to solve the orientation problem by analysis of the diffraction data in the absence of a model, usually by use of a self-rotation function.

Phases need to be derived for each of the measured reflections before electron density maps can be calculated. Estimates can be obtained by the traditional method of isomorphous replacement or by molecular replacement with a model structure, perhaps derived by cryo-EM. The noncrystallographic symmetry provides powerful phase constraints. In the case of isomorphous replacement studies the noncrystallographic constraints available mean that only a few percent of the data may be required to define the heavy atom positions, solve the orientation and position of the particle, and generate initial phases, whereas in the case of a cryo-EM starting point even a poor model can be sufficient to get started. Even poor phase estimates can be improved dramatically by the cyclic imposition of icosahedral symmetry. The procedure is relatively straightforward. The noncrystallographically related portions of the electron density map of the virus are averaged, and the regions not defined as virus are flattened to an average electron density value. This map is then back-transformed to generate calculated amplitudes and phases. The calculated structure factors are then recombined, suitably weighted, with the observed amplitudes and a new map is calculated. The procedure is then iterated to convergence.

At this stage, assuming that the nominal resolution is 4 Å or better, the electron density map should be of sufficient quality to interpret readily the course of the polypeptide chain and rapidly build a model, usually for the icosahederal asymmetric unit. Conventional crystallographic refinement techniques are then employed to refine the model against the observed data.

2. *Cryoelectron Microscopy*

i. The Method. Cryo-EM has developed steadily to such a point that it has several times attained resolutions of better than 10 Å for viral samples: for example, 9 Å for Semliki Forest virus (Mancini *et al.*, 2000), 8.5 Å for herpes simplex virus (Zhou *et al.*, 2000), 7.4 Å for hepatitis B virus (Bottcher *et al.*, 1997), 6.8 Å for rice dwarf virus (Zhou *et al.*, 2001), and, finally, more than 20 years after crystallography reached this resolution, 5.9 Å for TBSV (Heel *et al.*, 2000). Cryo-EM requires the collection of thousands of images of a structure, each of which is assumed to represent a projection of its three-dimensional electron density. If the orientation of each projected particle (which for a virus is often essentially random) can be determined, Fourier methods permit the facile reconstruction of the three-dimensional structure. As well as viruses, single-particle cryo-EM has been applied to structural studies of a wide range of macromolecular complexes, including ribosomes (Gabashvili *et al.*, 2000; Stark *et al.*, 2000), pore-forming toxins (Gilbert *et al.*, 1999; Orlova *et al.*, 2000), and chaperones (Schoehn *et al.*, 2000).

In cryo-EM a small amount of solution containing the macromolecular complex under study is placed on an electron microscope grid. The grid is blotted to leave only a thin film of sample and plunged rapidly into a bath of liquid ethane cooled by liquid nitrogen. The sample is thereby cooled at a rate of $>10^4$ degrees/s, which is more rapidly than ice crystals can form, leading to its vitrification. Fixed in a layer ($\sim$1000 Å thick) of solid amorphous water, the sample can be loaded into an electron microscope operating at liquid nitrogen or liquid helium temperatures and visualized by collecting images formed from electrons, commonly emitted from a field emission gun, passing through the electron optics of the microscope. The Fourier transform of these images of the translucent three-dimensional object is assumed to be a central section of its three-dimensional Fourier transform, perpendicular to the direction of the electron beam. Consequently, determining the orientation of the particle relative to the electron beam for a series of such images allows the central sections to be added in Fourier space; back-transformation then gives a three-dimensional map of the electron density for the particle. Another important observation is that any two projections of a three-dimensional object share a common line. Virus structure represents a special case in reconstruction of cryo-EM images because icosahedral symmetry yields 37 pairs of common lines present in every projection, which assist the determination of the particle orientation (Fuller *et al.*, 1996). Overall, the prior knowledge of symmetry constraints not only facilitates the process of image reconstruction but also, as for crystallography, dramatically enhances the quality of the reconstruction.

The projections obtained practically by imaging a macromolecular complex such as a virus in the electron microscope are unfortunately not simply projections of its Coulomb potential. In practice, the sample is embedded in ice and the contrast of the specimen relies on the difference in scattering power between the atoms in the vitreous water surrounding the complex and those in the proteins, lipids, and nucleic acids of which the specimen is composed and is further reduced by the finite ice thickness and the thermal properties of electrons. This amplitude contrast is therefore weak and some phase contrast is added by focusing the electron beam in a plane above the specimen (overfocus) or beneath it (underfocus). An unwanted corollary of this is that the information in the projected image is convoluted with the *contrast transfer function* (CTF) of the microscope. The CTF has phase and amplitude components. The phase is inverted in adjacent regions of the CTF, separated by nodes at which there is no information. The phase component can therefore be corrected by flipping the phases in alternate zones of the CTF, and combining reconstructions calculated from a range of different defocuses to ensure the data fill reciprocal space up to the resolution required. The amplitude component of the CTF is not well understood but has two practical effects. First, the higher resolution amplitudes are decreased because of an exponential decay that arises from limited temporal coherence of the electron beam, ice thickness, movement, and other factors (Mancini and Fuller, 2000). Second, the lower resolution amplitudes are conversely artificially strong, which is a poorly comprehended effect possibly arising from inelastic electron scatter and the thickness of the water layer (Mancini and Fuller, 2000). The amplitude components may be corrected by fitting the experimental map to an appropriate envelope function (Mancini and Fuller, 2000), scaling them to crystallographic data, or obtaining an amplitude profile by a different technique such as small-angle scattering (see below) (Schmid *et al.*, 1999).

ii. Fitting Prior (X-Ray) Structures into Cryoelectron Microscopy Maps. The combination of cryo-EM and X-ray crystallography has been the most fruitful hybrid technical approach in analyzing virus structure, as will be apparent from the discussion in Section III. Various methods for fitting crystallographic data to electron density maps from microscopy are in use. Often, a crystal structure can be placed by eye within the density. However, the use of a correlation coefficient-based criterion combined with iterated refinement of the rotation and translation of atoms within the map will maximize the accuracy of the fit and greatly increase the validity of the interpretation derived from it. This has been carried out in both real and reciprocal space (Grimes *et al.*, 1997; San Martin *et al.*, 2001; Rossmann, 2000). The usual method is to treat the whole molecule, or a subunit, or, at the least, a domain as a rigid body during this refinement process. The result of this incorporation of prior information is that the accuracy of the final model can far exceed that of the EM reconstruction. It has variously been estimated that a 22-Å microscopy map yielded positional accuracy for an atomic model of 2.2 Å (Rossmann, 2000) and a 24-Å map one of 4 Å (Grimes *et al.*, 1997). This is an area of active development and we expect that improved and more automatic procedures will soon be available. Rather than develop the theoretic underpinnings of the methods we simply provide some illustrations from the extant literature that illustrate the power and limitations of what has been done to date (see Section III). Essentially, it should be obvious that the problem is similar to the molecular replacement method in single-crystal X-ray analysis, with the simplification that the phase is a known quantity. What remain less well established are (1) a general criterion of uniqueness and (2) a measure of accuracy (this second limitation probably reflects the underlying lack of an error model for cryo-EM reconstructions).

iii. Resolution in Cryoelectron Microscopy Reconstructions. In X-ray crystallography the resolution of an analysis is essentially defined by the maximum Bragg angle of the measured data. However, the completeness with which data are measured at higher resolutions is often less than at lower resolution; thus if we consider an example in which diffraction intensities may be measured to 3.5 Å in a virus structure, the completeness (the extent to which spherical reciprocal space is filled at that resolution) may be far short of 100%. Redundancy arising from the icosahedral symmetry of the virus can in this case be invaluable because it means that measurement of one intensity yields 59 identical values at icosahedrally related points of the virion transform cell. The use of such noncrystallographic constraints can result in a greater precision in atomic modeling within the X-ray crystallographic electron density than the nominal resolution might suggest. This is not equivalent to a genuine increase in resolution; however, it does allow a clarity reminiscent of a much higher resolution map. The nature of the protein structure is also relevant to the quality of the atomic model that can be obtained from data of a particular resolution because a constrained fold such as a β sheet will lead to the side-chain positions being well determined whereas a region lacking well-defined secondary structure places fewer constraints on the side-chain direction and peptide plane and therefore will require higher resolution data for a confident determination of its structure. The oft-quoted major difference between X-ray crystallographic and EM data is that the former provides amplitudes but the phases must be calculated indirectly whereas the latter defines both amplitude and phase, being a real-space image. Apart from the fact that this hides a major problem in cryo-EM (a two-dimensional image contains two-dimensional phase information; phase information in the third dimension needs to be determined computationally by assigning projection angles and a phase origin to the image in some consistent but arbitrarily chosen coordinate system) it does mean that the resolution of a reconstruction could, in principle, be determined on the basis of either the phase or amplitude components of the data. Note, however, that whereas the X-ray resolution attainable is essentially determined from the beginning by the extent of the diffraction data set (assuming that sufficient data are measured to solve the phase problem), the EM resolution attainable is less predictable, being determined by the homogeneity of the specimen, the accuracy with which phase origin and projection angles can be determined, and the number of images used, as well as the imaging conditions. EM resolution is usually based on some measure of self-consistency, in either phase or amplitude. Typically the total data set is randomly halved and two reconstructions are computed for which the Fourier shell differential phase residual or amplitude correlation is then determined. This has the disadvantage that only half the data are present in the maps, which may significantly hobble the resolution. It would be better to compare a reconstruction from the whole data set with some objective reference such as a crystal structure, if an appropriate one exists (Mancini *et al.*, 2000). In that case measures well tested by crystallographers can be imported (J. Navaza, personal communication). However, at present, there remains the problem of agreeing on criteria on which phase residual or amplitude correlation between two maps define the resolution. There have been two schools of thought concerning the amplitude correlation method of resolution estimation. One defines the resolution as simply the point in reciprocal space at which the Fourier shell autocorrelation (FSC) of a reconstruction dips below 0.5 (50%) (Bottcher *et al.*, 1997; Gabashvili *et al.*, 2000). This criterion appears to hold the moral high ground because it is more conservative than the other method of resolution estimation, by which the cutoff is taken to be the point in reciprocal space where the Fourier shell autocorrelation of a reconstruction falls below 3σ, where σ is the standard deviation in the background (noise) in the data. The use of this second criterion has been strongly argued in a review (van Heel *et al.*, 2000), in which a treatment of the effects of symmetry on the FSC was presented (see also Orlova *et al.*, 1997).

This treatment of symmetry demonstrates that the threshold above which significant autocorrelation is estimated to lie must be revised upward in line with the point group symmetry of the structure (van Heel *et al.*, 2000; Orlova *et al.*, 1997). Interestingly this results in the correct significance level for an icosahedral object being similar to the 0.5 correlation coefficient FSC criterion.

A further issue that should ideally be addressed is that the resolution of a reconstruction is unlikely to be homogeneous. This issue has been explicitly treated in the reconstruction of Semliki Forest virus (SFV) to 9-Å resolution (Mancini *et al.*, 2000). In that work both phase residual and FSC methods of resolution estimation were used, and the autocorrelation functions for resolution estimation were computed in real space radial density shells to show that different regions of the virus were ordered to differing extents. The phase residual between the capsid region of the SFV reconstruction and an atomic model of the capsid fitted into it was 75° at 1/9 Å, and this was the ultimate basis for the "headline" resolution figure (Mancini *et al.*, 2000). However, the point at which the FSC dipped below 0.5 was ~10.4 Å for the capsid region, ~15 Å for the lipid bilayer region, and ~20 Å for the visible regions of the nucleic acid (Mancini *et al.*, 2000). The question of EM resolution remains contentious; perhaps increasing comparison of cryo-EM structures with X-ray models will provide a clear guide to best practice (van Heel *et al.*, 2000; Mancini *et al.*, 2000).

3. X-Ray and Light Scattering

Several forms of solution scattering have been applied to viral structure and assembly. Small-angle X-ray scattering (SAXS) has been used to obtain spherically-averaged amplitude profiles over the resolution range present in cryo-EM data. This means that the amplitude components of the CTF can be approximately corrected for by scaling the cryo-EM reconstruction of an object to its solution scattering profile (Schmid *et al.*, 1999). A development of this approach has been to use fitting of the amplitudes of a reconstruction of an object to its SAXS profile to determine parameters for functions mathematically describing the data distortions arising from the CTF, the effective envelope function of the object, and the background noise in the images (Saad *et al.*, 2001). This approach has, in particular, been applied to the herpes simplex virus capsid in determining a reconstruction to 8.5Å (Saad *et al.*, 2001; Zhou *et al.*, 2000). Scaling to SAXS profiles has also been applied in the study of ribosomes (Gabashvili *et al.*, 2000), although the publication of the crystal structures of bacterial 30S and 50S subunits means that this can now be accomplished with crystallographic data (Ban *et al.*, 2000; Wimberly *et al.*, 2000).

Dynamic light scattering, operating at wavelengths far longer than SAXS, provides information about the size distribution of a sample because the rate of diffusion is dependent on the size of a scattering object and this relationship can be quantified on the assumption that the object is spherical. Because virus assembly represents a process in which a number of proteins self-associate to form a much larger structure, there is a dramatic increase in the size of the objects present, which means that virus assembly can be monitored by light scattering to yield kinetic models (Zlotnick *et al.*, 1999, 2000).

4. Neutron Scattering

Small-angle neutron scattering (SANS) yields much weaker data than SAXS, but the ease with which, at low resolution, the contrast of a particular chemical species within the virus such as lipid, nucleic acid, or protein can be suppressed by adjusting the ratio of ^2H_2O:^1H_2O to match the mean scattering of that component means it can, for example, be useful in determining their radial distributions. However, it should be noted that methods have now been developed that yield this kind of information with improved resolution from cryo-EM maps (Spahn *et al.*, 2000, 2001) (and indeed X-ray crystallography maps; J. M. Diprose, J. M. Grimes, and D. I. Stuart, unpublished data).

The key use to which neutron scattering in general is currently put in biological systems is to reveal the hydration characteristics of a macromolecule or a system of which it forms a part (Byron and Gilbert, 2000). Here neutron scattering is understood, in a broad sense, to include neutron crystallography, lamellar diffraction, fiber diffraction, reflectivity, and incoherent neutron scattering, as well as SANS. However, particularly in the 1970s and early 1980s, neutron scattering (especially SANS) played a major role in revealing aspects of viral structure. In 1975, knowledge of viral structure and particularly that of the viral genome was limited; the use of neutron scattering with contrast variation to differentiate nucleic acid genome from protein coat and to show that the latter enclosed the former and enjoyed a differing relationship to it in different viruses and at different stages of the viral life cycle was a major contribution to this knowledge (Burns *et al.*, 1992; Devaux *et al.*, 1983; Jacrot *et al.*, 1977; Kruse *et al.*, 1982; Witz *et al.*, 1993). It became clear that some ordering of the nucleic acid in layers echoing the internal surface of the capsid was present in some viruses, for example, adenovirus, in which a 29-Å repeat was observed in its double-stranded (dsDNA) (Devaux *et al.*, 1983); this finding echoes direct observations made in other viruses on the basis of cryo-EM or crystallographic work (e.g., Gouet *et al.*, 1999; Hill *et al.*, 1999;

Tang *et al.*, 2001). Another SANS investigation showed changes in nucleic acid distribution in turnip yellow virus following freezing-induced expansion of the capsid (Witz *et al.*, 1993). Swelling was also defined by SANS for viruses such as TBSV and bromegrass mosaic virus (BMV) under the influence of divalent cations and pH. TBSV swelling appeared fully reversible, and involved uniform expansion of the protein and RNA components of the virus, the RNA being sandwiched between the outer capsid layer and a pendant domain of the capsid protein (Kruse *et al.*, 1982). Neither the RNA nor the inner capsid protein domain was visible in the crystal structures solved for either the compact (Olson *et al.*, 1983) or expanded (Robinson and Harrison, 1982) forms of the virus. This expansion derives from an opening of the capsid to yield pores, as detected in the ethidium bromide accessibility conferred by swelling of the genomic RNA. The same kind of swelling was evident in BMV, but appeared to be irreversible under regimens in which TBSV swelling was reversible.

The distribution of lipid within a virus can be investigated by the same kind of approach as that to nucleic acid, and this has been done with influenza virus and the surface antigen of hepatitis B virus (Cusack *et al.*, 1985; Sato *et al.*, 1995; Tomita *et al.*, 1999). In the former case this approach revealed a lipid bilayer within the virus (Cusack *et al.*, 1985); in the latter it seems unlikely that a bilayer is present, but instead lipids closely associate with protein (Sato *et al.*, 1995; Tomita *et al.*, 1999). The influenza work made use of mathematical models describing scattering from spherical shells in order to extract dimensions for components of the virus such as lipid bilayer, protein capsid, and surface glycoproteins; deviations between some data and the spherical models suggested significant contributions to scattering curves from the packing of glycoproteins on the virus surface (Cusack *et al.*, 1985). A similar mathematical approach to describing the structural aspects of viruses detected by neutron scattering was applied to alfalfa mosaic virus (AMV), which is a nonicosahedral virus with an ellipsoidal or rodlike shape (Cusack *et al.*, 1981, 1983; Oostergetel *et al.*, 1983). SANS data were also measured from a quasi-spherical subviral AMV particle; fitted smooth spherical models agreed poorly with the data but including a T=1 icosahedral lattice in the model greatly improved the fit. Like the influenza data (Cusack *et al.*, 1985), this suggests that SANS is capable of measuring significant features within the surface of a scattering body, as well as the overall shape and size of that body.

Neutron scattering in the form of diffraction from viral crystals has been used to investigate the structure of RNA within some viruses at ~16-Å resolution. This approach has been particularly useful where poorly ordered density can be seen within X-ray maps of viral capsids, but cannot be identified as either RNA or protein. Satellite tobacco necrosis virus (STNV) was used in one study, in which the structures formed by the viral genome could be identified and appeared to have partial icosahedral symmetry giving rise to tubular densities in icosahedrally averaged maps (Bentley *et al.*, 1987). This in turn allowed elements of the protein–nucleic acid interface formed between the interior of the capsid and the RNA genome to be delineated (Bentley *et al.*, 1987). A similar study was also performed on TBSV, which has triangulation T=3 and therefore three different environments occupied by its capsid protein (denoted A, B, and C). The maps derived at a number of contrasts for TBSV (Timmins *et al.*, 1994) suggested that the nucleic acid contacts were in this case limited to the 3-fold axes of the virion, occupied by the C-type subunits, and provided further insights into the locations of the RNA and the ordered and disordered protein domains relative to that inferred from the SANS study described above (Kruse *et al.*, 1982). The radical nonequivalence in RNA interactions between A and B subunits and the C subunit of the icosahedral asymmetric unit reflects differences in the organization of A and B over C in the capsid itself (Olson *et al.*, 1983; Robinson and Harrison, 1982).

5. *Nuclear Magnetic Resonance and Electron Spin Resonance*

Although structure determination by nuclear magnetic resonance (NMR) can at present be usefully applied only to relatively small proteins (and certainly not to intact viruses!) it can prove a highly valuable technique when dealing with important systems that possess high structural plasticity. A good example of this is the structure of RNA-binding proteins in retroviruses (de Guzman *et al.*, 1998), which are relatively lacking in tertiary structure, and the flexible retrovirus envelope protein gp41 (Turner and Summers, 1999). A notable hybrid use of NMR structures of proteins is as molecular replacement models for subsequent crystallographic experiments (Pauptit *et al.*, 2001); however, the lack of reliable side-chain conformational information means that only the backbone structure is useful. Furthermore, an NMR analysis usually provides a family of structures, so that whereas the topology and core structure of a molecule may be well determined, outlying loops or more flexible regions will not be (the NMR "structure" is an ensemble of separate backbone models that satisfy the short-range constraints determined from resonance peak assignments). This becomes more problematic for molecules consisting of two or more domains, where the interdomain angle is often not well determined. A review has set out factors that are likely to favor successful use of NMR models for crystallographic molecular replacement phasing (Pauptit *et al.*, 2001).

^{31}P solid-state NMR is highly sensitive to phospholipid and nucleic acid organization, allowing detection of dynamic aspects of capsid structure. For lipids, ^{31}P measurements can be used to infer whether lipids are in bilayer or other phases such as cubic or hexagonal (Bonev *et al.*, 2001; Cullis and de Kruijff, 1979; Yeagle, 1990). For nucleic acids, this approach was used to investigate the mobility of genomic RNA inside compacted and swollen forms of TBSV, showing that in the compact form the RNA was tightly bound to the inside of the capsid but in the swollen form was mobile (Kruse *et al.*, 1982; Munowitz *et al.*, 1980). In AMV, the rigidity of the viral RNA decreased as the temperature was raised and was associated with increased mobility of the N-terminal nucleic acid-binding portion of the coat protein (Kan *et al.*, 1987). Another study showed that the RNA in tobacco mosaic virus was immobilized by its interactions with the coat protein (Cross *et al.*, 1983).

Another spin-sensitive technique is electron-spin reonance (ESR) which, like NMR, can be used to detect the chemical and dynamic environment of specific spin labels. Early studies showed that spin-labeled influenza virus, Rauscher murine leukemia virus, and human red blood cell membrane gave similar spectra, indicating that all three possess lipid bilayers (Landsberger *et al.*, 1971, 1972). In another study it was shown that whereas Sendai virus lipids were bilamellar both within the virus and once extracted from it, hepatitis B virus lipids were bilamellar only when extracted but were closely associated with the protein in a different kind of organization within the virus itself (Satoh *et al.*, 2000). ESR was also used to investigate the way in which fluidity of phosphatidylcholine and phosphatidylethanolamine is modulated when Sendai virus lyses red blood cells, suggesting that they may have undergone a process of undifferentiation during the fusion process (Lyles and Landsberger, 1977).

6. *Fiber diffraction*

Both X-ray and neutron fiber diffraction (as well as electron microscopy) techniques have been applied to filamentous viruses, for which the prospect of three-dimensional crystals is poor. By combining neutron and X-ray fiber diffraction, NMR, circular dichroism, and Raman and infrared spectroscopies, an atomic model for the filamentous bacteriophage Pf1 has been derived (Liu and Day, 1994). Other studies concerning Pf1 have relied on purely X-ray fiber diffraction data, together with molecular modeling, to provide detailed filament structures (Pederson *et al.*, 2001; Welsh *et al.*, 1998a,b, 2000). Fiber diffraction was also used to solve the structure of the rodlike helical tobacco mosaic virus (TMV), where all of the coat protein and three genomic nucleotides bound to each subunit were visible at a resolution of 2.9 Å (Namba *et al.*, 1989; Stubbs, 1999; Stubbs *et al.*, 1977). A segment of TMV in the precursor stacked-disk conformation was previously solved separately by X-ray crystallography at 2.8-Å resolution (Bloomer *et al.*, 1978; Champness *et al.*, 1976; Klug, 1999). Historically significant viruses such as TMV have usually been subject to hybrid approaches, because discoveries about their structure and its mechanism have paralleled technical developments. Determining the structure of TMV itself from helical fiber diffraction data waited on analytical and computational developments for many years; and once established opened the way to solving a number of other viruses by similar methods (Pattanayek and Stubbs, 1992; Wang *et al.*, 1997; Wang and Stubbs, 1994).

7. *Mass Spectrometry*

Mass spectrometry can provide information about virus structure from three perspectives: the mass and sequences of viral components/subunits, the dynamic aspects of a virus, and the structure of the virus as a whole (Fuerstenau *et al.*, 2001; Siuzdak, 1998; Tito *et al.*, 2000). It has been shown that some viruses introduced into the gas phase of a mass spectrometer and collected from the detector retain their structural integrity and remain infectious (Siuzdak *et al.*, 1996). This means that mass spectrometry has the potential to be a preparative tool as well as an analytical one in sorting polydisperse samples for structural analysis. In another study, flock house virus (FHV) particles prepared from infected *Drosophila* cells and a baculovirus protein expression system were compared (Bothner *et al.*, 1999). Although these two forms of the same virus were crystallographically indistinguishable at 2.8 Å, proteolysis and chemical modification combined with mass spectrometry showed that the infectious virus derived from *Drosophila* was more proteolytically resistant than the virus-like particles (VLPs) derived from baculovirus, whereas the primary amino group reactivity of the recombinant baculovirus VLPs was greater than that of the true viruses (Bothner *et al.*, 1999). These results indicate greater solvent accessibility for the capsid in recombinant than in viral particles, which must arise within a more dynamic state because the crystallographic structures are the same. It is thought that this breathing of the VLP structure arises from sequence-specific deficiencies in the RNA packaged within it, because the only difference between viral and recombinant systems is that one contained viral RNA and the other contained randomly selected cellular RNA (Bothner *et al.*, 1999). A different way of testing chemical accessibility on viral surfaces in order to detect dynamic changes in structures is to perform hydrogen–deuterium

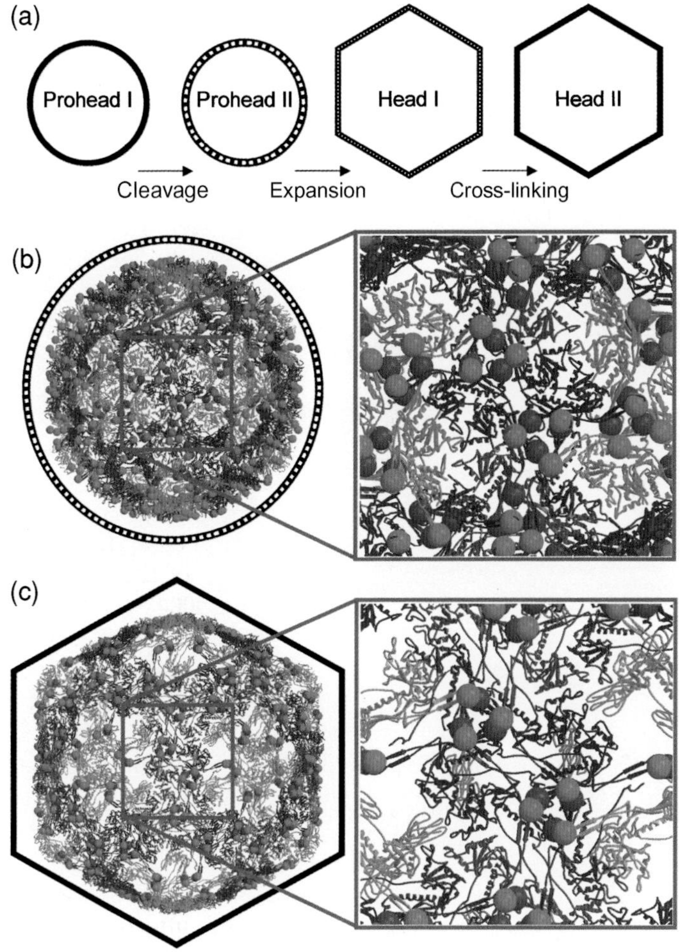

GILBERT *ET AL.*, FIG. 1. The maturation of bacteriophage HK97. (a) Schematic representation of the maturation pathway of bacteriophage HK97 (after Lata *et al.*, 2000). Proteolytic cleavage activates the nascent phage from prohead I to prohead II, whereupon it expands to form head I and is converted into head II by autocatalytic polymerization of the capsid protein. (b) Atomic model of the prohead II form of HK97 (Conway *et al.*, 2001), obtained by fitting the capsid protein crystal structure (Lata *et al.*, 2000) into a cryo-EM reconstruction. The C_α of amino acid Lys-169 is shown as an aquamarine sphere and that of Gln-356 is shown as a lilac sphere. These residues form a peptide bond in the head II form of the virus. (c) Crystal structure of head II, revealing the covalent interactions that bind the mature phage head together. Residues Lys-169 and Gln-356 are colored as in (c).

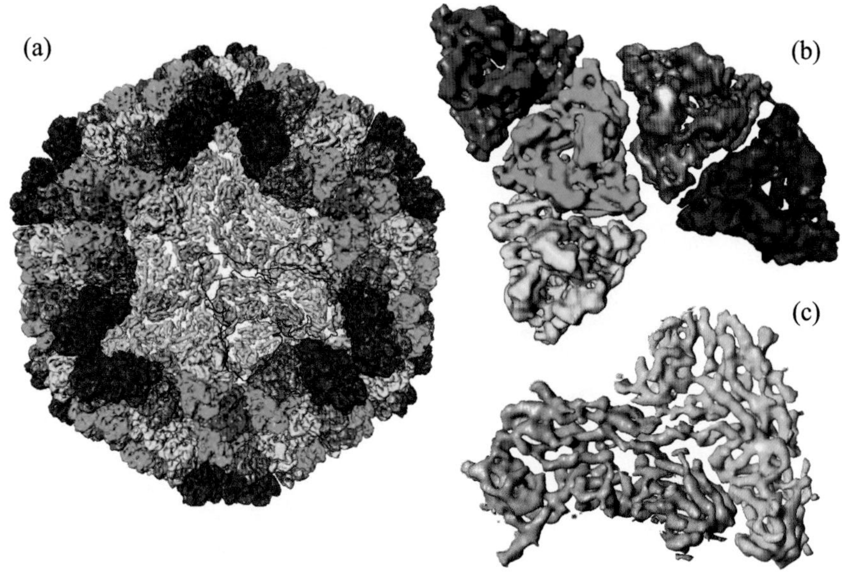

ZHOU AND CHIU, FIG. 7. Visualization of large virus particles via segmentation as illustrated by rice dwarf virus (RDV). (a) Full view of the entire RDV particle obtained by imposing icosahedral symmetry on an asymmetric unit segmented out from the original 3-D map. The subunits are segmented out and colored differently to facilitate visualization of intermolecular interactions. Each asymmetry unit (indicated by lines) contains four- and one-third trimers (in red, green, blue, yellow, and orange, respectively) of the outer shell protein P8 and two inner shell protein P3 monomers (in gray and aquamarine, respectively). In this display, 1 of the 20 triangular faces is removed to review the inner shell. (b and c) Close-up views of the unique subunits segmented out from the 3-D map, including the five P8 trimers on the outer shell (b) and the two P3 monomers on the inner shell (c) in each asymmetric unit.

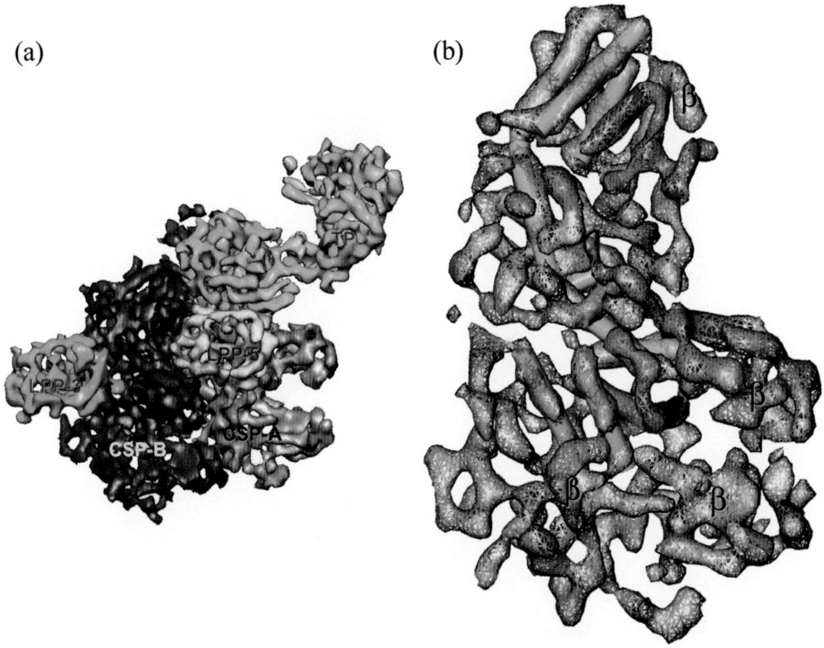

Zhou and Chiu, Fig. 9. Visualization of secondary structures in cytoplasmic polyhedrosis virus (CPV) capsid shell protein (Liang *et al.*, 2002; Zhou *et al.*, 2003). (a) An asymmetric unit extracted from the 8-Å structure of CPV. The structural components within the asymmetric unit are labeled, including one turret protein (TP, in light blue), two capsid shell protein (CSP) molecules (CSP-A, in red, and CSP-B, in gray), and two large protrusion protein molecules (the 5-fold proximal LPP-5 in yellow and 3-fold proximal LPP-3 in green). (b) CSP-A displayed at a higher density contour level, using wire frame representation, superimposed with visually identified helices, which were modeled as 5-Å-diameter cylinders. A total of 18 helices were identified in CSP-A, 14 (aquamarine) of which are structurally homologous to those in bluetongue virus VP3 and 4 (red) of which are unique to CPV. Several β sheet-rich regions were also identified and are indicated (β).

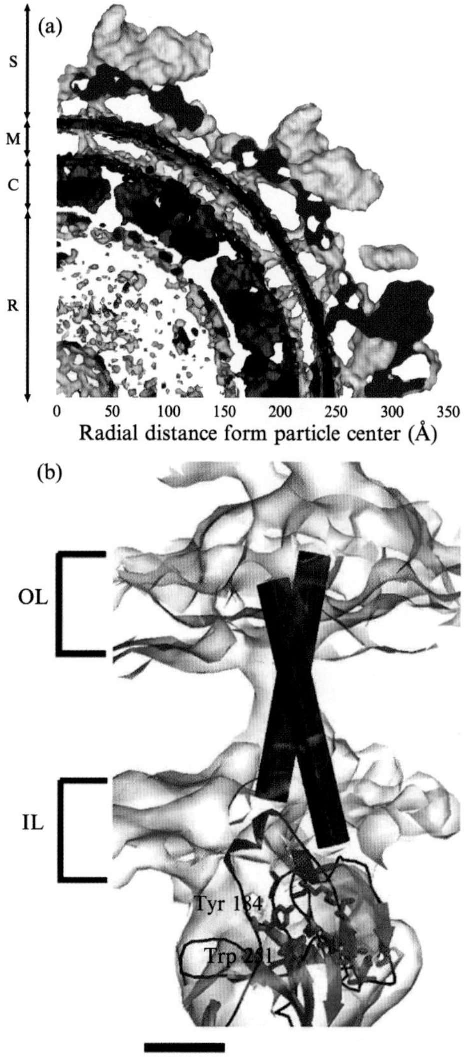

ZHOU AND CHIU, FIG. 10. Identification of transmembrane helices in Semliki Forest virus (SFV) E1 and E2 (Mancini *et al.*, 2000). (a) Shaded surface view of a central slab reveals the multiple shell organization of SFV. SFV is an envelope virus consisting of an RNA core (R) (yellow), a nucleocapsid made of 240 copies of protein C (C), which is enclosed by a lipid bilayer membrane (M) (inner and outer leaflets shown in blue) decorated with 80 glycoprotein spikes (S) (shown in white). (b) Transmembrane regions of the E1 and E2 densities interacting with the capsid are shown. The C protein is shown as a ribbon diagram. The positions of the inner leaflet (IL; radius, 213 Å) and the outer leaflet (OL; radius, 261 Å) of the membrane are marked. The putative α-helical transmembrane segments for E1 and E2 are depicted with a paired helical segment from a known protein structure, demonstrating that the dimensions and the topology of the density are consistent with a pair of helices. The transmembrane domains are seen as paired rods of density approximately 10 Å wide, separated at the top by approximately 10 Å and twisting about each other. Scale bar: 15 Å. Adapted from Mancini *et al.* (2000) with permission from the publisher and the authors.

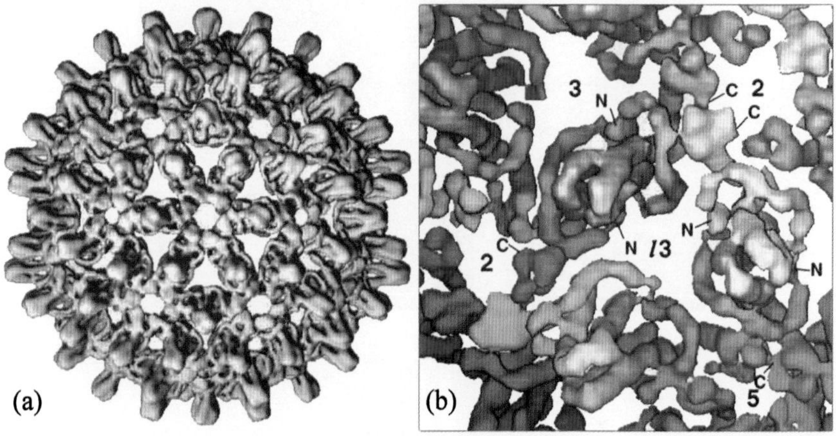

ZHOU AND CHIU, FIG. 11. Identification of the path of each polypeptide of the hepatitis B virus core protein (Böttcher *et al.*, 1997a). (a) Shaded surface representation of the T=4 HBV core structure determined at 7.4-Å resolution, viewed down a strict 2-fold axis. (b) Close-up view down a local 3-fold axis (*l*3). The positions of neighboring strict 2-fold, 3-fold, and 5-fold axes are marked. The density has been colored to indicate the four symmetrically independent monomers. The path of each polypeptide is revealed by using a higher density threshold for the display. The positions of the putative N and C termini of the polypeptide are indicated. [Adapted from Böttcher *et al.* (1997b) with permission from the author and the publisher.]

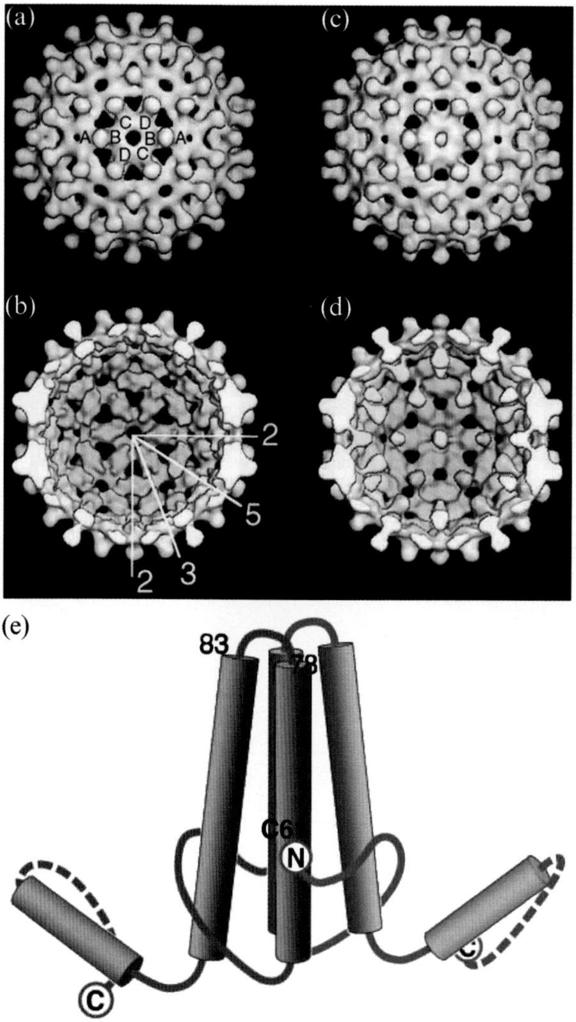

ZHOU AND CHIU, FIG. 12. Identification of specific amino acid residuals by difference imaging with chemical labels. (a and b) Three-dimensional structure of empty HBV Cp147 capsid at 17 Å. (c and d) Three-dimensional structure (20 Å) of the HBV Cp* 150 capsid with its C terminus labeled with undecagold label Au_{11}. (e) Model of the HBV capsid protein dimer with locations of the amino acids localized by difference imaging, including the C terminus by gold labeling, the N terminus by an octapeptide insertion (Conway *et al.*, 1998), and the loop covering residues 78–93. [Adapted from Zlotnick *et al.* (1997) and Conway *et al.* (1998) with permission from the publishers.]

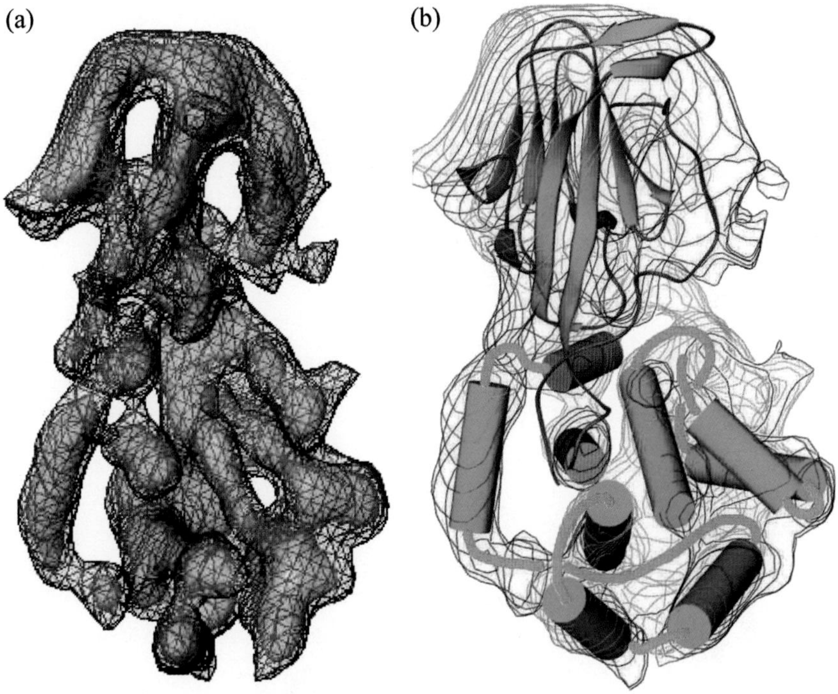

ZHOU AND CHIU, FIG. 13. Interpretation of subnanometer-resolution structure through integration of bioinformatics and secondary structure assignment (Zhou *et al.*, 2001). (a) A side view of the density map of a P8 monomer extracted from the 6.8-Å RDV structure. The higher density features are shown as shaded surfaces and their connectivities are revealed in the wire frame representation shown at a relatively lower density contour level. (b) Interpreted folds of the secondary structure elements in RDV P8. Each P8 contains a lower domain with nine helices and an upper β-barrel domain. The homologous bluetongue virus VP7 β-barrel fold was identified by *foldHunter* and shown as ribbon. [Picture courtesy of Dr. Matthew L. Baker.]

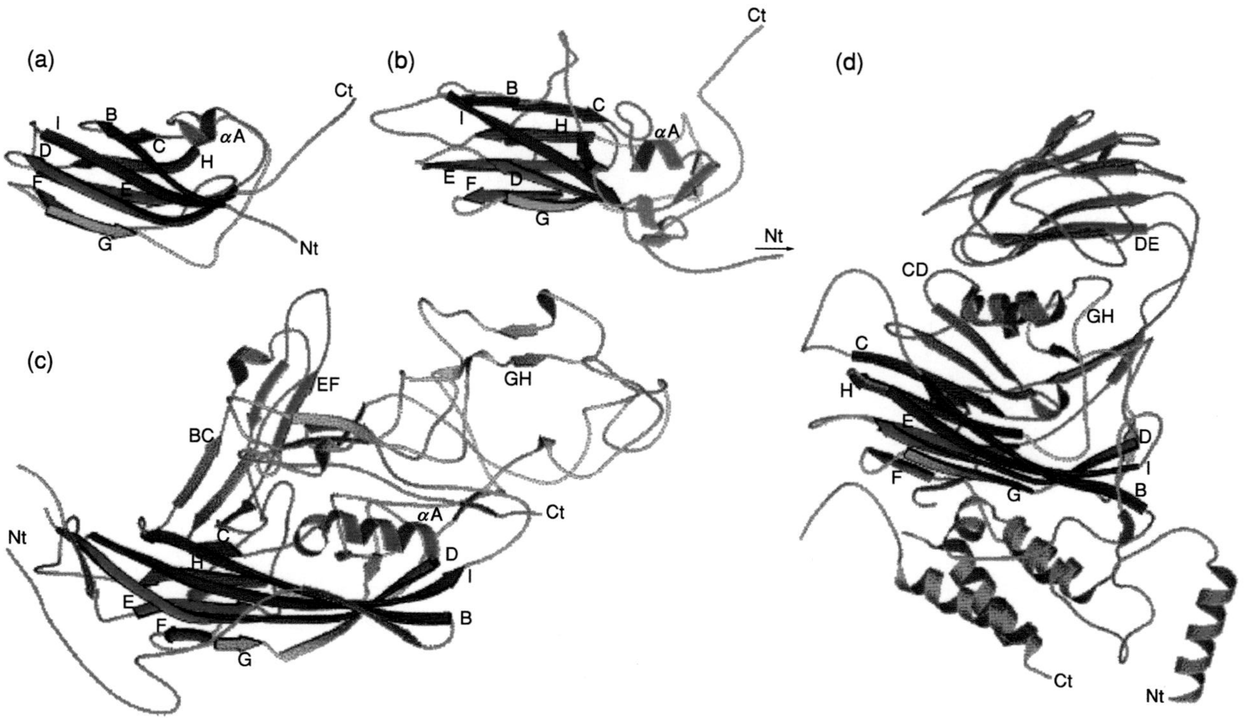

CHAPMAN AND LILJAS, FIG. 1. The canonical jelly-roll β barrel seen in many capsid structures, as exemplified by (a) satellite *Panicum panicum* mosaic virus (Ban and McPherson, 1995), (b) poliovirus VP3 (Filman *et al.*, 1989), (c) canine parvovirus (Filman *et al.*, 1989), and (d) *Nudaurelia capensis* ω virus (Munshi *et al.*, 1996). The view is approximately tangential to the capsid surface. As in many of the figures, the eight strands of the jelly roll are highlighted by darker colors from blue (βB) to red (βI) The visible N termini and C termini are marked with Nt and Ct, respectively, and the loops connecting the strands are denoted BC, CD, etc. All the ribbon drawings have been made with the program Molscript (Kraulis, 1991).

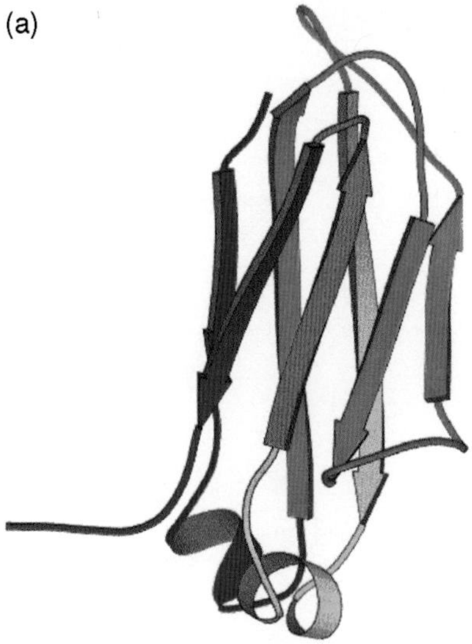

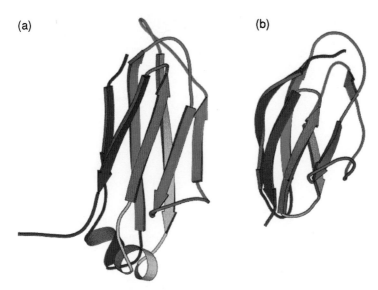

CHAPMAN AND LILJAS, FIG. 3. The immunoglobulin fold. (a) The constant domain of the light chain of an antibody; (b) the immunoglobulin domain of tick-borne encephalitis glycoprotein E (Rey *et al.*, 1995).

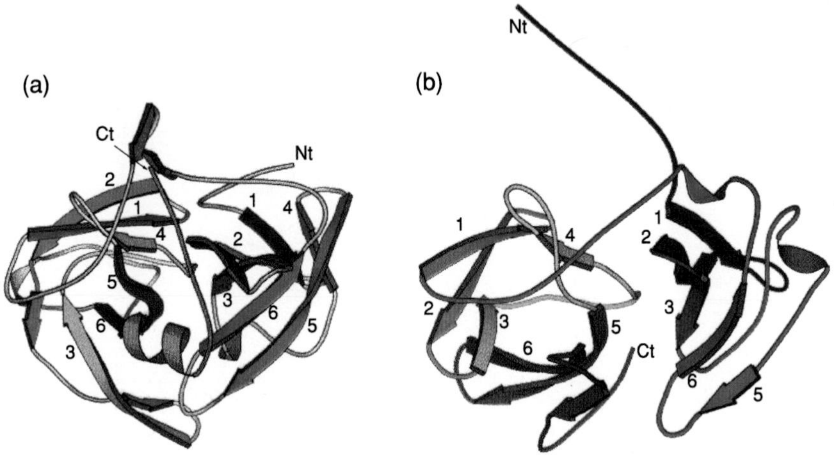

CHAPMAN AND LILJAS, FIG. 4. The serine protease fold. The serine proteases have two domains with the same topology. The six strands in the β barrel are denoted 1–6. The active site is in the cleft between these domains. (a) α-Lytic protease (Fujinaga *et al.*, 1985); (b) the fold of the Sindbis coat protein (Choi *et al.*, 1996). The corresponding strands in the two proteins are colored in the same way to simplify the comparison.

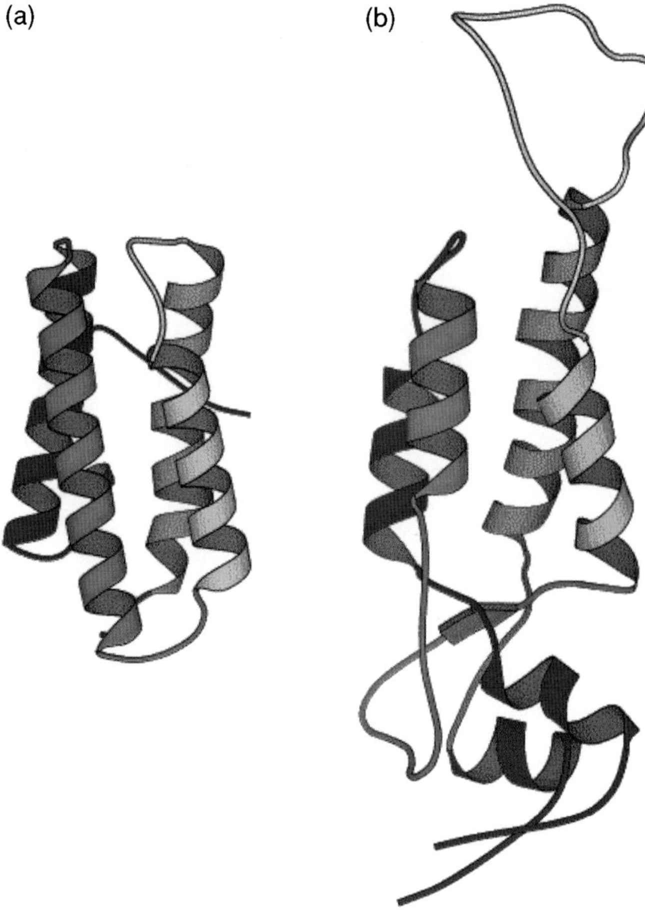

CHAPMAN AND LILJAS, FIG. 5. The four-helix bundle. (a) The prototypical conformation of myohemerythrin (Sheriff *et al.*, 1987); (b) the tobacco mosaic virus coat protein (Namba *et al.*, 1989).

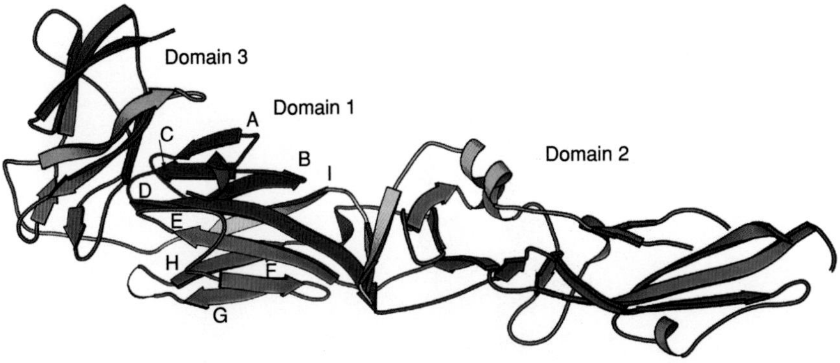

CHAPMAN AND LILJAS, FIG. 7. The structure of tick-borne encephalitis virus glycoprotein E (Rey *et al.*, 1995). The color scheme is blue to red from the N terminus to the C terminus. Domain 3 is also shown in Fig. 3b.

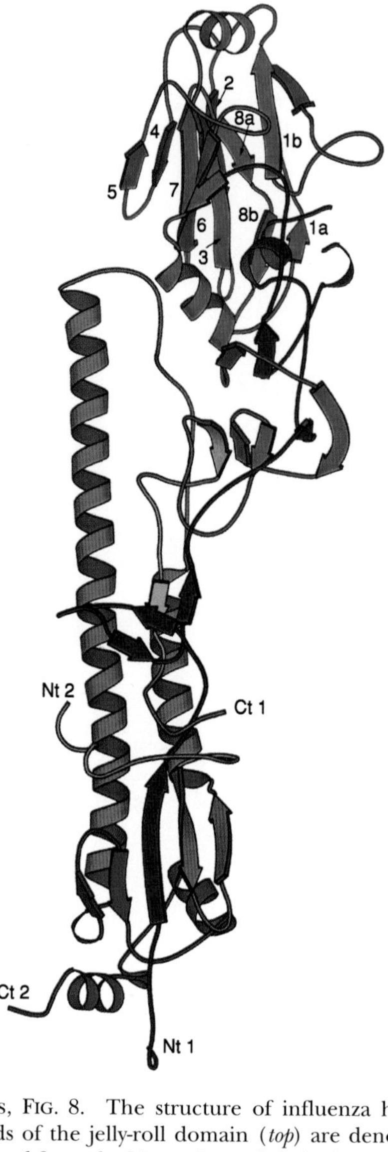

CHAPMAN AND LILJAS, FIG. 8. The structure of influenza hemagglutinin (Wilson *et al.*, 1981). The strands of the jelly-roll domain (*top*) are denoted 1 through 8. The color scheme is blue to red from the N terminus of chain 1 (Nt 1) to the C terminus of chain 2 (Ct 2), which is cleaved from the membrane anchor. The fusion peptide is at the N terminus of chain 2 (Nt 2). In the virus, the protein forms a trimer where the long helices are parallel to the 3-fold axis and form a stem.

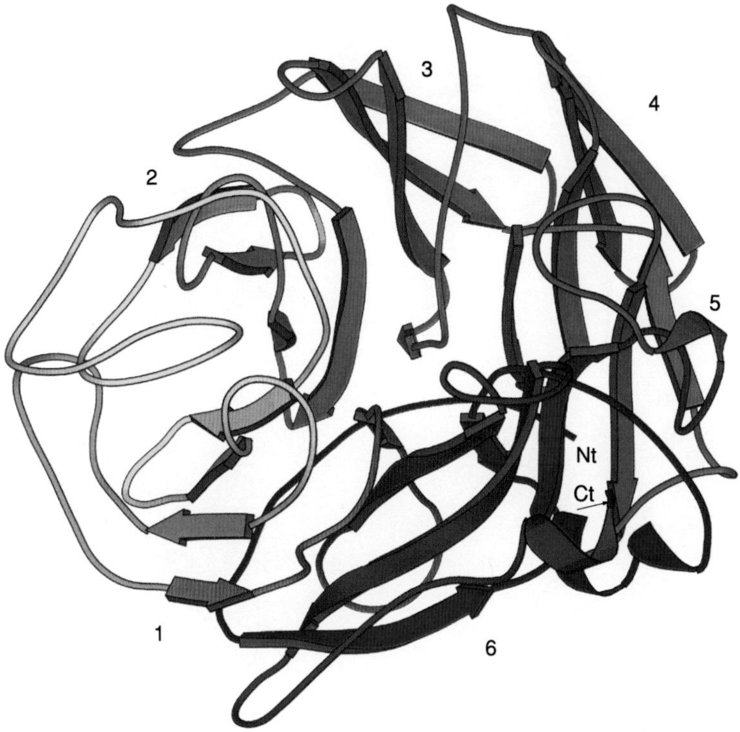

CHAPMAN AND LILJAS, FIG. 9. The structure of a monomer of the tetrameric influenza neuraminidase (Varghese *et al.*, 1983). The view is from the distal side. The six four-stranded sheets are denoted 1–6. The N-terminal strand (blue) is the outermost strand of sheet 6.

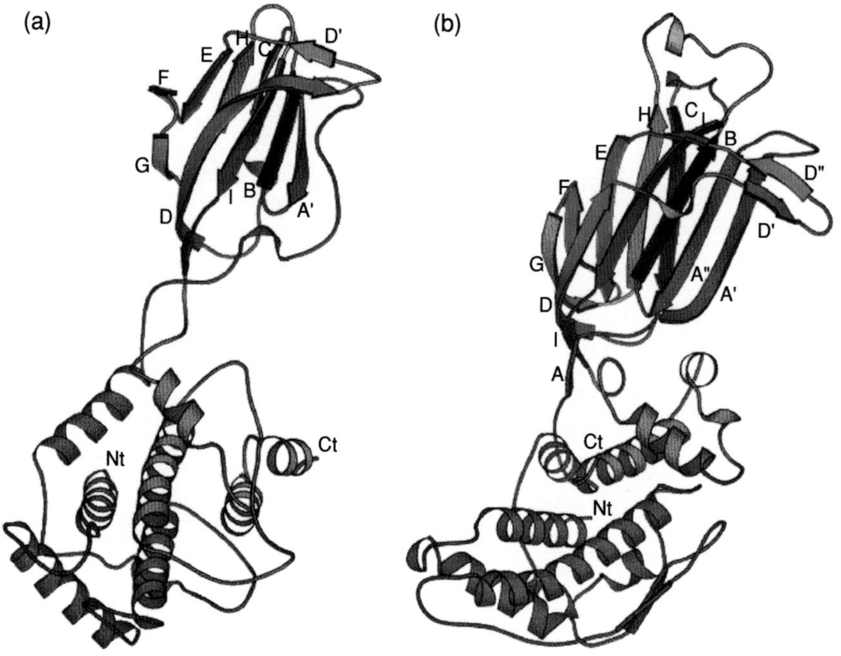

CHAPMAN AND LILJAS, FIG. 11. (a) The structures of orbivirus (bluetongue) VP7 (Grimes *et al.*, 1995) and (b) rotavirus VP6 (Mathieu *et al.*, 2001). The structures are embellished viral jelly rolls with large N- and C-terminal additions that collectively form a helical domain that attaches to the inner shell of the virus.

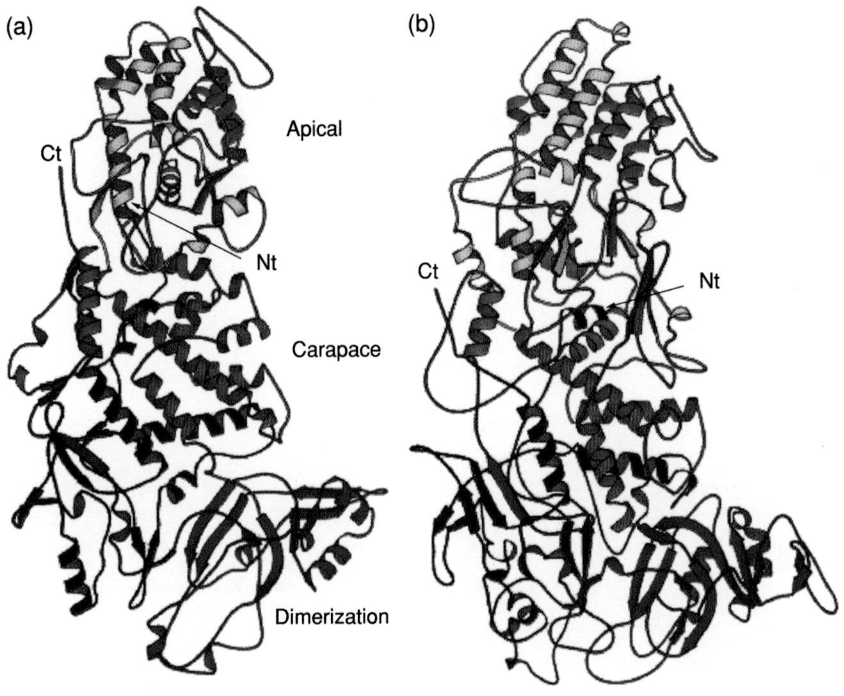

CHAPMAN AND LILJAS, FIG. 12. The shell-forming proteins of bluetongue virus and reovirus: (a) bluetongue VP3 protein (Grimes *et al.*, 1998); (b) reovirus $\lambda 1$ protein (Reinisch *et al.*, 2000). In the bluetongue VP3 protein, three domains (apical, carapace, and dimerization domains) have been identified. The secondary structure elements have been colored to emphasize the general structural similarity between the two proteins.

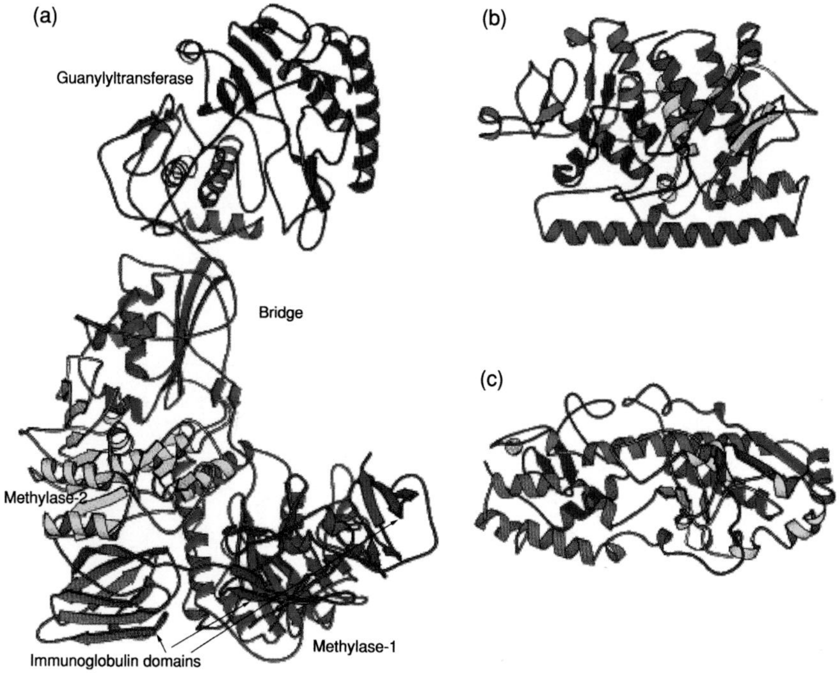

CHAPMAN AND LILJAS, FIG. 13. Reovirus proteins. (a) The $\lambda 2$ turret protein (Reinisch *et al.*, 2000). The individual domains are marked. (b) The $\sigma 2$ clamp protein (Reinisch *et al.*, 2000). The two halves of the main part of the protein have approximate 2-fold symmetry. The view is down the symmetry axis. (c) The $\sigma 3$ outer capsid protein (Olland *et al.*, 2001).

exchange on the virus before mass spectrometry (Tuma *et al.*, 2001; Wang *et al.*, 2001). The combination of proteolysis to test structural integrity and mass spectrometry to measure it has been termed "mass mapping" (Siuzdak, 1998). When a virus disassembles in the gas phase within the mass spectrometer, the variety of viral particles detected from whole capsid downward can demonstrate the hierarchy of the disassembly and so that of the intersubunit bonds present (Siuzdak, 1998; Tito *et al.*, 2000).

8. *Other Biophysical Techniques*

Atomic force microscopy has been used to study the morphology and growth of virus crystals, visualizing both the individual viruses within the lattice (Kuznetsov *et al.*, 2001; McPherson *et al.*, 2001) and the ultrastructural features of the virus surface (Malkin *et al.*, 1999). Optical tweezers, another optical technique, are used in the study of single-molecule dynamics and tensile properties; these have been successfully applied to the bacteriophage ϕ29 to measure the force with which the packaging motor of the phage must drive DNA into its capsid, reflecting the increasing internal pressure as more and more nucleic acid is encapsidated (Smith *et al.*, 2001). Understanding these energetic aspects of ubiquitous viral processes is essential if realistic mechanical models are to be conceived that relate to how the virus works as a self-replicating device. Fluorescence resonance energy transfer (FRET) is a powerful probe of the chemical environment of defined regions of proteins and as such has been used to reveal conformational changes in the human immunodeficiency virus (HIV) gp41 envelope protein on virus–cell interaction (Kliger *et al.*, 2000). Similarly, FRET has been used to try and resolve the positioning of HIV nucleocapsid zinc-finger motifs and show that their mutual arrangement is not random but statistically ordered (Mely *et al.*, 1994). Both fluorescence and force-sensitive techniques are growing in power, being increasingly able to work at one extreme at the level of single molecules and at the other with *in vivo* interactions at the cell surface. It therefore seems likely that a range of applications to virus structure–function analysis will be developed.

Finally, X-ray free electron lasers (XFELs) are expected to become operational by 2008 and there is much excitement at the potential they may have for imaging macromolecules with femtosecond X-ray pulses (Hajdu, 2000; Miao *et al.*, 2000, 2001). We consider that the published studies do not fully address the sources of error likely to arise in such analyses and that the case has been overstated (as has also been pointed out by Henderson, 2002). Nevertheless, viruses would be an ideal model system in the development of this method.

9. Mutagenesis

Perhaps the most widely applied method for probing protein function is the use of site-directed mutagenesis. This is no less a prime tool for virus structure analysis, and provides a way in which use of the techniques discussed above can be systematized. There are a large number of virus structure studies that have involved mutagenesis. Examples include the chemical modification and mutagenesis of the foot-and-mouth disease virus (FMDV), which ordered the antigenic receptor-binding loop invisible in native structures due to disorder (Logan *et al.*, 1993; Parry *et al.*, 1990); and use of glycosylation mutants of Sindbis virus matrix protein to identify regions it occupies in a reconstruction of the virus (Pletnev *et al.*, 2001). The glycosylation mutation method is a particularly eloquent way of identifying the position and, in favorable cases, the orientation of molecules in cryo-EM reconstructions at rather modest resolution (about 20 Å) and may be thought of as an evolution of the isomorphous replacement method developed for crystallography.

B. Common Combinations

The most common hybrid, as mentioned above, is that between cryo-EM and X-ray crystallography. The complementarity normally comes from the continuing supremacy of X-ray crystallography in providing the ultimate in resolution whereas EM excels in speed, easy analysis of complex systems, and application to heterogeneity. However, the relationship can work in a wide range of different ways, with either method capable of assisting the other. Cryo-EM frequently feeds crystallography by providing initial phases in the determination of a structure. In a small number of cases, this has involved the direct application of phases from a cryo-EM map to the experimental structure factor amplitudes, followed by cyclic averaging to improve the phase determination and bootstrap the resolution of the map from the EM model to the limits of the crystal (Jack *et al.*, 1975; McKenna *et al.*, 1992; Reinisch *et al.*, 2000; Wynne *et al.*, 1999). Other approaches have used a regular distribution of scattering centers within the envelope defined by the cryo-EM reconstruction to provide the starting phases (Prasad *et al.*, 1999; Wikoff *et al.*, 2000), whereas a third strategy—the sandwich strategy—is to model the structure of a known capsid component into its deduced position in an EM reconstruction of the virus and thereby derive a starting model (Grimes *et al.*, 1997, 1998). A variation on this theme is to make an educated guess as to the fold of a component of the virus and thereby derive a pseudo-atomic model for the target structure (Speir *et al.*, 1995). Perhaps cryo-EM at high resolution with fold identification methods might itself provide atomic models derived from maps that, of themselves, would not normally be classed as atomic resolution. The work on rice dwarf (Zhou *et al.*, 2001) and hepatitis B (Bottcher *et al.*, 1997) viruses suggests that there is scope for increasingly rich analysis of such intermediate-resolution maps, and indeed with the intense efforts devoted to the automatic interpretation of X-ray crystallography maps at 3-Å resolution or lower we can expect rapid progress in this important range of resolutions.

Both cryo-EM and more classic electron microscopy techniques can also be used to determine directly the packing of viruses or proteins in a crystal, where it cannot be deduced directly from the diffraction data. Crystal packing can be derived either from scanning electron microscopy (SEM) methods, by which the surface of a specimen is observed, or from higher resolution transmission electron microscopy (TEM), of which cryo-EM is a development. Methods for the preparation of crystals for TEM include freeze-etching, to highlight their surface features (Meining *et al.*, 1995), and fixation followed by thin sectioning (Bonami *et al.*, 1997; Dokland *et al.*, 1998). A particularly interesting combination of SEM and TEM has been applied to the alkaline-labile polyhedra in which viruses such as baculovirus and cytoplasmic polyhedrosis virus are occluded during viral replication and transmission. Whereas TEM has indicated the packing of the polyhedrin protein that forms the matrix of the polyhedron (Bonami *et al.*, 1997), SEM has shown the overall morphology of the polyhedron and revealed some details of viral spacing in the matrix (Qanungo *et al.*, 2000; Woo *et al.*, 1998). Another microscopy technique that can provide information about packing within a crystal is atomic force microscopy (AFM) (Kuznetsov Yu *et al.*, 1997). AFM has also been applied to the study of crystallization processes, revealing in one case, for two crystal forms of the same virus, particle size heterogeneity within an otherwise more-or-less flawless crystal form, and uniformity of virus size in a crystal with a high density of defects (Kuznetsov *et al.*, 2001); although flawed in a crystalline sense the latter is of use for structure determination whereas the former is not. In another case the structure of the growing edge of a crystal was analyzed, revealing cooperative transformations in the nascent crystal lattice as it developed in order and extent (Malkin *et al.*, 1999). In general, AFM has the potential to assist in the systematization of crystallization strategies, revealing relationships between different growth mechanisms, the appearance and development of defects, and the incorporation of impurities (McPherson *et al.*, 2001).

Crystallography can enhance cryo-EM analyses to provide more detailed insights to virus structure. In a strategy increasingly common in all cryo-EM work, the three-dimensional reconstruction can be fitted with the crystal structures of constituent proteins (see e.g., Belnap *et al.*, 2000a). Apart from analyzing viral capsid tectonics, this approach is a rich source of information in studying the complexes formed between viruses and their receptors, or neutralizing antibodies (see below).

It can also be useful to start with an "atomic" model in determining a cryo-EM reconstruction. In some cases, construction of an atomic model based on lower resolution reconstructions or reconstruction of a related virus has been used to align images (Mancini *et al.*, 2000), whereas in others it has been necessary to construct an "atomic" model *ex nihilo*. This has been especially useful when the likely structural principles of a capsid are apparent but there is no precedent for determining such a structure, as with prolate bacteriophages (Tao *et al.*, 1998) and the fascinating gemini viruses (Zhang *et al.*, 2001).

III. Hybrids

A. Structural Transformations in the Maturation of a DNA Bacteriophage

The homopolymeric head of DNA bacteriophage HK97 endures a number of conformational rearrangements in passing from a prohead structure to a mature phage containing DNA. *In vivo,* this may be intimately associated with the packaging of its nucleic acid through a portal/connector protein in one 5-fold vertex of the nascent capsid. Structural analysis of its maturation pathway was facilitated, however, by the omission of the portal protein such that the phage retained its icosahedral symmetry unbroken, but also failed to package DNA (Lata *et al.*, 2000). *In vitro,* an acid-stimulated series of transformations was followed by three techniques: time-resolved SAXS, cryo-EM, and X-ray crystallography.

The ground to be covered in order to follow sequential changes in HK97 structure during maturation was surveyed by measuring variations in the SAXS profile of a maturing viral population with time and obtaining cryo-EM reconstructions for a series of intermediates (Lata *et al.*, 2000). The initial preparation for these experiments consisted of the proteolytically activated prohead II (which is derived from prohead I) (Fig. 1a; see Color Insert). Representative scattering profiles were obtained for capsid conformations at the beginning and end of the maturation pathway, and then the gradual acid-induced change in the global scattering profile of a prohead II population was monitored while it matured as far as the stage (head I) immediately preceding mature head formation (head II) (Fig. 1a and b). Separating the different intermediates from each other in cryo-EM images of the maturing phage population proved problematic, but an objective scheme was devised (Lata *et al.*, 2000). The mature head is generated via autocatalytic peptide bond formation between subunits in the adolescent capsid, resulting in a remarkable and thus far unique chain-mail structure (Duda, 1998; Wikoff *et al.*, 2000). The mature head is 659 Å in its longest dimension (between 5-fold verticies), yet possesses a shell only 18 Å thick. The crystal structure of the mature head of HK97 was subsequently solved at 3.6-Å resolution, revealing a unique fold for a viral protein and a complex pattern of covalent cross-linking between and entwining capsid subunits, which generates the robust chain-mail architecture of the capsid (Fig. 1c).

A later, higher resolution, cryo-EM reconstruction of prohead II provided a matrix for fitting the crystal structure to derive a chemically informative atomic model (Fig. 1b). This indicated spatial separation in prohead II of the residues through which the virus is concatenated in the mature head and a possible mechanism for promoting this prohead-to-head transition as arising from the DNA packaging event (Conway *et al.*, 2001).

The structural analysis of HK97 has relied on the combined use of a number of techniques. The crystal structure determination of the capsid itself used the cryo-EM reconstruction to provide a phasing start for the measured diffraction data. The resolution of the cryo-EM map was 25 Å, and diffraction data were consequently measured over the range 200–3.45 Å to ensure sufficient overlap between the EM model and the crystal data. The cryo-EM model was then filled with equally spaced dummy atoms (scattering centers), which were scaled to the X-ray structure factors and then used to calculate phases between 200 and 50 Å. Cyclical 60-fold averaging derived from the noncrystallographic icosahedral symmetry of the virus was then applied, with five cycles being completed before each extension of the phases by one reciprocal lattice point. This procedure was iterated until the phases reached 3.6 Å, whereupon the interpretability of the map was assisted by the application of an isotropic B factor of -40 Å^2. Sharpening of map features by application of a negative B factor has been routinely used for crystallographic analyses of larger viruses (see, e.g., Grimes *et al.*, 1998); however, a much more dramatic application was in the cryo-EM reconstruction of the hepatitis B virus core at 7.4-Å resolution, -500 Å^2 (Bottcher *et al.*, 1997). The large negative B factor applied to EM reconstructions is necessary if the higher resolution terms in the map are to contribute sufficiently to produce the expected level of detail. As discussed in the introductory section on cryo-EM, complex imaging effects conspire to downweight the higher resolution terms compared with the lower resolution terms much more dramatically than for X-ray analysis (Mancini and Fuller, 2000).

B. Structure and Dynamics of Bacteriophage ϕ29

Hybrid methods have been central to describing the structure, maturation, and mechanistic dynamics of a tailed bacteriophage, ϕ29 (Simpson *et al.*, 2000; Tao *et al.*, 1998). This phage is somewhat similar to T4, but smaller and therefore more tractable. The attraction of systems such as ϕ29 and T4 is in part that the mechanisms associated with the viral life cycle are enacted through conformational changes in different distinctive limbs and devices of the virus structure, providing a series of model macromolecular machines. The mature phage consists of a prolate head from one vertex of which extends a complex tubular assembly responsible for the delivery of DNA to the host cell and, in a modified form, for its subsequent packaging into daughter virions (Fig. 2). The analysis of this virus has involved an ingenious combination of cryo-EM and X-ray crystallography.

In an analogous approach to that used for HK97, cryo-EM was used to mark out the series of structures adopted by ϕ29, from isometric particles (which are by-products of assembly not on the main maturation pathway) with and without appended fibers, to the mature phage and the emptied phage that has jettisoned its DNA (Tao *et al.*, 1998). The structure of the isometric fiberless particle was solved by the usual icosahedral reconstruction techniques (Tao *et al.*, 1998). Reconstructing the prolate prohead, mature phage, and emptied capsid made use of the isometric fiberless reconstruction, elongated by the equatorial addition of 10 hexamers (the isometric particle is T=3 and the prolate particles are T=3, elongation factor *Q*=5). This theoretical model was used in initial projection angle assignment of the prolate images, which were reconstructed imposing initially 52-point and finally just 5-fold symmetry coaxial with the prolate shell, to preserve the asymmetry of the portal and closed ends of the capsid. The structures resulting from iterated angle assignment and reconstruction possessed fibers attached at quasi-3-fold axes absent from the starting model, which strongly indicated that the particle orientations were being correctly determined.

One interesting aspect of ϕ29 is the presence of a structural RNA moiety (pRNA) in the prohead at the 5-fold vertex where packaging occurs (Simpson *et al.*, 2000). Although genetic studies have suggested that pRNA has a hexameric structure (Guo *et al.*, 1998; Zhang *et al.*, 1998), cryo-EM demonstrated it to be pentameric when a reconstruction without any symmetry imposed was calculated (Morais *et al.*, 2001; Simpson *et al.*, 2000). Modeling the theoretical structure of the pRNA pentamer into the relevant region of the prohead also gave a convincing fit to the density. However, the genetic evidence (Guo *et al.*, 1998; Zhang *et al.*, 1998) and

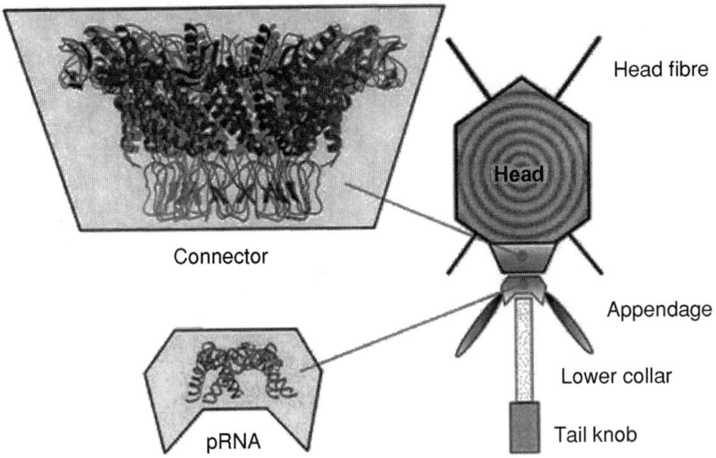

FIG. 2. Bacteriophage ϕ29. On the right is shown a schematic of the mature bacteriophage ϕ29, with the various limbs and devices it uses in effecting its life cycle. The phage head is filled with DNA, and extends from one vertex the dodecameric connector structure and pRNA, at the top of the long tubular structure through which the viral genome is injected into the host.

other studies such as the demonstration that tandem dimers of pRNA incorporate into phage proheads that can then package DNA (Mat-Arip *et al.*, 2001) remain unexplained if pRNA does form a simple pentamer. We propose here one explanation that could combine the structural and genetic data in a single model for the pRNA within the phage. Hexameric pRNA would have demonstrable 5-fold symmetry if five of its constituent molecules packed around the phage vertex while the sixth thrust outward like a spoke. In that case the spoke portion would be averaged out even in unsymmetrized cryo-EM reconstructions so that only a cyclical pentagon would be seen. If this arrangement were present at the ϕ29 vertex, it could play a material role in the mechanism of DNA packaging, either through being metastable and self-propagating around the circuit of pRNAs, or by supplying a lever interacting with the capsid itself or some other structure found at this end of the phage.

The pRNA and the capsid between them sandwich a dodecameric connector, the structure of which has been determined crystallographically (Simpson *et al.*, 2000) (Fig. 2). The connector has 12-fold rotational symmetry and its structure determination was a beautiful application of hybrid methods. Orthogonal views of the connector reconstructed from two-dimensional arrays by electron diffraction analysis had been published previously (Valpuesta *et al.*, 1999). These published images were used to generate a three-dimensional model that was positioned translationally and rotationally by cross-correlation with the diffraction data (Simpson *et al.*, 2000). Structure determination proceeded by cyclical 12-fold averaging and solvent flattening to improve the phases. The best map arising from this process was sufficient to locate two mercury sites at 4-Å resolution that could be combined with 12-fold noncrystallographic symmetry averaging to compute phases to 3.2 Å. The structure was refined at this resolution by iterative noncrystallographic symmetry averaging and solvent flattening.

As with HK97, both accessing and interpreting high-resolution information about ϕ29 involved hybrid methods. The resulting understanding of its structure has suggested one mechanism describing how its moving parts might operate in DNA packaging (Simpson *et al.*, 2000). The central insight is of a concentric triple symmetry mismatch at the packaging portal, consisting of DNA with 10-fold symmetry, capsid and pRNA with 5-fold symmetry, and the funnel-shaped connector with 12-fold symmetry. The authors of the connector structure propose that the head and $(pRNA)_5$ with attendant $(ATPase)_5$ act as a stator, the connector as a ball race, and the DNA as a movable central spindle (a movie depicting this model can viewed at http://news.uns.purdue.edu/UNS/mov/rossmann. motormovie.mov). As packaging proceeds a large internal pressure builds up within the prohead, as measured with optical tweezers (Smith *et al.*, 2001); this causes the packaging rate to slow and may store the necessary energy for ejecting the DNA into a host cell cytoplasm.

The next stages of the ϕ29 project, involving crystallographic analysis of the prolate head itself and further hybrid modeling of cryo-EM maps with viral components, will no doubt continue to fascinate. Furthermore, the goal of a structural dissection of T4 looms, succeeding a reconstruction of an isometric prohead (T=13) (Olson *et al.*, 2001) with the prolate structure (T=13, *Q*=21) according to the same philosophy applied to ϕ29, while, in parallel, crystal structures that will assist in interpretation of T4 reconstructions continue to be amassed (Kanamaru *et al.*, 2002; Kostyuchenko *et al.*, 1999; Leiman *et al.*, 2000; Rossmann, 2000; Strelkov *et al.*, 1996, 1998; Tao *et al.*, 1997).

C. *Structures of Membrane-Containing Isometric Viruses: Fusion Machines*

The membrane of enveloped viruses makes them highly attractive structural targets in seeking a general understanding of protein–membrane interactions and membrane fusion events. Perhaps the best structural information currently available for a membrane virus has been obtained for the alphaviruses, which possess two protein layers sandwiching a lipid bilayer (Fig. 3). The outer protein layer contains glycoproteins E1 and E2, which mediate fusion of viral and cellular membranes, and cell recognition, respectively. A third protein, E3, is derived, like E2, from a P62 precursor by proteolytic maturation; in most alphaviruses this is then lost, although in Semliki Forest virus (SFV) it remains associated with the outside of the virion. The inner capsid layer of the virus consists of the capsid (C) protein. Both the outer and inner layers have icosahedral T=4 symmetry (Cheng *et al.*, 1995; Mancini *et al.*, 2000). An icosahedral reconstruction of SFV has been obtained with a resolution of up to 9 Å (Mancini *et al.*, 2000) (see discussion in introductory section concerning resolution) (Fig. 3). To assist the initial alignment of the particles and to define the hand of the reconstruction use was made of an atomic model built by fitting the structure of the isolated capsid protein solved previously (Choi *et al.*, 1997) into an earlier 22-Å SFV map (Fuller *et al.*, 1995). The atomic model was summed with the EM map and this hybrid structure was used to assign projection angles to the cryo-EM data. When the resolution of the reconstruction reached 14 Å, however, the atomic model ceased to be of assistance and was discarded because of departures of the actual core structure from the ideal quasi-equivalence on which the atomic model for it had been constructed (Mancini *et al.*, 2000) (an alternative approach would have been to refine the atomic model during phase extension to higher resolutions as used for bluetongue virus; Grimes *et al.*, 1998). The reconstruction shows that the layer of protein above the bilayer consisting of E1 and E2 is formed of heterotrimeric [$(E1)_3(E2)_3$] spikes. Previous data had been taken to suggest that E1, the fusion protein, formed the spike erection and that E2 formed a surrounding skirt (Venien-Bryan and Fuller, 1994). However, solving the crystal structure of the isolated fusion protein E1 and a fine difference-imaging study subsequently demonstrated that the spike was largely E2 and the surrounding skirt a rather flat E1 (Lescar *et al.*, 2001; Pletnev *et al.*, 2001) (Fig. 3b).

Surprisingly, the crystallographic structure of SFV E1 was similar to that of the tick-borne encephalitis virus (TBEV) fusion envelope protein E, revealing an unexpected evolutionary relationship between spikey alphaviruses and the smooth flaviviruses such as TBEV (Lescar *et al.*, 2001). E1 could be fitted into the skirt regions of the outer protein layer in the cryo-EM map of SFV, but not the spikes themselves, indicating its location in the virus. This inferential model for the arrangement of E1 and E2 was confirmed by a careful difference-imaging study, comparing native Sindbis virus structure with mutants that have had glycosylation sequons removed individually (Pletnev *et al.*, 2001). Sindbis is an alphavirus with a structure similar to that of SFV. In this study difference maps between

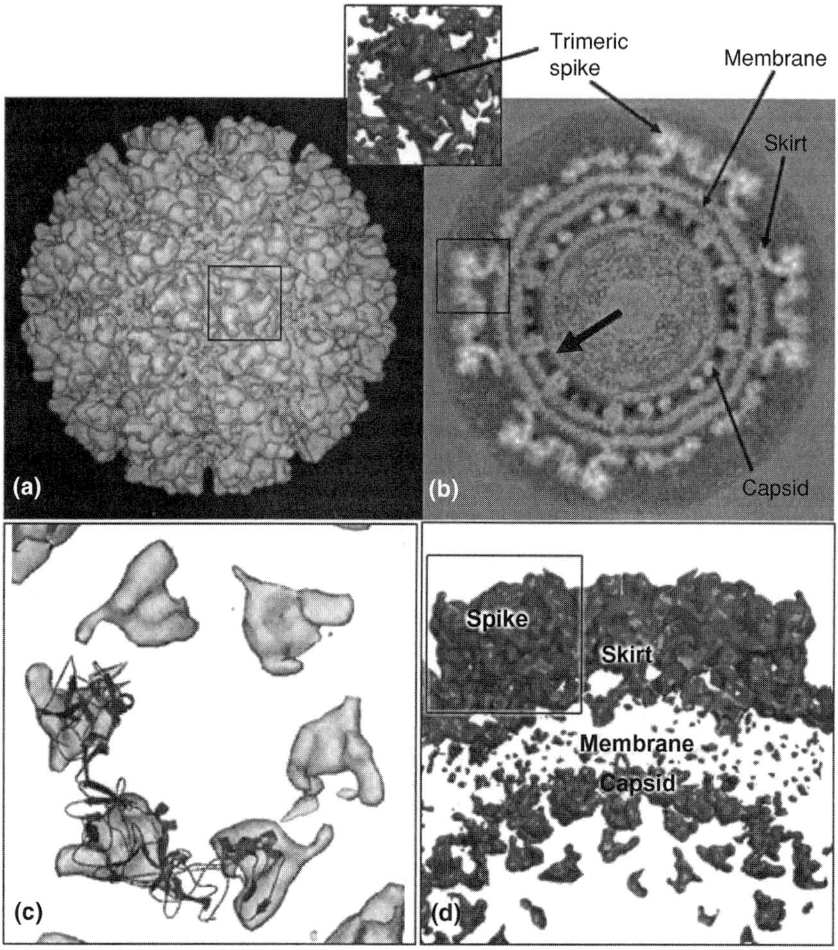

FIG. 3. A 9-Å resolution reconstruction of Semliki Forest virus. (a) Density isosurface view of the reconstruction down an icosahedral 2-fold axis. One trimeric spike is boxed, and shown close up in a similarly boxed inset panel between parts (a) and (b). (b) The central section of the SFV reconstruction, showing the inner capsid, membrane bilayer, and outer E1–E2–E3 layer composed primarily of trimeric spikes, one of which is again boxed. The thick arrow points along a 2-fold axis. (c) A close-up view of the capsid layer from the virion interior, with the capsid protein fitted into the reconstructed electron density. This view is along a 2-fold axis, as marked in (b). (d) A close-up surface-rendered view across the viral surface through a 2-fold axis, showing the layers visible in (b). Showing the density at a single contour level highlights the different extents to which the layers of the virus are ordered.

mutant and wild-type Sindbis viruses were calculated and the reconstructions interpreted at a series of real space radii to reveal the radial positioning of each glycosylation site in turn (Pletnev *et al.*, 2001). The misidentification of the density due to E1 highlights the difficulty in interpreting cryo-EM maps, even at 9-Å resolution, and demonstrates the incisive power of fitting higher resolution (atomic) structures into these maps. Interestingly, the evolutionary links between the alphaviruses and flaviviruses would have given the game away earlier, but this was revealed only by the crystallographic investigation.

These studies also provided information about the processes by which alphaviruses fuse with and bud from cells. Fusion by alphaviruses is initiated through the pH-triggered deployment of a fusion peptide from within the planar E1 molecule into the target membrane. However, because the spike is formed of E2 a dramatic rearrangement must occur during the fusion process, with movement of E2 to allow homotrimerization of E1 (Lescar *et al.*, 2001; Pletnev *et al.*, 2001). The result would be a fusion pore formed at an icosahedral 5-fold or local 6-fold symmetry axis, surrounded by five or six E1 trimers. Because E1 itself forms an icosahedral lattice, it can act to cloak the nascent virion with a protein coat of defined size, sandwiching the membrane with the preexistent core during budding (Pletnev *et al.*, 2001).

For hepatitis B virus (HBV) a 22-nm-diameter complex of lipid and the HBV surface antigen is observed during both expression of recombinant antigen and infection. In the mature ~40-nm HBV capsid this protein is part of the outer proteolipid layer. The surface antigen particle has so far eluded a structural analysis using imaging or crystallographic approaches. However, given the considerable interest surrounding HBV, a wide range of techniques have been targeted at it. For example, small-angle neutron scattering coupled with contrast variation has suggested that the surface antigen is located peripherally with respect to a lipid core (Sato *et al.*, 1995). Furthermore, electron spin resonance has been used to investigate the structure of the lipid within the particle (Satoh *et al.*, 2000). Spectra typical of a bilayer were obtained with vesicles formed using lipids extracted from the particle, and similarly with Sendai virus. However, the surface antigen particles themselves did not have a bilayer, but the lipid gave a signal typical of tight association with protein. In addition, lipids within the HBV particles were susceptible to phospholipase but could not be exchanged with a pool of free lipids. These results suggest that there is a tight association between surface antigen and lipid, with the lipid moieties accessible from the particle surface but, on average, located at lower radii than the protein components of the particle.

D. *Difference Imaging and an Internal Membrane*

A third example of a membrane virus that is now the subject of intense structural studies is the *Salmonella* bacteriophage PRD1. This virus is evolutionarily related to human adenovirus, but possesses a membrane within its protein capsid (Benson *et al.*, 1999; San Martin *et al.*, 2001). It therefore represents an ideal model system for both adenovirus (being smaller and more approachable) and membrane-containing viruses. In the latter case the prokaryotic nature of its host organism provides significant advantages over viruses infecting eurkaryotes, where the glycosylation of the outer virus proteins may raise major difficulties in obtaining crystals that diffract to high resolution. A quantum leap in our structural understanding of PRD1 is now tantalizingly close, with the achievement of crystals diffracting beyond 4-Å resolution (Bamford *et al.*, 2002). However, structural studies of both PRD1 and adenovirus have already made extensive use of fitting cryo-EM reconstructions with atomic coordinates, and the computation of difference maps, in order to dissect the organizational principles of these large viruses. Indeed, the first use of a combination of crystal structure information and a cryo-EM map was for adenovirus (Stewart *et al.*, 1991, 1993). Summarizing aspects of the picture of adenovirus structure that has emerged will lead us on to consider the parallel achievements with PRD1.

Adenovirus (Fig. 4, top) is a special example of failure to follow the rules of quasi-equivalence in that its complex capsid is formed mostly from a trimeric protein, hexon, which is built into a close-packed hexavalent net to form the rather flat faces of the virus, whereas closure of the shell relies on the presence of a different, pentameric, protein (penton) at the 5-fold vertices. A spike or fiber is extended from the 5-fold vertices and interacts with the cellular receptor of the virus during internalization (Chiu *et al.*, 1999; Stewart *et al.*, 1991, 1993). The capsid is organized on a pseudo $T=25$ geometry, comprising 240 trimers of the hexon protein and 12 pentamers of the penton protein, and has a maximum dimension of approximately 900 Å. Fitting the crystal structure of the hexon protein (Roberts *et al.*, 1986) into a cryo-EM reconstruction of the virus allowed the calculation of a difference map (Stewart *et al.*, 1993), which revealed the electron density arising not only from the penton base and fiber portions of the capsid, but also from two minor proteins (IIIa and VI) that are thought to be key elements in cementing the integrity of the adenovirus capsid. The structure of the adenovirus capsid limits the number of environments that hexon protein must occupy but necessarily results in weaker interactions at the limits of capsid facets (Burnett, 1985). The weaker regions of the capsid are supplemented with the minor proteins, with protein IIIa being an elongated tie riveting pairs of facets together and protein VI cementing the 5-fold rings of penton protein to the hexons surrounding them (Burnett, 1985; Stewart *et al.*, 1993).

More recently the crystal structures of the adenovirus fiber shaft and receptor-binding fiber head (van Raaij *et al.*, 1999) and of the head in complex with the coxsackie-adenovirus receptor molecule (CAR) (Bewley *et al.*, 1999) have been solved. The fiber shaft structure revealed a novel β-sheet triple spiral fold, which is perhaps particularly suited to forming rigid protein projections for interaction with and penetration through cell membranes by adenovirus; the structure of the complex that performs the same function in bacteriophage T4 has been solved, revealing a similar fold (Kanamaru *et al.*, 2002).

The first convincing evidence that PRD1 and adenovirus are related came from the X-ray structure of the PRD1 major capsid protein, P3 (Benson *et al.*, 1999), which is essentially a trimmed-down version of the core of adenovirus hexon protein (Athappilly *et al.*, 1994). Both PRD1 P3 and hexon core have two β jelly-roll domains that are oriented similarly with respect to each other and have the same fold topology (Benson *et al.*, 1999). In addition, the structure of PRD1 capsid as revealed by cryo-EM and image reconstruction is a pseudo T=25 lattice (Butcher *et al.*, 1995; Rydman *et al.*, 1999) (Fig. 4, bottom), the only other example of this architecture being adenovirus. Whereas PRD1 P3 corresponds to the adenovirus hexon, PRD1 P31 is thought to be the penton base equivalent (Benson *et al.*, 1999). Like adenovirus, PRD1 also has fibers protruding from its vertices (Rydman *et al.*, 1999) and a linear dsDNA genome. The larger size of adenovirus essentially arises from its major capsid protein being larger than PRD1 P3 (967 as opposed to 394 residues), and is reflected in adenovirus having a genome twice as large (Belnap and Steven, 2000). The radical difference in host range between *Salmonella* bacteriophage PRD1 and human adenovirus is presumably reflected in structural differences between the viruses, although these remain, for now, largely obscure.

In studies analogous to those performed on adenovirus, reconstructions of (1) wild-type PRD1, (2) a mutant that possesses a membrane but fails to package DNA, and (3) this mutant treated with detergent to obtain a P3 shell alone, were compared and fitted with the atomic structure of P3 (Martin *et al.*, 2001). The phage DNA was visible in the wild-type particle in layers, indicating that it adopts a statistically ordered state within the virion, as found in a number of other large viruses (Gouet *et al.*, 1999; Hill *et al.*, 1999; Reinisch *et al.*, 2000). There is cryo-EM evidence that in adenovirus too this is the case, consonant with solution scattering studies (Devaux *et al.*, 1983). The lipid bilayer membrane appeared to be angled

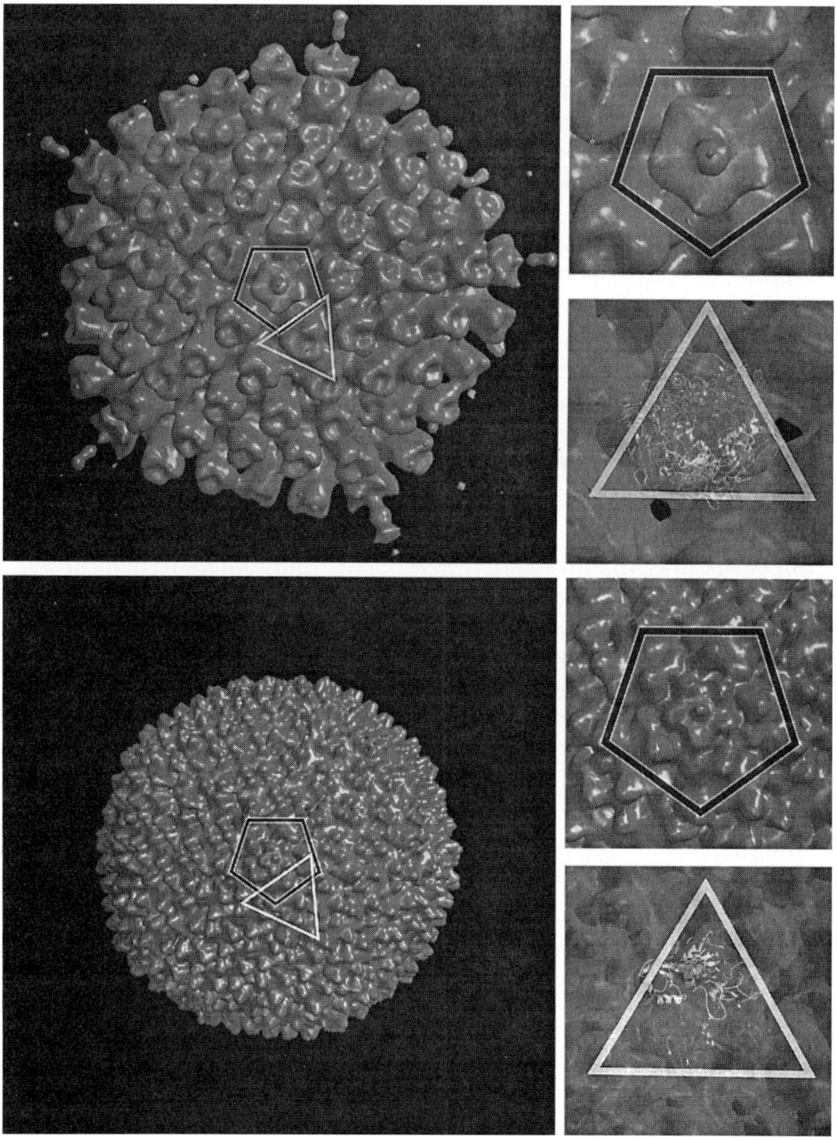

FIG. 4. Adenovirus and bacteriophage PRD1. *Top*: On the left, a density isosurface representation of adenovirus at 25-Å resolution is shown. The 5-fold axis, occupied by the protein penton, is marked with a pentagon. A trimer of hexon is marked with a triangle close by; arrays of hexon extend outward in all directions from the pentagonal vertex, forming the flat faces of the virus. On the right, a close-up of the 5-fold axis is shown (*top*) and below that a close-up of the hexon trimer with it crystal structure fitted (Athappilly *et al.*, 1994; Stewart *et al.*, 1991). *Bottom*: The *Sus1* mutant of PRD1 is shown

around the icosahedral capsid. Spectroscopic experiments have been used to try and define the structural characteristics of the lipid and nucleic acid within PRD1. Laser Raman spectroscopy indicated that the lipids were in a liquid crystalline phase and that the DNA was B form (Tuma *et al.*, 1996b). Intriguingly, these experiments also suggested that there might be structural cooperativity between the conformations of the membrane and the DNA, and that there are therefore close interactions between the PRD1 genome and its lipid bilayer (Tuma *et al.*, 1996b). PRD1 (dis)assembly was also investigated by Raman techniques, making use of temperature-driven disassembly of the phage capsid between 50 and 70°C (Tuma *et al.*, 1996a). These studies indicated that there may be a conformational change of part (~5%) of P3 from a β-sheet structure to an α-helical structure during assembly (Tuma *et al.*, 1996a). Although such conclusions relating to coupled folding and assembly of PRD1 rest on indirect data (and in this case are not supported by more thorough analysis; San Martin *et al.*, 2001), techniques of this kind have some use in addressing dynamic and short-lived structures adopted by macromolecular assemblies that are currently inaccessible by other means.

Fitting of P3 to the three PRD1 reconstructions revealed aspects of its interactions within the capsid and with the membrane and how these were changed in the virus possessing DNA compared with the mutant that failed to package its genome. It seems that the presence of DNA increases the number of protein–membrane contacts (San Martin *et al.*, 2001), which presumably relates to the increase in order within the membrane seen in a previous study on DNA packaging (Butcher *et al.*, 1995). A close analysis of the interfaces made between P3 trimers in the capsid suggested that there may be cementing proteins in PRD1 as there were in adenovirus (Stewart *et al.*, 1991, 1993). Given that cementing proteins were successfully identified in adenovirus at 35-Å resolution (Stewart *et al.*, 1993), it should be possible to identify the corresponding regions if they are present in the current, much higher resolution, reconstructions. [Note added in proof: This has now been achieved (San Martin, 2002).]

E. Structures of Large Viruses: Transcription Machines

Although some phenomenally large viruses (of diameter ~0.2 μm) have been studied by cryo-EM (Yan *et al.*, 2000), the largest structures for which

on the left, looking down a 5-fold axis marked by a pentagon. On the right, a close-up of the 5-fold (formed by protein P31) is shown (*top*), and below that is shown a close-up of the P3 trimer that is structurally homologous to adenovirus hexon and forms the bulk of the capsid in a similar way (Benson *et al.*, 1999; Butcher *et al.*, 1995).

a hybrid approach yielding atomic information has been obtained are the double-stranded RNA (dsRNA) viruses of the family *Reoviridae*. These viruses are constructed of between one and three capsid layers and possess segmented genomes that are transcribed into mRNA from within the virus core itself once the host cell is invaded. Reconstruction of a number of these viruses from cryo-EM data has been performed, providing representative structures for the three architectural styles adopted within the family: smooth with three complete layers [e.g., bluetongue virus (BTV); Grimes *et al.*, 1997], double-shelled with turrets at the icosahedral 5-folds surrounded by an incomplete second layer (e.g., orthoreovirus; Dryden *et al.*, 1998), and single-shelled with turrets at the 5-folds (e.g., cytoplasmic polyhedrosis virus [CPV]; Hill *et al.*, 1999; Zhang *et al.*, 1999) (Fig. 5). In addition, atomic structures have been published for BTV (Grimes *et al.*, 1998) and orthoreovirus (Reinisch *et al.*, 2000) core particles (lacking the outermost layer in each case). The determination of these atomic structures was in each case a major exercise in the application of hybrid techniques.

The first structures for triple-shelled reoviruses such as BTV and rotavirus were obtained using cryo-EM (Hewat *et al.*, 1992; Yeager *et al.*, 1994). Later, the X-ray crystal structure of a BTV structural protein that forms the middle layer of the triple-shelled mature virus (VP7) was solved (Grimes *et al.*, 1995), which made microscopy subsidiary to crystallography but of key interpretative value (Grimes *et al.*, 1997). The crystallographic structure of the BTV serotype 1 (BTV1) core (the double-shelled subparticle) was determined by fitting the VP7 crystallographic structure to a cryo-EM reconstruction as a phasing start (Grimes *et al.*, 1998). The atomic pseudo-structure of the VP7 (T=13) outer layer of the virus was positioned and oriented within the crystallographic unit cell by cross-correlation of calculated and measured diffraction data between 60 and 12 Å. This process of positional and orientational refinement was then repeated at 6-Å resolution. Phases were calculated from this model and then the icosahedral symmetry of the virus was harnessed to real space cyclic averaging combined with solvent flattening. This process refined the structure constrained by the known symmetry of the particle at the position already determined by translation and rotation searches. The polypeptide backbones of the subunits in the inner VP3 (T=2) layer were built into the 6-Å map at this stage and refined as rigid bodies at incrementally higher resolutions to 3.8 Å, whereupon real space cyclic averaging was resumed. The chain tracing for both the inner VP3 (T=2, 120-protein subunit) and the outer VP7 (T=13, 720-protein subunit) layers was then unambiguous (Grimes *et al.*, 1998). The data set was at that stage merged with a second that extended to 3.5-Å resolution, and

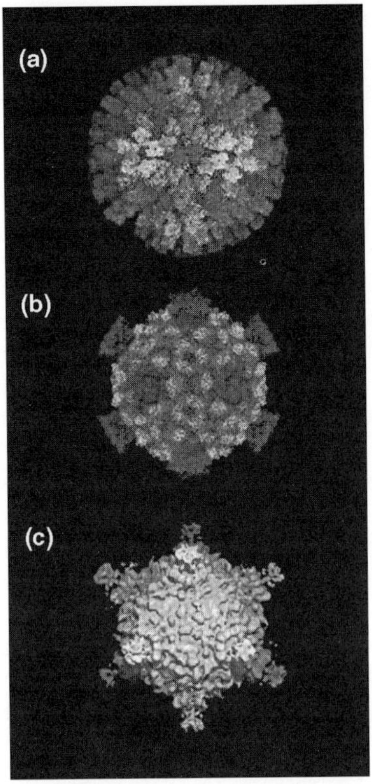

FIG. 5. Structural classification of *Reoviridae* architecture. (a) Bluetongue virus, a typical orbivirus (Grimes *et al.*, 1998). (b) Orthoreovirus, a typical reovirus (Reinisch *et al.*, 2000). (c) CPV, a typical cypovirus (Hill *et al.*, 1999).

sharpened by applying a *B* factor of −44.5 Å^2 [a similar *B* factor was used with HK97—see above; Wikoff *et al.*, 2000]. In this final stage of calculation, the position and orientation of the particle were again refined before calculation of a final map and atomic model.

The interior structure of the virus was analyzed in further experiments using data between 105 and 6.5 Å for BTV1 and 60–6.5 Å for a second serotype, BTV10 (in which the crystallographic cell axes all exceeded 1000 Å) (Gouet *et al.*, 1999). In both cases and in a cross-correlated map between the two, the dsRNA was visible as layers of density concentric with the capsid surface (indicating a role for the capsid in ordering the dsRNA) and spaced at ~30 Å. This 30-Å repeat within the core also gave rise to a powder ring arising from diffusely scattered X-rays in the diffraction patterns of the crystals. The identification of this density and powder ring as arising from dsRNA was confirmed by contrast-matching experiments in which the electron density of the dsRNA was approximately matched with cesium chloride, masking the signal from the powder ring. Fascinatingly, the powder ring increased in intensity when the cores were activated (i.e., induced into transcription), which suggests that the dsRNA may become more ordered in actively transcribing compared with quiescent cores, perhaps because of concerted movements of the nucleic acid through the transcription complexes close to the icosahedral 5-fold axes where the viral mRNA is synthesized and capped. The dsRNA is believed to exist in a spiral conformation around the fivefold vertices, permitting separate and simultaneous transcription of the 10 genome segments at different vertices (Gouet *et al.*, 1999). The density observed and modeled for the dsRNA in these structures is believed to account for ~80% of the BTV genome. More recent studies of crystals soaked with transcriptional substrates and substrate and product analogs have shed precise light on how the bluetongue core functions as a transcriptional device; substrates for transcription enter through pores in the capsid and the products, single-stranded mRNA molecules, exit through pores at its 5-fold vertices (Diprose *et al.*, 2001).

The significance of obtaining atomic information for large viruses such at BTV is found partly in the insights afforded by the observed interactions of proteins in each capsid layer and between layers, for example, the relative positions of symmetry axes in the separate layers, reveal the localized coherence through which such a large structure is assembled. Examples of the insights in this line derived from BTV include (1) a novel pseudo T=2 structure for the inner VP3 layer that is formally excluded by classic quasi-equivalence due to the hand of proteins but is accommodated by conformational switching of the VP3 subunits to form a relationship termed geometric quasi-equivalence; (2) a T=13 layer that adheres to classic quasi-equivalence to an extraordinary degree as a result of proteins VP7 making rather narrow and oily contacts with each other; and (3) the coincidence of symmetry between VP3 and VP7 layers at the icosahedral 3-fold axes, which indicates these to be the points of nucleation for the VP7 T=13 layer on a preformed VP3 T=2 layer, supported by the lessening strength of interaction apparent as one moves away from the 3-fold axis, suggestive of a hierarchical method for ordering T=13 growth.

It is also interesting to discuss work on rotavirus, which has a similar architecture to BTV. The crystal structure of VP6 (the rotavirus equivalent of BTV VP7) has been solved and fitted into a low-resolution cryo-EM reconstruction to reveal information about the contacts between viral subunits and layers (Mathieu *et al.*, 2001). The same conclusion was made concerning nucleation of the T=13 layer at the 3-fold axes and hierarchical outgrowth over the preformed T=2 layer according to the current most-favored interaction (Grimes *et al.*, 1998; Mathieu *et al.*, 2001). However, where the rotavirus and BTV systems apparently do differ is in the extent of the contacts made between proteins in the T=13 layer. In the rotavirus system the contacts between proteins are more extensive and not focused in a narrow basal band, and are apparently electrostatic and not hydrophobic. Furthermore, the rotaviral system shows substantial oligomeric polymorphism absent in BTV. BTV VP7 trimers are apparently incapable of forming higher order structures alone, but require as a template the VP3 T=2 layer. However, purified preparations of the equivalent rotavirus protein form spherical bodies at low pH and tubular structures at intermediate pH (Lepault *et al.*, 2001). The structural phases of the rotavirus system were mapped out with respect to pH using SAXS, and negative stain as well as cryo-EM (Lepault *et al.*, 2001). Helical reconstructions were performed on the tubular oligomers, which were of two diameters, but the spherical objects varied greatly in size so that a reconstruction of even one representative subclass was not possible. The conformation of the rotavirus protein in the tubes was apparently the same as in the crystal structure and the cryo-EM reconstruction of the double-layered core particle (Lepault *et al.*, 2001; Mathieu *et al.*, 2001). The fact that lowered pH induces self-organization supports the idea that electrostatic, not hydrophobic, contacts are predominant in the rotaviral system, whereas the formation of widely varying spherical assemblies in the lower pH ranges supports the idea that the inner T=2 layer is responsible for determining the size of the viral particle (Grimes *et al.*, 1998; Lepault *et al.*, 2001). Another example of polymorphism in the organization of a reoviral protein is African horse sickness virus (AHSV) VP7 (Basak *et al.*, 1996). AHSV is of the same architectural form as BTV and rotavirus, and like BTV is a member of the genus *Orbivirus.* In infected cells AHSV VP7 spontaneously forms small, flat hexagonal crystals, which have not been observed with any other orbiviruses (Burroughs *et al.*, 1994; Chuma *et al.*, 1992). The surface of AHSV VP7 has electrostatic properties different from those of BTV VP7 (Basak *et al.*, 1996); the base of BTV VP7 that interacts with the 180-subunit layer beneath is more hydrophobic than the equivalent region of AHSV, whereas the upper part, which would contact the outermost layer of the virus formed of VP2 and VP5, is more hydrophilic in BTV VP7 than in AHSV VP7 (Basak *et al.*, 1996). Presumably these differences are reflected in complementary changes in the interacting proteins, but equally may be the source of the unusual *in vivo* crystallization of AHSV VP7.

Although the VP7 protein of BTV shows essentially no conformational variation across the 13 quasi-equivalent copies in the core asymmetric unit

(Grimes *et al.*, 1998) the molecule is, nonetheless, capable of dramatic conformation rearrangements, as demonstrated by the crystallographic analysis of a different crystal form and a cleavage product (Basak *et al.*, 1997). This analysis suggested that the subunits of the trimer, which are "wound up" on the native core, can readily undergo a concerted unwinding. It has been proposed (Basak *et al.*, 1997) that such changes are relevant to the biological properties of the virus, perhaps being utilized during cell entry; however, at present, in the absence of direct evidence, this remains speculation.

A third virus that is structurally similar to BTV is rice dwarf virus (RDV) for which a cryo-EM reconstruction at a nominal resolution of 6.8 Å has been obtained (Zhou *et al.*, 2001). Bioinformatics coupled with novel methods in electron density map analysis were used to define the organization of secondary structure elements within the density and identify the primary sequences that form them (Zhou *et al.*, 2001). This involved identifying tubular density corresponding to helices in assigning structure *de novo* and comparing regions of the viral structural proteins with domains of known folds (Zhou *et al.*, 2001). The organization of RDV revealed was similar to that of BTV.

A study that compared cryo-EM reconstructions of different forms of orthoreovirus has defined the various structural targets this virus presents, as it did with HK97 and ϕ29. The intact virus (triple layered), a protease-treated infectious derivative, and the turreted core particle were visualized. Like BTV, orthoreovirus encapsidates 10 genome segments within its core (which has a diameter of ~700 Å), the structure of which was determined at 3.6-Å resolution by X-ray crystallography (Reinisch *et al.*, 2000). There are a number of exact parallels in both the technical approach made to and the results derived from the BTV and orthoreovirus structures. The orthoreovirus core surface is composed of 120 copies of a protein λ1 [homologous to BTV VP3 (T=2)], with λ2 forming the turrets and σ2 occupying three symmetrically distinct regions within the icosahedral asymmetric unit and acting as a clamp. A clamping protein was also seen with CPV (Hill *et al.*, 1999) and is presumably required to provide stability to the core surface, which in BTV derives from the T=13 (VP7) layer (Grimes *et al.*, 1998). The internal surface of the λ2 turrets provides the enzymatic guanyltransferase and methylase capping activity of the core, such that mRNA transcribed from the viral dsRNA beneath the 5-fold vertex receives a 5′ 7-methyl-GDP cap as eukaryotic mRNAs do. In BTV this function is performed by the transcription complex suspended from the interior surface of the inner capsid layer adjacent to the 5-fold vertices. This cap has the dual purpose of increasing the stability of the mRNA and targeting it to the cellular translation machinery. *S*-adenosylmethionine (SAM) cofactorially provides the methyl group for the cap, which is added to a GDP prebound at the 5′ end of the mRNA. The cofactor is thereby converted to *S*-Adenosylhomocysteine (SAH): the orthoreovirus crystals were soaked with SAH in order to locate the methylase domain and SAM-binding site. The arrangement of guanylyltransferase and methylase activities revealed is vectorial, resulting in a spatiotemporal chemical ordering as the 5′ terminus of the mRNA passes into and through the turret. Within the core, concentric layers of genomic dsRNA were visible as previously observed in BTV and other reoviruses (Gouet *et al.*, 1999; Hill *et al.*, 1999; Reinisch *et al.*, 2000; Zhang *et al.*, 1999) in accord with the suggestion that such ordering is functionally significant (Gouet *et al.*, 1999).

For orthoreovirus, diffraction data collected between 65- and 3.6-Å resolution were sharpened by application of a *B* factor of -55 Å^2, similar to that used in the BTV work (Grimes *et al.*, 1998). A 27-Å resolution cryo-EM reconstruction was used as a phasing start, simplified to a binary mask in an effort to escape the CTF errors. The position and orientation of the viral particle were essentially determined by the crystallographic space group. Phase extension was carried out incrementally by one reciprocal lattice point during cyclical averaging combined with solvent flattening. To extend from 4 to 3.6 Å, local averaging additional to the icosahedral noncrystallographic symmetry was applied to the asymmetric unit, while refinement was assisted by inserting a partial model for $\sigma 2$ from one symmetry environment into another. A platinum heavy atom derivative confirmed the chain tracing and facilitated extension to 3.6 Å. SAH-binding sites were located, using data sets from SAH-soaked crystals between 60 and 4 Å. The dsRNA was visualized by including additional 200- to 50-Å resolution diffraction data.

The outer $\sigma 3$ protein surface of the orthoreovirus was studied by an alternative combination of cryo-EM and crystallography (Olland *et al.*, 2001). The structure of the $\sigma 3$ monomer was determined at 1.8-Å resolution and then fitted into a low-resolution cryo-EM map on the basis of a combination of intuition, cross-correlation calculation, and immunological evidence (Olland *et al.*, 2001). The fit suggested the location of a fifth orthoreoviral protein, $\mu 1$, which forms an interface with $\sigma 3$ within a heterododecameric structure mutually exclusive to its crystallographic dimerization interface, suggesting a playoff between a homodimer and heterodimer during viral assembly.

No crystallographic structure is currently available for CPV-type single-shelled dsRNA virus capsids. Two cryo-EM reconstructions have been performed, which reveal much the same architecture as the orthoreovirus core: a single layer with turrets at the 5-fold vertices, statistically ordered layers of dsRNA, and a clamping protein at 2-fold axes (Hill *et al.*, 1999;

Zhang *et al.*, 1999). The layers of dsRNA have also been detected by SAXS (Harvey *et al.*, 1981). As with BTV and orthoreovirus, a 120-subunit surface forms the first (and here only) layer, a capsid geometry that was previously unknown and that appears to be ubiquitous among reoviruses. The virus is occluded within a crystal of the viral protein polyhedrin (Payne and Mertens, 1983), which dissolves in the alkaline pH of the insect gut, releasing infectious viruses. Occlusion within a robust crystal lattice provides a communal solution to the need for protection provided in other reoviruses by outer viral layers.

F. *Structural Analyses of Human Immunodeficiency Virus*

Human immunodeficiency virus (HIV) is the most studied and clinically important retrovirus. No single technique has been able to provide a structure for HIV, so that studies of its structure are chimeric rather than hybrid. Two HIV particles have received particular attention—the immature particle that buds from the host cell membrane, and the mature virus. Both particles have a similar radial organization, with differences in the structure of the individual layers between immature and mature forms (Fig. 6). Unfortunately it appears that HIV particles, whether native virus or recombinant virus-like particles, are heterogeneous in size and shape. It has been proposed that this is because the local geometric relationships of capsid subunits from which the virus is assembled fail to define a unique global particle symmetry (Wilk and Fuller, 1999). HIV has an outer membrane that contains the envelope proteins gp41 and gp120 (derived from a gp160 precursor by proteolysis) (Fig. 6). The structure of gp120 complexed with its receptor CD4 and a neutralizing antibody was solved by X-ray crystallography (Kwong *et al.*, 1998). The core structure of the gp41 ectodomain has also been solved by crystallography (Chan *et al.*, 1997; Tan *et al.*, 1997; Weissenhorn *et al.*, 1997), whereas the structure of a larger portion of this domain from the simian immunodeficiency virus (SIV) gp41 was solved by NMR (Caffrey *et al.*, 1998).

Beneath the membrane lie proteins derived, by proteolysis, from a Gag precursor. The outer protein, matrix, has been solved crystallographically for both SIV and HIV (Hill *et al.*, 1996; Rao *et al.*, 1995) (Fig. 6). Within this lies the capsid protein, which forms the characteristic conical capsid of the mature virus. Capsid has been solved in complex with an Fab fragment, as a head-to-tail dimer (Berthet-Colominas *et al.*, 1999), and structures for capsid protein fragments have also been solved: for the N-terminal core domain by NMR first (Gitti *et al.*, 1996), and later crystallographically (Gamble *et al.*, 1996). This N-terminal domain

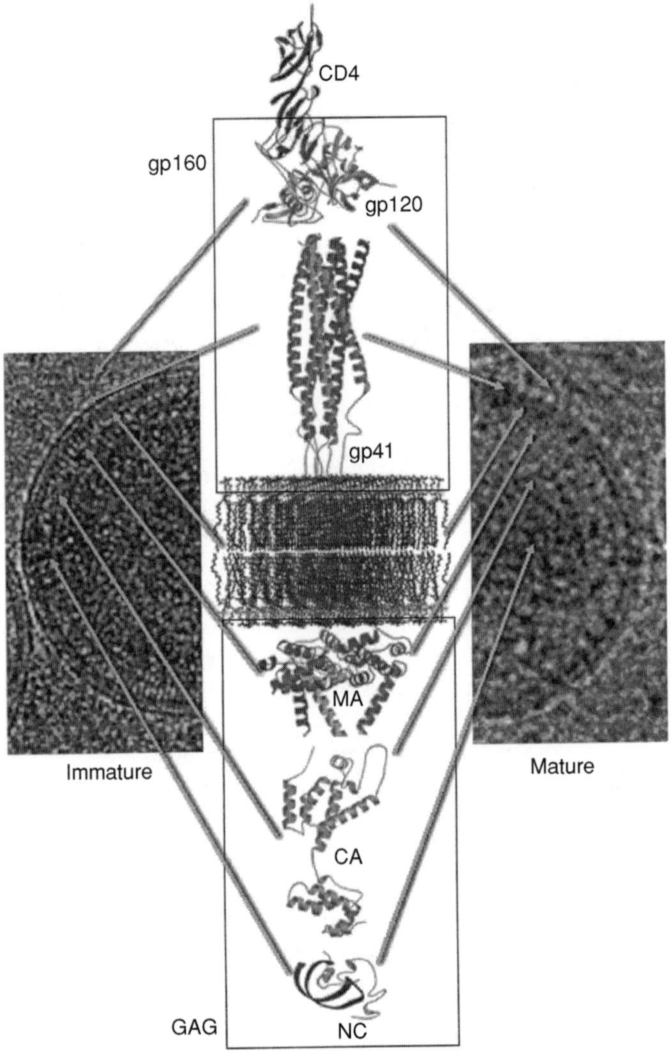

FIG. 6. Arrangement of proteins in immature and mature HIV. The central column shows representations of different HIV structural proteins in the order they occur radially from the outside (top) to the inside of the virus. Gp120 is shown complexed with its receptor CD4 from a crystal structure Protein Data Bank (PDB) ref 1GC1; (Kwong *et al.*, 1998), gp41 from a theoretical model developed using NMR constraints for SIV protein (PDB ref 1IF3), matrix protein (MA) from the X-ray crystal structure of its trimeric form (PDB ref 1HIW; Hill *et al.*, 1996), capsid protein (CA) from an X-ray structure (PDB ref 1E6J; Berthet-Colominas *et al.* 1999) and nucleocapsid protein (NC) from an NMR structure in which it was complexed with RNA (PDB ref 1A1Y; de Guzman *et al.*, 1998). On either side are shown cryo-electron micrographs of immature and mature forms, note that the CA protein has retreated within to form a conical core containing the NC in the natural form (Wilk *et al.*, 2001).

structure was also solved in crystals of full-length capsid protein in an antibody complex in which it was the only visible capsid protein density (Momany *et al.*, 1996). The C-terminal domain of capsid protein has also been solved crystallographically in isolation (Gamble *et al.*, 1997). Within the capsid the viral genome is complexed with a nucleocapsid protein, which has been solved by NMR alone (Morellet *et al.*, 1992; Summers *et al.*, 1992) and in complex with viral RNA (Amarasinghe *et al.*, 2000; de Guzman *et al.*, 1998) (Fig. 6). The structure of nucleocapsid protein is reminiscent of the RNA-binding proteins found, for example, in the ribosome that exhibits long extended folds relatively free of tertiary and secondary structure (Ban *et al.*, 2000; Wimberly *et al.*, 2000). Such a conformation assists the assembly and orchestration of RNA helices, which will be as essential in a replicating machine such as HIV, as it is in a translating machine like the ribosome.

The envelope glycoproteins gp120 and gp41 are believed to be arranged in a trimer of gp120–gp41 heterodimers (Kwong *et al.*, 2000). Current models of this structure are assembled on the basis of criteria such as carbohydrate exposure, positioning of conserved residues, epitope mapping, and steric constraints derived from the known structure of a neutralized gp120–antibody complex (Kwong *et al.*, 1998, 2000; Wyatt *et al.*, 1998). In solution gp120, the cell-binding domain, is monomeric, which suggests that its homotrimerization is the result of either induced conformational changes or association with the ectodomain of gp41, the fusion domain responsible for HIV internalization. gp41 possesses an architecture similar to that of the fusion proteins of influenza virus and Moloney murine leukemia virus (Caffrey *et al.*, 1998; Chan *et al.*, 1997; Tan *et al.*, 1997; Weissenhorn *et al.*, 1997). These are three examples of "class I" fusion proteins, which are similar to the cellular SNAREs (Mayer, 1999). Their α-helical coiled-coil arrangement in a spike should be contrasted with the sheet-rich flat fusion proteins that orchestrate "class II" fusion events in viruses such as SFV and tick-borne encephalitis virus (TBEV) (Ferlenghi *et al.*, 2001; Lescar *et al.*, 2001). The observation of threefold symmetry for both the matrix protein and gp41 suggests the possibility of specific symmetry-matched associations between the envelope and matrix layers on the inner surface of the viral membrane.

Gag has three consituent proteins, linked by spacer peptides. These are the matrix protein associated with the membrane, the capsid protein forming the internal capsid of the virus, and the nucleocapsid protein binding the viral RNA. A classic electron microscopy study of the interaction of actin with HIV revealed the radial organization of the proteins in the immature capsid, where it binds the nucleocapsid domain of the Gag structural polyprotein (Wilk *et al.*, 1999). A further study has shown the radial organization of these proteins in immature virus, using cryo-EM and deletion mutagenesis (Wilk *et al.*, 2001). Images were collected of *in vitro*-assembled Gag particles and also live immature HIV. The data were analyzed in terms of the radial density variations of the images and the repeating patterns of electron density present around the particle circumference. The viral membrane is the outermost layer. The matrix protein is myristoylated and through this is bound to the membrane (Hill *et al.*, 1996), although the protein itself is not membrane inserted. The capsid protein is joined to the matrix protein by a linker region and consists of two major domains forming the next layer of the particle. Finally, the innermost layer is formed of the nucleocapsid protein (Fig. 6). Because the same radial arrangement of Gag is present in *in vitro*-assembled particles lacking a membrane and in those that have a membrane, as well as in authentic immature HIV, the bilayer seems to have a limited role in organizing viral assembly (Wilk *et al.*, 2001).

Both SIV matrix (Rao *et al.*, 1995) and HIV matrix (Hill *et al.*, 1996) crystallize as trimers. The arrangements of these (similar) trimers suggested a possible organization of the matrix protein layer (and thereby Gag in the immature particle) in a hexameric net with the crystallographic trimer present at the junction between two hexamers (Rao *et al.*, 1995). This arrangement would yield a center-to-center spacing between matrix domains of 66 Å. Such a model would predict the 33-Å circumferential spacing of the matrix domain repeat observed in the cryo-EM study of immature virus and virus-like particles (Wilk *et al.*, 2001). The hexagonal arrangement of both matrix and capsid protein domains in Gag of immature viruses has also received support from classic electron microscopy techniques (Nermut *et al.*, 1998).

An interesting observation is that although the matrix protein forms trimers, the C-terminal domain of the capsid protein dimerizes both in solution and crystallographically (Gamble *et al.*, 1997), whereas a different, N-to-C domain (head-to-tail) interface is present in the crystals of full-length capsid protein with Fab (Berthet-Colominas *et al.*, 1999). These observations suggest there is substantial plasticity in the topology of capsid protein dimers, which seems to make repeated use of a particular selection of molecular interfaces that may facilitate the conformational changes in the virus structure accompanying maturation (Berthet-Colominas *et al.*, 1999). However, the antiparallel head-to-tail conformation of the full-length capsid protein dimer (Berthet-Colominas *et al.*, 1999) is not likely to be directly relevant to capsid assembly, which is much more likely to be based on parallel dimers. In summary, it is conceivable that both dimeric and trimeric interactions occur in the immature particle. Basing the construction of the matrix and capsid protein layers in the immature virion on different symmetries would increase the tensile strength of the Gag protein coat of these particles.

In mature HIV the capsid, matrix, and nucleocapsid domains of Gag are separated into autonomous proteins. The matrix protein remains bound to the quasi-spherical viral membrane. The core, which was concentric with the membrane, retreats within the particle and forms a cone-shaped structure containing the nucleocapsid protein complexed with RNA. Structural analyses of capsid protein assemblies have revealed a hexagonal net similar to that present in the matrix protein crystals (Li *et al.*, 2000; Rao *et al.*, 1995). It is thought that this represents the associational principle of the capsid. *In vitro*-assembled tubes of capsid protein possessed helical symmetry and were thereby reconstructed (Li *et al.*, 2000). The resulting electron density was fitted with the N- and C-terminal domains of the capsid protein, to yield a model for its arrangement in the cones in a hexagonal lattice. The size distribution of *in vitro*-assembled cones formed from a capsid–nucleocapsid fusion protein had previously been analyzed, suggesting that certain cone apex angles were allowed and others disallowed (Ganser *et al.*, 1999). Consequently, the local hexameric symmetry appears to impose global constraints on capsid assembly. The observed structures have been successfully modeled as a fullerene cone or fullerene sphere. A fullerene is created from a hexagonal net by the inclusion of pentagonal defects to close the structure. The conical or spherical nature of the final structure would be determined by the distribution of the pentagons. Approximately 5% of mature HIV capsids appear tubular rather than conical (Welker *et al.*, 2000); it will be interesting to compare the structural nature of such *ex vivo* tubes with those formed *in vitro* (Li *et al.*, 2000).

G. *Structural Insights from Virus Complexes*

With one exception (Smith *et al.*, 1996), all the studies of complexes of viruses with other proteins are hybrids of cryo-EM for the complex and crystallography for its components. The fitting of atomic models within cryo-EM density maps has revealed the relative positioning of structural components of viruses, and their receptors or antibodies, and has afforded insights into the mechanisms of cell invasion, viral uncoating, and immune recognition.

The picornavirus family, which includes rhinovirus, poliovirus, and foot-and-mouth disease virus, has provided the test bed for structural studies on virus–antibody and virus–receptor complexes. The crystal structure of human rhinovirus 14 (HRV14) revealed a canyon encircling the 5-fold vertex of the capsid (Rossmann *et al.*, 1985). Neutralizing immunogenic sites on the capsid surface were located around but not within this canyon, and the residues within the canyon were more conserved than those on its rim (Rossmann *et al.*, 1985). Consequently it was proposed that the viral receptor might bind deep in the canyon, while neutralizing antibodies bound to exposed surface features. This was intuitively satisfying for two reasons: first, the bulkiness of the antibody might well prevent it entering the canyon whereas a slender receptor molecule could easily insinuate itself there; second, the separation of residues involved in receptor binding from those forming immunogenic sites would allow resistance mutations in the viral coat without compromising cell-binding affinity (Rossmann *et al.*, 1985).

Cryo-EM reconstructions of virus–receptor complexes have subsequently confirmed that one class of rhinovirus receptors, namely intercellular adhesion molecule 1 (ICAMI), does bind in the canyon (Olson *et al.*, 1993). In poliovirus the receptor (CD155) also binds into the canyon around the icosahedral 5-fold axis (Belnap *et al.*, 2000b; He *et al.*, 2000). In the work on both rhinovirus– and poliovirus–receptor complexes, determining the positions of the receptor molecule glycans has proved of great assistance in confirming the fit of an atomic model to the cryo-EM density. Methods similar to those described above for alphaviruses were used to show the sugar side-chain density (He *et al.*, 2000; Kolatkar *et al.*, 1999; Xiao *et al.*, 2001) (Fig. 7). This allowed positioning of crystal structures or homology-based models of the receptor with likely glycosylation sequons adjacent to the glycan density. In the study of the poliovirus–CD155 interaction, the glycosylation sequons were simply placed adjacent to protrusions from the receptor electron density, which were assumed to arise from glycans (Belnap *et al.*, 2000b). In rhinovirus–receptor (Kolatkar *et al.*, 1999), poliovirus–receptor (Belnap *et al.*, 2000b; He *et al.*, 2000), and coxsackievirus–receptor (Xiao *et al.*, 2001) structures, the footprint of the receptor on the virus is similar, although the angle of the receptor molecule to the viral surface is different in each case.

The structure of foot-and-mouth disease virus (FMDV) (Acharya *et al.*, 1989) demonstrated that surface receptor interactions could use a quite different mode of attachment. This was developed further by visualizing FMDV–receptor interactions crystallographically (Fry *et al.*, 1999b). Furthermore, in the case of the binding of antibodies by viruses, in the single case for which the crystal structure of a virus–antibody complex has been accomplished, residues from the antibody paratope were found inserted into the canyon (Smith *et al.*, 1996). Thus, although residues within the canyon are more conserved than those outside, and selected

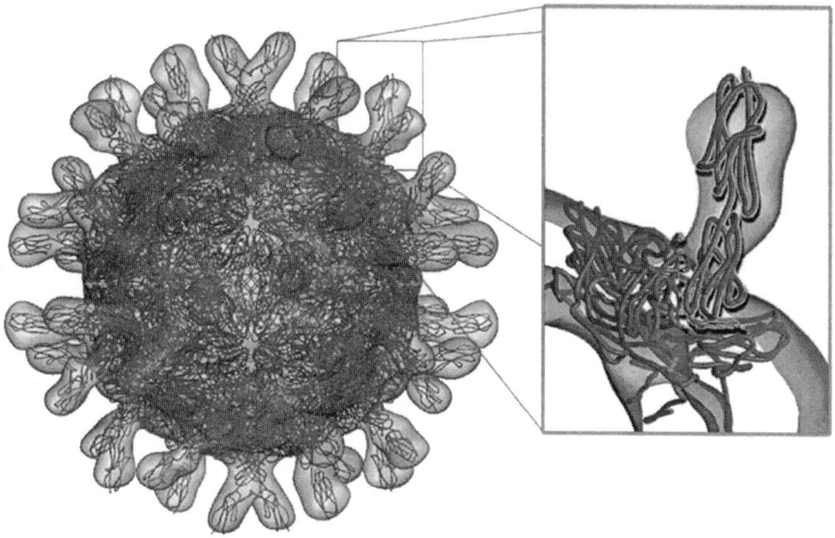

FIG. 7. Views of the interaction between human rhinovirus-16 and ICAM. The main body of the figure shows an isosurface representation of the human rhinovirus-16 ICAM complex reconstruction, fitted with the HRV-16 atomic coordinates and the ICAM on the basis of alignment of glycosylation sequons with the difference density arising from sugars when a reconstruction of a deglycosylated HRV-16 ICAM reconstruction was substracted from this one (Kolatkar *et al.*, 1999). A close-up of a single HRV 16 icosahedral asymmetric unit in interaction with ICAM is shown, with the ICAM highlighted by an outerglow. This work used a combination of cryo-EM reconstruction and the fitting of components to reveal the binding geometry and footprint of a viral receptor and is deposited in the Research Collaboratory for Structural Bioinformatics (RCSB) PDB, ref 1D3E.

neutralizing antibody escape mutations lie outside the canyon (Xiao *et al.*, 2001), antibodies are not in fact excluded from the receptor-binding sites of the virus. Indeed, it has been pointed out that overlap between antibody and receptor binding might indirectly give rise to host cell tropism through antibody-directed selection of mutants (Baranowski *et al.*, 2001). This would neatly invert the canyon hypothesis, and suggest that receptor binding in a canyon might be driven by another functional requirement.

This theme has been developed by studies of receptor–virus interactions for low-density lipoprotein receptor (LDLR), the receptor for a minor receptor group rhinovirus HRV-2 (Hewat *et al.*, 2000), and the glycolipid globoside bound to human parvovirus (Chipman *et al.*, 1996). Both of these receptors are small and globular and bind at different positions on the viral surface. In the case of LDLR, binding is at a star-shaped dome sitting on the 5-fold vertex of HRV-2, rather than in the canyon further "south" as for ICAM1 (Hewat *et al.*, 2000). In the case of globoside, the binding occurs at the 3-fold axis, in a depression, although the parvovirus has a canyon around its 5-fold vertex (Chipman *et al.*, 1996). The dichotomy in the binding mode of receptors is reflected in the difference in their mechanisms of uncoating: those viruses that have long receptors inserted into the surface of their capsids within the canyon are destabilized for delivery of their genome into the host cell by being levered open; on the other hand, viruses with receptors that do not insert into a canyon are endosomatically taken into the cell, where the capsids are destabilized by low pH, giving genome release. The case details for the exquisitely pH-sensitive FMDV are instructive. FMDV binds via a flexible loop containing an Arg-Gly-Asp sequence to an integrin, $\alpha_v\beta_6$ (Jackson *et al.*, 2000). In the original crystal structure of FMDV, this receptor-binding loop was not visible because it was disordered (Acharya *et al.*, 1989), presumably as a result of being made dynamically available to passing receptor molecules. The loop was observed, however, in virus in which a disulfide bond at the base of the integrin-binding loop was reduced, allowing it to collapse down onto the capsid surface (Logan *et al.*, 1993). The receptor-binding loop is also highly immunogenic, and the interaction of FMDV with neutralizing antibodies has been studied by cryo-EM (Hewat *et al.*, 1997; Verdaguer *et al.*, 1999). The cryo-EM density could in each case be fitted with the crystal structure of the virus and the Fab fragment of the antibody. The Fab structures had been solved in complex with peptides corresponding to their epitope, the receptor-binding loop. Therefore, fitting of the Fab–loop crystal structure into the cryo-EM density revealed the structure of the loop as it would be with the antibody bound on the viral surface. The orientation of the loop was different for each antibody, and it appears to act as a hinged structure (Hewat *et al.*, 1997; Verdaguer *et al.*, 1999), as proposed earlier on the basis of subtle structural and serological data (Parry *et al.*, 1990).

Other studies have focused on rhinovirus–antibody complexes, using cryo-EM and crystallography (Smith *et al.*, 1993a,b). These studies showed that, to achieve bivalent binding, the antibody rotated about an elbow axis, whereas the viral capsid was unaffected by antibody binding. This might have seemed surprising; however, because the antigenic and cell-binding sites around the canyon overlap the neutralizing effects of the antibodies are likely to arise from prevention of cell binding [this is also true for FMDV (Hewat *et al.*, 1997; Verdaguer *et al.*, 1999)]. The same antibody–virus interaction was studied crystallographically in a structure resolved at 4 Å which remains the only crystallographic study of a whole virus–protein ligand complex (Smith *et al.*, 1996). The Fab fragment was bound to

HRV14 and the complex purified before crystallization. The structure was solved by molecular replacement, using the HRV14 and Fab17-IA structures fitted into the cryo-EM density of the HRV14–Fab17-IA reconstruction. An envelope was calculated on the basis of this model and used during phase extension through rounds of cyclic averaging. The structure possessed 20-fold noncrystallographic symmetry. Model phases were extended one reciprocal lattice point at a time, with at least six cycles of noncrystallographic averaging at each point. The initial 4-Å map showed that the Fab fragments were ~4 Å away from their original positions and that the Fab constant domain density was diffuse and uninterpretable; this domain was therefore removed from the model. Phase refinement was completed by ~45 cycles of noncrystallographic symmetry averaging (Smith *et al.*, 1996). This structure showed the Fab to penetrate deep within the canyon, which had not been expected from the cryo-EM analysis.

Bivalent antibody binding has also been studied for HRV2 for a weakly neutralizing antibody, which bound across the icosahedral 2-fold axis and did not obstruct the canyon (Hewat and Blaas, 1996). This study showed that bivalent binding does not entail strong neutralization; it seems logical to believe that the weak neutralization of this antibody arose from its binding away from the canyon where the HRV2 receptor binds. Studying the interaction between the calicivirus rabbit hemorrhagic disease virus (RHDV) and a neutralizing antibody addressed the problem of partial occupancy of the antibody on the virus due to steric hindrance between adjacently bound antibodies by fitting of crystal structures to the density (Thouvenin *et al.*, 1997). Partial occupancy was also a problem in a study of antibody-mediated inhibition of the transcriptase activity of rotavirus studied by cryo-EM and crystal structure fitting (Thouvenin *et al.*, 2001). This interesting study analyzed the interaction of three different antibodies with the rotavirus double-layered particle (DLP). Two of the antibodies have no effect on transcription whereas the third inhibits it. Uniquely, the transcription-inhibiting antibody bound across two protomers of the outer trimeric molecule. It was therefore suggested that the inhibitory action of this antibody arose from the prevention of a conformational change in the VP7-equivalent protein associated with mRNA exit from the DLP, because its binding did not block the exit channel of the mRNA, nor did it cause a conformational change in the protein visible at the (albeit modest) resolution of the reconstruction of 23 Å. The questions of steric hindrance and partial occupancy, which are likely to be recurrent problems in cryo-EM analyses, have been addressed in an interesting review article (Thouvenin and Hewat, 2000).

IV. Conclusion

The structural biology of viruses not only benefits from a wide range of perspectives from techniques that reveal both static and dynamic aspects of the virus, but also invites cross-talk between these techniques. Approaches such as crystallography frequently yield the atomic secrets of the virion more readily by using a cryo-EM reconstruction as a starting point; such crystallographic analysis may only provide one structure of the range of conformers adopted by the virus during its life cycle, and cryo-EM can set such atomic models in context; conversely, our understanding of low-resolution cryo-EM maps can be massively assisted by atomic data on components of the structure being studied, and more fundamentally the process of deriving reconstructions can use preexisting or theoretical atomic models. The theme throughout these methodological interactions is that cryo-EM provides shape, while crystallography gives us chemistry: the play of chemistry through shapes is a powerful combination.

References

Acharya, R., Fry, E., Stuart, D., Fox, G., Rowlands, D., and Brown, F. (1989). *Nature* **337,** 709–716.

Amarasinghe, G. K., De Guzman, R. N., Turner, R. B., Chancellor, K. J., Wu, Z. R., and Summers, M. F. (2000). *J. Mol. Biol.* **301,** 491–511.

Athappilly, F. K., Murali, R., Rux, J. J., Cai, Z., and Burnett, R. M. (1994). *J. Mol. Biol.* **242,** 430–455.

Bamford, J.K., Cockburn, J., Diprose, J., Grimes, J.M., Sutton, G., Stuart, D.I., and Bamford, D.H. (2002) *J. Struct. Biol.* **2,** 103–112.

Ban, N., Nissen, P., Hansen, J., Moore, P. B., and Steitz, T. A. (2000). *Science* **289,** 905–920.

Baranowski, E., Ruiz-Jarabo, C. M., and Domingo, E. (2001). *Science* **292,** 1102–1105.

Basak, A. K., Gouet, P., Grimes, J., Roy, P., and Stuart, D. (1996). *J. Virol.* **70,** 3797–3806.

Basak, A. K., Grimes, J. M., Gouet, P., Roy, P., and Stuart, D. I. (1997). *Structure* **5,** 871–883.

Belnap, D. M., and Steven, A. C. (2000). *Trends Microbiol.* **8,** 91–93.

Belnap, D. M., Filman, D. J., Trus, B. L., Cheng, N., Booy, F. P., Conway, J. F., Curry, S., Hiremath, C. N., Tsang, S. K., Steven, A. C., and Hogle, J. M. (2000a). *J. Virol.* **74,** 1342–1354.

Belnap, D. M., McDermott, B. M. Jr., Filman, D. J., Cheng, N., Trus, B. L., Zuccola, H. J., Racaniello, V. R., Hogle, J. M., and Steven, A. C. (2000b). *Proc. Natl. Acad. Sci. USA* **97,** 73–78.

Benson, S. D., Bamford, J. K., Bamford, D. H., and Burnett, R. M. (1999). *Cell* **98,** 825–833.

Bentley, G. A., Lewit-Bentley, A., Liljas, L., Skoglund, U., Roth, M., and Unge, T. (1987). *J. Mol. Biol.* **194,** 129–141.

Berthet-Colominas, C., Monaco, S., Novelli, A., Sibai, G., Mallet, F., and Cusack, S. (1999). *EMBO J.* **18,** 1124–1136.

Bewley, M. C., Springer, K., Zhang, Y. B., Freimuth, P., and Flanagan, J. M. (1999). *Science* **286,** 1579–1583.

Bloomer, A. C., Champness, J. N., Bricogne, G., Staden, R., and Klug, A. (1978). *Nature* **276,** 362–368.

Bonami, J. R., Aubert, H., Mari, J., Poulos, B. T., and Lightner, D. V. (1997). *J. Struct. Biol.* **120,** 134–145.

Bonev, B. B., Gilbert, R. J., Andrew, P. W., Byron, O., and Watts, A. (2001). *J. Biol. Chem.* **276,** 5714–5719.

Bothner, B., Schneemann, A., Marshall, D., Reddy, V., Johnson, J. E., and Siuzdak, G. (1999). *Nat. Struct. Biol.* **6,** 114–116.

Bottcher, B., Wynne, S. A., and Crowther, R. A. (1997). *Nature* **386,** 88–91.

Burnett, R. M. (1985). *J. Mol. Biol.* **185,** 125–143.

Burns, N. R., Saibil, H. R., White, N. S., Pardon, J. F., Timmins, P. A., Richardson, S. M., Richards, B. M., Adams, S. E., Kingsman, S. M., and Kingsman, A. J. (1992). *EMBO J.* **11,** 1155–1164.

Burroughs, J. N., O'Hara, R. S., Smale, C. J., Hamblin, C., Walton, A., Armstrong, R., and Mertens, P. P. (1994). *J. Gen. Virol.* **75,** 1849–1857.

Butcher, S. J., Bamford, D. H., and Fuller, S. D. (1995). *EMBO J.* **14,** 6078–6086.

Byron, O., and Gilbert, R. J. (2000). *Curr. Opin. Biotechnol.* **11,** 72–80.

Caffrey, M., Cai, M., Kaufman, J., Stahl, S. J., Wingfield, P. T., Covell, D. G., Gronenborn, A. M., and Clore, G. M. (1998). *EMBO J.* **17,** 4572–4584.

Caspar, D.L.D., and Klug, A. (1962). *Cold Spring Harb. Symp. Quant. Biol.* **27,** 1–24.

Champness, J. N., Bloomer, A. C., Bricogne, G., Butler, P. G., and Klug, A. (1976). *Nature* **259,** 20–24.

Chan, D. C., Fass, D., Berger, J. M., and Kim, P. S. (1997). *Cell* **89,** 263–273.

Cheng, R. H., Kuhn, R. J., Olson, N. H., Rossmann, M. G., Choi, H. K., Smith, T. J., and Baker, T. S. (1995). *Cell* **80,** 621–630.

Chipman, P. R., Agbandje McKenna, M., Kajigaya, S., Brown, K. E., Young, N. S., Baker, T. S., and Rossmann, M. G. (1996). *Proc. Natl. Acad. Sci. USA* **93,** 7502–7506.

Chiu, C. Y., Mathias, P., Nemerow, G. R., and Stewart, P. L. (1999). *J. Virol.* **73,** 6759–6768.

Choi, H. K., Lu, G., Lee, S., Wengler, G., and Rossmann, M. G. (1997). *Proteins* **27,** 345–359.

Chuma, T., Le Blois, H., Sanchez-Vizcaino, J. M., Diaz-Laviada, M., and Roy, P. (1992). *J. Gen. Virol.* **73,** 925–931.

Conway, J. F., Wikoff, W. R., Cheng, N., Duda, R. L., Hendrix, R. W., Johnson, J. E., and Steven, A. C. (2001). *Science* **292,** 744–748.

Crick, F., and Watson, J. (1956). *Nature* **177,** 473–475.

Cross, T. A., Opella, S. J., Stubbs, G., and Caspar, D. L. (1983). *J. Mol. Biol.* **170,** 1037–1043.

Cullis, P. R., and de Kruijff, B. (1979). *Biochim. Biophys. Acta* **559,** 399–420.

Cusack, S., Miller, A., Krijgsman, P. C., and Mellema, J. E. (1981). *J. Mol. Biol.* **145,** 525–539.

Cusack, S., Oostergetel, G. T., Krijgsman, P. C., and Mellema, J. E. (1983). *J. Mol. Biol.* **171,** 139–155.

Cusack, S., Ruigrok, R. W., Krygsman, P. C., and Mellema, J. E. (1985). *J. Mol. Biol.* **186,** 565–582.

Devaux, C., Timmins, P. A., and Berthet-Colominas, C. (1983). *J. Mol. Biol.* **167,** 119–132.

Diprose, J. M., Burroughs, J. N., Sutton, G. C., Goldsmith, A., Gouet, P., Malby, R., Overton, I., Zientara, S., Mertens, P. P., Stuart, D. I., and Grimes, J. M. (2001). *EMBO J.* **20,** 7229–7239.

Dokland, T., McKenna, R., Sherman, D. M., Bowman, B. R., Bean, W. F., and Rossmann, M. G. (1998). *Acta Crystallogr. D Biol. Crystallogr.* **54,** 878–890.

Dryden, K. A., Farsetta, D. L., Wang, G., Keegan, J. M., Fields, B. N., Baker, T. S., and Nibert, M. L. (1998). *Virology* **245,** 33–46.

Duda, R. L. (1998). *Cell* **94,** 55–60.

Ferlenghi, I., Clarke, M., Ruttan, T., Allison, S. L., Schalich, J., Heinz, F. X., Harrison, S. C., Rey, F. A., and Fuller, S. D. (2001). *Mol. Cell* **7,** 593–602.

Fry, E. E., Grimes, J., and Stuart, D. I. (1999a). *Mol. Biotechnol.* **12,** 13–23.

Fry, E. E., Lea, S. M., Jackson, T., Newman, J. W., Ellard, F. M., Blakemore, W. E., Abu-Ghazaleh, R., Samuel, A., King, A. M., and Stuart, D. I. (1999b). *EMBO J.* **18,** 543–554.

Fuerstenau, S. D., Benner, W. H., Thomas, J. J., Brugidou, C., Bothner, B., and Siuzdak, G. (2001). *Angew. Chem. Int. Ed. Engl.* **40,** 541–544.

Fuller, S. D., Berriman, J. A., Butcher, S. J., and Gowen, B. E. (1995). *Cell* **81,** 715–725.

Fuller, S. D., Butcher, S. J., Cheng, R. H., and Baker, T. S. (1996). *J. Struct. Biol.* **116,** 48–55.

Gabashvili, I. S., Agrawal, R. K., Spahn, C. M., Grassucci, R. A., Svergun, D. I., Frank, J., and Penczek, P. (2000). *Cell* **100,** 537–549.

Gamble, T. R., Vajdos, F. F., Yoo, S., Worthylake, D. K., Houseweart, M., Sundquist, W. I., and Hill, C. P. (1996). *Cell* **87,** 1285–1294.

Gamble, T. R., Yoo, S., Vajdos, F. F., von Schwedler, U. K., Worthylake, D. K., Wang, H., McCutcheon, J. P., Sundquist, W. I., and Hill, C. P. (1997). *Science* **278,** 849–853.

Ganser, B. K., Li, S., Klishko, V. Y., Finch, J. T., and Sundquist, W. I. (1999). *Science* **283,** 80–83.

Gilbert, R. J., Jimenez, J. L., Chen, S., Tickle, I. J., Rossjohn, J., Parker, M., Andrew, P. W., and Saibil, H. R. (1999). *Cell* **97,** 647–655.

Gitti, R. K., Lee, B. M., Walker, J., Summers, M. F., Yoo, S., and Sundquist, W. I. (1996). *Science* **273,** 231–235.

Gouet, P., Diprose, J. M., Grimes, J. M., Malby, R., Burroughs, J. N., Zientara, S., Stuart, D. I., and Mertens, P. P. (1999). *Cell* **97,** 481–490.

Grimes, J., Basak, A. K., Roy, P., and Stuart, D. (1995). *Nature* **373,** 167–170.

Grimes, J. M., Jakana, J., Ghosh, M., Basak, A. K., Roy, P., Chiu, W., Stuart, D. I., and Prasad, B. V. (1997). *Structure* **5,** 885–893.

Grimes, J. M., Burroughs, J. N., Gouet, P., Diprose, J. M., Malby, R., Zientara, S., Mertens, P. P., and Stuart, D. I. (1998). *Nature* **395,** 470–478.

Guo, P., Zhang, C., Chen, C., Garver, K., and Trottier, M. (1998). *Mol. Cell* **2,** 149–155.

de Guzman, R. N., Wu, Z. R., Stalling, C. C., Pappalardo, L., Borer, P. N., and Summers, M. F. (1998). *Science* **279,** 384–388.

Hajdu, J. (2000). *Curr. Opin. Struct. Biol.* **10,** 569–573.

Harvey, J. D., Bellamy, A. R., Earnshaw, W. C., and Schutt, C. (1981). *Virology* **112,** 240–249.

He, Y., Bowman, V. D., Mueller, S., Bator, C. M., Bella, J., Peng, X., Baker, T. S., Wimmer, E., Kuhn, R. J., and Rossmann, M. G. (2000). *Proc. Natl. Acad. Sci. USA* **97,** 79–84.

Henderson, R. (2002). *Nature* **415,** 833.

Hewat, E. A., and Blaas, D. (1996). *EMBO J.* **15,** 1515–1523.

Hewat, E. A., Booth, T. F., Loudon, P. T., and Roy, P. (1992). *Virology* **189,** 10–20.

Hewat, E. A., Verdaguer, N., Fita, I., Blakemore, W., Brookes, S., King, A., Newman, J., Domingo, E., Mateu, M. G., and Stuart, D. I. (1997). *EMBO J.* **16,** 1492–1500.

Hewat, E. A., Neumann, E., Conway, J. F., Moser, R., Ronacher, B., Marlovits, T. C., and Blaas, D. (2000). *EMBO J.* **19,** 6317–6325.

Hill, C. L., Booth, T. F., Prasad, B. V., Grimes, J. M., Mertens, P. P., Sutton, G. C., and Stuart, D. I. (1999). *Nat. Struct. Biol.* **6,** 565–568.

Hill, C. P., Worthylake, D., Bancroft, D. P., Christensen, A. M., and Sundquist, W. I. (1996). *Proc. Natl. Acad. Sci. USA* **93,** 3099–3104.

Jack, A., Harrison, S. C., and Crowther, R. A. (1975). *J. Mol. Biol.* **97,** 163–172.

Jackson, T., Sheppard, D., Denyer, M., Blakemore, W., and King, A. M. (2000). *J. Virol.* **74,** 4949–4956.

Jacrot, B., Chauvin, C., and Witz, J. (1977). *Nature* **266,** 417–421.

Johnson, J. E. (1999). Principles of virus structure. *In* "Encyclopaedia of Virology" (A. Granoff and R. G. Webster, Eds.), 2nd Ed., pp. 1946–1956. Academic Press, London.

Kan, J. H., Cremers, A. F., Haasnoot, C. A., and Hilbers, C. W. (1987). *Eur. J. Biochem.* **168,** 635–639.

Kanamaru, S., Leiman, P. G., Kostyuchenko, V. A., Chipman, P. R., Mesyanzhinov, V. V., Arisaka, F., and Rossmann, M. G. (2002). *Nature* **415,** 553–557.

Kliger, Y., Peisajovich, S. G., Blumenthal, R., and Shai, Y. (2000). *J. Mol. Biol.* **301,** 905–914.

Klug, A. (1999). *Philos. Trans. R. Soc. Biol. Sci.* **354,** 531–535.

Kolatkar, P. R., Bella, J., Olson, N. H., Bator, C. M., Baker, T. S., and Rossmann, M. G. (1999). *EMBO J.* **18,** 6249–6259.

Kostyuchenko, V. A., Navruzbekov, G. A., Kurochkina, L. P., Strelkov, S. V., Mesyanzhinov, V. V., and Rossmann, M. G. (1999). *Structure* **7,** 1213–1222.

Kruse, J., Kruse, K. M., Witz, J., Chauvin, C., Jacrot, B., and Tardieu, A. (1982). *J. Mol. Biol.* **162,** 393–414.

Kuznetsov Yu, G., Malkin, A. J., Land, T. A., DeYoreo, J. J., Barba, A. P., Konnert, J., and McPherson, A. (1997). *Biophys. J.* **72,** 2357–2364.

Kuznetsov, Y. G., Larson, S. B., Day, J., Greenwood, A., and McPherson, A. (2001). *Virology* **284,** 223–234.

Kwong, P. D., Wyatt, R., Robinson, J., Sweet, R. W., Sodroski, J., and Hendrickson, W. A. (1998). *Nature* **393,** 648–659.

Kwong, P. D., Wyatt, R., Sattentau, Q. J., Sodroski, J., and Hendrickson, W. A. (2000). *J. Virol.* **74,** 1961–1972.

Landsberger, F. R., Lenard, J., Paxton, J., and Compans, R. W. (1971). *Proc. Natl. Acad. Sci. USA* **68,** 2579–2583.

Landsberger, F. R., Compans, R. W., Paxton, J., and Lenard, J. (1972). *J. Supramol. Struct.* **1,** 50–54.

Lata, R., Conway, J. F., Cheng, N., Duda, R. L., Hendrix, R. W., Wikoff, W. R., Johnson, J. E., Tsuruta, H., and Steven, A. C. (2000). *Cell* **100,** 253–263.

Leiman, P. G., Kostyuchenko, V. A., Shneider, M. M., Kurochkina, L. P., Mesyanzhinov, V. V., and Rossmann, M. G. (2000). *J. Mol. Biol.* **301,** 975–985.

Lepault, J., Petitpas, I., Erk, I., Navaza, J., Bigot, D., Dona, M., Vachette, P., Cohen, J., and Rey, F. A. (2001). *EMBO J.* **20,** 1498–1507.

Lescar, J., Rossel, A., Wien, M. W., Navaza, J., Fuller, S. D., Wengler, G., Wengler, G., and Rey, R. A. (2001). *Cell* **105,** 137–148.

Leslie, A. G. W. (1992). "Joint CCP4 + ESF EAMCB Newsletter on Protein Crystallography," **26,** http://www.mrc-lmb.cam.ac.uk/harry/mosflm/mosflm_user_guide.html.

Li, S., Hill, C. P., Sundquist, W. I., and Finch, J. T. (2000). *Nature* **407,** 409–413.
Liddington, R. C., Yan, Y., Moulai, J., Sahli, R., Benjamin, T. L., and Harrison, S. C. (1991). *Nature* **354,** 278–284.
Liu, D. J., and Day, L. A. (1994). *Science* **265,** 671–674.
Logan, D., Abu-Ghazaleh, R., Blakemore, W., Curry, S., Jackson, T., King, A., Lea, S., Lewis, R., Newman, J., and Parry, N., *et al.* (1993). *Nature* **362,** 566–568.
Luger, K., Mader, A. W., Richmond, R. K., Sargent, D. F., and Richmond, T. J. (1997). *Nature* **389,** 251–260.
Lyles, D. S., and Landsberger, F. R. (1977). *Proc. Natl. Acad. Sci. USA* **74,** 1918–1922.
Malkin, A. J., Kuznetsov, Y. G., Lucas, R. W., and McPherson, A. (1999). *J. Struct. Biol.* **127,** 35–43.
Mancini, E. J., and Fuller, S. D. (2000). *Acta Crystallogr. D Biol. Crystallogr.* **56,** 1278–1287.
Mancini, E. J., Clarke, M., Gowen, B. E., Rutten, T., and Fuller, S. D. (2000). *Mol. Cell* **5,** 255–266.
Martin, C. S., Burnett, R. M., de Haas, F., Heinkel, R., Rutten, T., Fuller, S. D., Butcher, S. J., and Bamford, D. H. (2001). *Structure* **9,** 917–930.
Mat-Arip, Y., Garver, K., Chen, C., Sheng, S., Shao, Z., and Guo, P. (2001). *J. Biol. Chem.* **276,** 32575–32584.
Mathieu, M., Petitpas, I., Navaza, J., Lepault, J., Kohli, E., Pothier, P., Prasad, B. V. V., Cohen, J., and Rey, F. A. (2001). *EMBO J.* **20,** 1485–1497.
Mayer, A. (1999). *Curr. Opin. Cell. Biol.* **11,** 447–452.
McKenna, R., Xia, D., Willingmann, P., Ilag, L. L., and Rossmann, M. G. (1992). *Acta Crystallogr. B* **48,** 499–511.
McPherson, A., Malkin, A. J., Kuznetsov, Y. G., and Plomp, M. (2001). *Acta Crystallogr. D Biol. Crystallogr.* **57,** 1053–1060.
Meining, W., Bacher, A., Bachmann, L., Schmid, C., Weinkauf, S., Huber, R., and Nar, H. (1995). *J. Mol. Biol.* **253,** 208–218.
Mely, Y., Jullian, N., Morellet, N., De Rocquigny, H., Dong, C. Z., Piemont, E., Roques, B. P., and Gerard, D. (1994). *Biochemistry* **33,** 12085–12091.
Miao, J., Kirz, J., and Sayre, D. (2000). *Acta Crystallogr. D Biol. Crystallogr.* **56,** 1312–1315.
Miao, J., Hodgson, K. O., and Sayre, D. (2001). *Proc. Natl. Acad. Sci. USA* **98,** 6641–6645.
Momany, C., Kovari, L. C., Prongay, A. J., Keller, W., Gitti, R. K., Lee, B. M., Gorbalenya, A. E., Tong, L., McClure, J., Ehrlich, L. S., Summers, M. F., Carter, C., and Rossmann, M. G. (1996). *Nat. Struct. Biol.* **3,** 763–770.
Morais, M. C., Tao, Y., Olson, N. H., Grimes, S., Jardine, P. J., Anderson, D. L., Baker, T. S., and Rossmann, M. G. (2001). *J. Struct. Biol.* **135,** 38–46.
Morellet, N., Jullian, N., De Rocquigny, H., Maigret, B., Darlix, J. L., and Roques, B. P. (1992). *EMBO J.* **11,** 3059–3065.
Munowitz, M. G., Dobson, C. M., Griffin, R. G., and Harrison, S. C. (1980). *J. Mol. Biol.* **141,** 327–333.
Namba, K., Pattanayek, R., and Stubbs, G. (1989). *J. Mol. Biol.* **208,** 307–325.
Nermut, M. V., Hockley, D. J., Bron, P., Thomas, D., Zhang, W. H., and Jones, I. M. (1998). *J. Struct. Biol.* **123,** 143–149.
Olland, A. M., Jane-Valbuena, J., Schiff, L. A., Nibert, M. L., and Harrison, S. C. (2001). *EMBO J.* **20,** 979–989.
Olson, A. J., Bricogne, G., and Harrison, S. C. (1983). *J. Mol. Biol.* **171,** 61–93.
Olson, N. H., Kolatkar, P. R., Oliveira, M. A., Cheng, R. H., Greve, J. M., McClelland, A., Baker, T. S., and Rossmann, M. G. (1993). *Proc. Natl. Acad. Sci. USA* **90,** 507–511.
Olson, N. H., Gingery, M., Eiserling, F. A., and Baker, T. S. (2001). *Virology* **279,** 385–391.

Oostergetel, G. T., Mellema, J. E., and Cusack, S. (1983). *J. Mol. Biol.* **171,** 157–173.
Orlova, E. V., Dube, P., Harris, J. R., Beckman, E., Zemlin, F., Markl, J., and van Heel, M. (1997). *J. Mol. Biol.* **271,** 417–437.
Orlova, E. V., Rahman, M. A., Gowen, B., Volynski, K. E., Ashton, A. C., Manser, C., van Heel, M., and Ushkaryoy, Y. A. (2000). *Nat. Struct. Biol.* **7,** 48–53.
Otwinowski, Z., and Minor, W. (1997). *Methods Enzymol.* **276,** 307–326.
Parry, N., Fox, G., Rowlands, D., Brown, F., Fry, E., Acharya, R., Logan, D., and Stuart, D. (1990). *Nature* **347,** 569–572.
Pattanayek, R., and Stubbs, G. (1992). *J. Mol. Biol.* **228,** 516–528.
Pauptit, R. A., Dennis, C. A., Derbyshire, D. J., Breeze, A. L., Weston, S. A., Rowsell, S., and Murshudov, G. N. (2001). *Acta Crystallogr. D Biol. Crystallogr.* **57,** 1397–1404.
Payne, C. C., and Mertens, P. P. C. (1983). Cytoplasmic polyhedrosis viruses. *In* "The Reoviridae" (W. F. Joklik, Ed.), pp. 425–504. Plenum Press, New York.
Pederson, D. M., Welsh, L. C., Marvin, D. A., Sampson, M., Perham, R. N., Yu, M., and Slater, M. R. (2001). *J. Mol. Biol.* **309,** 401–421.
Pletnev, S. V., Zhang, W., Mukhopadyay, S., Fisher, B. R., Hernandez, R., Brown, D. T., Baker, T. S., Rossmann, M. G., and Kuhn, R. J. (2001). *Cell* **105,** 127–136.
Prasad, B. V., Hardy, M. E., Dokland, T., Bella, J., Rossmann, M. G., and Estes, M. K. (1999). *Science* **286,** 287–290.
Qanungo, K. R., Kundu, S. C., and Ghosh, A. K. (2000). *Acta Virol.* **44,** 349–357.
Rao, Z., Belyaey, A. S., Fry, E., Roy, P., Jones, I. M., and Stuart, D. I. (1995). *Nature* **378,** 743–747.
Ravantti, J. J., and Bamford, D. H. (1999). *J. Struct. Biol.* **125,** 216–222.
Reinisch, K. M., Nibert, M. L., and Harrison, S. C. (2000). *Nature* **404,** 960–967.
Roberts, M. M., White, J. L., Grutter, M. G., and Burnett, R. M. (1986). *Science* **232,** 1148–1151.
Robinson, I. K., and Harrison, S. C. (1982). *Nature* **297,** 563–568.
Rossmann, M. G. (1999). *J. Synchrotron Radiat.* **6,** 816–821.
Rossmann, M. G. (2000). *Acta Crystallogr. D Biol. Crystallogr.* **56,** 1341–1349.
Rossmann, M. G. Arnold, E. Erickson, J. W. Frankenberger, E. A. Griffith, J. P. Hecht, H. J. Johnson, J. E. Kamer, G. Luo, M. Mosser, A. G., *et al.* (1985). *Nature* **317,** 145–153.
Rydman, P. S., Caldentey, J., Butcher, S. J., Fuller, S. D., Rutten, T., and Bamford, D. H. (1999). *J. Mol. Biol.* **291,** 575–587.
Saad, A., Ludtke, S. J., Jakana, J., Rixon, F. J., Tsuruta, H., and Chiu, W. (2001). *J. Struct. Biol.* **133,** 32–42.
San Martin, C., Huiskonen, J. T., Bamford, J. K., Butcher, S. J., Fuller, S. D., Bamford, D. H., and Burnett, R. M. (2002). *Nat. Struct. Biol.* **10,** 756–763.
Sato, M., Sato Miyamoto, Y., Kameyama, K., Ishikawa, N., Imai, M., Ito, Y., and Takagi, T. (1995). *J. Biochem. (Tokyo)* **118,** 1297–1302.
Satoh, O., Imai, H., Yoneyama, T., Miyamura, T., Utsumi, H., Inoue, K., and Umeda, M. (2000). *J. Biochem. (Tokyo)* **127,** 543–550.
Schmid, M. F., Sherman, M. B., Matsudaira, P., Tsuruta, H., and Chiu, W. (1999). *J. Struct. Biol.* **128,** 51–57.
Schoehn, G., Hayes, M., Cliff, M., Clarke, A. R., and Saibil, H. R. (2000). *J. Mol. Biol.* **301,** 323–332.
Simpson, A. A., Tao, Y., Leiman, P. G., Badasso, M. O., He, Y., Jardine, P. J., Olson, N. H., Morais, M. C., Grimes, S., Anderson, D. L., Baker, T. S., and Rossmann, M. G. (2000). *Nature* **408,** 745–750.
Siuzdak, G. (1998). *J. Mass. Spectrom.* **33,** 203–211.

Siuzdak, G., Bothner, B., Yeager, M., Brugidou, C., Fauquet, C. M., Hoey, K., and Chang, C. M. (1996). *Chem. Biol.* **3,** 45–48.

Smith, D. E., Tans, S. J., Smith, S. B., Grimes, S., Anderson, D. L., and Bustamante, C. (2001). *Nature* **413,** 748–752.

Smith, T. J., Olson, N. H., Cheng, R. H., Chase, E. S., and Baker, T. S. (1993a). *Proc. Natl. Acad. Sci. USA* **90,** 7015–7018.

Smith, T. J., Olson, N. H., Cheng, R. H., Liu, H., Chase, E. S., Lee, W. M., Leippe, D. M., Mosser, A. G., Rueckert, R. R., and Baker, T. S. (1993b). *J. Virol.* **67,** 1148–1158.

Smith, T. J., Chase, E. S., Schmidt, T. J., Olson, N. H., and Baker, T. S. (1996). *Nature* **383,** 350–354.

Spahn, C. M., Penczek, P. A., Leith, A., and Frank, J. (2000). *Structure* **15,** 937–948.

Spahn, C. M., Beckmann, R., Eswar, N., Penczek, P. A., Sali, A., Blobel, G., and Frank, J. (2001). *Cell* **107,** 373–386.

Speir, J. A., Munshi, S., Wang, G., Baker, T. S., and Johnson, J. E. (1995). *Structure* **3,** 63–78.

Stark, H., Rodnina, M. V., Wieden, H. J., van Heel, M., and Wintermeyer, W. (2000). *Cell* **100,** 301–309.

Stewart, P. L., Burnett, R. M., Cyrklaff, M., and Fuller, S. D. (1991). *Cell* **67,** 145–154.

Stewart, P. L., Fuller, S. D., and Burnett, R. M. (1993). *EMBO J.* **12,** 2589–2599.

Strelkov, S. V., Tao, Y., Rossmann, M. G., Kurochkina, L. P., Shneider, M. M., and Mesyanzhinov, V. V. (1996). *Virology* **219,** 190–194.

Strelkov, S. V., Tao, Y., Shneider, M. M., Mesyanzhinov, V. V., and Rossmann, M. G. (1998). *Acta Crystallogr. D Biol. Crystallogr.* **54,** 805–816.

Stubbs, G. (1999). *Philos. Trans. R. Soc. B Biol. Sci.* **354,** 551–557.

Stubbs, G., Warren, S., and Holmes, K. (1977). *Nature* **267,** 216–212.

Suck, D., Rayment, I., Johnson, J. E., and Rossmann, M. G. (1978). *Virology* **85,** 187–197.

Summers, M. F. Henderson, L. E. Chance, M. R. Bess, J. W., Jr. South, T. L. Blake, P. R. Sagi, I. Perez Alvarado, G. Sowder, R. C., III Hare, D. R., *et al.* (1992). *Protein Sci.* **1,** 563–574.

Tan, K., Liu, J., Wang, J., Shen, S., and Lu, M. (1997). *Proc. Natl. Acad. Sci. USA* **94,** 12303–12308.

Tang, L., Johnson, K. N., Bal, L. A., Lin, T., Yeager, M., and Johnson, J. E. (2001). *Nat. Struct. Biol.* **8,** 77–83.

Tao, Y., Strelkoy, S. V., Mesyanzhinov, V. V., and Rossmann, M. G. (1997). *Structure* **5,** 789–798.

Tao, Y., Olson, N. H., Xu, W., Anderson, D. L., Rossmann, M. G., and Baker, T. S. (1998). *Cell* **95,** 431–437.

Thouvenin, E., and Hewat, E. (2000). *Acta Crystallogr. D Biol. Crystallogr.* **56,** 1350–1357.

Thouvenin, E., Laurent, S., Madelaine, M. F., Rasschaert, D., Vautherot, J. F., and Hewat, E. A. (1997). *J. Mol. Biol.* **270,** 238–246.

Thouvenin, E., Schoehn, G., Rey, F., Petitpas, I., Mathieu, M., Vaney, M. C., Cohen, J., Kohli, E., Pothier, P., and Hewat, E. (2001). *J. Mol. Biol.* **307,** 161–172.

Timmins, P. A., Wild, D., and Witz, J. (1994). *Structure* **2,** 1191–1201.

Tito, M. A., Tars, K., Valegard, K., Hajdu, J., and Robinson, C. V. (2000). *J. Am. Chem. Soc.* **122,** 3550–3551.

Tomita, M., Hasegawa, T., Tsukihara, T., Miyajima, S., Nagao, M., and Sato, M. (1999). *J. Biochem. (Tokyo)* **125,** 916–922.

Tuma, R., Bamford, J. H., Bamford, D. H., Russell, M. P., and Thomas, G. J., Jr. (1996a). *J. Mol. Biol.* **257,** 87–101.

Tuma, R., Bamford, J. H., Bamford, D. H., and Thomas, G. J. Jr. (1996b). *J. Mol. Biol.* **257,** 102–115.
Tuma, R., Coward, L. U., Kirk, M. C., Barnes, S., and Prevelige, P. E., Jr. (2001). *J. Mol. Biol.* **306,** 389–396.
Turner, B. G., and Summers, M. F. (1999). *J. Mol. Biol.* **285,** 1–32.
Valpuesta, J. M., Fernandez, J. J., Carazo, J. M., and Carrascosa, J. L. (1999). *Structure* **7,** 289–296.
van Heel, M., Gowen, B., Matadeen, R., Orlova, E. V., Finn, R., Pape, T., Cohen, D., Stark, H., Schmidt, R., Schatz, M., and Patwardan, A. (2000). *Q. Rev. Biophys.* **33,** 307–369.
van Raaij, M. J., Mitraki, A., Lavigne, G., and Cusack, S. (1999). *Nature* **401,** 935–938.
Venien-Bryan, C., and Fuller, S. D. (1994). *J. Mol. Biol.* **236,** 572–583.
Verdaguer, N., Schoehn, G., Ochoa, W. F., Fita, I., Brookes, S., King, A., Domingo, E., Mateu, M. G., Stuart, D., and Hewat, E. A. (1999). *Virology* **255,** 260–268.
Wang, H., and Stubbs, G. (1994). *J. Mol. Biol.* **239,** 371–384.
Wang, H., Culver, J. N., and Stubbs, G. (1997). *J. Mol. Biol.* **269,** 769–779.
Wang, L., Lane, L. C., and Smith, D. L. (2001). *Protein Sci.* **10,** 1234–1243.
Weissenhorn, W., Dessen, A., Harrison, S. C., Skehel, J. J., and Wiley, D. C. (1997). *Nature* **387,** 426–430.
Welker, R., Hohenberg, H., Tessmer, U., Huckhagel, C., and Krausslich, H. G. (2000). *J. Virol.* **74,** 1168–1177.
Welsh, L. C., Marvin, D. A., and Perham, R. N. (1998a). *J. Mol. Biol.* **284,** 1265–1271.
Welsh, L. C., Symmons, M. F., Sturtevant, J. M., Marvin, D. A., and Perham, R. N. (1998b). *J. Mol. Biol.* **283,** 155–177.
Welsh, L. C., Symmons, M. F., and Marvin, D. A. (2000). *Acta Crystallogr. D Biol. Crystallogr.* **56,** 137–150.
Wikoff, W. R., Liljas, L., Duda, R. L., Tsuruta, H., Hendrix, R. W., and Johnson, J. E. (2000). *Science* **289,** 2129–2133.
Wilk, T., and Fuller, S. D. (1999). *Curr. Opin. Struct. Biol.* **9,** 231–243.
Wilk, T., Gowen, B., and Fuller, S. D. (1999). *J. Virol.* **73,** 1931–1940.
Wilk, T., Gross, I., Gowen, B. E., Rutten, T., de Haas, F., Welker, R., Krausslich, H. G., Boulanger, P., and Fuller, S. D. (2001). *J. Virol.* **75,** 759–771.
Wimberly, B. T., Brodersen, D. E., Clemons, W. M., Jr., Morgan Warren, R. J., Carter, A. P., Vonrhein, C., Hartsch, T., and Ramakrishnan, V. (2000). *Nature* **407,** 327–339.
Winkler, F. K., Schutt, C. E., Harrison, S. C., and Bricogne, G. (1977). *Nature* **265,** 509–513.
Witz, J., Timmins, P. A., and Adrian, M. (1993). *Proteins* **17,** 223–231.
Woo, S. D., Kim, W. J., Kim, H. S., Jin, B. R., Lee, Y. H., and Kang, S. K. (1998). *Arch. Virol.* **143,** 1209–1214.
Wyatt, R., Kwong, P. D., Desjardins, E., Sweet, R. W., Robinson, J., Hendrickson, W. A., and Sodroski, J. G. (1998). *Nature* **393,** 705–711.
Wynne, S. A., Crowther, R. A., and Leslie, A. G. (1999). *Mol. Cell* **3,** 771–780.
Xiao, C., Bator, C. M., Bowman, V. D., Rieder, E., He, Y., Hebert, B., Bella, J., Baker, T. S., Wimmer, E., Kuhn, R. J., and Rossmann, M. G. (2001). *J. Virol.* **75,** 2444–2451.
Yan, X., Olson, N. H., Van Etten, J. L., Bergoin, M., Rossmann, M. G., and Baker, T. S. (2000). *Nat. Struct. Biol.* **7,** 101–103.
Yeager, M., Berriman, J. A., Baker, T. S., and Bellamy, A. R. (1994). *EMBO J.* **13,** 1011–1018.
Yeagle, P. L. (1990). *Biol. Magn. Res.* **9,** 1–50.
Zhang, F., Lemieux, S., Wu, X., St Arnaud, D., McMurray, C. T., Major, F., and Anderson, D. (1998). *Mol. Cell* **2,** 141–147.

Zhang, H., Zhang, J., Yu, X., Lu, X., Zhang, Q., Jakana, J., Chen, D. H., Zhang, X., and Zhou, Z. H. (1999). *J. Virol.* **73,** 1624–1629.

Zhang, W., Olson, N. H., Baker, T. S., Faulkner, L., Agbandje-McKenna, M., Boulton, M. I., Davies, J. W., and McKenna, R. (2001). *Virology* **279,** 471–477.

Zhao, W., Ji, W. G., and Rowlands, J. A. (2001). *Med. Phys.* **28,** 2039–2049.

Zhou, Z. H., Dougherty, M., Jakana, J., He, J., Rixon, F. J., and Chiu, W. (2000). *Science* **288,** 877–880.

Zhou, Z. H., Baker, M. L., Jiang, W., Dougherty, M., Jakana, J., Dong, G., Lu, G., and Chiu, W. (2001). *Nat. Struct. Biol.* **8,** 868–873.

Zlotnick, A., Johnson, J. M., Wingfield, P. W., Stahl, S. J., and Endres, D. (1999). *Biochemistry* **38,** 14644–14652.

Zlotnick, A., Aldrich, R., Johnson, J. M., Ceres, P., and Young, M. J. (2000). *Virology* **277,** 450–456.

DETERMINATION OF ICOSAHEDRAL VIRUS STRUCTURES BY ELECTRON CRYOMICROSCOPY AT SUBNANOMETER RESOLUTION

By Z. HONG ZHOU* AND WAH CHIU†

*Department of Pathology and Laboratory Medicine, University of Texas–Houston Medical School, Houston, Texas 77030, and †National Center for Macromolecular Imaging, Verna and Marrs McLean Department of Biochemistry and Molecular Biology, Baylor College of Medicine, Houston, Texas 77030

I. Introduction 93
II. Electron Cryomicroscopy 94
A. XTheoretical Considerations of Electron Imaging 94
B. Choice of Instrument 97
C. Number of Particle Images Needed for a Three-Dimensional Reconstruction 101
III. Overview of Methods for Subnanometer-resolution Reconstructions 101
A. CTF and *B* Factor Corrections 102
B. Orientation and Center Determination 105
IV. Example of Data Collection, Evaluation, and Processing 105
A. Imaging 105
B. Digitization and Particle Selection 107
C. Image Screening 109
D. Orientation and Center Determination by the Focal Pair Method 110
E. Three-Dimensional Reconstruction by Fourier–Bessel Synthesis 113
F. Assessment of Effective Resolution 114
G. Practical Use of Software Packages 115
V. Visualization and Structure Interpretation 117
A. Three-Dimensional Visualization Methods 117
B. Visualization of Secondary Structure Elements 118
C. Derivation of Folds 120
D. Toward Near-Atomic Resolution: Three-Dimensional Modeling 121
VI. Conclusion 122
References 122

I. Introduction

X-Ray crystallography is the method of choice for revealing atomic structures of large macromolecules and viruses. As shown in various examples in this volume, electron cryomicroscopy has emerged rapidly and has become a parallel technique to reveal additional information about virus structures, even in the situation in which the crystal structure of the virus may have already been obtained. The information that can be extracted from a hybrid approach of X-ray crystallography and electron cryomicroscopy is

ADVANCES IN PROTEIN CHEMISTRY, Vol. 64

0065-3233/03 $35.00

well illustrated in the article by Gilbert, Grimes, and Stuart (this volume). An exhaustive survey of electron cryomicroscopy applications related to viral assembly and virus cell entry has also been reviewed (Baker *et al.*, 1999).

In most of the applications of virus structure determination to date, electron cryomicroscopy has been used as a low-resolution technique (15–30 Å). Several studies have shown the feasibility of resolving the three-dimensional (3-D) structures of icosahedral virus particles at subnanometer (7- to 9-Å) resolution, where no crystal structures of the viruses or their components were previously known (Böttcher *et al.*, 1997a; Conway *et al.*, 1997; Trus *et al.*, 1997; Zhou *et al.*, 2000, 2001, 2003). Together with other biochemical and bioinformatics analyses, long helices inside the protein components of the viruses have been identified and their structural folds deduced. These developments represent a significant step forward because such *de novo* structure determinations have revealed new structural features and uncovered novel protein folds. Moreover, in the favorable event when a high-resolution model of individual protein(s) is available, the secondary structure elements identified inside the subunits of macromolecular assemblies by the electron cryomicroscopy approach provide internal markers to guide the fit of the atomic structures of the components when establishing a pseudo-atomic model of the entire assembly. These capabilities will make the hybrid approach more accurate, reliable, and powerful.

Because the methodology of obtaining subnanometer structures by electron cryomicroscopy is not yet a common practice, we focus in this article on reviewing this emerging technique, relating our own experiences and approaches. Readers interested in the general methods of virus reconstruction may refer to previous reviews on basic physical principles and procedures for low-resolution icosahedral particle reconstruction (Baker *et al.*, 1999; Thuman-Commike and Chiu, 2000) in addition to the original literature describing image-processing algorithms (Crowther, 1971; Conway and Steven, 1999; Zhou *et al.*, 1998; Fuller *et al.*, 1996; Baker and Cheng, 1996).

II. Electron Cryomicroscopy

A. *Theoretical Considerations of Electron Imaging*

A 3-D structure of an object viewed through an electron microscope is described in terms of the 3-D Coulomb potential function within the object. The image recorded in an electron microscope is a convolution of the projected potential function of the object with the contrast transfer function (CTF) of the electron microscope (EM) (Erickson and Klug, 1970) and various other experimental factors (Jiang and Chiu, 2001). Because the electrons used here have wavelengths smaller than 0.2 Å, the depth of field [see Eq. (5) below] is large enough so that the top and bottom parts of the particle can be thought of as having the same focus. The deconvolution of the image can be conveniently carried out in Fourier space. In addition, the effect of the potential field of thin biological samples on the incident electrons is small (i.e., the so-called weak phase approximation), which allowed the relatively convenient formulation of the image formation theory (Erickson and Klug, 1970; Thon, 1971). Under this formulation, the Fourier intensity, $I(s)$, of an electron image as a function of spatial frequency, s, is expressed as (Ludtke *et al.*, 2001)

$$I(s) = F^2(s)\mathrm{CTF}^2(s)E^2(s) + N^2(s) \tag{1}$$

where

$$\mathrm{CTF}(s) = -k[\sqrt{1 - Q^2}\,\sin(\gamma) + Q\,\cos(\gamma)] \tag{2}$$

and

$$\gamma = -2\pi\left(\frac{C_s\lambda^3 s^4}{4} - \frac{\Delta Z\lambda s^2}{2}\right) \tag{3}$$

and k is a scaling factor relating the arbitrary scale of structure factor, $\overline{F}(s, \theta)$, to the specific electron dose and film sensitivity. C_s is the spherical aberration coefficient of the objective lens, λ is the electron wavelength, and ΔZ is the objective lens defocus and is positive for underfocus in the convention used here. The image contrast contains two terms, one for phase contrast, $\sin(\gamma)$, and the other for amplitude contrast, $\cos(\gamma)$. Q is the fractional amplitude contrast, and its value depends on the electron energy and specimen thickness. Under a given experimental setting, all the parameters in Eq. (3) are constant except ΔZ, which is set by the microscope operator. Figure 1 shows the CTF curves in a 300-kV electron microscope for underfocus values of 0.5 μm (short-dashed curve) and 3.0 μm (long-dashed curve). At a low-resolution range, the values of CTF are smaller with a small defocus setting than those with a high defocus setting. Conversely, at a higher resolution range, there are fewer oscillations with a small defocus setting than with a high defocus setting. The modulation of Fourier amplitudes due to the CTF thus affects the contrast at various spatial frequencies, depending on the defocus setting of the micrograph.

According to Equation (3), when ΔZ is negative (overfocus) or has a small positive (underfocus) setting, γ will have a negative value.

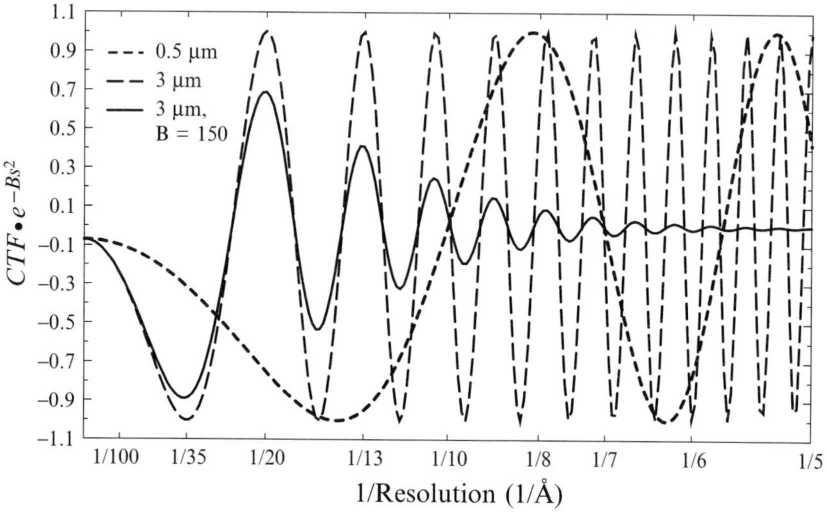

FIG. 1. CTF and E function simulation. EM imaging parameters: electron energy = 300 keV; spherical aberration coefficient C_s = 1.6 mm; Q = 7%. No envelope decay (B = 0) was used for the dotted (underfocus = 0.5 μm) and dashed (underfocus = 3.0 μm) curves.

Consequently the sine and cosine terms in Eq. (2) would have opposite signs at low frequency (i.e., small s), which leads to a poor low-resolution image contrast that is difficult to interpret for image processing. Therefore, micrographs are usually recorded with significantly underfocused settings (i.e., at least ~1–4 μm, depending on the voltage of the electrons used in imaging).

The cumulative envelope function, $E(s)$, can be complex and is attributable to a number of instrumental and experimental effects, such as spatial and temporal coherence and specimen motion. It has been shown that in practice a simple Gaussian function with width B adequately describes the cumulative envelope function (Saad *et al.*, 2001):

$$E(s) = e^{-Bs^2} \tag{4}$$

Under this definition, the E function is characterized by an experimental B factor, which can be estimated experimentally (Saad *et al.*, 2001). Note that the cumulative B factor applied in the final reconstruction is a composite of experimental and computational causes. The computational B factor is attributable to additional blurring effects such as inaccuracy in determining the orientation of particles, which could also be described by a Gaussian function type. The dampening of the image contrast by the function of a typical cumulative *B* factor (150 Å^2) is steep at a resolution beyond 10 Å (Fig. 1, solid curve). Note that this dampening function is mathematically analogous to the crystallographic temperature factor. However, the physical causes of these dampening functions are different. Moreover, the *B* factor as defined here is smaller by a factor of four than that used in the crystallographic formulation adopted in some studies (Böttcher *et al.*, 1997a).

B. *Choice of Instrument*

Modern transmission electron cryomicroscopes are capable of routinely recording inorganic crystal lattice images beyond 2.5-Å resolution. Unfortunately, because of the constraint of radiation damage, it is not possible to use the same electron optical conditions to image ice-embedded virus particles. With the best imaging procedure, it is straighforward to obtain an image of ice-embedded virus particles with a detectable contrast signal of up to 7–9 Å. Indeed, some of the best resolution structures published so far were recorded with a 20-year-old instrument (Table I). The desirable specifications of an electron cryomicroscope for imaging virus particles in the 7- to 9-Å resolution range include a field emission gun, an electron voltage in the range of 200–400 kV, and a cryospecimen holder with a low-dose kit.

The primary reason for favoring the field emission gun is the high spatial coherence of its illuminating beam (Zhou and Chiu, 1993). The envelope function due to the partial spatial coherence of the electron beam is a function of defocus value and angular source size (Zhou and Chiu, 1993; Chiu, 1978). The field emission gun would allow the use of a smaller source size and large defocus value so that the low-resolution image contrast is high while the high-resolution contrast is still present. Without the field emission gun, it would be necessary to use a smaller defocus value to avoid the severe high resolution image contrast dampening due to the partial spatial coherence of electron source illumination (Zhou *et al.*, 2000, 2001). In this case, the low-resolution contrast would be rather low, making it difficult to recognize particle images. The use of a large defocus value at realistic angular source size can lead to dampening the image contrast at high resolution, as illustrated by the solid curve in Fig. 1.

There are several advantages of choosing higher voltages to record images. The first is the smaller chromatic aberration effect on the images; and the second is the larger depth of field [see Eq. (5) below]. As pointed out above, the formulation of the currently used virus reconstruction

TABLE I
Comparisons of Different Strategies Used in subnanometer Reconstructions

Virus name, particle diameter	Voltage (kV)	Electron gun	Focal pair/ series	Range of underfocus (μm)	*B* factor[a] (Å^2)	Refinement method	No. of particles included in the final map	Effective resolution (Å)	Ref.
Herpesvirus capsid, 1250 Å	400	LaB_6	Focal pair, discard far-from-focus images	0.2–2.0	180	20 projections, common -line based	5860	8.5	Zhou *et al.*, 2000, 2001
Cytoplasmic polyhedrosis virus, 850 Å[b]	400	LaB_6	Focal pair, discard far-from-focus images	0.3–2.8	180	20 projections, common-line based	4532	8.0	Liang *et al.*, 2002; Zhou *et al.*, 2003
Rice dwarf virus, 780 Å	400	LaB_6	Focal pair, discard far-from-focus images	0.3–2.2	150	20 projections, common-line based	3261	6.8	Zhou *et al.*, 2001
Semlike: Forest virus, 700 Å	200	FEG	Single focus	0.98–7.6	15[c]	PFT	5276	9[d] or 10.5	Mancini *et al.*, 2000
P22 phage, 680 Å	400	LaB_6	Focal pair	0.5–2.0	50–250	20 projections, common-line based	8723 and 5000	8.5 and 9.5	Jiang *et al.*, 2001a, 2003

Papillomavirus capsid, 600 Å	200	FEG	Merged focal series	1.4–2.7	0	PFT	209	9[d]	Trus *et al.*, 1997
Hepatitis B virus core, 360 Å	120	FEG	Merged focal pair	0.86 and 2.2	~100[c]	PFT	600 particle pairs	9[d]	Conway *et al.*, 1997
Hepatitis B virus core, 360 Å	200	FEG	Single focus	1.3–1.5	125	6 projections, common-line based	6384	7.4	Böttcher *et al.*, 1997a
Tomato bushy stunt virus, 300 Å	200	FEG	Sinogram	Unknown	Unknown	Projection matching between class averages and projections	~6000	5.9[d]	van Heel *et al.*, 2000

Abbreviation: PFT, Polar fourier transform; FEG, Field emission gun.

[a]There is a factor-of-four difference between the definitions of the *B* factor used by the MRC group (Böttcher *et al.*, 1997a) and those used by others (Conway *et al.*, 1997; Zhou *et al.*, 2000, 2001, 2003). The *B* factors shown have been adjusted to follow the definition of Eq. (4) described in text.

[b]Diameter measured including the poorly ordered A spike. When excluding the A spike, cytoplasmic polyhedrosis virus has a diameter of 710 Å.

[c]The low-resolution features have also been additionally down-weighted in these studies.

[d]Resolution assessment based on the 3σ or other criteria that are less stringent than the 0.5 FSC or 45° phase difference criteria.

methods are all based on the assumption that the top and bottom parts of the particle have equivalent focus. However, as can be derived from Eq. (3), the difference in the defocus-related phase modifications between the top and bottom surfaces of a specimen with a thickness t is

$$\Delta\gamma = \pi \cdot t \cdot \lambda \cdot s^2 \tag{5}$$

Thus, if $\pi/2$ is allowed to be the maximum tolerable phase error, there is a constraint on the maximum thickness [commonly referred to as the depth of field, that is, $t = 1/(2\lambda \cdot s^2)$] for a given accelerating voltage and at a given target resolution. For instance, the maximum thickness of a particle for a 7-Å resolution structure must be less than 1500 Å with a 400-kV instrument. Figure 2 gives the maximum thickness of a particle for 100–400 kV as a function of spatial frequency. Therefore, there is a compelling reason to use as high a voltage as possible if high-resolution studies with large virus particles are to be pursued.

The cryoholder is used to keep the specimen in a frozen hydrated state in addition to reducing radiation damage. Several types of cryoholders operated at either liquid nitrogen or liquid helium temperature are now

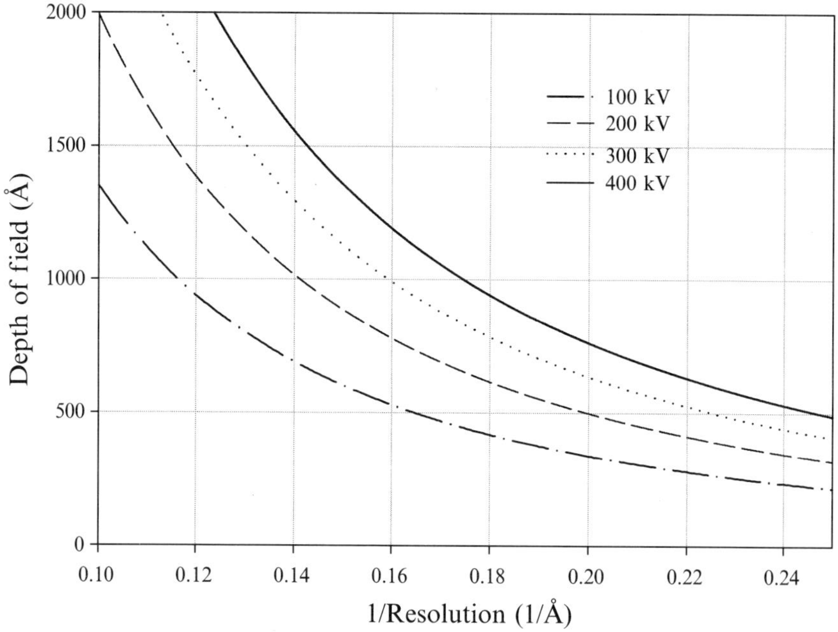

FIG. 2. Depth of fields at 100, 200, 300, and 400 kV as a function of reciprocal resolution.

available and are sufficiently stable for recording data between 7 and 9 Å from ice-embedded virus particles. The low-dose option is now available in all electron cryomicroscopes. Their modes of operation vary depending on the instruments but are generally operationally effective. For subnanometer resolution, the cumulative electron dose should be less than 15 electrons/Å^2 at liquid nitrogen specimen temperature (Schmid *et al.*, 1993). The reduction in radiation damage at liquid helium temperature is no more than a factor of 1.5 to 2, relative to the liquid nitrogen temperature (Chiu *et al.*, 1981). Therefore, it is essential to closely monitor the data collection procedure to avoid overdosing the specimen. Images suffering from radiation damage appear fuzzy or lack detailed features even though their incoherently averaged Fourier transforms may reveal excellent CTF rings. In more severely overdosed micrographs, bubbles can be seen around particle images.

C. Number of Particle Images Needed for a Three-Dimensional Reconstruction

On the basis of geometric consideration alone, the number N of particle images of different views needed for reconstructing a 3-D structure of a large asymmetric complex is dependent on targeted resolution d and particle diameter or specimen thickness t, that is, $N = \pi t/d$ (Crowther *et al.*, 1970a). Because of the redundancy of 60 copies of asymmetric units in an icosahedron, N is reduced 60-fold. For example, a 1250-Å-diameter icosahedral particle requires as few as 17 evenly spaced views to compute a 4-Å map. However, the signal/noise (S/N) ratio is low in low-dose images, particularly those recorded at close-to-focus conditions. The number of views required for a 3-D reconstruction is thus dependent not on particle size, but on the targeted resolution and many other experimental factors, including the S/N ratio in the low-dose image, the structural integrity of the particles and the computational accuracy of the particle orientation parameters determined. In practice, 4000–10,000 particles have been used to reconstruct a 7- to 9-Å map (see Table I).

III. Overview of Methods for Subnanometer-Resolution Reconstructions

Several structures of icosahedral particles with diameters in the range of 300–1250 Å have been reported to reach a subnanometer resolution to date. Different methods of data collection and analyses were used. Table I summarizes the characteristics of the data acquisition and data processing in the order of the structural complexity or particle diameters. All these studies were carried out with 120- to 400-kV electron cryomicroscopes with liquid nitrogen specimen cryoholders. Most were equipped with a field emission gun to obtain the maximum spatial coherence of the illuminating electrons. The methods differ primarily in the corrections of the CTF and cumulative *B* factor and in the methods of refining the orientation and center parameters of particle images.

A. *CTF and B Factor Corrections*

The defocus values for electron micrographs can be readily estimated on the basis of the CTF rings visible in the incoherently averaged Fourier transforms of individual particle images (Zhou *et al.*, 1994, 1996). This method has become a routine practice universally adapted for the initial evaluation and determination of CTF parameters as defined in Eqs. (2) and (3). So far, the determination of the cumulative *B* factors for 3-D reconstruction has been somewhat *ad hoc.* The cumulative *B* factor used in these studies is determined either by trial and error with the initial value derived from previous results, or from the incoherently averaged Fourier transforms of particle images. Different approaches have been adopted to make corrections for the CTF and the *E* function of the micrographs. They differ in the steps where these corrections are made and in whether or how the *E* function is corrected.

Zhou and Chiu made the CTF and *B* factor corrections at the image level and the details of corrections differed at different stages of data processing. The defocus and other parameters were estimated by fitting of the functions with the profile of incoherently averaged Fourier intensities of particle images (Zhou *et al.*, 1994, 1996; Ludtke *et al.*, 1999). In some of their studies, they utilized the X-ray solution scattering intensity of the viral particles to provide a one-dimensional profile of the structure factor of the virus particle (Fig. 3) for the estimation of the *B* factors of the micrographs (Saad *et al.*, 2001). Immediately before the merging of Fourier data of all the particles, the Fourier amplitude of each particle image was corrected by dividing the CTF and the Gaussian function, using a modified Wiener type of filter function by which the data points near the CTF zeros were excluded to avoid amplifying the noise in these regions (Zhou *et al.*, 1994, 1999, 2000, 2001). However, in the initial and other intermediate reconstructions generated during the orientation and center refinement, only CTF correction without the *B* factor correction was performed. For orientation and center refinement, particles from each micrograph were refined by minimizing their difference with model

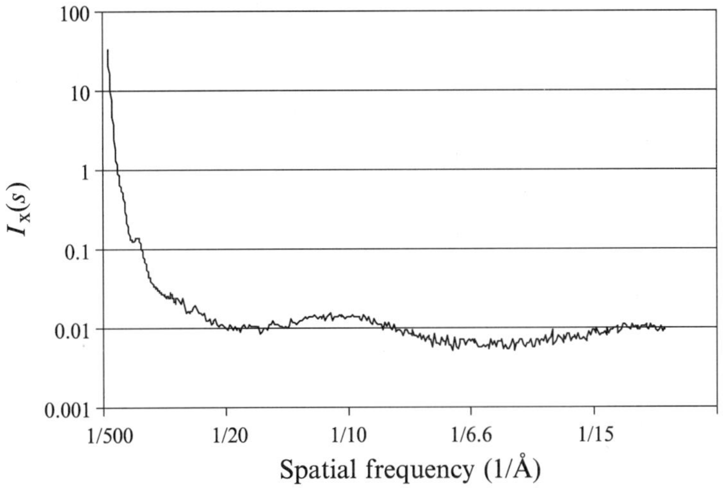

FIG. 3. X-ray solution scattering intensity of a suspension of herpes simplex virus 1 B capsids recorded at the Stanford SLAC beam line. [Adapted from Saad *et al.* (2001) with permission from the publisher and the author.]

projections, which were multiplied by a CTF function determined for the micrograph. The final reconstruction was synthesized from the Fourier data of particle images corrected by CTF and *B* factor.

A slightly different approach of CTF and *E* function correction was used by Conway, Trus, and Steven (Conway *et al.*, 1997; Trus *et al.*, 1997; Conway and Steven, 1999). For each micrograph an *E* function was empirically determined by fitting to a Gaussian curve with four parameters. The CTF, also determined from incoherently averaged particle images, and the *E* functions were then corrected in the particle averages generated by combining the Fourier transforms of micrographs in a focal pair/series, using a formulation with a Wiener-like filter (Conway *et al.*, 1997; Trus *et al.*, 1997; Conway and Steven, 1999). Three-dimensional maps were subsequently generated from these CTF-corrected particle averages. In their approach, the low-resolution terms (within the first peak of CTF) were left uncorrected, presumably to down-weight the low-resolution feature in the 3-D reconstruction.

van Heel and colleagues favored performing the CTF correction directly on the micrographs at the beginning of data processing (van Heel *et al.*, 2000). In addition, the CTF correction was performed separately in different areas of each micrograph because of defocus variation across the micrograph in their data (van Heel *et al.*, 2000). In their case, phase reversal was corrected without amplitude scaling to avoid amplifying the noise present in the image.

Böttcher and Crowther made the CTF correction at the 3-D map level (Böttcher and Crowther, 1996; Böttcher *et al.*, 1997b). They first determined a medium-resolution 3-D map from the particles in a single micrograph. The CTF parameter of the micrograph was determined by cross-correlating its CTF-uncorrected 3-D reconstruction with a CTF-corrected model. The correlation coefficient as a function of spatial frequency has the same positive and negative oscillation patterns as the CTF of the micrograph. Thus the positions of CTF zeros can be identified by finding the locations where the correlation coefficient reaches zero. Many of these CTF-uncorrected maps with the determined CTF parameters were then merged to generate a CTF-corrected map by least-squares fitting. The map was further scaled with a Gaussian function with a cumulative *B* factor based on other studies (Böttcher *et al.*, 1997a,b). The rationale for using this approach was that relatively noise-free uncorrected 3-D maps could be used to determine the CTF parameters more accurately. However, this approach is impractical for large viruses, for which an insufficient number of particles can be obtained on a single micrograph at an appropriate magnification to generate a CTF-uncorrected map with subnanometer resolution data.

The defocus accuracy achieved on the basis of the incoherent averaging of particle Fourier transforms (Zhou *et al.*, 1996) may not be sufficient to correct the CTF for micrographs with relatively large defocus values because of the close proximity of CTF oscillations in the high spatial frequency region. For such micrographs, it is desirable to refine the estimated defocus values by taking advantage of an existing model (Böttcher *et al.*, 1997a). For larger virus particles, for which it is impractical to obtain a 3-D reconstruction from a single micrograph, we would calculate the average Fourier ring correlation (FRC) function between a particle image of unknown defocus with its corresponding projections computed from the latest CTF-corrected model. The average of the FRC functions of all particles in the same micrograph should yield a relative noise-free curve to locate the CTF zeros. The experimental *B* factor of each micrograph can also be determined. Initially, it may simply be assumed that the structural factor profile in the subnanometer region is a flat curve to obtain a rough *B* factor estimate for each micrograph by fitting to the profile of the incoherently averaged Fourier transforms of particle images. A more plausible approach is to use generic models such as atomic models of related protein complexes or of particles of similar dimension and shape to approximate the structural factor profile of the particle for *B* factor estimation.

B. *Orientation and Center Determination*

van Heel and co-workers used a real space common-lines (i.e., sinograms) algorithm to determine and refine the center and orientation of particle images, as applied to asymmetric particle reconstruction (van Heel *et al.*, 2000). In other laboratories, two completely different methods of orientation estimation and refinement have been employed: the Fourier common-lines method (Crowther *et al.*, 1970b) and the polar Fourier transform (PFT) method (Baker and Cheng, 1996). The Fourier common-lines method can be used without an initial model and has often been employed at the beginning of a project. After an initial model becomes available, either of these methods can be used to carry out particle parameter refinement iteratively by minimizing the differences between the raw particle image and a set of computed projections of the latest reconstruction. The parameters of minimization are different in the two methods. The Fourier common-lines method is based on the phase residual differences, whereas the PFT method is based on the correlation matching. The PFT method (Baker and Cheng, 1996) has the flexibility to circularly mask either the inner or outer components in real space and has been used in structural determination up to 9 Å (Conway *et al.*, 1997; Trus *et al.*, 1997; Mancini *et al.*, 2000). On the other hand, the Fourier common-lines method is capable of refining the positional parameters and Euler angles simultaneously (Zhou *et al.*, 1998) and has been successfully used to reach 6- to 7-Å resolutions (Böttcher *et al.*, 1997a; Zhou *et al.*, 2001).

IV. Example of Data Collection, Evaluation, and Processing

Using the procedure illustrated in Fig. 4, we have determined several subnanometer-resolution structures of icosahedral particles (Zhou *et al.*, 2000, 2001, 2003; Viang *et al.*, 2001a, 2003; Liang *et al.*, 2002). In this section, we discuss each of the steps in our procedure and mention the specific modular programs (in *italics*) employed.

A. *Imaging*

Most of the subnanometer-resolution studies used field emission gun cryomicroscopes because of the advantages mentioned above (Table I). When the LaB_6 gun is used, the spatial coherence of the instrument is relatively poor. Therefore, a high-resolution image must be recorded with a small defocus value to minimize the dampening due to the partial coherence envelope function (Zhou and Chiu, 1993). Consequently, these images have low contrasts for their low-resolution features, making it difficult to locate the particles and to determine their initial orientation

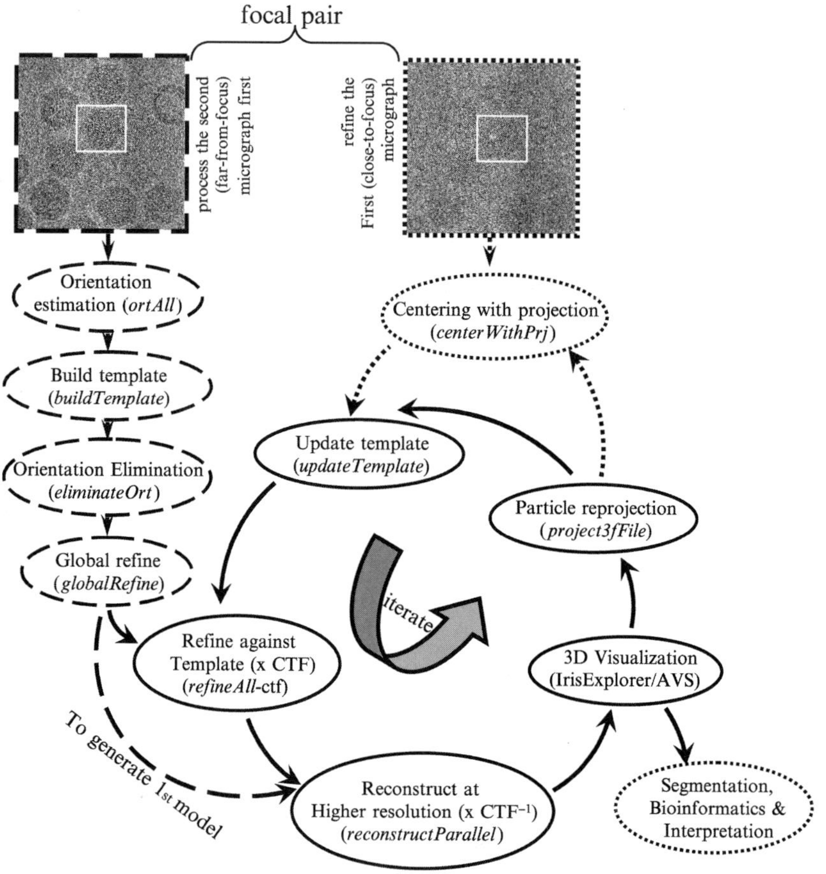

FIG. 4. Flow chart of data processing, using the focal pair approach (Zhou *et al.*, 1995, 1998; Liang *et al.*, 2002). In this approach, the far-from-focus micrograph (*top left*) is used to assist in orientation determination of particles in the close-to-focus micrograph (*top right*). Data-processing steps are listed, inside dashed-line ovals (for far-from-focus particles), dotted-line ovals (for close-to-focus particles), or solid-line ovals (for both far-from-focus and close-to-focus particles). The procedure begins with far-from-focus particles, first by obtaining a set of possible orientation estimates for each particle, using self common-line phase residual minimization, establishment of a template set of particles, followed by the elimination of incorrect orientation estimates, and then global simultaneous orientation and center refinement, using particle images (Zhou *et al.*, 1998), and finally 3-D Fourier merging and inversion by Fourier–Bessel synthesis. The 3-D merging and orientation refinement steps are iterated for several cycles, each time with the update of the template set through the computation of 2-D projections from the latest 3-D reconstructions. The center of each close-to-focus particle image is first determined by cross-correlating the particle image with a computed projection along the orientation estimated from the corresponding far-from-focus particle (*Centering with Projection*). The orientation and center parameters are

parameters confidently. An experimental solution is to use the so-called focal pair method [Figs. 4 and 5a–d; see detailed description elsewhere (Liang *et al.*, 2002)]. In this method, the first micrograph is taken with a small defocus value (typically with an underfocus value such that the first CTF zero occurs at about 10 Å) to maximize the high-resolution signals and to increase the separation of neighboring CTF rings in the subnanometer spatial frequency range. The second, far-from-focus, micrograph is taken from the same specimen area at relatively higher defocus value (typically targeting the first CTF zero at about 20–30 Å, depending on the morphology of the particles imaged) to maximize the image contrast in the low spatial frequency range. The high contrast of low-resolution features of the far-from-focus micrograph is essential for the initial estimation of particle position and orientation. The particle orientation parameters estimated from this micrograph are then used as the starting point to refine those for the corresponding particles in the first, close-to-focus micrograph. In the final reconstruction, only the particles from the close-to-focus micrographs are included.

Although the use of the focal pair method doubles the task of image acquisition and their subsequent processing, it offers the most reliable means to eliminate false orientation assignments, which often occur for close-to-focus particle images. It should be pointed out that the focal pair approach was not used with the structures determined with microscopes equipped with a field emission gun, by which the micrographs were recorded with relatively large underfocus values. However, when extending structural analysis beyond 7–9 Å, the focal pair approach may be necessary even with the use of a field emission gun. This is because adjacent CTF rings present in the micrographs with large defocus values may be too close to one another in the high spatial frequency range to achieve accurate CTF determination and correction.

B. *Digitization and Particle Selection*

Before digitization, micrographs are first examined visually to discard those with apparent image astigmatism or drift. So far, all the image data used in subnanometer-resolution reconstructions have been recorded on

---

refined iteratively toward higher resolution against a template set of a list of CTF-multiplied, evenly spaced projections made from the current best model. Usually about 20 projections are used so that the Fourier transforms are fully sampled. In each iteration, the resolution of the reconstruction is gradually improved by including image data at a higher spatial frequency range. Note that the final 3-D reconstruction includes only particle images from the close-to-focus micrographs.

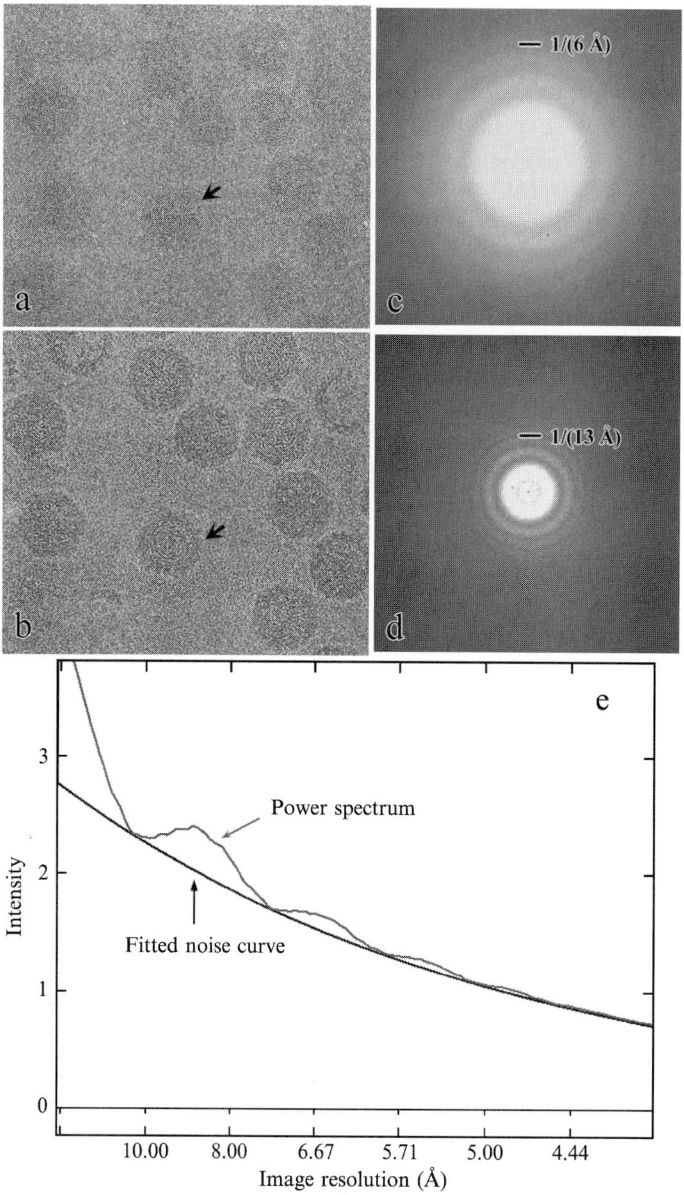

FIG. 5. Imaging and image quality assessment by incoherent averaging of Fourier transforms. (a and b) A focal pair of 400-keV electron micrographs of ice-embedded rice dwarf virus (RDV) was recorded in a JEOL4000 electron cryomicroscope with a Gatan cryoholder operated at -170°C. Images were taken at $\times$50,000 magnification with an electron dose less than 20 electrons/Å^2. One corresponding RDV particle is

photographic films and, consequently, must be digitized in a high-resolution film scanner with minimal modulation transfer function. The digitization step on the specimen scale should be approximately three to four times finer than the expected resolution of the 3-D reconstruction. The step size limits of 7.5–10 μm/pixel of currently available high-speed film scanners thus require that the micrographs be taken at relatively high magnification ($\times$50,000–60,000). At such a magnification, the specimen exposure dose limit of 10–15 electrons/Å^2 is sufficient to produce an acceptable optical density on the photographic film.

The digitized micrographs are subject to further graphical inspection for particle selection. This step is typically performed on a graphics workstation with a large memory because each micrograph is rather large ($\sim$0.5 GB for each micrograph scanned at 1.4 Å/pixel on the specimen scale). In practice, particles are picked either manually or semiautomatically and saved to image particle files, using graphic tools such as *boxMrc* (Liang *et al.*, 2002) and *EMAN* (Ludtke *et al.*, 1999). These tools are also able to automatically match corresponding particles on focal pair micrographs.

C. *Image Screening*

Before embarking on extensive data processing, a number of screening steps are routinely performed to select particles appropriate for subnanometer-resolution reconstructions. Examining the power spectrum or incoherently averaged Fourier transforms of particle images is the first step to assess the quality of images as shown in Fig. 5c and d (Zhou *et al.*, 1994, 1996; Ludtke *et al.*, 1999). The most frequently encountered image defects are subtle beam-induced movement (such as charging) and specimen/cryostage drifting, which manifest as nonisotopic or incomplete CTF rings in the power spectrum of particle images (Thuman-Commike and Chiu, 2000). In some micrographs, specimen charging is localized to a small region, resulting in different power spectra for particle images at different locations on the same micrograph.

A potentially difficult problem is ice thickness variations across a micrograph, which would result in defocus difference at different regions

highlighted by arrows. (c and d) Incoherently averaged Fourier transforms of individual RDV particle images selected from micrographs shown in (a) and (b), respectively. Each spectrum was obtained by incoherently averaging the Fourier transforms of about 100 particle images. (e) Circularly averaged Fourier intensity as a function of resolution from the power spectrum in (c). The profile indicates that the close-to-focus micrograph has detectable contrast beyond 5-Å resolution.

on the same micrograph. Changes in ice thickness are evident from variations in the particle contrast. These variations can affect the accuracy in determinating the effective CTF of the micrograph and further complicate the problems imposed by the depth of field limit for large particles (Fig. 2).

D. Orientation and Center Determination by the Focal Pair Method

As mentioned above, we used the focal pair method in the orientation and center determination for subnanometer-resolution structure determination. In this method, the orientation parameters obtained from the far-from-focus particle images are used as the starting point for processing the corresponding close-to-focus particles in a focal pair. The center parameters of each of the close-to-focus images are determined by focal-pair matching during particle selection and refined by cross-correlation with computed projections of a preliminary 3-D model calculated from the far-from-focus micrograph. These center and orientation parameters are further refined gradually, using template-based refinement by including image data at higher spatial frequency range until the resolution of the map can no longer be improved. Only the close-to-focus images are included to compute the final high-resolution reconstruction. Some of these key steps are described in more detail below.

1. Initial Estimation of Orientation and Center Parameters

The first task is orientation and center estimation, or orientation search (*ortAll*), for all particles selected from far-from-focus micrographs. In this program, each of the selected particles is first premasked with a Gaussian mask to exclude noise from the corners and to reduce Fourier artifacts (Frank, 1979). The size of the mask is typically slightly larger than the particle diameter. Artifacts that occur when no mask is used typically manifest as a cross at the origin of the Fourier transform, due to the noncontinuity in densities at opposing edges of the particle. The particle center, defined as the projection of the origin of the icosahedral axes in the image plane, is approximated by locating the peak in the correlation image between the premasked particle and a reference image. The reference image can simply be the particle image itself rotated by 180°, or a rotationally averaged image sum of many particle images or computed projections generated by *azimuthalAvg*. The estimated center of each particle is used to remask the particle.

For each masked particle, a list (~30) of the most likely angular and center parameter estimates is obtained by sorting all possible orientations according to their corresponding self common-line phase residuals calculated in a brute force manner at 1° intervals for the three angular parameters (θ, ϕ, and ω) over the entire icosahedral asymmetric unit (Crowther, 1971). Four slightly different self common-line based formulations are evaluated in this brute force calculation (Crowther, 1971; Prasad *et al.*, 1988; Fuller, 1987; Baker *et al.*, 1988), only using particle Fourier data within the spatial frequency range of 1/35–1/300 Å^{-1}. Except for the Crowther formulation, these functions employ statistical measures such as the Student *t* test (Prasad *et al.*, 1988) and the χ^2 distribution (Fuller, 1987) to reduce estimation errors resulting from self common-line degeneracies. Self common-line degeneracy arises when the direction of view coincides with any of the 5-, 2-, and 3-fold symmetry axes, leading to fewer than 37 pairs of self common-lines in the Fourier transform of a particle image and smaller phase residuals if not down-weighted properly (Crowther, 1971).

2. *Orientation Elimination and Selection*

The Orientation Elimination program (*eliminateOrt*) sorts the lists of possible orientation and center estimates according to the cross common-lines phase residuals between the particle and a template set of particles or computed projections. The template set can be obtained by *buildTemplate* (when no preexisting model available) or *project3fFile* (when a preliminary model available) and consists of a group of either particle images with refined orientation and center parameters or projections computed from the best reconstruction available. During the execution of the *eliminateOrt* program, two cycles of coarse refinements are first performed, typically using particle Fourier data in the spatial frequency range of 1/300–1/30 Å^{-1}, for each listed set of center and orientation parameters before their ranking of cross common-lines phase residuals. For each particle, if the set of orientation and center parameters with the smallest cross common-line phase residual is smaller than a user-supplied cutoff, the particle with this set of orientation and center parameters is kept, or otherwise is "eliminated," for further processing. The phase residual cutoff is typically about 45–55°, depending on the low-resolution contrast (a function of defocus) of the particle images. The use of cross common-lines and a template set with more than one particle projection in *eliminateOrt* effectively eliminates the "inaccuracy near symmetry axes" problem associated with self common-line degeneracy that is often encountered in self common-lines based procedures (such as *ortAll*). The result of the *eliminateOrt* procedure is a subset of particles selected for further refinement.

3. *Orientation and Center Refinement*

Orientation and center parameters are refined by minimizing the cross common-line phase residuals between each particle and all other particles (*globalRefine*), and/or between the particle and a set (~20) of model projections (*refineAll*). These two approaches are known as Global Refine and Refine against Template, respectively. Both procedures carry out simultaneous orientation and center refinement, using Fourier common-lines. The computation time of *refineAll* is linear with the number of particles (~10 s/particle in a PC with a single 2-GHz Intel Pentium IV CPU). On the other hand, the computation time of the global refinement procedure increases quadratically with the number of particles and thus can be prohibitive for refining a large set of particles (Zhou *et al.*, 1998). However, global refinement does provide a useful alternative in two situations: the first being the initial stage of reconstruction when a template or a model is not yet available; the second being the refinement of close-to-focus images in the same defocus group, such as particles from the same micrograph.

In practice, the major portion of refinement is carried out by the iterative template-based refinement procedure (*refineAll*), which uses the CTF-corrected 3-D map as a template to refine all particles in the entire data set. The projections from a CTF-corrected model are first multiplied by the CTF of the particle image before refinement. This approach avoids the CTF correction on the particle images, thus avoiding possible noise amplification due to division by zero or by small CTF values near CTF zeros. Note that additional particles with incorrect parameters may be eliminated in this step because data at higher resolution are included in the refinement to provide more discriminating power. Once an adequate number of particles is accumulated on the refined list of orientation and center parameters, these particles are merged to calculate a new reconstruction, which can then be used to generate the current template set (*project3fFile* and *updateTemplate*). Typically it is necessary to iterate the above-described steps of orientation and center estimation, elimination, and refinement for three to five cycles until no more particles can be added to the list of refined particles.

4. *Center and Orientation Refinement for Close-to-Focus Images*

The orientation parameters obtained from the far-from-focus particle images in the focal pair are used as the starting point for processing the close-to-focus particles. Because of the poor S/N ratio of low-resolution features, the center and orientation parameters of the close-to-focus particles cannot be reliably determined by directly using the orientation estimation and selection steps described above. Typically, the center of each close-to-focus particle image in the entire data set is determined by focal pair matching and refined by locating the peak in the cross-correlation image (*centerWithPrj*) of the particle image and a projection computed in the same orientation as determined from its corresponding far-from-focus particles (*project3fFile*). Further refinement of the center and orientation parameters of all these particles is carried out by gradually (at an increment of 1–3 Å each iteration) extending toward the targeted resolution, using *refineAll* (or *globalRefine*) and 3-D reconstruction (see Section IV.E, below) programs.

E. Three-Dimensional Reconstruction by Fourier–Bessel Synthesis

The last major tasks of data processing are data merging and reconstruction by 3-D Fourier inversion, using the Reconstruct program (Fig. 4, *reconstructParallel*). Instead of performing a direct 3-D inverse Fourier transformation, Fourier–Bessel synthesis is used to avoid an explicit interpolation in the 3-D Fourier space, thus allowing irregularly spaced Fourier transforms to be conveniently merged and averaged for 3-D Fourier inversion (Crowther, 1971). The program for the Fourier–Bessel synthesis as incorporated in our package was recoded on the basis of various previous implementations (Crowther, 1971; Prasad *et al.*, 1988; Baker *et al.*, 1988) to improve accuracy and to allow a much more efficient merging of large numbers of particles for higher resolution reconstruction. This program consists of four major computational steps. The first step prepares the normal matrices by applying one 5-fold, one 3-fold, and two 2-fold symmetry rotations to the Fourier transform of each particle to generate 60 Fourier "planes" in 3-D Fourier space for each particle. Before merging the Fourier planes of all the particles, the Fourier values (both phases and amplitudes) are corrected for CTF (Section III.A) (Zhou *et al.*, 1999). The Fourier values near the CTF zeros (e.g., $|CTF| < 0.15$) are simply discarded, thus avoiding potential problems of noise amplification. One normal matrix is prepared for each annulus in the polar coordinates of the 3-D Fourier space. The normal matrices are then solved by a linear algebra method for the discrete Fourier–Bessel transform values. This is followed by the summation of Bessel functions of different orders and the final step of Fourier–Bessel synthesis to compute a 3-D density map.

The computation time for the 3-D reconstruction depends mainly on the normal matrix preparation step and the Fourier–Bessel synthesis step. The time for the normal matrix calculation increases roughly linearly with the particle numbers and the targeted resolution. The Fourier–Bessel synthesis step is independent of the number of particles but its computation time increases roughly cubically with the targeted resolution. Thus, although the execution time of this procedure could be rather lengthy [e.g., it took a full day on a 24-processor SGI Origin to calculate an 8.5-Å herpesvirus capsid map (5)], it is substantially faster than other real space reconstruction methods (e.g., the back projection reconstruction), whose computation time increases cubically with the targeted resolution and linearly with the number of particles. In addition, we have ported our Fourier–Bessel reconstruction program to run as parallel processes on both the SGI Iris shared memory platform and the Linux cluster distributed memory platform.

F. Assessment of Effective Resolution

The quality of 3-D reconstructions should be assessed at various steps of the data processing to ensure data convergence and to measure the effective resolution of the final 3-D map. An obvious measurement that should be performed is to calculate the cross-correlation coefficient of the computed 3-D reconstruction and the true structure of the object. However, because the true structure is the very entity that is being reconstructed and, thus, is unknown in most situations, it is impossible to perform such a comparison. Instead, a widely used approach is to estimate how well two independent reconstructions agree with one another by calculating the differential phase residues (DPR) (Fig. 6a) and the Fourier shell correlation (FSC) coefficients (Fig. 6b) between the two maps (Böttcher *et al.*, 1997a; Zhou *et al.*, 1994; van Heel, 1987). The spatial frequency, where the FSC reaches 0.5 or the DPR increases to 45°, is commonly considered to be the effective resolution of the reconstructions; therefore, the term "resolution" does not really mean resolvability so much as reproducibility.

To monitor the progress of the reconstruction, the resolution test would be performed as a function of particle numbers. As shown in Fig. 6, the resolution would be expected to improve as the number of particles increases. Thus such a plot could allow the prediction of the number of particles needed to achieve a certain resolution. The required number of particles varied somewhat in practice (Table I), probably because of variations in the robustness of icosahedral symmetry in the particle, the image quality, as well as the effectiveness of the data processing procedures used. It should be pointed out that the FSC should approach unity at low resolution and generally decrease rapidly closer to the resolution limit for perfectly icosahedrally ordered particles. The presence

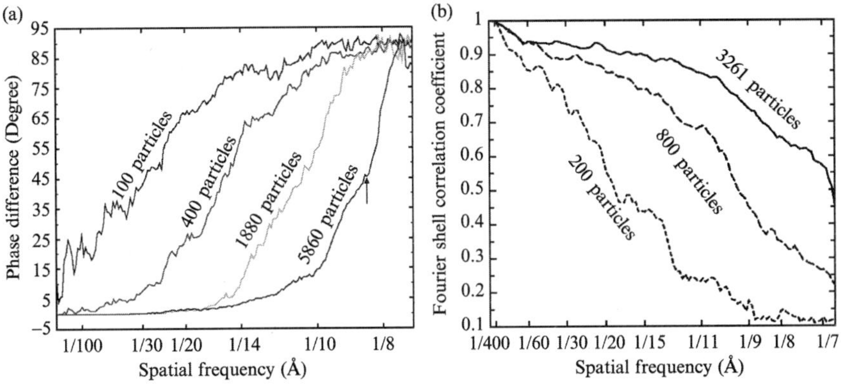

FIG. 6. Resolution assessments. (a) Differential phase residual method based on the 45° phase difference criterion used in the 8.5-Å resolution (indicated by arrow) structure of the herpesvirus capsid (Zhou *et al.*, 2000). (b) Fourier shell correlation method based on the 0.5 Fourier shell correlation coefficient criterion used in the assessment of effective resolution of the rice dwarf virus (RDV) structure at 6.8 Å (Zhou *et al.*, 2001). The imperfect FSC value (<1.0) is partly due to the presence of nonicosahedrally ordered dsRNA genomes within the RDV virions. [Adapted from Zhou *et al.* (2000, 2001), with permissions from the publishers.]

of nonicosahedrally ordered materials such as nucleic acid may lead to a less than unity FSC at low resolution.

G. *Practical Use of Software Packages*

The data set for structure determination of an icosahedral particle of a moderate size (300–1200 Å in diameter) at subnanometer resolution typically needs 30–200 focal pairs of micrographs containing several thousands of particle images (Table I). For example, the number of particles increased about 40 times, from 159 to 5900, to improve the herpesvirus capsid reconstruction from 25 Å (Zhou *et al.*, 1994) to 8.5 Å (Zhou *et al.*, 2000). Table II shows the size of the data sets for various particle diameters for four targeted resolutions. Notably, the amount of data involved for a near atomic resolution reconstruction of a 1200-Å particle is nearly 1 terabytes (TB) and is almost 1 million times more than that needed for a low-resolution small particle reconstruction.

To engage in the investigation of structures at subnanometer resolution, inexperienced users often encounter two challenges: the first is to manage the large number of particle images efficiently and the second is to carry

TABLE II
Estimated Data Size, in Gigabytes, for Icosahedral Reconstruction of Different Particle Diameters at Different Resolutions[a]

Diameter	Resolution (Å)			
(Å)	24	12	8	4[b]
300	0.008	0.21	2.8	54.4
700	0.048	0.96	12.6	248
1200	0.13	2.8	34.1	671
2000	0.41	7.7	89.4	1773

[a]Image data are stored as floating point numbers. A working disk space consumption twice that of the image data size is assumed in this estimate. This estimate is based on using the focal pair method, so roughly 60% of the image data are used only in the initial stage of data processing and discarded in the final reconstruction. Only the individual selected particle images, not the original micrographs containing these particles, are included in this estimate.

[b]A 5-fold decrease in the image signal-to-noise ratio is assumed from 8 to 4 Å.

out data processing with available computing resources. The traditional mechanism of data management through a flat or tagged text file, or by using image headers, is error prone and difficult for data tracking and, thus, is inadequate for processing the substantially large data set needed for high-resolution studies. The scheme described above involves multiple computer programs. Each step requires the appropriate input of some user parameters and involves the proper selection of particle image files. The most repetitive procedure is the iterative refinement. Generally, it takes more than 10 iterative cycles to reach a 7- to 9-Å structure. In an effort to avoid any human handling errors of the data and to maximize the efficiency of using computing resources, the aforementioned programs have been incorporated into two software packages, which make the data management and data processing relatively easier and more convenient.

The first package is called IMIRS, which was originally coded for UNIX platforms with the refinement and reconstructions both parallelized to run on multiprocessor computer servers (Zhou *et al.*, 1998; Johnson *et al.*, 1997). This package has been ported to the widely available Microsoft Windows platforms and integrated with an SQL image database (http://hub.med.uth.tmc.edu/~hong/IMIRS) (Liang *et al.*, 2002). The client/server architecture of the IMIRS design divides the image processing task into a front-end component and a back-end or server component. An intuitive graphic user interface (GUI) provides novice users an easy access to the many data processing steps (as shown in Fig. 4) and constitutes the front end. The database server constitutes the back end, with which many different front-end client applications communicate over the Internet. The database server maintains referential integrity, security, and logs, and ensures that operations can be painlessly recovered in the event of user mistakes and system failures. The data management tools provided in an SQL database are designed and optimized for applications that require not only large amounts of data, but also many simultaneous users across distributed computing environments. Thus, the use of relational image databases optimizes the tedious and error-prone data management tasks in 3-D reconstructions, allowing multiple users and many computers to work collaboratively and more independently, an inevitable situation encountered in projects targeting at atomic resolution.

The second package is called SAVR (http://ncmi.bcm.tmc.edu/~wjiang) (Jiang *et al.*, 2001a). This package has been designed to "glue" together the most CPU-intensive and iterative steps (or modules) in the IMIRS (Liang *et al.*, 2002) and the EMAN (Ludtke *et al.*, 1999) packages, using the scripting language Python. SAVR is portable across various UNIX- flavored platforms and has been parallelized to run on both shared and distributed memory platforms. SAVR also allows the incorporation of new algorithms and facilitates the management of the increasingly large data sets needed to achieve higher resolution reconstructions. More importantly, this package will allow users to perform checks on the results during various steps of the refinement to ensure the process is heading in the right direction. The package automatically e-mails the user with processing notification and data summaries at various stages. This software has been applied to solve two structures of icosahedral particles at subnanometer resolution (Jiang *et al.*, 2001a).

V. Visualization and Structure Interpretation

A. *Three-Dimensional Visualization Methods*

The 3-D visualization of electron densities of a virus structure is accomplished mostly with commercial volume and surface-rendering software packages with customized modules or programs. The most commonly used packages include Iris Explorer (NAG, Oxford, UK), which is available on both UNIX (SGI Irix and Linux) and Microsoft Windows platforms, and AVS (Advanced Visual Systems, Waltham, MA), which is available on the UNIX platform. Many enhancements have been added to both packages to accommodate specialized needs to visualize the large density volumes of virus structures, including import and export modules, segmentation tools, as well as different ways of color coding the rendering (Spencer *et al.*, 1997; Sheehan *et al.*, 1996; Dougherty and Chiu, 1998, 2000).

Because surface rendering of large density volume data can be computationally prohibitive, it is necessary to dissect or segment individual structural components out of the entire virus reconstruction so that subnanometer-resolution features can be conveniently identified and compared (Zhou *et al.*, 2000, 2001) (Fig. 7; see Color Insert). Segmentation of individual components also allows nonicosahedral averaging of structurally similar components to enhance the *S/N* ratio (He *et al.*, 2001). Only structural components within an icosahedral asymmetric unit are structurally unique and necessary for detailed examination. When desired, the entire viral particle can be generated by performing the 5-3-2 symmetry operations from the asymmetric unit (Fig. 7).

B. *Visualization of Secondary Structure Elements*

A distinctive feature of 3-D maps at subnanometer resolution is the resolution of molecular internal structural features representative of secondary structure elements (α helices and β sheets). The identification of these features, however, is not straightforward using surface representations alone and requires the combined use of different visualization techniques. Although it has been a common practice to display virus structures with shaded surface views, using a density cutoff or contour level equivalent to the molecular weight of the virus, such views do not readily reveal internal secondary structural features. At 7- to 9-Å resolution, long α helices appear as straight rods of densities ~5–6 Å in diameter when displayed at higher contour levels (e.g., above 2σ), whereas large β sheets appear as continuous surfaces at lower counter levels (Fig. 8). It should be cautioned, however, that small β sheets containing two or three β strands have a dimension similar to that of α helices and are difficult to distinguish at this resolution.

Therefore, a substantial portion of the secondary structure elements can be readily identified in 3-D maps calculated to 6- to 8-Å resolution by interactive visual inspection of the density volume, using 3-D volume-rendering techniques (Fig. 9; see Color Insert). Examination of the density range distribution can also be used as a means to verify secondary structure elements identified on the basis of shaded surface representations. For example, the gray density displays of density sections extracted from a 3-D volume are often useful to verify the assignment based on information such as the relative densities of the identified α helices and β

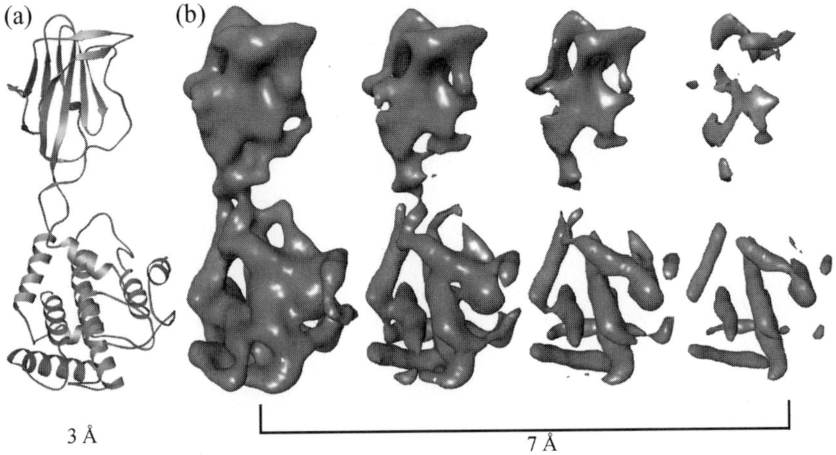

FIG. 8. Illustration of helices and β sheets at 7 Å by a simulated density map. The atomic model of a VP7 monomer of bluetongue virus (Grimes *et al.*, 1998) was obtained from the Protein Data Bank and rendered as ribbons (a). The same model was then Gaussian filtered to 7 Å to generate a density map, which is displayed as shaded surfaces, from left to right, using gradually increasing contour levels (b). [Courtesy of Dr. Matthew L. Baker.]

sheets, and distances and twist angles between adjacent helices inside a helix bundle.

Interpretation of the membrane proteins in an envelope virus can be assisted by the identification of transmembrane α helices. Membrane proteins represent a special class of proteins because of the predominant presence of transmembrane helices connected by extramembrane loops and domains. For example, even at 10.5-Å resolution, a pair of transmembrane helices could be identified in the Semliki Forest virus E1 and E2 proteins (Fig. 10; see Color Insert).

Automatic pattern recognition tools such as *helixHunter* have been developed to perform the tedious manual identification of helices longer than 2.5 turns (Jiang *et al.*, 2001b). *helixHunter* uses a cylinder of 5 Å in diameter as a generic helix template and carries out an exhaustive 3-D cross-correlation search in a 3-D density map to identify the locations of helices. This objective and automatic approach involves a multistep process of cross-correlation, density segmentation, quantification, and helix identification as well as an explicit description of the helices. For visualization purposes, final helices can be annotated as cylinders (such as those modeled in Figs. 9 and 10), which can be described by six parameters (three for center, two for orientation, and one for length).

C. Derivation of Folds

The fold of a protein can be described by its overall topology, which includes the spatial arrangement and the connectivity of the secondary structure elements of the polypeptide chain into the tertiary structure (Lo Conte *et al.*, 2000; Orengo *et al.*, 1997). In a large protein, a single chain may form one or several distinct domains, each with a particular fold. However, α helices and β sheets identified either manually or automatically do not contain any information for mapping the identified secondary structure elements to their amino sequences in the resolution regimen from 7–9 Å. When the fold is relatively simple, it is possible to assign such information by correlating existing biochemical and structural information. The map of the hepatitis B virus core protein has a simple fold with a pair of bundled helices and represents the first time that the fold of a protein has been derived by image analysis of single particles (Böttcher *et al.*, 1997a) (Fig. 11; see Color Insert). The putative amino and carboxyl termini were identified by integrating its known biochemical and immunological information. In this particular case, undecagold-labeled cysteine was engineered onto its C terminus (Zlotnich *et al.*, 1997). In addition, difference imaging was used to locate a peptide of 10 amino acids inserted onto the N terminus of the core protein (Conway *et al.*, 1998). The availability of a large body of biochemical mutagenesis data together with chemical labeling studies made it possible to establish a fold model of the core protein (Fig. 12; see Color Insert) (Böttcher *et al.*, 1997a; Conway *et al.*, 1998). The predicted fold is in good agreement with that subsequently determined by X-ray crystallography at 3.3 Å (Wynne *et al.*, 1999).

For more complex structures, it is possible to combine electron cryomicroscopy structures with sequence-based secondary structure predictions to interpret the observed secondary structure elements. In the outer shell protein P8 of rice dwarf virus (RDV), where nine helices were predicted in the domain formed by the N and C termini, it was possible to match the lengths of the helices identified in the 3-D density map to those predicted from a consensus secondary structure analyses (Fig. 13a; see Color Insert). The connections between the helical densities can be seen in the lower domain of P8, allowing us to establish a rough backbone model for the lower domain of P8 (Zhou *et al.*, 2001).

To assess the accuracy of such a prediction, *DejaVu* (Kleywegt and Jones, 1997) and *COSEC* (Mizuguchi and Lio, 1995) can be used to perform spatial fold recognition. A successful match in the helix arrangement between the *helixHunter* results of a structure in the Protein Data Bank suggested a possible fold homolog. Six of the nine helices within the lower domain of RDV P8 subunit were matched to bluetongue virus VP7 (centroid RMS <5 Å) even though these proteins have only remote or no sequence similarity (<20%).

In addition, an individual protein or portions of a protein within a macromolecular complex may have a homologous structure. Once a homologous fold has been identified, it is necessary to place this structure back into the entire complex. *foldHunter,* a template-based cross-correlation tool, automatically searches all possible rotations and translations for the best fit of the homologous structure to the density map of the macromolecule (Jiang *et al.*, 2001b). In RDV P8, it was possible to identify the middle sequence segment to be structurally similar to the β-sheet domain of VP7 of the bluetongue virus. Figure 13b shows the localization of the putative jelly-roll β-sandwich fold of bluetongue virus in the density in the upper domain of RDV P8. By combining this fold identification result with the earlier *helixHunter* results, a model for the entire structure of P8 can be derived (Fig. 13b).

D. *Toward Near-Atomic Resolution: Three-Dimensional modeling*

Electron cryomicroscopy has been used to determine near-atomic resolution structures of two-dimensional crystals (Henderson *et al.*, 1990; Nogales *et al.*, 1998) and, therefore, in principle, may possibly be used to reach the same resolution in determining the structures of icosahedral virus particles. Modern electron cryomicroscopes are capable of recording images containing data at this resolution. A trivial limiting factor currently facing us is the digitization of data. The availability of a high-throughput scanner or of a high-resolution CCD camera will be helpful in this regard to generate image data in a digital form within a reasonable amount of time. A nontrivial limitation seems to lie in the development of a more powerful image-processing algorithm for improved accuracy in the refinement of orientation and center parameters and in the determination of CTF, particularly in those micrographs with residual astigmatisms. Finally, for larger particles, the correction for curvature of the Ewald sphere due to the depth of field limitation must be taken into account (DeRosier, 2000; Jensen, 2001). In any case, tens of thousands of particle images will need to be processed (Saad *et al.*, 2001). Thus, a powerful, efficient, user-friendly, and data management-capable software package is certainly required to accomplish the massive task of data processing for near-atomic resolution reconstructions.

VI. Conclusion

Structural determination of icosahedral viruses at subnanometer resolution by electron cryomicroscopy has become routine about 15 years after the first cryomicroscope virus structure at about 40 Å was reported. With the current approach of combining these structures and other bioinformatics analyses, it is possible to derive a pseudo-atomic model of an entire virion in some favorable cases. This capability has transformed electron cryomicroscopy to a method of choice for rapidly solving the structures of viruses and their complexes with other factors. By combining the electron cryomicroscopy structures of a virus with the X-ray crystal structures of its components, it is possible to describe structure changes during viral assembly and define the interactions between the virus and its cellular factors. Most important, the prospect is indeed promising for obtaining 3-D reconstructions of viruses approaching atomic resolution by using electron cryomicroscopy alone. While virologists enjoy the imaging power that electron cryomicroscopy offers for understanding the structure–function relationships of viral assembly and infection, the advancement of this technique toward near-atomic resolution will also benefit other efforts to improve the resolution of structural determination of macromolecular complexes with less or no symmetry.

Acknowledgments

The research activities described here have been supported in part by grants from the NIH (P41RR02250, AI43656, and AI38469 to W.C., AI46420 and CA94809 to Z.H.Z.), the Welch Foundation (Q1242 to W.C. and AU-1492 to Z.H.Z.), the March of Dimes Birth Defects Foundation (5-FY99-852 to Z.H.Z.), and the American Heart Association (0240216N to Z.H.Z.), Z.H.Z. is a Pew Scholar in Biomedical Sciences. We thank Dr. Michael Schmid, Dr. Wen Jiang, Jacob Zachariah, and Sanket Shah for their comments on this manuscript.

References

Baker, T. S., and Cheng, R. H. (1996). *J. Struct. Biol.* **116,** 120–130.

Baker, T. S., Drak, J., and Bina, M. (1988). *Proc. Natl. Acad. Sci. USA* **85,** 422–426.

Baker, T. S., Olson, N. H., and Fuller, S. D. (1999). *Microbiol. Mol. Biol. Rev.* **63,** 862–922.

Böttcher, B., and Crowther, R. A. (1996). *Structure* **4,** 387–394.

Böttcher, B., Wynne, S. A., and Crowther, R. A. (1997a). *Nature* **386,** 88–91.

Böttcher, B., Wynne, S. A., and Crowther, R. A. (1997b). *Hitachi Instr. News Electron Microsc. Ed.* **32,** 3–8.

Chiu, W. (1978). *Scanning Electron Microsc.* **1,** 569–580.

Chiu, W., Knapek, E., Jeng, T. W., and Dietrick, I. (1981). *Ultramicroscopy* **6,** 291–296.

Conway, J. F., and Steven, A. C. (1999). *J. Struct. Biol.* **128,** 106–118.

Conway, J. F., Cheng, N., Zlotnick, A., Wingfield, P. T., Stahl, S. J., and Steven, A. C. (1997). *Nature* **386,** 91–94.

Conway, J. F., Cheng, N., Zlotnick, A., Stahl, S. J., Wingfield, P. T., and Steven, A. C. (1998). *Proc. Natl. Acad. Sci. USA* **95,** 14622–14627.

Crowther, R. A. (1971). *Philos. Trans. R. Soc. Lond. B Biol. Sci.* **261,** 221–230.

Crowther, R. A., Amos, L. A., Finch, J. T., DeRosier, D. J., and Klug, A. (1970a). *Nature* **226,** 421–425.

Crowther, R. A., DeRosier, D. J., and Klug, A. (1970b). *Proc. R. Soc. Lond. B Biol. Sci.* **317,** 319–340.

DeRosier, D. J. (2000). *Ultramicroscopy* **81,** 83–98.

Dougherty, M., and Chiu, W. (2000). *Microsc. Microanal.* **6,** 282–283.

Dougherty, M. T., and Chiu, W. (1998). *In* "Microscopy and Microanalysis 1998" (G. W. Bailey, K. B. Alexander, W. G. Jerome, M. G. Bond, and J. J. McCarthy, Eds.), pp. 452–453. Springer-Verlag, Atlanta, GA.

Erickson, H. P., and Klug, A. (1970). *Philos. Trans. R. Soc. Lond. B. Biol. Sci.* **261,** 105–118.

Frank, J. (1979). *J. Microsc.* **117,** 25–38.

Fuller, S. D. (1987). *Cell* **48,** 923–934.

Fuller, S. D., Butcher, S. J., Cheng, R. H., and Baker, T. S. (1996). *J. Struct. Biol.* **116,** 48–55.

Grimes, J. M., Burroughs, J. N., Gouet, P., Diprose, J. M., Malby, R *et al.* (1998). *Nature* **395,** 470–478.

He, J., Schmid, M. F., Zhou, Z. H., Rixon, F., and Chiu, W. (2001). *J. Mol. Biol.* **309,** 903–914.

Henderson, R., Baldwin, J. M., Ceska, T. A., Zemlin, F., Beckmann, E., and Downing, K. H. (1990). *J. Mol. Biol.* **213,** 899–929.

Jensen, G. J. (2001). *J. Struct. Biol.* **133,** 143–155.

Jiang, W., and Chiu, W. (2001). *Microsc. Microanal.* **7,** 329–334.

Jiang, W., Li, Z., Zhang, Z., Booth, C. R., Baker, M. L., and Chiu, W. (2001a). *J. Struct. Biol.* **136,** 214–225.

Jiang, W., Baker, M. L., Ludtke, S. J., and Chiu, W. (2001b). *J. Mol. Biol.* **308,** 1033–1044.

Jiang, W., Li, Z., Zhang, Z., Baker, M. L., Prevelige, P. E., and Chiu, W. (2003). *Nat. struct. Biol.* **10,** 131–135.

Johnson, O., Govindan, V., Park, Y., and Zhou, Z. H. (1997). *In* "Proceedings of the 4th International Conference in High Performance Computing," pp. 517–521. IEEE Computer Society Press, Los Alamitos, California.

Kleywegt, G. J., and Jones, T. A. (1997). "Detecting Folding Motifs and Similarities in Protein Structures," pp. 525–545. Academic Press, London.

Liang, Y., Ke, E. Y., and Zhou, Z. H. (2002). *J. Struct. Biol.* **137,** 292–304.

Lo Conte, L., Ailey, B., Hubbard, T. J., Brenner, S. E., Murzin, A. G., and Chothia, C. (2000). *Nucleic Acids Res.* **28,** 257–259.

Ludtke, S. J., Baldwin, P. R., and Chiu, W. (1999). *J. Struct. Biol.* **128,** 82–97.

Ludtke, S. J., Jakana, J., Song, J.- L., Chuang, D., and Chiu, W. (2001). *J. Mol. Biol.* **314,** 253–262.

Mancini, E. J., Clarke, M., Gowen, B. E., Rutten, T., and Fuller, S. D. (2000). *Mol. Cell* **5,** 255–266.

Mizuguchi, K., and Go, N. (1995). *Protein Eng.* **8,** 353–362.

Nogales, E., Wolf, S. G., and Downing, K. H. (1998). *Nature* **391,** 199–203.

Orengo, C. A., Michie, A. D., Jones, S., Jones, D. T., Swindells, M. B., and Thornton, J. M. (1997). *Structure* **5,** 1093–1108.

Prasad, B. V.V., Wang, G. J., Clerx, J. P.M., and Chiu, W. (1988). *J. Mol. Biol.* **199,** 269–275.

Saad, A., Ludtke, S. J., Jakana, J., Rixon, F. J., Tsuruta, H., and Chiu, W. (2001). *J. Struct. Biol.* **133,** 32–42.

Schmid, M. F., Jakana, J., Matsudaira, P., and Chiu, W. (1993). *J. Mol. Biol.* **230,** 384–386.

Sheehan, B., Fuller, S. D., Pique, M. E., and Yeager, M. (1996). *J. Struct. Biol.* **116,** 99–106.

Spencer, S., Sgro, J., Dryden, K., Baker, T., and Nibert, M. (1997). *J. Struct. Biol.* **120,** 11–21.

Thon, F. (1971). *In* "Electron Microscopy in Material Sciences," pp. 572–625. Academic Press, New York.

Thuman-Commike, P. A., and Chiu, W. (2000). *Micron* **31,** 687–711.

Trus, B. L., Roden, R. B., Greenstone, H. L., Vrhel, M., Schiller, J., and Booy, F. P. (1997). *Nat. Struct. Biol.* **4,** 411–418.

van Heel, M. (1987). *Ultramicroscopy* **21,** 95–100.

van Heel, M., Gowen, B., Matadeen, R., Orlova, E. V., Finn, R *et al.* (2000). *O. Rev. Biophys.* **33,** 307–369.

Wynne, S. A., Crowther, R. A., and Leslie, A. G. (1999). *Mol. Cell* **3,** 771–780.

Zhou, Z. H., and Chiu, W. (1993). *Ultramicroscopy* **49,** 407–416.

Zhou, Z. H., Prasad, B. V., Jakana, J., Rixon, F. J., and Chiu, W. (1994). *J. Mol. Biol.* **242,** 456–469.

Zhou, Z. H., He, J., Jakana, J., Tatman, J. D., Rixon, F. J., and Chiu, W. (1995). *Nat. Struct. Biol.* **2,** 1026–1030.

Zhou, Z. H., Hardt, S., Wang, B., Sherman, M. B., Jakana, J., and Chiu, W. (1996). *J. Struct. Biol.* **116,** 216–222.

Zhou, Z. H., Chiu, W., Haskell, K., Spears, H.Jr., Jakana, J *et al.* (1998). *Biophys. J.* **74,** 576–588.

Zhou, Z. H., Chen, D. H., Jakana, J., Rixon, F. J., and Chiu, W. (1999). *J. Virol.* **73,** 3210–3218.

Zhou, Z. H., Dougherty, M., Jakana, J., He, J., Rixon, F. J., and Chiu, W. (2000). *Science* **288,** 877–880.

Zhou, Z. H., Baker, M. L., Jiang, W., Dougherty, M., Jakana, J *et al.* (2001). *Nat. Struct. Biol.* **8,** 868–873.

Zhou, Z. H., Zhang, H., Jakana, J., Lu, X.-Y., and Zhang, J.-Q. (2003). *Structure* **11,** in press.

Zlotnick, A., Cheng, N., Stahl, S. J., Conway, J. F., Steven, A. C., and Wingfield, P. T. (1997). *Proc. Natl. Acad. Sci. USA* **94,** 9556–9561.

STRUCTURAL FOLDS OF VIRAL PROTEINS

By MICHAEL S. CHAPMAN* AND LARS LILJAS[†]

*Department of Chemistry and Biochemistry and Institute of Molecular Biophysics, Florida State University, Tallahassee, Florida 32306, and [†]Department of Cell and Molecular Biology, Uppsala University, Biomedical Center, 751 24 Uppsala, Sweden

I.	Introduction	125
II.	Terminology	126
III.	Virus Families	126
IV.	Determination of Structural Fold	126
V.	Prototypical Viral Folds	128
	A. Jelly-Roll β Barrel	128
	B. The Immunoglobulin Fold	131
	C. The Serine Protease Fold	132
	D. Four-Helix Bundle	132
VI.	Survey through the Virus Families	132
	A. Positive-Strand RNA Viruses and Satellite Viruses	139
	B. Negative-Strand RNA Viruses	158
	C. Double-Stranded RNA Viruses from the Reoviridae Family	163
	D. Single-Stranded DNA Viruses	167
	E. Double-Stranded DNA Viruses	171
	F. Reverse-Transcribing Viruses	178
VII.	Common Themes	184
VIII.	Phylogenetic Relationships	185
	References	187

I. Introduction

Viruses vary greatly. "Simple" viruses make parsimonious use of a few gene products, gaining additional functionality from host proteins. Others are complicated machines that encode many of the structural and enzymatic proteins needed for their own replication. This review, first, focuses on proteins whose structure or function is primarily associated with viruses. It does not cover the myriad of enzymes encoded by larger viruses that are homologous to cellular proteins. Second, it is restricted to those proteins whose structures are known at near atomic resolution. This will include the building blocks of symmetrical virus capsids and cores, and key membrane proteins of some enveloped viruses, and a few enzymes with function unique to viruses.

The topology of the folds, how they are suited to their function, and present prevalent ideas about their possible evolutionary relationships are

ADVANCES IN PROTEIN CHEMISTRY, Vol. 64

0065-3233/03 $35.00

described. Through tables annotating structures this article attempts to give some perspective on how much about virus structure is known, and how much remains to be discovered. Structural data are expanding exponentially, and reviews can no longer be truly comprehensive. We have made representative selections, but it is likely that there are also omissions by oversight.

II. Terminology

The terminology of Harrison *et al.* (1996) is followed, updated according to current preferences. *Subunits* are distinct protein chains containing one or more integral (globular) *domains* that have single peptide connections to neighboring domains. Sometimes "domain" is used when these criteria are not all fully met, sometimes "subdomain." This review is concerned with the tertiary structure of domains/subdomains and how it relates to higher orders of structure. *Protomers* are collections of several nonidentical proteins that form the smallest unique part of a symmetric assembly. *Nucleocapsids* or "cores" are the protein–nucleic acid complexes packaged in viruses. *Capsids* are the assembled protein moiety that surrounds the nucleic acid in nucleocapsids. *Envelopes* of lipid plus associated proteins and carbohydrates surround the nucleocapsids of some viruses. Capsid, nucleocapsid, and envelope proteins are all considered here.

III. Virus Families

The number of known virus families is currently greater than 90 and increasing at 2 per year (van Regenmortel *et al.*, 2000). They vary greatly in size, morphology, genome type, and proteome (Murphy and Kingsbury, 1996). Morphologically, they may be isometric, spherical, rodlike, filamentous, or pleomorphic. Examples of many of these morphologies are found in lipid-enveloped or naked forms. Thus, there are a wide variety of architectures, and it is to be expected that there will be a wide variety of folds used in the protein-building blocks.

IV. Determination of Structural Fold

The predominant technique is X-ray crystallography, followed by fiber diffraction and nuclear magnetic resonance (NMR) spectroscopy. A few structures have been predicted, on the basis of homology, but, by definition, these are not predicting new folds, but relationships to known ones, and are not considered further in this article. Viruses are often composed of large complexes of large proteins, often beyond the realm of NMR spectroscopy. The discussion of folds includes NMR-determined domain structures but not short peptide fragments. Structure determination of large assemblies through crystal or fiber diffraction methods is possible only because of the symmetry that is common in virus particles, but not universal, leading to a heavy bias in structural information to the isometric, rodlike, and filamentous viruses.

Many of the structural proteins are components of the protein capsids/cores that surround the nucleic acid, or are components of lipid envelopes. Each has its own challenges for structure determination. Membrane proteins are generally difficult to crystallize or to study by NMR in monodisperse solutions. Domains that are not integral to the membrane can of course be cleaved off for structural studies. For the most part, the capsids and core proteins have been studied as assemblies isolated from natural sources or reconstituted/expressed to mimic those of the infectious virus. By the standards of diffraction methods, these are large complexes, the smallest of which contain >1 MDa of protein.

The importance of symmetry to structure determinations can be explained to the layman in two ways. First, crystals of these large complexes are not of the quality of typical proteins, and the diffraction data have experimental errors greater by a factor of 3 to 5. By averaging the images of symmetry-related parts, the signal-to-noise ratio of electron density of maps can be improved, in principle >8-fold (Arnold and Rossmann, 1986), to the point at which they can be interpreted. Virus crystallographers actually use symmetry in a more powerful way to solve the greatest challenge in protein crystallography: the phase problem. Electron density maps are calculated by Fourier transformation of the diffraction amplitudes. Each of the many thousands of sine waves contributing to the map needs not only an amplitude, but also a starting point, or phase. The experimental methods of phase determination applicable to proteins are extremely difficult to apply to viral assemblies. Symmetry averaging is the saving grace. It is applied in an iterative process through which constraints of the symmetry are used to refine precise phases from a crude starting point (Chapman *et al.*, 1998; Rossmann, 1995). The power of the symmetry-based phase refinement is such that effectively no bias remains of an atomic model used to initiate the phasing. It also generates maps that are among those of the highest quality possible for a given resolution.

Cryoelectron microscopy can similarly take advantage of the symmetry. It has been most effective in revealing the (low-resolution) morphology of viruses that could not be crystallized. It has begun achieving the resolutions necessary for determination of protein fold. One of the early successes has been the core of hepatitis B (Böttcher *et al.*, 1997; Conway *et al.*, 1997), which was subsequently also determined by X-ray diffraction (Wynne *et al.*, 1999). Electron microscopy has been used to align previously known high-resolution structures in low-resolution images of assemblies (Baker and Johnson, 1996). It has not, on its own, generated any of the current high-resolution Protein Data Bank virus entries, but this is likely to change soon.

Filamentous viruses have been studied mostly by fiber diffraction (Namba *et al.*, 1989). Fibers can be considered to be crystals in which there is rotational disorder about the helical axis. This has the effect of smearing the discrete spots of crystalline diffraction into layer lines. There is a resulting loss of information, but sufficient experimental observations can be made for highly symmetric virus structures so that a unique high-resolution structure is obtainable (Makowski, 1991).

V. Prototypical Viral Folds

In this section the folds of domains found repeatedly in different viruses are described. The discussion is of the canonical form with details of the differences between different viruses following in the survey of viral taxa.

A. *Jelly-Roll β Barrel*

At one time, seemingly all capsids contained a jelly-roll β-barrel fold, leading to the synonym "virus capsid β barrel," with other synonyms being (distorted) "eight-stranded antiparallel β-barrel" and "β sandwich." Two four(+)-stranded sheets face one another (Fig. 1 [see Color Insert]; and Figure 2). The end strands almost form a closed circular barrel structure. However, except in rare cases noted, there is no hydrogen bonding between the end strands of facing sheets (as there is in some nonviral barrels), leading some to prefer the term "sandwich" rather than the more popular "barrel." All strands run antiparallel to their neighbors. Topologically, there is a single tight turn between strands E and F, which then leads to the pairing of strands D and G, C and H, and B and I. These then roll up, Swiss cake style, so that the BI and DG pairs combine to form the BIDG sheet, and the CH and EF pairs combine to form the CHEF sheet. An excellent comparative illustration of the strand pairing in structures known in 1989 appeared in Fig. 5 of Rossmann and Johnson (1989), to which should be added Fig. 8 of Xie and Chapman (1996), to show the embellishments possible in a large jelly-roll domain.

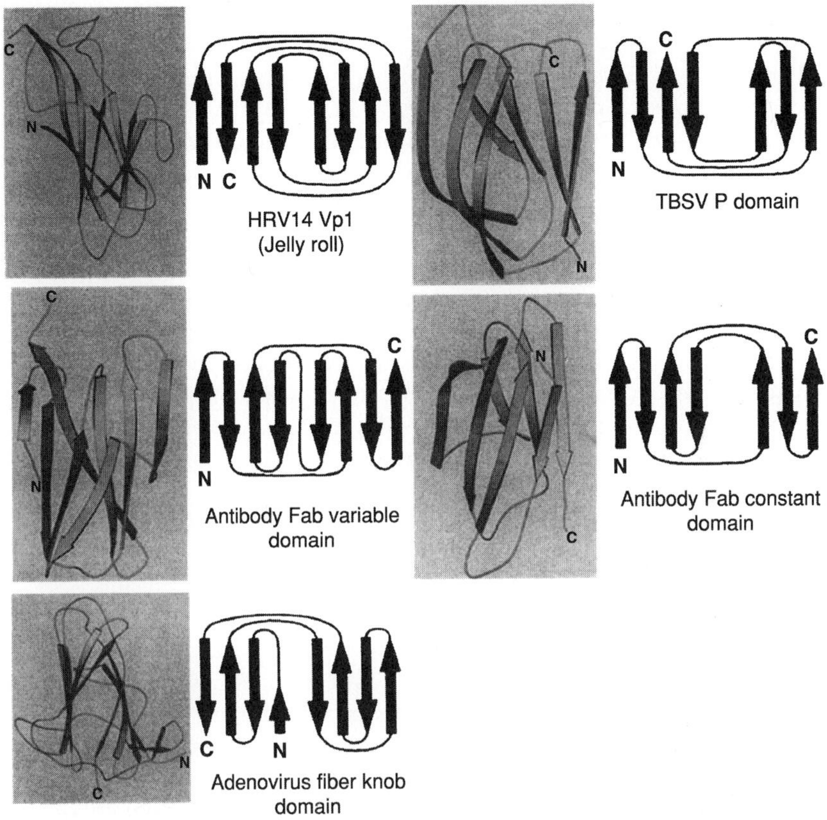

FIG. 2. Comparison of some common β-barrel topologies seen in virus and other structures. [Adapted from Xia *et al.* (1994) with permission from Elsevier Science.]

In the prevalent (but not sole) configuration, the domains are oriented within the capsid so that all strands are roughly tangential to the viral capsid. Strands B, D, F, and H point toward 5-fold or quasi-6-fold axes, where both sheets are oriented so that the strands are stacked on one another radially, with strands G and F innermost and strands B and C outermost. β Sheets have a right-handed helical twist looking along the peptides, so progressing from the 5-fold/quasi-6-fold axes, the sheets flatten to a tangential orientation with the BIDG sheet forming the inner surface of the capsid and the CHEF sheet farther from the viral center. Even within this prevalent configuration, differences between the plant and picornaviruses include 19° rotations and 2.5-Å translations of the whole domain. Some quite different orientations in other viruses, such as in the adenovirus hexon (Roberts *et al.*, 1986), are discussed later.

Domains in different viruses differ in several ways. There may be one or two additional strands. Strand A is found in some viruses, but in different locations, sometimes as a direct extension of sheet BIDG, sometimes domain swapped so that it hydrogen bonds with a neighboring subunit. Within the core barrel, strand lengths differ. For T=1 capsids βB ranges from 9 to 17 amino acids, βC from 3 to 10, βD from 10 to 25, and so on (Xie and Chapman, 1996).

The biggest variations are in the loops—in size and additional secondary structural elements (or complete domains) that they contain. In the nonenveloped vertebrate viruses, it is the external loops that contain the antigenic sites (Rossmann *et al.*, 1985; Tsao *et al.*, 1991), with epitopes formed where several loops come together. The nomenclature "BC" is used to describe the loop between strands B and C. Generally, the BC, HI, DE, and FG loops that are close to the 5-fold/quasi-6-fold axes tend to be short, whereas the CD, EF, and GH loops tend to be longer. Whereas most of the jelly-roll β barrels are about 180 amino acids, they go up in size to 584 amino acids in parvoviruses with large insertions in the loops (Tsao *et al.*, 1991).

With few exceptions, the jelly-roll proteins forming capsids have arms at their N termini and often also at their C termini. These arms are often partly disordered, and the disordered segments might be important for interaction with the nucleic acid molecule. Ordered parts of these arms are used to regulate subunit packing and particle stability.

Several hypotheses have been proposed for the prevalence of this domain type. Harrison and colleagues (1996) sketched the domain schematically in a way that is appealing in its simplicity and emphasizes its approximate trapezoidal cross-section. This is a natural building-block shape for various icosahedral assemblies. The small single-stranded RNA (ssRNA) capsid structures determined through the 1980s showed compact β-barrel domains interacting directly with those of neighboring subunits. However, with the structures of canine parvovirus (CPV) and bacteriophage ϕX174 (McKenna *et al.*, 1992a; Tsao *et al.*, 1991), it became clear that increased size of T=1 capsids was achieved not by larger barrels, but by adding to the loops. It is, then, the loops—not the trapezoidal barrel surfaces—that in CPV form the primary subunit interfaces (Xie and Chapman, 1996). Another explanation for the domain prevalence was suggested by the early picornavirus structures that revealed a hole in the middle of the barrel. When the pocket is filled with drugs or natural factors, the protein is stabilized, preventing conformational changes necessary for uncoating (Filman *et al.*, 1989; Hadfield *et al.*, 1997; Kim *et al.*,

1989; Smith *et al.*, 1986; Tsang *et al.*, 2000). This could present a mechanism for controlling uncoating. However, later came examples such as Mengovirus (Luo *et al.*, 1987), in which the side chains completely filled the barrel. The true reason for the prevalence of the domain is likely now obscured by the diversity that has emerged between viruses since a point in early evolution when one of these (or another) factor presumably gave a strong selective advantage for the barrel domain.

The domain is found in icosahedral viruses of different sizes and T numbers. In some of the T=3 plant viruses, such as southern bean mosaic virus (SBMV) (Abad-Zapatero *et al.*, 1980), the capsid is composed of 180 identical subunits. In the pseudo-T=3 capsids of picornaviruses, the corresponding part of the capsid is composed of 60 copies of each of 3 different proteins. Although there is less residual sequence homology between VP1, VP2, and VP3 (Palmenberg, 1989), there is strong structural homology (Rossmann, 1987) suggesting that the three proteins in picornaviruses evolved by gene triplication. The laboratory of J. Johnson has strengthened this argument with structures of "missing links"—plant comoviruses with two, or nepoviruses with three of these domains within single covalent peptides. Cowpea mosaic virus (CPMV) and bean pod mottle virus (BPMV) have two capsid proteins, one of which has two domains (Chen *et al.*, 1989; Lin *et al.*, 1999), whereas the tobacco ringspot virus capsid has 60 copies of a single protein, which contains 3 barrel domains (Chandrasekar and Johnson, 1998).

B. *The Immunoglobulin Fold*

Here we start describing folds that are primarily associated with nonviral proteins, but are found in some viral structures. The immunoglobulin fold (Fig. 3; see Color Insert) is found in antibodies, cellular adhesion, and many other molecules. One viral example is one of the domains of a flavivirus envelope glycoprotein (Rey *et al.*, 1995). The canonical immunoglobulin fold has a seven-stranded antiparallel β-barrel core to which two additional short strands can be added in the variable immunoglobulin domain. The barrel can be split into two sheets within which there is good hydrogen bonding. The three- and four-member sheets face each other and are commonly linked by a disulfide bond. The linear topology of the domain is Greek key. All of the strands are antiparallel, so there are no cross-over connections that pass over the sheet. Loops that connect strands are both intrasheet and also cross between the sheets.

C. *The Serine Protease Fold*

The serine protease fold was first seen in 1967 in Blow's, D. structure of α-chymotrypsin, and in many protease structures since (Matthews *et al.*, 1967). It is found in a number of viral proteins including some proteases (Bazan and Fletterick, 1988; Matthews, D. A. *et al.*, 1994) and the alphavirus core protein (Choi *et al.*, 1991). Once again it is an antiparallel Greek key β barrel, this time with six strands (Fig. 4; see Color Insert). Although viral capsids, immunoglobulins, and serine proteases all contain antiparallel β barrels, the order of the strands and the connections between them are different. In the serine and viral cysteine proteases the protein is composed of two copies of the same domain fold that presumably evolved long ago by gene duplication. The active site is in the cleft between with the catalytic triad residues contributed by the $\beta 3$–$\beta 4$ and the $\beta 5$–$\beta 6$ loops of the N-terminal domain, and by the $\beta 3$–$\beta 4$ loop of the C-terminal domain.

D. *Four-Helix Bundle*

The four-helix bundle is a common motif in which (usually) antiparallel α helices are packed side by side. It is found in myohemerythrin, various cytochromes, and a number of other proteins. A viral example is the coat protein of tobacco mosaic virus (TMV) (Bloomer *et al.*, 1978). TMV represents the most common type, in which the helix axes are nearly antiparallel, off by 18°, coiled with a left-handed superhelical twist (Fig. 5; see Color Insert). The slight misalignment of the individual helix axes allows the side chains to interdigitate efficiently, burying internal hydrophobic side chains.

VI. Survey Through the Virus Families

Table I through Table VII place in some context the extent of our understanding of virus structure. There are a remarkable number of structures that have led to insights of a general virological nature and specific to the virus type. Tables I–VII also make clear how far we have to go. About most virus families we know nothing in atomic detail. For others we have determined structures for one or two of several critical proteins. That said, the discussion starts with one of the best-characterized groups.

Table I

Taxonomy of Known Viral Structure: *RNA Genome, Single Positive-Sense Strand*[a,b]

Order / ***Family***—Distinguishing features / ***Subfamily*** / Fold:	*Genus* / Description	*Examples* / Elements of known structure	Host / Representative PDB id
Leviviridae—Nonenveloped; 180 copies of 1 protein in a 290-Å T=3 capsid			
Fold:	(Atypical) β meander with α helices over an antiparallel β sheet		
	Levivirus	*Enterobacteria phage MS2*	Bacteria
		Capsids (Golmohammadi *et al.*, 1993)	2ms2
		Bacteriophage fr	Bacteria
		Capsids (Liljas *et al.*, 1994)	1frs
		Bacteriophage GA	Bacteria
		Capsids (Ni *et al.*, 1996; Tars *et al.*, 1997)	1gav, 1una
		Caulobacter phage PP7	Bacteria
		Capsids (Tars *et al.*, 2000)	1dwn
	Allolevivirus	*Enterobacteria phage Qβ*	Bacteria
		Capsids (Golmohammadi *et al.*, 1996)	1qbe
Picornaviridae—Nonenveloped; T=1 capsid with 60 copies of 4 proteins (VP1–VP4); VP1–3 arranged to form a pseudo-T=3, ~300-Å-diameter capsid. A 2.4-kDa genome-linked protein VPg and an RNA polymerase are of unknown structure. Genome is ~7.8 kb.			
Fold:	Capsid, jelly-roll β sandwich; 3C protease, trypsin-like; leader protease, papain-like		
	Enterovirus	*Poliovirus 1, 3*	Vertebrates
		Capsid; VP1–VP4 (Hiremath *et al.*, 1995; Hogle *et al.*, 1985; Lentz *et al.*, 1997)	2plv, 1eah, 1pvc
		Coxsackievirus B3, A9	Vertebrates
		Capsid; VP1–VP4 (Hendry *et al.*, 1999; Muckelbauer *et al.*, 1996)	1cov, 1d4m
		Bovine enterovirus	Vertebrates
		Capsid; VP1–VP4 (Smyth *et al.*, 1993)	1bev
		Echovirus 1, 11	Vertebrates

(*continues*)

TABLE I (Continued)

Order / ***Family***—Distinguishing features / ***Subfamily***	*Genus*	*Examples* / Elements of known structure	Host / Representative PDB id
Fold:	Description		
		Capsid; VP1–VP4 (Filman *et al.*, 1998; Stuart *et al.*, 2002)	1 evl, 1h8t
	Rhinovirus	*Human rhinoviruses 1A, 2, 3, 14, 16, 50*	Vertebrates
		Capsid; VP1–VP4 (Blanc *et al.*, 2002; Kim *et al.*, 1989; Oliveira *et al.*, 1993; Rossmann *et al.*, 1985; Verdaguer *et al.*, 2000; Zhao *et al.*, 1996); 3C protease (D.A. Matthews, *et al.*, 1994)	1 r1a, 1 fpn, 1 rhi, 4rhv, 1 aym, 1 cqq (protease)
	Hepatovirus	*Hepatitis A virus*	Vertebrates
		3c protease (Bergmann *et al.*, 1997)	1 hav
	Cardiovirus	*Mengovirus*	Vertebrates
		Capsid; VP1–VP4 (Luo *et al.*, 1987)	2mev
		Theiler's murine encephalomyelitis virus	Vertebrates
		Capsid; VP1–VP4 (Grant *et al.*, 1992; Luo *et al.*, 1992)	1tme, 1tmf
	Aphthovirus	*Foot-and-mouth disease virus*	Vertebrates
		Capsid; VP1–VP4 (Acharya *et al.*, 1989; Lea *et al.*, 1994); leader protease (Guarne *et al.*, 1998)	1bbt, 1fmd (capsid), 1qmy (protease)
Dicistroviridae—Nonenveloped; T=1 capsid with 60 copies of 4 proteins (VP1–VP4); VP1–VP3 arranged to form a 305-Å-diameter pseudo-T=3 capsid			
	Cripavirus	*Cricket paralysis virus*	Invertebrates
		Capsid; VP1–VP4 (Tate *et al.*, 1999)	1b35
Comoviridae—Coat protein(s) form a ~320-Å-diameter T=1 capsid. Protein(s) contains three homologous domains, arranged with pseudo-T=3 icosahedral symmetry. Structures for three of four genera			
Fold:	Capsid, jelly-roll β sandwich. One protein with two domains, the other with one		
	Comovirus	*Cowpea mosaic virus*	Plants
		Capsid (Lin *et al.*, 1999)	1ny7
		Bean pod mottle virus	Plants
		Capsid (Chen *et al.*, 1989)	1bmv

		Red clover mottle virus	Plants
		Capsid (Lin *et al.*, 2000)	—
Fold:	Capsid, jelly-roll β sandwich. One protein with three jelly-roll domains		
	Nepovirus	*Tobacco ringspot virus*	Plants
		Capsid (Chandrasekar and Johnson, 1998)	1a6c
Caliciviridae—~340-Å capsid of 180 proteins surrounding ~7.5-kb genome; several nonstructural proteins, nonenveloped. Structures for one of four genera			
Fold:	Capsid, jelly-roll β sandwich, protruding domain with two subdomains with antiparallel β structure, one of which has the fold of EF-Tu domain 2		
	Norovirus	*Norwalk virus*	Vertebrates
		Capsid (Prasad *et al.*, 1999)	1ihm
Nodaviridae—Nonenveloped; 180 subunits in a ~355-Å-diameter T=3 capsid			
Fold:	Capsid—jelly-roll β sandwich		
	Alphanodavirus	*Nodamura virus*	Invertebrates
		Capsid (Zlotnick *et al.*, 1997)	1nov
		Flock house virus	Invertebrates
		Capsid (Fisher and Johnson, 1993)	1fhv
		Black beetle Virus	Invertebrates
		Capsid (Wery *et al.*, 1994)	2bbv
		Pariacoto virus	Invertebrates
		Capsid (Tang *et al.*, 2001)	1f8v
	Betanodavirus	(No structural information)	Vertebrates
Tetraviridae—Nonenveloped; 240 subunits in a ~430-Å-diameter T=4 capsid			
Fold:	Capsid, jelly-roll β sandwich, protruding domain with immunoglobulin fold		
	Betatetravirus	(No structural information)	Invertebrates
	Omegatetravirus	*Nudaurelia capensis* ω *virus*	Invertebrates
		Capsid (Munshi *et al.*, 1996)	—
Unclassified	*Sobemovirus*	Nonenveloped, 180 × single subunit in ~330-Å-diameter, T=3 icosahedral capsid	

(*continues*)

TABLE I (*Continued*)

Order / ***Family***—Distinguishing features / ***Subfamily*** / Fold:	*Genus* / Description	*Examples* / Elements of known structure	Host / Representative PDB id
Fold:	Capsid, jelly-roll β sandwich		
		Southern bean mosaic virus	Plants
		Capsid (Abad-Zapatero *et al.*, 1980)	4sbv
		Sesbania mosaic virus	Plants
		Capsid (Gopinath *et al.*, 1994)	1smv
		Rice yellow mottle virus	Plants
		Capsid (Qu *et al.*, 2000)	1f2n
Luteoviridae—180 copies of subunit in 320- to 360-Å-diameter T=3 icosahedral capsid			
Fold:	Capsid S (shell) domain, jelly-roll β sandwich; projecting P domain, jelly-roll β sandwich. Structures for three of eight genera		
	Tombusvirus	*Tomato bushy stunt virus*	Plants
		Capsid (Harrison *et al.*, 1978)	2tbv
	Carmovirus	*Turnip crinkle virus*	Plants
		Capsid (Hogle *et al.*, 1986)	
		Carnation mottle virus	Plants
		Capsid (Morgunova *et al.*, 1994)	1opo
Fold:	Capsid domain, jelly-roll β sandwich, no protruding domain		
	Necrovirus	*Tobacco necrosis virus*	Plants
		Capsid (Oda *et al.*, 2000)	1c8n
Nidovirales			
Flaviviridae—Spherical, enveloped, 450- to 600-Å-diameter, nonsegmented genome; 9.5–12.5 kb. Contains core protein plus two or three membrane proteins including a small transmembrane protein and a major surface glycoprotein			

Fold:	Flavivirus glycoprotein: domain 1 is an antiparallel β barrel that is quite different from the viral capsid jelly roll or immunoglobulin fold. Domain 2 contains several β sheets and is of complicated topology. Domain 3 is an immunoglobulin antiparallel β fold. Structures from two of three families		
	Flavivirus	*Tick-borne encephalitis virus*	Verebrates/Insect
		Glycoprotein (Rey *et al.*, 1995) (also yellow fever)	1svb
	Hepacivirus	*Hepatitis C virus*	Vertebrates
		NS3 proteinase (helicase) (Love *et al.*, 1996), NS5B RNA polymerase (Bressanelli *et al.*, 1999; Lesburg *et al.*, 1999)	1a1q (NS3), 1c2p, 1csj (NS5B)
Togaviridae—10- to 12-kb genome within ~400-Å-diameter T=4 icosahedral nucleocapsid; envelope with peplomers in T=4 symmetry and several glycoproteins. Total diameter is about 700 Å			
Fold:	Core protein, serine protease-like		
	Alphavirus	*Sindbis virus*	Vertebrates
		Core protein (Choi *et al.*, 1991)	2snv
		Semliki Forest virus	Vertebrates
		Core protein (Choi *et al.*, 1997)	1vcp
	Rubivirus	*Rubella virus*	Vertebrates
		(No structural information)	
Unclassified	*Rodlike helical virus*		
Fold:	All α helix; four-helix up–down bundle. There are additional short helices at the N and C termini. Structures for one of four genera		
	Tobamovirus	*Tobacco mosaic virus*	Plants
		Intact virus, protein (Bhyravbhatla *et al.*, 1998; Namba and Stubbs, 1986)	2tmv, 1e17
		Ribgrass mosaic virus	Plants
		Intact virus (Wang *et al.*, 1997a)	1rmv
		Cucumber green mosaic mottle virus	Plants
		Intact virus (Wang and Stubbs, 1994)	1cgm
Bromoviridae—Icosahedral capsids, details vary (see below). Structures for three of five genera			
Fold:	Capsid protein, jelly-roll β sandwich		
	Alfamovirus	60 subunits in a 220-Å-diameter T=1 icosahedral capsid	
		Alfalfa mosaic virus	Plants
		Capsid (Kumar *et al.*, 1997)	—

(*continues*)

TABLE I (*Continued*)

Order / Family / Subfamily / Fold	Genus / Description	Examples / Elements of known structure	Host / Representative PDB id
Order			
Family—Distinguishing features			
Subfamily	*Genus*	*Examples*	Host
		Elements of known structure	Representative PDB id
Fold:	Description		
	Bromovirus—180 subunits in a 290-Å-diameter T=3 icosahedral capsid		
		Cowpea chlorotic mottle virus	Plants
		Capsid (Speir *et al.*, 1995)	1cwp
		Brome mosaic virus	Plants
		Capsid (Lucas *et al.*, 2002)	1js9
	Cucumovirus—180 subunits in a 305-Å-diameter T=3 icosahedral capsid		
		Cucumber mosaic virus	Plants
		Capsid (Smith *et al.*, 2000)	1f15
Tymoviridae	*Tymovirus*—180 copies of subunit in ~320-Å-diameter T=3 icosahedral capsid. Structures for 1 of 13 families		
Fold:	Capsid protein, jelly-roll β sandwich		
		Turnip yellow mosaic virus	Plants
		Capsid (Canady *et al.*, 1996)	1auy
		Physalis mottle virus	Plants
		Capsid (Krishna *et al.*, 1999)	1qjz
		Desmodium yellow mottle virus	Plants
		Capsid (Larson *et al.*, 2000)	1ddl

[a]Nineteen named families, 22 unnamed.

[b]Assembled from several secondary sources: the 7th ICVT report (2000; van Regenmortel *et al.*, 2000), Viper (Reddy, 1999), the Protein Data Bank (Berman *et al.*, 2000), SwissProt (Bairoch and Apweiler, 2000), and Murphy and Kingsbury (1996).

TABLE II
Taxonomy of Known Viral Structure: Satellite Viruses[a]

Order Satellites (otherwise not classified)		
	Examples	Host
	Elements of known structure	Representative PDB id
Nonenveloped; 60 copies of 1 protein in a 170- to 200-Å T=1 capsid		
Fold:	Capsid protein, jelly-roll β sandwich	
	Satellite tobacco necrosis virus	Plants
	Capsids (Jones and Liljas, 1984; Liljas *et al.*, 1982)	2stv
	Satellite panicum mosaic virus	Plants
	Capsids (Ban and McPherson, 1995)	1stm
	Satellite Tobacco mosaic virus	Plants
	Capsids (Larson *et al.*, 1993b)	1a34

[a]Assembled from several sources: Bairoch and Apweiler (2000); Berman *et al.* (2000); Murphy and Kingsbury (1996); Reddy (1999); van Regenmortel *et al.* (2000).

A. *Positive-Strand RNA Viruses and Satellite Viruses*

1. *Bacteriophages: Leviviridae Family*

The capsid subunit tertiary structure in the *Leviviridae* family is different from any other viral fold. It is dominated by a β meander, a five-stranded antiparallel β sheet that is tangential to the spherical capsid (Golmohammadi *et al.*, 1993; Valegård *et al.*, 1990) (Fig. 6). The long C-terminal strands of 2-fold related subunits abut each other edge on, doubling the extent of the β sheet if two subunits are combined. Dimer interactions are further solidified by intercalation of the N- and C-terminal subdomains of the adjacent subunits in what some might term "domain swapping" (Bennett *et al.*, 1995). Before the start of the β meander, there is an N-terminal β ribbon. At the C terminus there are two short α helices that would be continuous except for a short region of extended chain. Packing alternately in the subunit dimer are the β ribbon from subunit A, the α helices from B, the helices from A, and finally the ribbon of B. All lie on top of the β meander on the outer surface of the virus. Loops are small, but form the majority of the other intersubunit contacts. Although morphologically this may resemble an "open-faced sandwich" fold, the topological order of secondary structure elements is quite different. The dimer is superficially similar to the peptide-binding part of the

TABLE III

Taxonomy of Known Viral Structure: RNA Genome, Single Negative-Sense Strand[a,b]

Order / *Family*—Distinguishing features / *Subfamily* / Fold:	*Genus* / Description	*Examples* / Elements of known structure	Host / Representative PDB id
Mononegavirales—Four families, enveloped			
Filoviridae—Pleomorphic (various shapes), 800 Å in diameter, up to 14 μm in length, nonsegmented genome in a rigid helical 500-Å nucleocapsid			
Fold:	gp2 ectodomain contains trimeric coiled coil. Matrix protein has two sandwich domains, each with 3 + 3 antiparallel strands		
	Ebola-like viruses	*Ebola virus* gp2 membrane fusion glycoprotein (Malashkevich *et al.*, 1999; Weissenhorn *et al.*, 1998); matrix protein (Dessen *et al.*, 2000)	Vertebrates 1ebo, 2ebo (fusion), 1es6 (matrix)
	Marburg-like viruses	*Marburg virus*	Vertebrate
Paramyxoviridae—Pleomorphic (various shapes) with dimensions 300–10,000 nm, enveloped, ~15-kb genome in helical nucleocapsids of ~800 × ~15 nm, includes well-known viruses, e.g., mumps			
Fold:	Fusion protein contains trimeric coiled coil, different from influenza HA, with coiled coil in opposite direction. Hemagglutinin–neuraminidase has β-propeller topology like NA from influenza A		
	Rubulavirus	*Simian parainfluenza virus 5* Fragment of fusion protein F (Baker *et al.*, 1999)	Vertebrates 1svf
		Newcastle disease virus Hemagglutinin–neuraminidase (HN) (Crennell *et al.*, 2000); fusion (F) protein (Chen *et al.*, 2001)	Vertebrates 1e8u, 1g5g
Rhabdoviridae			
Fold:	Matrix protein has a five-stranded curved antiparallel sheet with a C-terminal hairpin and two helices on one side		
	Vesiculovirus	*Vesicular stomatitis virus Indiana* Matrix protein M (Baker *et al.*, 1999)	Vertebrates 1lg7

One other family
Unnamed orders

Orthomyxoviridae—Pleomorphic, often spherical, 800–1200 Å in diameter, enveloped, peplomers containing hemagglutinin and neuraminidase, segmented genome of six to eight segments totaling 10–14 kb. In addition to proteins below, virions contain three polymerase proteins, nucleocapsid protein (NP), up to two nonstructural proteins. Structures from influenza A, representing one of six families

Folds:

HA: Five domains: (1) N-terminal receptor-binding domain is β jelly roll; (2) hairpin of two α helices forms coiled coil in assembled trimer; (3) membrane-proximal β sheet; (4 and 5) transmembrane and internal domains of unknown structure

NA: (1) N-terminal membrane attachment domain of unknown structure; (2) propeller of six blades, each a four-stranded parallel β sheet

M_1: First two domains are four-helix bundles; the first is a classic antiparallel bundle, and the second is a crossed-helix bundle. The C-terminal domain is of unknown structure

M_2: External (N-terminal) and internal minidomains of unknown structure with an α-helical transmembrane domain that tetramerizes to form channel

Influenzavirus A	*Influenza A virus* Hemagglutinin (HA) (Ha *et al.*, 2002; Harrison, 1990; Wilson *et al.*, 1981) and its fusion peptide (Han *et al.*, 2001); neuraminidase (NA) (Colman *et al.*, 1983); matrix protein M_1 (Sha and Luo, 1997); M_2 (Kovacs *et al.*, 2000).	Vertebrates 1ha0 (HA_0); 1ibn/1ibo (fusion peptide); 1nnf (NA); 1aa7 (M_1)

Several other genera, including *Influenzavirus B, C*

---

[a]Total of 10 families.

[b]Assembled from several sources: Bairoch and Apweiler (2000); Berman *et al.* (2000); Murphy and Kingsbury (1996); Reddy (1999); van Regenmortel *et al.* (2000).

TABLE IV

Taxonomy of Known Viral Structure: RNA Genome, Double Stranded[a,b]

Family—Distinguishing features Subfamily / Genus	Examples Elements of known structure	Host Representative PDB id
Fold: Description		
Reoviridae—Nearly spherical, diameter 600–800 Å, nonenveloped, two or three concentric icosahedral protein shells. Outer shell structure depends on genus. Virions contain the transcriptional machinery		
Folds: VP7 (orbivirus numbering): jelly-roll antiparallel β barrel with large additions to the N- and C-terminal ends that together form a large α-helical domain		
VP3: The apical domain is α/β; the carapace domain is two-layered with α helices in each; the dimerization domain has two β sheets		
Orbivirus	*Bluetongue virus*	Vertebrates
	VP7 (Grimes *et al.*, 1995), internal capsid particle of VP3 and VP7 (Grimes *et al.*, 1998)	1bvp, 2btv
	African horse sickness virus	Vertebrates
	VP7 top domain (Basak *et al.*, 1996)	1ahs
Rotavirus	*Rotavirus*	Vertebrates
	VP6 (homologous to *Orbivirus* VP7) (Mathieu *et al.*, 2001); VP4 outer capsid spike protein (Dormitzer *et al.*, 2002)	1qhd 1kqr
Orthoreovirus	*Reovirus*	Vertebrates
	Internal capsid particle (Reinisch *et al.*, 2000), $\sigma3$ (Olland *et al.*, 2001), $\mu1\sigma3$ complex (Liemann *et al.*, 2002), $\sigma1$ protein (Chappell *et al.*, 2002)	1ej6 (particle), 1 fn9 ($\sigma3$), 1jmu ($\mu1\sigma3$ complex), 1kke ($\sigma1$)
Six other genera		Vertebrates, invertebrates, plants
Six other families		

[a]Total of seven families.

[b]Assembled from several sources: Bairoch and Apweiler (2000); Berman *et al.* (2000); Murphy and Kingsbury (1996); Reddy (1999); van Regenmortel *et al.* (2000).

Table V

Taxonomy of Known Viral Structure: DNA Genome, Single Stranded[a]

Order	Family—Distinguishing features / Subfamily	Genus	Example / Elements of known structure	Host / Representative PDB id
	Fold:	Description		
	Inoviridae—6.4-kb ssDNA genome is enclosed in a flexible helical filament of ~2800 copies of the major coat protein pVIII. Proteins in smaller copy number cap the ends			
	Fold:	Coat protein pVIII, single α helix packed in a helical array; ssDNA-binding protein, antiparallel β sheet		
		Inovirus	*Coliphage strains M13, fd, Pf1, Xf* ssDNA-binding protein pV (Skinner *et al.*, 1994), major coat protein pVIII (Marvin *et al.*, 1994; McDonnell *et al.*, 1993; Nambudripad *et al.*, 1991a)	Bacteria 1vqb (pV); 1ifi, 1ifm, 2ifo, 1pfi (major coat protein pVIII)
		Plectrovirus	*Acholeplasma phage L51*	Mycoplasm
	Microviridae—nonenveloped; icosahedral with diameter of 260 Å plus spikes at the 12 5-folds that are 32 Å tall and 70 Å in diameter; 60 copies of J, F, and G proteins, 12 of H; B (60 copies) and D (240 copies) scaffolding proteins in procapsids			
	Fold:	F and G proteins are viral jelly-roll antiparallel β barrels. J is small. D scaffolding protein has a stable core of six α helices and loops that differ in conformation at different sites. Structures from one of four genera		
		Microvirus	*Coliphage ϕX174* Capsid (McKenna *et al.*, 1992a), procapsid (Dokland *et al.*, 1997, 1999)	Bacteria 2bpa, 1cd3
			Coliphage G4 Capsid (McKenna *et al.*, 1996)	Bacteria 1gff

(*continues*)

TABLE V (*Continued*)

Order	Family—Distinguishing features / Subfamily	Genus	Example / Elements of known structure	Host / Representative PDB id
	Fold:	Description		
	Parvoviridae—Nonenveloped; T=1 capsids with 60 near-equivalent proteins that are a mixture mostly of VP2 or VP3 (which is shorter by an N-terminal ~20 residues), or occasionally VP1 (longer than VP2 by ~150 N-terminal residues)			
	Fold:	Capsid proteins, antiparallel β barrel with longer loops than in other viral folds		
	Parvovirinae	*Parvovirus*	*Canine parvovirus* (*CPV*); *feline panleukopenia virus* (*FPV*); *mouse minute virus* (*MVM*); *porcine parvovirus* Capsid VP3 (Agbandje *et al.*, 1993; Agbandje-McKenna *et al.*, 1998; Simpson *et al.*, 2002; Tsao *et al.*, 1991)	Mammals 4dpv (CPV), 1fpv, 1mvm, 1krv
		Erythrovirus	*B19 virus*	Vertebrates
		Dependovirus	*Adeno-associated virus* Capsid (Xie *et al.*, 2002), nuclease domain of Rep (replication protein) (Hickman *et al.*, 2002)	Vertebrates 1lp3, 1m55
	Densovirinae	*Densovirus*	*Wax moth densovirus* Capsid (Simpson *et al.*, 1998)	Insects 1 dnv
		Two other genera		
Three other families				

[a]Assembled from several sources: Bairoch and Apweiler (2000); Berman *et al.* (2000); Murphy and Kingsbury (1996); Reddy (1999); van Regenmortel *et al.* (2000).

TABLE VI
Taxonomy of Known Viral Structure: DNA Genome, Double Stranded[a]

Order ***Family***—Distinguishing features ***Subfamily*** Fold:	*Genus* Description	*Examples* Elements of known structure	Host Representative PDB id
Myoviridae—Isometric or elongated heads and a tail complex consisting of central tube, contractile sheath, collar, base plate, six short spikes, and six long fibers			
Fold:	gp9: trimer of three domains: (1) N-terminal elongated domain with a helix forming a coil coil; (2) middle domain with a seven-stranded β sandwich; and (3) C-terminal domain with an eight-stranded β sandwich of topology partly similar to common jelly roll gp11: trimer of three domains, small helical N-terminal domain and C-terminal antiparallel sheet, finger domain seven-stranded antiparallel β roll with one helix. Fibritin: mostly long helix interrupted by loops, in trimer three helices form coiled coil. gp5: trimer of three domains, an N-terminal OB-fold domain, a lysozyme domain, and a β-helix domain gp27: a trimeric four-domain protein with two similar sheet domains that form a hexagonal cylinder in the trimer		
	T4-like viruses	*Coliphage T4* Baseplate protein gp9 (Kostyuchenko *et al.*, 1999), baseplate protein gp11 (Leiman *et al.*, 2000), fibritin (Tao *et al.*, 1997), baseplate proteins gp5 and gp27 (Kanamaru *et al.*, 2002); enzymes, regulatory proteins, etc.	Bacteria 1qex (gp9), 1el6 (gp11), 1aa0 (fibritin), 1k28 (gp5–gp27 complex) 16 PDB entries
	Six other genera		
Siphoviridae—Icosahedral heads about 660 Å in diameter, with a tail. Structures from one of six genera			
Fold:	HK97, capsid protein has two unique domains, both with an antiparallel sheet and helices, an extended N-terminal arm, and an extended hairpin		

(*continues*)

TABLE VI *(Continued)*

Order / Family—Distinguishing features / Subfamily / Fold:	Genus / Description	Examples / Elements of known structure	Host / Representative PDB id
	λ-like viruses	*Coliphage λ* Proteins gp W and gpFII (Maxwell *et al.*, 2001, 2002) and head protein D (Yang *et al.*, 2000)	Bacteria 1hyw, 1koh, 1c5e
		Coliphage HK97 Capsid (Wikoff *et al.*, 2000)	Bacteria fh6
Podoviridae—Isometric heads with short tail and six short fibers			
Folds:	Trimeric tailspike: N-terminal head-binding domain has a three-stranded and a five-stranded antiparallel sheet, C terminus is mainly a parallel β helix, some regions form separate domains in trimer Connector protein: Three domains: (1) a central domain formed by three long helices; (2) a domain formed by three layers of sheet and two helices; and (3) a small domain by β strands and a small helix		
	P22-like viruses	*Enterobacteriophage P22* Tailspike protein (gp9, endorhamnosidase) (Steinbacher *et al.*, 1994, 1997)	Bacteria 1tyu, 1lkt
	ϕ29-like viruses	*Phage ϕ29* head–tail connector protein (Simpson *et al.*, 2000; Guasch *et al.*, 2002)	Bacteria 1ijg, 1h5w
	T7-like viruses	*Coliphage T7*	Bacteria
Tectiviridae—Icosahedral phages usually lacking a tail but with an internal membrane			
Fold:	P3 protein has two jelly-roll domains		
	PRD1-like viruses	*Enterobacteria phage PRD1* P3 major coat protein (Benson *et al.*, 1999)	Bacteria 1cjd

Herpesviridae—Pleomorphic, 1500–2000 Å in diameter, envelope surrounding amorphous tegument surrounding 1000-Å icosahedral nucleocapsid containing a 120-to 240-kbp genome wound on a spool. About 30 structural proteins, including 11 envelope glycoproteins and ~30 nonstructural proteins

Fold: BCRF1 is also known as viral interleukin 10 because of its homology

Alpha herpesvirinae	*Simplexvirus*	*Human herpesvirus 1, herpes simplex virus* Envelope glycoprotein D (Carfi *et al.*, 2001); VP16, transcription regulation (Liu *et al.*, 1999); protease (Hoog *et al.*, 1997); processivity factor (Zuccola *et al.*, 2000); thymidine kinase (Vogt *et al.*, 2000); U12, uracil-DNA glycosylase (Savva *et al.*, 1995)	Vertebrates 1jma (gD in complex with receptor), 16vp (VP16), 1at3 (protease), 1dml (processivity), 1e2h (TK), 1lau (U12)
Gammaherpesvirinae	*Lymphocryptovirus*	*Human herpesvirus 4 = Epstein-Barr virus* Nuclear antigen 1, DNA-binding/dimerization fragment; BCRF1 (Zdanov *et al.*, 1997); gp42 C-type lectin (Mullen *et al.*, 2002)	Vertebrates 1vhi (nuclear antigen 1), 1vlk (BCRF1), 1kg0 (gp42 in complex with HLA-DR1)

Adenoviridae—Nonenveloped, icosahedral, 800- to 1100-Å-diameter capsids containing 240 hexon capsomers and 12 vertex penton capsomers from which protrudes a fiber of up to 350 Å. At least 12 structural proteins and many nonstructural. Genome is ~37 kbp of dsDNA

Fold: Hexon protein: 2 anti-parallel viral β barrels with extended loops that form the "tower" region

Fiber: shaft plus head domain. Shaft is a novel triple β-spiral fibrous fold. Head is an antiparallel β barrel with a novel Greek key-based strand topology

ssDNA-binding protein: unique fold that is mostly α, with some β

	Mastadenovirus	*Human adenovirus 2, 3, 5, and 12* Type 2 hexon (Athappilly *et al.*, 1994); type 5 hexon (Rux and Burnett, 2000); fiber head type 2 (van Raaij *et al.*, 1999a), type 3 (Durmort *et al.*, 2001), type 5 (Xia *et al.*, 1994), and type 12 (Bewley *et al.*, 1999); fiber type 2 (van Raaij *et al.*, 1999b); single-stranded DNA-binding protein (Kanellopoulos *et al.*, 1996)	Vertebrates 1dhx, 1rux (hexon types 2 and 5), 1qhv, 1h7z, 1knb, 1nob (knob types 2, 3, 5, and 12), 1qiu (fiber), 1adu (ssDNA-binding protein)

(continues)

TABLE VI *(Continued)*

Order / ***Family***—Distinguishing features / ***Subfamily*** / Fold:	*Genus* / Description	*Examples* / Elements of known structure	Host / Representative PDB id
	Aviadenovirus	*Fowl adenovirus 1*	Vertebrates
Polyomaviridae—Nonenveloped, icosahedral, ~450 Å in diameter with 360 VP1, 6 in each identical unit. Three capsid proteins, VP1–VP3			
Fold:	VP1 is an embellished viral jelly roll, oriented radially instead of tangentially as in the (+)ssRNA viruses		
	Polyomavirus	*Murine polyomavirus*	Vertebrates
		VP1 (Stehle *et al.*, 1994)	1vps
		Simian virus 40 (SV40)	Vertebrates
		VP1 (Liddington *et al.*, 1991; Stehle *et al.*, 1996), DNA-binding domain of T antigen	1sva (VP1), 2tbd (T antigen)
Papillomaviridae—Nonenveloped, icosahedral, similar to *Polyomaviridae*			
Fold:	L1 is homologous to VP1 of polyomavirus		
	Papillomavirus	*Human papillomavirus 16*	Vertebrates
		Recombinant capsid-like structure containing L1 (Chen *et al.*, 2000)	1dzl
14 other families—unknown structure, including ***poxviridae***, and ***Baculoviridae*** that infect bacteria, archaea, algae, invertebrates, and invertebrates			

[a]Assembled from several sources: Bairoch and Apweiler (2000); Berman *et al.* (2000); Murphy and Kingsbury (1996); Reddy (1999); van Regenmortel *et al.* (2000).

Table VII

Taxonomy of Known Viral Structure: DNA and RNA Reverse-Transcribing Viruses[a]

Order ***Family***—Distinguishing features ***Subfamily***	*Genus*	*Examples* Elements of known structure	Host Representative PDB id
Fold:	Description		
Hepadnaviridae—Spherical, 400–480 Å in diameter, sometimes pleomorphic. Envelope contains surface antigen (HbsAg) and three glycoproteins (gp27, gp36, and gp42) and surrounds a T=3, 270- to 350- Å nucleocapsid (180 core antigen proteins, HbcAg), containing 3.2 kbp, mostly dsDNA			
	Othohepadnavirus	*Hepatitis B virus*	Vertebrates
		Core protein (Wynne *et al.*, 1999)	1qgt
	Avihepadnavirus	*Duck hepatitis B virus*	Vertebrates
Fold:	Core protein mostly helical, two helices form four-helix bundle in dimer		
Retroviridae—Spherical, 800–1000 Å in diameter, enveloped with peplomers surrounding icosahedral capsid, surrounding helical nucleocapsid of genus-specific shape (cone, rod, sphere, etc.). Genome is homodimer of 2×7–11 kb of ssRNA. Virions contain two envelope glycoproteins, three to six internal structural proteins, matrix, capsid, and nucleocapsid proteins			
Folds:	Mo-MuLV surface glycoprotein: No homology to HIV gp120; proximal domain is similar to immunoglobulin β barrel; distal domain is α helical		
	HIV proteins		
	CA capsid protein: N-terminal domain, α-helical coiled coil (unlike canonical jelly roll); C-terminal domain, four antiparallel helices		
	MA (matrix protein): Five-helix globular region plus three-stranded mixed β sheet		
	NC (nucleocapsid protein): Zinc fingers		
	TM (gp41) (transmembrane ectodomain): Homotrimer forms central coiled coil with helices of 49 residues. This is flanked by antiparallel helices of 40 residues		

(*continues*)

TABLE VII (*Continued*)

Order / ***Family***—Distinguishing features / ***Subfamily*** / Fold:	*Genus* / Description	*Examples* / Elements of known structure	Host / Representative PDB id
	SU (surface glycprotein) (gp120): No homology to MuLV SU or other proteins. Inner β-sandwich, bridging β sheet and double β-barrel outer domain; protease, antiparallel β, like other aspartyl proteases; integrase, N-terminal domain is helical, containing helix–turn–helix motif; core domain, RNase H-like (mixed β with helices on either side)		
	RT (reverse transcriptase): "Palm" subdomain is antiparallel β; "fingers" subdomain is mixed β/α; "connection" is mixed β with two α helices; "thumb" is a four-helix bundle; C-terminal domain is RNase H-like (mixed β with helices either side)		
	Nef: Helix–turn–helix motif plus antiparallel β sheet		
	Gammaretrovirus	Moloney murine leukemia virus (*Mo-MuLV*) Transmembrane protein fragment (Fass *et al.*, 1996), surface glycoprotein (Fass *et al.*, 1997a), reverse transcriptase (Georgiadis *et al.*, 1995), Ncp10 (Schuler *et al.*, 1999)	Vertebrates 1mof (TM), 1aol (SU), 1mml (RT), 1a6b (Ncp10)
	Deltaretrovirus	*Bovine leukemia virus, human T-lymphotropic virus 1 (HTLV-1)* HTLV-1 matrix protein (Christensen *et al.*, 1996), HTLV-1 TM protein (Kobe *et al.*, 1999)	Vertebrates 1jvr (HTLV matrix), 1mgl (TM)
	Lentivirus	*Human immunodeficiency virus 1* (*HIV-1*)	Vertebrates

		CA (p24) fragment (Gamble *et al.*, 1996, 1997; Gitti *et al.*, 1996; Momany *et al.*, 1996); MA (Hill *et al.*, 1996; Massiah *et al.*, 1996); NC (Morellet *et al.*, 1992; Summers *et al.*, 1992); TM ectodomain (Caffrey *et al.*, 1997; Chan *et al.*, 1997; Tan *et al.*, 1997; Weissenhorn *et al.*, 1997); SU (gp120) (Kwong *et al.*, 1998); aspartyl protease (Miller *et al.*, 1989); integrase (Cai *et al.*, 1997; Dyda *et al.*, 1994); reverse transcriptase (Kohlstaedt *et al.*, 1992); fragments of Nef (Grzesiek *et al.*, 1996; Lee *et al.*, 1996) and Tat (Battiste *et al.*, 1996)	1ak4, 1a8o, 1gds, 1afv (CA), 1hiw, 2hmx (MA), 1bj6, 1aaf (NC), 1aik, 1szt, 1env, 2ezo (TM), 1gcl (SU)
		Equine infectious anemia virus (EIAV)	Vertebrates
		Capsid protein (p26) (Jin *et al.*, 1999), protease; MA (Hatanaka *et al.*, 2002)	2eia (capsid), 1fmb (protease), 1hek (MA)
		Visna virus	Vertebrates
		TM protein fragment (Malashkevich *et al.*, 2001)	1jek
	Avian type C retroviruses	*Avian sarcoma virus*	Vertebrates
		Capsid (Campos-Olivas *et al.*, 2000), protease (Jaskolski *et al.*, 1990), integrase–core domain (Bujacz *et al.*, 1996)	1dld (CA), 1rsp (protease), 1asu (integrase)
	Three other genera		
Three other families			

[a]Assembled from several sources: Bairoch and Apweiler (2000); Berman *et al.* (2000); Murphy and Kingsbury (1996); Reddy (1999); van Regenmortel *et al.* (2000).

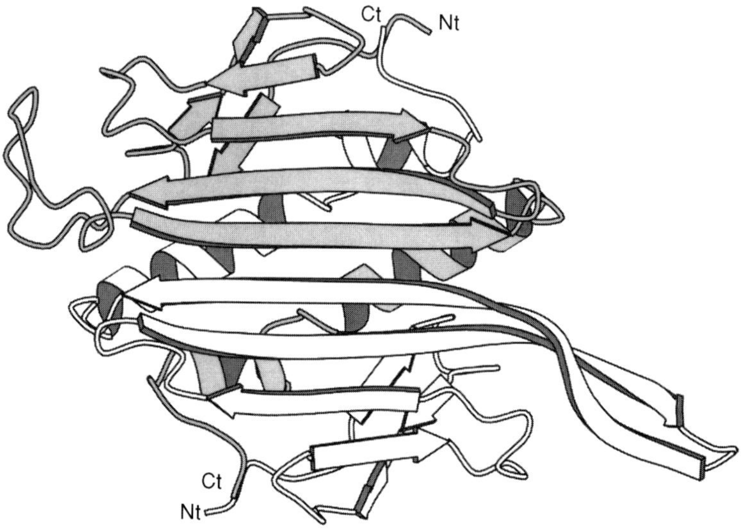

FIG. 6. The fold of the *Leviviridae* capsid proteins. A dimer (at the quasi-2-fold) of the coat protein of MS2 viewed from the inside of the particle is shown (Golmohammadi *et al.*, 1993). The subunits differ in shade. The N terminus of one subunit is close to the C terminus of the other subunit in the dimer.

histocompatibility antigens (Bjorkman *et al.*, 1987), but the topology is different and there is no binding pocket between the two helices.

The capsid proteins of the *Leviviridae* family are all similar. The main difference is that the dimers of Qβ (Golmohammadi *et al.*, 1996) and PP7 (Tars *et al.*, 2000) are both linked by disulfides between the loops close to the 5-fold and quasi-two-fold axes. In these viruses, there are fewer dimer–dimer contacts, and the capsids are less stable when the disulfides are not formed (Cielens *et al.*, 2000).

2. *Icosahedral Nonenveloped Plant and Insect Viruses: Comoviridae, Nodaviridae, Tetraviridae, Luteoviridae, and Bromoviridae*

All known single positive-sense strand [(+)ssRNA] plant virus capsids are built from jelly-roll β-barrel building blocks. They represent the most classic expression of the canonical domain form, with few of the embellishments seen in the animal virus homologs described later. Perhaps there was not the selective pressure to decorate the surface with loops of variable sequence, because immune surveillance was not as stringent as in animal hosts. The result is that the plant loops between the β strands are consistently shorter.

One interesting exception among the plants is found in the *Luteoviridae* family. In both tomato bushy stunt virus (TBSV) (Harrison *et al.*, 1978) and turnip crinkle virus (TCV) (Hogle *et al.*, 1986) each capsid protein is comprised of two (S and P) jelly-roll domains. The S (shell) domain occupies the usual spot in the capsid assembly. The N-terminal P (projecting) domain is a stripped-down jelly roll with tight turns of zero to four residues that projects outward from the capsid surface. The relative orientations of S and P domains differ slightly between TBSV and TCV, but in both cases pairs of domains from neighboring subunits dimerize, presumably stabilizing the assembly and forming pronounced protrusions on the viral surface. Although the P domain is a jelly roll, it is a significant deviation from the canonical viral fold. It is as if half way through the second roll, the C-terminal strand is truncated so that each sheet ends up with three jelly-roll strands instead of four (Fig. 2). A fourth strand is added to one of the sheets with an up–down hairpin.

Of the insect virus families, the capsid proteins of members of the *Nodaviridae* family are similar to those of plant viruses such an SBMV and TBSV, which have the same T=3 arrangement of subunits (Hosur *et al.*, 1987). The capsid protein of the T=4 virus *Nudaurelia capensis* ω (NωV) (Fig. 1d; see Color Insert), belonging to the *Tetraviridae* family, also has a jelly-roll topology, but the EF loop contains a complete domain of the immunoglobulin type c topology that is located on the outside surface of the capsid (Munshi *et al.*, 1996). The N and C termini form a separate helical domain on the inside of the protein shell.

3. *Plant Satellite Viruses*

Satellite viruses are those that are dependent for their own replication on some (catalytic) activity encoded in another "helper" virus that coinfects the host cell. The structures of three plant ssRNA satellite viruses represent some of the highest resolutions known and have been comparatively reviewed (Ban *et al.*, 1995). The structures of satellite tobacco mosaic virus (STMV) (Larson *et al.*, 1993a,b), satellite tobacco necrosis virus (STNV) (Jones and Liljas, 1984; Liljas *et al.*, 1982), and satellite panicum mosaic virus (SPMV) (Ban and McPherson, 1995) have T=1 capsids composed of 60 identical copies of unembellished jelly-roll β barrels constructed of only 155 to 195 amino acids (Fig. 1a; see Color Insert). What is remarkable is how little the assembly context of these domains is conserved. The same end always points toward the 5-fold axis, but the domains are rotated to different extents around the 5-fold axis. Furthermore, between STNV and the others, there is a 70° rotation of the barrel about its long axis. Contacts across the dimer interface are different: end-to-end (with respect to the barrel in STNV), compared with side-to-side in SPMV and STMV, involving only loops in STMV, but also βB in SPMV. Such disparity may be an evolutionary consequence of these viruses being satellites, perhaps with the opportunity to exchange genetic information with their (quite different) helper viruses (Ban *et al.*, 1995).

4. *Caliciviruses*

The only capsid structure known from an animal T=3 virus is that of Norwalk virus, a member of the *Caliciviridae* family (Prasad *et al.*, 1999). Similar to plant viruses such an TBSV and the insect NωV (Munshi *et al.*, 1996), it has an S domain with jelly-roll topology and a protruding domain. Also in Norwalk virus, the protruding domain forms dimer contacts, but the topology of the domain is not found in other viral proteins. It has two subdomains. One continuous segment of 95 amino acids forms a β-barrel domain with the same topology as domain 2 of elongation factor Tu. The other subdomain is also formed mainly by β structure. In the known cases of jelly-roll capsid proteins with inserted protruding domains, the extra domain is a β domain, but in these three cases the topology of the inserted domain is different.

5. *Picornaviruses*

The capsid is composed of 60 copies of three classic jelly-roll β barrels (VP1, VP2, and VP3), and a small VP4 on the inside surface of the capsid. The proteins vary in length, between about 230 and 300 amino acids. The corresponding proteins of different viruses are more similar to one another than are VP1–VP3 within one species (Rossmann, 1987). VP1 and VP3 have N-terminal extensions to the barrel that meander away to form contacts with neighboring subunits (Hogle *et al.*, 1985; Rossmann *et al.*, 1985). The VP2 extension forms an additional β ribbon on the RNA side of the capsid. Various secondary structural elements are inserted within the loops. Like SBMV, there is a helix in the CD loops of all of the capsid proteins of poliovirus, rhinovirus, and so on. In the picornaviruses there is a short helix either breaking or preceding βB.

The most interesting features are the longest of the loops that differ between VP1, VP2, and VP3, and give each species and type of virus its unique characteristics. The BC loop of VP1, near the 5-fold axis, contains dominating neutralizing immunogenic (NIm) sites in both rhino- and polioviruses (Hogle *et al.*, 1985; Minor *et al.*, 1986; Rossmann *et al.*, 1985; Sherry *et al.*, 1986). The VP1 GH loop in foot-and-mouth disease virus (FMDV) is antigenic (Acharya *et al.*, 1989), and forms part of a protrusion in many of the picornaviruses. Surface-exposed loops appear to be more readily recognized by neutralizing antibodies, and are the sites of sequence variability between serotypes, presumably as viruses evolve to evade immune detection (Rossmann *et al.*, 1985). Between the VP1 BC and GH loops of poliovirus and rhinoviruses (at opposite ends of the barrel) lies a depression in which there is surface access to the opening of the β barrel. The depressions of adjacent subunits connect, leading to a ring about the 5-fold axis that was predicted, and then demonstrated, to be the site of rhino- and poliovirus attachment to cellular receptors (Belnap *et al.*, 2000; Chapman and Rossmann, 1993a; He *et al.*, 2000; Olson *et al.*, 1993; Rossmann and Palmenberg, 1988). The location of the cellular receptor-binding site in a depression allows sequence conservation of the interface in an area that is less accessible to immune surveillance (Rossmann and Palmenberg, 1988).

Although many features are conserved between picornaviral VP1s, not all are. For example, the longer CD loop of Mengovirus combined with the shorter BC loop relative to rhinoviruses change the canyon into a pit, which is also the putative receptor-binding site (Chapman *et al.*, 1990). In coxsackievirus B3 the DE loop of nine residues is longer, and the BC loop of four residues is shorter, which, combined with a shorter VP2 GH loop, leads to a slightly different virus surface topology (Muckelbauer *et al.*, 1995).

VP2 is slightly shorter than VP1 and has a shorter BC loop, but a long GH loop that forms the antigenic "puff" in rhinoviruses. VP3 has the shortest loops that do not form large surface features. However, βB is broken roughly halfway with a hairpin loop extending outward, known as the "knob." In rhinoviruses there is a short helix on the outward path of the knob. Mengovirus has no structure in the knob. Poliovirus is even less structured, with the chain following the same path, but without much β strand before the knob. It is worth noting that cricket paralysis virus, an insect virus belonging to the *Dicistroviridae* family and structurally similar to picornaviruses, has a VP3 protein with a feature resembling the knob, but in this case the extending loop is found in the CD loop (Tate *et al.*, 1999).

FMDV is the outlier of picornavirus structures (Fry *et al.*, 1990), with capsid proteins that are 20% shorter than in the other viruses. The VP1 loops near the 5-fold axis are "sheared off," so that there is not the pronounced 5-fold protrusion and canyon of rhinoviruses and polioviruses (Acharya *et al.*, 1989). This leaves a longer VP1 GH loop as the prominent surface feature, which is highly antigenic, the site of the RGD receptor attachment sequence, but disordered in structure unless a disulfide is reduced (Acharya *et al.*, 1989; Fox *et al.*, 1989; Lea *et al.*, 1994; Rowlands *et al.*, 1994). FMDV VP2 is more similar to its homologs, except that the GH loop puff is 50 residues shorter than in poliovirus, and its space is occupied partly by the longer VP1 GH loop.

The structure of human rhinovirus 16 (HRV16) was pursued at higher resolution (2.15 Å) than most, and revealed interesting structures around the 5-fold axis (Hadfield *et al.*, 1997). These were composed in part of residues that had been disordered in prior rhinoviral crystal structures, but had been seen in type 3 poliovirus (Filman *et al.*, 1989). N termini of five symmetry-equivalent VP3 come together to form a five-stranded parallel β barrel, plugging a gap between the five VP1 jelly-roll domains. Each of 5 VP4 N termini contributes a β hairpin to a 10-stranded antiparallel β barrel closer to the virus center. In both rhino- and polioviruses VP4 is N-terminally myristoylated and these elements of structure are thought to be intimately involved with conformational transitions that occur on cell entry and uncoating.

6. *Enveloped Single Positive-Sense Strand RNA Viruses*

i. Flavivirus. Flaviviruses are among the smallest (~500 Å) that have a lipid bilayer. The family includes yellow fever virus, hepatitis C virus, and tick-borne encephalitis virus, for which there is some structural information. They contain three structural proteins. The 500-amino acid surface glycoprotein "E" trimerizes irreversibly at low pH, a process that is thought to be central to membrane fusion. The structure of a 400-residue tryptic fragment has been determined (Rey *et al.*, 1995) to have a structure different from hemagglutinin, the protein that has homologous function in influenza A (see below). Protein E is a 150-Å-long molecule that lies parallel with the membrane surface, anchored by two transmembrane sequences near the C terminus that has been cleaved off in the tryptic fragment.

Domain 1 is at the N terminus, but spatially is the central of three domains. It is an antiparallel β barrel that is superficially reminiscent of the viral capsid fold, but actually has completely different topology (Fig. 7; see Color Insert). In the jelly roll, the β strands alternate between the two opposing sheets of the barrel. In E protein they zigzag up and down the same sheet, with the following exception. Strand B, which in chemical sequence falls between strands A and C of the outermost sheet, actually forms the end strand of the inner sheet. So, the strand order is ACDEF for the outer sheet, and BIHG for the inner. Strands F and G are hydrogen bonded, but there is no direct continuation of the sheets across the AB strand gap.

In the strictest sense, "domain" 2 is really a subdomain, connected to domain 1 by three polypeptide chains. It is really a globular extension formed by loops DE and HI of domain 1 coming together to form β-sheet structures. The core of this is formed by a β ribbon (within the domain

1 DE loop) that forms the center of the proximal five-stranded sheet. The ribbon continues past this sheet with a sharp twist, restarting as part of the second sheet. The second, three-stranded sheet is completed by a looping round of the chain to form a third strand. The sheet is held in place by disulfide bonds. This distal part of the domain is completed with another β hairpin coming from the HI loop of domain 1, the β hairpin packing face to face with the three-stranded sheet. On the way out to the distal sheet of domain 2 from domain 1, the chains of the HI loop have been split apart to form two strands and one strand that bracket the DE ribbon in the proximal domain 2 sheet.

Domain 3 is simpler. It is a classic immunoglobulin fold (Fig. 3; see Color Insert) connected to domain 1 by a single polypeptide chain. This fold contains Greek key motifs and is a seven-stranded antiparallel β barrel with hydrogen bonding that that is broken into a three-stranded sheet packed against a four-stranded sheet. There are three short three-residue β strands that form an additional small sheet.

ii. Togaviruses. Togaviruses are also small enveloped viruses, many of which are insect-borne. The genus *Rubivirus* contains rubella virus, but it is the *Alphavirus* genus that is partially characterized structurally. Alphaviruses have a 410-Å-diameter icosahedral nucleocapsid core that becomes surrounded by a 40-Å-thick lipid layer containing two or three (E_1–E_3) membrane glycoproteins. The transcribed polyprotein p130 contains the core protein followed by E_3, E_2, and E_1. The core protein is first released, probably autocatalytically, leaving a protease inactive core protein and an E_3–E_1 polyprotein that is cleaved by other enzymes (Strauss and Strauss, 1986). Although there was sequence similarity, and site-directed mutagenesis of the putative catalytic triad (Boege *et al.*, 1981; Hahn *et al.*, 1985; Hahn and Strauss, 1990; Melancon and Garoff, 1987), it was a surprise to the virus structure community that the structures of Sindbis core protein (SCP) and the Semliki Forest virus core protein formed icosahedral structures out of a serine protease-like domain (Choi *et al.*, 1991, 1997).

The fold of the serine protease domain-type was described in Section V.C. SCP is more like α-lytic protease (Fujinaga *et al.*, 1985) than α-chymotrypsin (Matthews *et al.*, 1967), but with loops that are even shorter (Fig. 4; see Color Insert). Unlike the other proteases, there are no disulfide bonds. The structure of the C terminus is completely different from either of the other two proteases, and it leaves the final three amino acids in the active site. These superimpose on the structure of a peptide inhibitor determined as a complex with α-lytic protease (Bone *et al.*, 1987). This indicates that the N-terminal product of the autocatalytic lysis of p130 is never expelled from the active site, remaining as an inhibitor, and accounting for the lack of protease activity in the mature core protein.

Whether the core was T=3 or T=4 had been a contentious issue, in part because it is possible to prepare, *in vitro*, core particles of smaller diameter (330 Å) than *in vivo* (410 Å) (Coombs and Brown, 1987a,b; Enzmann and Weiland, 1979; Fuller, 1987; Fuller and Argos, 1987; Horzinek and Mussgay, 1969). With determination of the domain fold, it was possible to build models of the entire core, and to show that the T=3 proposed assembly would leave unreasonable gaps between the protomers (Choi *et al.*, 1991).

iii. Helical Tobamaviruses. Tobacco mosaic virus (TMV) and its relatives ribgrass mosaic virus and cucumber green mottle mosaic virus form completely different assemblies from what has been discussed hitherto. These are helical viruses in which the capsid protein winds around the genomic RNA in a helix. The structures have been determined either by fiber diffraction of oriented viruses, or by single-crystal diffraction of disk aggregates that have 17-fold symmetry (Bhyravbhatla *et al.*, 1998; Bloomer *et al.*, 1978; Namba *et al.*, 1989; Wang and Stubbs, 1994; Wang *et al.*, 1997a). The domain is a classic four-helix bundle (Fig. 5; see Color Insert) with small embellishments. Domains are packed so that the helix axes extend radially away from the RNA. The N terminus of the domain is tucked inside the protein, and the C terminus is exposed on the outside. Embellishments include short extra helices at the N and C termini of the bundle. The high-resolution structure (Bhyravbhatla *et al.*, 1998) shows two tiny two- or three-stranded β sheets in which the strands are only one or two residues long. Within the primary structure, the strands that comprise these sheets are before and after the N-terminal helix, between the middle two helices, and on either side of the C-terminal helix.

B. *Negative-Strand RNA Viruses*

Negative-strand RNA viruses are enveloped and larger than those considered so far. Genomes are packaged into a helical nucleoprotein complex. Envelopes contain one or two glycoproteins for which structures are available for hemagglutinin (HA) (Wilson *et al.*, 1981) and neuraminidase (NA) (Colman *et al.*, 1983) from influenza A (Table III).

1. *Influenza A*

i. Hemagglutinin. Hemagglutinin has been studied not only as a viral protein, but also as a paradigm of membrane fusion. In addition to the primary literature, there are excellent summaries in Brändén and Tooze (1998) and Harrison (2001). The originally synthesized entity, HA_0, starts as a 567-amino acid protein. A 16-residue signal sequence is removed after localization in the endoplasmic reticulum (ER). During infection HA_0 is cleaved at residue 330, but the resulting HA_1 and HA_2 remain associated through a disulfide linkage. It is the 20 residues in the newly formed N terminus of HA_2 that are responsible for the membrane fusion. The crystal structure is of the natural trimer, but 47 C-terminal HA_2 residues were cleaved off before crystallization, including the membrane anchor.

Dominant in HA_1 is an antiparallel jelly-roll β-barrel domain that has a similar topology to the common viral capsid domain (Fig. 8; see Color Insert). It is relatively distorted in the placement of the strands and with breaks in neighboring strands 1 and 8. Between strands 8a and 8b there is just a β bulge and a change in direction of about 60°. The break in the corresponding position between strands 1a and 1b inserts a large loop that adds an additional antiparallel strand before looping around to return to strand 1b. The domain is 145 residues in length, about the same as the compact barrels of the satellite RNA viruses. Thus, the other loops are relatively short turns, with the exception of that between strands 3 and 4, which contains a helix that forms part of the receptor-binding site (see below). The domain starts about 100 Å from where HA_2 is anchored to the membrane, and extends to about 140 Å. It is in a pocket at the most distal part, near the end of the barrel, which is the site of attachment of the cellular receptor as implicated through binding of sialic acid, the terminal moiety of the receptor (Weis *et al.*, 1988). The N-terminal region of 63 residues extends from near the membrane (to which it was previously attached by the signal sequence), outward about 100 Å to the barrel. It meanders alongside the path of HA_2 (to be described below), forming part of the stem on which the barrel "head" is placed. The N-terminal region is without globular domains, but with a β ribbon and with four strands that combine with those of the C-terminal HA_1 linker and HA_2 to form small β sheets. The 70-residue C-terminal linker backtracks from the barrel domain, similarly without globular structure, but with small β structures, and again meandering alongside the path of HA_2. The C terminus of HA_1, and therefore the point of proteolytic cleavage between HA_1 and HA_2, is two-thirds of the way back from the barrel domain to the membrane just external to a β sheet at the base of the stem.

The N terminus of HA_2 leads directly into two strands on one side of this five-stranded antiparallel sheet that forms the base of the stem. The middle strand is the N terminus of HA_1. The two strands on the other side come from residues near the C terminus of HA_2. The dominant structural feature of HA_2 is a hairpin of two long α helices that form the core of the fibrous stem. In the pH-neutral form, a 19-residue helix, N25 Å long, extends away from the membrane continuing from the end strand of the membrane-proximal β sheet. The path then extends to form the end strand of a β structure with the C-terminal region of HA_1, before turning to return toward the membrane. This it does in spectacular fashion with a single 52-amino acid, 76-Å helix before the chain turns to form the final two strands of the membrane-proximal sheet. In the natural assembly, long helices from the three subunits form a coiled coil that winds slowly as a left-handed triple helix. The shorter helices near the HA_2 N terminus flank the central long helices.

Hemagglutinin is central in the membrane fusion that occurs after endocytosis and at acidic pH. It was not possible to study the large conformational changes in the intact molecule, but in a proteolytic fragment that contains residues 38 to 175 of HA_2 and the N-terminal 27 residues of HA_1 (Bullough *et al.*, 1994). These studies provided the details of a large conformational change that had been implicated earlier (Skehel *et al.*, 1982). Two HA_2 helices align and 20 intervening loop residues become helix, extending the long helix N terminally by 26 residues. One effect is to propel the HA_2 N-terminal membrane fusion region 100 Å outward beyond the jelly-roll domain. Near the C-terminal end of the long helix a turn is introduced, so that the helix residues 106–112 run antiparallel to the long helix in the low-pH form. This moves two strands of the proximal sheet outward by 40 Å toward the jelly-roll domain. The joint effect of these large conformational changes is to bring together the receptor-binding region on the tip of the HA_1 jelly roll, the HA_2 C-terminal anchoring region, and the HA_2 N-terminal fusion region. The pH-neutral structure enables the virus to reach 140 Å out to attach to the membrane and then, with the conformational change, brings the fusing membranes into close proximity. The low-pH form is more stable, but cannot be formed until the HA_1–HA_2 cleavage is made in the mature trimer, and then the conformational change is not possible until the pH is lowered.

The structure of the HA_2 N-terminal fusion peptide has been probed by a combination of NMR and electron paramagnetic resonance (EPR) in detergent micelles that mimic the lipid bilayer, at both acid and neutral pHs (Han *et al.*, 2001). At both acidic and neutral pH the structure is predominantly helical with a kink where it rises most prominently to the presumptive membrane surface. At lower pH the kink is stronger, there is additional 3_{10} helix, and two charged residues are rotated out of the membrane plane. The stronger kink likely allows the peptide to become more deeply immersed, perhaps disrupting the membrane and facilitating fusion.

ii. Neuraminidase. Neuraminidase is a membrane glycoprotein enzyme that cleaves sialic acid, helping progeny influenzaviruses leave without reinfecting the host cell. Like hemagglutinin, each neuraminidase molecule consists of a globular head on a stalk, extending out to about 120 Å from the membrane. It is a proteolytic fragment containing the tetramer of the head domain, whose structure has been determined (Varghese *et al.*, 1983). The head part of each subunit consists of 400 residues, connected by 70 N-terminal residues to the membrane signal sequence. Each head domain has a particularly elegant all-β structure consisting of six up–down four-stranded antiparallel sheets, arranged with pseudo-6-fold symmetry so that the sheets resemble the blades of a propeller (Fig. 9; see Color Insert). The sheets have greater twist than usual, so that the first and last strands are nearly orthogonal. Topologically, the structure is simple, with 6×4-stranded sheets in a row. There is a circular permutation (presumably to strengthen the folded configuration), so that the N and C ends of the domain are not between sheets. Rather, the polypeptide starts as the end strand of sheet 6 before continuing to strand 1 (the inner strand) of sheet 1. Thus, the propeller structure is closed by hydrogen bonding between strands of sheet 6 that are the N and C termini of the domain. The propeller is splayed out on the stem side, but the connections between the strands are all short loops and tight turns. On the distal side, some of the connections need to cross over from an outer strand of one sheet to the inner strand of the next propeller blade, but all are relatively long loops. Between the long external loops is located the enzyme active site.

iii. M_1. M_1 mediates encapsidation of nucleoprotein cores into the envelope and interacts with both RNA and the membrane. The structure of a fragment (residues 2–158 of 252) shows two 4-helix bundle domains (Sha and Luo, 1997). The N-terminal domain is a classic up–down bundle. The second is a variant in which diagonally opposite helices are connected, a "crossed helix bundle," so that each helix has one neighbor that is parallel, one that is antiparallel. There is a presumed third domain of unknown structure. Hydrophobic residues on one surface of the N-terminal domain are buried at an interface to the second domain, but with a conformational change could become exposed to a membrane surface. Positively charged residues are on the surface of the second domain and could interact with RNA.

iv. M_2. M_2 is an ion channel and the target for the drug amantidine. By blocking ion transport, the drug has two effects: (1) inhibition of acid-induced dissociation of the genome from the matrix protein, and

(2) premature acid-induced conformational changes in HA during maturation in the Golgi vesicles, due to inhibition of proton efflux. There is no atomic resolution structural information about the 97-residue protein, but a middle 19-residue segment was predicted to form a transmembrane helix, and to tetramerize, forming a parallel bundle (Holsinger and Lamb, 1991; Sugrue and Hay, 1991) of 4 amphiphilic helices. The channel, blocked by amantidine, would run down the middle. The essential elements of the four-helix bundle have been confirmed, and details of the helix orientation have been provided by spectroscopy (Kukol *et al.*, 1999) and new techniques of solid-state NMR (Kovacs *et al.*, 2000).

2. *Paramyxovirus Fusion Protein and Hemagglutinin–Neuraminidase*

Structures have become available within this important family that includes major pathogens such as measles and mumps. Both examples have high homology to previously known related structures. The fusion protein of simian parainfluenza virus contains the same coiled-coil topology as in influenza A HA, and homologous proteins in Ebola virus and retroviruses (see below) (Baker *et al.*, 1999). The globular part of the protein with both hemagglutinin and neuraminidase activity is similar to influenza A neuraminidase with a six-blade propeller β-sheet topology (Crennell *et al.*, 2000). The fusion protein (F) of Newcastle disease virus has a head region with a twisted β structure plus an immunoglobulin domain, but the F protein differs from influenza A hemagglutinin in that the central coiled coil is oriented in the opposite direction with respect to the membrane (Chen *et al.*, 2001).

3. *Ebola Virus Matrix Protein and Glycoprotein*

The structure of the Ebola virus matrix protein (VP40) has been determined (Dessen *et al.*, 2000). It has two domains of the same, unique topology (Fig. 10) consisting of sandwiches made from two three-stranded antiparallel β sheets. The topology bears no resemblance to other viral matrix proteins of known structure. Indeed, differences in the sequence suggest that it might be different from the closest related of other families, including *Paramyxoviridae* and *Rhabdoviridae* (Dessen *et al.*, 2000).

The structure of a core 74-amino acid fragment of the ectodomain of the envelope glycoprotein gp2 has been determined (Malashkevich *et al.*, 1999; Weissenhorn *et al.*, 1998). In the homotrimer, three symmetry-related α helices form a coiled coil. The chain folds back on itself, so that surrounding and antiparallel to the coiled coil are three outer

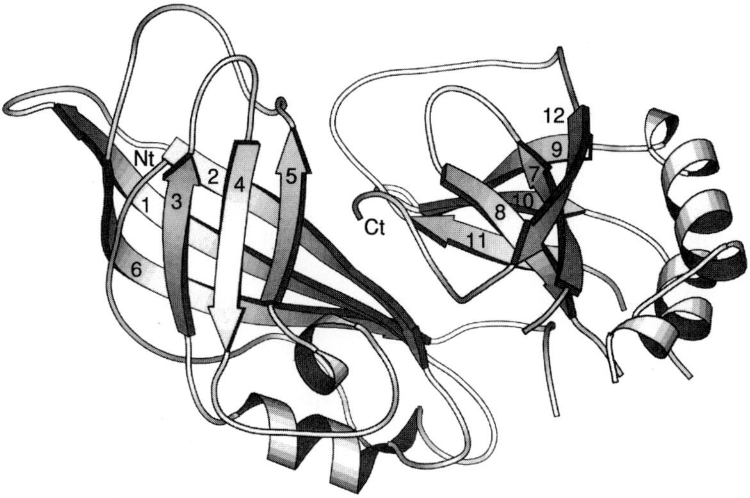

FIG. 10. The Ebola virus matrix protein (Dessen *et al.*, 2000). The two domains have the same topology. Some loops in the second domains are disordered and not included in the drawing.

symmetry-related helices. This overall topology is exactly the same as found in membrane fusion proteins in the *Orthomyxoviridae* (including influenza A), *Paramyxoviridae,* and many of the retroviruses. This is remarkable because of the low sequence identity (less than ~20%) and the unrelatedness of some of these viruses (Malashkevich *et al.*, 1999; Weissenhorn *et al.*, 1998).

C. Double-Stranded RNA Viruses from the Reoviridae Family

1. Genus Orbivirus

At a diameter of 800 Å, orbiviruses are among the largest viruses characterized to date. They are nonenveloped and have four major capsid proteins. An outer layer of VP2 and VP5 is removed on cell entry, and the intracellular cores characterized structurally are composed of an internal scaffolding built from 120 copies of VP3, which is then decorated with 260 capsomers, each containing 3 copies of the 38-kDa VP7, and arranged on a T=13 lattice.

The first structure was of the bluetongue virus VP7 protein (Grimes *et al.*, 1995), which has two domains. The central part of the sequence (residues 121–249) forms a β sandwich (or distorted barrel) most like hemagglutinin (see above), but also similar to the (+)ssRNA capsid structures (Fig. 11; see Color Insert). The structural similarity to HA is similar to that between STNV and SBMV (Grimes *et al.*, 1995). The sides of the barrel are far from closed, and the A′BIDG sheet has longer strands than CHEF. As a relatively short jelly roll, it is to be expected that the loops are relatively short. The other "domain" is composed of both the N-terminal and C-terminal regions, totaling 221 residues. It is α helical with nine helices. The core is a four-helix antiparallel bundle, two contributed by each of the N- and C-terminal regions. There is no homology to other known domain structures. It is the larger α-helical base domain that interacts with the VP3 scaffold. The β-barrel domains are oriented with the barrel axis radial with the BC, DE, FG, and HI loop pointing outward. The connection between the domains is loose. Indeed, in the natural homotrimers, there is a twist so that each jelly roll sits atop the α-helical domain of a neighboring subunit.

Structure determination of the entire core of diameter 700 Å, 54 MDa, and about 1000 proteins was a monumental feat (Grimes *et al.*, 1998; Johnson and Reddy, 1998). The prime interests in the formation of such a complex assembly, and the implications for in-virus transcription are discussed elsewhere in this volume. Of interest here is that VP3 had a novel fold (Grimes *et al.*, 1998). There are slight differences between 60 "A" subunits and "60" B subunits that are in a slightly different packing context. (These differences are denoted by adding relevant information for B subunits in parentheses.)

The molecule has been described in terms of three domains: apical, carapace, and dimerization (Fig. 12; see Color Insert). The dimerization domain, residues 699–854, is close to the C terminus, and is the most distinct domain. At its core are two β sheets that are approximately orthogonal to each other and comprise a total of 13 (14) strands. The smaller of these sheets forms the dimer interface. The larger sheet packs against a sheet at the bottom of the carapace domain, forming a β sandwich. In addition to the β structure, this domain has five helices. The carapace and apical domains are primarily helix, with another 22 helices, about one-third of which are 3_{10} helices rather than α helices. The apical domain (residues 298–587) is mixed α/β with 10 (11) strands and 11 (10) helices. The carapace domain is flat and platelike. It has two layers that are primarily helical with helices that run in approximately orthogonal directions in the two layers, and are long—up to 28 residues. At the base of the domain is a region of β structure that forms the sandwich with the dimerization domain. The carapace domain consists of residues 7–297, 588–698, and 855–901, and thus has two peptide connections to each of the two neighboring domains.

2. Genus Rotavirus

The structure of rotavirus VP6 (Mathieu *et al.*, 2001) is highly homologous to *Orbivirus* VP7 (Grimes *et al.*, 1995) (see previous discussion). In this 2-Å resolution structure, some subtle embellishments to the basic jelly-roll domain are clear (Fig. 11; see Color Insert). First, the N-terminal addition to the BIDG sheet is not just an A′ strand, but an up–down pair, so that the strand order is A′ A″BIDG. Second, the I strand is particularly long, and as its C-terminal end it hydrogen bonds to the bent ends of the BIDG sheet, closing one of the ends. Thus, there is an additional small β structure, AIDG, with A being separated from the A′ strand of the other sheet by some extended chain. Finally, in *Orbivirus*, there was a strand D′ inserted between the D and E strands. In *Rotavirus*, there is an additional insertion of 13 residues, some of which provide a second β strand so that the insertion is a D′D″ ribbon.

3. Genus Orthoreovirus

The internal capsid particle (ICP) has been determined for a reovirus from the *Orthoreovirus* genus (Reinisch *et al.*, 2000). Some differences from the *Orbivirus* ICP were expected, but the level of similarity has been a point of contention. *Orthoreovirus* $\lambda 1$, residues 240–1275, and *orbivirus* VP3 share little sequence similarity (<9%) in the core region. However, they have the same general shape, similar layers of orthogonal helices, similar β structures at or near domain I (*Orthoreovirus*) or the dimerization domain (*Orbivirus*) (Fig. 12; see Color Insert). They both form a similar 120-protein structures within the ICP. However, these proteins show different domain boundaries and different pivot points through which the A- and B-type subunits are related in each virus. Similarity of topology has been argued. Harrison and colleagues reported that essentially no secondary structure or topology could be superimposed (Reinisch *et al.*, 2000). Stuart's group has reevaluated this, suggesting that 63% of the protein was alignable (Bamford *et al.*, 2001). Figure 12 clearly illustrates the similarity in topology, although the orientations of secondary structure elements vary considerably. The first 240 residues are an addition unique to $\lambda 1$ and are disordered in half of the copies. In the other half, they lie on the inside surface forming a network of interactions. Residues 181 to 208 form a Cys_2His_2 zinc finger.

The $\sigma 2$ protein is a clamp attached to the $\lambda 1$ shell. The protein has three distinct binding sites on the shell, one of which is across a 2-fold axis. There are thus 150 copies of this protein. The 417 residues form a globular domain consisting mostly of helices (Fig. 13; see Color Insert). Most of the structure is formed by a repeated motif of about 150 residues. The motif starts with a strand and is followed by three helices, the last two of which are antiparallel. A loop region containing a two-stranded antiparallel sheet comes next, and the motif ends with a strand antiparallel to the first and then a helix. In the C-terminal 120 residues, a long helix is packed along the whole subunit.

The $\lambda 2$ proteins of *Orthoreovirus* and VP7 of *Orbivirus* form structures on the surface of the ICPs that were expected to be entirely different, and so they are. Pentamers of $\lambda 2$ lie at the vertices, with channels through the middle large enough to allow the transcribing RNA to exit the ICP. In *Orthoreovirus* the proteins form a turret with several RNA-processing enzyme activities. Each $\lambda 2$ subunit is elongated with two lobes of unequal size, and is aligned with its long axis 45° to the channel (Fig. 13; see Color Insert). The first of seven domains starts at the base of the turret, and is a 385-residue guanylyltransferase with a novel fold. It is a predominantly antiparallel β sheet in the core with surrounding loops and helices. The next domain, formed with the next ~50 residues and a later segment of ~100 residues is thought to be a structural linker. Open-faced sandwiches of symmetry-related subunits pack with four-stranded antiparallel β sheets on the inside of the turret and α helices are on the external surface. The next two regions (residues 434–691 and 804–1022) are seen by homology to form methyltransferases, known as methylase-1 and methylase-2 domains. These presumably accept RNA substrate from the guanylyltransferase site (in an unknown order) for O and N methylation. Methylase-2 has the closest homology to the consensus *S*-adenosyl-L-methionine (SAM)-binding domain. This is a mixed, but nearly parallel, β-sandwich structure in which α helices layer either side of a seven-stranded sheet. The strand topology is (1×, 1×, −3×, −1×, −2×, 1). This means that the N terminus starts in the middle of the sheet, with the path adding strands to the "right" with cross-over helix connections below the sheet. The fourth strand neighbors the first with subsequent strands added "leftward" and with the two remaining helices on "top" of the sheet. The last strand in the chemical sequence is antiparallel to all of the others and is placed between strands 5 and 6 in the sheet. To this consensus topology is inserted in methylase-2: (1) a small three-stranded up–down antiparallel sheet in the cross-connection between strands 5 and 6, and (2) two helices that lie on a face of this sheet, inserted into the turn between strands 6 and 7. Methylase-1 does not have these embellishments, but is a greater variant on the consensus fold. The consensus strand 3 is missing, but, instead, an additional strand is at the N terminus in sequence, but physically at the other end of the sheet. This changes the topology to (+5, 1×, −2×, −1×, −2×, 1). Finally, the subunit ends with 250 residues in 3 immunoglobulin folds: V-like, C-like, and a truncated V-like domain with two three-stranded sheets. It is to this region that the cell attachment protein, $\sigma 1$, would be bound in the complete virus.

$\sigma 3$ is one of the external structural proteins removed on cell entry, and is therefore not part of the internal capsid particle studied in both *Reoviridae.* It forms a two-lobed structure (determined at high resolution) with neither domain showing a previously known fold (Olland *et al.*, 2001). The protein starts with an α helix, and then an up–down antiparallel four-stranded sheet that ends with a CCHC zinc finger motif (Fig. 13; see Color Insert). The long helix B then makes the first of three polypeptide connections to the second lobe or "domain." The second lobe is two-layered with B plus five other α helices lying on top of a β layer made up of three small sheets. After helix C, the path runs to the end strands of a three-stranded β sheet in which the cross-over is helix D, and for which the middle strand is the C terminus. Next is a four-stranded antiparallel sheet with an insertion between the first two strands that includes one of the strands of the last β structure, and helix E. Between this and the final β sheet the path returns temporarily to lobe 1 for helices F and G.

The $\sigma 3$ protein is associated with the $\mu 1$ protein, and a complex of these proteins has been determined (Liemann *et al.*, 2002). $\mu 1$ forms a T=13 layer in the reovirus particles but is not present in the crystal structure of the capsid (Reinisch *et al.*, 2000). The protein is similar to *Orbivirus* VP7 in that it has a jelly-roll domain positioned external to a base formed mostly by helices. The jelly-roll domain contains a few extra strands, and in the sequence it is flanked by the residues forming the helical region. Although this is true also for the corresponding region in VP7 of bluetongue virus, the fold in this part of the protein is different.

D. *Single-Stranded DNA Viruses*

1. *Parvoviruses*

Parvoviruses are small T=1 icosahedral viruses about 250 Å in diameter. They contain three capsid proteins, VP1–VP3, that are variants of each other, resulting from alternative start codons and RNA splicing mechanisms. The basic building block, an ~540 residue classic jelly roll, is common to VP1, VP2, and VP3. Whether it is VP2 or VP3 that dominates the mature capsid depends on the species, but VP2 has an additional ~20 amino acids at the N terminus that are of unknown structure (Berns, 1996). VP1 is a minority protein that has an additional ~150 amino acids (cf. VP2). It is thought that the jelly-roll cores of VP1–VP3 can largely substitute for each other in the capsid assembly. There is no evidence for the systematic distribution of VP1, VP2, and VP3, other than that disordered density is consistent with a single chain passing up some of the 5-fold channels in full, but not empty, capsids (Wu and Rossmann, 1993; Xie and Chapman, 1996). This is consistent with a sometime external location for the VP1-unique N-terminal 150 residues and with the fact that only 1 of 5 proteins surrounding each 5-fold axis can adopt this configuration. Several structures are available for the assembled capsid, which shows the jelly-roll cores, but not the N-terminal unique regions (Agbandje *et al.*, 1993; Agbandje-McKenna *et al.*, 1998; Simpson *et al.*, 1998; Simpson *et al.*, 2002; Tsao *et al.*, 1991; Xie *et al.*, 2002).

Although the cores of VP1–VP3 are classic jelly rolls (Fig. 1c; see Color Insert) they are more than three times longer than the most compact of the RNA virus barrels. This led to the one-time, incorrect conclusion that parvoviruses were pseudo-T=3 viruses with three domains, like the comoviruses (Murphy and Kingsbury, 1996). The long chain length really reflects a single jelly-roll domain with especially long loops that is packed into a true T=1 assembly. The first long loop (35 residues) is "loop 1" at the BC location, where the insertion is longer than in rhinoviruses, and is mainly a long β-ribbon hairpin. (As in the picornaviruses, the loops at this end of the barrel, which point toward the 5-fold axis, are shorter than at the other end.) Even though they are otherwise so dissimilar, it is remarkable that αA, the helix in the short connection between strands C and D, seen in picornaviruses and other viruses, is conserved in parvoviruses. Between the D and E strands, there is a small insertion of a β ribbon that keeps the points of the barrels apart at the 5-fold axis. The ribbon runs parallel to the 5-fold axis and, superficially, it looks like there is a 10-stranded antiparallel barrel formed by contributions from the neighboring subunits, but between the subunits the strands do not come close enough for hydrogen binding. The next large insertion is at "loop 2" (72 residues) between strands E and F. It also forms a β ribbon that packs against loop 1. At its end as it returns to βF is another conserved helix, αB. The biggest insertion (221 amino acids) is what in vertebrate parvoviruses is called both "loop 3" and "loop 4," which together correspond to the GH loop of picornaviruses where there are also insertions, albeit shorter by an order of magnitude. Most of the solved parvovirus structures share >50% sequence identity (Chapman and Rossmann, 1993b) and, as exemplified by canine parvovirus (Xie and Chapman, 1996), have similar but highly convoluted GH loop structures, containing several elements of β structure, some in which strands of adjacent subunits come together. As might be expected from prior studies of the picornaviruses, it is the long exposed loops 1–4 that are the sites of antigenic recognition and high sequence variability between the species and strains (Chapman and Rossmann, 1993a,b). In addition, the long loops 3 and 4 form the bulk of protrusions near the 3-fold axes of parvoviruses. The chains of 3-fold related subunits intertwine and form the bulk of the interactions responsible for stabilizing the assembled form. Indeed, regions of adjacent chains are hooked over each other, so that the folding can be completed only during assembly into the capsid (Xie and Chapman, 1996).

Two structures represent parvoviruses that are less similar. Wax moth densovirus, representing invertebrate parvoviruses, shares only ~10% sequence identity with other parvoviruses of known structure (Chapman and Rossmann, 1993b). The β barrel is mostly conserved, but there are other structural differences (Simpson *et al.*, 1998). Strand βA in canine parvovirus (CPV) folds back against βB as in plant viruses such as TBSV. In wax moth parvovirus, βA is also associated with βB, but in a domain-swapped way in which, extending back toward the N terminus from βB, the chain continues in the same direction to become strand A of a 2-fold-related subunit. The barrel of the wax moth form is shifted about 10 Å with respect to the vertebrate form, the tertiary structure of the long loops are quite different, and the invertebrate form lacks most of the GH loop and therefore the prominent surface protrusions near the 3-fold axes.

Adeno-associated virus 2 (AAV-2) represents the dependoviruses, so named because they are dependent for replication on a helper virus, often adenovirus. Their sequences have about 20% of capsid amino acids identical to the autonomous parvoviruses of known structure (Chapman and Rossmann, 1993b). AAV structure is more similar to the canine parvovirus-like viruses than is the densovirus structure, with a superimposable β barrel, and an equally long GH loop. Likewise, the GH loop has additional elements of secondary structure, but these are quite different from those in the other parvoviruses, and there is little correspondence between the GH loop structures of AAV and other parvoviruses (Xie *et al.*, 2002). Subloops extend radially, intertwining with those of 3-fold related neighbors to form a set of three separate (antigenic) peaks surrounding each 3-fold and giving AAV more prominent surface features than other parvoviruses. Cellular receptor binding appears to be on the sides of the peaks, in the valleys between adjacent symmetry-related peaks (Xie *et al.*, 2002).

2. *Microviridae Bacteriophages*

The structures of ssDNA bacteriophages ϕX174 and G4 have been reported in mature and provirus forms (Dokland *et al.*, 1997; McKenna *et al.*, 1992b, 1996). The mature viruses are T=1 with 60 copies each of the F, G, and (small J) proteins, and 12 copies of the H protein. Both the F and G proteins are classic viral jelly-roll structures. It is the F protein that occupies the positions homologous to the (+)ssRNA capsids. At 430 residues, the F protein is closest in size to the parvoviral capsids, and achieves its size through large loop insertions, primarily in the EF and HI loops. It is these loops that form the primary inter subunit contacts rather than direct barrel contacts as in the (+)ssRNA viruses. The G proteins form large pentameric protrusions that decorate the surface of the F-protein assembly. The barrels of the G proteins are oriented with strands radial with respect to the virus center. The H protein is not seen in the mature virus crystal structure. The structure of the procapsid has now been refined (Dokland *et al.*, 1999), and there are several interesting findings relevant to assembly. The external "D" scaffolding protein forms a T=4 lattice that is therefore mismatched with the T=1 structure of F and G proteins. It forms a shell surrounding, but only loosely associated with, the F-protein shell, filling and leveling the surface between the G-protein spikes. The D protein has a compact globular structure and has 7 α helices connected by short loops, one of which has a β strand. About half the internal "B" scaffolding protein is seen, and it is also helical.

3. *Inoviridae Bacteriophages*

The *Inoviridae* are typified by *Escherichia coli* phages M13, fd, and f1 that share >98% sequence identity and are collectively known as Ff phages, as well as some more distantly related viruses such as *Pseudomonas* phage Pf3. Their 6.4-kb ssDNA genome is circular, and the virus has a flexible rod morphology with variable length (e.g., 500 Å) and diameter of ~65 Å (Makowski and Russel, 1997). It is therefore completely different from the viruses discussed to this point. Two of the eight gene products have known atomic structure. A single-stranded DNA-binding protein (pV) has a five-stranded antiparallel β-sheet fold with two extending β-ribbon strand pairs (Skinner *et al.*, 1994). The overall shape leaves a cleft in which DNA could be contained. The major coat protein (pVIII) is also known. About 2800 copies are used in a helical array that covers the length of the DNA. Smaller numbers of other proteins cap the ends, but these are of unknown atomic structure. Neutron and X-ray diffraction of fibers (Nambudripad *et al.*, 1991a,b) showed that the small protein of 55 amino acids forms a single slightly curved α helix of 40 residues (62 Å). These are packed to form a cylinder that is 65 Å wide and 20 Å thick. The C-terminal end of pVIII interacts with the inner nucleic acid. The α-helix axis points about 20° from the fiber axis, tilted to form a right-handed super helix, and pointing out slightly, so that the N terminus is on the surface. The virus assembles directly from membrane-attached coat proteins with an interesting conformational change. In the membrane, the helix is broken with a right-angle turn at Tyr-24. The N-terminal part is amphiphilic, lying on the periplasmic surface. The remaining ~45 Å of the helix points through the membrane and beyond to the cytoplasm (McDonnell *et al.*, 1993). As pVIII is incorporated into the virus, the helix is straightened, extending its length by 16 Å, the increase in length of the fiber for each layer of pVIII that is added.

E. Double-stranded DNA viruses

1. Polyomaviruses

The family *Polyomaviridae* includes both polyomavirus and simian virus 40 (SV40), whose structures are both known at atomic resolution (Liddington *et al.*, 1991; Stehle *et al.*, 1994, 1996). These viruses had been grouped within the *Papovavirus* family with other tumor-inducing papillomaviruses that have been studied by electron microscopy, but not at atomic resolution (Baker *et al.*, 1991; Trus *et al.*, 1997). The families have now been recategorized separately (van Regenmortel *et al.*, 2000). They share similar genetic structure, but little sequence similarity between capsid proteins, and show some differences in overall dimensions (Baker *et al.*, 1991).

It is the structure of the major structural protein VP1 that was best characterized (see below) (Liddington *et al.*, 1991). The minor proteins VP2 and VP3 share a C terminus, but VP2 has a myristoylated N-terminal addition of ~100 residues. These proteins are mostly internal in the capsid, but the C terminus binds to VP1.

The greatest interest in the structures of the polyomaviruses was in the implications for capsid assembly. This is detailed elsewhere in the volume, but a brief summary is given here. Early electron micrographs were interpreted as consistent with a T=7 assembly, and with the Caspar–Klug rules of quasi-equivalence (Caspar and Klug, 1962). There were thought to be 12 capsomers, each encircling a 5-fold, and 60 encircling pseudo-6-folds, totaling 420 subunits. Diffraction at a low 22.5 Å showed that at the pseudo-6-fold positions there were also pentamers, giving 6 subunits per unique repeat, and a total of 360 copies of the major capsid protein, VP1 (Rayment *et al.*, 1982). This presented a problem. A pentameric unit was surrounded by six neighbors. Clearly the subunit contacts could not be the same.

An explanation came with the atomic structures (Garcia and Liddington, 1997; Liddington *et al.*, 1991; Stehle *et al.*, 1994, 1996). It extended our ideas of quasi-equivalence that previously had been implicitly limited to considering the interactions between relatively rigid globular domains. In the polyomaviruses the contacts between capsomers are not barrel to barrel, but are mediated by a long flexible C-terminal arm that reaches to and interacts with a neighboring subunit in another capsomer. The details of the interactions between one C-terminal arm and the neighboring subunit are identical, even though the packing environments of different subunits vary.

How does the fold of the protein help in this assembly function? The core of the domain is a jelly-roll barrel with relatively few embellishments, but a different orientation from the (+)ssRNA viruses. The axis of the barrel is nearly radial, parallel to the 5-fold axis (Fig. 14; see Color Insert). Both sheets of the barrel have additional strands added. Some of the strands are particularly long, and the N-terminal half of βG hydrogen bonds to strand F of the CHEF sheet of a 5-fold-related adjacent subunit, providing stability to the pentamer. An addition at the other edge of the BIDG sheet is also interesting. To the N terminus of strand B is an additional strand, but instead of forming an A strand antiparallel to B, it is parallel, with the sheet cross-over connection made by an α helix. Furthermore, it is not the immediate neighbor of B, but strand J″ that is inserted from the C-terminal arm of a neighboring subunit, so that the entire sheet is antiparallel β. The effect of this is to lock the C-terminal arm in place with hydrogen bonding to two strands of the β sheet in a neighboring subunit.

2. *Papillomavirus*

Papillomavirus is difficult to produce in amounts suitable for structural studies, but the particles, like polyoma and SV40, appear to be built up of 72 pentamers. Recombinant expression of the main capsid protein, L1, in bacteria leads to capsids with 12 pentamers, and it has been possible to study these by crystallography (Chen *et al.*, 2000). The C-terminal segment of the chain that is exchanged between pentamers in polyomavirus is in the papillomavirus recombinant capsids forming a projection, but the chain returns to the jellyroll from where it emanates. In the native T=7 particles, this region is probably exchanged between pentamers like in polyomavirus.

3. *Adenovirus*

Although adenovirus is nonenveloped, it is an example of a larger virus (~150 MDa) with more complex organization (Burnett, 1997). There are 11 structural proteins, 7 of which are capsid proteins, and 4 of which are core proteins. Some of the structures are known at high resolution (see below). The locations of some others have been determined by one of the first combined electron microscopic and X-ray studies. This was *a tour de force* in which difference images were analyzed between various natural and recombinant assemblies that lacked specific components (e.g., Stewart *et al.*, 1991). Adenovirus is assembled through various intermediates that carry names of historical significance. "Groups of nine" make up much of the icosahedral repeating unit. These are made up of nine hexons, named for their hexagonal packing in the group of nine, and in spite of the fact that they are each trimers of polypeptide II. The presence of 3-fold, but not 6-fold, symmetry eliminated the possibility that adenovirus was a T=25 virus, and established it as an exception to the Caspar–Klug rules (Caspar and Klug, 1962; Crowther and Amos, 1971). At the base of the 12 × 5-fold vertices are assemblies of five polypeptides III, known as pentons. To these are attached long fibers, consisting of trimers of a 582-residue polypeptide IV, that give electron micrographs their characteristic "spiked" look, and extend outward 450 Å from the virus center.

Polypeptide II, that as trimers form the hexon, has been determined to 1.8-Å resolution (Athappilly *et al.*, 1994). It is a protein of 967 amino acids that forms two side-by-side viral jelly-roll domains named P1 and P2 (Fig. 15; see Color Insert). The two domains are oriented around the 3-fold trimer axis to make pseudo-6-fold symmetry (giving the capsid pseudo-T=25 symmetry). The P1 domain looks particularly open because the BIDG strands are longer than the CHEF strands, and because the two sheets flare apart. The jelly-roll domains are held together by a third "PC" connecting domain, which has at its core a three-stranded antiparallel sheet. Two strands come from the C terminus of P2. The other strand follows the final "I" strand of domain P1. The N terminus of P1 (before βB) folds over the PC domain before looping off to interact with both of the other subunits of the trimer. As to be expected of jelly-roll domains of ~530 and ~290 residues, some of the loops are extended and follow convoluted paths. The barrels are oriented radially with respect to the center of the virus, and, in contrast to the (+)ssRNA viruses, it is the loops at the BC, DE, FG, and HI ends that are long, whereas loops at the other ends are short. The FG "l_4" loop of P2 is the most structured, with a turn and an α helix following βF. There is then a 70-residue hairpin loop that interacts with loop l_2 of P1, the first and last 12 residues forming a β ribbon. The largest loop is the FG l_2 loop of P1 that totals nearly 200 amino acids. As in the P2 FG loop, it starts with a turn of an α helix, but is then followed by ~50 residues of extended chain before the β ribbon of the loop starts. The ribbon has two short segments of 5 and 5 residue pairs before the path meanders back to strand G through 3 α helices in an ~100-residue segment. The P1 DE l_1 loop is about 90 residues. Its path is convoluted, but includes two β ribbons, strands C and D with six residue pairs, and strands E and F with seven residue pairs. It is not obvious from Fig. 15a, showing the monomer, that loops l_1, l_2, and l_4 intertwine with those of their 3-fold related neighbors to form the "tower" region of hexons. Thus, in the most superficial of ways, adenovirus resembles parvoviruses in that the long loops are convoluted, contain small elements of β structure as ribbons, and are intertwined to cement trimeric associations.

The other adenovirus protein known at atomic resolution is the fiber. The first structure obtained was of a C-terminal fragment containing the distal "knob" (Xia *et al.*, 1994) as the natural homotrimer. The subunit fold is an eight-stranded antiparallel β barrel (Fig. 15; see Color Insert), but of topology and strand order that were new and unlike the canonical viral fold (Fig. 2). The R sheet, containing strands DIHG, is a classic Greek key. The V sheet contains strands (FE)ABCJ. The authors consider strands E and F to be a small part of the R sheet DG loop. However, if the short F and E strands are included in V, the sheet starts with a Greek key of strands EABC, modified to insert the D strand of the R sheet in the CE loop. The F strand is an up–down hairpin addition to one side of the sheet. The J strand is an addition on the other side of the V sheet, after insertion of all other R-sheet strands. The two sheets pack face to face with an angle of about 30° between the strands. The R sheet faces out with hydrophilic residues toward the cellular receptor (hence the "R" designation). In the trimeric assembly, the R sheets form a three-bladed propeller. Conserved, putative receptor-binding residues are found within the depression at the center, shielded from antibody binding as in the picornaviruses.

Parts of the stem were seen in a later structure determination of a different expressed construct of the fiber (van Raaij *et al.*, 1999b). As had been predicted, the shaft forms a triple helix, but the strand conformation had not been anticipated correctly. There are 22 near repeats of a 15-residue sequence motif in adenovirus types 2 and 5 (Fig. 15b). These total 66 for the assembled trimer, of which 4 were visualized in the structure. The 15-residue motif forms one long β strand roughly parallel to the fiber axis, a hairpin turn closed by a single hydrogen bond followed by a shorter strand running about 45° from anti parallel. There is only one hydrogen bond within this hairpin, but then the chain repeats, forming four hydrogen bonds with the next instance. In addition, the shaft is wound as a triple helix, and there are interchain hydrogen bonds, three to each of two neighboring motifs. The combination of extensive burial of hydrophobic surface and hydrogen bonding is thought to endow rigidity on the shaft. Extrapolating from what was visualized by repeating the same motifs, the shaft would be about 300Å long, the observed length.

4. *Double-stranded DNA Bacteriophages: T4, HK97, P22, and* ϕ*29*

i. The HK97 Head. The dsDNA bacteriophages are complicated structures composed of heads, tails, and fibers. The structure is known for the icosahedral head of the λ-like phage HK97, where the tail part normally present is missing (Wikoff *et al.*, 2000). The head structure represents one of several steps in phage assembly. It is a T=7 arrangement of 420 fragments of the coat protein, from which the N-terminal 103 amino acids have been cleaved in the maturation process. In contrast to the all-pentamer arrangement of subunits in the T=7 polyoma or SV40 capsids, the shell is composed of 60 hexamers and 12 pentamers of identical subunits. The coat protein has a fold not found in any other viral protein (Fig. 16; see Color Insert). It has two domains, an extended N-terminal arm and an extensive hairpin loop (E loop). One domain, the A domain, forms the 5-fold and 6-fold contacts. It has a central six-stranded sheet and two helices that form most of the contacts and the outer surface of the domain. The P domain has the shape of a rectangular box and has three antiparallel strands that form one side of the domain. The innermost strand has an unusual kink that interrupts the twist of the sheet and makes it relatively flat. The other side of the domain consists of a helix. The helix is interrupted by a long loop that extends over the rest of the domain. The extended E loop links the subunits covalently through an asparagine–lysine cross-link between residues 169 and 363 in subunits related by the 5-fold or quasi-6-fold axis. The cross-links between the subunits are not within the pentamer or hexamer but forms covalently linked rings of five or six subunits around the pentamers or hexamers, and by icosahedral symmetry these rings become interlinked into a closed chain-mail structure. These covalent links allow the protein shell of the head to be unusually thin but still stable.

ii. The PRD1 Capsid Protein. PRD1 is an unusual DNA bacteriophage lacking a tail but having a membrane inside a protein coat. The protein capsid has a pseudo-T=25 lattice like adenovirus. The major capsid protein, P3, is trimeric and occupies four hexavalent positions in the same way as the adenovirus hexon protein. The crystal structure of the P3 trimer shows that it has two jelly-roll domains and a conformation that is similar to the hexon protein (Fig. 15b; see Color Insert) (Benson *et al.*, 1999). The long loops found in the hexon protein are almost absent, and the connecting domain is missing, making the distance between the jelly rolls smaller.

iii. The ϕ*29 Motor Protein.* In the tailed phages, the DNA is injected into the head, using a special mechanism that involves a connector protein. The connector occupies a pentagonal vertex of the head. In phage ϕ29, the connector is a dodecamer of the gp10 protein with a molecular mass of 36 kDa. This is an example of symmetry mismatch that probably is of importance for the function; the connector has 12-fold symmetry and binds to a position of 5-fold symmetry in the head. The crystal structure shows that the protein is elongated and has three domains (Fig. 17; see Color Insert). The top domain is formed by three small antiparallel sheets and two helices (Simpson *et al.*, 2000). The central domain is formed by three long helices (α1, α3, and α5). A small sheet and a helix form the small bottom domain. All domains interact in the connector with their equivalent in other subunits to form cylindrical regions with different radii. The bottom domain forms a narrow ring that protrudes from the head, while the top domain forms a wide ring inside the head.

iv. The P22 Tailspike Protein. The *Salmonella* phage P22 is a phage with a relatively short tail. To this tail, six spikes formed by the tailspike protein (gp9) are attached a short tail. The tailspikes are responsible for receptor binding, but they also have endoglycosidase activity. The tailspikes are trimeric and have two parts, an N-terminal head-binding domain and a C-terminal part. The structure of the main C-terminal part is unusual (Steinbacher *et al.*, 1994). It is elongated and formed mostly by 13 turns of a β helix (Fig. 18a). β Helices figure prominently in the extended appendages of bacteriophages, but there are fundamental differences, such as single versus triple helix. In P22, the single helix has the shape of a flattened cylinder with one convex and one concave side. At a few positions in the helix, extra residues are inserted in the turns between the β strands, forming protruding subdomains. The C terminus contains two sheets. This part of the protein interacts closely with the corresponding parts of the other subunits in the trimer. The structure of the head-binding domain has been determined separately (Steinbacher *et al.*, 1997). It has two antiparallel sheets of three and five strands (Fig. 18b).

v. T4 Fibritin and Baseplate Proteins. The tail of T4 and many other phages connects the baseplate to the head, and forms a hollow tube through which DNA is injected. gp5–gp27 forms the central hub of the bacteriophage T4 baseplate. The trimeric structure, which fits well into a cryoelectron microscopy (cryo-EM) image reconstruction of the baseplate, shows a central β helix (Kanamaru *et al.*, 2002). gp27 extends the amino end, forming a hollow cylinder that would be associated with the phage tail tube through which the DNA is ejected. The complex thus forms the

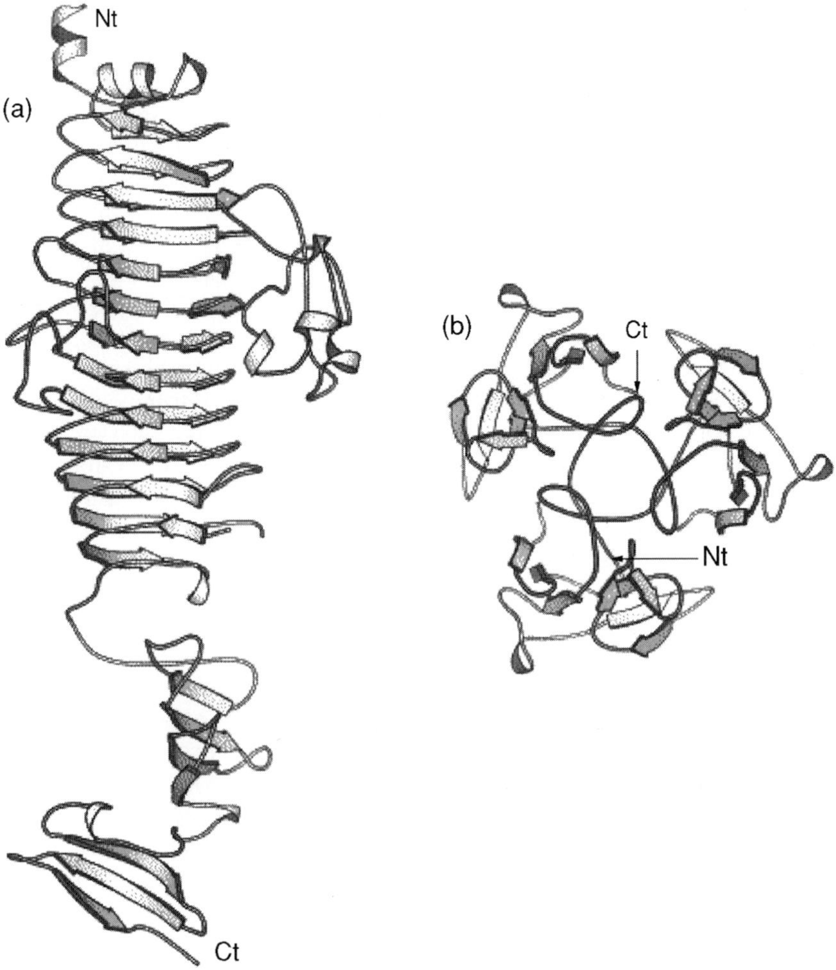

FIG. 18. The P22 tailspike protein. (a) The main domain (Steinbacher *et al.*, 1994). The protein is trimeric, and the 3-fold axis is parallel to the long axis of the monomer. The C-terminal sheets and the connecting loops link the subunits together. (b) A trimer of the N-terminal head-binding domain seen down the 3-fold axis (Steinbacher *et al.*, 1997).

tip of the syringe that enters and opens the outer cell membrane and peptidoglycan layer of the host *E. coli.*

The baseplate becomes attached to the host cell through both long and short fibers attached to the baseplate. gp9 connects long fibers that recognize the *E. coli* receptor, while gp11 connects the short fibers whose interaction with *E. coli* make attachment irreversible. Fibritin assists in the assembly of the long fibers, their attachment to the baseplate, and their retraction in unfavorable environments. The structures of fibritin, gp9, and gp11 have been determined. Several fragments of fibritin have been studied. All contain a 30-residue C-terminal domain that contains a β hairpin and a C-terminal helix. Three hairpins associate in the trimer with two intrachain hydrogen bonds and a salt bridge to form a propeller (Tao *et al.*, 1997). The full-length protein was predicted to have 12 helical regions, of which 3 are seen in one of the fragment structures, forming a trimeric coiled coil of head-to-tail helices linked by short loops. The stability of the C-terminal associations and lability of the coiled-coil regions are thought to be critical for the functioning of fibritin. gp11 forms a trimer of three domains each, with a coiled coil at its center formed by symmetry-equivalent N-terminal domains (Leiman *et al.*, 2000). The C-terminal domain forms a β annulus and probably functions in trimerization, as does the C terminus of fibritin (see above). The middle domain forms fingers extending from the annulus, outside the coiled coil. gp9 is again trimeric and three-domain, with the domains in a row parallel to the 3-fold axis (Kostyuchenko *et al.*, 1999). Closest to the tail emanating from the bacteriophage head, and farthest from the baseplate, is the N-terminal domain that is associated with the long fiber. It starts as extended chain, and then there are six turns of α helix that form a trimeric coiled coil. The middle domain is a seven-stranded β sandwich of previously unseen topology. The C-terminal domain is an eight-stranded β jelly roll, superficially similar to those found in the (+)ssRNA capsids, although the strand connectivity differs. It is this domain that is associated with the baseplate. The connection between the middle and C-terminal domains is long, with the covalently connected domains partially swapped to increase the interactions with 3-fold-related neighbors. The connection forms a 3-fold symmetric β annulus with each chain divided into two segments, the first six residues interacting with one neighboring subunit and the next five with the other neighbor.

F. *Reverse-Transcribing Viruses*

1. *Hepadnaviridae*

Hepatitis B virus is enveloped, but it has been possible to study the inner DNA-containing capsid after removal of the lipid. The capsid protein or core antigen forms a mixture of T=3 and T=4 particles when expressed in

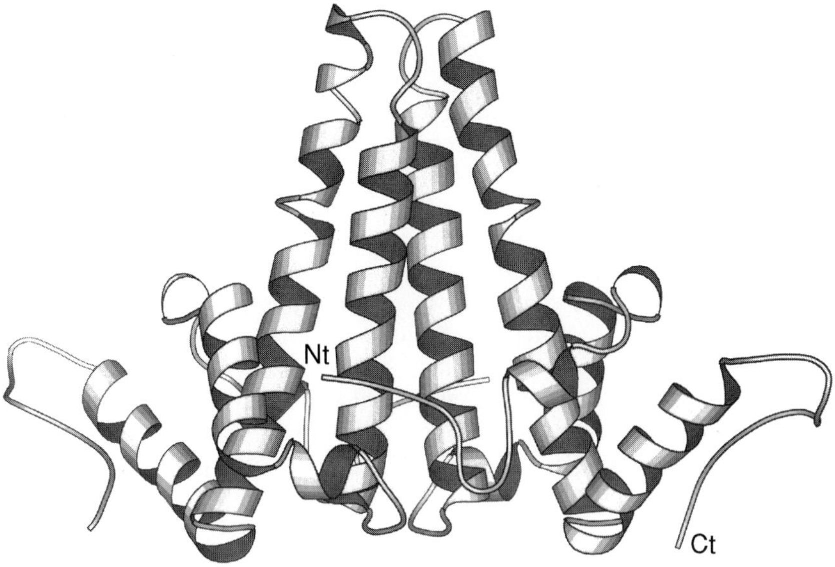

FIG. 19. The dimer of the capsid protein of hepatitis B virus (Wynne *et al.*, 1999).

a bacterial system. When the protein is truncated at its C terminus, however, a fragment of 149 amino acids forms T=4 capsids that have been studied by crystallography (Wynne *et al.*, 1999). The coat protein forms strongly interacting dimers with a fold that is not observed in other viruses. The protein is helical (Fig. 19). Two antiparallel helices from each subunit form a spike on the surface of the capsid. This four-helical bundle does not show the type of packing found in typical four-helix bundles, the contacts mostly formed by one of the helices. The rest of the subunit is formed by three short helices.

2. *Retroviridae*

The retroviruses have been the subject of intense study because of the emergence of human immunodeficiency virus (HIV) and acquired immunodeficiency syndrome (AIDS), and this study has led to a structural understanding of complicated viruses that has been unfolding rapidly. In the immature virus, two copies of (+)ssRNA are surrounded by a spherical immature capsid that contains about 2000 copies of the Gag polyprotein (the precursor of the structural proteins), and up to a few hundred copies of larger polyproteins, the most ubiquitous being Gag–Pol, a polyprotein that in addition contains precursors of the proteinase (PR), reverse transcriptase (RT), and integrase (IN) (Weldon and Hunter, 1997). Both immature "procapsid" and the mature "core" are surrounded by a bilayer host-derived membrane that contains a membrane-spanning protein (TM), anchoring a surface glycoprotein (SU). During maturation the Gag polyprotein is cleaved to form (1) a matrix protein (MA) that is associated with the inner surface of the membrane, and forms an icosahedrally symmetric (or sometimes pleomorphic) assembly; (2) a capsid protein (CA) that forms various rod- and cone-shaped assemblies; (3) a nucleocapsid protein (NC) that associates with the RNA; and (4) other proteins (p1, p2, and p6) of unknown function. Gag–Pol is similarly cleaved to form the structural proteins and PR, RT, and IN. Atomic structures are now known for many of the viral components, many of which are covered in a 1999 review (Turner and Summers, 1999). A more complete list of structures is given in Table VII, but in the following sections, as with the other viruses, the focus 15 on structural proteins, to the exclusion of several enzymes and RNA-binding proteins.

i. Gag Structural Proteins: Capsid, Matrix, and Nucleocapsid. In HIV the intact core is cone shaped. Rod like and other assemblies have been found in other retroviruses, but none have been amenable to high-resolution structure determination, in part because the cores are not as symmetric as icosahedral or helical particles. The capsid protein (p24 for HIV; p27 and p30 in others) has been studied as separate N- and C-terminal domains, because the purified subunit tends to form many aggregates. In one exception, an antibody Fab–CA complex was crystallized, but the C-terminal domain was too disordered to visualize (Momany *et al.*, 1996). Structures have come from both NMR (Gitti *et al.*, 1996) and X-ray crystallography (Gamble *et al.*, 1996, 1997; Momany *et al.*, 1996). The domains are quite unlike the canonical jelly-roll viral fold, or any other virus structure, and completely different from numerous prior predictions. The N-terminal core domain has seven α helices, five of which (A, B, C, D, and G) are configured in a coiled coil (Fig. 20) (Gitti *et al.*, 1996; Khorasanizadeh *et al.*, 1999; Momany *et al.*, 1996). The helices are antiparallel with respect to their 3-D neighbors, except for A and C. The α helices are at various angles (0 to 40°) with respect to the superhelical axis that is at the center of the coil of helices that winds in a left-handed sense. The C-terminal oligomerization domain structure has been determined from two slightly different constructs that both show dimers with an N-terminal strand followed by four antiparallel helices (Fig. 20). The linker between domains is flexible and has not been characterized. Thus the assembly of the two domains, and the assembly of these subunits into entire capsids, is a matter for modeling, consistent with electron

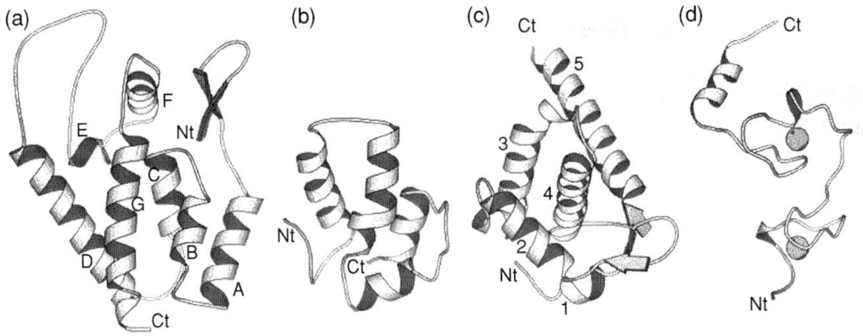

FIG. 20. Structurally characterized retroviral Gag proteins. (a) Capsid protein, N-terminal part (Gamble *et al.*, 1996); (b) capsid protein, C-terminal oligomerization domain (Gamble *et al.*, 1997); (c) matrix protein (Hill *et al.*, 1996); (d) nucleocapsid protein (De Guzman *et al.*, 1998).

micrographs of p24 assemblies, and with contacts between molecules observed in crystals (e.g., Jin *et al.*, 1999; Momany *et al.*, 1996).

The matrix protein (MA) forms an often icosahedrally symmetric assembly on the inner surface of the membrane. Structures of heterologously expressed subunits have been determined by both NMR and crystallography for HIV-1 and several other retroviruses (e.g., Hill *et al.*, 1996; Massiah *et al.*, 1994, 1996; S. Matthews *et al.*, 1994). Much of the fold resembles interferon γ (S. Matthews *et al.*, 1994). Near the N terminus there is a short helix, followed by the first two of three strands of a mixed β sheet (Fig. 20). There then follows helices 2 to 5 in a globular fold. The fourth helix is in the center and hydrophobic. The other helices surround it and are amphipathic. The final strand of the small β sheet (three strands of four residues) comes before the final C-terminal helix. MA is N-terminally myristoylated in most retroviruses, although the structure has not been visualized. The sheet, and residues immediately before and after, has a highly basic surface with eight conserved lysines a arginia in the strands that are thought to interact with the negatively charged phospholipid surface. Although there is not yet direct evidence that it is physiologically relevant, in the crystals the basic surfaces of the sheets sit side by side to form a large common surface in a trimer.

The nucleocapsid protein (NC) has many nucleic acid-associated functions, such as packaging, transcription initiation, and stabilization of proviral DNA (Turner and Summers, 1999, and references therein). The parts that have been characterized structurally are the zinc fingers, which are common nucleic acid-binding motifs (Berg, 1986). Most retroviral NCs contain one, and usually two, zinc fingers of the CCHC type, in which the ligands of a zinc ion include three cysteines and one histidine in a peptide of the following consensus sequence: $CX_2CX_4HX_4C$. NMR characterization of NC for HIV and other retroviruses has included synthetic motifs (Omichinski *et al.*, 1991; Summers *et al.*, 1990) and complete protein (Morellet *et al.*, 1992; Summers *et al.*, 1992). The NC zinc fingers are truncated forms of the classic motif, in which a β ribbon is followed by a loop and an approximately three-turn helix that is parallel to the ribbon (Brändén and Tooze, 1998). The cysteine and histidine ligands come from the first β strand, the turn between the strands, and two from the last turn of the helix. It is the 12-residue segment of loop and helix between the middle two ligands that normally interacts with the major groove of dsDNA. One of the NC structures is a complex with a tetraloop ssRNA from the Ψ-region recognition sequence (De Guzman *et al.*, 1998). As a single-stranded interaction, variations from the classic (CCHH) motif might be expected. Between the middle two zinc ligands there are only 4 residues instead of 12, and the mode of binding is completely different (Fig. 20). It is an amino-terminal 3_{10} helix before the first zinc finger that packs into the major groove of the RNA stem. The zinc fingers interact with the single-stranded region through specific base hydrogen bonds, and through packing of two exposed bases into hydrophobic pockets formed by zinc finger side chains.

ii. Envelope Proteins: Transmembrane and Surface Glycoprotein gp120. Both TM and SU proteins are derived from the Env precursor polyprotein gp160, and emerge from the ER and Golgi apparatus, posttranslationally modified and as a $(TM\text{–}SU)_3$ trimer, as in the mature virus. SU is responsible for binding to the primary cellular receptor (CD4 for HIV). TM acts both as an anchor for SU, and its N-terminal region is responsible for membrane fusion. There are alternative hypotheses for the fusion mechanism, which is not yet well understood, but one of these involves a "spring-loaded" conformational change, and is based on at least superficial sequence and structural homology with influenza hemagglutinin (HA), although there are important differences (Turner and Summers, 1999).

With TM, it is fragments of the external domain whose structures have been characterized by both NMR and crystallography (Caffrey *et al.*, 1997; Chan *et al.*, 1997; Tan *et al.*, 1997; Weissenhorn *et al.*, 1997). The transmembrane domain, and an internal region that interacts with MA, are of unknown structure. The ~30-residue N-terminal fusogenic region is also of unknown structure. Then there is a 49-residue α helix that forms a coiled coil in the trimer, a 30-residue loop that is more mobile, and then a

40-residue helix at the C terminus of this domain that packs on the outside of the coiled coil, antiparallel to the inner helices. Thus, there are many similarities with the structure of the influenza HA stem that suggest to many, in spite of important differences, that a conformational change may project the fusogenic region toward the membrane of a potential host cell.

The surface glycoprotein (SU, gp120 in HIV) is the protein responsible for recognizing the primary cellular receptor before cell entry. Unlike other proteins that share homology either within the retroviruses, or even with other families of viruses, the surface glycoproteins have sequence and structure that are dependent on the receptor, and differ between members of the retrovirus family.

HIV gp120 is a protein that likely undergoes several conformational changes: (1) on binding the primary cellular receptor (CD4 for HIV) to allow a second interaction with a chemokine receptor, that (2) leads to a change communicated to TM that triggers membrane fusion. The structure that has been determined (Kwong *et al.*, 1998) was modified extensively for crystallization by removing heterogeneity arising from glycosylation and variable conformation. Removed was 90% of the glycosylation, 52 N-terminal and 19 C-terminal residues, as well as loops of 64 and 29 residues at the presumptive distal side of the molecule, all done while preserving CD4 and antibody interactions (Binley *et al.*, 1998). The gp120 structure was solved in complex with parts of CD4 and an antibody that blocks interaction with the cytokine receptor.

The topological fold of HIV gp120 is complicated, and with little precedent in terms of domain homology with prior structures (Fig. 21; see Color Insert) (Kwong *et al.*, 1998). The molecule is heart shaped with three domains at each of the two lobes, and with a bridging domain at the "V," but there are multiple chain connections between these domains. The N terminus is at the top left of Fig. 21 (inner domain), and starts with a β strand and α-helix 1 running immediately down to the first two strands of the bridging sheet (Fig. 21, bottom left). The chain then returns to nearly complete the inner domain with $\beta 4$ and $\beta 8$ forming a ribbon between the entering and exiting chains. $\beta 5$ and $\beta 7$ form two of the strands of a sheet lying over and orthogonal to $\beta 1$, to which is hydrogen bonded a fragmentary $\beta 6$, part of the loop between $\beta 5$ and $\beta 7$. The final strand of the $\beta 5/7/25$ sheet is contributed by the C terminus, after it returns from the outer domain, and completing the five-stranded β sandwich. The outer domain is composed of two β barrels that share a common core and strands that would be common except for an intervening break from β structure. The proximal barrel (Fig. 21, top right) is six stranded, mixed direction, and contains both long and short strands, as well as helix 2. Its strands are near both the N- and C-terminal ends of the domain. The distal barrel (Fig. 21, bottom right) contains the first strand of the domain (β9) and the middle ones, forming a seven-stranded antiparallel barrel. Between the last two strands of this barrel (β19 and β22) is inserted an up–down strand pair of the sheet of the bridging domain. There were many interesting revelations from the structure, including an unglycosylated CD4-binding site that was remote from the secondary receptor-binding site, further implicating conformational changes in the function of the molecule (Kwong *et al.*, 1998).

The other surface glycoprotein of known structure is that of murine leukemia virus (Fass *et al.*, 1997b). The presumptive proximal region, formed by both N- and C-terminal portions, forms two antiparallel sheets that stack 40° to one another, to form a distorted (unclosed) barrel-like configuration. Topologically, this is a modest variation on the variable domain of an immunoglobulin fold. The loops between strands 3 and 4 and between strands 6 and 7 form a partially helical region that constitutes the distal domain of the structure. It is formed from sequence regions A and B, two of the three regions that differ among various type C murine leukemia viruses that have different cellular tropisms (Fass *et al.*, 1997b). The lack of similarity between moloney murine leukemia virus (Mo-MuLV) and HIV glycoproteins is consistent with the hypothesis that different proteins have been marshaled for viral interactions with different cellular receptors.

VII. Common Themes

Viruses are at once surprising in the diversity of approaches to satisfy the same end, and surprising in the range of viruses that share homologous proteins. Perhaps most surprising is the use of the same fold in a variety of contexts. The jelly-roll antiparallel β barrel seems ubiquitous, yet it is used in a variety of ways. Once thought to be a particularly well-shaped building block for capsid assembly, in influenzavirus hemagglutinin and bluetongue virus it is present without forming the major protein–protein contacts in the shell. The presence of the jelly roll in all these viruses is probably not due to a current unknown common function of this domain. Probably, it was inherited from ancestral viruses, but the primary function has long since diverged. When it indeed forms a protein shell by itself we find it used in a number of different orientations, tangentially as in the case of (+)ssRNA viruses or radially as in adenovirus. However, even within the tangential group, we find positions adjusted by rotations about the 5-fold, and differences in canting, with the barrels of picornaviruses canted up toward the 5-folds in picornaviruses, relative to the plant viruses. It appears that the jelly roll does not have any unique properties suitable for formation of icosahedral capsids, but rather that this simple fold has been easily adapted for new interactions in the evolution of viruses.

There are homologies between other viral proteins. Thus, although there are important differences, there are at least superficial similarities in the coiled-coil trimeric structure of the stem of influenza hemagglutinin and HIV TM, perhaps indicating a similar spring-loaded mechanism for bringing the viral and host fusing membranes together (Harrison, 2001; Turner and Summers, 1999). Just as with the icosahedral capsids, several nonhomologous folds can achieve the same function. Tick-borne encephalitis virus glycoprotein E mediates a pH-induced fusion mechanism, just like influenza hemagglutinin, but its structure is completely different (Rey *et al.*, 1995).

Perhaps the most spectacular homology seen to date is the similarity between the hexon protein of a mammalian virus, adenovirus, and the P3 coat protein of a bacteriophage PRD1, both containing two jelly-roll domains (Athappilly *et al.*, 1994; Benson *et al.*, 1999). This only goes to show that as our understanding of other viral proteins expands, so will the homologies that will likely become apparent.

VIII. Phylogenetic Relationships

Phylogeny has some, loose relation to the taxonomy that has framed our discussion. Virus phylogeny is particularly challenging, because of repeated exchange of genetic material between viruses and other viruses or their hosts, and because with such small genomes, there may be few opportunities to cross-check proposed phylogenies (Murphy and Kingsbury, 1996). Past natural bases for classification have included host type (pro-versus eukaryote, etc.), genome type (RNA versus DNA, single-stranded versus double, reverse-transcribing or not, etc.). It was also appreciated that viruses have been in existence far longer than many higher forms of life.

Structure has figured prominently in virus phylogeny. Structure is more conserved than sequence, and structural similarity can therefore suggest a relation that is impossible to find in the amino acid sequences of the proteins. Different capsid morphologies between genomically related helical TMV and icosahedral plant viruses had suggested that capsid structure might not be fundamental to virus evolution. Nevertheless, when it was found that animal viruses shared the same capsid fold as plant viruses (Hogle *et al.*, 1985; Rossmann *et al.*, 1985), it was then easy to extrapolate and assume that all small unenveloped icosahedral ssRNA viruses might share what was increasingly called the "viral fold." The structure of MS2 (Valegård *et al.*, 1990) showed the errors of that. Then, lest anyone assume that genome type was fundamental, the first ssDNA virus structures (Tsao *et al.*, 1991) showed that protein fold could be conserved more than genome type.

Several of the potential complexities of viral evolution were raised with the structure determination of Sindbis core protein (Choi *et al.*, 1991). Although other genes have similarity to their homologs in other positive-strand ssRNA viruses (Haseloff *et al.*, 1984), the core protein bears absolutely no homology to the jelly-roll structures of the capsid proteins of other (+)ssRNA viruses. Clearly, there has been some mix- and-matching of genomes. Intriguingly, the homology of SCP to picornaviral 3C and comoviral cysteine proteases is stronger than to other serine proteases (Choi *et al.*, 1991). [The cysteine of the viral proteases is thought to replace the serine nucleophile in a serine protease-like fold (Bazan and Fletterick, 1988).] Thus, it is possible that the *Alphavirus* core evolved by recruiting an existing viral gene product to serve a completely new purpose (Choi *et al.*, 1991).

The potential complications of inferring phylogeny from molecular data have also been emphasized by studies of Ebola virus. One protein, the fusion glycoprotein gp2, showed structural homology to viruses that were quite unrelated (Malashkevich *et al.*, 1999; Weissenhorn *et al.*, 1998), whereas another, the matrix protein, showed no homology even to its closest relatives (Dessen *et al.*, 2000).

With increasing sequence data, there are now several bases on which to construct phylogenies. However, there is not yet a satisfactory phylogenetic classification that encompasses all viral families. Three-dimensional structure will continue to play a pivotal role as new viral families are explored. Such comparative studies are not just of academic importance. Beginning to understand the mechanisms of viral evolution is an important part of understanding how new viruses and their associated diseases emerge (Condit, 2001).

Acknowledgments

M.S.C. was supported by the American Cancer Society (RPG-99-365-01-GMC) and National Institutes of Health (RO1-GM66875), and L.L. was supported by the Swedish Natural Science Research Council. M.S.C. wishes to thank Michael Rossmann, his postdoctoral mentor, for introducing him to virus structure, and several colleagues in the years since whose discussions have maintained this interest: Jack Johnson, Lee Makowski, and Don Caspar.

References

Abad-Zapatero, C., Abdel-Meguid, S. S., Johnson, J. E., Leslie, A. G. W., Rayment, I., and Rossmann, M. G. (1980). *Nature* **286,** 33–39.

Acharya, R., Fry, E., Stuart, D., Fox, G., Rowlands, D., and Brown, F. (1989). *Nature* **337,** 709–716.

Agbandje, M., McKenna, R., Rossmann, M. G., Strassheim, M. L., and Parrish, C. R. (1993). *Proteins* **16,** 155–171.

Agbandje-McKenna, M., Llamas-Saiz, A. L., Wang, F., Tattersall, P., and Rossmann, M. G. (1998). *Structure* **6,** 1369–1381.

Arnold, E., and Rossmann, M. G. (1986). *Proc. Natl. Acad. Sci. USA* **83,** 5489–5493.

Athappilly, F. K., Murali, R., Rux, J. J., Cai, Z., and Burnett, R. M. (1994). *J. Mol. Biol.* **242,** 430–455.

Bairoch, A., and Apweiler, R. (2000). *Nucleic Acids Res.* **28,** 45–48.

Baker, K. A., Dutch, R. E., Lamb, R. A., and Jardetzky, T. S. (1999). *Mol. Cell* **3,** 309–319.

Baker, T. S., and Johnson, J. E. (1996). *Curr. Opin. Struct. Biol.* **6,** 585–594.

Baker, T. S., Newcomb, W. W., Olson, N. H., Cowsert, L. M., Olson, C., and Brown, J. C. (1991). *Biophys. J.* **60,** 1445–1456.

Bamford, D. H., Gilbert, R. J., Grimes, J. M., and Stuart, D. I. (2001). *Curr. Opin. Struct. Biol.* **11,** 107–113.

Ban, N., and McPherson, A. (1995). *Nat. Struct. Biol.* **2,** 882–890.

Ban, N., Larson, S. B., and McPherson, A. (1995). *Virology* **214,** 571–583.

Basak, A. K., Gouet, P., Grimes, J., Roy, P., and Stuart, D. (1996). *J. Virol.* **70,** 3797–3806.

Battiste, J. L., Mao, H., Rao, N. S., Tan, R., Muhandiram, D. R., Kay, L. E., Frankel, A. D., and Williamson, J. R. (1996). *Science* **273,** 1547–1551.

Bazan, J. F., and Fletterick, R. J. (1988). *Proc. Natl. Acad. Sci. USA* **85,** 7872–7876.

Belnap, D. M., McDermott, B. M., Jr., Filman, D. J., Cheng, N., Trus, B. L., Zuccola, H. J., Racaniello, V. R., Hogle, J. M., and Steven, A. C. (2000). *Proc. Natl. Acad. Sci. USA* **97,** 73–78.

Bennett, M. J., Schlunegger, M. P., and Eisenberg, D. (1995). *Protein Sci.* **4,** 2455–2468.

Benson, S. D., Bamford, J. K., Bamford, D. H., and Burnett, R. M. (1999). *Cell* **98,** 825–833.

Berg, J. M. (1986). *Science* **232,** 485–487.

Bergmann, E. M., Mosimann, S. C., Chernaia, M. M., Malcolm, B. A., and James, M. N. (1997). *J. Virol.* **71,** 2436–2448.

Berman, H. M., Westbrook, J., Feng, Z., Gilliland, G., Bhat, T. N., Weissig, H., Shindyalov, I. N., and Bourne, P. E. (2000). *Nucleic Acids Res.* **28,** 235–242.

Berns, K. I. (1996). *In* "Virology" (B. N. Fields, D. M. Knipe, and P. M. Howley, Eds.), 3rd Ed., pp. 1017–1041. Raven, Philadelphia.

Bewley, M. C., Springer, K., Zhang, Y. B., Freimuth, P., and Flanagan, J. M. (1999). *Science* **286,** 1579–1583.

Bhyravbhatla, B., Watowich, S. J., and Caspar, D. L. (1998). *Biophys. J.* **74,** 604–615.

Binley, J. M., Wyatt, R., Desjardins, E., Kwong, P. D., Hendrickson, W., Moore, J. P., and Sodroski, J. (1998). *AIDS Res. Hum. Retroviruses.* **14,** 191–198.

Bjorkman, P. J., Saper, M. A., Samraoui, B., Bennett, W. S., Strominger, J. L., and Wiley, D. C. (1987). *Nature* **329,** 506–512.

Blanc, E., Giranda, V., Alexander, R. S., Pevear, D. C., Grorke, J., Gattis, J., and Chapman, M. S. (2002). *Structure* (in press).

Bloomer, A. C., Champness, J. N., Bricogne, G., Staden, R., and Klug, A. (1978). *Nature* **276,** 362–368.

Boege, U., Wengler, G., and Wittmann-Liebold, B. (1981). *Virology* **113,** 293–303.

Bone, R., Shenvi, A. B., Kettner, C. A., and Agard, D. A. (1987). *Biochemistry* **26,** 7609–7614.

Böttcher, B., Wynne, S. A., and Crowther, R. A. (1997). *Nature* **386,** 88–91.

Brändén, C.-I., and Tooze, J. (1998). "Introduction to Protein Structure." Garland, New York.

Bressanelli, S., Tomei, L., Roussel, A., Incitti, I., Vitale, R. L., Mathieu, M., De Francesco, R., and Rey, F. A. (1999). *Proc. Natl. Acad. Sci. USA* **96,** 13034–13039.

Bujacz, G., Jaskolski, M., Alexandratos, J., Wlodawer, A., Merkel, G., Katz, R. A., and Skalka, A. M. (1996). *Structure* **4,** 89–96.

Bullough, P. A., Hughson, F. M., Skehel, J. J., and Wiley, D. C. (1994). *Nature* **371,** 37–43.

Burnett, R. M. (1997). *In* "Structural Biology of Viruses," (W. Chiu, R. M. Burnett, and R. L. Garcia, Eds.), pp. 209–239. Oxford University Press, New York.

Caffrey, M., Cai, M., Kaufman, J., Stahl, S. J., Wingfield, P. T., Gronenborn, A. M., and Clore, G. M. (1997). *J. Mol. Biol.* **271,** 819–826.

Cai, M., Zheng, R., Caffrey, M., Craigie, R., Clore, G. M., and Gronenborn, A. M. (1997). *Nat. Struct. Biol.* **4,** 567–577.

Campos-Olivas, R., Newman, J. L., and Summers, M. F. (2000). *J. Mol. Biol.* **296,** 633–649.

Canady, M. A., Larson, S. B., Day, J., and McPherson, A. (1996). *Nat. Struct. Biol.* **3,** 771–781.

Carfi, A., Willis, S. H., Whitbeck, J. C., Krummenacher, C., Cohen, G. H., Eisenberg, R. J., and Wiley, D. C. (2001). *Mol. Cell.* **8,** 169–179.

Caspar, D. L. D., and Klug, A. (1962). *Cold Spring Harb. Symp. Quant. Biol.* **27,** 1–24.

Chan, D. C., Fass, D., Berger, J. M., and Kim, P. S. (1997). *Cell* **89,** 263–273.

Chandrasekar, V., and Johnson, J. E. (1998). *Structure* **6,** 157–171.

Chapman, M. S., and Rossmann, M. G. (1993a). *Virology* **195,** 745–765.

Chapman, M. S., and Rossmann, M. G. (1993b). *Virology* **194,** 491–508.

Chapman, M. S., Giranda, V. L., and Rossmann, M. G. (1990). *Semin. Virol.* **1,** 413–427.

Chapman, M. S., Blanc, E., Johnson, J. E., McKenna, R., Munshi, S., Rossmann, M. G., and Tsao, J. (1998). *In* "Direct Methods for Solving Macromolecular Structures" (S. Fortier, Ed.), pp. 433–442. Kluwer, Dortrecht, The Netherlands.

Chappell, J. D., Prota, A. E., Dermody, T. S., and Stehle, T. (2002). *EMBO J.* **21,** 1–11.

Chen, L., Gorman, J. J., McKimm-Breschkin, J., Lawrence, L. J., Tulloch, P. A., Smith, B. J., Colman, P. M., and Lawrence, M. C. (2001). *Structure (Camb)* **9,** 255–266.

Chen, X. S., Garcea, R. L., Goldberg, I., Casini, G., and Harrison, S. C. (2000). *Mol. Cell* **5,** 557–567.

Chen, Z., Stauffacher, C., Li, Y., Schmidt, T., Bomu, W., Kamer, G., Shanks, M., Lomonsoff, G., and Johnson, J. E. (1989). *Science* **245,** 154–159.

Choi, H. K., Tong, L., Minor, W., Dumas, P. A. B. U., Rossmann, M. G., and Wengler, G. (1991). *Nature* **354,** 37–43.

Choi, H. K., Lee, S., Zhang, Y. P., McKinney, B. R., Wengler, G., Rossmann, M. G., and Kuhn, R. J. (1996). *J. Mol. Biol.* **262,** 151–167.

Choi, H. K., Lu, G., Lee, S., Wengler, G., and Rossmann, M. G. (1997). *Proteins* **27,** 345–359.

Christensen, A. M., Massiah, M. A., Turner, B. G., Sundquist, W. I., and Summers, M. F. (1996). *J. Mol. Biol.* **264,** 1117–1131.

Cielens, I., Ose, V., Petrovskis, I., Strelnikova, A., Renhofa, R., Kozlovska, T., and Pumpens, P. (2000). *FEBS Lett.* **482,** 261–264.
Colman, P. M., Laver, W. G., and Varghese, J. N. (1983). *Nature* **303,** 41–47.
Condit, R. C. (2001). *In* "Virology" (B. N. Fields, D. M. Knipe, and P. M. Howley, Eds.), 4th Ed., pp. 19–52. Lippincott Williams & Wilkins, Philadelphia.
Conway, J. F., Cheng, N., Zlotnick, A., Wingfield, P. T., Stahl, S. J., and Steven, A. C. (1997). *Nature* **386,** 91–94.
Coombs, K., and Brown, D. T. (1987a). *J. Mol. Biol.* **195,** 359–371.
Coombs, K., and Brown, D. T. (1987b). *Virus Res.* **7,** 131–149.
Crennell, S., Takimoto, T., Portner, A., and Taylor, G. (2000). *Nat. Struct. Biol.* **7,** 1068–1074.
Crowther, R. A., and Amos, L. A. (1971). *J. Mol. Biol.* **60,** 123–130.
De Guzman, R. N., Wu, Z. R., Stalling, C. C., Pappalardo, L., Borer, P. N., and Summers, M. F. (1998). *Science* **279,** 384–388.
Dessen, A., Volchkov, V., Dolnik, O., Klenk, H. D., and Weissenhorn, W. (2000). *EMBO J.* **19,** 4228–4236.
Dokland, T., McKenna, R., Ilag, L. L., Bowman, B. R., Incardona, N. L., Fane, B. A. A., and Rossmann, M. G. (1997). *Nature* **389,** 308–313.
Dokland, T., Bernal, R. A., Burch, A., Pletnev, S., Fane, B. A., and Rossmann, M. G. (1999). *J. Mol. Biol.* **288,** 595–608.
Dormitzer, P. R., Sun, Z. Y., Wagner, G., and Harrison, S. C. (2002). *EMBO J.* **21,** 885–897.
Durmort, C., Stehlin, C., Schoehn, G., Mitraki, A., Drouet, E., Cusack, S., and Burmeister, W. P. (2001). *Virology* **285,** 302–312.
Dyda, F., Hickman, A. B., Jenkins, T. M., Engelman, A., Craigie, R., and Davies, D. R. (1994). *Science* **266,** 1981–1986.
Enzmann, P. J., and Weiland, F. (1979). *Virology* **95,** 501–510.
Fass, D., Harrison, S. C., and Kim, P. S. (1996). *Nat. Struct. Biol.* **3,** 465–469.
Fass, D., Davey, R. A., Hamson, C. A., Kim, P. S., Cunningham, J. M., and Berger, J. M. (1997a). *Science* **277,** 1662–1666.
Fass, D., Davey, R. A., Hamson, C. A., Kim, P. S., Cunningham, J. M., and Berger, J. M. (1997b). *Science* **277,** 1662–1666.
Filman, D. J., Syed, R., Chow, M., Macadam, A. J., Minor, P. D., and Hogle, J. M. (1989). *EMBO J.* **8,** 1567–1579.
Filman, D. J., Wien, M. W., Cunningham, J. A., Bergelson, J. M., and Hogle, J. M. (1998). *Acta Crystallogr. D Biol. Crystallogr.* **54,** 1261–1272.
Fisher, A. J., and Johnson, J. E. (1993). *Nature* **361,** 176–179.
Fox, G., Parry, N., Barnett, P. V., McGinn, B., Rowlands, D. J., and Brown, F. (1989). *J. Gen. Virol.* **70,** 625–637.
Fry, E., Logan, D., Acharya, R., Fox, G., Rowlands, D., Brown, F., and Stuart, D. (1990). *Semin. Virol.* **1,** 439–451.
Fujinaga, M., Delbaere, L. T., Brayer, G. D., and James, M. N. (1985). *J. Mol. Biol.* **184,** 479–502.
Fuller, S. D. (1987). *Cell* **48,** 923–934.
Fuller, S. D., and Argos, P. (1987). *EMBO J.* **6,** 1099–1105.
Gamble, T. R., Vajdos, F. F., Yoo, S., Worthylake, D. K., Houseweart, M., Sundquist, W. I., and Hill, C. P. (1996). *Cell* **87,** 1285–1294.
Gamble, T. R., Yoo, S., Vajdos, F. F., von Schwedler, U. K., Worthylake, D. K., Wang, H., McCutcheon, J. P., Sundquist, W. I., and Hill, C. P. (1997). *Science* **278,** 849–853.

Garcia, R. L., and Liddington, R. C. (1997). *In* "Structural Biology of Viruses" (W. Chiu, R. M. Burnett, and R. L. Garcia, Eds.), pp. 187–208. Oxford University Press, New York.

Georgiadis, M. M., Jessen, S. M., Ogata, C. M., Telesnitsky, A., Goff, S. P., and Hendrickson, W. A. (1995). *Structure* **3,** 879–892.

Gitti, R. K., Lee, B. M., Walker, J., Summers, M. F., Yoo, S., and Sundquist, W. I. (1996). *Science* **273,** 231–235.

Golmohammadi, R., Valegard, K., Fridborg, K., and Liljas, L. (1993). *J. Mol. Biol.* **234,** 620–639.

Golmohammadi, R., Fridborg, K., Bundule, M., Valegard, K., and Liljas, L. (1996). *Structure* **4,** 543–554.

Gopinath, K., Sundareshan, S., Bhuvaneswari, M., Karande, A., Murthy, M. R., Nayudu, M. V., and Savithri, H. S. (1994). *Indian J. Biochem. Biophys.* **31,** 322–328.

Grant, R. A., Filman, D. J., Fujinami, R. S., Icenogle, J. P., and Hogle, J. M. (1992). *Proc. Natl. Acad. Sci. USA* **89,** 2061–2065.

Grimes, J., Basak, A. K., Roy, P., and Stuart, D. (1995). *Nature* **373,** 167–170.

Grimes, J. M., Burroughs, J. N., Gouet, P., Diprose, J. M., Malby, R., Zientara, S., Mertens, P. P., and Stuart, D. I. (1998). *Nature* **395,** 470–478.

Grzesiek, S., Bax, A., Clore, G. M., Gronenborn, A. M., Hu, J. S., Kaufman, J., Palmer, I., Stahl, S. J., and Wingfield, P. T. (1996). *Nat. Struct. Biol.* **3,** 340–345.

Guarne, A., Tormo, J., Kirchweger, R., Pfistermueller, D., Fita, I., and Skern, T. (1998). *EMBO J.* **17,** 7469–7479.

Guasch, A., Pous, J., Ibarra, B., Gomis-Ruth, F. X., Valpuesta, J. M., Sousa, N., Carrascosa, J. L., and Coll, M. (2002). *J. Mol. Biol.* **315,** 663–676.

Ha, Y., Stevens, D. J., Skehel, J. J., and Wiley, D. C. (2002). *EMBO J.* **21,** 865–875.

Hadfield, A. T., Lee, W. M., Zhao, R., Oliveira, M. A., Minor, I., Rueckert, R. R., and Rossmann, M. G. (1997). *Structure* **5,** 427–441.

Hahn, C. S., and Strauss, J. H. (1990). *J. Virol.* **64,** 3069–3073.

Hahn, C. S., Strauss, E. G., and Strauss, J. H. (1985). *Proc. Natl. Acad. Sci. USA* **82,** 4648–4652.

Han, X., Bushweller, J. H., Cafiso, D. S., and Tamm, L. K. (2001). *Nat. Struct. Biol.* **8,** 715–720.

Harrison, S. C. (1990). *In* "Virology" (B. N. Fields and D. M. Knipe, Eds.), pp. 37–61. Raven Press, New York.

Harrison, S. C. (2001). *In* "Virology" (B. N. Fields, D. M. Knipe, and P. M. Howley, Eds.), 4th Ed. pp. 53–86. Lippincott Williams & Wilkins, Philadelphia.

Harrison, S. C., Olson, A., Schutt, C. E., Winkler, F. K., and Bricogne, G. (1978). *Nature* **276,** 368–373.

Harrison, S. C., Skehel, J. J., and Wiley, D. C. (1996). *In* "Virology" (B. N. Fields, D. M. Knipe, and P. M. Howley, Eds.), 3rd Ed., pp. 59–99. Raven Press, New York.

Haseloff, J., Goelet, P., Zimmern, D., Ahlquist, P., Dasgupta, R., and Kaesberg, P. (1984). *Proc. Natl. Acad. Sci. USA* **81,** 4358–4362.

Hatanaka, H., Iourin, O., Rao, Z., Fry, E., Kingsman, A., and Stuart, D. I. (2002). *J. Virol.* **76,** 1876–1883.

He, Y., Bowman, V. D., Mueller, S., Bator, C. M., Bella, J., Peng, X., Baker, T. S., Wimmer, E., Kuhn, R. J., and Rossmann, M. G. (2000). *Proc. Natl. Acad. Sci. USA* **97,** 79–84.

Hendry, E., Hatanaka, H., Fry, E., Smyth, M., Tate, J., Stanway, G., Santti, J., Maaronen, M., Hyypia, T., and Stuart, D. (1999). *Structure Fold Des.* **7,** 1527–1538.

Hickman, A., Ronning, D., Kotin, R., and Dyda, F. (2002). *Mol. Cell* **10,** 327.

Hill, C. P., Worthylake, D., Bancroft, D. P., Christensen, A. M., and Sundquist, W. I. (1996). *Proc. Natl. Acad. Sci. USA* **93,** 3099–3104.

Hiremath, C. N., Grant, R. A., Filman, D. J., and Hogle, J. M. (1995). *Acta Crystallogr. D Biol. Crystallogr.* **51,** 473–489.

Hogle, J. M., Chow, M., and Filman, D. J. (1985). *Science* **229,** 1358–1365.

Hogle, J. M., Maeda, A., and Harrison, S. C. (1986). *J. Mol. Biol.* **191,** 625–638.

Holsinger, L. J., and Lamb, R. A. (1991). *Virology* **183,** 32–43.

Hoog, S. S., Smith, W. W., Qiu, X., Janson, C. A., Hellmig, B., McQueney, M. S., O'Donnell, K., O'Shannessy, D., DiLella, A. G., Debouck, C., and Abdel-Meguid, S. S. (1997). *Biochemistry* **36,** 14023–14029.

Horzinek, M., and Mussgay, M. (1969). *J. Virol.* **4,** 514–520.

Hosur, M. V., Schmidt, T., Tucker, R. C., Johnson, J. E., Gallagher, T. M., Selling, B. H., and Rueckert, R. R. (1987). *Proteins* **2,** 167–176.

Jaskolski, M., Miller, M., Rao, J. K., Leis, J., and Wlodawer, A. (1990). *Biochemistry* **29,** 5889–5898.

Jin, Z., Jin, L., Peterson, D. L., and Lawson, C. L. (1999). *J. Mol. Biol.* **286,** 83–93.

Johnson, J. E., and Reddy, V. S. (1998). *Nat. Struct. Biol.* **5,** 849–854.

Jones, T. A., and Liljas, L. (1984). *Acta Crystallogr.* **177,** 737–767.

Kanamaru, S., Leiman, P. G., Kostyuchenko, V. A., Chipman, P. R., Mesyanzhinov, V. V., Arisaka, F., and Rossmann, M. G. (2002). *Nature* **415,** 553–557.

Kanellopoulos, P. N., Tsernoglou, D., van der Vliet, P. C., and Tucker, P. A. (1996). *J. Mol. Biol.* **257,** 1–8.

Khorasanizadeh, S., Campos-Olivas, R., and Summers, M. F. (1999). *J. Mol. Biol.* **291,** 491–505.

Kim, S., Smith, T. J., Chapman, M. S., Rossmann, M. G., Pevear, D. C., Dutko, F. J., Felock, P. J., Diana, G. D., and McKinlay, M. A. (1989). *J. Mol. Biol.* **210,** 91–111.

Kobe, B., Center, R. J., Kemp, B. E., and Poumbourios, P. (1999). *Proc. Natl. Acad. Sci. USA* **96,** 4319–4324.

Kohlstaedt, L. A., Wang, J., Friedman, J. M., Rice, P. A., and Steitz, T. A. (1992). *Science* **256,** 1783–1790.

Kostyuchenko, V. A., Navruzbekov, G. A., Kurochkina, L. P., Strelkov, S. V., Mesyanzhinov, V. V., and Rossmann, M. G. (1999). *Structure Fold Des.* **7,** 1213–1222.

Kovacs, F. A., Denny, J. K., Song, Z., Quine, J. R., and Cross, T. A. (2000). *J. Mol. Biol.* **295,** 117–125.

Kraulis, P. (1991). *J. Appl. Crystallogr.* **24,** 946–950.

Krishna, S. S., Hiremath, C. N., Munshi, S. K., Prahadeeswaran, D., Sastri, M., Savithri, H. S., and Murthy, M. R. (1999). *J. Mol. Biol.* **289,** 919–934.

Kukol, A., Adams, P. D., Rice, L. M., Brunger, A. T., and Arkin, T. I. (1999). *J. Mol. Biol.* **286,** 951–962.

Kumar, A., Reddy, V. S., Yusibov, V., Chipman, P. R., Hata, Y., Fita, I., Fukuyama, K., Rossmann, M. G., Loesch-Fries, L. S., Baker, T. S., and Johnson, J. E. (1997). *J. Virol.* **71,** 7911–7916.

Kwong, P. D., Wyatt, R., Robinson, J., Sweet, R. W., Sodroski, J., and Hendrickson, W. A. (1998). *Nature* **393,** 648–659.

Larson, S. B., Koszelak, S., Day, J., Greenwood, A., Dodds, J. A., and McPherson, A. (1993a). *Nature* **361,** 179–182.

Larson, S. B., Koszelak, S., Day, J., Greenwood, A., Dodds, J. A., and McPherson, A. (1993b). *J. Mol. Biol.* **231,** 375–391.

Larson, S. B., Day, J., Canady, M. A., Greenwood, A., and McPherson, A. (2000). *J. Mol. Biol.* **301,** 625–642.

Lea, S. Hernandez, J., Blakemore, W., Brocchi, E., Curry, S. Domingo, E., Fry, E. Abu-Ghazaleh, R., King, A., Newman, J. *et al.*(1994). *Structure* **2,** 123–139.

Lee, C. H., Saksela, K., Mirza, U. A., Chait, B. T., and Kuriyan, J. (1996). *Cell* **85,** 931–942.

Leiman, P. G., Kostyuchenko, V. A., Shneider, M. M., Kurochkina, L. P., Mesyanzhinov, V. V., and Rossmann, M. G. (2000). *J. Mol. Biol.* **301,** 975–985.

Lentz, K. N., Smith, A. D., Geisler, S. C., Cox, S., Buontempo, P., Skelton, A., DeMartino, J., Rozhon, E., Schwartz, J., Girijavallabhan, V., O'Connel, J., and Arnold, E. (1997). *Structure* **5,** 961–978.

Lesburg, C. A., Cable, M. B., Ferrari, E., Hong, Z., Mannarino, A. F., and Weber, P. C. (1999). *Nat. Struct. Biol.* **6,** 937–943.

Liddington, R. C., Yan, Y., Moulai, J., Sahli, R., Benjamin, T. L., and Harrison, S. C. (1991). *Nature* **354,** 278–284.

Liemann, S., Chandran, K., Baker, T. S., Nibert, M. L., and Harrison, S. C. (2002). *Cell* **108,** 283–295.

Liljas, L., Unge, T., Jones, T. A., Fridborg, K., Lovgren, S., Skoglund, U., and Strandberg, B. (1982). *J. Mol. Biol.* **159,** 93–108.

Liljas, L., Fridborg, K., Valegard, K., Bundule, M., and Pumpens, P. (1994). *J. Mol. Biol.* **244,** 279–290.

Lin, T., Chen, Z., Usha, R., Stauffacher, C. V., Dai, J. B., Schmidt, T., and Johnson, J. E. (1999). *Virology* **265,** 20–34.

Lin, T., Clark, A. J., Chen, Z., Shanks, M., Dai, J. B., Li, Y., Schmidt, T., Oxelfelt, P., Lomonossoff, G. P., and Johnson, J. E. (2000). *J. Virol.* **74,** 493–504.

Liu, Y., Gong, W., Huang, C. C., Herr, W., and Cheng, X. (1999). *Genes Dev.* **13,** 1692–1703.

Love, R. A., Parge, H. E., Wickersham, J. A., Hostomsky, Z., Habuka, N., Moomaw, E. W., Adachi, T., and Hostomska, Z. (1996). *Cell* **87,** 331–342.

Lucas, R. W., Larson, S. B., and McPherson, A. (2002). *J. Mol. Biol.* **317,** 95–108.

Luo, M., Vriend, G., Kamer, G., Minor, I., Arnold, E., Rossmann, M. G., Boege, U., Scraba, D. G., Duke, G. M., and Palmenberg, A. C. (1987). *Science* **235,** 182–191.

Luo, M., He, C., Toth, K. S., Zhang, C. X., and Lipton, H. L. (1992). *Proc. Natl. Acad. Sci. USA* **89,** 2409–2413.

Makowski, L. (1991). *Acta Crystallogr. A* **47,** 562–567.

Makowski, L., and Russel, M. (1997). *In* "Structural Biology of Viruses" (W. Chiu, R. M. Burnett, and R. L. Garcia, Eds.), pp. 352–380. Oxford University Press, New York.

Malashkevich, V. N., Schneider, B. J., McNally, M. L., Milhollen, M. A., Pang, J. X., and Kim, P. S. (1999). *Proc. Natl. Acad. Sci. USA* **96,** 2662–2667.

Malashkevich, V. N., Singh, M., and Kim, P. S. (2001). *Proc. Natl. Acad. Sci. USA* **98,** 8502–8506.

Marvin, D. A., Hale, R. D., Nave, C., and Helmer-Citterich, M. (1994). *J. Mol. Biol.* **235,** 260–286.

Massiah, M. A., Starich, M. R., Paschall, C., Summers, M. F., Christensen, A. M., and Sundquist, W. I. (1994). *J. Mol. Biol.* **244,** 198–223.

Massiah, M. A., Worthylake, D., Christensen, A. M., Sundquist, W. I., Hill, C. P., and Summers, M. F. (1996). *Protein Sci.* **5,** 2391–2398.

Mathieu, M., Petitpas, I., Navaza, J., Lepault, J., Kohli, E., Pothier, P., Prasad, B. V., Cohen, J., and Rey, F. A. (2001). *EMBO J.* **20,** 1485–1497.

Matthews, B. W., Sigler, P. B., Henderson, R., and Blow, D. M. (1967). *Nature* **214,** 652–656.

Matthews, D. A., Smith, W. W., Ferre, R. A., Condon, B., Budahazi, G., Slasson, W., Villafranca, J. E., Janson, C. A., McElroy, H. E., Gribskov, C. L., and Worland, S. (1994). *Cell* **77,** 761–771.

Matthews, S., Barlow, P., Boyd, J., Barton, G., Russell, R., Mills, H., Cunningham, M., Meyers, N., Burns, N., Kingsman, S., Kingsman, A., and Campbell, I. (1994). *Nature* **370,** 666–668.

Maxwell, K. L., Yee, A. A., Booth, V., Arrowsmith, C. H., Gold, M., and Davidson, A. R. (2001). *J. Mol. Biol.* **308,** 9–14.

Maxwell, K. L., Yee, A. A., Arrowsmith, C. H., Gold, M., and Davidson, A. R. (2002). *J. Mol. Biol* **318,** 1395–1404.

McDonnell, P. A., Shon, K., Kim, Y., and Opella, S. J. (1993). *J. Mol. Biol.* **233,** 447–463.

McKenna, R., Xia, D., Willingham, P., Ilag, L. L., Krishnaswarmy, S., Rossmann, M. G., Olson, N. H., Baker, T. S., and Incardona, N. (1992a). *Nature* **355,** 137–143.

McKenna, R., Xia, D., Willingmann, P., Ilag, L., and Rossmann, M. G. (1992b). *Acta Crystallogr. B* **48,** 499–511.

McKenna, R., Bowman, B. R., Ilag, L. L., Rossmann, M. G., and Fane, B. A. (1996). *J. Mol. Biol.* **256,** 736–750.

Melancon, P., and Garoff, H. (1987). *J. Virol.* **61,** 1301–1309.

Miller, M., Jaskolski, M., Rao, J. K., Leis, J., and Wlodawer, A. (1989). *Nature* **337,** 576–579.

Minor, P. D., Ferguson, M., Evans, D. M. A., Almond, J. W., and Icenogle, J. P. (1986). *J. Gen. Virol.* **67,** 1283–1291.

Momany, C., Kovari, L. C., Prongay, A. J., Keller, W., Gitti, R. K., Lee, B. M., Gorbalenya, A. E., Tong, L., McClure, J., Ehrlich, L. S., Summers, M. F., Carter, C., and Rossmann, M. G. (1996). *Nat. Struct. Biol.* **3,** 763–770.

Morellet, N., Jullian, N., De Rocquingy, H., Maigret, B., Darlix, J.-L., and Roques, B. P. (1992). *EMBO J.* **11,** 3059–3065.

Morgunova, E., Dauter, Z., Fry, E., Stuart, D. I., Stel'mashchuk, V., Mikhailov, A. M., Wilson, K. S., and Vainshtein, B. K. (1994). *FEBS Lett.* **338,** 267–271.

Muckelbauer, J. K., Kremer, M., Minor, I., Diana, G., Dutko, F. J., Groarke, J., Pevera, D. C., and Rossmann, M. G. (1995). *Structure* **3,** 653–667.

Muckelbauer, J. K., Kremer, M., Minor, I., Tong, L., Zlotnick, A., Johnson, J. E., and Rossmann, M. G. (1996). *Acta Crystallogr. D Biol. Crystallogr.* **51,** 871–887.

Mullen, M. M., Haan, K. M., Longnecker, R., and Jardetzky, T. S. (2002). *Mol. Cell* **9,** 375–385.

Munshi, S., Liljas, L., Cavarelli, J., Bomu, W., McKinney, B., Reddy, V., and Johnson, J. E. (1996). *J. Mol. Biol.* **261,** 1–10.

Murphy, F. A., and Kingsbury, D. W. (1996). *In* "Virology" (B. N. Fields and D. M. Knipe *et al.*, Eds.), 3rd Ed., pp. 15–57. Raven Press, New York.

Namba, K., and Stubbs, G. (1986). *Science* **231,** 1401–1406.

Namba, K., Pattanayek, R., and Stubbs, G. (1989). *J. Mol. Biol.* **208,** 307–325.

Nambudripad, R., Stark, W., Opella, S. J., and Makowski, L. (1991a). *Science* **252,** 1305–1308.

Nambudripad, R., Stark, W., Opella, S. J., and Makowski, L. (1991b). *J. Mol. Biol.* **220,** 359–379.

Ni, C. Z., White, C. A., Mitchell, R. S., Wickersham, J., Kodandapani, R., Peabody, D. S., and Ely, K. R. (1996). *Protein Sci.* **5,** 2485–2493.

Oda, Y., Saeki, K., Takahashi, Y., Maeda, T., Naitow, H., Tsukihara, T., and Fukuyama, K. (2000). *J. Mol. Biol.* **300,** 153–169.

Oliveira, M. A., Zhao, R., Lee, W. M., Kremer, M., Minor, I., Diana, G. D., Pevear, D. C., Dutko, F. J., McKinlay, M. A., and Rossmann, M. G. (1993). *Structure* **1,** 51–68.

Olland, A. M., Jane-Valbuena, J., Schiff, L. A., Nibert, M. L., and Harrison, S. C. (2001). *EMBO J.* **20,** 979–989.

Olson, N., Kolatkar, P., Oliveira, M. A., Cheng, R. H., Greve, J. M., McClelland, A., Baker, T. S., and Rossmann, M. G. (1993). *Proc. Natl. Acad. Sci. USA* **90,** 507–511.

Omichinski, J. G., Clore, G. M., Sakaguchi, K., Appella, E., and Gronenborn, A. M. (1991). *FEBS Lett.* **292,** 25–30.

Palmenberg, A. C. (1989). *In* "Molecular Aspects of Picornavirus Infection and Detection" (B. L. Semler and E. Ehrenfeld, Eds.), pp. 211–241. American Society for Microbiology, Washington, D.C.

Prasad, B. V., Hardy, M. E., Dokland, T., Bella, J., Rossmann, M. G., and Estes, M. K. (1999). *Science* **286,** 287–290.

Qu, C., Liljas, L., Opalka, N., Brugidou, C., Yeager, M., Beachy, R. N., Fauquet, C. M., Johnson, J. E., and Lin, T. (2000). *Structure Fold Des.* **8,** 1095–1103.

Rayment, I., Baker, T. S., Caspar, D. L. D., and Murakami, W. T. (1982). *Nature* **295,** 110–115.

Reddy, V. S. (1999). The Scripps Research Institute NIH research resource: Multiscale Modeling Tools for Structural Biology (MMTSB), RR12255.

Reinisch, K. M., Nibert, M. L., and Harrison, S. C. (2000). *Nature* **404,** 960–967.

Rey, F. A., Heinz, F. X., Mandl, C., Kunz, C., and Harrison, S. C. (1995). *Nature* **375,** 291–298.

Roberts, M. M., White, J. L., Grutter, M. G., and Burnett, R. M. (1986). *Science* **232,** 1148–1151.

Rossmann, M. G. (1987). *Bioessays* **7,** 99–103.

Rossmann, M. G. (1995). *Curr. Opin. Struct. Biol.* **5,** 650–655.

Rossmann, M. G., and Johnson, J. E. (1989). *Annu. Rev. Biochem.* **58,** 533–573.

Rossmann, M. G., and Palmenberg, A. C. (1988). *Virology* **164,** 373–382.

Rossmann, M. G., Arnold, E., Erickson, J. W., Frankenberger, E. A., Griffith, J. P., Hecht, H., Johnson, J. E., Kamer, G., Luo, M., Mosser, A., Rueckert, R., Sherry, B., and Vriend, G. (1985). *Nature* **317,** 145–153.

Rowlands, D., Logan, D., Abu-Ghazaleh, R., Blakemore, W., Curry, S. Jackson, T., King, A., Lea, S., Lewis, R., Newman, J., *et al.* (1994). *Arch. Virol. Suppl.* **9,** 51–58.

Rux, J. J., and Burnett, R. M. (2000). *Mol. Ther.* **1,** 18–30.

Savva, R., McAuley-Hecht, K., Brown, T., and Pearl, L. (1995). *Nature* **373,** 487–493.

Schuler, W., Dong, C., Wecker, K., and Roques, B. P. (1999). *Biochemistry* **38,** 12984–12994.

Sha, B., and Luo, M. (1997). *Nat. Struct. Biol.* **4,** 239–244.

Sheriff, S., Hendrickson, W. A., and Smith, J. L. (1987). *J. Mol. Biol.* **197,** 273–296.

Sherry, B., Mosser, A. G., Colonno, R. J., and Rueckert, R. R. (1986). *J. Virol.* **57,** 246–257.

Simpson, A. A., Chipman, P. R., Baker, T. S., Tijssen, P., and Rossmann, M. G. (1998). *Structure* **6,** 1355–1367.

Simpson, A. A., Tao, Y., Leiman, P. G., Badasso, M. O., He, Y., Jardine, P. J., Olson, N. H., Morais, M. C., Grimes, S., Anderson, D. L., Baker, T. S., and Rossmann, M. G. (2000). *Nature* **408,** 745–750.

Simpson, A. A., Hebert, B., Sullivan, G. M., Parrish, C. R., Zadori, Z., Tijssen, P., and Rossmann, M. G. (2002). *J. Mol. Biol.* **315,** 1189–1198.

Skehel, J. J., Bayley, P. M., Brown, E. B., *et al.* (1982). *Proc. Natl. Acad. Sci. USA* **79,** 968–972.

Skinner, M. M., Zhang, H., Leschnitzer, D. H., Guan, Y., Bellamy, H., Sweet, R. M., Gray, C. W., Konings, R. N., Wang, A. H., and Terwilliger, T. C. (1994). *Proc. Natl. Acad. Sci. USA* **91,** 2071–2075.
Smith, T. J., Kremer, M. J., Luo, M., Vriend, G., Arnold, E., Kamer, G., Rossmann, M., McKinlay, M., Diana, G., and Otto, M. J. (1986). *Science* **233,** 1286–1293.
Smith, T. J., Chase, E., Schmidt, T., and Perry, K. L. (2000). *J. Virol.* **74,** 7578–7586.
Smyth, M., Hall, J., Fry, E., Stuart, D., Stanway, G., and Hyypia, T. (1993). *J. Mol. Biol.* **230,** 667–669.
Speir, J., Munshi, S., Wang, G., Baker, T., and Johnson, J. (1995). *Structure* **3,** 63–77.
Stehle, T., and Harrison, S. C. (1996). *Structure* **4,** 183–194.
Stehle, T., Yan, Y., and Harrison, S. C. (1994). *Nature* **369,** 160–163.
Stehle, T., Gamblin, S. J., Yan, Y., and Harrison, S. C. (1996). *Structure* **4,** 165–182.
Steinbacher, S., Seckler, R., Miller, S., Steipe, B., Huber, R., and Reinemer, P. (1994). *Science* **265,** 383–386.
Steinbacher, S., Miller, S., Baxa, U., Budisa, N., Weintraub, A., Seckler, R., and Huber, R. (1997). *J. Mol. Biol.* **267,** 865–880.
Stewart, P. L., Burnett, R. M., Cyrklaff, M., and Fuller, S. D. (1991). *Cell* **67,** 145–154.
Strauss, E. G., and Strauss, J. H. (1986). *In* "The Togaviridae and Flaviviridae" (S. Schlesinger and M. J. Schlesinger, Eds.), pp. 35–90. Plenum, New York.
Stuart, A. D., McKee, T. A., Williams, P. A., Harley, C., Shen, S., Stuart, D. I., Brown, T. D., and Lea, S. M. (2002). *J. Virol.* **76,** 7694–7704.
Sugrue, R. J., and Hay, A. J. (1991). *Virology* **180,** 617–624.
Summers, M. F., South, T. L., Kim, B., and Hare, D. R. (1990). *Biochemistry* **29,** 329–340.
Summers, M. F., Henderson, L. E., Chance, M. R., Bess, J. W. J., South, T. L., Blake, P. R., Sagi, I., Perez-Alvarado, G., Sowder, R. C. I., Hare, D. R., and Arthur, L. O. (1992). *Protein Sci.* **1,** 563–574.
Tan, K., Liu, J., Wang, J., Shen, S., and Lu, M. (1997). *Proc. Natl. Acad. Sci. USA* **94,** 12303–12308.
Tang, L., Johnson, K. N., Ball, L. A., Lin, T., Yeager, M., and Johnson, J. E. (2001). *Nat. Struct. Biol.* **8,** 77–83.
Tao, Y., Strelkov, S. V., Mesyanzhinov, V. V., and Rossmann, M. G. (1997). *Structure* **5,** 789–798.
Tars, K., Bundule, M., Fridborg, K., and Liljas, L. (1997). *J. Mol. Biol.* **271,** 759–773.
Tars, K., Fridborg, K., Bundule, M., and Liljas, L. (2000). *Acta Crystallogr. D Biol. Crystallogr.* **56,** 398–405.
Tate, J., Liljas, L., Scotti, P., Christian, P., Lin, T., and Johnson, J. E. (1999). *Nat. Struct. Biol.* **6,** 765–774.
Trus, B. L., Roden, R. B. S., Greenstone, H. L., Vrhel, M., Schiller, J. T., and Booy, F. P. (1997). *Nature Struct. Biol.* **4,** 413–420.
Tsang, S. K., Danthi, P., Chow, M., and Hogle, J. M. (2000). *J. Mol. Biol.* **296,** 335–340.
Tsao, J., Chapman, M. S., Agbandje, M., Keller, W., Smith, K., Wu, H., Luo, M., Smith, T. J., Rossmann, M. G., Compans, R. W., and Parrish, C. (1991). *Science* **251,** 1456–1464.
Turner, B. G., and Summers, M. F. (1999). *J. Mol. Biol.* **285,** 1–32.
Valegård, K., Liljas, L., Fridborg, K., and Unge, T. (1990). *Nature* **344,** 36–41.
van Raaij, M. J., Louis, N., Chroboczek, J., and Cusack, S. (1999a). *Virology* **262,** 333–343.
van Raaij, M. J., Mitraki, A., Lavigne, G., and Cusack, S. (1999b). *Nature* **401,** 935–938.
van Regenmortel, M. H. V., Fauquet, C. M., Bishop, D. H. L., Carstens, E. B., Estes, M. K., Lemon, S. M., Maniloff, J., Mayo, M. A., McGeoch, D. J., Pringle, C. R., and Wickner, R. B. (2000). "Virus Taxonomy: The Classification and Nomenclature of

Viruses. The Seventh Report of the International Committee on Taxonomy of Viruses.'' Academic Press, San Diego, CA.
Varghese, J. N., Laver, W. G., and Colman, P. M. (1983). *Nature* **303,** 35–40.
Verdaguer, N., Blaas, D., and Fita, I. (2000). *J. Mol. Biol.* **300,** 1179–1194.
Vogt, J., Perozzo, R., Pautsch, A., Prota, A., Schelling, P., Pilger, B., Folkers, G., Scapozza, L., and Schulz, G. E. (2000). *Proteins* **41,** 545–553.
Wang, H., and Stubbs, G. (1994). *J. Mol. Biol.* **239,** 371–384.
Wang, H., Culver, J. N., and Stubbs, G. (1997a). *J. Mol. Biol.* **269,** 769–779.
Wang, H., Culver, J. N., and Stubbs, G. (1997b). *J. Mol. Biol.* **269,** 769–779.
Weis, W., Brown, J. H., Cusak, S., Paulson, J. C., Skehel, J. J., and Wiley, D. C. (1988). *Nature* **333,** 426–431.
Weissenhorn, W., Dessen, A., Harrison, S. C., Skehel, J. J., and Wiley, D. C. (1997). *Nature* **387,** 426–430.
Weissenhorn, W., Carfi, A., Lee, K. H., Skehel, J. J., and Wiley, D. C. (1998). *Mol. Cell* **2,** 605–616.
Weldon, R. A., and Hunter, E. (1997). *In* ''Structural Biology of Viruses,'' (W. Chiu, R. M. Burnett, and R. L. Garcia, Eds.), pp. 381–410. Oxford University Press, New York.
Wery, J.-P., Reddy, V., Hosur, M. V., and Johnson, J. E. (1994). *J. Mol. Biol.* **235,** 565–586.
Wikoff, W. R., Liljas, L., Duda, R. L., Tsuruta, H., Hendrix, R. W., and Johnson, J. E. (2000). *Science* **289,** 2129–2133.
Wilson, I. A., Skehel, J. J., and Wiley, D. C. (1981). *Nature* **289,** 373–378.
Wu, H., and Rossmann, M. G. (1993). *J. Mol. Biol.* **233,** 231–244.
Wynne, S. A., Crowther, R. A., and Leslie, A. G. (1999). *Mol. Cell* **3,** 771–780.
Xia, D., Henry, L. J., Gerard, R. D., and Deisenhofer, J. (1994). *Structure* **2,** 1259–1270.
Xie, Q., and Chapman, M. S. (1996). *J. Mol. Biol.* **264,** 497–520.
Xie, Q., Bu, W., Bhatia, S., Hare, J., Somasundaram, T., Azzi, A., and Chapman, M. S. (2002). *Proc. Natl. Acad. Sci. USA* **99,** 10405–10410.
Yang, F., Forrer, P., Dauter, Z., Conway, J. F., Cheng, N., Cerritelli, M. E., Steven, A. C., Pluckthun, A., and Wlodawer, A. (2000). *Nat. Struct. Biol.* **7,** 230–237.
Zdanov, A., Schalk-Hihi, C., Menon, S., Moore, K. W., and Wlodawer, A. (1997). *J. Mol. Biol.* **268,** 460–467.
Zhao, R., Pevear, D. C., Kremer, M. J., Giranda, V. L., Kofron, J. A., Kuhn, R. J., and Rossmann, M. G. (1996). *Structure* **4,** 1205–1220.
Zlotnick, A., Natarajan, P., Munshi, S., and Johnson, J. (1997). *Acta Crystallogr.* **53,** 738–746.
Zuccola, H. J., Filman, D. J., Coen, D. M., and Hogle, J. M. (2000). *Mol. Cell* **5,** 267–278.

VIRUS PARTICLE DYNAMICS

By JOHN E. JOHNSON

Department of Molecular Biology, Scripps Research Institute, La Jolla, California 92037

I. Introduction 197
II. Particle Fluctuations and Infectivity 199
A. Antigenic Epitopes and VP4 of Polio Virus Fluctuate from Inside to Outside the Particle 199
B. Study of Particle Fluctuations by Proteolysis and Mass Spectrometry 200
C. Effect of the Viral Genome on Particle Dynamics 202
D. Plant Virus Capsids: Less Dynamic Than Animal Virus Capsids 203
III. Large-Scale Reversible Quaternary Structure Changes in Viruses 203
A. Reversible Swelling in T=3 Plant Viruses 203
B. Biological Role of Large-Scale Changes of Morphology in Plant Viruses 205
C. Large-Scale Quaternary Structure Changes in Single-Stranded RNA Animal Viruses 205
IV. Large-Scale Irreversible Quaternary Structure Changes in Double-Stranded DNA Bacteriophage 209
A. Introduction to Tailed Double-Stranded DNA Phage and HK97 209
B. Structure of the HK97 Subunit 212
C. Quaternary Structure of the HK97 Particle 213
D. Prohead-to-Head Transition of HK97 213
V. Conclusions 216
References 216

I. Introduction

Biological molecules are characterized by dynamic behavior that is not obvious from studies with crystallography or electron microscopy. Gregorio Weber made this point in a review article in 1975 with the comment; "Indeed the protein molecule resulting from the X-ray crystallographic observations is a "platonic" protein well removed in its perfection from the kicking and screaming "stochastic" molecule that we infer must exist in solution." Not surprisingly, it is now clear that assemblies of subunits as they are deployed in virions are also dynamic. Many of us were, however, seduced by the appearance of refined molecular models into the assumption that virions are as stable as a billiard ball. The myth is propagated by electron cryomicroscopy reconstructions that portray the virus as almost stonelike in appearance (Fig. 1). Today, however, there are data from many sources indicating that an animal virus particle in solution is also a "kicking and screaming stochastic collection of protein subunits" that apparently varies widely

ADVANCES IN PROTEIN CHEMISTRY, Vol. 64

0065-3233/03 $35.00

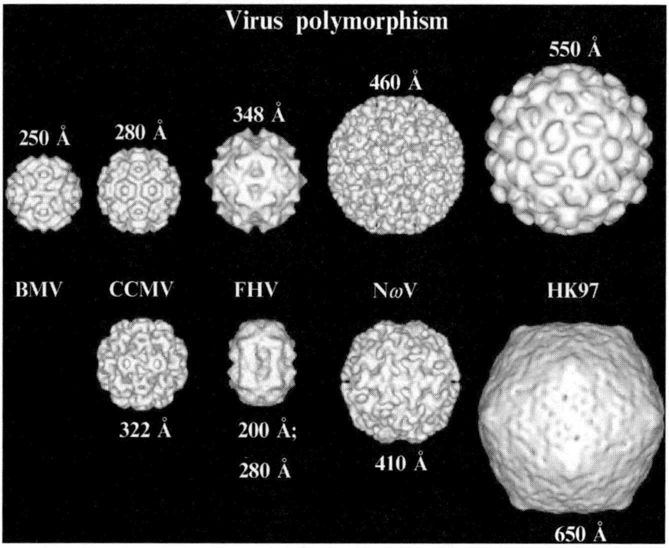

FIG. 1. Particle reconstructions based on electron cryomicroscopy images. The particles are representative of the different assembly forms that different viral subunits can take given various expression systems, mutations, and physical conditions. *From the left:* First—a 120-subunit particle of brome mosaic virus (BMV) formed when the subunits are expressed in a yeast system (Krol *et al.*, 1999). Normally BMV forms T=3 particles with a morphology similar to CCMV. Second—two forms of cowpea chlorotic mottle virus (CCMV) observed under different physical conditions. *Top:* The native form of the virus at pH 6 in the presence of divalent metal ions. *Bottom:* A metal-free form at pH 7 (Speir *et al.*, 1995). Third—two forms of flock house virus (FHV). Native FHV (*top*) (Cheng *et al.*, 1994) and a particle obtained when FHV subunits, with residues 1–31 genetically removed (*bottom*), were expressed in a baculovirus system. (Dong *et al.*, 1998). Fourth—the procapsid form (*top*) and capsid form (*bottom*) of *Nudaurelia capensis* ω virus (NωV). Both forms were isolated from a baculovirus assembly system, the procapsid at pH 7.6 and the capsid at pH 5.0. The particles isolated at pH 7.6 can be converted to the pH 5.0 form by lowering the pH. At pH 5.0 an autocatalytic cleavage occurs that prevents the particles from expanding to the pH 7.6 form even at high pH (Canady *et al.*, 2000). Fifth—the procapsid form (*top*) and capsid form (*bottom*) of subunits of the dsDNA bacteriophage HK97 expressed in *E. coli.* The expansion of procapsid to capsid is normally triggered by the packaging of dsDNA, but with the expressed particles it can be induced by lowering the pH to pH 4.0 (Conway *et al.*, 1995).

about the equilibrium structure that is frozen in space and time by the crystal lattice. It is fortunate that crystalline viruses tend to conform to a minimal energy structure in the lattice that quenches the dynamic properties of the particle. This allows observation of portions of the particle that are now known to fluctuate significantly. Extreme examples demonstrate that portions of the subunit polypeptide that are clearly inside the virions in the crystal structure are exposed to the outer surface intermittently when the virus is free in solution. This review concentrates first on the fluctuating dynamic properties of viruses and then discusses the large-scale dynamics often associated with particle maturation. The dynamics of viral proteins associated with membrane fusion are discussed in the article by Fass (this volume).

II. Particle Fluctuations and Infectivity

A. *Antigenic Epitopes and VP4 of Polio Virus Fluctuate from Inside to Outside the Particle*

The first data suggesting that there were fluctuating dynamic aspects of protein subunits associated with an animal virus emerged in 1985. Chow, Baltimore, and coworkers mapped the antigenic properties of the VP1 subunit of poliovirus using polypeptide fragments that spanned the sequence of the viral subunit (Chow *et al.*, 1985). Antibodies raised against these peptides were tested for their ability to neutralize poliovirus infection and those that did were deemed to be antibodies to antigenic epitopes of the virus. When the structure of poliovirus was determined most of these peptide sequences were found to be on the surface of the particle as anticipated (Hogle *et al.*, 1985). One sequence, however, was near the N terminus of the VP1 subunit and this was clearly internal in the X-ray structure. Subsequently a group investigating neutralizing epitopes of poliovirus discovered the same region and demonstrated that in time more antibodies derived from the polypeptide in this region would bind to the particle, suggesting that it was intermittently exposed and could be trapped on the outside of the particle by the antibodies (Roivainen *et al.*, 1991, 1993). In addition, it was demonstrated that these fluctuations were required for infectivity of poliovirus and that the exposure of an amphipathic helical region near the N terminus of VP1 allowed the virus to bind to liposomes and that this was an essential step in the infection process (Fricks and Hogle, 1990). It was also known that the VP4 polypeptide, which lies on the inside surface of the capsid in the X-ray structure, was released early in infection when the particle was still intact, supporting the role of particle dynamics in virus infection.

B. *Study of Particle Fluctuations by Proteolysis and Mass Spectrometry*

Although these results were widely known in the mid 1990s, it was still surprising when a more quantitative method was reported for examining particle fluctuations. Bothner *et al.* had set out to study the susceptibility of the flock house viral (FHV) surface to proteases with the anticipation that antigenic sites would be more readily cleaved by a protease because of their demonstrated greater mobility in X-ray structures (Bothner *et al.*, 1998). The experimental protocol was straightforward, with virus particles exposed to a protease in a time course of 15-min intervals followed by an analysis of the products by matrix-assisted laser desorption/ionization mass spectrometry (MALDI–MS). Given the known sequence of the capsid protein, the specificity of the protease, and the precise mass of the products, the locations of cleavage sites were unambiguously determined (Fig. 2). The previous published results not withstanding, the authors of the article were incredulous when they discovered that the most susceptible regions of the insect flock house virus (FHV) to protease were the N and C termini, both inferred to be internal in the X-ray structure (Fisher and Johnson, 1993; Wery *et al.*, 1994). Following a variety of control experiments that proved the observations were representative of the general population of particles and not the result of proteolysis of a few broken particles the results were reported (Bothner *et al.*, 1998).

The work on FHV stimulated definitive studies of human rhinovirus 14 (HRV14) that demonstrated fluctuations that exposed VP4 and the N terminus of VP1 (both internal in the X-ray structure) in particles incubated under normal buffer conditions (Lewis *et al.*, 1998). The results were consistent with the earlier studies of poliovirus but were quantitated by mass spectrometry. The authors extended the study and removed all questions of possible artifacts by performing a second experiment in which the virus was first incubated with a drug (WIN 52084) known to inhibit virus infection by stabilizing the capsid and preventing uncoating of the RNA. Preincubation with the drug totally inhibited the exposure of VP4 and the N terminus of VP1. Indeed, particles with drug bound were able to incubate indefinitely in the protease without any proteolysis. The result was significant at a practical level because it demonstrated a potential method to screen for antiviral compounds to animal viruses. The experiments described require little material and can be fully automated to search for inhibition of capsid dynamics.

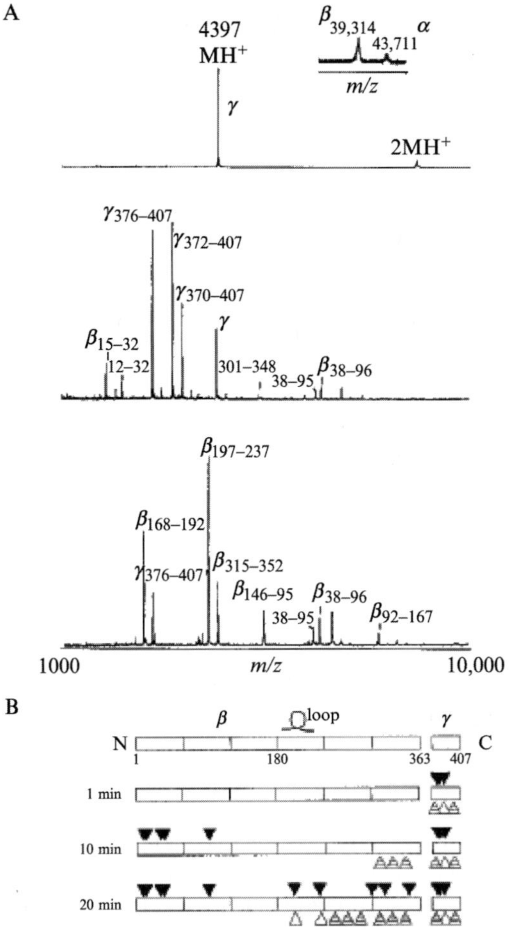

FIG. 2. (A) MALDI–MS data generated from the trypsin digest of FHV. *Top:* Native FHV. The capsid protein undergoes an autocatalyzed cleavage event in most of its subunits during maturation. The precursor α protein and products, β protein and γ peptide, were detected. *Middle:* Fifteen minutes after the addition of trypsin the γ peptide and three fragments were observed along with proteolytic fragments from β protein. *Bottom:* After 24 h of exposure the γ peptide exists only as the uncleavable fragment 376–407. The ion of highest intensity contains the loop region of the β protein present on the viral capsid surface. All digests were performed with FHV at 1.0 mg/ml and at 25°C. (B) Proteolytic cleavage sites mapped to the FHV capsid protein. The kinetics of the proteolysis reaction are demonstrated in this time course experiment. The γ peptide and the N and C terminis of β protein are domains that are localized internally. Cleavages localized to the capsid surface (loop domain) are not initially present. The cleavage sites of trypsin, Lys-C, and Glu-C, are represented by solid, open, and striped arrowheads, respectively (Bothner *et al.*, 1998).

C. *Effect of the Viral Genome on Particle Dynamics*

The same procedure proved to be a sensitive measure for the role of RNA in virus particle dynamics. Assembly studies of FHV utilized a baculovirus system to express the capsid protein gene of the virus. Expression of this gene in SF21 cells leads to the spontaneous assembly of the subunits to form virus-like particles (Schneemann *et al.*, 1993). These particles have the same capsid structure as wild-type virus (Fisher *et al.*, 1993), but lack the viral genome because only the gene for the coat protein is present. The subunits package a variety of cellular RNA molecules to neutralize the charge of the basic amino termini of the subunits. The VLPs were crystallized and the structure was determined to 2.5-Å resolution. The crystals and the structure of the VLPs were isomorphous with the authentic virus to the resolution of the data. Every atom in the final refined model of the VLP was within experimental error of being located in the same position as atoms in the authentic virus. As an extension of the crystallographic study the two types of particles were analyzed with the proteolysis protocol described above. Remarkably, in spite of their virtually identical structure, the termini of the VLPs were digested seven times faster than authentic virus particles, indicating a much higher level of dynamic fluctuations in the VLPs (Bothner *et al.*, 1999). The result has striking significance regarding the evolution of the viral RNA. The assembly of FHV and of the nodaviruses studied by crystallography depends on their RNA genome in regulating capsid assembly (Fig. 3; see Color Insert). The RNA interacts at subunit interfaces to control the subunit dihedral angle contacts and therefore the formation of the shell. It is clear that the subunits can recruit cellular RNA to perform this function in the VLPs and that an authentic appearing capsid results. These capsids, however, are much more dynamic than the virions, suggesting that viral RNA evolution not only optimizes the genetic fitness of the virus, but also the chemical stability of the particle.

Particle dynamics and fluctuations appear to be a "built-in" feature of the animal virus particles studied to date. Inhibition of the dynamic behavior of a number of genera of picornaviruses through the binding of small molecules into a hydrophobic pocket within the VP1 subunit prevents infection by inhibiting uncoating. Computational and experimental studies indicate that these drugs increase the compressibility of the viral capsid, providing an entropically driven stabilization of the particle (Phelps and Post, 1995). A remarkable body of research has been performed on picornavirus infectivity and its dependence on drug binding in this pocket and a rich array of drug-resistant and drug-dependent mutants have been characterized over the years (McKinlay, 1993). Drug-resistant mutants generally blocked binding of the molecules into the pocket by altering residues adjacent to the pocket to large side-chain amino acids (Heinz *et al.*, 1989). Drug-dependent mutants were apparently destabilized to such an extent that they were dynamic and could uncoat even in the presence of the drug, but, in the absence of the drug, they were so unstable that they spontaneously lost the VP4 polypeptide at 37°C, rendering them non-infectious (Mosser and Rueckert, 1993). Mutations causing drug dependence are not localized to specific sites and apparently generally destabilize the particle.

D. Plant Virus Capsids: Less Dynamic Than Animal Virus Capsids

There is no evidence that plant viruses enter their hosts through a receptor-mediated pathway. Generally a chewing insect functions as the vector and places the virus in the damaged cell to initiate infection. Plant viruses or their genomes then move through the plant systemically, through the intercellular connections call desmodesmota. Because the dynamic capsid is not required to assist in cell entry, plant viruses studied in their native form do not display the type of fluctuations described in the animal virus capsid. Comoviruses are a group of plant viruses in the picornavirus supergroup and they have a capsid strikingly similar to the picornavirus capsids in animal viruses (Lomonossoff and Johnson, 1991). These particles, however, will not bind the drugs that bind to picorna animal viruses and the reason is obvious when the structure of cowpea mosaic virus, the type member, is examined (Lin *et al.*, 1999). The entire hydrophobic pocket of the analogous VP1 domain in the plant virus is filled with large side chains, effectively creating a natural stabilization of the particle by occupying the site with its own amino acids. The effect is to make the particle less dynamic as demonstrated by the lack of proteolysis with trypsin when experiments were performed under conditions used to study the dynamics of FHV and HRV14.

III. Large-Scale Reversible Quaternary Structure Changes in Viruses

A. Reversible Swelling in T=3 Plant Viruses

Large-scale fluctuations of an animal virus particle are important for its ability to infect cells, but another type of particle dynamic behavior has been known for decades and was first identified in a plant virus. Brome mosaic virus (BMV), a single-stranded RNA (ssRNA) virus with T=3 icosahedral symmetry, was found to swell from a diameter of 260 to 300 Å when the particles were treated with EDTA and exposed to neutral pH (Incardona and Kaesberg, 1964). The swelling was reversible, with the particle converting to the compact form in the presence of divalent metal ions or pH values of 5 or less. This behavior was studied in detail by Bancroft and co-workers (1969) with another member of the brome mosaic virus group, cowpea chlorotic mottle virus (CCMV). High-resolution titrations of CCMV demonstrated that the swelling occurred as the result of deprotonation of acidic residues postulated to bind divalent metal ions when they were present (Durham *et al.*, 1977; Pfeiffer and Durham, 1977). These so-called carboxyl clusters would have altered pK_a values because of their close proximity to one another and were protonated at pH 5 to 6 instead of at their normal value of pH 4.5. Swelling resulted from the electrostatic repulsion of the negatively charged carboxyl groups at pH 7. Interaction with nucleic acid stops the particles from disintegrating. If the salt concentration is raised, the particles do disassemble. The swelling phenomena were found in the sobemovirus group and the tombusvirus group in addition to the brome mosaic virus group. The first plant virus structures determined, tomato bushy stunt virus (TBSV) (Harrison *et al.*, 1978) and southern bean mosaic virus (SBMV) (Abad-Zapatero *et al.*, 1980), revealed the divalent metal ion-binding sites at subunit interfaces exactly as predicted by Bancroft and others. A low-resolution structure of the swollen form of TBSV was eventually determined and the nature of the swelling was seen for the first time (Robinson and Harrison, 1982). The virus particles expand by opening at the quasi-3-fold symmetry axes due to the charge repulsion. There was evidence that interactions at the 2-fold symmetry axes retained much of the character as in the compact form, but the quasi-2-fold axes changed significantly by adding the so-called β-A strand to the eight-stranded β-sandwich making the 2-fold and quasi-2-fold symmetry axes more similar in the swollen form whereas they are distinctly different in the compact form.

The structure of CCMV revealed a similar cluster of carboxyl groups, also at subunit interfaces, although the organization of the CCMV capsid differs significantly from that of TBSV. An image reconstruction of swollen CCMV with electron cryomicroscopy data revealed that the particles also opened at the quasi-3-fold symmetry axes due to charge repulsion. Indeed, it was possible to use the model of the subunits from the 3.2-Å X-ray structure to rationalize the density of the 25-Å reconstruction (Speir *et al.*, 1995). The CCMV capsid has the shape of a truncated icosahedron, with icosahedral and quasi-2-fold contacts appearing similar to each other. This similarity is maintained in the swollen form although both dimer contacts are significantly different from those in the compact form. It is striking how the hexamer and pentamer interactions are retained during the expansion, with the majority of the alterations occurring at the quasi-3-fold and dimer contacts (Fig. 4; see Color Insert).

B. *Biological Role of Large-Scale Changes of Morphology in Plant Viruses*

There has been considerable debate through the years on the biological role of particle swelling in plant viruses and there is still no clear consensus. The first biological studies suggested that swelling was required for the particles to uncoat and the hypothesis was advanced that the low Ca^{2+} concentration inside the cell leads to particle swelling and cotranslational disassembly (Durham *et al.*, 1977). It was shown that a mutant of CCMV that does not swell is fully infectious, casting some question on the role of swelling in the native virus (Albert *et al.*, 1997). A new hypothesis for the role of metal ion binding was put forward in studies of rice yellow mottle virus (RYMV), a sobemovirus, in which the authors indicate that metal binding may contribute to the pathology of infection by leaching Ca^{2+} from pit membranes, permitting virus to migrate through them (Opalka *et al.*, 1998). Such a mechanism emphasizes the role of chelation at the metal-binding sites and not the importance of a structural change, although both may be important. Like histological studies of other plant virus infections, the study of RYMV infection shows that virus assembly and particle formation occur in crystalline arrays and that assembly probably takes place in vacuoles. If these vacuoles were at pH 5.5 or lower, like many intracellular compartments, the assembled particles would exist in the compact form even in the absence of divalent metal ions. The swollen form may still be an important transient intermediate that facilitates particle assembly as described below for animal viruses. When the membranes of the vacuoles break because of the mechanical force of the growing crystal, the pH is raised, giving the particles a high affinity for divalent metal ions. The virus then becomes a chelating agent, weakening the pit membranes by removing structurally important calcium. Given the evidence leading to different roles for chelation and particle expansion, it is still not clear whether there is a single role for this phenomenon in all the plant viruses where it has been found.

C. *Large-Scale Quaternary Structure Changes in Single-Stranded RNA Animal Viruses*

Swelling that is dependent on divalent metal ions and pH is a common phenomenon in T=3 plant viruses, but it has not been observed in the picorna-like plant viruses or T=3 RNA animal viruses. The T=3 nodaviruses

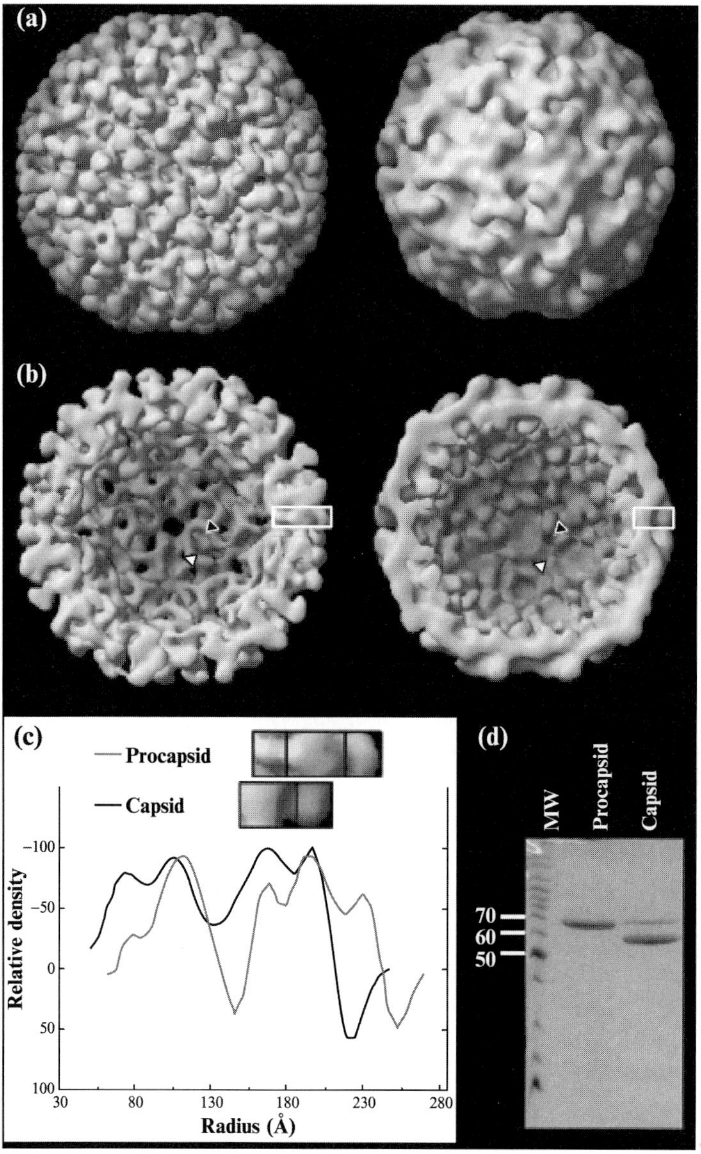

FIG. 5. Three-dimensional, surface-shaded (a) full particle and (b) sectioned views of the procapsid (*left*) and capsid (*right*) of NωV VLPs viewed down a 2-fold symmetry axis, with 3-fold (solid arrowhead) and quasi-3-fold (open arrowhead) axes marked in (b). The procapsid is larger, rounded, and porous, while the mature capsid has a smaller, angular, solid shell. (c) The radial density plots reveal that the capsid protein

bind divalent metal ions with carboxyl clusters at locations in the quaternary structure similar to those in SBMV and TBSV plant viruses, but removal of divalent ions and elevated pH do not change the particle size. This behavior is paradoxical in that the nodaviruses undergo the polypeptide fluctuations described above, but do not undergo large-scale size changes when they are perturbed as described. Picornaviruses undergo large-scale reorganization on binding receptors and this has been documented in studies related to cell entry.

The first ssRNA animal virus capsid characterized in two dramatically different forms was a T=4 insect virus, *Nudaurelia capensis* ω virus (NωV). Authentic virus was characterized only in a compact, 410-Å-diameter form, but expression of the capsid protein gene in a baculovirus system generates a particle that can exist at pH 7.0 with a 450-Å diameter and as a compact form at pH 5.0 (Fig. 5) that is indistinguishable from authentic particles (Canady *et al.*, 2000). Solution X-ray scattering was used to define a titration curve (Fig. 6) for the transition, which is sharply defined between pH values of 6.5 and 5.5 (Canady *et al.*, 2001). The biological role of this transition is not definitive. It has been observed only in an artificial expression/assembly system and it must be determined whether it is just an interesting artifact or of consequence in the virus life cycle. As discussed below, the complex DNA bacteriophages go through well-established assembly intermediates in a coordinated manner because of the complexity of the final capsids. NωV and other tetraviruses are the only known examples of nonenveloped T=4 ssRNA viruses. Because these particles require subunits to exist in four structurally unique environments within the icosahedral asymmetric unit, it may be necessary for them to initially assemble in a procapsid form. A working hypothesis involves assembly in a vacuole, as discussed for RYMV above. In such an environment the procapsid would be a transient intermediate with the low-pH form the end product. The spontaneous transition does not occur in the expression/assembly baculovirus system because the vacuoles do not exist without the other viral gene products expressed. Assembly occurs in the cytoplasm in the expression system near neutral pH and if particles are purified at this pH, assembly is arrested at the procapsid stage. If

shell (black line) spanned 62 Å with two domains, whereas the procapsid (gray line) had a thickness of 83 Å and comprised three domains. A cross-section of each map is shown above the plot, the domains delineated and placed in register with the radii to which they correspond. (d) An SDS–polyacrylamide gel showing that the procapsid sample contains the 70-kDa uncleaved coat protein, while the capsid contains mostly 62- and 8-kDa (not visible) coat protein fragments, which result from autoproteolysis on lowering of the pH to pH 5.0.

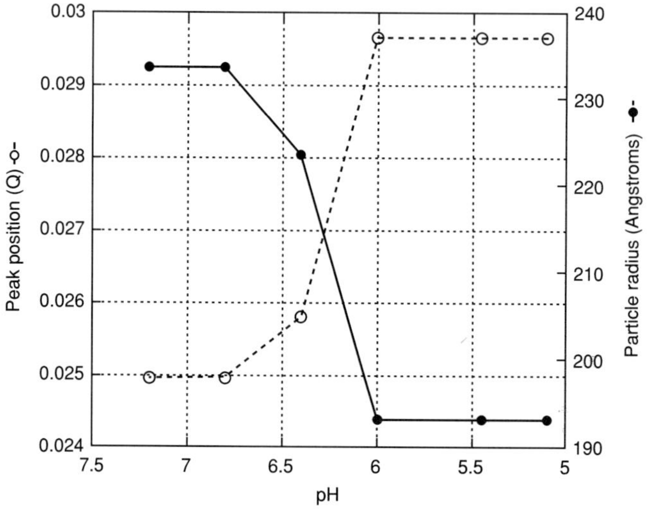

FIG. 6. A plot of the characteristic peak positions (dashed line) in a solution X-ray scattering experiment ($Q = 4\pi \sin \theta/\lambda$, where θ is half the X-ray scattering angle and λ is 1.38 Å, the wavelength of the X-rays) used to monitor the change in NωV particle size as a function of pH. The spherically averaged particle radius is shown by the solid line. The plot illustrates the highly cooperative pH dependence of the transition (Canady *et al.*, 2001).

this hypothesis is correct, it may be expected that the pH of wild-type virus could be raised, thus generating the expanded putative procapsid. This does not occur because the low-pH form of the virus undergoes an autoproteolysis in which cleavage (Fig. 5) occurs at residue 571 (Agrawal and Johnson, 1995). The cleaved polypeptide (residues 572–644) remains associated with the particle (Munshi *et al.*, 1996), but the cleavage "locks" the capsid state and it cannot be reversed (Canady *et al.*, 2001). The autoproteolysis occurs in particles produced in the expression/assembly system when they are lowered to pH 5 and in authentic virus. It was shown that the transition is reversible if the cleavage is inhibited by mutating Asn-570 to glutamine, although there is an obvious hysteresis when the forward and reverse conversions are compared (Taylor *et al.*, 2002).

An argument for the role of the intermediate in assembly emerges by comparing the details of the procapsid and capsid structures. The procapsid displays a high degree of quasi-equivalence in the environments of the individual subunits. In particular, contacts in different categories of dimer interactions are strikingly similar (Fig. 7; see Color Insert). This suggests that an initial assembly product may generate the greatest degree of equivalence between subunits so, as they assemble, there is only minimal need for molecular switching to categorize the four subunits into the proper T=4 surface lattice. This particle, while displaying near equivalence between quasi-equivalent subunits, is not as stable as the capsid. Following assembly of the procapsid, the quaternary structure is programmed to differentiate quasi-equivalent dimer contacts, leading to nonequivalence between dimer interactions and strong trimer interactions that lead to greater stability. The "seed" for this transition may lie on the interior surface of the capsid, where the only distinct trimer contacts are visible in the procapsid density (Canady *et al.*, 2000). Figure 5 shows that the trimer density is clearly present at icosahedral and quasi-3-fold symmetry axes in the procapsid and that the trefoils of density are not visible in the capsid. The structural flexibility of this system, as a function of pH, makes it likely that a continuum of cryo-EM structures will be determined that will allow a detailed mapping of subunit interfaces of different intermediates in the maturation.

IV. Large-Scale Irreversible Quaternary Structure Changes in Double-Stranded DNA Bacteriophage

Many double-stranded DNA (dsDNA) viruses undergo a maturation that involves large-scale reorganization of the capsid proteins. In most cases the particle transformation increases the virion diameter by 20% or more and changes its shape from round to polyhedral. Initially it was assumed that the transformation was limited to dsDNA bacteriophage, but it is now clear that members of the *Herpesviridae* and *Adenoviridae* undergo similar maturation steps. This section discusses in detail the structural features of the transition in HK97 and it is believed that these observation are relevant for other phages and may be a model for *Herpesviridae*, although it is clear that the details of the proteins involved in the capsid are quite different between the phage and animal viruses.

A. *Introduction to Tailed Double-Stranded DNA Phage and HK97*

The tailed, dsDNA bacteriophages are among the most studied dynamic systems in biology. Their assembly is a marvel of genetic, biochemical, and mechanical control (Hendrix and Duda, 1998). Particle morphogenesis is characterized by the formation of a metastable procapsid that, DNA packaging, forms a mature capsid. Typically, the particle diameter changes from roughly 450 to 650 Å during this irreversible transition while the protein composition remains the same. Atomic resolution detail has become available for the capsid form of a virus in this group and its structure, combined with static and time-resolved cryo-EM studies of the procapsid and capsid, has led to an understanding of the remarkable transitions in the particle development. A more extended description of the assembly and maturation of HK97 is provided here, as it is currently the best structurally characterized system that undergoes large-scale particle dynamics. The capsid atomic model for the λ-like bacteriophage HK97 was determined by crystallography (Wikoff *et al.*, 2000) at 3.5 Å. A 12-Å resolution cryo-EM reconstruction of the procapsid was convincingly interpreted in terms of the capsid subunit, permitting a detailed description of the protein reorganization during the maturation process (Conway *et al.*, 2001).

HK97 is a temperate, λ-like coliophage with a 40-kbp dsDNA genome. Near one end of the genome are three contiguous genes comprising less than 4 kbp that encode the proteins found in the naturally occurring capsid (Duda *et al.*, 1995b). Two are present in whole or in part throughout morphogenesis; the third is a transient component found only in the initial assembly product (Duda *et al.*, 1995a). Figure 8 illustrates the components and maturation of natural and expressed capsids. The initially assembled particle has T=7 quasi-symmetry containing 415 identical proteins (gene *5*) with one pentamer replaced by 12 portal proteins (gene *3*) and roughly 50 copies of a putative protease (gene *4*) packaged within the particle (Conway *et al.*, 1995). On assembly the protease immediately digests the first 103 amino acids from the head protein and then digests itself, with all the polypeptide fragments exiting the capsid and leaving a particle composed of a 31-kDa head protein and a portal, now competent to package DNA. The 40-kbp DNA genome is next inserted into the head through the portal and this triggers the remarkable reorganization of the particle quaternary structure and a partial refolding of the subunits. The overall shape of the particle changes from a corrugated, round shell with protuberances at the pentamer and hexamer axes to an icosahedrally shaped particle with thin walls and smooth, flat faces connecting the 5-fold axes (Conway *et al.*, 1995). Immediately after expansion, but at a slower rate, an autocatalytic cross-linking of the subunits occurs in which the side chains of Lys-169 and Asn-356 are joined with the release of NH_4^+ (Fig. 8). This reaction chemically joins subunits to each other, forming hexamers and pentamers, and also physically concatenates hexamers and pentamers to each other, suggesting a "chain-mail" association of proteins that resembles a chain-linked fence. Either following or simultaneously with the cross-linking, the tail assembly is added to the portal to complete the formation of the mature particle.

A breakthrough in the study of HK97 was achieved when an *Escherichia coli* expression system was developed in which gene *4* and gene *5* products alone were shown to assemble and mature in a manner identical with the authentic virus (Xie and Hendrix, 1995). Expansion was triggered by

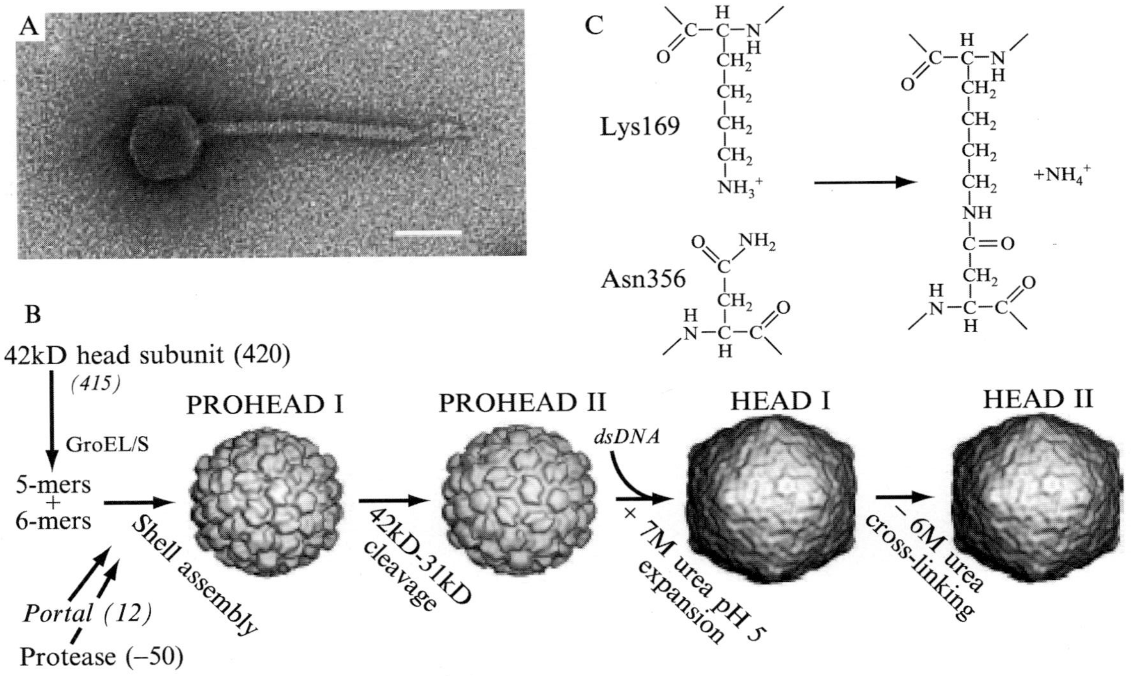

FIG. 8. HK97 assembly and maturation. (A) Negatively stained electron micrograph of a mature dsDNA-filled capsid, with noncontractile tail and accessory proteins. (B) Steps in capsid assembly and maturation (see text; *in vitro* conditions are in boldface, and *in vivo* conditions or components that differ from the *in vitro* conditions are in italic) (Conway *et al.*, 1995). (C) Chemistry of the cross-linking reaction.

treating the prohead II particles with any of a variety of denaturants (Duda *et al.*, 1995a). Each of the particles shown in Fig. 8 was isolated in milligram quantities from the expression system by (1) selecting mutations in the gene *4* protease (stops at prohead I), (2) by preventing the exposure of the cleaved prohead II to expansion inducer (prohead II), and (3) by using a mutant of gene *5* that prevents cross-linking (head I). All these particles were the subject of moderate-resolution cryo-EM investigation (Conway *et al.*, 1995) that paved the way for a detailed structural study of the dynamics of this system by crystallography and EM

The homogeneity, stability, and symmetry of the head II particles allowed the production of high-quality crystals that diffracted X-rays to 3.5-Å resolution (Wikoff *et al.*, 1998, 1999). Data were collected and the phase problem was solved by using the cryo-EM density as a low-resolution model for phases (Wikoff *et al.*, 2000). Phases to high resolution were computed with the 60-fold symmetry of the capsid for real space averaging and the phase extension procedure. The electron density map was readily interpreted and a model was constructed of one complete gp5 subunit. An early step in capsid maturation is the cleavage of 103 residues from the N terminus of the capsid protein, and thus the sequence numbering of mature gp5 begins with Ser-104. Clearly defined electron density was present for residues 104 to 383, with the main chain being visible up to residue 384 (Fig. 9; see Color Insert). The bacteriophage capsid is constructed from pentamers and hexamers, with seven unique copies of the polypeptide chain in the icosahedral asymmetric unit (T=7). The density for these chains is similar overall except for details at the subunit contacts and the conformations of extensions from the central domains. The model for one subunit was adjusted to the electron density to fit the other six subunits in the icosahedral asymmetric unit.

B. *Structure of the HK97 Subunit*

gp5 has no similarity to any previously determined capsid protein, and represents a new category of subunit fold. It contains 28% α helix and 32% β strand, and is organized into two compact and spatially distinct domains that are not contiguous in sequence. The axial or A domain is near the 5-fold and quasi-6-fold symmetry axes and the peripheral or P domain, with two elongated extensions, fills the region between adjacent quasi- and/or icosahedral 3-fold axes. The defined domains are not unusual in structure; the novel form and dynamics of the subunit derive from the two long extensions from these domains, an N-terminal arm and an extended loop, and their acrobatic quaternary associations, described below. The icosahedral asymmetric unit, color coded by domain and with the extensions colored in green and yellow, is shown in Fig. 9.

C. Quaternary Structure of the HK97 Particle

The HK97 mature capsid has the shape of an icosahedron, with a maximum diameter of 659 Å near the 5-fold axes (Fig. 10; see Color Insert). The radial thickness of the shell is only 18 Å or less, giving the large particle the appearance of a balloon (Fig. 10). The HK97 capsid is thinner than that of plant and animal viruses constructed with a β barrel, which typically have a mean thickness of 40 Å or more. The capsid subunits are organized as hexamers and pentamers and are arranged with nearly ideal T=7 quasi-symmetry (Fig 10), an arrangement not previously observed in a virus crystal structure. The hexamers display near-perfect 6-fold symmetry. Each polypeptide chain lies almost completely within the hexamer or pentamer boundaries defining the morphological units (outlined by the cage connecting quasi- and/or icosahedral 3-fold symmetry axes in Fig. 10). The positions where the chain extends slightly over the geometric boundaries are crucial for the cross-linking. An icosahedron organized with a T=7 lattice is enantiomorphic; the hand of the HK97 capsid, determined from the crystal structure, is *levo*.The closest example to this arrangement of subunits is found in the papovaviruses, in which the morphological units are arranged on a T=7 *dextro* lattice, but with pentamers substituted in the hexavalent lattice positions (Liddington *et al.*, 1991). The atomic model of the head II form of HK97 together with a 12-Å cryo-EM structure of the prohead II form of the particles allow a detailed analysis of the transition from prohead II.

D. Prohead-to-Head Transition of HK97

The prohead-to-head transition in HK97 and other bacteriophages is irreversible. The basis of the transition is an exothermic switching from a local energy minimum for the particle to a global minimum. The changes in structure are dramatic and were interpreted by modeling the head II subunit into the 12-Å prohead reconstruction. Remarkably, most of the HK97 subunit was fitted into the prohead density with a high degree of fidelity (Conway *et al.*, 2001). There were distinctive features in the EM density that functioned as fiducials for positioning the head II subunit structure with precision and a high degree of confidence (Fig. 11; see Color Insert). A striking feature of the prohead surface density is the breakdown of the 6-fold symmetry that is obvious in the head. Each hexavalent site in the capsid is a sheared set of trapezoids that appear as a dimer of trimers instead of the expected hexamer. Figure 12 (see Color Insert), shows that this feature is closely modeled by the subunit. Regions of the subunit that do not behave as a rigid body are the N-terminal extension (residues 104–135) and the E loop composed of residues 150–180. Both of these regions are not accounted for by the procapsid density and are either folded differently from the head tertiary structure or are more dynamic in the prohead form.

The change in subunit interactions and locations during the transition is summarized in Table I showing the large-scale reorganization in the transition. It is clear that the subunits totally reorganize, mostly as rigid bodies, during the transition while maintaining particle integrity. Extrapolating the extent of these changes to other bacteriophages readily explains the exposure of new antigenic sites in T4 phage during the expansion process in this virus (Steven, 1991). The expansion can viewed as comparable to other processive events in biology including the dynamics of muscle, RNA replication, and protein translation. Virus particle transitions, however, are novel in that all the other dynamic actions investigated have at least one component of the system that is a linear polymer that slides through a second component. The movements in virus particle transitions are not "anchored" directly by an element of the system. Phage transitions probably nucleate at a point on the particle and then propagate from this position through the capsid like a spherical wave. Such a transition was observed in phage T4 mutants in which viral

TABLE I
Change in Subunit Interactions and Locations during Transition

			Cross-link distance	
Subunit	Distance[a]	Rotation[b]	Prohead II[c]	Head II
A	53	24	32	9
B	45	39	31	7
C	45	36	34	8
D	45	20	39	9
E	42	36	37	8
F	51	37	39	8
G	58	22	33	9

[a]Distance between the subunit center of mass of prohead II and head II.
[b]The rotation angle between the prohead II and head II subunit.
[c]Distance between C_α atoms of Lys-169 in the indicated subunit and Asn-356 in the subunit that is cross-linked (head II) or will be cross-linked (prohead II).

subunits formed extended cylinders that underwent expansion as a "wave" propagating down the cylinder. This implies that icosahedral symmetry breaks down during the transition; however, the process enables the maintenance of particle integrity by preserving all but the particular class of interactions that are changing at one moment in time. Thus, in a generalized sense, there are stable polymers (in two distinct states) that adjoin the point of actual transition and this must account for the maintenance of particle integrity.

A striking feature of the head II particle when compared with the prohead is the degree of buried surface of the subunits. Table II shows that there is nearly a 40% increase in buried surface area between the prohead and head. This provides a significant driving force for the transition and explains why the head is at the energy minimum compared with the prohead. It is likely that with the interactions of the N-terminal polypeptides cleaved off between prohead I and prohead II add significant additional buried surface area in prohead I as do interactions with the protease. Thus, the remodeling by the protease, among other possible effects, reduces the buried surface area of the prohead form and explains why it is metastable. Examining the charge distribution within the capsid suggests a mechanism for overcoming the energy barrier of the local minimum. Contrary to expectation, the interior of the prohead is negatively charged. This is in sharp contrast to most RNA viruses, where the interior is basic. The acidic interior, when interacting with the negative charged DNA that naturally triggers the transition, is probably enough of a perturbation to dislodge the procapsid from its local minimum and initiate the remarkable expansion that is programmed into the structure.

TABLE II
Buried Surface Area between Prohead and Head

	Prohead II		Head II[a]	
	Pentamer	Hexamer	Pentamer	Hexamer
Intracapsomer[b]	15,665	17,694	21,720	23,277
Intercapsomer[c]	13,385	14,696	21,030	24,950

[a]By this measure, intra- and intercapsomer interactions contribute almost equally to capsid stabilization (Ross *et al.*, 1985).

[b]Buried surface areas were calculated between adjacent pairs of subunits within a capsomer.

[c]Buried surface areas were calculated between all subunits of a given capsomer and contacting subunits of neighboring capsomers.

V. Conclusions

Particle dynamics are a critical component of animal viruses and appear to fall into two broad categories: fluctuations about an equilibrium point and large-scale dynamics that lead to a change in particle morphology. The former are essential for virus interactions with a cell and its uncoating. The latter are necessary for complex virus structures that cannot directly assemble into the final functional form.

Fluctuating regions appear to interact with membranes in non-enveloped animal viruses and the fluctuations are preferable to permanently exposed hydrophobic peptides, as this prevents the peptides from causing virus aggregation. It is likely that the fluctuations are generally exaggerated when the particle binds to the cellular receptor, as is the case with polio-and rhinovirus in the exposure of VP4 and the N terminus of VP1. The mechanism of this change in exposure frequency is almost certainly associated with the "pocket" in the VP1 subunits in picornaviruses. It is not clear what the mechanism for this change might be in nodaviruses, but this is under investigation.

Large-scale quaternary structure changes probably reflect the multistep nature of assembling a complicated particle. The process requires at least two stages, the first creating an association of subunits that minimizes the difference between the solution state and assembled state of the subunits. This assembly is necessarily plastic and allows the subunits the mobility to seek an overall stable configuration. Although favorable for assembly, this state is not sufficiently stable to maintain particle integrity under harsh conditions. When the defining architecture is achieved, the second, stable and final morphology is achieved by the transition. The systems described have much to offer as organizational models for understanding large-scale dynamics at the chemical level.

References

Abad-Zapatero, C., Abdel-Meguid, S., Johnson, J., Leslie, A., Rayment, I., Rossmann, M., Suck, D., and Tsukihara, T. (1980). *Nature* **286,** 33–39.

Agrawal, D., and Johnson, J. (1995). *Virology* **207,** 89–97.

Albert, F. G., Fox, J. M., and Young, M. J. (1997). *J. Virol.* **71,** 4296–4299.

Bancroft, J. B., Bracker, C. E., and Wagner, G. W. (1969). *Virology* **38,** 324–335.

Bothner, B., Dong, X. F., Bibbs, L., Johnson, J. E., and Siuzdak, G. (1998). *J. Biol. Chem.* **273,** 673–676.

Bothner, B., Schneemann, A., Marshall, D., Reddy, V., Johnson, J. E., and Siuzdak, G. (1999). *Nat. Struct. Biol.* **6,** 114–116.

Canady, M. A., Tihova, M., Hanzlik, T. N., Johnson, J. E., and Yeager, M. (2000). *J. Mol. Biol.* **299,** 573–584.

Canady, M., Tsuruta, H., and Johnson, J. (2001). *J. Mol. Biol.* **311,** 803–814.

Cheng, R., Reddy, V., Olson, N., Fisher, A., Baker, T., and Johnson, J. (1994). *Structure* **2,** 271–282.
Chow, M., Yabrov, R., Bittle, J., Hogle, J., and Baltimore, D. (1985). *Proc. Natl. Acad. Sci. USA* **82,** 910–914.
Conway, J., Wikoff, W., Cheng, N., Duda, R., Hendrix, R., Johnson, J., and Steven, A. (2001). *Science* **292,** 744–748.
Conway, J. F., Duda, R. L., Cheng, N., Hendrix, R. W., and Steven, A. C. (1995). *J. Mol. Biol.* **253,** 86–99.
Dong, X., Natarajan, P., Tihova, M., Johnson, J., and Schneemann, A. (1998). *J. Virol.* **72,** 6024–6033.
Duda, R. L., Hempel, J., Michel, H., Shabanowitz, J., Hunt, D., and Hendrix, R. W. (1995a). *J. Mol. Biol.* **247,** 618–635.
Duda, R. L., Martincic, K., and Hendrix, R. W. (1995b). *J. Mol. Biol.* **247,** 636–647.
Durham, A. C., Hendry, D. A., and Von Wechmar, M. B. (1977). *Virology* **77,** 524–533.
Fisher, A., and Johnson, J. (1993). *Nature* **361,** 176–179.
Fisher, A., McKinney, B., Schneemann, A., Rueckert, R., and Johnson, J. (1993). *J. Virol.* **67,** 2950–2953.
Fricks, C. E., and Hogle, J. M. (1990). *J. Virol.* **64,** 1934–1945.
Harrison, S. C., Olson, A. J., Schutt, C. E., Winkler, F. K., and Bricogne, G. (1978). *Nature* **276,** 368–373.
Heinz, B. A., Rueckert, R. R., Shepard, D. A., Dutko, F. J., McKinlay, M. A., Fancher, M., Rossmann, M. G., Badger, J., and Smith, T. J. (1989). *J. Virol.* **63,** 2476–2485.
Hendrix, R. W., and Duda, R. L. (1998). *Adv. Virus Res.* **50,** 235–288.
Hogle, J. M., Chow, M., and Filman, D. J. (1985). *Science.* **229,** 1358–1365.
Incardona, N., and Kaesberg, P. (1964). *Biophys. J.* **4,** 11–22.
Krol, M. A., Olson, N. H., Tate, J., Johnson, J. E., Baker, T. S., and Ahlquist, P. (1999). *Proc. Natl. Acad. Sci. USA* **96,** 13650–13655.
Lewis, J. K., Bothner, B., Smith, T. J., and Siuzdak, G. (1998). *Proc. Natl. Acad. Sci. USA* **95,** 6774–6778.
Liddington, R. C., Yan, Y., Moulai, J., Sahli, R., Benjamin, T. L., and Harrison, S. C. (1991). *Nature* **354,** 278–284.
Lin, T., Chen, Z., Usha, R., Stauffacher, C. V., Dai, J. B., Schmidt, T., and Johnson, J. E. (1999). *Virology* **265,** 20–34.
Lomonossoff, G., and Johnson, J. (1991). *Prog. Biophys. Mol. Biol.* **55,** 107–137.
McKinlay, M. A. (1993). *Scand. J. Infect. Dis. Suppl.* **88,** 109–115.
Mosser, A. G., and Rueckert, R. R. (1993). *J. Virol.* **67,** 1246–1254.
Munshi, S., Liljas, L., Cavarelli, W., Bomu, W., McKinney, B., Reddy, V., and Johnson, J. (1996). *J. Mol. Biol.* **261,** 1–10.
Opalka, N., Brugidou, C., Bonneau, C., Nicole, M., Beachy, R. N., Yeager, M., and Fauquet, C. (1998). *Proc. Natl. Acad. Sci. USA* **95,** 3323–3328.
Pfeiffer, P., and Durham, A. C. (1977). *Virology* **81,** 419–432.
Phelps, D. K., and Post, C. B. (1995). *J. Mol. Biol.* **25,** 544–551.
Robinson, I. K., and Harrison, S. C. (1982). *Nature* **297,** 563–568.
Roivainen, M., Narvanen, A., Korkolainen, M., Huhtala, M. L., and Hovi, T. (1991). *Virology* **180,** 99–107.
Roivainen, M., Piirainen, L., Rysa, T., Narvanen, A., and Hovi, T. (1993). *Virology* **195,** 762–765.
Ross, P., Black, L., Bisher, M., and Steven, A. (1985). *J. Mol. Biol.* **183,** 353–362.
Schneemann, A., Dasgupta, R., Johnson, J., and Rueckert, R. (1993). *J. Virol.* **67,** 2756–2763.

Speir, J. A., Munshi, S., Wang, G., Baker, T. S., and Johnson, J. E. (1995). *Structure* **3,** 63–78.

Steven, A. C., Bauer, A. C., Bisher, M. E., Robey, F. A., and Black, L. W. (1991). *J. Struct. Biol.* **106,** 236–242.

Tang, L., Johnson, K., Ball, L., Lin, T., Yeager, M., and Johnson, J. (2001). *Nat. Struct. Biol.* **8,** 77–83.

Taylor, D., Krishna, N., Canady, M., Schneemann, A., and Johnson, J. (2002). *J. Virol.* **76,** 9972–9980.

Weber, G. (1975). *Adv. Protein Chem.* **29,** 1–83.

Wery, J., Reddy, V., Hosur, M., and Johnson, J. (1994). *J. Mol. Biol.* **235,** 565–586.

Wikoff, W., Duda, R., Hendrix, R., and Johnson, J. (1998). *Virology* **243,** 113–118.

Wikoff, W., Liljas, L., Duda, R., Tsuruta, H., Hendrix, R., and Johnson, J. (2000). *Science* **289,** 2129–2133.

Wikoff, W. R., Duda, R. L., Hendrix, R. W., and Johnson, J. E. (1999). *Acta Crystallogr. D Biol. Crystallogr.* **55,** 763–771.

Xie, Z., and Hendrix, R. W. (1995). *J. Mol. Biol.* **253,** 74–85.

VIRAL GENOME ORGANIZATION

By B. V. VENKATARAM PRASAD* AND PETER E. PREVELIGE, JR.†

*Department of Biochemistry and Molecular Biology, Keck Center for Computational Biology, Baylor College of Medicine, Houston, Texas 77030, and †Department of Microbiology, University of Alabama at Birmingham, Birmingham, Alabama 35294

I. Introduction 219
Structural Techniques 220
II. Single-Stranded RNA Viruses 221
A. Conformation of RNA in Icosahedral Single-Stranded RNA Viruses 222
B. Structural Organization of the Entire Genome 224
C. Genome Packaging 225
D. Genome Release 226
E. Specific Recognition of the Genome in Single-Stranded RNA Viruses 227
F. Genome Structure in Helical Single-Stranded RNA Viruses 228
III. Double-Stranded RNA Viruses 229
A. General Genomic and Capsid Features 230
B. Unique Organization of Capsid Layer That Surrounds the Genome 231
C. Endogenous Transcription and Exit Pathway of the Transcripts 231
D. Endogenous Transcription is a Highly Dynamic Process 232
E. Structural Organization of the Genome 232
F. Genome Replication and Packaging 234
IV. Single-Stranded DNA Viruses 235
A. Microviruses 236
B. Model for Genome Entry 237
C. Ordered DNA in the Capsid Structure 237
D. Genome Release 238
E. The *Parvoviridae* 238
V. Double-Stranded DNA Viruses 240
A. The Double-Stranded DNA Bacteriophages 240
B. Structural Organization of the Genome 241
C. Genome Packaging 242
D. DNA Release and Entry 243
E. The *Herpesviridae* 244
F. The Polyomaviruses 245
G. Adenovirus 245
VI. Conclusions 246
References 248

I. Introduction

Viruses are macromolecular assemblies designed to contain and protect the genome, and deliver it to a specific host cell. Viruses come in various sizes, shapes, and forms. Some are large and some small, some are rodlike

ADVANCES IN
PROTEIN CHEMISTRY, Vol. 64

0065-3233/03 $35.00

and some are spherical, and some have lipid envelopes. Viruses are also distinguished on the basis of the chemical nature of the genome that they contain: single-stranded or double-stranded RNA or DNA. Some viruses have a single segment of nucleic acid whereas others have multiple segments. Although the details may vary from one virus to another, in all the viruses, the genome is enclosed within a proteinacious capsid. Often the size of the virus is proportional to the size of the genome. Given the biophysical characteristics of the nucleic acids, such as the diameter of the strand, the partial specific volume, and the molecular mass, we see that often the genome is compacted into a significantly smaller volume inside the virus. Such a compaction or condensation raises interesting questions: What is the conformation and the structural organization of the compacted genome? Given the polyanionic nature of the nucleic acid, how are charges neutralized? How is the condensed form later on decondensed or unraveled to allow the normal functions of the genome once inside the host cell? The process of nucleic acid condensation and decondensation is not unique to viruses; it is in fact a fundamental cellular feature, for example, chromatin condensation.

In some viruses, such as segmented double-stranded RNA (dsRNA) viruses, understanding of the structural organization of the genome assumes a greater importance as their genomes are intimately involved in several enzymatic reactions inside the capsid layers. Although the conformation of the encapsidated genome is of interest, there are other interesting aspects related to the genome encapsidation. These include the following: How do viruses specifically incorporate their own genes, distinguishing them from other nucleic acids in the cellular milieu? Is the genome encapsidation concurrent with the capsid assembly or is it threaded into a preformed capsid? How is the genome released? The underlying mechanisms for these aspects evidently vary from one family of viruses to the other. Although the focus of this article is mainly on the encapsidated conformation of the genome, we briefly touch on these issues as well. Several excellent reviews on this subject have been published periodically since 1980 (Casjens, 1985; Johnson and Rueckert, 1997; Rossmann *et al.*, 1983), hence we focus on more recent developments in the field. Specifically for the purpose of this review, we have classified the viruses into four groups: single-stranded RNA (ssRNA), dsRNA, ssDNA, and dsDNA viruses.

Structural Techniques

Beginning with the structures of three small icosahedral plant viruses in the early 1980s (Abad-Zapatero, 1980; Harrison *et al.*, 1978; Liljas *et al.*, 1982), over the last two decades X-ray crystallography has been successfully applied to study a variety of larger and more complex icosahedral viruses (Grimes *et al.*, 1998; Reinisch *et al.*, 2000; Wikoff *et al.*, 2000). The closely related technique of X-ray fiber diffraction has been used to study viruses that have helical symmetry (Namba *et al.*, 1989). In the last three decades, building on the foundation laid by Klug and colleagues in the early 1970s (Crowther, 1971; Crowther *et al.*, 1970), owing to spectacular advances in specimen preparation, electron imaging, and computer image reconstructions, three-dimensional electron microscopy (cryo-EM) has evolved either as an independent or as a complementary technique to X-ray diffraction to study viruses at high resolution (Adrian *et al.*, 1984; Bottcher *et al.*, 1997b; Conway *et al.*, 1997; Zhou *et al.*, 2000) (see reviews by Baumeister and Steven, 2000; Chiu *et al.*, 1999). All these techniques, to obtain three-dimensional structural information on such large macromolecular assemblies as viruses, rely implicitly on the symmetry of the capsid. As a result, the structural organization of the encapsidated genome is amenable to these structural techniques only when the genome follows the capsid symmetry. There are several examples in which the entire genome or a significant portion of it is observed to follow the capsid symmetry and visualized in the structural analysis. Although, in general, precise structural information in terms of the genome sequence is not obtained, these analyses have provided important insights into the structural properties of the genome, which are the main of focus of this review. In addition to the X-ray crystallographic and cryo-EM structural techniques, other diffraction techniques such as neutron diffraction (Bentley *et al.*, 1987), low-angle X-ray scattering (Earnshaw *et al.*, 1976; Harvey *et al.*, 1981; Jack and Harrison, 1975; Tsuruta *et al.*, 1998), and spectroscopic techniques (Thomas, 1999) have been useful in understanding the conformational properties of the encapsidated genome.

II. Single-Stranded RNA Viruses

The single-stranded RNA viruses constitute a large group of viruses, which include viruses of plant, bacterial, human, and animal origin. These viruses are generally divided into two groups; those that contain positive-strand ssRNA in their genomes, and others that contain negative-strand ssRNA genome. Both spherical or icosahedral, and helical or rod-shaped, organization are common among ssRNA viruses. In positive-strand ssRNA viruses, the naked genome is encapsidated; however, in certain plus-strand ssRNA viruses, such as picornaviruses, the genomic RNA is covalently linked to a small protein called VPg, which is used in priming RNA synthesis on genome release into host cells (Flint *et al.*, 2000; Kitamura *et al.*, 1981). In contrast, in the negative-strand ssRNA viruses, the genome is bound to nucleocapsid proteins and other accessory proteins such as RNA-dependent RNA polymerase required for initiation of mRNA synthesis. Virion-associated polymerase activity is a necessity not only for negative-strand ssRNA viruses but also for dsRNA viruses, as discussed in the next section, because the host cells lack the necessary enzymes to translate their genomes. Most well-characterized negative-strand ssRNA viruses, such as influenzavirus, vesicular stomatitis virus, arenavirus, and bunyavirus, are enveloped with possibly helical nucleocapsids. Although X-ray crystallographic structures of some of the viral proteins of these ssRNA viruses have been determined, in terms of overall structural organization they remain poorly characterized. To date, all high-resolution structural analyses of ssRNA-containing viruses have been of positive-strand ssRNA viruses.

A. *Conformation of RNA in Icosahedral Single-Stranded RNA Viruses*

Many of the ssRNA viruses have icosahedral symmetry. X-ray crystallographic structures of several of these viruses have been determined. In some of these structures, a significant portion of the genome is ordered.

1. *Satellite Single-Stranded RNA Viruses*

The satellite ssRNA viruses are a group of ssRNA icosahedral viruses, which are the satellites to certain plant viruses (Pritsch and Mayo, 1989). Satellite tobacco necrosis virus (STNV), a satellite virus to tobacco necrosis virus, was in fact one of the first icosahedral virus structures to be determined by X-ray crystallography (Liljas *et al.*, 1982). In more recent years, two other satellite plant viruses, satellite tobacco mosaic virus (STMV) (Larson *et al.*, 1993, 1998) and satellite panicum mosaic virus (SPMV) (Ban and McPherson, 1995), have been determined. These are perhaps the simplest and smallest icosahedral viruses whose structures have been determined by X-ray crystallography. Of relevance to our discussion is the structure of STMV determined to 1.8-Å resolution, which shows extraordinary details about genome organization (Larson *et al.*, 1998; Larson and McPherson, 2001).

The capsid of STMV, ~170 Å in diameter, with 60 monomeric subunits, is organized on a T=1 icosahedral lattice. The capsid encloses an ssRNA genome of 1059 bases. The 159-amino acid-long capsid protein exhibits a canonical antiparallel 8-stranded β-sandwich fold, with the N-terminal 36 residues in highly extended conformation facing the interior of capsid. About 45% of the genome is seen in the electron density map along with several water molecules. Portions of the ssRNA genome in a double-helical conformation, predominantly in the A form, are closely associated with the interior face of the capsid protein dimers at all the icosahedral 2-fold axes (Fig. 1a; see Color Insert). The RNA helices are oriented with their helical axes perpendicular to the icosahedral 2-fold axes. Each of the observed 30 RNA segments consist of 9 base pairs plus an unpaired base in each strand, thus accounting for 600 of the 1059 bases of the genomic RNA.

The observed protein–RNA interactions, as expected, are nonspecific. However, there is a high degree of steric complementarity between the interior surface of the dimer and the RNA helix. The interacting surface of the dimer is saddle shaped, similar to TATA-binding protein, cradling the bent conformation of RNA. A significant portion of the protein–RNA interactions is water-mediated hydrogen bond interactions involving side-chain atoms of the capsid protein and the oxygen atoms from the phosphate and sugars on the RNA. The high-resolution structure of STMV, in which water molecules are clearly identified, underscores the importance of water molecules not only in stabilizing the protein–protein interfaces but also in mediating protein–RNA contacts.

2. *T=3 Single-Stranded RNA Viruses*

Although X-ray crystallographic structures of several T=3 ssRNA viruses have been determined, structures of viruses in the family *Nodaviridae* have provided excellent opportunity to visualize the genome organization at high resolution (Fisher and Johnson, 1993; Tang *et al.*, 2001). How does the genome organization in the viruses with T=3 icosahedral symmetry, compare with that seen in T=1 satellite ssRNA viruses? The first visualization of ordered RNA in *Nodaviridae* was in the structure of flock house virus (Fisher and Johnson, 1993). These insect viruses, which are about ∼325 Å in diameter, have two plus-strand RNA molecules in their genomes. The capsid consists of 180 molecules of a single gene product. In the X-ray structure of pariacoto virus (PaV), another member of this family, a stunningly larger portion of the ordered RNA forming a dodecahedral shell underneath the T=3 capsid shell was visualized (Tang *et al.*, 2001). Similar to the helical segments at the icosahedral 2-fold axes in the T=1 STMV structure, in PaV also, the segments of genomic RNA in the double-helical conformation interact with the dimeric subunits of the capsid protein (Fig. 1b; see Color Insert). Each double-helical segment at an icosahedral 2-fold axis consists of 25 base pairs, compared with 12 base pairs in the related flock house virus structure, thus accounting for 1500 of 4322 nucleotides of the viral genome.

3. *Pseudo-T=3 Single-Stranded RNA Viruses*

Some of the ssRNA viruses such as comoviruses and picornaviruses can be classified under the pseudo-T=3 ssRNA category (Rossmann and Johnson, 1989). In contrast to strict T=3 viruses, the β-barrels that tile the icosahedral lattice are contributed by chemically distinct capsid protein subunits. In comoviruses, 60 copies of 2 capsid proteins, L and S, form the capsid (Chen *et al.*, 1989). The L protein has two domains, each with a β-barrel motif, whereas the S protein has one β-barrel motif. In the picornaviruses, 60 copies VP1, VP2, and VP3, each with a β-barrel motif, form the viral capsid (Acharya *et al.*, 1989; Hogle *et al.*, 1985; Rossmann *et al.*, 1985). In addition to these proteins, the capsid also contains 60 copies of another internally located protein, VP4. VP4 and VP2 are the maturation-dependent autocatalytic cleavage products of VP0 (Jacobson *et al.*, 1970). Although a significant portion of VP4 is disordered, the ordered portion of VP4, in close association with VP1 and VP2, surrounds the 5-fold axis. The arrangement of the 180 β-barrels in pseudo-T=3 structures is similar to that observed in the strict T=3 icosahedral structures. In the X-ray crystallographic structure of a comovirus, about 20% of the genomic RNA is ordered (Chen *et al.*, 1989). In fact, this structure was the first example in which ordered RNA was observed in an X-ray structure. In contrast to STMV, and nodaviruses, the observed RNA in comovirus is located at the 3-fold axes, and it is single stranded. The ordered RNA interacts exclusively with the L subunits, which surround the icosahedral 3-fold axis. In this structure, as expected, there are no sequence-specific interactions. It is interesting to note that, although structures of several picornaviruses have been determined to date, ordered RNA is not observed in any of them.

B. *Structural Organization of the Entire Genome*

Although a significant portion of the genome is observed in some of the ssRNA viruses, the chemical identity of each of the bases is lost because of the icosahedral averaging. Thus, it is difficult to decipher whether the observed duplexes, for example in STMV and nodaviruses, represent nearby segments linked by short loops or are formed from distant segments of RNA along the gene sequence. Larson *et al.* (1998) have argued in favor of a helix–loop packaging process in which each duplex is derived from local contiguous segments separated by a loop, and these duplexes are connected by passing sequentially from one icosahedral 2-fold axis to the other. The other possibility is that pairing of distant stretches of the RNA sequences forms each duplex. With such a model the packaged RNA would be severely constrained, posing problems in constructing an acceptable motif. The helix–loop packaging process, in contrast, allows for short, efficient, and more systematic connections between the duplexes and the single-stranded RNA loops. Similar packing of the RNA in comoviruses is proposed, in which the observed trefoil-shaped single-stranded RNAs associated with the icosahedral triads are joined by loops (Johnson and Rueckert, 1997). The reason why the loops become invisible in the structural analysis is likely because each loop has a different conformation, and also because the loops may not be precisely in tune with the icosahedral symmetry.

C. *Genome Packaging*

How does RNA become encapsidated? Is the RNA packaged into a preformed capsid or is the encapsidation concomitant with capsid assembly? In general, structural studies on the ssRNA viruses appear to favor a model in which genome encapsidation is concomitant with capsid assembly (Harrison, 1989; Rossmann and Erickson, 1985). From analysis of the buried surface area, and the nature of the intersubunit contacts, for STMV, Larson *et al.* have proposed a model in which the assembly is initiated by dimers of the capsid protein that associate with the double-helical segments of the genomic RNA (Larson *et al.*, 1998; Larson and McPherson, 2001). These dimers with the associated genomic RNA form trimers of dimers, which then interact to form an intact icosahedral capsid with pentameric vertices. In this process, it is not clear yet whether RNA plays an active or a passive role. It is possible that RNA by assuming discrete conformations will present an array of binding sites to recruit the dimers and direct their assembly in a cooperative manner, thus playing an active role. It is also possible that the capsid protein dimers, because of their intrinsic affinity to RNA, interact with local RNA conformations and drive the assembly and encapsidation primarily by protein–protein interactions, inducing appropriate conformational changes in the RNA during this processes. In such a process, the basic N-terminal arm of the capsid may play a crucial role. Similar cocondensation models in which the basic N-terminal arm plays an important role have been proposed for T=3 plant viruses. Whereas in tombusviruses such as tomato bushy stunt virus (TBSV) and turnip crinkle virus, the initial protein–RNA complex has been suggested to involve trimers of dimers (Harrison, 1989; Wei and Morris, 1991), in sobemoviruses such as southern bean mosaic virus (SBMV), pentamers of dimers have been proposed (Rossmann and Erickson, 1985).

The X-ray crystallographic structures of nodaviruses have unraveled a unique role for the double-helical segments of the encapsidated RNA in controlling capsid assembly (Fisher and Johnson, 1993). In the formation of T=3 capsids, the capsid protein dimer exists in two distinct, flat and bent, conformations. The conformation of the dimer at icosahedral 2-fold axes is flat (C–C interactions), whereas at the local 2-fold axes the conformation (B–A interactions) is bent. In contrast to other T=3 viruses such as SBMV and TBSV, in which the flat and bent conformations are controlled by disorder-to-order switching of the N-terminal arm of the capsid protein, in nodaviruses the ordered RNA helical segment present only at the icosahedral 2-fold axis functions as a wedge in stabilizing the flat conformation. This RNA segment interacts with a 44-amino acid polypeptide fragment that is produced by the autocatalytic cleavage of the capsid protein. Increased stability of the capsid following maturation-dependent autocatalytic cleavage is likely due to this RNA–protein interaction. The role of RNA in the maturation-dependent stabilization of the capsid is also indicated in picornaviruses (Basavappa *et al.*, 1994; Bishop and Anderson, 1993; Curry *et al.*, 1995; Jacobson *et al.*, 1970).

In some of the ssRNA viruses like STMV, and nodaviruses, empty particles devoid of the genomic RNA are rarely seen. This observation underscores the importance of RNA in the capsid assembly and substantiates the model in which the RNA and the capsid proteins cocondense. The role of RNA in nucleating the assembly or controlling the assembly process is further supported by biochemical and structural studies on several other ssRNA viruses such as brome mosaic virus (Sacher and Ahlquist, 1989), cowpea chlorotic mottle virus (Dasgupta and Kaesberg, 1982), sobemoviruses (Erickson *et al.*, 1985), and tombusviruses (Sorger *et al.*, 1986) (also see review by Rossmann and Erickson, 1985). However, in several other ssRNA viruses including comoviruses, tymoviruses (Argos and Johnson, 1984; Chen *et al.*, 1989), and caliciviruses (Prasad *et al.*, 1994, 1999) the capsid protein forms empty capsids both *in vitro* and *in vivo*. Interestingly, in these viruses the N-terminal arm is not basic. Does this mean that RNA is encapsidated into preformed capsids in these viruses? For some of these viruses, we cannot rule out the cocondensation process without further studies, and genome encapsidation into preformed or partially assembled capsids is a distinct possibility.

D. Genome Release

During its infection cycle, for a productive infection to occur, the virus must release the encapsidated genome at an appropriate location inside the host cell. The genome release is often associated with the disassembly process. Studies on nodaviruses and picornaviruses have provided some mechanistic insights into this process (Johnson and Rueckert, 1997). In both nodaviruses and picornaviruses, the maturation-dependent autoproteolysis that generates the γ peptide and VP4, respectively, correlates with viral infectivity (Gallagher and Rueckert, 1988; Lee *et al.*, 1993; Zlotnick *et al.*, 1994). Several studies on picornaviruses have shown that abrogation of cleavage results in a loss of infectivity (Bishop and Anderson, 1993; Compton *et al.*, 1990; Lee *et al.*, 1993). Although the mechanism of how infectivity is correlated to autoproteolysis is not entirely clear, studies on poliovirus and flock house virus have provided some evidence that suggests that autoproteolysis is required for release of the genome at late stages of cell entry. In picornaviruses, during the process of receptor-mediated cell entry, the capsid undergoes conformational changes resulting in the externalization of the N-terminal portion of VP1 and myristilic acid-associated VP4. A similar phenomenon is envisaged for nodaviruses, in which externalization of the γ peptide is implicated in forming the fusion pore and assisting translocation of the genome across the membrane barrier (Bong *et al.*, 1999; Johnson and Rueckert, 1997; Schneemann *et al.*, 1998).

E. *Specific Recognition of the Genome in Single-Stranded RNA Viruses*

Every virus has the capability of specifically identifying its genome for packaging. Several biochemical studies have long recognized the existence of a packaging site in a viral genome that facilitates specific recognition. Such a recognition event must occur once and before the packaging of the entire genome occurs. Whereas the capsid proteins interact with the rest of the genome in a nonspecific manner, interactions with the packaging site must involve sequence-specific interactions and perhaps high-affinity binding. Structural insights into this specific recognition of the packaging signal have been obtained from studies on icosahedral MS2 bacteriophage and a helical virus tobacco mosaic virus (TMV) (see below). The capsid of MS2 bacteriophage is composed of 180 molecules of a single protein of 129 residues, arranged as 90 dimers on a T=3 icosahedral lattice (Valegard *et al.*, 1990). Although the tertiary structure of the capsid protein shows a marked difference from the canonical 8-stranded β-barrel structure that is commonly found in other T=3 ssRNA viruses, the general arrangement of the capsid subunits into 30 CC dimers with exact 2-fold symmetry and 60 AB dimers with quasi-2-fold symmetry is similar to that seen in other T=3 structures. The crystal structures of a recombinant MS2 capsid with a 19-mer RNA stem–loop, and with other RNA variants, have been determined

(Grahn *et al.*, 1999; Stockley *et al.*, 1995; Valegard *et al.*, 1994). It is to be noted that the packaging site in all the known ssRNA viruses is a stem–loop structure. The crystal structures of the recombinant MS2 capsid and operator sequences elegantly demonstrate that a stem–loop structure provides a scaffold in which operator sequences in an appropriate conformation are presented for specific recognition by the capsid protein. In contrast to the RNA–protein interactions seen in STMV, and nodaviruses, in the MS2–operator complex structure there are clear sequence-specific interactions in which Asn-87 of the MS2 capsid protein plays an important role in specifically recognizing the MS2 packaging site.

F. *Genome Structure in Helical Single-Stranded RNA Viruses*

The best studied example of a helical virus with an ssRNA genome is tobacco mosaic virus (TMV) belonging to family *Tobamoviridae.* TMV is perhaps the simplest virus both in terms of overall organization and genome organization. The structure of TMV has been determined to 2.8-Å resolution in X-ray fiber diffraction studies (Namba *et al.*, 1989). The structure of TMV, assembly, genome packaging, and other related aspects have been reviewed extensively (Caspar and Namba, 1990; Klug, 1999). We provide here a brief discussion for the sake of completeness. In contrast to icosahedral viruses, in which only a portion of the genome is ordered, in TMV the entire genome is well ordered, with the same helical symmetry as the capsid protein. TMV is a rod-shaped virus 3000 Å long and 180 Å in diameter, with a central hole 40 Å in diameter. The capsid protein of TMV forms a one-start, right-handed helix of pitch 23 Å, with $16^1/_3$ subunits in each turn. These subunits wrap around the genomic RNA, such that the RNA lies inside a groove, at a radius of 40 Å, between successive helical turns. Each subunit interacts with three nucleotides. The subunit structure is predominantly α helical. The core of the structure consists of a right-handed four antiparallel α-helix bundle. One of the helices from the bundle makes extensive contacts with the genomic RNA.

The protein–RNA interactions are in general nonspecific and predominantly ionic, occurring between the phosphate groups of the RNA and the basic residues of the protein. One of the interesting observations regarding protein–RNA interactions is an anomalous repulsive interaction that is seen between the carboxylate group of Asp-116 and a phosphate group of the RNA. Namba *et al.*, considering that TMV, and in general any viral assembly, must assemble and disassemble during its infectious cycle, have argued that such repulsive interactions (including carboxylate–carboxylate

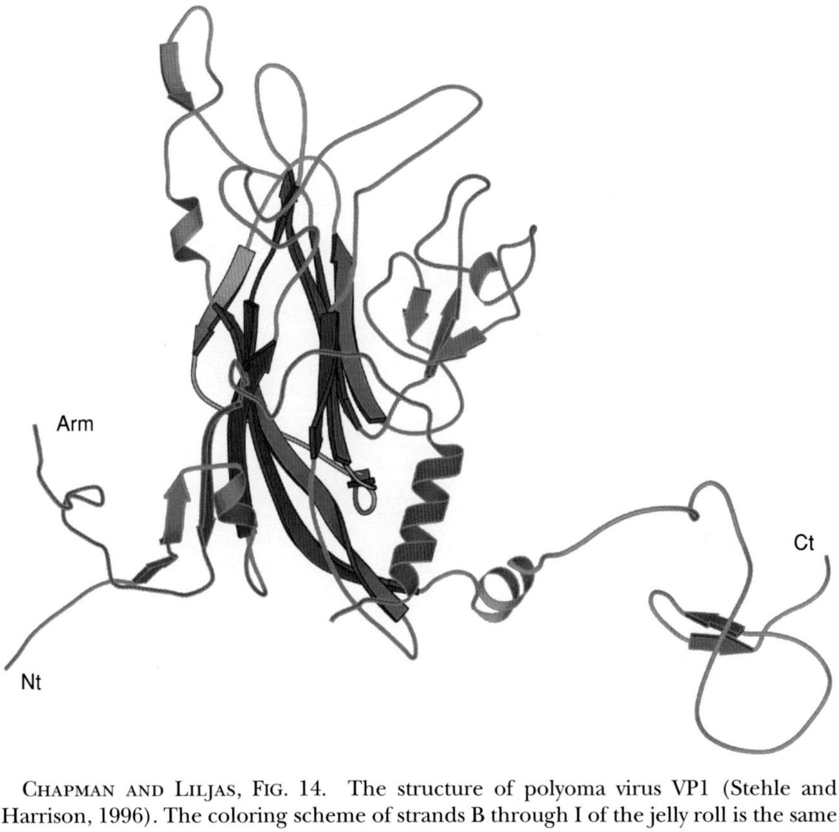

CHAPMAN AND LILJAS, FIG. 14. The structure of polyoma virus VP1 (Stehle and Harrison, 1996). The coloring scheme of strands B through I of the jelly roll is the same as in Fig. 1. The C-terminal arm of a neighboring subunit (purple) is inserted in an extension of the BIDG sheet of the viral jelly-roll domain.

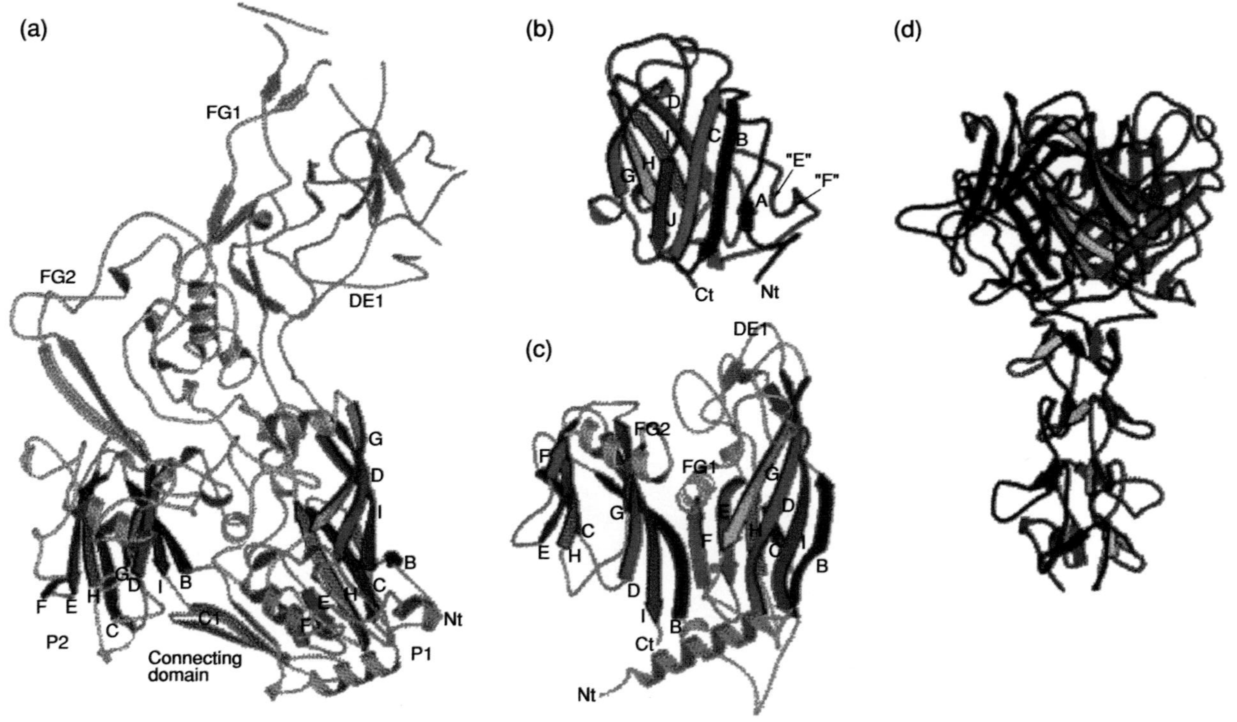

Chapman and Liljas, Fig. 15. (a) The structure of the adenovirus hexon polypeptide II (Athappilly *et al.*, 1994). The P1 (*right*) and P2 (*left*) domains are antiparallel eight-stranded β barrels. The connecting domain and intertwined loops hold them together. The coloring of both jelly-roll domains is as in Fig. 1. (b) The P3 major capsid protein of bacteriophage PRD1 (Benson *et al.*, 1999) with the same coloring scheme. (c) The knob domain of the adenovirus fiber (van Raaij *et al.*, 1999b). (d) Part of the trimeric adenovirus type 2 fiber, with 4 of the 15-residue repeats (total, 22 in this strain).

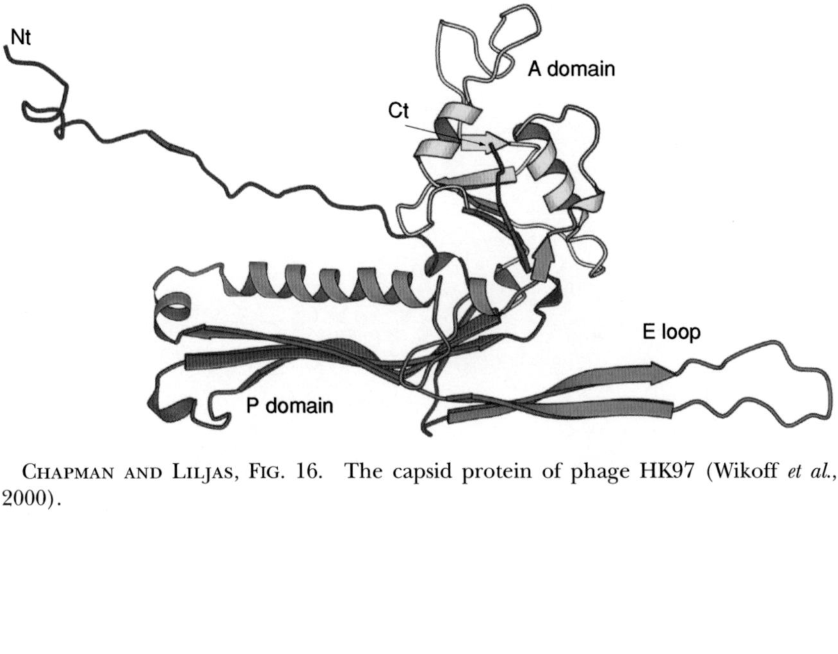

CHAPMAN AND LILJAS, FIG. 16. The capsid protein of phage HK97 (Wikoff *et al.*, 2000).

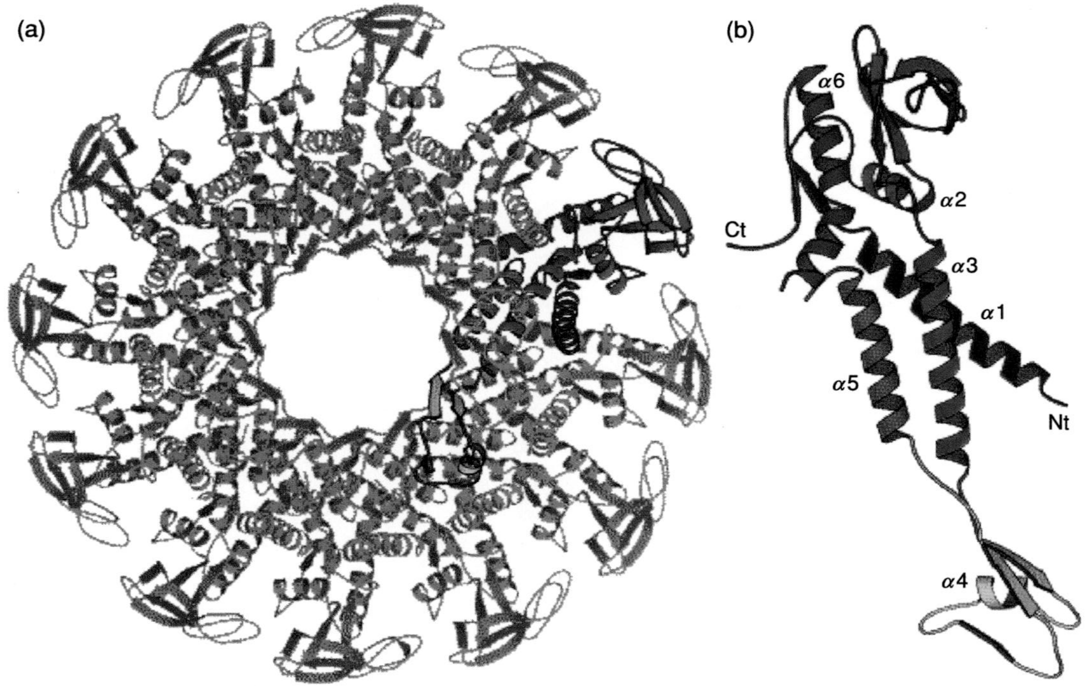

CHAPMAN AND LILJAS, FIG. 17. The ϕ29 connector protein (Simpson *et al.*, 2000). (a) The connector, with 1 of the 12 subunits highlighted. (b) Enlargement of one of the subunits. The 11 N-terminal and 23 C-terminal residues are disordered, as well as 18 residues between helices 5 and 6.

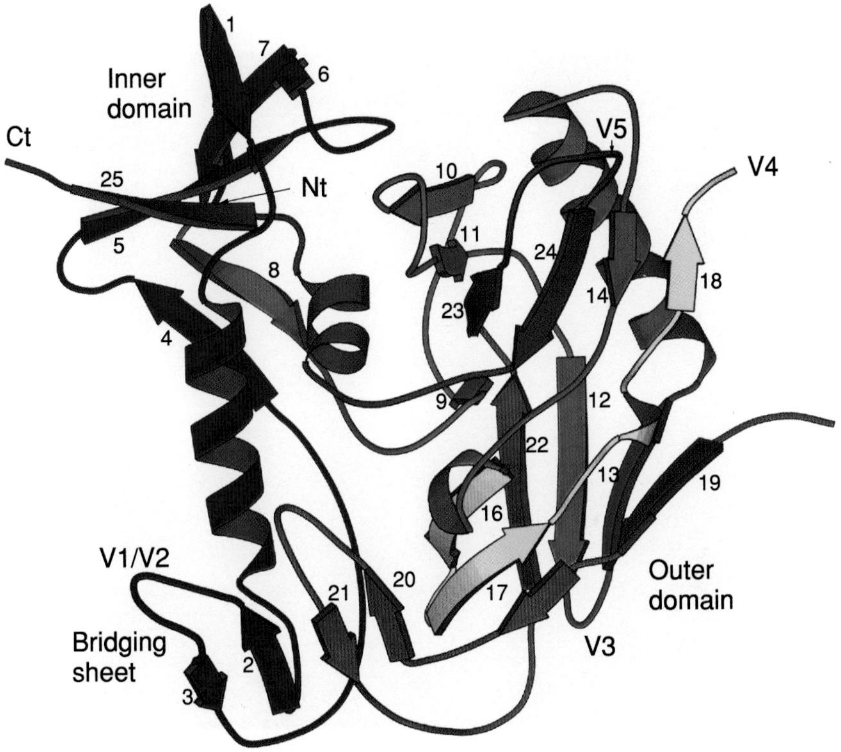

CHAPMAN AND LILJAS, FIG. 21. Structure of HIV gp120 (Kwong *et al.*, 1998). Strands 1 through 25 are marked, as well as the positions of the variable loops that were deleted from the protein construct used for the structural study.

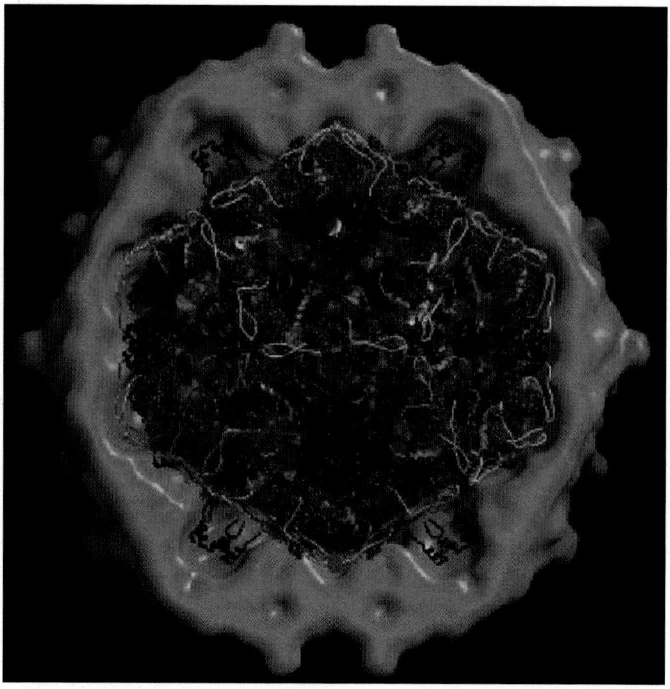

JOHNSON, FIG. 3. A composite of the capsid of paricoto virus (a nodavirus related to flock house virus) produced by electron cryomicroscopy and the packaged nucleic acid produced from the X-ray structure. The regions of the capsid protein that interact with the RNA are also shown from the X-ray structure (Tang *et al.*, 2001).

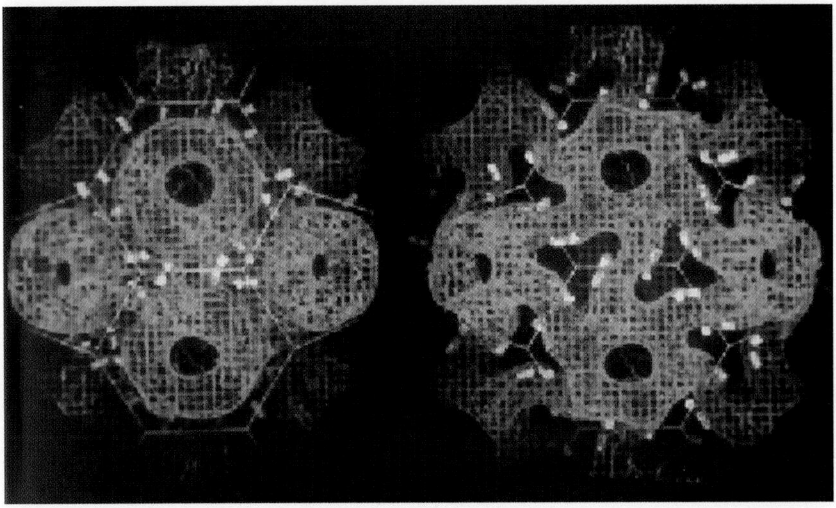

JOHNSON, FIG. 4. *Left:* Density for the cryo-EM reconstruction of CCMV (turquoise) with the model from the 3.2-Å resolution X-ray structure superimposed. *Right:* The cryo-EM reconstruction of the swollen form of CCMV generated by removing divalent metal ions and raising the pH to pH 7.0. The subunit models of the X-ray structure are fitted to the cryo-EM density with high fidelity, indicating that the subunits are moving largely as rigid units during the expansion (Speir *et al.*, 1995).

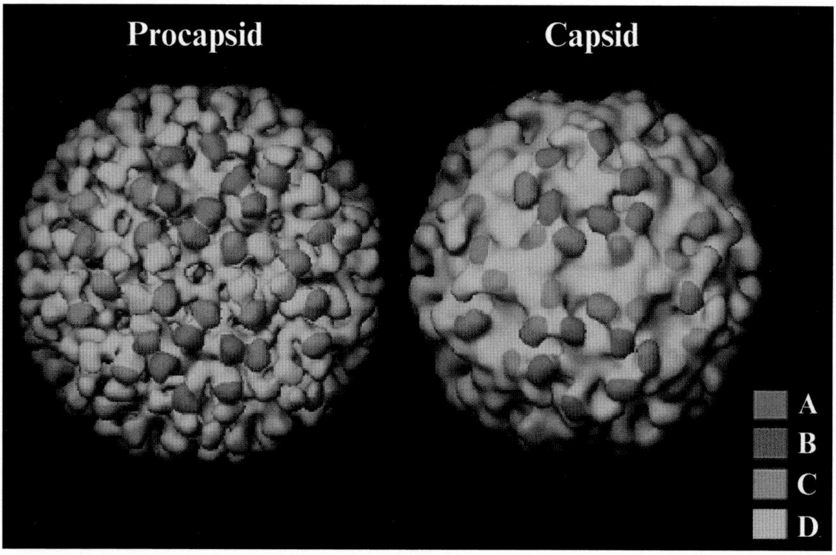

JOHNSON, FIG. 7. A color-coded representation of the outer domain of NωV subunits depicted in cryo-EM reconstructions of the procapsid and capsid. As expected, the X-ray model from the authentic virus fit the cryo-EM reconstruction of capsid with high fidelity, allowing the assignment of the density. The X-ray model of each subunit was then placed in the density of the procapsid, allowing the assignment of the density to individual subunits in that particle. It is clear from the color coding that the outer domains are dimeric in the procapsid and differentiate into trimers in the capsid.

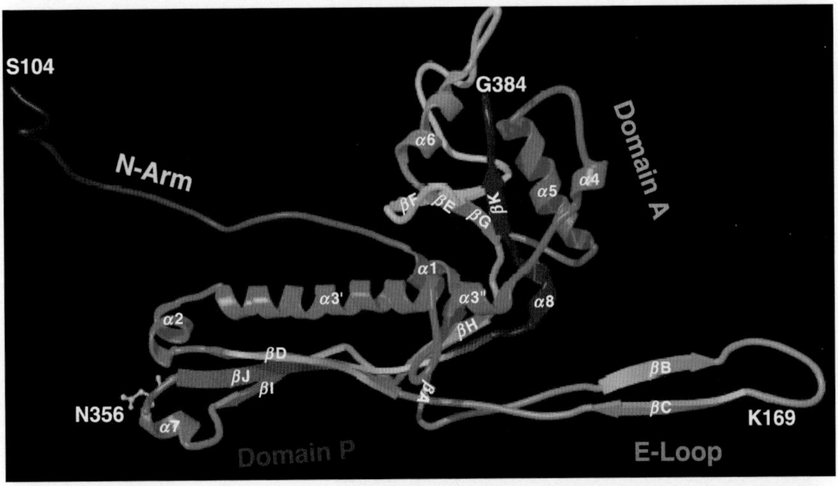

JOHNSON, FIG. 9. Structure of one gp5 subunit, color ramped from the NH_2 terminus (violet) to the COOH terminus (red) (label colors correspond to the domain colors in Fig. 3). The head II NH_2 terminus becomes Ser-104 by maturational proteolysis in the prohead I-to-II transition. The subunit is organized into A and P domains, plus the extended N arm (violet) and E loop (cyan). Lys-169, on the E loop, forms an isopeptide bond with Asn-356 on a neighboring subunit (Fig. 8).

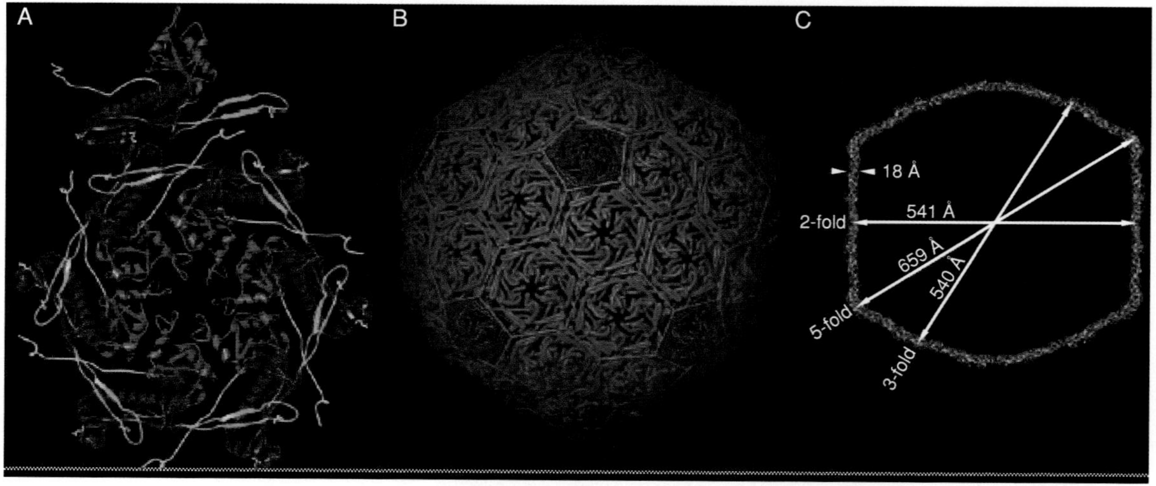

JOHNSON, FIG. 10. Capsid organization. (A) The capsid asymmetric unit (A domain, blue; P domain, red; N arm, yellow; E loop, green). The capsid is a T=7 *levo* arrangement of 420 subunits, organized into hexamers (one shown) and pentamers (one pentamer subunit shown). The subunits wrap around each other in an intricate arrangement. Cross-links cannot form between subunits within the asymmetric unit, because the cross-linking residues (Lys-169 and Asn-356, in white) are not in close proximity. (B) The complete capsid from the particle exterior (each subunit backbone is a smoothed tube). The hexamers (green) are flat, with most of the particle curvature at the concave pentamer (magenta), producing the distinctive icosahedral capsid shape. A T=7 *levo* cage (gray) indicates the quasi-symmetry axes. The pentagon and hexagon vertices are icosahedral or quasi-3-fold axes, with icosahedral or quasi-2-fold axes equidistant between them. (C) Cross-section through the unusually thin empty capsid, which despite its large size (659 Å along the 5-fold) is only 18 Å thick. Icosahedral symmetry axes are indicated.

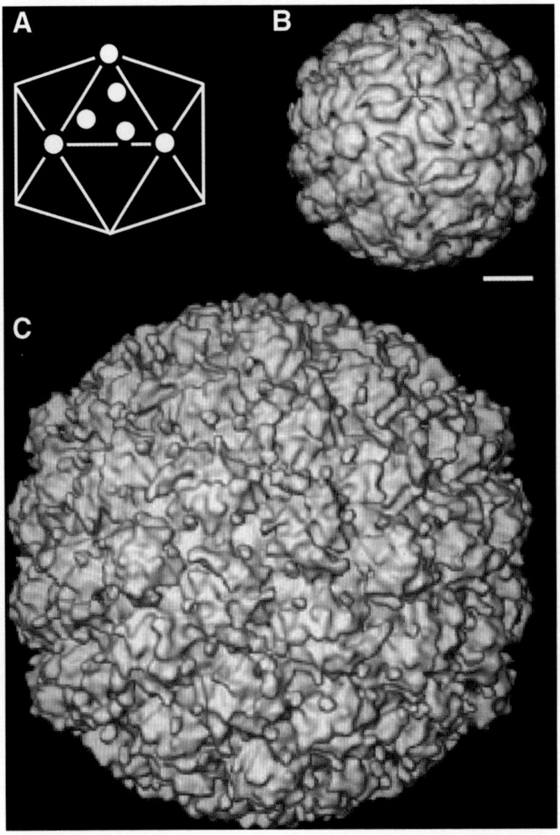

JOHNSON, FIG. 11. HK97 prohead II at 12-Å resolution as viewed along a 2-fold axis. (A) Diagram showing placement of capsomers (hexamers and pentamers) on an icosahedral surface lattice, triangulation number T=7 *levo.* (B) Prohead II at 25-Å resolution (Conway *et al.*, 2001).(C) Exterior view of prohead II, with one hexamer colored in red and blue, corresponding to its two trimers related by a 30-Å "shear" dislocation, and the pentamer subunit in green to complete the asymmetric unit. The contour level corresponds to 100% of expected mass. Bar: 100 Å.

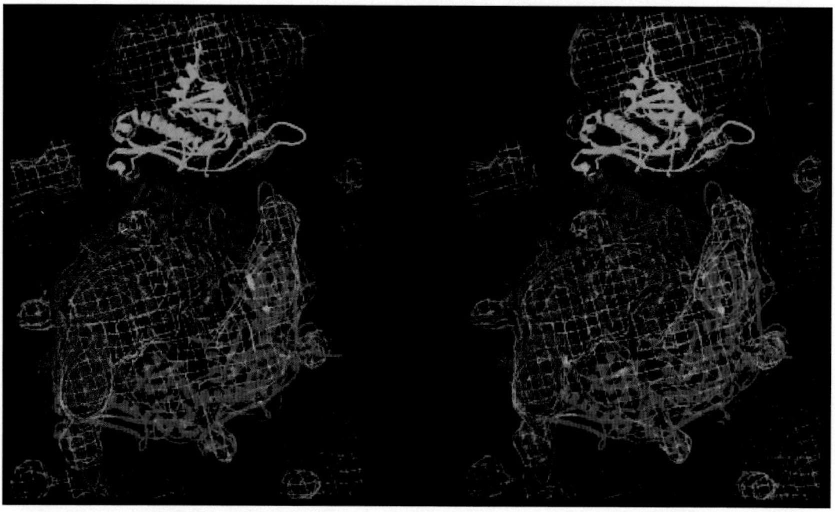

Johnson, Fig. 12. Stereo view of the asymmetric unit of prohead II, consisting of a pentamer subunit and a hexamer, colored as in Fig. 11B; the pseudo-atomic model is enclosed within the density map. The E loop forms a well-defined knob for each subunit; the angle between the loop and domain P was adjusted at the "hinge." The N arm was adjusted as a rigid body, hinged at about Arg-130.

PRASAD AND PREVELIGE, FIG. 1. Stereo views of the encapsidated genome in the X-ray structures of (a) STMV (Larson *et al.*, 1998) and (b) PaV (Tang *et al.*, 2001). Only the genome portion is shown for clarity. The views are along the icosahedral 2-fold axis.

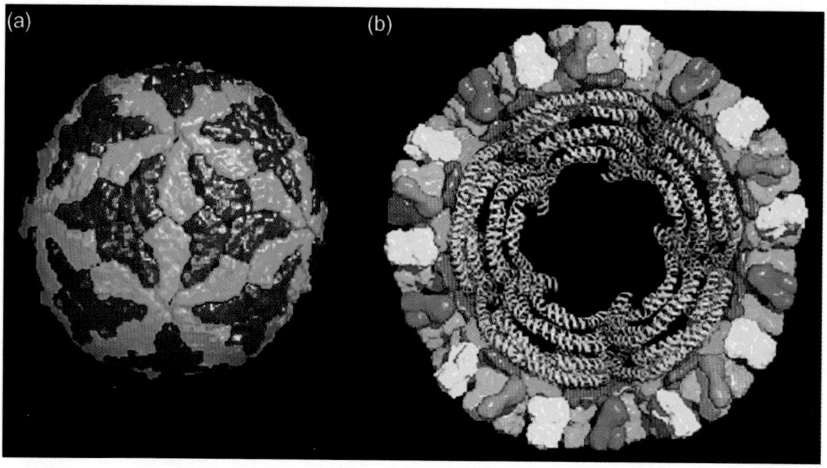

PRASAD AND PREVELIGE, FIG. 2. X-ray structure of the bluetongue virus (BTV) core (Grimes *et al.*, 1998). (a) The "T=2" layer in the BTV core structure. This layer is composed of 120 subunits, with 2 subunits (shown in green and red) in the icosahedral asymmetric units. (b) An equatorial cross-section from the X-ray structure of bluetongue virus core, showing the concentric layers of dsRNA density (in cyan).

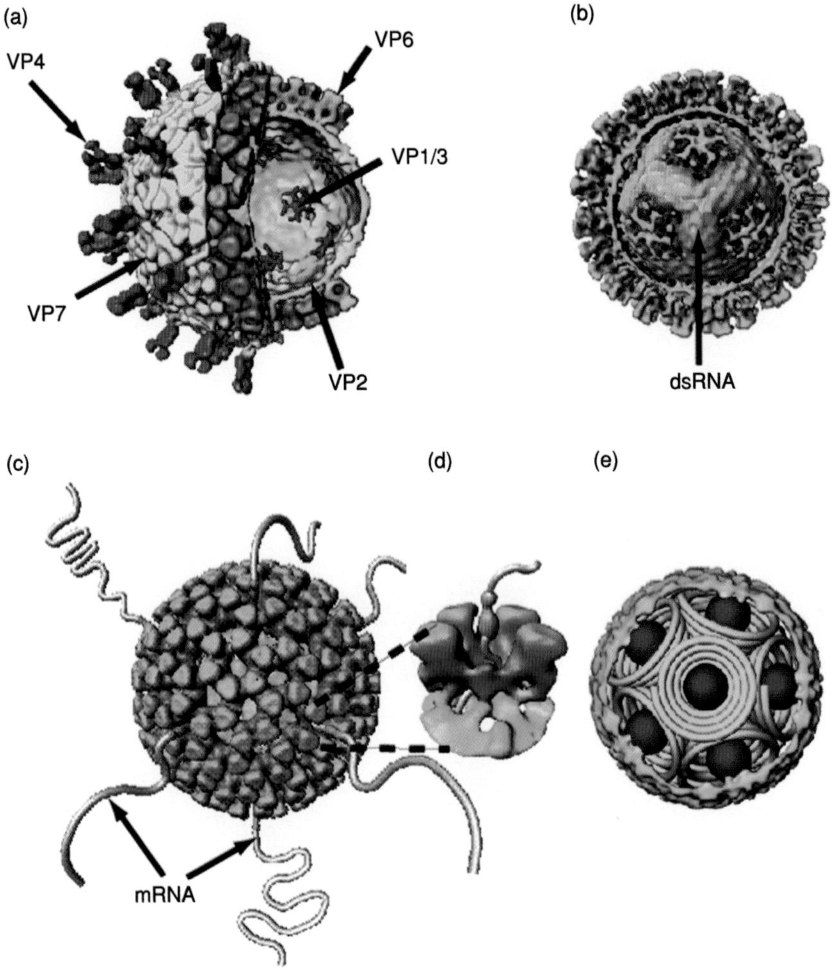

PRASAD AND PREVELIGE, FIG. 3. Summary of the architectural features of rotavirus, a prototypical member of the *Reoviridae* family. (a) Cutaway of the mature virion structure, showing the locations of the various structural proteins. (b) Cutaway of the transcriptionally competent double-layered particle, showing the RNA core. (c). Structure of actively transcribing rotavirus double-layered particles. (d) Close-up of the exit pathway of the mRNA during transcription (Lawton *et al.*, 1997). (e) Model for the organization of the RNA segments (Pesavento *et al.*, 2001).

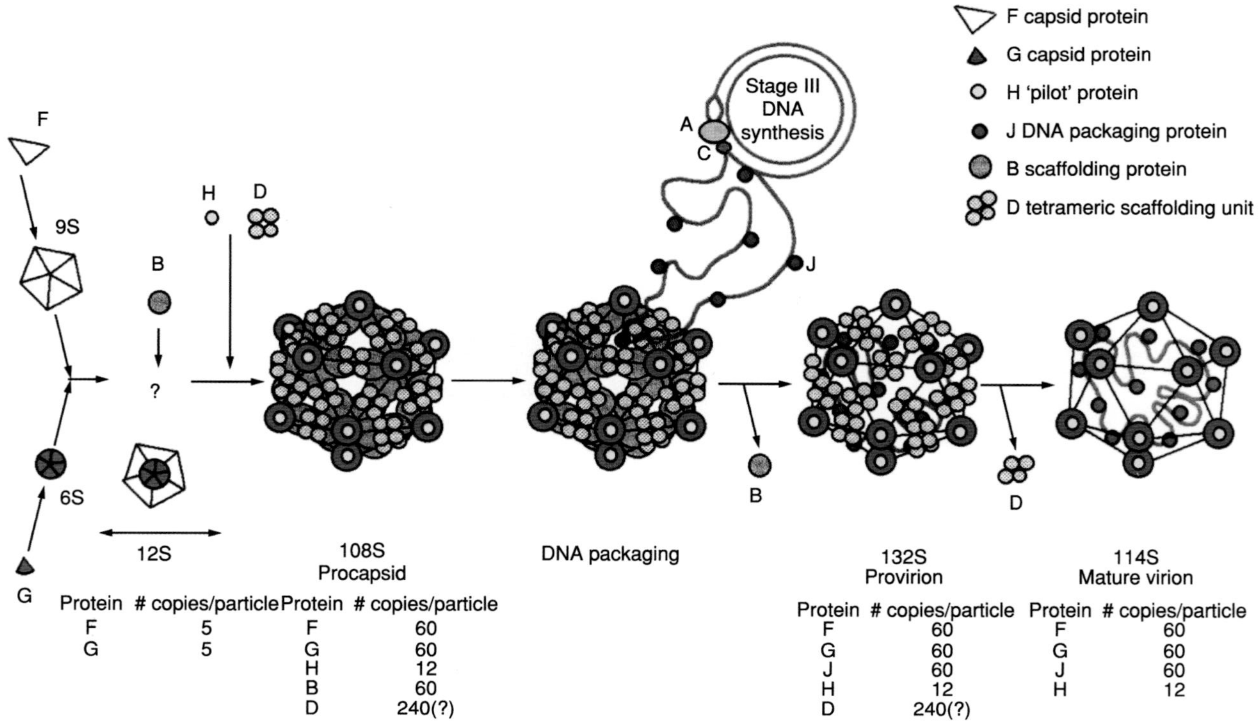

PRASAD AND PREVELIGE, FIG. 4. Schematic representation of the ϕX174 assembly pathway (Ilag *et al.*, 1995).

interactions) may confer a metastable nature to the TMV (Culver *et al.*, 1995; Namba *et al.*, 1989). Such an interaction may be required to maintain an energy balance and potentially be a trigger for driving viral disassembly (Stubbs, 1999).

The assembly process in TMV has been studied extensively, and it is perhaps the best characterized example of cocondensation of capsid protein and RNA. Briefly, the viral RNA interacts with a 20S aggregate, a two-turn helix of the capsid protein, and the assembly proceeds by addition of 20S aggregates through a highly cooperative process (Butler, 1999; Caspar and Namba, 1990; Klug, 1999). It is suggested that a disorder-to-order transition of a loop in the capsid protein, induced by the binding of RNA, may play an important role in this process (Culver *et al.*, 1995; Namba *et al.*, 1989). In the X-ray structure of the 20S aggregate, this particular loop is disordered, whereas in the TMV structure it is ordered. A remarkable feature of the TMV assembly is that the 5′ end of the RNA is pulled through the central hole of the growing TMV rod.

In TMV, the specific recognition of the viral RNA is facilitated by an initiation sequence, AAGAAGUCG, which forms the loop portion of the stem–loop structure (Butler, 1999; Zimmern, 1977). Although from the structure of the fully assembled virus it is difficult to assess how TMV protein initially recognizes this sequence, the high-resolution structure does indeed provide some insights into this process (Namba *et al.*, 1989). Although the three RNA-binding sites in each subunit can accommodate any base, one of the binding sites is particularly suitable for G, and allows favorable hydrogen bond interactions. The repetition of the sequence with every third nucleotide being G thus may provide a strong discrimination for the higher affinity binding of the packaging signal over the rest of the sequence in which the XXG motif does not occur in phase at a statistically significant frequency.

III. Double-Stranded RNA Viruses

The dsRNA viruses are ubiquitous in nature and infect hosts that range from bacteria and fungi to species throughout the plant and animal kingdom (van Regenmortel *et al.*, 2000). Although some of the principles that we have seen with ssRNA viruses are likely to be utilized in the genome organization and packaging in dsRNA viruses, these viruses present a unique set of conditions for genome organization. Because the host cells do not have the enzymatic machinery to convert dsRNA into a translatable mRNA molecule, these viruses must provide a mechanism to synthesize mRNA from the genomic dsRNA. The dsRNA viruses have evolved to carry out genome transcription within the intact particles, using an endogenous transcription apparatus that is an integral part of the virus structure. It is to the advantage of the virus to carry out transcription inside a confined environment not only to avoid any degradation of the genome by cellular nucleases but also to prevent unfavorable antiviral interferon synthesis, particularly in mammalian hosts, which is triggered by increased concentrations of dsRNA. The structural organization of the virus and its genome therefore must be conducive to endogenous enzymatic activities required for transcription.

A. *General Genomic and Capsid Features*

The dsRNA viruses are classified into five major groups (Lawton *et al.*, 2000; van Regenmortel *et al.*, 2000). With the exception of members of the *Totiviridae*, all the dsRNA viruses have multiple segments of dsRNA in their genomes. Members of the *Reoviridae*, a large family of dsRNA viruses that infect a wide variety of hosts including plants, animals, and humans, and cause mild and life-threatening illness, have 10–12 unique dsRNA segments in their genomes (Fields *et al.*, 1996). Generally, the dsRNA segments in members of the *Reoviridae* are monocistronic. Polycistronic segments are common in other groups of dsRNA viruses. All the well-characterized dsRNA viruses have icosahedral capsids, and except for ϕ6, a prototypical bacterial virus in the family *Cystoviridae*, they are nonenveloped. Perhaps necessitated by the general requirement for cell entry and a specialized requirement for endogenous transcription, the capsids of these viruses, with few exceptions, consist of multiple layers.

Structures of several dsRNA viruses including L-A virus (*Totiviridae* family) (Caston *et al.*, 1997) with a single segment, infectious bursal disease virus (*Birnaviridae* family) (Bottcher *et al.*, 1997a) with two segments, ϕ6 (*Cystoviridae*) (Butcher *et al.*, 1997) with three segments, and several members of the *Reoviridae* representing various genera including rotavirus (Prasad and Estes, 2000; Tihova *et al.*, 2001), bluetongue virus (BTV) (Grimes *et al.*, 1997), orthoreovirus (Dryden *et al.*, 1993), aquareovirus (Nason *et al.*, 2000), rice dwarf virus (Zhou *et al.*, 2001), and cypovirus (Hill *et al.*, 1999), have been analyzed by cryo-EM techniques. X-ray structures of L-A virus (J. E. Johnson, personal communication), and transcriptionally competent cores of bluetongue virus (Grimes *et al.*, 1998), orthoreovirus (Reinisch *et al.*, 2000), and rice dwarf virus (J. E. Johnson, personal communication), have been determined to near 3-Å resolution. In these viruses, despite noticeable differences, with the exception of cypovirus and L-A virus, the outer capsid layer generally is based on T=13 icosahedral symmetry.

B. *Unique Organization of Capsid Layer That Surrounds the Genome*

In all the structurally characterized dsRNA viruses, the innermost layer that surrounds the genome has a unique icosahedral organization with 120 subunits (reviewed in Lawton *et al.*, 2000). Such a structural organization, also referred to as "T=2" icosahedral organization (Grimes *et al.*, 1998), has not been found in any other type of virus (Hill *et al.*, 1999) (Fig. 2a; see Color Insert). In several of these viruses, the protein that forms this innermost layer is an RNA-binding protein and may play a role in the structural organization and the endogenous transcription of the underlying genome (Bisaillon and Lemay, 1997; Harrison *et al.*, 1999; Labbe *et al.*, 1994; Loudon and Roy, 1991). In rotavirus, biochemical and structural studies using mature virions and recombinant virus particles have indicated that viral polymerase and the capping enzyme are incorporated as a heterodimer anchored to the inside surface of the T=2 capsid layer, at each of the twelve 5-fold vertices (Prasad *et al.*, 1996). A similar structural organization of the transcription enzymes anchored to the inner surface of the T=2 layer has also been seen in orthoreovirus (Dryden *et al.*, 1998), BTV (Gouet *et al.*, 1999), cypovirus (Hill *et al.*, 1999; Zhang *et al.*, 1999), and cystovirus (Butcher *et al.*, 1997). However, one principal difference between rotavirus or BTV, and orthoreovirus or acquareovirus is that in the latter two, the capping enzyme, although located at the 5-fold axes, is external to the innermost layer, forming a distinct turret-like feature. Although cypovirus, which is architecturally similar to orthoreovirus core with pronounced turrets at the 5-fold axes, and L-A virus, are exceptions to having multilayered capsids, the single capsid layer in these two viruses exhibits T=2 icosahedral organization. Thus the T=2 organization appears to be highly conserved in all the dsRNA viruses. In addition to highly conserved structural organization, the available structural data also indicate that the polypeptide folds of the proteins that form this layer, despite lacking any noticeable sequence similarity, are similar. It is possible that this unique T=2 organization of the innermost capsid layer in these viruses has evolved to serve a dual purpose: to properly position the transcription enzyme complex and to organize the genome to facilitate endogenous transcription.

C. *Endogenous Transcription and Exit Pathway of the Transcripts*

In the several members of *Reoviridae*, following cell entry the mature virion, which generally is transcriptionally incompetent, is converted into a transcriptionally competent unit by the removal of outer capsid layer(s). In rotavirus, for instance, the mature particles have three concentric capsid layers (Prasad and Estes, 2000) (Fig. 3; see Color Insert). Following cell entry, the outer layer is removed and the resulting double-layered particle becomes transcriptionally active (Cohen, 1977; Estes, 2001). The observation that the transcription enzymes, particularly the polymerase, are anchored to the inner surface surrounded by the genome(which is likely the case for all the dsRNA viruses), suggests that the RNA template moves around this complex during transcription (Dryden *et al.*, 1998; Prasad *et al.*, 1996). Three-dimensional cryo-EM reconstructions of actively transcribing rotavirus particles have shown that nascent transcripts exit through a system of channels at the 5-fold axes, consistent with the observation that the initiation of transcription occurs inside the particles in the vicinity of the 5-fold axes (Lawton *et al.*, 1997) (Fig. 3c and d; see Color Insert). In orthoreoviruses also, using both conventional and cryo-EM techniques, the nascent transcripts have been shown to exit through the turrets at the 5-fold axes (Bartlett *et al.*, 1974; Gillies *et al.*, 1971; Yeager *et al.*, 1996). X-ray crystallographic studies on BTV cores in complex with various precursors of transcription also indicate that the exit pathway is through the channels at the 5-fold axes (Diprose *et al.*, 2001).

D. Endogenous Transcription is a Highly Dynamic Process

Kinetic studies on cypovirus, which contains 10 segments, strongly suggest that an independently functioning transcription enzyme complex transcribes each genome segment, and that all the genome segments are transcribed simultaneously (Smith and Furuichi, 1982). Biochemical studies on orthoreoviruses and structural studies on rotaviruses also support such independent and simultaneous transcription of the genome segments (Banerjee and Shatkin, 1970; Bartlett *et al.*, 1974; Gillies *et al.*, 1971; Skehel and Joklik, 1969). These segments are synthesized at the same rate and they accumulate in molar quantities inversely proportional to their length. In *in vitro* experiments, transcriptionally competent particles continue to transcribe as long as the precursors last, indicating thereby that these particles are capable of repeated cycles of transcription. During each cycle of transcription the template must be unwound, separated, rejoined, and rewound for further cycles of transcription. Taken together these data suggest that genome transcription in dsRNA viruses is a highly dynamic process.

E. Structural Organization of the Genome

Although the precise organization of the genome that can facilitate repeated cycles of transcription remains to be elucidated, structural studies have provided some useful hints. Cryo-EM studies on rotavirus

(Pesavento *et al.*, 2001; Prasad *et al.*, 1996) and X-ray crystallographic studies on BTV (Gouet *et al.*, 1999) and orthoreovirus cores (Reinisch *et al.*, 2000) have shown that a significant portion of the genome is statistically ordered and manifests as concentric layers of density in the icosahedrally averaged structures of these viruses (Figs. 2b and 3b; see Color Insert). Earlier, low-angle X-ray scattering studies on orthoreoviruses also indicated that the dsRNA genome is tightly packed as parallel helices in a semicrystalline array (Harvey *et al.*, 1981). On the basis of the volume occupied by the RNA in the BTV core structure, and the molecular weight of the BTV genome, the concentration of the encapsidated genome is estimated to be about 400 mg/ml (Gouet *et al.*, 1999). At such concentrations, if it is assumed that the dsRNA behaves like dsDNA (Livolant and Leforestier, 1996), then dsRNA is likely to exhibit local hexagonal packing with an interstrand spacing of about ~30–32 Å. Such spacing translates to 26- to 28-Å separation between the RNA layers, consistent with what is observed in these structures. More recently, cryo-EM studies on rotaviruses examined under various chemical conditions have shown that the genome in this virus can be condensed to a radius of 180 Å from an original radius of 210 Å by treating the particles with high pH in the presence of ammonium ions (Pesavento *et al.*, 2001). When these pH-treated particles are brought back to physiological pH, the genome expands to the original radius. This study demonstrates the remarkable stability of the capsid, and resilience of the genome, which may be required attributes to carry out continuous transcription in a confined environment.

A plausible model for the structural organization of the genome that emerges from the above-mentioned biochemical and structural information on dsRNA viruses is that each dsRNA segment is spooled around a transcription enzyme complex at the 5-fold axes inside the innermost capsid layer (Gouet *et al.*, 1999; Pesavento *et al.*, 2001; Prasad *et al.*, 1996). This model allows a capsid to contain up to 12 independent transcription complexes, each with an individual dsRNA segment attached for concurrent transcription. Such a model is also consistent with the observation that no dsRNA virus containing more than 12 segments has ever been isolated. A stylized version of this model is shown in Fig. 3e. Each dsRNA segment is depicted as an inverted cone at the 5-fold vertex surrounding a transcription enzyme complex. In addition to allowing for simultaneous and independent transcription of the dsRNA segments, this model also provides a simple mechanistic explanation for the ability of the genome to undergo reversible expansion and condensation. The isometric and concentric condensation is achieved simply by reducing the interstrand separation in each of these cones.

F. Genome Replication and Packaging

Equally fascinating and perhaps more mysterious is the question of how a correct set of dsRNA segments is packaged inside each particle. The plus-strand RNA transcripts that exit from the transcribing particles, in addition to directing the synthesis of viral proteins, function as templates for negative-strand synthesis. Thus, replication and transcription can be thought of as complementary processes both involving the viral polymerase (reviewed in Lawton *et al.*, 2000). Lawton *et al.* have suggested that the association of one transcription enzyme complex with one genome segment, as envisioned inside the virus particles, may initially begin with the assembly process (Lawton *et al.*, 2000). The polymerase responsible for replicating and packaging a particular segment may remain associated with it during the particle assembly, and be responsible for transcribing that same segment later during the endogenous transcription process. In some of the well-studied members of the *Reoviridae*, such as rotavirus, orthoreovirus, and BTV, although the mechanism is presently unclear, it is evident that the entire process of genome replication, packaging, and perhaps segment assortment is choreographed by some of the virus-encoded nonstructural proteins.

Envisioning a model for how the genome segments are packaged inside the particles in dsRNA viruses is more complicated than in the ssRNA viruses discussed earlier, not only because of the multiple segments but also because of the requirement for duplex formation inside a confined environment. In none of the dsRNA viruses has free dsRNA been found in infected cells. *In vitro* biochemical studies together with structural studies of $\phi6$ (three dsRNA segments) clearly indicate a packaging model in which the three mRNA segments are sequentially incorporated into a preformed core in a process coupled to sequential conformational changes within the core (reviewed in Mindich, 1999). Subsequent to mRNA incorporation, the core undergoes a dramatic expansion and activates the endogenous polymerase for negative-strand synthesis and duplex formation (Butcher *et al.*, 1997).

Development of such an *in vitro* packaging and replication system has not been possible for any members of the *Reoviridae*, despite success in producing recombinant proteins and empty virus-like particles. As shown by cryo-EM analysis, unlike empty cores of $\phi6$, empty recombinant core particles of rotavirus, BTV, or orthoreovirus are identical in size to native particles. Another critical difference between the $\phi6$ system and reoviruses is that the packaging protein P4 of $\phi6$ is an integral part of the virion structure located at each of the 5-fold axes as a hexamer (de Haas *et al.*, 1999). This protein, an NTPase, is suggested to package RNA through the 5-fold vertices (de Haas *et al.*, 1999; Juuti *et al.*, 1998). However, in the

Reoviridae, such a packaging protein as an integral part of virion structure has not been observed. Instead, in these viruses, virus-encoded nonstructural proteins are suggested to be involved (Patton and Spencer, 2000). The X-ray structure of a nonstructural protein, NSP2 of rotavirus, implicated in replication/packaging has been determined (Jayaram *et al.*, 2002). This protein exhibits NTPase, ssRNA-binding, and nucleic acid helix-destabilizing activities (Schuck *et al.*, 2001; Taraporewala *et al.*, 1999; Taraporewala and Patton, 2001). Temperature-sensitive mutants of NSP2 fail to replicate the genome and produce mostly empty particles, implicating NSP2 in genome replication and packaging (Chen *et al.*, 1990; Ramig and Petrie, 1984). *In vivo* studies have shown that NSP2, in association with viral RNA and polymerase, localizes to viroplasms of the infected cells where these process occur (Aponte *et al.*, 1996; Kattoura *et al.*, 1994; Petrie *et al.*, 1984). On the basis of these biochemical data, it is hypothesized that NSP2 functions as a molecular motor using the energy derived from NTP hydrolysis to package the dsRNA. The existing biochemical data suggest that NSP2 may be functionally homologous to NS2 in BTV (Fillmore *et al.*, 2002; Taraporewala *et al.*, 2001; Zhao *et al.*, 1994) and σNS of orthoreovirus (Gillian and Nibert, 1998). Given the differences between ϕ6 and members of the *Reoviridae,* it remains to be seen whether the model of packaging RNA into preformed cores is applicable to dsRNA viruses with larger numbers of segments. On the basis of the existing biochemical and structural data on *Reoviridae* members, alternative packaging models assisted by nonstructural proteins including coassembly of core proteins and genome segments remain a distinct possibility.

IV. Single-Stranded DNA Viruses

Viruses with ssDNA in their genomes are classified into three groups: microviruses, which have a circularized ssDNA genome; parvoviruses, which have a linear ssDNA genome; and geminiviruses, which have two circular genomes (van Regenmortel *et al.*, 2000). Because of the ssDNA genome, the replication strategies in these viruses differ considerably from those in RNA viruses. Most notably, replication and packaging, particularly in animal ssDNA viruses, occur in the nucleus of the host cell. In addition to crossing the initial cell membrane barrier, the genomes of these viruses must find their way into the nucleus. In this section, we focus mainly on the microvirus and parvovirus groups, as X-ray crystallographic structures are available for some of the representative members of only these groups of ssDNA viruses. Geminiviruses, so called because of a geminate capsid consisting of two incomplete icosahedra (T=1) with a total of 22 pentameric capsomers, are indeed interesting. Although an elegant cryo-EM reconstruction of maize streak virus in this family has been determined (Zhang *et al.*, 2001), as yet there is no high-resolution structure of any member of this group of economically important plant viruses.

A. *Microviruses*

A well-characterized prototype of the microvirus group of ssDNA viruses is ϕX174, an icosahedral bacterial virus. The ϕX174 system represents the first virus system for which not only the complete sequence was determined but also the *in vitro* genome synthesis and packaging were successfully achieved (Aoyama *et al.*, 1983; Aoyama and Hayashi, 1982; Sanger *et al.*, 1977). The genome of this virus consists of a positive-sense circular DNA with 5386 nucleotides. The assembly of this virus proceeds in several stages, which includes the formation of a procapsid (see the article by Fane and Prevelige in this volume). Formation of an empty procapsid, into which the genome is inserted, is a common phenomenon in bacteriophages, as elaborated in Section V on dsDNA viruses (Fig. 4; see Color Insert). In ϕX174, the assembly is initiated by the formation of pentameric structures of two types of capsid proteins, F and G (reviewed in Hayashi *et al.*, 1988). These pentamers associate with H protein, and two scaffolding proteins, B and D, to form an icosahedral procapsid structure (Mukai *et al.*, 1979). The procapsid consist of 60 copies each of F, G, and B proteins, 240 copies of D protein, and 12 copies of H protein (Dokland *et al.*, 1997). The viral DNA along with 60 copies of J protein is inserted into this procapsid, facilitated by virus-encoded packaging accessory proteins A and C (Aoyama and Hayashi, 1982). During this process, one of the scaffolding proteins, B, is eliminated from the capsid, and the procapsid enters a penultimate provirion stage. In the final stages of maturation the remaining scaffolding protein is also eliminated from the capsid. The final maturation stage is thought to be triggered by the increased levels of divalent cations during cell lysis.

Capsid Maturation

The X-ray structure of mature ϕX174 (McKenna *et al.*, 1992), and cryo-EM structures of procapsid and provirion (Ilag *et al.*, 1995), have provided some insights into the nature of conformational changes during capsid maturation, the role of the scaffolding proteins, and also the entry path of the DNA. These studies have shown that the G protein, which forms the spikes at the icosahedral 5-fold axes, is not perturbed during maturation, whereas the F protein, which forms the capsid, undergoes significant conformational changes. Both F and G proteins have an eight-stranded antiparallel β-barrel structure commonly seen in other icosahedral capsid proteins. In the procapsid structure, the F and G pentamers are not in close association. The scaffolding proteins "glue" the F and G pentamers to form a stable procapsid. The presence of holes of suitable size for entry of DNA in the vicinity of the icosahedral 3-fold axes, and their absence in the mature virions, suggest that the ssDNA may enter through these pores. Following DNA entry, and exit of the scaffolding B protein, the F protein undergoes conformational changes accompanied by a radially inward shift to establish new contacts and form a protective coat for the encapsidated genome (Dokland *et al.*, 1998). Although maturation-dependent conformational changes are seen in other bacteriophages, the shrinkage of the capsid contrasts with dsDNA bacteriophages in which distinct expansion is observed during maturation (see Section V on dsDNA viruses).

B. *Model for Genome Entry*

How does circular DNA enter the icosahedral procapsid? The entry of DNA is assisted by three virus-encoded proteins, A, C, and J, of which proteins A and C are not part of the virion structure either at the procapsid or at the maturation stage. Earlier studies on ϕX174 have shown that genome packaging into procapsids is concurrent with DNA replication (Dressler *et al.*, 1978; Hayashi, 1978). On the basis of these and other studies, as reviewed by Casjens (1985), a model for DNA packaging has been proposed. In this model, the virus-encoded A protein, a nicking enzyme, nicks the viral strand from the circular dsDNA replicative intermediate, and remains covalently associated with the 5′ end providing 3′-hydroxyl end to prime unidirectional leading-strand synthesis. The A protein, along with the dsDNA replicative intermediate and host cell replication enzymes, then binds to the procapsid, perhaps assisted by the C protein, to form a preinitiation complex, which facilitates insertion of the viral strand as replication progresses (Aoyama and Hayashi, 1986). The X-ray structure of the procapsid and mutational analysis, have identified the possible location for binding of the preinitiation complex to be in a depression within the capsid protein that skirts around the icosahedral 2-fold axes (Dokland *et al.*, 1997, 1998).

C. *Ordered DNA in the Capsid Structure*

Along with the DNA, the virus-encoded J protein also enters the procapsid. What is the role of this protein in assembly/genome encapsidation? The J protein, although not necessary for DNA replication, is required for DNA packaging (Hamatake *et al.*, 1985). Its demonstrated ability to bind to both single- and double-stranded DNA, and the highly basic nature of this protein, support the view that in addition to being involved in charge neutralization it may be involved in translocating the DNA into the procapsid and also in protecting the DNA from nucleases (Hamatake *et al.*, 1985). In the X-ray structure of mature ϕX174, only a part of the C-terminal region, which is not as basic, is ordered (McKenna *et al.*, 1992). The rest of the protein possibly in association with the genome is disordered. In the X-ray structure of mature ϕX174, a small portion of the ordered DNA (11 nucleotides per icosahedral asymmetric units) is also observed. This ordered DNA is seen in the cavities between the F proteins, and near the ordered portion of the J protein.

D. *Genome Release*

Early electron microscopic studies on ϕX174 suggest that DNA ejection is through the G-protein spikes at the 5-fold vertices (Mano *et al.*, 1982). The X-ray structure of the mature capsid indicates that the pentameric G protein spikes create a hydrophilic hole along the 5-fold axes. Also present inside this hole is a putative Ca^{2+}-binding site and some disordered density. This disordered density is tentatively interpreted as being due to the H pilot protein, consistent with the observation that along with DNA H protein is injected into *Escherichia coli.* Binding of Ca^{2+} to the channel during the entry processes may trigger the release of both H protein and DNA. The hydrophilic channel, the interior of which is lined by acidic residues, may serve as an electrostatic focusing device for the smooth exit of negatively charged DNA. It is not clear whether the ejection of H protein during genome release is functionally similar to the externalization of γ peptide in nodaviruses or of picornavirus VP4 during their genome release.

E. *The* Parvoviridae

Parvoviridae is a family of small icosahedral, nonenveloped animal viruses that contain a linear ssDNA genome of approximately 5000 bases with short unique terminal palindromic sequences that fold back on themselves to form hairpin duplexes (van Regenmortel *et al.*, 2000). They are further classified into two subfamilies, *Parvovirinae* and *Densovirinae,* and various genera depending on host specificity, strand specificity of the genomic ssDNA, and on whether helper viruses are required for productive infection (dependoviruses). Members of the *Parvoviridae* generally have 60 copies of 2 to 4 proteins, VP1, VP2, VP3, and VP4, which are alternative forms of the same gene product, differing only at their N termini.

The X-ray structures of several members in this family have been determined including canine parvovirus [CPV (Tsao *et al.*, 1991; Wu and Rossmann, 1993)], murine parvovirus [MPV (Agbandje-McKenna *et al.*, 1998)], minute virus of mice [MVM (Agbandje-McKenna *et al.*, 1998)] in the subfamily *Parvovirinae*, and an insect parvovirus in the subfamily *Densovirinae* (Simpson *et al.*, 1998). Although these viruses are architecturally similar, with a T=1 icosahedral capsid and their capsid proteins exhibiting an eight-stranded antiparallel β-barrel fold, there are significant conformational differences in the capsid protein. Confining our discussion to genome organization, in all these viruses a significant amount of ordered DNA is observed. The maximum amount of ordered DNA, representing about 32% of the genome (23 nucleotides per icosahedral asymmetric unit), is seen in the MVM structure. The ordered DNA is located at the interior surface of the capsid in a cleft between the 5-fold-related subunits. The conformation of the DNA is rather unusual with bases pointing out toward the protein-binding site, and backbone phosphates, coordinated by putative divalent cations, pointing inward with respect to the loop conformation of the DNA (Chapman and Rossmann, 1995). Thus there are no interactions between the phosphate groups and the basic residues of the capsid protein that are observed in parvovirus structure. Comparison of the empty and full canine parvovirus structures has indicated that the capsid protein undergoes modest conformational changes on DNA binding.

In the specific recognition of DNA, the functional groups on the bases are usually involved; perhaps the ordered structure of DNA in parvovirus is suggestive of some specific interactions. The statistical matching of the electron density profile with possible stretches of 11 nucleotides along the sequence of the genome resulted in ~40 sites that the capsid protein can bind (Chapman and Rossmann, 1995). In terms of genome packaging, the significance of such mildly specific interactions is not immediately obvious. Considering that the replicative intermediate must be a dsDNA, the process of replication and perhaps subsequent packaging bears a resemblance to gene replication and packaging in ϕX174. However, in these viruses with linear ssDNA, the hairpin duplexes formed by the palindromic sequences in the genomic DNA, in contrast to the circular replication intermediates in ϕX174 (rolling circle versus rolling hairpin), provide a basis for replication (Cotmore and Tattersall, 1996). In this process, similar to the A and C proteins of ϕX174, the nonstructural proteins of parvovirus are implicated in replication. In MVM, NS1 is found covalently bound to the 5′ end of the progeny strand, similar to the A protein of ϕX174 (Christensen *et al.*, 1995; Cotmore *et al.*, 1995). In addition to nonstructural proteins, autonomous parvoviruses rely on the host proteins during replication. In dependoviruses, which are not replication competent, replication depends on helper viruses such as herpesvirus or adenovirus.

The empty particles of parvovirus are formed during infection with kinetics that suggest them to be packaging precursors (Casjens, 1985). However, the structures of parvoviruses do not provide sufficient hints to ascertain the possibility that the genome is inserted into preformed empty particles. The parvovirus structure has a channel at the 5-fold axes that is to some extent filled with the glycine-rich N-terminal residues of the capsid protein VP2, a canyon that surrounds the 5-fold axes, and a small depression at the icosahedral 2-fold axes. It is proposed that the channel at the 5-fold axis is used for externalizing the N-terminal ends for cleavage during the cell entry process, analogous to externalization of picornavirus VP4 or nodavirus γ peptide (Agbandje-McKenna *et al.*, 1998). Such a cleavage in parvovirus appears to be necessary for efficient translocation of the capsid from plasma membrane to nucleus. Whether the 5-fold channel is used for releasing the genome is not clear.

V. Double-Stranded DNA Viruses

The families of dsDNA-containing viruses include bacterial, plant, and animal viruses. Because of early work done in defining their life cycles, the bacteriophages perhaps represent the best understood dsDNA viruses. Studies have suggested striking similarities between the life cycle of bacteriophages and that of herpesviruses. A common feature of these two systems is the packaging of the DNA to extremely high density through a portal complex situated at a single vertex into a preformed capsid, or procapsid. In contrast, some dsDNA viruses, such as the *Papovaviridae*, are thought to package their dsDNA genome by cocondensation, and for others, such as adenovirus, the verdict is still out.

A. *The Double-Stranded DNA Bacteriophages*

The Capsid Structure

The dsDNA-containing bacteriophages are icosahedral virions composed of a single major capsid protein, sometimes although not always decorated with accessory proteins. They can be prolate, such as is the case with T4 and ϕ29, or isometric, such as is the case for λ, P22, and T7, which all display T=7 lattices. To date, the only structure of a dsDNA-containing bacteriophage that has been solved to atomic resolution is that of HK97 (Wikoff *et al.*, 2000). The fold of the HK97 coat protein represents a new fold. It is a mixed α/β protein (28% α helix, 32% β sheet) organized into distinct axial and peripheral domains. Of particular note is the presence of an N-terminal arm and the E loop, both of which represent large excursions from the otherwise compact subunit. The E loop in particular invades an adjacent subunit during maturation and autocatalytically forms an intersubunit isopeptide bond between its lysine residue 169 and Asn-356 of the adjacent subunit. The topology in this arrangement is such that the entire capsid becomes cross-linked into a chain-mail arrangement. This allows an extremely thin-shelled capsid to be structurally stable. Although the cross-linking is not conserved among the dsDNA bacteriophages, given the similarity in life cycle it would not be surprising if the protein fold were conserved.

B. *Structural Organization of the Genome*

The organization of the dsDNA within phage capsids has long intrigued researchers in part because it serves as a model for eukaryotic chromosome condensation. Early X-ray diffraction experiments displayed a strong reflection with a 24-Å spacing, indicative of close parallel packing of the DNA, and 12-, 8-, and 3.4-Å reflections consistent with the bulk of the DNA being B-form (Earnshaw *et al.*, 1976; Earnshaw and Harrison, 1977; North and Rich, 1961). However, despite considerable effort, it has been difficult to determine the detailed packaging of the DNA within the phage head, and a wide variety of models have been proposed. Among these are the ball of string, coaxial spool, liquid crystal, spiral fold, and folded toroid (Black, 1989; Earnshaw *et al.*, 1978a,b; Harrison, 1983; Hud, 1995; Lepault *et al.*, 1987; Richards *et al.*, 1973). It is also unclear whether the DNA packaging arrangement is identical in all capsids within a population, or whether it varies from particle to particle.

Toroidal morphology is observed in electron micrographs of the DNA released from gently lysed phage heads, suggesting the DNA within the capsid may be packed with a similar arrangement (Earnshaw *et al.*, 1978b; Kosturko *et al.*, 1979; Richards *et al.*, 1973). A variety of biophysical studies lend support to the coaxial spool model; ripples modulating the 25-Å diffraction band suggest coaxial winding (Earnshaw and Harrison, 1977), as do comparisons of electron micrographs of intact phage with models of projection images

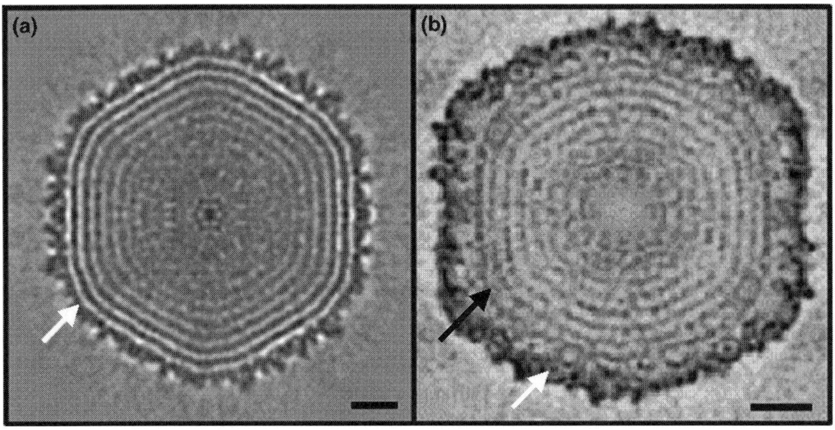

FIG. 5. Structural organization of the genome in dsDNA viruses revealed by cryo-EM techniques. Cross-sectional views from the cryo-EM reconstructions of (a) P22 (Zhang *et al.*, 2000) and (b) PRD1 (Martin *et al.*, 2001). Concentric rings of DNA, with a spacing of 25 Å, are seen inside the capsid layers (indicated by white arrows in each). PRD1 virus has a lipid bilayer (black arrow) between the genome and the capsid layer. Scale bars: 100 Å.

derived from different packaging arrangements (Cerritelli *et al.*, 1997) (and see the article by Cerritalli *et al.*, this volume). Studies on bacteriophage P22 provide a similar picture (Zhang *et al.*, 2000), that of concentric rings of DNA arising from coaxial spooling (Fig. 5a). In T7, the DNA is coiled about an axis extending through the portal, this also appears to be the case for bacteriophage P22 (Zhang *et al.*, 2000). Earlier work using oriented samples suggested that the spool axis would lie perpendicular (Earnshaw *et al.*, 1978b) or at 43° (Kosturko *et al.*, 1979) relative to the phage axis.

C. Genome Packaging

For the dsDNA-containing phages, the mechanism of DNA recognition and packaging is well described. The phage genome is replicated as a concatemer and packaging proceeds unidirectionally and processively along the concatemer from a sequence-defined initial cut site termed a *pac* site (Black, 1989). Whereas the initial cut is sequence specific (the exception being bacteriophage T4; Kalinski and Black, 1986), subsequent cuts can occur either at the reoccurrence of the same sequence, or the second (and subsequent cut sites) can be defined by DNA length in a process called headful cutting. Examples of the former are λ (Feiss *et al.*, 1977; Hohn, 1983) and T7, whereas examples of the latter include P22

(Casjens *et al.*, 1987; Tye *et al.*, 1974) and T4 (Kalinski and Black, 1986). The number of genome equivalents packaged from a single concatemer varies from phage type to phage type but is typically approximately three.

The packaging sequence on the DNA is recognized by a group of phage-encoded enzymes called "terminases." The terminases are generally two protein complexes (with the exception of ϕ29, the terminase of which is a single protein subunit). The small subunit (~20 kDa) is responsible for recognizing the DNA, while the large subunit (~70 kDa) recognizes the portal complex in the prohead. The terminase complex has ATPase activity, and DNA packaging is an energy-dependent process (Black, 1989). Estimates are that it requires a molecule of ATP for every base pair packaged (Guo *et al.*, 1987a).

The portal complex is composed of 12 molecules of a phage-encoded protein (Bazinet and King, 1985). This dodecamer is located at a 5-fold vertex of the icosahedral capsid, and it was suggested early on that the function of this symmetry mismatch might be to enable smooth rotation during DNA packaging (Hendrix, 1978). Although there is limited sequence homology between portal proteins, there is considerable morphological homology. The crystal structure of the portal complex from ϕ29 has been reported (Guasch *et al.*, 2002; Simpson *et al.*, 2000), and there is reason to believe that other portal complexes will be similar (Moore and Prevelige, 2001). The ϕ29 portal complex appears to be unique, however, in that a small (174-nucleotide) RNA molecule termed a pRNA is required for packaging (Guo *et al.*, 1987b). This RNA molecule, as either a pentamer or hexamer, forms a structural element of the portal complex. It appears that the DNA translocates into the procapsid through the central channel of the portal protein, and current models favor the idea that the portal protein does in fact rotate in the procapsid during packaging; thus it comprises a molecular motor (Simpson *et al.*, 2000). Single-molecule packaging experiments have estimated the stall force at approximately 50 pN, making it one of the strongest motors examined to date (Smith *et al.*, 2001). Interestingly, the packaging rate deceases as the capsid becomes filled, suggesting a build-up in internal force, and a pressure of 6 MPa has been estimated for the packaged DNA.

D. DNA Release and Entry

Given the amount of pressure stored in the packaged DNA it is tempting to speculate that this might provide the driving force for entry of the DNA into the host cell (Smith *et al.*, 2001). (Phages are not internalized during infection, but rather inject their DNA into the host.) However, evidence in bacteriophage T7 suggests the possibility of a more complex mechanism. Careful measurements of the rate of DNA entry show that it is constant across the entire length of the genome, an observation inconsistent with the idea that pressure alone drives translocation. Biochemical and genetic evidence suggests that approximately 850 base pairs of DNA can enter the host cell unassisted and that transcription by the *E. coli* and T7 RNA polymerases then powers continued translocation (Garcia and Molineux, 1996; Molineux, 2001; Struthers-Schlinke *et al.*, 2000). Experiments with λ and T5 suggest that transcription-independent mechanisms may be at work in these phages, with the energy perhaps supplied by the stored energy in the highly condensed DNA (Filali Maltouf and Labedan, 1983, 1985).

E. The Herpesviridae

In many respects the *Herpesviridae* resemble the dsDNA bacteriophages. The capsids are icosahedral T=16 lattices composed of a single major coat protein. Although no crystallographic data are available, cryo-EM image reconstructions have been carried out to 8.5-Å resolution (Zhou *et al.*, 2000). At this resolution, it is possible to fit in α helices and approximately 17% of the total protein is helical. Like phages, the first assembly intermediate is a procapsid, devoid of nucleic acid, into which the DNA is packaged (Newcomb *et al.*, 1999, 2000). The structure of the packaged DNA has been investigated by thin-section and negative stain electron microscopy and more recently by cryoelectron microscopy (Booy *et al.*, 1991). Capsids containing DNA (C capsids) displayed a 25-Å striation or fingerprint pattern reflective of tightly packed DNA that was not seen in empty (A capsid) particles. The density appeared uniform and contained no strongly contrasted features, results consistent with the DNA being nonicosahedrally distributed. The geometry of the packaged DNA remains unsolved.

The occurrence of a procapsid form into which the DNA was packaged suggested the possibility of a portal protein complex, analogous to that found in bacteriophages, and indeed one has been identified. The protein, pUL6, is located at a single vertex of the procapsid, forms dodecameric rings with a central channel, and is required for DNA packaging into the procapsid (Newcomb *et al.*, 2001). Seven proteins are required for stable packaging of DNA into the procapsid. There is good evidence to suggest that two of these, the 81-kDa UL15 and UL28 proteins, comprise the terminase complex (Sheaffer *et al.*, 2001). The similarity in life cycle and packaging components suggests the basic mechanism of DNA packaging is conserved between the herpesviruses and bacteriophages.

F. The Polyomaviruses

The capsids of the polyomaviruses are also T=7 icosahedra. Strikingly, they are composed of 360 subunits arranged as all pentamers rather than an the expected mixture of quasi-equivalent pentamers and hexamers (Belnap *et al.*, 1996; Liddington *et al.*, 1991; Rayment *et al.*, 1982). To accomplish this striking departure from quasi-equivalent packing, stable pentameric building blocks are tied together by differently ordered C-terminal arms (Liddington *et al.*, 1991). Whereas the capsids of the dsDNA phages and the *Herpesviridae* have naked, highly condensed dsDNA contained within, the DNA within polyomaviruses is both highly condensed and complexed with cellular histones in the form of a minichromosome. The assembly pathway is differentiated from that of the dsDNA phages and *Herpesviridae* in that the minichromosome and capsid proteins appear to copolymerize. Interestingly, the histone H1 is found in the cellular minichromosome but not in the virion, suggesting it is selectively stripped during assembly (reviewed in Garcia and Liddington, 1997).

In cryo-EM-based reconstructions of the virion, the chromosome appears as an inner core, separated from the capsid wall by a gap of 0.5–1.5 nm (Baker *et al.*, 1988). This gap is surprising because of the clear biochemical evidence of an interaction between the virus structural proteins and the minichromosome. It is possible that flexibility in the binding domains results in the appearance of gaps in the reconstructions. No strong features are observed in the chromosome core, suggesting that it is not packed with icosahedral symmetry, although the data do not rule out other symmetries. The geometry of the packing of the chromatin within the core remains an unsolved problem.

G. Adenovirus

The adenovirus capsid is an icosahedral capsid that can be described as pseudo-T=25. The 12 pentameric vertices are each composed of 5 molecules of the penton bases, carrying a trimeric fiber. The 20 faces are composed of the 4 molecules of hexon protein, which itself is a trimer of the 967-residue-long polypeptide II. The crystal structure of the hexon has been solved and it reveals that the trimeric subunits are intimately intertwined, thus accounting for their observed stability. The capsid also contains four minor proteins, which are thought to act as cementing proteins stabilizing the entire capsid.

Adenovirus contains a 35-kb dsDNA complexed with four adenovirus-encoded proteins (V, VII, m, and terminal), perhaps in a nucleosome-like arrangement (Burnett, 1997). The arrangement has been controversial, as it has been suggested that the nucleoprotein is arranged both with and without overall icosahedral symmetry. Cryo-EM data support the notion that there is perhaps a small degree of icosahedral symmetry enforced by the capsid lattice, but that there is no well-structured overall icosahedral symmetry within the core (Stewart *et al.*, 1991).

A striking example of an evolutionary relationship between adenovirus and a bacteriophage has been recognized. The bacteriophage PRD1 is an icosahedral bacteriophage that contains an internal membrane. The architecture of the virion is similar to that of adenovirus, a pseudo-T=25 capsid in which 240 molecules of a homotrimer of a 43-kDa protein called P3 are assembled into faces and the pentameric protein P31, in concert with a trimer of a two-protein receptor-binding complex, form a vertex structure similar to adenovirus spikes (Butcher *et al.*, 1995). The crystal structure of the PRD1 P3 protein revealed that the unusual fold and intimate interactions between subunits observed in adenovirus hexon are reiterated in PRD1 P3 (Benson *et al.*, 1999). The structural similarity between these two viruses, as well as life cycle similarities, provide strong evidence of an evolutionary relationship (Hendrix, 1999). Cryoelectron micrographs of central sections of the PRD1 virion reveal five concentric rings, corresponding to DNA, which display a 13-Å width and 25-Å spacing as seen in the dsDNA bacteriophages and herpesviruses (Martin *et al.*, 2001) (Fig. 5b).

VI. Conclusions

X-ray crystallography and cryo-EM techniques have provided substantial structural information about the conformation of the encapsidated genome in various icosahedral viruses. The number of instances in which we have seen ordered genomes has steadily increased, and in three of the virus structures (STMV, PaV, and BTV) more than 50% of the genome is visualized. However, the tobacco mosaic virus remains the only virus for which the entire genome is accessible to structural techniques. In all these virus structures, the discernible regions of the genome are in direct contact with the capsid, thus acquiring the capsid symmetry and becoming structurally visible. In those cases in which the nucleic acid interacts less intimately with the capsid and thereby fails to adopt the capsid symmetry, it has proved more difficult to obtain structural information. Likewise, it has proved difficult to obtain information about the connectivity of the nucleic acid in those cases in which only ordered patches are seen. Whether these patches are unique in their sequence or physicochemical properties remains to be resolved. Further studies involving development of *in vitro* encapsidation experiments coupled with mutational analysis, such as are being carried out in nodaviruses, may provide some answers to questions concerning why there is a variability in the amount of genome observed between closely related viruses, such as FHV and PaV, and whether that variability is due to differences in the genome and/or capsid protein sequences.

A growing consensus regarding genome encapsidation in simple ssRNA viruses is that capsid assembly is concomitant with genome encapsidation. Within this broad model, whether the RNA first collapses with protein subunits subsequently binding to collapsed RNA, or whether the polymerization of the capsid protein drives the collapse of the RNA, remains an open question. In many dsDNA bacterial viruses, encapsidation of the genome into preformed capsid appears to be a common phenomenon, and this theme appears to be reiterated in the herpesvirus family. In each of the four different types of viruses discussed in this article, we can see that studies on bacterial viruses have had a profound influence on our understanding of genome organization, recognition, encapsidation, and release. How far these parallels will extrapolate to other, nonbacterial viruses remains to be seen. Whether the general principles of these processes as applied to ssDNA ϕX174 hold true for parvoviruses, whether what we have learned from dsDNA bacterial viruses is applicable to viruses such as herpesviruses and adenoviruses, and whether ϕ6 virus can be a model for other multisegmented dsRNA viruses are interesting questions that need further structural and biochemical study.

The dsRNA viruses, because of the multisegmented nature of the genome and the requirement for endogenous transcription, pose several interesting mechanistic questions regarding genome organization, transcription, replication, and packaging. Although attractive plausible models have been put forward, further extensive structural and biochemical studies are required to comprehend various aspects of genome-associated processes in these viruses.

Exciting developments in X-ray crystallography and cryo-EM techniques have made them increasingly accessible to study larger systems at higher resolutions. Near-atomic resolution structures of more complex and larger viruses such as adenovirus and herpesvirus and even viruses that are not entirely icosahedral are likely to be possible by taking advantage of the complementarity of these two techniques. Although increasingly sophisticated structural studies are likely to resolve outstanding questions about genome organization, questions about dynamic processes such as encapsidation or release will require both structural and biochemical studies.

Using cryo-EM to obtain structural information about assembly/encapsidation intermediates and about capsids that are in the process of releasing their genomes, combined with the atomic resolution structures of the capsids or the relevant capsid proteins derived from X-ray crystallography, may prove powerful in gaining further mechanistic insights. The cryo-EM technique may also be effectively employed to retrieve structural information about the genome or a portion that does not obey capsid symmetry, by using single-particle image analysis, or tomographic approaches. The feasibility of such an application of the cryo-EM technique has been elegantly demonstrated in studies on bacterial virus ϕ29 (Morais *et al.*, 2001; Tao *et al.*, 1998). In addition to existing structural methodologies, novel strategies may have to be developed for studying such processes as genome release. Genome release is often associated with virus entry and subsequent virus uncoating, processes that involve virus receptors and cellular membranes. To gain a more realistic understanding, imaging virus particles in the context of the lipid membranes, perhaps in the form of vesicles, may have to be developed (J. Hogle, personal communication). Thus, there is realistic optimism that exciting new information and novel insights into the various processes relating to the viral genome will be forthcoming in the near future.

Acknowledgments

This work was supported by grants AI36040 (B.V.V.P.) and GM47980 (P.P.) from the NIH and a grant from R. W. Welch Foundation (B.V.V.P.). We thank R. Chen, H. Jayaram, Z. Zhang, R. Bernal, R. Burnett, and J. Grimes for help in preparing the figures.

References

Abad-Zapatero, C., Abdel-Meguid, S. S., Johnson, J. E., Leslie, A. G. W., Rayment, I., Rossmann, M. G., Suck, D., and Tsukihara, T. (1980). Structure of southern bean mosaic virus at 2.8 Å resolution. *Nature* **286,** 33–39.

Acharya, R., Fry, E., Stuart, D., Fox, G., Rowlands, D., and Brown, F. (1989). The three-dimensional structure of foot-and-mouth disease virus at 2.9 Å resolution. *Nature* **337,** 709–716.

Adrian, M., Dubochet, J., Lepault, J., and McDowall, A. W. (1984). Cryo-electron microscopy of viruses. *Nature* **308,** 32–36.

Agbandje-McKenna, M., Llamas-Saiz, A. L., Wang, F., Tattersall, P., and Rossmann, M. G. (1998). Functional implications of the structure of the murine parvovirus, minute virus of mice. *Structure* **6,** 1369–1381.

Aoyama, A., Hamatake, R. K., and Hayashi, M. (1983). In vitro synthesis of bacteriophage ϕX174 by purified components. *Proc. Natl. Acad. Sci. USA* **80,** 4195–4199.
Aoyama, A., and Hayashi, M. (1982). In vitro packaging of plasmid DNAs into ϕX174 bacteriophage capsid. *Nature* **297,** 704–706.
Aoyama, A., and Hayashi, M. (1986). Synthesis of bacteriophage ϕX174 in vitro: Mechanism of switch from DNA replication to DNA packaging. *Cell* **47,** 99–106.
Aponte, C., Poncet, D., and Cohen, J. (1996). Recovery and characterization of a replicase complex in rotavirus-infected cells by using a monoclonal antibody against NSP2. *J. Virol.* **70,** 985–991.
Argos, P., and Johnson, J. E. (1984). "Chemical Stability in Simple Plant Viruses," Vol. 1, John Wiley & Sons, New York.
Baker, T. S., Drak, J., and Bina, M. (1988). Reconstruction of the three-dimensional structure of simian virus 40 and visualization of the chromatin core. *Proc. Natl. Acad. Sci. USA* **85,** 422–426.
Ban, N., and McPherson, A. (1995). The structure of satellite panicum mosaic virus at 1.9 Å resolution. *Nat. Struct. Biol.* **2,** 882–890.
Banerjee, A. K., and Shatkin, A. J. (1970). Transcription in vitro by reovirus-associated ribonucleic acid-dependent polymerase. *J. Virol.* **6,** 1–11.
Bartlett, N. M., Gillies, S. C., Bullivant, S., and Bellamy, A. R. (1974). Electron microscopy study of reovirus reaction cores. *J. Virol.* **14,** 315–326.
Basavappa, R., Syed, R., Flore, O., Icenogle, J. P., Filman, D. J., and Hogle, J. M. (1994). Role and mechanism of the maturation cleavage of VP0 in poliovirus assembly: Structure of the empty capsid assembly intermediate at 2.9 Å resolution. *Protein Sci.* **3,** 1651–1669.
Baumeister, W., and Steven, A. C. (2000). Macromolecular electron microscopy in the era of structural genomics. *Trends Biochem. Sci.* **25,** 624–631.
Bazinet, C., and King, J. (1985). The DNA translocating vertex of dsDNA bacteriophage. *Annu. Rev. Microbiol.* **39,** 109–129.
Belnap, D. M., Olson, N. H., Cladel, N. M., Newcomb, W. W., Brown, J. C., Kreider, J. W., Christensen, N. D., and Baker, T. S. (1996). Conserved features in papillomavirus and polyomavirus capsids. *J. Mol. Biol.* **259,** 249–263.
Benson, S. D., Bamford, J. K., Bamford, D. H., and Burnett, R. M. (1999). Viral evolution revealed by bacteriophage PRD1 and human adenovirus coat protein structures. *Cell* **98,** 825–833.
Bentley, G. A., Lewit-Bentley, A., Liljas, L., Skoglund, U., Roth, M., and Unge, T. (1987). Structure of RNA in satellite tobacco necrosis virus: A low resolution neutron diffraction study using $^1H_2O/^2H_2O$ solvent contrast variation. *J. Mol. Biol.* **194,** 129–141.
Bisaillon, M., and Lemay, G. (1997). Molecular dissection of the reovirus λ1 protein nucleic acids binding site. *Virus Res.* **51,** 231–237.
Bishop, N. E., and Anderson, D. A. (1993). RNA-dependent cleavage of VP0 capsid protein in provirions of hepatitis A virus. *Virology* **197,** 616–623.
Black, L. W. (1989). DNA packaging in dsDNA bacteriophages. *Annu. Rev. Microbiol.* **43,** 267–292.
Bong, D. T., Steinem, C., Janshoff, A., Johnson, J. E., and Reza Ghadiri, M. (1999). A highly membrane-active peptide in flock house virus: Implications for the mechanism of nodavirus infection. *Chem. Biol.* **6,** 473–481.

Booy, F. P., Newcomb, W. W., Trus, B. L., Brown, J. C., Baker, T. S., and Steven, A. C. (1991). Liquid-crystalline, phage-like packing of encapsidated DNA in herpes simplex virus. *Cell* **64,** 1007–1015.

Bottcher, B., Kiselev, N. A., Stel'Mashchuk, V. Y., Perevozchikova, N. A., Borisov, A. V., and Crowther, R. A. (1997a). Three-dimensional structure of infectious bursal disease virus determined by electron cryomicroscopy. *J. Virol.* **71,** 325–330.

Bottcher, B., Wynne, S. A., and Crowther, R. A. (1997b). Determination of the fold of the core protein of hepatitis B virus by electron cryomicroscopy [see comments]. *Nature* **386,** 88–91.

Burnett, R. M. (1997). "The Structure of Adenovirus." Oxford University Press, New York.

Butcher, S. J., Bamford, D. H., and Fuller, S. D. (1995). DNA packaging orders the membrane of bacteriophage PRD1. *EMBO J.* **14,** 6078–6086.

Butcher, S. J., Dokland, T., Ojala, P. M., Bamford, D. H., and Fuller, S. D. (1997). Intermediates in the assembly pathway of the double-stranded RNA virus $\phi 6$. *EMBO J.* **16,** 4477–4487.

Butler, P. J. (1999). Self-assembly of tobacco mosaic virus: The role of an intermediate aggregate in generating both specificity and speed. *Philos. Trans. R. Soc. Lond. B Biol. Sci.* **354,** 537–550.

Casjens, S. (1985). "Nucleic Acid Packaging by Viruses." Jones and Bartlett Publishers, Boston.

Casjens, S., Huang, W. M., Hayden, M., and Parr, R. (1987). Initiation of bacteriophage P22 DNA packaging series: Analysis of a mutant that alters the DNA target specificity of the packaging apparatus. *J. Mol. Biol.* **194,** 411–422.

Caspar, D. L., and Namba, K. (1990). Switching in the self-assembly of tobacco mosaic virus. *Adv. Biophys.* **26,** 157–185.

Caston, J. R., Trus, B. L., Booy, F. P., Wickner, R. B., Wall, J. S., and Steven, A. C. (1997). Structure of L-A virus: A specialized compartment for the transcription and replication of double-stranded RNA. *J. Cell Biol.* **138,** 975–985.

Cerritelli, M. E., Cheng, N., Rosenberg, A. H., McPherson, C. E., Booy, F. P., and Steven, A. C. (1997). Encapsidated conformation of bacteriophage T7 DNA. *Cell* **91,** 271–280.

Chapman, M. S., and Rossmann, M. G. (1995). Single-stranded DNA–protein interactions in canine parvovirus. *Structure* **3,** 151–162.

Chen, D., Gombold, J. L., and Ramig, R. F. (1990). Intracellular RNA synthesis directed by temperature-sensitive mutants of simian rotavirus SA11. *Virology* **178,** 143–151.

Chen, Z. G., Stauffacher, C., Li, Y., Schmidt, T., Bomu, W., Kamer, G., Shanks, M., Lomonossoff, G., and Johnson, J. E. (1989). Protein–RNA interactions in an icosahedral virus at 3.0 Å resolution. *Science* **245,** 154–159.

Chiu, W., McGough, A., Sherman, M. B., and Schmid, M. F. (1999). High-resolution electron cryomicroscopy of macromolecular assemblies. *Trends Cell Biol.* **9,** 154–159 .

Christensen, J., Cotmore, S. F., and Tattersall, P. (1995). Minute virus of mice transcriptional activator protein NS1 binds directly to the transactivation region of the viral P38 promoter in a strictly ATP-dependent manner. *J. Virol.* **69,** 5422–5430.

Cohen, J. (1977). Ribonucleic acid polymerase activity associated with purified calf rotavirus. *J. Gen. Virol.* **36,** 395–402.

Compton, S. R., Nelsen, B., and Kirkegaard, K. (1990). Temperature-sensitive poliovirus mutant fails to cleave VP0 and accumulates provirions. *J. Virol.* **64,** 4067–4075.

Conway, J. F., Cheng, N., Zlotnick, A., Wingfield, P. T., Stahl, S. J., and Steven, A. C. (1997). Visualization of a 4-helix bundle in the hepatitis B virus capsid by cryo-electron microscopy [see comments]. *Nature* **386,** 91–94.

Cotmore, S. F., and Tattersall, P. (1996). "Parvovirus DNA Replication." Cold Spring Harbor Laboratory Press, Cold Spring Harbor, NY.

Cotmore, S. F., Christensen, J., Nuesch, J. P., and Tattersall, P. (1995). The NS1 polypeptide of the murine parvovirus minute virus of mice binds to DNA sequences containing the motif $[ACCA]_{2-3}$. *J. Virol.* **69,** 1652–1660.

Crowther, R. A. (1971). Procedures for three-dimensional reconstruction of spherical viruses by Fourier synthesis from electron micrographs. *Philos. Trans. R. Soc. Lond. B Biol. Sci.* **261,** 221–230.

Crowther, R. A., Amos, L. A., Finch, J. T., De Rosier, D. J., and Klug, A. (1970). Three dimensional reconstructions of spherical viruses by Fourier synthesis from electron micrographs. *Nature* **226,** 421–425.

Culver, J. N., Dawson, W. O., Plonk, K., and Stubbs, G. (1995). Site-directed mutagenesis confirms the involvement of carboxylate groups in the disassembly of tobacco mosaic virus. *Virology* **206,** 724–730.

Curry, S., Abrams, C. C., Fry, E., Crowther, J. C., Belsham, G. J., Stuart, D. I., and King, A. M. (1995). Viral RNA modulates the acid sensitivity of foot-and-mouth disease virus capsids. *J. Virol.* **69,** 430–438.

Dasgupta, R., and Kaesberg, P. (1982). Complete nucleotide sequences of the coat protein messenger RNAs of brome mosaic virus and cowpea chlorotic mottle virus. *Nucleic Acids Res.* **10,** 703–713.

de Haas, F., Paatero, A. O., Mindich, L., Bamford, D. H., and Fuller, S. D. (1999). A symmetry mismatch at the site of RNA packaging in the polymerase complex of dsRNA bacteriophage ϕ6. *J. Mol. Biol.* **294,** 357–372.

Diprose, J. M., Burroughs, J. N., Sutton, G. C., Goldsmith, A., Gouet, P., Malby, R., Overton, I., Zientara, S., Mertens, P. P., Stuart, D. I., and Grimes, J. M. (2001). Translocation portals for the substrates and products of a viral transcription complex: The bluetongue virus core. *EMBO J.* **20,** 7229–7239.

Dokland, T., McKenna, R., Ilag, L. L., Bowman, B. R., Incardona, N. L., Fane, B. A., and Rossmann, M. G. (1997). Structure of a viral procapsid with molecular scaffolding. *Nature* **389,** 308–313.

Dokland, T., McKenna, R., Sherman, D. M., Bowman, B. R., Bean, W. F., and Rossmann, M. G. (1998). Structure determination of the ϕX174 closed procapsid. *Acta Crystallogr. D Biol. Crystallogr.* **54,** 878–890.

Dressler, D., Hourcade, D., Koths, K., and Sims, J. (1978). "The DNA Replication Cycle of the Isometric Phages." Cold Spring Harbor Laboratory Press, Cold Spring Harbor, NY.

Dryden, K. A., Farsetta, D. L., Wang, G., Keegan, J. M., Fields, B. N., Baker, T. S., and Nibert, M. L. (1998). Internal/structures containing transcriptase-related proteins in top component particles of mammalian orthoreovirus. *Virology* **245,** 33–46.

Dryden, K. A., Wang, G., Yeager, M., Nibert, M. L., Coombs, K. M., Furlong, D. B., Fields, B. N., and Baker, T. S. (1993). Early steps in reovirus infection are associated with dramatic changes in supramolecular structure and protein conformation: Analysis of virions and subviral particles by cryoelectron microscopy and image reconstruction. *J. Cell Biol.* **122,** 1023–1041.

Earnshaw, W., Casjens, S., and Harrison, S. C. (1976). Assembly of the head of bacteriophage P22: X-ray diffraction from heads, proheads and related structures. *J. Mol. Biol.* **104,** 387–410.

Earnshaw, W. C., and Harrison, S. C. (1977). DNA arrangement in isometric phage heads. *Nature* **268,** 598–602.

Earnshaw, W. C., King, J., and Eiserling, F. A. (1978a). The size of the bacteriophage T4 head in solution with comments about the dimension of virus particles as visualized by electron microscopy. *J. Mol. Biol.* **122,** 247–253.

Earnshaw, W. C., King, J., Harrison, S. C., and Eiserling, F. A. (1978b). The structural organization of DNA packaged within the heads of T4 wild-type, isometric and giant bacteriophages. *Cell* **14,** 559–568.

Erickson, J. W., Silva, A. M., Murthy, M. R., Fita, I., and Rossmann, M. G. (1985). The structure of a T=1 icosahedral empty particle from southern bean mosaic virus. *Science* **229,** 625–629.

Estes, M. K. (2001). Rotaviruses and their replication. *In* ''Fields Virology'' (D. M. Knipe and P. M. Howley, Eds.), pp. 1747–1785. Lippincott Williams & Wilkins, Philadelphia.

Feiss, M., Fisher, R. A., Crayton, M. A., and Egner, C. (1977). Packaging of the bacteriophage λ chromosome: Effect of chromosome length. *Virology* **77,** 281–293.

Fields, B. N. (1996). Reoviridae. *In* ''Virology'' (B. N. Fields, D. M. Knipe, R. M. Channock, M. S. Hirsch, J. L. Melnick, T. P. Monath, and B. Roizman, Eds.), pp. 1553–1556. Raven Press, New York.

Filali Maltouf, A., and Labedan, B. (1983). Host cell metabolic energy is not required for injection of bacteriophage T5 DNA. *J. Bacteriol.* **153,** 124–133.

Filali Maltouf, A. K., and Labedan, B. (1985). The energetics of the injection process of bacteriophage λ DNA and the role of the ptsM/pel-encoded protein. *Biochem. Biophys. Res. Commun.* **130,** 1093–1101.

Fillmore, G. C., Lin, H., and Li, J. K. (2002). Localization of the single-stranded RNA-binding domains of bluetongue virus nonstructural protein NS2. *J. Virol.* **76,** 499–506.

Fisher, A. J., and Johnson, J. E. (1993). Ordered duplex RNA controls capsid architecture in an icosahedral animal virus. *Nature* **361,** 176–179.

Flint, S. J., Enquist, L. W., Krug, R. M., Racaniello, V. R., and Skalka, A. M. (2000). ''Principles of Virology.''ASM Press, Washington, D.C.

Gallagher, T. M., and Rueckert, R. R. (1988). Assembly-dependent maturation cleavage in provirions of a small icosahedral insect ribovirus. *J. Virol.* **62,** 3399–3406.

Garcia, L. R., and Liddington, R. C. (1997). ''Structural Biology of Polyomaviruses.'' Oxford University Press, New York.

Garcia, L. R., and Molineux, I. J. (1996). Transcription-independent DNA translocation of bacteriophage T7 DNA into. *Escherichia coli. J. Bacteriol.* **178,** 6921–6929.

Gillian, A. L., and Nibert, M. L. (1998). Amino terminus of reovirus nonstructural protein sigma NS is important for ssRNA binding and nucleoprotein complex formation. *Virology* **240,** 1–11.

Gillies, S., Bullivant, S., and Bellamy, A. R. (1971). Viral RNA polymerases: Electron microscopy of reovirus reaction cores. *Science* **174,** 694–696.

Gouet, P., Diprose, J. M., Grimes, J. M., Malby, R., Burroughs, J. N., Zientara, S., Stuart, D. I., and Mertens, P. P. (1999). The highly ordered double-stranded RNA genome of bluetongue virus revealed by crystallography. *Cell* **97,** 481–490.

Grahn, E., Stonehouse, N. J., Murray, J. B., van den Worm, S., Valegard, K., Fridborg, K., Stockley, P. G., and Liljas, L. (1999). Crystallographic studies of RNA hairpins in complexes with recombinant MS2 capsids: Implications for binding requirements. *RNA* **5,** 131–138.

Grimes, J. M., Burroughs, J. N., Gouet, P., Diprose, J. M., Malby, R., Zientara, S., Mertens, P. C. P., and Stuart, D. I. (1998). The atomic structure of the bluetongue virus core. *Nature* **395,** 470–478.

Grimes, J. M., Jakana, J., Ghosh, M., Basak, A. K., Roy, P., Chiu, W., Stuart, D. I., and Prasad, B. V. (1997). An atomic model of the outer layer of the bluetongue virus core derived from X-ray crystallography and electron cryomicroscopy. *Structure* **5,** 885–893.

Guasch, A., Pous, J., Ibarra, B., Gomis-Ruth, F. X., Valpuesta, J. M., Sousa, N., Carrascosa, J. L., and Coll, M. (2002). Detailed architecture of a DNA translocating machine: The high-resolution structure of the bacteriophage ϕ29 connector particle. *J. Mol. Biol.* **315,** 663–676.

Guo, P., Peterson, C., and Anderson, D. (1987a). Prohead and DNA-gp3-dependent ATPase activity of the DNA packaging protein gp16 of bacteriophage ϕ29. *J. Mol. Biol.* **197,** 229–236.

Guo, P. X., Erickson, S., and Anderson, D. (1987b). A small viral RNA is required for in vitro packaging of bacteriophage ϕ29 DNA. *Science* **236,** 690–694.

Hamatake, R. K., Aoyama, A., and Hayashi, M. (1985). The J gene of bacteriophage ϕX174: In vitro analysis of J protein function. *J. Virol.* **54,** 345–350.

Harrison, S. C. (1983). Packaging of DNA into bacteriophage heads: A model. *J. Mol. Biol.* **171,** 577–580.

Harrison, S. C. (1989). What do viruses look like? *Harvey Lect.* **85,** 127–152.

Harrison, S. C., Olson, A., Schutt, C. E., Winkler, F. K., and Bricogne, G. (1978). Tomato bushy stunt virus at 2.9 Å resolution. *Nature* **276,** 368–373.

Harrison, S. J., Farsetta, D. L., Kim, J., Noble, S., Broering, T. J., and Nibert, M. L. (1999). Mammalian reovirus L3 gene sequences and evidence for a distinct amino-terminal region of the λ1 protein. *Virology* **258,** 54–64.

Harvey, J. D., Bellamy, A. R., Earnshaw, W. C., and Schutt, C. (1981). Biophysical studies of reovirus type 3. IV. Low-angle X-ray diffraction studies. *Virology* **112,** 240–249.

Hayashi, M. (1978). "Morphogenesis of Isometric Phages."Cold Spring Harbor Laboratory Press, Cold Spring Harbor, NY.

Hayashi, M., Aoyama, A., Richrdson, D. L., Jr., and Hayashi, M. N. (1988). "Biology of Bacteriophage ϕX174," **Vol. 2.** Plenum, New York.

Hendrix, R. W. (1978). Symmetry mismatch and DNA packaging in large bacteriophages. *Proc. Natl. Acad. Sci. USA* **75,** 4779–4783.

Hendrix, R. W. (1999). Evolution: the long evolutionary reach of viruses. *Curr. Biol.* **9,** R914–R917.

Hill, C. L., Booth, T. F., Prasad, B. V., Grimes, J. M., Mertens, P. P., Sutton, G. C., and Stuart, D. I. (1999). The structure of a cypovirus and the functional organization of dsRNA viruses. *Nat. Struct. Biol.* **6,** 565–568.

Hogle, J. M., Chow, M., and Filman, D. J. (1985). Three-dimensional structure of poliovirus at 2.9 Å resolution. *Science* **229,** 1358–1365.

Hohn, B. (1983). DNA sequences necessary for packaging of bacteriophage λ DNA. *Proc. Natl. Acad. Sci. USA* **80,** 7456–7460.

Hud, N. V. (1995). Double-stranded DNA organization in bacteriophage heads: An alternative toroid-based model. *Biophys. J.* **69,** 1355–1362.

Ilag, L. L., Olson, N. H., Dokland, T., Music, C. L., Cheng, R. H., Bowen, Z., McKenna, R., Rossmann, M. G., Baker, T. S., and Incardona, N. L. (1995). DNA packaging intermediates of bacteriophage ϕX174. *Structure* **3,** 353–363.

Jack, A., and Harrison, S. C. (1975). On the interpretation of small-angle X-ray solution scattering from spherical viruses. *J. Mol. Biol.* **99,** 15–25.

Jacobson, M. F., Asso, J., and Baltimore, D. (1970). Further evidence on the formation of poliovirus proteins. *J. Mol. Biol.* **49,** 657–669.

Jayaram, H., Taraporewala, Z., Patton, J. T., and Prasad, B. V. V. (2002). X-ray structure of a rotavirus protein involved in genome packaging and replication exhbits a HIT-like fold. *Nature* **417,** 311–315.
Johnson, J. E., and Rueckert, R. R. (1997). "Packaging and Release of Viral Genomes." Oxford University Press, Oxford.
Juuti, J. T., Bamford, D. H., Tuma, R., and Thomas, G. J., Jr. (1998). Structure and NTPase activity of the RNA-translocating protein (P4) of bacteriophage ϕ6. *J. Mol. Biol.* **279,** 347–359.
Kalinski, A., and Black, L. W. (1986). End structure and mechanism of packaging of bacteriophage T4 DNA. *J. Virol.* **58,** 951–954.
Kattoura, M., Chen, X., and Patton, J. (1994). The rotavirus RNA-binding protein NS35 (NSP2) forms 10S multimers and interacts with the viral RNA polymerase. *Virology* **202,** 803–813.
Kitamura, N., Semler, B. L., Rothberg, P. G., Larsen, G. R., Adler, C. J., Dorner, A. J., Emini, E. A., Hanecak, R., Lee, J. J., van der Werf, S., *et al.* (1981). Primary structure, gene organization and polypeptide expression of poliovirus RNA. *Nature* **291,** 547–553.
Klug, A. (1999). The tobacco mosaic virus particle: Structure and assembly. *Philos. Trans. R. Soc. Lond. B Biol. Sci.* **354,** 531–553.
Kosturko, L. D., Hogan, M., and Dattagupta, N. (1979). Structure of DNA within three isometric bacteriophages. *Cell* **16,** 515–522.
Labbe, M., Baudoux, P., Charpilienne, A., Poncet, D., and Cohen, J. (1994). Identification of the nucleic acid binding domain of the rotavirus VP2 protein. *J. Gen. Virol.* **75,** 3423–3430.
Larson, S. B., Day, J., Greenwood, A., and McPherson, A. (1998). Refined structure of satellite tobacco mosaic virus at 1.8 Å resolution. *J. Mol. Biol.* **277,** 37–59.
Larson, S. B., Koszelak, S., Day, J., Greenwood, A., Dodds, J. A., and McPherson, A. (1993). Double-helical RNA in satellite tobacco mosaic virus. *Nature* **361,** 179–182.
Larson, S. B., and McPherson, A. (2001). Satellite tobacco mosaic virus RNA: Structure and implications for assembly. *Curr. Opin. Struct. Biol.* **11,** 59–65.
Lawton, J. A., Estes, M. K., and Prasad, B. V. (1997). Three-dimensional visualization of mRNA release from actively transcribing rotavirus particles. *Nat. Struct. Biol.* **4,** 118–121.
Lawton, J. A., Estes, M. K., and Prasad, B. V. (2000). Mechanism of genome transcription in segmented dsRNA viruses. *Adv. Virus Res.* **55,** 185–229.
Lee, W. M., Monroe, S. S., and Rueckert, R. R. (1993). Role of maturation cleavage in infectivity of picornaviruses: Activation of an infectosome. *J. Virol.* **67,** 2110–2122.
Lepault, J., Dubochet, J., Baschong, W., and Kellenberger, E. (1987). Organization of double-stranded DNA in bacteriophages: A study by cryo-electron microscopy of vitrified samples. *EMBO J.* **6,** 1507–1512.
Liddington, R. C., Yan, Y., Moulai, J., Sahli, R., Benjamin, T. L., and Harrison, S. C. (1991). Structure of simian virus 40 at 3.8-Å resolution. *Nature* **354,** 278–284.
Liljas, L., Unge, T., Jones, T. A., Fridborg, K., Lovgren, S., Skoglund, U., and Strandberg, B. (1982). Structure of satellite tobacco necrosis virus at 3.0 Å resolution. *J. Mol. Biol.* **159,** 93–108.
Livolant, F., and Leforestier, A. (1996). Condensed phases of DNA: Structures and phase transitions. *Prog. Polym. Sci.* **21,** 1115–1164.
Loudon, P. T., and Roy, P. (1991). Assembly of five bluetongue virus proteins expressed by recombinant baculoviruses: Inclusion of the largest protein VP1 in the core and virus-like proteins. *Virology* **180,** 798–802.

Mano, Y., Kawabe, T., Komano, T., and Yazaki, K. (1982). *Agric. Biol. Chem.* **46,** 2041–2049.

Martin, C. S., Burnett, R. M., de Haas, F., Heinkel, R., Rutten, T., Fuller, S. D., Butcher, S. J., and Bamford, D. H. (2001). Combined EM/X-ray imaging yields a quasi-atomic model of the adenovirus-related bacteriophage PRD1 and shows key capsid and membrane interactions. *Structure* **9,** 917–930.

McKenna, R., Xia, D., Willingmann, P., Ilag, L. L., Krishnaswamy, S., Rossmann, M. G., Olson, N. H., Baker, T. S., and Incardona, N. L. (1992). Atomic structure of single-stranded DNA bacteriophage ϕX174 and its functional implications. *Nature* **355,** 137–143.

Mindich, L. (1999). Precise packaging of the three genomic segments of the double-stranded-RNA bacteriophage ϕ6. *Microbiol. Mol. Biol. Rev.* **63,** 149–160.

Molineux, I. J. (2001). No syringes please, ejection of phage T7 DNA from the virion is enzyme driven. *Mol. Microbiol.* **40,** 1–8.

Moore, S. D., and Prevelige, P. E., Jr. (2001). Structural transformations accompanying the assembly of bacteriophage P22 portal protein rings in vitro. *J. Biol. Chem.* **276,** 6779–6788.

Morais, M. C., Tao, Y., Olson, N. H., Grimes, S., Jardine, P. J., Anderson, D. L., Baker, T. S., and Rossmann, M. G. (2001). Cryoelectron-microscopy image reconstruction of symmetry mismatches in bacteriophage ϕ29. *J. Struct. Biol.* **135,** 38–46.

Mukai, R., Hamatake, R. K., and Hayashi, M. (1979). Isolation and identification of bacteriophage ϕX174 prohead. *Proc. Natl. Acad. Sci. USA* **76,** 4877–4881.

Namba, K., Pattanayek, R., and Stubbs, G. (1989). Visualization of protein–nucleic acid interactions in a virus: Refined structure of intact tobacco mosaic virus at 2.9 Å resolution by X-ray fiber diffraction. *J. Mol. Biol.* **208,** 307–325.

Nason, E. L., Samal, S. K., and Venkataram Prasad, B. V. (2000). Trypsin-induced structural transformation in aquareovirus. *J. Virol.* **74,** 6546–6555.

Newcomb, W. W., Homa, F. L., Thomsen, D. R., Trus, B. L., Cheng, N., Steven, A., Booy, F., and Brown, J. C. (1999). Assembly of the herpes simplex virus procapsid from purified components and identification of small complexes containing the major capsid and scaffolding proteins. *J. Virol.* **73,** 4239–4250.

Newcomb, W. W., Juhas, R. M., Thomsen, D. R., Homa, F. L., Burch, A. D., Weller, S. K., and Brown, J. C. (2001). The UL6 gene product forms the portal for entry of DNA into the herpes simplex virus capsid. *J. Virol.* **75,** 10923–10932.

Newcomb, W. W., Trus, B. L., Cheng, N., Steven, A. C., Sheaffer, A. K., Tenney, D. J., Weller, S. K., and Brown, J. C. (2000). Isolation of herpes simplex virus procapsids from cells infected with a protease-deficient mutant virus. *J. Virol.* **74,** 1663–1673.

North, A. C. T., and Rich, A. (1961). X-ray diffraction studies of bacterial viruses. *Nature* **191,** 1242–1245.

Patton, J. T., and Spencer, E. (2000). Genome replication and packaging of segmented double-stranded RNA viruses. *Virology* **277,** 217–225.

Pesavento, J. B., Lawton, J. A., Estes, M. E., and Venkataram Prasad, B. V. (2001). The reversible condensation and expansion of the rotavirus genome. *Proc. Natl. Acad. Sci. USA* **98,** 1381–1386.

Petrie, B. L., Greenberg, H. B., Graham, D. Y., and Estes, M. K. (1984). Ultrastructural localization of rotavirus antigens using colloidal gold. *Virus Res.* **1,** 133–152.

Prasad, B. V., Rothnagel, R., Zeng, C. Q., Jakana, J., Lawton, J. A., Chiu, W., and Estes, M. K. (1996). Visualization of ordered genomic RNA and localization of transcriptional complexes in rotavirus. *Nature* **382,** 471–473.

Prasad, B. V. V., and Estes, M. K. (2000). "Electron Cryomicroscopy and Computer Image Processing Techniques: Use in Structure–Function Studies of Rotavirus." Humana Press, Totowa, NJ.

Prasad, B. V. V., Hardy, M. E., Dokland, T., Bella, J., Rossmann, M. G., and Estes, M. K. (1999). X-ray crystallographic structure of the Norwalk virus capsid. *Science* **286,** 287–290.

Prasad, B. V. V., Rothnagel, R., Jiang, X., and Estes, M. K. (1994). Three-dimensional structure of baculovirus-expressed Norwalk virus capsids. *J. Virol.* **68,** 5117–5125.

Pritsch, C., and Mayo, M. A. (1989). Staellites of plant viruses. *In* "Plant Viruses" (C. Mandahar, Ed.), pp. 289–321. CRC Press, Boca Raton, FL.

Ramig, R., and Petrie, B. L. (1984). Characterization of temperature-sensitive mutants of simian rotavirus SA11: Protein synthesis and morphogenesis. *J. Virol.* **49,** 665–673.

Rayment, I., Baker, T. S., Caspar, D. L., and Murakami, W. T. (1982). Polyoma virus capsid structure at 22.5 Å resolution. *Nature* **295,** 110–115.

Reinisch, K. M., Nibert, M. L., and Harrison, S. C. (2000). Structure of the reovirus core at 3.6 Å resolution. *Nature* **404,** 960–967.

Richards, K. E., Williams, R. C., and Calendar, R. (1973). Mode of DNA packing within bacteriophage heads. *J. Mol. Biol.* **78,** 255–259.

Rossmann, M. G., Abad-Zapatero, C., Erickson, J. W., and Savithri, H. S. (1983). RNA–protein interactions in some small plant viruses. *J. Biomol. Struct. Dyn.* **1,** 565–579.

Rossmann, M. G., and Erickson, J. W. (1985). "Structure and Assembly of Icosahedral Shells." Jones and Bartlett Publishers, Boston.

Rossmann, M. G., Arnold, E., Erickson, J. W., Frankenberger, E. A., Griffith, J. P., Hecht, H. J., Johnson, J. E., Kamer, G., Luo, M., Mosser, A. G., *et al.* (1985). Structure of a human common cold virus and functional relationship to other picornaviruses. *Nature* **317,** 145–153.

Rossmann, M. G., and Johnson, J. E. (1989). Icosahedral RNA virus structure. *Annu. Rev. Biochem.* **58,** 533–573.

Sacher, R., and Ahlquist, P. (1989). Effects of deletions in the N-terminal basic arm of brome mosaic virus coat protein on RNA packaging and systemic infection. *J. Virol.* **63,** 4545–4552.

Sanger, F., Air, G. M., Barrell, B. G., Brown, N. L., Coulson, A. R., Fiddes, C. A., Hutchison, C. A., Slocombe, P. M., and Smith, M. (1977). Nucliotide sequence of bacteriophage ϕX174 DNA. *Nature* **265,** 687–695.

Schneemann, A., Reddy, V., and Johnson, J. E. (1998). The structure and function of nodavirus particles: A paradigm for understanding chemical biology. *Adv. Virus Res.* **50,** 381–446.

Schuck, P., Taraporewala, Z., McPhie, P., and Patton, J. T. (2001). Rotavirus nonstructural protein NSP2 self-assembles into octamers that undergo ligand-induced conformational changes. *J. Biol. Chem.* **276,** 9679–9687.

Sheaffer, A. K., Newcomb, W. W., Gao, M., Yu, D., Weller, S. K., Brown, J. C., and Tenney, D. J. (2001). Herpes simplex virus DNA cleavage and packaging proteins associate with the procapsid prior to its maturation. *J. Virol.* **75,** 687–698.

Simpson, A. A., Chipman, P. R., Baker, T. S., Tijssen, P., and Rossmann, M. G. (1998). The structure of an insect parvovirus (*Galleria mellonella* densovirus) at 3.7 Å resolution. *Structure* **6,** 1355–1367.

Simpson, A. A., Tao, Y., Leiman, P. G., Badasso, M. O., He, Y., Jardine, P. J., Olson, N. H., Morais, M. C., Grimes, S., Anderson, D. L., *et al.* (2000). Structure of the bacteriophage ϕ29 DNA packaging-motor. *Nature* **408,** 745–750.

Skehel, J. J., and Joklik, W. K. (1969). Studies on the in vitro transcription of reovirus RNA catalyzed by reovirus cores. *Virology* **39,** 822–831.

Smith, D. E., Tans, S. J., Smith, S. B., Grimes, S., Anderson, D. L., and Bustamante, C. (2001). The bacteriophage straight ϕ29 portal motor can package DNA against a large internal force. *Nature* **413,** 748–752.

Smith, R. E., and Furuichi, Y. (1982). The double-stranded RNA genome segments of cytoplasmic polyhedrosis virus are independently transcribed. *J. Virol.* **41,** 326–329.

Sorger, P. K., Stockley, P. G., and Harrison, S. C. (1986). Structure and assembly of turnip crinkle virus. II. Mechanism of reassembly in vitro. *J. Mol. Biol.* **191,** 639–658.

Stewart, P. L., Burnett, R. M., Cyrklaff, M., and Fuller, S. D. (1991). Image reconstruction reveals the complex molecular organization of adenovirus. *Cell* **67,** 145–154.

Stockley, P. G., Stonehouse, N. J., Murray, J. B., Goodman, S. T., Talbot, S. J., Adams, C. J., Liljas, L., and Valegard, K. (1995). Probing sequence-specific RNA recognition by the bacteriophage MS2 coat protein. *Nucleic Acids Res.* **23,** 2512–2518.

Struthers-Schlinke, J. S., Robins, W. P., Kemp, P., and Molineux, I. J. (2000). The internal head protein gp16 controls DNA ejection from the bacteriophage T7 virion. *J. Mol. Biol.* **301,** 35–45.

Stubbs, G. (1999). Tobacco mosaic virus particle structure and the initiation of disassembly. *Philos. Trans. R. Soc. Lond. B Biol. Sci.* **354,** 551–557.

Tang, L., Johnson, K. N., Ball, L. A., Lin, T., Yeager, M., and Johnson, J. E. (2001). The structure of pariacoto virus reveals a dodecahedral cage of duplex RNA. *Nat. Struct. Biol.* **8,** 77–83.

Tao, Y., Olson, N. H., Xu, W., Anderson, D. L., Rossmann, M. G., and Baker, T. S. (1998). Assembly of a tailed bacterial virus and its genome release studied in three dimensions. *Cell* **95,** 431–437.

Taraporewala, Z., Chen, D., and Patton, J. (2001). Multimers of the bluetongue virus nonstructural protein, NS2, possess nucleotidyl phosphatase activity: Similarities between NS2 and rotavirus NSP2. *Virology* **280,** 221–231.

Taraporewala, Z., Chen, D., and Patton, J. T. (1999). Multimers Formed by the rotavirus nonstructural protein NSP2 bind to RNA and have nucleoside triphosphatase activity. *J. Virol.* **73,** 9934–9943.

Taraporewala, Z., and Patton, J. (2001). Identification and characterization of the helix-destabilizing activity of rotavirus nonstructural protein NSP2. *J. Virol.* **75,** 4519–4527.

Thomas, G. J. Jr. (1999). Raman spectroscopy of protein and nucleic acid assemblies. *Annu. Rev. Biophys. Biomol. Struct.* **28,** 1–27.

Tihova, M., Dryden, K. A., Bellamy, A. R., Greenberg, H. B., and Yeager, M. (2001). Localization of membrane permeabilization and receptor binding sites on the VP4 hemagglutinin of rotavirus: Implications for cell entry. *J. Mol. Biol.* **314,** 985–992.

Tsao, J., Chapman, M. S., Agbandje, M., Keller, W., Smith, K., Wu, H., Luo, M., Smith, T. J., Rossmann, M. G., Compans, R. W., *et al.* (1991). The three-dimensional structure of canine parvovirus and its functional implications. *Science* **251,** 1456–1464.

Tsuruta, H., Reddy, V. S., Wikoff, W. R., and Johnson, J. E. (1998). Imaging RNA and dynamic protein segments with low-resolution virus crystallography: Experimental design, data processing and implications of electron density maps. *J. Mol. Biol.* **284,** 1439–1452.

Tye, B. K., Huberman, J. A., and Botstein, D. (1974). Non-random circular permutation of phage P22 DNA. *J. Mol. Biol.* **85,** 501–528.

Valegard, K., Liljas, L., Fridborg, K., and Unge, T. (1990). The three-dimensional structure of the bacterial virus MS2. *Nature* **345,** 36–41.

Valegard, K., Murray, J. B., Stockley, P. G., Stonehouse, N. J., and Liljas, L. (1994). Crystal structure of an RNA bacteriophage coat protein–operator complex. *Nature* **371,** 623–626.

van Regenmortel, M. H. V., Fauquet, C. M., Bishop, D. H. L., Carstens, E. B., Estes, M. K., Lemon, S. M., Maniloff, J., Mayo, M. A., McGeoch, D. J., Pringle, C. H., and Wickner, R. B. (2000). "Virus Taxonomy: The Classification and Nomenclature of Viruses." Academic Press, San Diego, CA.

Wei, N., and Morris, T. J. (1991). Interactions between viral coat protein and a specific binding region on turnip crinkle virus RNA. *J. Mol. Biol.* **222,** 437–443.

Wikoff, W. R., Liljas, L., Duda, R. L., Tsuruta, H., Hendrix, R. W., and Johnson, J. E. (2000). Topologically linked protein rings in the bacteriophage HK97 capsid. *Science* **289,** 2129–2133.

Wu, H., and Rossmann, M. G. (1993). The canine parvovirus empty capsid structure. *J. Mol. Biol.* **233,** 231–244.

Yeager, M., Weiner, S., and Coombs, K. M. (1996). Transcriptionally active reovirus core particle visualized by electron cryo-microscopy and image reconstruction. *Biophys. J.* **70,** A116.

Zhang, H., Zhang, J., Yu, X., Lu, X., Zhang, Q., Jakana, J., Chen, D. H., Zhang, X., and Zhou, Z. H. (1999). Visualization of protein–RNA interactions in cytoplasmic polyhedrosis virus. *J. Virol.* **73,** 1624–1629.

Zhang, W., Olson, N. H., Baker, T. S., Faulkner, L., Agbandje-McKenna, M., Boulton, M. I., Davies, J. W., and McKenna, R. (2001). Structure of the maize streak virus geminate particle. *Virology* **279,** 471–477.

Zhang, Z., Greene, B., Thuman-Commike, P. A., Jakana, J., Prevelige, P. E. Jr., King, J., and Chiu, W. (2000). Visualization of the maturation transition in bacteriophage P22 by electron cryomicroscopy. *J. Mol. Biol.* **297,** 615–626.

Zhao, Y., Thomas, C., Bremer, C., and Roy, P. (1994). Deletion and mutational analyses of bluetongue virus NS2 protein indicate that the amino but not the carboxy terminus of the protein is critical for RNA–protein interactions. *J. Virol.* **68,** 2179–2185.

Zhou, Z. H., Baker, M. L., Jiang, W., Dougherty, M., Jakana, J., Dong, G., Lu, G., and Chiu, W. (2001). Electron cryomicroscopy and bioinformatics suggest protein fold models for rice dwarf virus. *Nat. Struct. Biol.* **8,** 868–873.

Zhou, Z. H., Dougherty, M., Jakana, J., He, J., Rixon, F. J., and Chiu, W. (2000). Seeing the herpesvirus capsid at 8.5 Å. *Science* **288,** 877–880.

Zimmern, D. (1977). The nucleotide sequence at the origin for assembly on tobacco mosaic virus RNA. *Cell* **11,** 463–482.

Zlotnick, A., Reddy, V. S., Dasgupta, R., Schneemann, A., Ray, W. J., Jr., Rueckert, R. R., and Johnson, J. E. (1994). Capsid assembly in a family of animal viruses primes an autoproteolytic maturation that depends on a single aspartic acid residue. *J. Biol. Chem.* **269,** 13680–13684.

MECHANISM OF SCAFFOLDING-ASSISTED VIRAL ASSEMBLY

By BENTLEY A. FANE* AND PETER E. PREVELIGE, JR.†

*Department of Veterinary Sciences and Microbiology, University of Arizona, Tucson, Arizona 85721, and †Department of Microbiology, University of Alabama at Birmingham, Birmingham, Alabama 35294

I. Introduction 259
II. ϕX174 Morphogenesis 261
III. Prescaffolding Stages: Coat Proteins and Chaperones 261
IV. The ϕX174 Internal Scaffolding Protein 263
V. Genetic Data for Scaffolding Protein Flexibility: ϕX174 and *Herpesviridae*.... 264
VI. Structural Data for Scaffolding Protein Flexibility: ϕX174, P22, and *Herpesviridae* 266
VII. So What's All This Fuss over These C Termini? 267
A. The ϕX174 Internal Scaffolding Protein 267
B. The *Herpesviridae* Scaffolding Protein 268
VIII. Internal Scaffolding Protein Function in One and Two Scaffolding Protein Systems: ϕX174 versus P22 and Herpesviruses 269
IX. The Assembly Pathway of Bacteriophage P22 270
X. The Role of the P22 Scaffolding Protein 272
XI. Functional Domains of the P22 Scaffolding Protein 274
A. Autoregulation of Synthesis 274
B. Oligomerization 275
C. Recruitment of Minor Proteins 276
D. Coat Protein Binding 277
E. Exit from Procapsids 279
XII. Physical Chemistry of the P22 Scaffolding Protein 280
A. Secondary Structure and Unfolding Studies 280
B. Small-Angle X-Ray Scattering Studies 281
XIII. The Mechanism of Scaffolding-Assisted Assembly 281
XIV. External Scaffolding Proteins 283
XV. The ϕX174 External Scaffolding Protein 284
XVI. P4 Sid Protein 290
XVII. Herpesvirus Triplex Proteins 292
XVIII. Scaffolding-Like Functions 293
References 295

I. Introduction

The existence of a class of molecules that came to be known as "scaffolding proteins" was recognized in 1974 by Jonathan King and Sherwood Casjens on the basis of their studies of the morphogenetic pathway of the *Salmonella* bacteriophage P22 (Casjens and King, 1974; King

0065-3233/03 $35.00

and Casjens, 1974), and those of other groups on bacteriophage T4 and T7 (Kellenberger *et al.*, 1968; Studier and Maizel, 1969; Showe and Black, 1973). Three characteristics of scaffolding protein were articulated: hundreds of molecules are necessary for the formation of one product, the catalyst molecule becomes a stable component of the immediate product, and dissociation of the complex is a complicated triggered reaction. Perhaps most importantly, their results indicated that proteins critical for supramolecular assembly may not be found as part of the mature structure. In subsequent years, the existence of scaffolding proteins has been documented in a variety of both bacterial and animal viruses. The critical functions of scaffolding proteins in virus assembly have become clear, and scaffolding-like functions have been identified in assembly-related proteins that may not meet the strict definition of scaffolding proteins.

There appear to be two essential functions of scaffolding protein during viral morphogenesis: (1) facilitating the nucleation of assembly and (2) subsequently mediating the reaction through completion. The assembly of supramolecular structures generally involves an unfavorable nucleation step, after which continued assembly becomes more favorable (Erickson and Pantaloni, 1981). There are two factors that may contribute to the existence of the nucleation barrier. The first is the statistical improbability of getting a sufficient number of molecules together in space and time to promote assembly (Erickson and Pantaloni, 1981). The second is the requirement for an effector molecule to promote a conformational change from an un-associable to an associable conformation (Caspar, 1980). One role of scaffolding proteins may be to increase the effective concentration of the assembly active molecules and thereby lower the nucleation barrier. Lowering the nucleation barrier is frequently coupled to an induced conformational change in the protein subunit. It is worth noting that the existence of a nucleation barrier, and the use of an auxiliary molecule to lower that barrier, provide a convenient biological mechanism to control assembly both temporally and spatially. After nucleating assembly, the second role of scaffolding proteins is to provide form-determining information. There is ample evidence that viral coat proteins are capable of assembling into a variety of morphological forms ranging from sheets and cylinders to icosahedral capsids of altered T number to octahedra (Earnshaw and King, 1978; Salunke *et al.*, 1989; Thuman-Commike *et al.*, 1998). These structures are thought to maintain similar local bonding interactions but to arise from subtle alterations in the relative disposition of the subunits. As is discussed in this review, the presence or absence of scaffolding protein can alter the disposition of the subunits within the lattice, and thereby ensure the proper form of the assembled particle.

In this review, we select two well-described phage systems, the *Escherichia coli* phage ϕX174 and the *Salmonella* phage P22, to illustrate the key functions of scaffolding proteins while drawing parallels to other viral systems. At this time, the functions of scaffolding proteins are best described in phage systems because of the well-developed genetics of phage biology. However, lessons learned from these simple systems readily translate to other, less tractable, viral systems.

II. ϕX174 Morphogenesis

With developed genetics, biochemistry, and biophysics the *Microviridae* system is well suited for the study of virion morphogenesis. Six assembly intermediates can be purified (Farber, 1976; Hayashi *et al.*, 1988; Ekechukwu and Fane, 1995) and the atomic structures of the ϕX174 virion and procapsid, containing both the internal and external scaffolding proteins, have been solved (McKenna *et al.*, 1992, 1994, 1996; Dokland *et al.*, 1997, 1999). While crystal structures provide a wealth of information, allowing the results of genetic experiments to be interpreted within a structural context, data are limited to the particle crystallized. Transient or less stable interactions between proteins and the morphogenetic functions of the elucidated structures are not always apparent (Ilag *et al.*, 1995; Dokland *et al.*, 1997; Burch and Fane, 2000b).

III. Prescaffolding Stages: Coat Proteins and Chaperones

As illustrated in Fig. 1, the first ϕX174 morphogenetic intermediates are the 9S and 6S particles, respective pentamers of viral coat and spike proteins. Several lines of evidence suggest that these particles self-assemble without the aid of scaffolding or host cell proteins, such as groEL and groES (Sternberg, 1973; Georgopoulos and Hohn, 1978; Hendrix and Tsui, 1978). In cells infected with nonsense and temperature-sensitive alleles of the internal scaffolding protein, protein B, 9S, and 6S particles accumulate (Tonegawa and Hayashi, 1970; Ekechukwu and Fane, 1995). Furthermore, the atomic structures of *Microviridae* capsids reveal extensive 5-fold-related contacts, suggesting a self-assembly mechanism (McKenna *et al.*, 1992, 1996; Dokland *et al.*, 1997, 1999).

Chaperone independence distinguishes the ϕX174 coat protein from those of larger bacteriophages (Gordon *et al.*, 1994; Ding *et al.*, 1995; Hanninen *et al.*, 1997; Nakonechny and Teschke, 1998). The *groE* genes were first defined as host cell mutants that failed to support the growth of

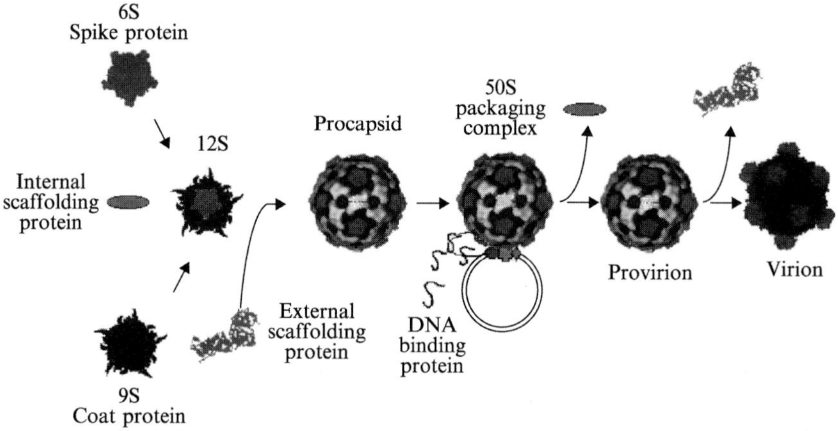

FIG. 1. The ϕX174 morphogenetic pathway. The first ϕX174 morphogenetic intermediates are the 9S and 6S particles, respective pentamers of viral coat and spike proteins. In the first internal scaffolding-mediated reaction, the internal scaffolding protein binds to the underside of the 9S particle, inducing a conformation switch that allows the intermediate to interact with the spike and external scaffolding proteins. Twelve of these pentameric intermediates associate to form the procapsid, into which the single-stranded DNA is packaged along with the DNA-binding protein. Single-stranded DNA biosynthesis and packaging are concurrent processes. The DNA replication/packaging complex associates with the procapsid, forming the 50S complex. During DNA packaging, the internal scaffolding protein is extruded from the structure. The penultimate intermediate, the viral procapsid, is infectious. Dissociation of the external scaffolding protein yields the mature virion.

several double-stranded DNA bacteriophages (Sternberg, 1973). Although extensive searches have been conducted with ϕX174, mutations in molecular chaperones have never been recovered. However, mutations in other genes, such as the *rep* helicase, are abundant (Tessman and Peterson, 1976) (M. Hayashi, personal communication). Chaperone independence may be a consequence of the T=1 capsid. Perhaps the ability of a protein to assume quasi-equivalent positions makes it more susceptible to aggregation. However, it should be noted that interaction of the P22 coat protein with groEL/ES machinery was demonstrated by the rescue of known folding mutants by groE/ES overexpression (Gordon *et al.*, 1994). Similar experiments have not been conducted in the ϕX174 system.

Preliminary evidence from some eukaryotic systems suggests that chaperones may be required for proper coat protein folding. Whereas the expression of functional eukaryotic proteins in bacteria has not been widely successful, eukaryotic expression systems have produced assembly-competent proteins (McKenna *et al.*, 1999; Newcomb *et al.*, 1999). However, whether eukaryotic expression systems provide essential chaperone and/or transport functions remains to be determined. It should be noted that the ability of eukaryotic proteins to fold into an assembly-competent state in bacteria does not necessarily indicate a chaperone-independent pathway. Nothing illustrates this point better than research conducted with *Papovaviridae* coat proteins (Wrobel *et al.*, 2000; Chen *et al.*, 2001). In eukaryotic cells, the polyoma coat protein coprecipitates with hsc70. In prokaryotic cells, it coprecipitates with dnaK, an hsc70 analog (Cripe *et al.*, 1995).

IV. The ϕX174 Internal Scaffolding Protein

In cells infected with null, temperature-sensitive, or cold-sensitive mutations of the ϕX174 B gene, coat protein pentamers (9S particles) accumulate. Pentamers formed in the absence of functional scaffolding protein do not differ from pentamers formed in its presence. On shift to permissive temperatures in *tsB*- and *csB*-infected cells, 9S particles are efficiently chased into virions (Siden and Hayashi, 1974; Ekechukwu and Fane, 1995). However, pentamers formed in the absence of B protein will self-associate, forming aberrant, heterogeneous particles (Tonegawa and Hayashi, 1970; Siden and Hayashi, 1974). This behavior has been observed in other assembly systems, such as P22 and herpes simplex virus type 1 (HSV-1) (Earnshaw and King, 1978; Desai *et al.*, 1994; Matusick-Kumar *et al.*, 1994, 1995). This aggregation is specifically caused by the lack of functional B protein, as opposed to general blocks in procapsid formation. For example, if procapsid morphogenesis is blocked by mutations in either the major spike or external scaffolding proteins, coat protein aggregates are not observed. The formation of aberrant structures with curved surfaces in the absence of viral scaffolding proteins in several viral systems suggests that icosahedral coat proteins have the inherent ability to produce structures with curvatures; however, internal scaffolding proteins are required for fidelity and/or the prevention of premature aggregates. At which level these two phenomena are related remains obscure.

After coat protein pentamer formation, the ϕX174 internal scaffolding protein binds to the underside of the pentamer and induces a conformational change in the particle. This change inhibits premature aggregation, and produces an assembly-competent state. B-protein binding is both necessary and sufficient to allow future interactions with the external scaffolding and major spike proteins. Therefore, the most dramatic consequences of the conformational switch are probably expressed on the outer surface, where these interactions will occur. Elucidating the exact structural changes resulting from conformational switching has proved difficult. Ideally it would be preferable to compare the atomic structure of a naive pentamer with that of a pentamer in the procapsid. However, assembly-naive 9S particles aggregate *in vitro* (R. McKenna; personal communication), complicating crystallization.

The results of second-site genetic analyses have offered some insights into the nature of the conformational switch (Fane and Hayashi, 1991; Ekechukwu and Fane, 1995). Although morphogenesis does not continue past the first B protein-mediated reaction in cells infected with the cold-sensitive B protein used in these studies, two lines of evidence suggest that the *cs* protein retains some level of function, indicating a defect in conformational switching. First, 9S particles do not aggregate *in vivo*, suggesting that the *csB* protein still inhibits the inappropriate aggregation of coat protein pentamers. Second, the mutant is rescued by substitutions located on the outer surface of the coat protein, not at the scaffolding–coat protein interface. The mutations are located within three distinct sequences of considerable homology, all found in loop regions of the protein, as opposed to the β-barrel core. These sequences may play a key role, perhaps as hinges, in mediating pentamer conformational switches. Coat protein mutations affecting other stages of morphogenesis, such as external–scaffolding protein interactions, packaging complex recognition, B-protein specificity, and provirion-to-virion transition, have also been isolated (Fane *et al.*, 1993; Ekechukwu *et al.*, 1995; Jennings and Fane, 1997; Burch *et al.*, 1999; Hafenstein and Fane, 2002). All these mutations are found within the loop regions of the atomic structure, suggesting that once folded the contribution of the β-barrel core to morphogenesis is minimal.

V. Genetic Data for Scaffolding Protein Flexibility: ϕX174 and *Herpesviridae*

Genetic studies of scaffolding function have, in general, been impeded by a dearth of scaffolding protein missense mutations that confer defects in morphogenesis. Of course, in eukaryotic systems, genetic analyses are much more difficult to conduct. However, with advances in recombinant DNA technologies, excellent forays in this area are being made (Desai and Person, 1999; Warner *et al.*, 2000). However, the dearth of such mutations in phage systems requires an explanation. If most substitutions are detrimental, few conditional lethal mutants will be recovered. If wild-type protein is provided *in trans*, as has been done in ϕX174, then dominant lethality would be the most common phenotype. In this model the mutated B protein would retain enough function to enter the morphogenetic pathway, and effectively compete with the wild-type protein supplied *in trans*. The location of gene B in an overlapping reading frame (Sanger *et al.*, 1978) may have also hindered the ability to isolate mutants. Proteins encoded by the overlapping genes could also be affected, creating double or triple mutants. Yet, it is difficult to rationalize the existence of such genetic intolerance within an otherwise dynamic evolutionary system. Alternatively, the lack of missense mutations could indicate that the proteins are highly tolerant of amino acid substitutions. This has been intuited by sequence, genetic, and structural analyses in many systems (Miller *et al.*, 1979; Dikerson and Geis, 1983; Krebs *et al.*, 1983; Bashford *et al.*, 1987; Fane and King, 1987; McKenna *et al.*, 1992, 1996; Chapman and Rossmann, 1993; Jennings and Fane, 1997). This alternate hypothesis is well supported in the ϕX174 and *Herpesviridae* systems.

The internal scaffolding proteins of ϕX174 and the related bacteriophages α3 and G4 have been cloned and expressed *in vivo*. These cloned genes were assayed for the ability to cross-complement ϕX174, G4, and α3 *nullB* mutants. Surprisingly, despite only 30% homology, the proteins, with one exception (see below), were capable of cross-complementation, yielding not only procapsids but mature virions (Burch *et al.*, 1999). However, in all cases, morphogenesis was more efficient when directed by the indigenous protein. In similar experiments, the bovine herpesvirus 1 proteinase and scaffolding proteins were shown to support the formation of hybrid HSV-1 B capsids, despite only 41% sequence identity (Haanes *et al.*, 1995). In essence, the "foreign" scaffolding proteins used in these systems can be regarded as "multiple mutants." In the ϕX174 system, for example, a scaffolding protein in which 70% of the amino acids are altered is functional. The difficulty in obtaining single amino acid substitutions with discernable phenotypes becomes apparent.

This cross-functional phenomenon may also indicate that scaffolding proteins are inherently flexible. Considering the dynamics of viral assembly, some inherent flexibility is probably required. Internal scaffolding proteins must first assume a structure that directs the assembly of coat protein pentamers (ϕX174) or monomers (P22) into a rigid capsid. Afterward, these proteins must assume an alternative structure, one that allows for extrusion from an internal location. In P22 and ϕX174, this structure must be compact enough to exit through 20 to 30-Å pores (Prasad *et al.*, 1993; Ilag *et al.*, 1995).

VI. Structural Data for Scaffolding Protein Flexibility: ϕX174, P22, and *Herpesviridae*

Evidence for inherent flexibility comes from the ϕX174 procapsid atomic structure. In the procapsid, the internal scaffolding protein binds to a cleft formed between α-helix 2 and the β barrel of the coat protein (McKenna *et al.*, 1996; Dokland *et al.*, 1997, 1999). Binding appears to be mediated by the last 24 amino acids of the scaffolding protein, which is the only region of the protein exhibiting a high degree of conservation among the related phages used in the cross-complementation studies (see above). These interactions are primarily aromatic and comprise the most intimate contacts between B and coat proteins. The C terminus is also the most ordered part of the protein. The first 60 amino acids of the protein yield primarily diffuse density, suggesting that interactions made by this region are variable and/or nonspecific. In addition, the amino termini of the cross-functional scaffolding proteins are highly divergent. Therefore, it is unlikely that coat–scaffolding interactions in this region are governed primarily by specific side chains.

The image of the ϕX174 internal scaffolding protein in the crystal structure is similar to the images of the HSV-1 and P22 scaffolding proteins in cryoimage reconstructions: local icosahedral symmetry. Unlike ϕX174, a strict coat: scaffolding protein stoichiometry is not observed in these larger capsids. In fact, biophysical data indicate that there are two classes of the P22 scaffolding, which are readily distinguishable by kinetic and calorimetric data (Parker *et al.*, 2001). Although the P22 protein does not form an internal icosahedral lattice, interactions with defined areas of the coat protein shell create local regions of icosahedral order, which can be visualized by comparing cryoimage reconstructions of P22 procapsids with and without scaffolding proteins (Thuman-Commike *et al.*, 1999). The difference map reveals scaffolding density associated with four distinct quasi-equivalent coat subunits. Although this density is ordered in close proximity to the overlaying T=7 coat protein lattice, it quickly dissolves as it moves inward. As discussed below, the results of structural and biochemical experiments strongly suggest that the scaffolding density represents the C terminus of the protein (Parker *et al.*, 1998; Tuma *et al.*, 1998). As with the P22 procapsid, cryoimage reconstruction of the HSV-1 procapsid demonstrates that the bulk of the HSV scaffolding protein is not icosahedrally ordered (Trus *et al.*, 1996; Zhou *et al.*, 1998). However, difference maps constructed from HSV-1 procapsids with and without maturation protease function (Newcomb *et al.*, 2000) have uncovered local icosahedral symmetry (Zhou *et al.*, 1998). The HSV-1 maturation protease releases the C-terminal 25 amino acids of the scaffolding protein. Again, ordered scaffolding density is seen only in association with four quasi-equivalent coat proteins. This density extends inwards for 40 Å before dissolving into an unordered density core, and, with good reason, is attributed to the C terminus of the protein (see below).

VII. So What's All This Fuss over These C Termini?

A. *The φX174 Internal Scaffolding Protein*

Data from three diverse systems, φX174, P22, and HSV-1, suggest that the C termini of the scaffolding proteins play the most critical role in coat protein recognition. The importance of this region in the φX174 system is reinforced by both genetic and biochemical data. As stated above, the scaffolding proteins of the related phages G4, φX174, and α3 are able to cross-complement (Burch and Fane, 2000b). However, there was one instance in which cross-complementation was not observed. The φX174 protein, provided *in trans*, cannot participate in the formation of the G4 procapsid in G4 *nullB* infections. Characterization of the G4 morphogenetic pathway under these conditions revealed an accumulation of major coat and spike protein pentamers, indicating that morphogenesis was arrested before the first internal scaffolding protein-mediated reaction. In addition, the presence of the φX174 B protein did not inhibit wild-type G4 morphogenesis. Taken together, these data suggest that the φX174 B protein does not recognize or otherwise interact with the G4 coat protein. However, two G4 mutants (*φXB-utilizers*) that can productively utilize the φX174 B protein were isolated and characterized. Both confer substitutions in regions of the G4 coat protein that contact the C-terminal half of the B protein. In both instances, the substitutions create local coat protein sequences that are more φX174-like. One of these substitutions is located directly within the aromatic B protein-binding cleft. It confers a Ser → Phe substitution and most likely reflects the importance of aromatic interactions in coat protein recognition. Interestingly, one of these gain-of-function mutations confers a loss of function: the ability to productively utilize the α3 scaffolding protein.

To further investigate the importance of C-terminal interactions, chimeric B genes were constructed and the proteins were expressed *in vivo* (Burch and Fane, 2000b). The atomic structure of the internal scaffolding protein was used as a guide in determining the junction points within the chimeric genes. The most logical junction appeared to be between amino acids 60 and 70, sequence that contains no secondary structures and bridges the structurally defined C terminus with the less-defined N-terminal density. The ϕXG4 B protein complements G4 *nullB* mutants, demonstrating that the inability of the ϕX174 B protein to interact with the G4 coat protein is a function of the C terminus, or lack thereof. In addition, when the C terminus of any chimeric scaffolding protein was of the same origin as the viral coat protein used in the experiment, complementation was the most efficient. However, several of the chimeric proteins demonstrated decreased complementation efficiencies under some conditions, when compared with wild-type proteins. This observation suggests that the termini do not function as totally separate and independent domains. Although the amino acids involved in specific coat protein interactions are located in the C terminus of the proteins, nonspecific contacts, which may be mediated by the unordered N terminus, cannot be entirely disregarded. For the most part, the function(s) of the N terminus remain obscure. The procapsid atomic structure suggests that the first 10 amino acids form an α helix that self-associates across the 2-fold axis of symmetry. However, proteins lacking this α helix have wild-type phenotypes (Burch and Fane, 2000b) (C. Novak and B. Fane, unpublished). These contradictory observations demonstrate why dogma should not be declared on the basis of structural or genetic data alone.

B. The Herpesviridae *Scaffolding Protein*

While the C terminus of the B protein can be directly visualized in the ϕX174 atomic structure, the results of genetic and biochemical experiments in the P22 and herpesvirus systems support the assignment of density to the C termini in cryoreconstructions. As stated above, the HSV-1 scaffolding protein can support the formation of bovine herpesvirus 1 procapsids (Haanes *et al.*, 1995). However, the importance of the C terminus in herpesvirus scaffolding–coat interactions was elucidated in experiments in which cross-complementation was not observed, unless chimeric proteins were utilized. For example, chimeric varicella-zoster virus (VZV) and HSV-1 scaffolding proteins cross-function only when the C terminus of the scaffolding protein is of the same origin as the viral coat protein (Preston *et al.*, 1997). In a more detailed study, replacing the last 12 amino acids of the cytomegalovirus (CMV) scaffolding proteins with the HSV-1 sequence was both necessary and sufficient to promote the HSV-1 procapsid assembly *in vitro* (Oien *et al.*, 1997). In addition, the failure of the CMV and HSV-1 scaffolding proteins to coprecipitate suggests that scaffolding dimerization domains are found upstream of the maturation protease cleavage site (for a more detailed discussion of dimerization see Section XI.B). The interactions of the C terminus have also been studied with glutathione fusion proteins (Hong *et al.*, 1996). These coprecipitation experiments gave similar results to the chimeric protein studies and further demonstrated the importance of an α helix and a critical phenylalanine residue in coat–scaffolding interactions. Conservative substitutions for the phenylalanine abrogated all coat–scaffolding interactions, and small peptides designed to mimic the structure of the α helix act as competitive inhibitors. The importance of aromatic residues is reminiscent of studies conducted with bacteriophage G4 (Burch *et al.*, 1999). The C-terminal regions of the ϕ29 (Lee and Guo, 1995) and P22 scaffolding proteins also mediate coat protein binding (see Section XI.D).

VIII. Internal Scaffolding Protein Function in One and Two Scaffolding Protein Systems: ϕX174 versus P22 and Herpesviruses

Although the ϕX174 internal scaffolding protein appears to be mechanistically similar to HSV-1 and P22 proteins in coat protein recognition, there are some critical differences that affect later morphogenetic stages. First, ϕX174 assembles via coat protein pentamers and internal scaffolding monomers. In contrast, the active forms of the P22 and HSV-1 scaffolding proteins are oligomers (Parker *et al.*, 1997a; Newcomb *et al.*, 1999). In addition, capsomers are a structural, not morphogenetic, phenomenon in these more complex capsids. Furthermore, ϕX174 morphogenesis relies on an external scaffolding protein (protein D) to assemble pentameric intermediates into the procapsid. However, prior interaction with the internal scaffolding protein is required for subsequent coat protein interactions with the external scaffolding and spike proteins (Siden and Hayashi, 1974). In the absence of a functional D protein, 12S particles accumulate (Tonegawa and Hayashi, 1970). These particles appear to have 5-fold rotational symmetry (Hayashi *et al.*, 1988). Three chemically distinct 12S particles have been isolated (Hayashi *et al.*, 1988; Fane and Hayashi, 1991) and all three contain the F and G proteins. They differ in the incorporation of the minor vertex and internal scaffolding proteins. The presence or absence of the B protein, a substrate for the ompT protease, is an artifact of purification procedures (Richardson *et al.*, 1988; Dalphin *et al.*, 1992). Although the incorporation of the H protein remains obscure, its absence does not affect the formation of capsid-like structures *in vivo* (Spindler and Hayashi, 1979).

While the 12S particle exhibits the biochemical properties traditionally associated with morphogenetic intermediates, that is, the ability to be chased into large particles in temperature-shift experiments with *tsD* mutants, it is not clear whether this particle represents a true morphogenetic intermediate or the product of an off-pathway but reversible reaction (Hayashi *et al.*, 1988). The procapsid atomic structure supports the latter possibility. There are apparently no contacts made between the F and G proteins within the structure. The spike pentamers are tethered to the underlying capsid proteins via the external scaffolding protein, which is not a component of the 12S particle. The 12S particle formed in *tsD* infections is most likely not the degradation product of a fully formed procapsid. However, in cells infected with *csD* alleles at restrictive temperatures, fragile procapsids are produced. During DNA packaging, these procapsids dissociate into 12S-like particles that cannot be chased into larger structures (Ekechukwu and Fane, 1995). Then what is the true assembly intermediate? The atomic structure suggests that it should contain not only the internal scaffolding, major spike, and capsid proteins but also 20 copies of the external scaffolding protein. This may be the fleeting 18S particle (Fane and Hayashi, 1991; Burch *et al.*, 1999; B. Fane, unpublished).

In addition, if pentamer association is primarily mediated by the external scaffolding protein, why does ϕX174 even have an internal species? Indeed, it is unique among viruses that assemble via pentameric intermediates, such as polyomaviruses and polioviruses, in that it requires a scaffolding protein, let alone two. The functions of the external scaffolding protein are discussed later in this review. As for the evolution of the internal scaffolding protein, perhaps ϕX174 employs this fidelity protein simply because it is free. Unlike large double-stranded DNA (dsDNA) bacteriophages, ϕX174 infection does not result in a complete shutdown of host cell functions. In fact, DNA replication and transcription are almost wholly dependent on the host (Hayashi *et al.*, 1988). For lysis to occur, the host must enter mitosis (Bernhardt *et al.*, 2000). Considering the relative amount of energy to complete at least a portion of the cell cycle, the energy exhausted in translating the internal scaffolding protein is trivial. As for transcription and DNA replication, evolution has placed the gene in an overlapping reading frame.

IX. The Assembly Pathway of Bacteriophage P22

The assembly pathway of the *Salmonella typhimurium* bacteriophage P22 is typical of the dsDNA-containing bacteriophages (Fig. 2). In a series of genetic and biochemical experiments, the laboratories of Jonathan King and David Botstein demonstrated that the first identifiable structural intermediate is a "procapsid" (Botstein *et al.*, 1973; King *et al.*, 1973). The P22 procapsid is composed of an outer shell of 420 molecules of the 47-kDa coat protein (the product of gene 5) arranged with T=7 symmetry

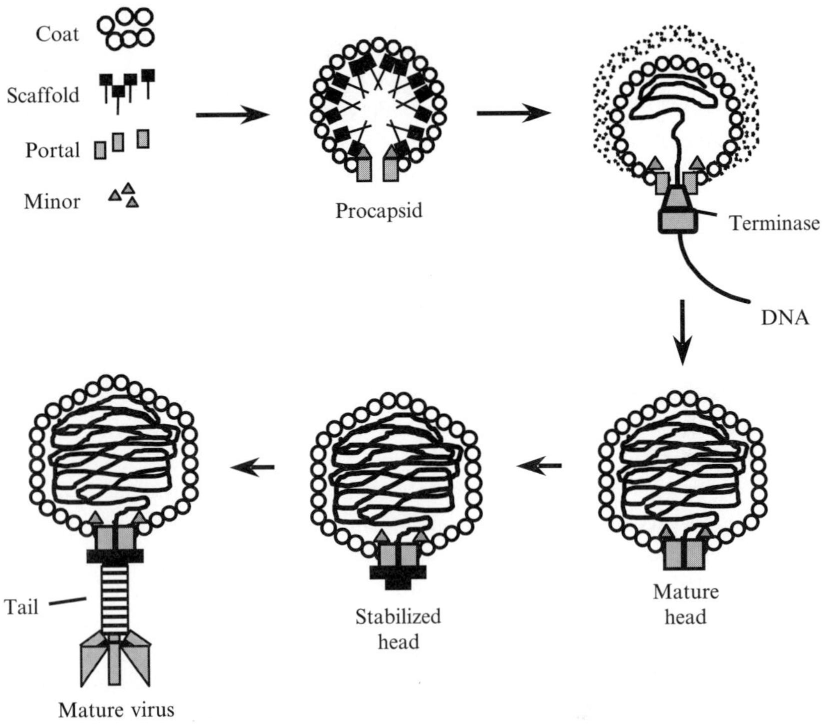

FIG. 2. The P22 morphogenetic pathway. Four hundred and fifteen molecules of the 42-kDa coat protein and 300 molecules of the 33-kDa scaffolding protein copolymerize to form a capsid whose radius is 255 Å. Twelve molecules of the 88-kDa portal protein as well as 12–20 molecules of the ejection proteins gp16 and gp20 are also incorporated at this stage. A *pac* site on the concatemeric DNA is recognized by the gp3 protein of the terminase complex, which in concert with the gp2 subunit docks at the portal vertex and packages the DNA into the prohead in an ATP-dependent reaction. DNA packaging results in a structural transformation in which the T=7 lattice becomes 10% larger, angular, and stable. Following DNA cleavage, the portal vertex is closed by the consecutive addition of gp4, gp10, and gp26. Tail attachment renders the phage infectious.

(Casjens, 1979; Prasad *et al.*, 1993). The procapsid does not contain nucleic acid but instead contains a core composed of ~300 molecules of the 33-kDa scaffolding protein (encoded by gene 8) (Casjens and King, 1974; King and Casjens, 1974; King *et al.*, 1976). Unlike the ϕX174 procapsid, a strict coat: scaffolding protein stoichiometry is not observed. Biochemical and genetic studies demonstrated that in addition to scaffolding protein the procapsid also contains approximately 12 copies of the portal protein (the product of gene 1), and 12–20 copies each of the pilot and ejection proteins (the products of genes 7, 16, and 20), required for productive infection (Botstein *et al.*, 1973; King *et al.*, 1973; Israel, 1977). In addition to promoting the fidelity of coat protein assembly, the P22 scaffolding protein may also mediate the incorporation of these minor capsid proteins.

The procapsid is the packaging precursor. The P22 genome is replicated as a concatemer and packaged by a headfull mechanism (Botstein and Levine, 1968; Tye and Botstein, 1974; Tye *et al.*, 1974a,b). A two-protein terminase complex (the products of genes 2 and 3) recognizes a *pac* sequence on the DNA and initiates packaging (Poteete and Botstein, 1979; Casjens and Huang, 1982; Casjens and Hayden, 1988). One 5-fold symmetrical vertex of the icosahedron is differentiated from the other 11 by the presence of a dodecameric portal protein complex; DNA is packaged through this portal vertex in an ATP-dependent reaction (Bazinet and King, 1985; Hartweig *et al.*, 1986; Bazinet *et al.*, 1988). DNA packaging results in an approximately 10% expansion of the T=7 lattice, a pronounced increase in stability, and the egress of the scaffolding protein (Casjens and King, 1974; King and Casjens, 1974; Earnshaw *et al.*, 1976; King *et al.*, 1976). In P22 and the *Bacillus subtilis* phage ϕ29, the scaffolding protein exits intact and can be recycled in further rounds of assembly (Casjens and King, 1974; King and Casjens, 1974; Nelson *et al.*, 1976). In most other dsDNA-containing bacteriophage and in the *Herpesviridae* cleavage of the scaffolding protein by a virally encoded protease facilitates its removal (Newcomb *et al.*, 2000).

X. The Role of the P22 Scaffolding Protein

In wild-type P22 infections, approximately 30 min postinfection, the cells burst and release on average 500 newly formed virus particles. In cells infected with mutants that block DNA packaging (e.g., those with mutations in the terminase proteins), the cells accumulate large numbers of procapsids within this time frame. In contrast, when cells are infected with mutants that exclusively prevent the synthesis of functional scaffolding protein, few particles are produced (Casjens and King, 1974). If infections are allowed to proceed long past the normal lysis time, particles are seen to accumulate. Rather than a nearly homogeneous population of well-formed procapsids, these particles fall into three classes: spirals, big shells, and small shells. The big shells are T=7 shells and the small shells are T=4 shells. The morphology of the spirals, which constitute ~45% of the population by mass, suggests that they have an incorrect radius of curvature perhaps due to improper position of the pentameric vertices (Earnshaw and King, 1978). From these simple observations, two of the key functions of scaffolding protein can be discerned. Scaffolding protein promotes assembly, somehow lowering the concentration of coat protein required for polymerization as evinced by the slower *in vivo* kinetics of aberrant particle formation, and ensures the fidelity of form determination. A variety of structural approaches have been taken to understand the role of the P22 scaffolding protein in form determination. Small-angle X-ray scattering (SAXS) experiments with procapsids indicated that the scaffolding protein density can be modeled as a thick shell or solid ball contained within the procapsid. The scaffolding density extends from near the center of the particle to a diameter of approximately 215 Å (Earnshaw *et al.*, 1976). This arrangement within the procapsid is in stark contrast to the ϕX174 crystal structure. Three-dimensional cryoelectron image reconstructions of P22 proheads revealed that the scaffolding protein was not packed with overall icosahedral symmetry (Thuman-Commike *et al.*, 1996). Rather, scaffolding protein interacts with the coat protein in a defined way only at the inside edge of the coat protein shell and the order falls off rapidly, presumably because of flexibility in the scaffolding protein (see Section VI). This result suggests that there are specific binding sites for the scaffolding protein on the coat protein subunit, which counters the argument that scaffolding protein forms a micellar core that is subsequently tiled by the coat protein. Further evidence against this model comes from sedimentation experiments under assembly conditions (see Section XI.B). A similar situation is seen in the *Herpesviridae*, where the scaffolding appears to be icosahedrally ordered only at the points of contact with the coat protein shell (see Section VI).

Thus scaffolding proteins may impart form-determining information through local interactions, for it is difficult to imagine how icosahedral information can pass through randomly oriented molecules. Cryo-EM image reconstruction has revealed that scaffolding protein interacts with four of the seven quasi-equivalent protein subunits within the T=7 lattice (Fig. 5) (Thuman-Commike *et al.*, 1996). An inherent asymmetry is observed in this interaction with scaffolding protein, appearing to interact with two of the three subunits that form a local 3-fold interaction. This theme is reiterated in the *Herpesviridae* (see Sections V and VI).

In addition to catalyzing high-fidelity assembly, scaffolding protein is also involved in recruiting the other proteins required for infectivity to the procapsid. Procapsid-like particles assembled in the absence of scaffolding protein do not incorporate either the portal protein, or the minor proteins gp7, gp16, and gp20, which are required for infectivity but not for the assembly of a stable DNA-containing phage particle (Earnshaw and King, 1978). Additional evidence for recruitment functions comes from studies by Greene and King (1996), in which P22 mutant scaffolding proteins were found to assemble procapsid-like particles that lacked the portal and/or minor head proteins. Although these findings suggest that the P22 scaffolding protein directly interacts with the portal and minor proteins, no such interactions have been demonstrated biochemically. The only physical scaffolding–portal interaction demonstrated is a relatively weak interaction between the phage $\phi 29$ scaffold and portal proteins (Guo *et al.*, 1991).

XI. Functional Domains of the P22 Scaffolding Protein

The scaffolding protein is a multifunctional protein, and a number of domains have been identified by mutational analysis (Fig. 3). The N-terminal region of the protein appears to be involved in translational autoregulation, the central region involved in oligomerization and exit during packaging, and the C-terminal region is involved in coat protein binding.

A. *Autoregulation of Synthesis*

The idea that proteins required as assembly cofactors might regulate their own synthesis is an attractive one; when these molecules are freely available synthesis could be down regulated and when their presence is required they could be upregulated. In the case of P22 scaffolding protein, during normal infections procapsids mature to phage releasing scaffolding protein to recycle in the process. Mutations that block DNA packaging accumulate intracellular procapsids, and the total amount of intracellular

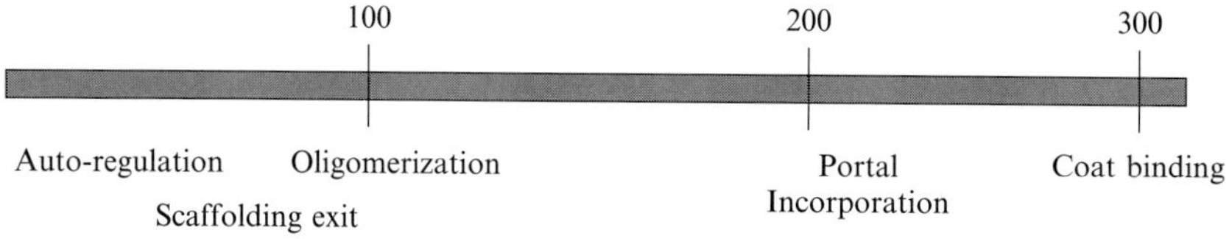

FIG. 3. Domain structure of the 303-amino acid-long P22 scaffolding protein as determined by mutational analysis. The N-terminal one-third of the protein is involved in posttranscriptional autoregulation of synthesis. An oligomerization domain, involved in dimerization and tetramerization, whose boundaries are not well defined is located in the central region of the protein. Mutations that ablate portal and minor protein incorporation are located in the region between residues 210 and 250, and the C-terminal region forms a tetratricopeptide-like fold that interacts with the coat protein.

scaffolding protein increases to approximately five times that seen in wild-type infections (King *et al.*, 1978; Casjens *et al.*, 1985). The bulk of this scaffolding protein is found contained within procapsids. Thus, scaffolding protein synthesis is upregulated in response to increased demand. Conversely, if mutations in the coat protein prevent the formation of procapsids, the amount of intracellular scaffolding protein is decreased in a manner suggesting that the ratio of scaffolding to coat protein is maintained (Casjens *et al.*, 1985). Thus, the presence of free intracellular scaffolding protein appears to repress the synthesis of new scaffolding protein whereas scaffolding protein sequestered within procapsids does not inhibit translation. Strong support for this hypothesis was obtained when the addition of purified scaffolding protein to a coupled *in vitro* DNA-dependent transcription/translation system programmed to synthesize P22 structural proteins resulted in a 15-fold decrease in the levels of scaffolding protein synthesis (Wyckoff and Casjens, 1985). These experiments, and experiments in which the production and lifetime of functional scaffolding protein mRNA was measured (Casjens and Adams, 1985), provided evidence that the autoregulation is a posttranscriptional event. Using a series of amber mutations that produced protein fragments of decreasing length, the inhibitory activity of the scaffolding protein was localized to the N-terminal one-third of the scaffolding protein (King *et al.*, 1978; Casjens and Adams, 1985; Casjens *et al.*, 1985; Wyckoff and Casjens, 1985). The molecular mechanism of scaffolding autoregulation is unknown but translational inhibition through scaffolding–mRNA binding is an attractive hypothesis.

B. *Oligomerization*

In an effort to determine whether scaffolding protein formed cores that would serve as a template surface on which to polymerize coat protein, a careful series of centrifugation experiments was performed. While these studies demonstrated that scaffolding protein did not form cores, they suggested that it might undergo some oligomerization. Analytical ultracentrifugation experiments under assembly conditions subsequently revealed that scaffolding protein forms dimers and tetramers in solution (Parker *et al.*, 1997a). The λ (Ziegelhoffer *et al.*, 1992) and herpesvirus internal scaffolding proteins also dimerize, as do the ϕX174 and P4 external scaffolding proteins (see Sections XIV and XVI). The dissociation constant for dimerization is 75 μM and for tetramerization it is 400 μM. Evidence of a physiological role for oligomerization comes from *in vitro* assembly experiments. Kinetic analysis of the rate of assembly as a function of scaffolding protein concentration suggested that assembly displayed a kinetic order of between 1.7 and 3.5 with regard to scaffolding protein (Prevelige *et al.*, 1993). Because the conditions accessible for *in vitro* assembly allow only approximately 15% of the scaffolding protein to be dimeric, it was difficult to demonstrate directly a role for oligomers in assembly. However, the isolation of a naturally occurring mutant (R74C/L177I) that spontaneously formed disulfide-cross-linked dimers allowed for the direct testing of dimers in assembly reactions. The rate of assembly with covalently dimeric scaffolding protein was significantly faster than with an equivalent concentration of wild-type scaffolding protein, suggesting an active role for dimers in assembly (Parker *et al.*, 1997a). The rate differential between covalently dimeric and wild-type scaffolding proteins decreased with increasing scaffolding protein concentration, as would be expected if the wild-type proteins were being driven to dimerize by mass action. Kinetic analysis of assembly with the covalently dimeric scaffolding protein revealed an order of 1.5, suggesting the dimers and perhaps tetramers may be involved in the rate-limiting step of assembly (M. H. Parker and P. E. Prevelige, unpublished observations). Reduction of the R74C disulfide bond led to a reduction of the assembly rate below that obtained for wild-type reactions, indicating a somewhat deleterious effect of the point mutation. Under reducing conditions the R74C/L177I mutant displayed a reduced tendency to form dimers; perhaps this reduced K_a accounts for the deleterious effect.

Although the regions of the scaffolding protein responsible for oligomerization remain undefined, there are some mutational data that bear on the question. The fact that the R74C mutation leads to the spontaneous formation of disulfide-cross-linked dimers suggests that the dimers are likely symmetrical with amino acid 74 located near the dimer interface. A scaffolding protein mutant in which the N-terminal 140 amino acids were deleted was capable of promoting assembly. Analytical ultracentrifugation experiments indicated that this protein was capable of dimerization but not tetramerization, suggesting that the domain driving tetramerization may lie closer to the N terminus of the protein (Parker *et al.*, 1997b). In accord with this suggestion, chemical cross-linking studies suggest that the tetramer is nonsymmetrical (M. H. Parker and P. E. Prevelige, unpublished observations).

C. *Recruitment of Minor Proteins*

Although not strictly required for assembly proper, several minor proteins ultimately required for infectivity are incorporated during procapsid assembly. These are the portal protein and the ejection proteins gp7, gp16, and gp20. The portal protein forms the conduit, located at a unique icosahedral vertex, through which DNA is packaged and subsequently exits during infection (Bazinet and King, 1985; Hartweig *et al.*, 1986; Bazinet *et al.*, 1988). The ejection proteins are required to deliver an infectious DNA molecule into the host cell (Israel, 1977; Bryant and King, 1984). Indications for the role of scaffolding protein in their incorporation were first obtained from the observation that procapsid-like particles assembled in the absence of scaffolding protein did not contain either the portal or pilot proteins (Earnshaw and King, 1978). The isolation of a temperature-sensitive scaffolding protein (S242F), which could drive the assembly of procapsids lacking both portal and pilot proteins, and the isolation of a Y214W mutant, which incorporated pilot proteins but not the portal protein, suggested the central region of the scaffolding protein was involved in these functions (Greene and King, 1996). Further support for the role of the central region in minor protein incorporation is the fact that fragments of scaffolding protein starting at residue 141 can assemble procapsid-like particles that still incorporate portal as do deletions from the N terminus up to residue 228. Deletion of nine additional residues ablates portal incorporation (S. Casjens, personal communication).

However, whether the interaction is direct or indirect remains an open question. Although there is biochemical evidence suggesting a direct interaction of one pilot protein (gp16) with the coat protein (Thomas and Prevelige, 1991), attempts to biochemically detect interactions between the scaffolding and the minor and portal proteins have met with failure (P. E. Prevelige, unpublished data).

D. Coat Protein Binding

A domain contributing to the binding of scaffolding protein to coat protein has been identified by deletion analysis. Deletion of the C-terminal 11 amino acids results in scaffolding protein that is incapable of either promoting assembly, or binding to coat protein subunits. However, on the basis of Raman, circular dichroic (CD), and nuclear magnetic resonance (NMR) analysis, the overall secondary structure of the protein appears largely unchanged except for the loss of some α helicity (Tuma *et al.*, 1998). In *in vitro* assembly reactions, truncated and full-length scaffolding subunits form mixed dimers, indicating that their dimerization domains remain intact and accordingly inhibit assembly (Parker *et al.*, 1998). The removal of the C-terminal 11 amino acids results in a loss of α helicity corresponding to approximately 30 amino acids. The three-dimensional structure of the C-terminal coat protein-binding domain (residues 238–303) was determined by NMR (Sun *et al.*, 2000). In these experiments, a fragment of the scaffolding protein corresponding to the carboxy-terminal 66 amino acids was studied. This fragment was still capable of binding to coat protein subunits and of inducing its polymerization. However, the fragment caused a much higher incidence of incorrect assembly than did the wild-type protein or the larger fragment 141–30. This result is in accord with the results with other truncation mutants, which suggest that the intact protein is required to confer the maximum fidelity of assembly over a broad range of conditions.

The structure of the C-terminal peptide revealed that a relatively flexible region spanning 27 amino acids, from residues 240 to 267, was followed by a well-defined helix–loop–helix motif spanning residues 268–303 (Fig. 4; see Color Insert). The two helices (helix I residues 268–283; helix II, residues 289–303) were amphipathic in nature. Hydrophobic residues on each helix are packed together to form a hydrophobic core. The packing of the two helices against one another explains the observation that removal of the C-terminal 11 amino acids destabilizes approximately 30 helical residues. The connecting loop consists of five residues (residues 284–288), of which the first four form a type I β turn. The fifth residue extends the loop, allowing the necessary freedom for the helices to associate along their lengths. One striking feature of the structure is the high density of charged residues on the outside of the coat protein-binding domain. Five basic residues (R293, K294, K296, K298, and K300) are located on one side of the outer face of the 12-residue C-terminal helix. The highly charged nature of this surface provides an explanation for the observed salt sensitivity of the scaffolding–coat protein interaction. The unique characteristics of the last amino acids are reminiscent of the ϕX174 coat–internal scaffolding protein interactions. However, in that system the interactions were mediated by hydrophobic and aromatic side chains.

Scaffolding protein can be stripped from procapsids, leaving an intact shell of coat protein in which the lattice is nearly identical to that of the scaffolding-containing procapsid (Thuman-Commike *et al.*, 1999). Surprisingly, when scaffolding protein is added back to the empty procapsid shells, it is capable of reentry and binding (Greene and King, 1994). This provides a mechanism to measure the thermodynamics of binding independently of the contributions from assembly. Isothermal titration calorimetry provided direct evidence of two classes of scaffolding protein within the procapsid, a tightly bound class and a weakly bound class (Parker *et al.*, 2001). These two classes roughly correlated with the two rates of scaffolding reentry seen in turbidity-based kinetic assays (Greene and King, 1994). The apparent K_d for binding to the high-affinity sites was

100–300 n*M* and was almost entirely enthalpy driven between 10 and 37°C. A more negative than expected ΔC_p suggested the possibility of a conformational change on binding or bridging water molecules (Parker *et al.*, 2001). Raman spectroscopy has suggested that binding of the scaffolding protein to the coat protein results in an increase in the amount of α-helical secondary structure. This conformational change may represent a component of the release mechanism that allows scaffolding exit during DNA packaging (Tuma *et al.*, 1996).

E. Exit from Procapsids

Although scaffolding-containing procapsids are stable structures in which the scaffolding protein can remain bound for prolonged periods of time, they are programmed to release the scaffolding protein intact during DNA packaging. Exit presumably occurs through ~25-Å holes present at the center of the hexameric capsomers (Prasad *et al.*, 1993b). Whether it is necessary to package some fraction of the DNA to drive egress, as is the case for bacteriophages T4 (Jardine *et al.*, 1998), λ (Hohn, 1983), T3 (Shibata *et al.*, 1987a,b), and T7 (Masker and Serwer, 1982), or whether only DNA–terminase complex docking at the portal vertex is required, is unknown. However, circumstantial evidence suggests the former (Poteete *et al.*, 1979). In the case of bacteriophage λ there appear to be specific sites of interaction of the DNA with the capsid. Although this has not been determined for P22, it seems likely. Parker and Prevelige (1998) proposed that scaffolding protein was bound through charge–charge interactions in which the positive charges on the basic C-terminal domain of the scaffolding protein interacted with negatively charged patches on the coat protein subunits. During DNA packaging, the negative charges on the DNA would serve to drive the charged patches away from each other, driving expansion and destroying the scaffolding protein-binding site. On the basis of high-resolution structural data, a similar model has been proposed for the role of DNA packaging in driving the expansion of the bacteriophage HK97 (Conway *et al.*, 2001). The NMR structure of the scaffolding protein represented the unbound state, and it is also possible that a conformational change in the scaffolding protein itself contributes to its release.

In accord with an active role for scaffolding protein in release, mutational studies have identified regions involved in release of the scaffolding protein. Deletions of the region from residues 1 to 58 allows for the production of infectious virions, whereas deletion of the region from residues 1 to 63 causes assembly to stop at the procapsid step. This suggests that the region from residues 58 to 63 is involved in scaffolding exit. Further support for this comes from the observation that mutation R74C results in a temperature-sensitive phenotype in which assembly is halted at the procapsid stage (Greene and King, 1996). Biochemical characterization suggests that the reason for this is failure of the scaffolding protein to exit. [This mutation was originally described as a leucine-to-isoleucine mutation at residue 177; however, the phage contained both L177I and R74C, and in subsequent experiments it was determined that the R74C mutation conferred the phenotype (B. Greene and J. King, personal communication).] Mutations in herpesvirus scaffolding protein that block scaffolding exit also prevent DNA packaging and maturation.

XII. Physical Chemistry of the P22 Scaffolding Protein

A. *Secondary Structure and Unfolding Studies*

To date the only internal scaffolding proteins for which there are high-resolution structural data is the N-terminal domain of the CMV scaffolding protein (Qiu *et al.*, 1996; Shieh *et al.*, 1996; Tong *et al.*, 1996), the C-terminal domain of the bacteriophage P22 scaffolding protein (Sun *et al.*, 2000), and the ϕX174 B protein (Dokland *et al.*, 1997, 1999). However, the secondary structure of P22 scaffolding protein has been examined by both circular dichroism (CD) (Teschke *et al.*, 1993) and Raman spectroscopy (Thomas *et al.*, 1982). Both techniques indicate that the protein is largely α helical. The CD studies suggest helicity on the order of 23%, with the rest being distributed between β sheet and turn (37%) and random coil (40%). The Raman studies also suggest substantial helicity.

Stability studies of the scaffolding protein using both temperature and guanidine hydrochloride as perturbants revealed that the protein is only marginally stable in solution. Approximately 13% of the α-helical structure is lost at 30°C, a temperature at which the phage is fully viable (Greene and King, 1999a,b). Similarly, a loss in helicity is seen on the addition of as little as 0.25 *M* guanidine hydrochloric. The unfolding curves measured by CD and fluorescence did not coincide, indicating multidomain unfolding. It required higher temperature or higher guanidine hydrochloric concentration to induce a change in fluorescence indicating that the single tryptophan, which resides at position 134, is in a relatively more stable domain. Proteolytic digestion in the presence of increasing concentrations of guanidine hydrochloric demonstrated that the C-terminal domain is the first unfolding domain, a result consistent with the observation that scaffolding protein can be stripped from procapsid by mild (0.5 *M*) guanidine hydrochloric treatment (Fuller and King, 1981).

Although the spectroscopic studies indicate that scaffolding has elements of secondary structure, it appears to lack a well-folded globular structure. Hydrodynamic studies suggest that the molecule is highly extended, having an axial ratio of approximately 10:1 (Fuller and King, 1982; Parker *et al.*, 1997a). Calorimetric studies revealed no evidence of a cooperative melting transition for scaffolding protein (Galisteo and King, 1993), and hydrogen–deuterium exchange studies provided no evidence of an exchange-protected core; the only protected region was the coat protein-binding domain (Tuma *et al.*, 1998). Consistent with the idea that the molecule lacks a well-folded stable core is the observation that scaffolding protein can bind up to 12 molecules of 1, 1′-bis(4-amilino)-naphthalene-5, 5′-disulfonic acid (bisANS), a compound frequently taken as an indicator of a molten globule state (Teschke *et al.*, 1993).

B. *Small-Angle X-Ray Scattering Studies*

Sedimentation velocity studies indicated that the scaffolding protein is a highly elongated molecule. Attempts to crystallize the scaffolding protein have been unsuccessful, perhaps because of the absence of a well-folded core. SAXS experiments were performed to obtain information about the structure of the protein in solution, using the covalent dimer. Modeling of the scattering data suggested that the dimeric scaffolding protein is a Y-shaped molecule approximately 155 Å in length, the coat-binding domains form the arms of the Y and are splayed out to about 48 Å (R. Tuma, H. Tsuruta, and P. E. Prevelige, unpublished data).

XIII. The Mechanism of Scaffolding-Assisted Assembly

Any model for the mechanism of scaffolding-assisted assembly must accommodate its role both in promoting assembly and ensuring proper form determination. In the assembly of the multishelled bluetongue virus, it has been suggested that the inner core forms a template surface on which the outer shell polymerizes (see Section XVIII). Although such a mechanism is attractive it appears not to be the case as no preformed cores of scaffolding protein could be identified under assembly conditions (Prevelige *et al.*, 1988). Initiation of assembly is a critical control point and it is possible that scaffolding protein might play a role in proper initiation but not be required for continued polymerization. However, *in vitro* experiments under conditions of limited scaffolding protein demonstrated that a minimum of approximately 120 scaffolding protein molecules is required for assembly, suggesting that scaffolding protein is required throughout the assembly reaction (Prevelige *et al.*, 1988). It should be noted that these experiments, and experiments in which assembly was performed with a 10× excess of scaffolding protein, suggest that, despite tight control, the ratio of scaffolding to coat protein per se is not a critical element in ensuring assembly fidelity.

Given that scaffolding protein is required for continued polymerization there are two mechanisms by which scaffolding protein might promote assembly. Scaffolding binding might switch the coat protein subunit from an inactive, unassociable state to one competent for assembly. Alternatively, scaffolding protein might act as an entropy sink. In this model, each of the two coat-binding domains of a scaffolding protein dimer binds one coat protein molecule, resulting in the formation of a heterotetramer (two each of scaffolding and coat subunits) in which the effective concentration of coat protein is greatly increased. Thus, the binding energy for scaffolding dimerization is being used to increase the effective concentration of the coat protein. Of course, both mechanisms could be operative at the same time.

The fact that truncated scaffolding protein molecules such as fragment 240–303 used for NMR can activate the coat protein for assembly yet only weakly dimerize suggests that scaffolding binding can activate the coat protein for assembly. However, the breakdown of fidelity observed suggests that the information required for form determination is not simply locally transmitted (Parker *et al.*, 1998). Fidelity of assembly seems to be correlated with the dimerization affinity. On the basis of a series of N-terminal deletion mutants (wild type, Δ1–140, Δ1–237), which show decreased ability to dimerize, fidelity seems to correlate with dimerization potential.

A suggestion for the mechanism of the control of form determination comes from the location of the scaffolding protein within the procapsid (Fig. 5; see Color Insert). The cryo-EM image analysis suggests that scaffolding protein is dimer clustered at local 3-fold sites (Thuman-Commike *et al.*, 1999). The first steps in assembly may be the formation of a 15-coat protein subunit complex in which an initiating pentamer of coat protein subunits subsequently binds 10 additional coat protein monomers; binding of scaffolding dimers to these subunits would provide a platform on which to build. In the next step, two coat protein molecules are added, assisted by the formation of a scaffolding protein tetramer. Unassisted addition of two more coat proteins results in the formation of a pentameric vertex surrounded by five hexameric coat protein capsomers. While this could represent the fundamental building block for assembly, with 12 of these structures assembling into a T=7 procapsid, SAXS studies conducted during the process of assembly were unable to detect any such structures (R. Tuma, H. Tsuruta, and P. E. Prevelige, unpublished data). Therefore, it seems likely that growth proceeds in a continuous process.

In this case, the role of scaffolding protein would be to transiently stabilize unstable coat subunit additions, allowing assembly to proceed. It is also possible that this stabilization servers to energetically steer the assembly pathway. For example, in the absence of scaffolding protein T=4 structures are built, and the pentameric and hexameric capsomers in these T=4 particles are virtually identical to those seen in the T=7 form (Thuman-Commike *et al.*, 1998). However, to form a T=4 capsid requires the *c* coat protein subunits (Fig. 5) to form a trimer cluster, which would require trimeric clustering of the scaffolding protein. It is possible that scaffolding protein functions to prevent the formation of the trimer cluster and thereby steer the assembly toward T=7 particles.

In these models, the role of scaffolding protein is to tie together and stabilize particular arrangements of coat protein subunits. In a sense, scaffolding protein may be viewed as a bungee cord, in which the hooks represent the coat protein-binding domain and the length of the cord defines the range of interactions that may be stabilized. In this way, a relatively flexible protein that is not icosahedrally arranged in the final structure could control icosahedral assembly.

XIV. External Scaffolding Proteins

If using a strict definition of a scaffolding protein, one found associated with assembly intermediates but not in the mature virion, external scaffolding proteins such as the ϕX174 D and P4 Sid proteins are rare. However, some structural proteins, such as the T4 Soc, alphavirus glycoproteins, and *Herpesviridae* triplex proteins, bear a strong functional and/or structural resemblance (Steven *et al.*, 1992; Trus *et al.*, 1996; Iwasaki *et al.*, 2000; Zhou *et al.*, 2000; Olson *et al.*, 2001; Pletnev *et al.*, 2001). Although the decoration protein Soc does not mediate morphogenesis, it stabilizes the final structure of the T4 head (Steven *et al.*, 1992). Its arrangement on the capsid surface (Iwasaki *et al.*, 2000; Olson *et al.*, 2001), bridging adjacent capsomers, is reminiscent of the ϕX174 external scaffolding protein (Dokland *et al.*, 1997, 1999). Alphavirus cores do not achieve their ordered T=4 symmetry until they have interacted with membrane-bound glycoproteins (Pletnev *et al.*, 2001). Similarities between external scaffolding and triplex proteins are even more pronounced (see Section XVI). Perhaps "classic" external scaffolding proteins represent only one extreme of a protein group, whose diverse members mediate morphogenesis from the exterior of the capsid and/or stabilize the structure of the final product.

XV. The ϕX174 External Scaffolding Protein

The ϕX174 external scaffolding protein (protein D) performs many of the functions typically associated with internal species in one-scaffolding-protein systems: the organization of assembly precursors into a procapsid and the stabilization of that structure. However, its function is physically and temporally dependent on the internal scaffolding protein, which induces the conformational changes in capsid pentamers to prevent their premature association. In the procapsid crystal structure, 20 D proteins are associated with each pentameric capsomer. Remarkably, there is little or no contact between capsid pentamers. The structure is primarily held together by 2-fold-related contacts between D proteins.

As illustrated in Fig. 6; see Color Insert, the four D subunits (D1, D2, D3, and D4) per asymmetric unit are arranged as two similar, but not identical, asymmetric dimers (D1D2 and D3D4). These dimeric subassemblies may be a component of the tetramer structures, as visualized by negative staining, in solutions of purified D protein (R. McKenna, personal communication). However, these particles have closed point group symmetry, a significant departure from the appearance of the D subunits in the asymmetric unit. It has not yet been determined whether these tetrads consist of four or eight proteins. These particles sediment at 4S and behave like assembly intermediates in pulse-chase experiments (Tonegawa and Hayashi, 1970).

The "canonical monomer" in the crystal structure is composed of seven α helices separated by loop regions. However, there is considerable structural variation between the subunits. This variation bears no resemblance to quasi-equivalence. The subunits are not arranged in a T=4 lattice. Structural data suggest that this unique arrangement is mediated, in part, by Gly-61 in α-helix 3. One monomer in each dimer is bent 30° at this site. The kink may be needed to switch the second monomer into a nonsticky conformation. Without this flexibility, D proteins might assemble into an indefinitely growing helical bundle. The organization of the genome also suggests a critical role for this amino acid. The Gly-61 and gene E start codons overlap in all the ϕX174-like phages (Godson *et al.*, 1978; Sanger *et al.*, 1978; Kodaira *et al.*, 1992). However, a limited number of site-directed mutants could be generated at this site and propagated in cells overexpressing the wild-type protein. The results of preliminary experiments indicate that substitutions confer dominant lethal phenotypes in coinfections with wild-type ϕX174, suggesting morphogenetic defects occurring after monomers interact to form dimers (A. D. Burch and B. A. Fane, unpublished data). The wild-type protein becomes sequestered in heterogeneous dimers with the mutant Gly-61 protein. These dimers, in turn, cannot form the external lattice because of the inability to bend 30° at the Gly-61 site.

In addition to Gly-61, individual subunit structures are influenced by unique sets of interactions made with the underlying coat and neighboring D proteins within and across asymmetric units. For example, α-helix 5′ forms β structure in subunit D2, where it participates in interdimer contacts with D3, and forms helical structure in D3, where it participates in D4 intradimer contacts. In subunit D4, it mediates scaffolding contacts across the 2-fold axis symmetry and forms loop structure. α-Helix 7 forms only in the D4 subunit, where it mediates the most extensive coat protein interactions found in the entire lattice.

Although the atomic structure of the ϕX174 procapsid has yielded a wealth of information, it does not reveal all of the contacts in which the external scaffolding protein participates. During crystallization, the procapsid matured and the coat protein assumed a conformation similar to its structure in the mature virion. These maturation events were elucidated by comparing the X-ray model with the cryoimage model, which most likely represents the native state of this metastable intermediate. There are two dramatic differences between the two models: (1) In the X-ray model, the coat protein has moved inward radially, away from the external scaffolding protein lattice, but it has now fully dissociated from it; and (2) a large coat protein α helix occupies the 3-fold axes of symmetry, which in the cryo-EM image is free of density and thus contains a 30-Å pore. In addition, it is likely that a component of the external scaffolding lattice associates with the packaging machinery, presumably at the 2-fold axis of symmetry (Ekechukwu *et al.*, 1995; Burch and Fane, 2000b).

The results of genetic experiments and experiments conducted with inhibitory cross-species and chimeric scaffolding proteins have helped elucidate some of these unseen contacts. The primary sequences of the ϕX174 and related bacteriophage α3 are 70% conserved (Sanger *et al.*, 1978; Kodaira *et al.*, 1992). Divergent sequences are localized to the N and C termini of the proteins. These sequences constitute α-helices 1 and 7 as loop 6 in the atomic structure. The ϕX174 and α3 external scaffolding genes have been cloned and expressed *in vivo* and do not cross-complement. However, the expression of foreign scaffolding proteins blocks wild-type morphogenesis, suggesting inhibitory foreign-indigenous scaffolding interactions (Burch and Fane, 2000a). The ability of foreign scaffolding proteins to inhibit morphogenesis is most likely due to the formation of cross-species dimers. The majority of the intra- and interdimer contacts are mediated by the middle region of the protein, α-helices 2–6, which are strongly conserved.

To determine whether one or both termini conferred inhibitory effects, chimeric genes have been expressed *in vivo*. In the chimeras, the first α helices were interchanged. Expression inhibits morphogenesis of wild-type ϕX174 and α3 in a somewhat species-specific manner. In these experiments, the chimeric proteins are expressed from plasmids whereas the wild-type proteins are expressed from the phage genome. Efficient inhibition is governed by the identity of the first α helix in the chimeric protein. The chimera that contains α-helix 1 from ϕX174, for example, strongly inhibits ϕX174 morphogenesis; α3 morphogenesis is only modestly affected. Inhibition is most likely due to dimerization between the wild-type and chimeric proteins, which would not be affected because the dimerization is mediated by α-helices 2–6. Although the chimeric proteins retain enough function to inhibit morphogenesis, neither chimera can complement *nullD* mutants of either virus.

The relative levels and species specificity of inhibition conferred by each of the chimeras merits further explanation. Although the phenomenon is symmetrical, depending on which phage, ϕX174 or α3, was used in the assay, for purposes of clarity the discussion focuses exclusively on ϕX174. The weak inhibition conferred by the α3–ϕX chimera is achieved only when the chimeric gene is maximally induced. Under these conditions, inhibition in plating assays ranges from 10^{-1} to 10^{-3}. The strong inhibitory phenomenon is observed when cloned ϕX–α3 is barely induced. Even under those conditions, plating efficiencies drop below 10^{-6}. Although plating efficiencies below 10^{-6} are also achieved when expressing the wild-type foreign α3 protein, the clone gene must be maximally induced. These data suggest a temporal mechanism in which the initial recognition of the coat protein is mediated by α-helix 1 of the external scaffolding protein. The presence of the proper first α-helix, of the same origin as the viral coat protein, facilitates the incorporation of the chimeric protein into external lattice, acting as a vehicle for the incorporation of inhibitory foreign sequences, the unconserved amino acids found in loop 6 and α-helix 7.

ϕX174 intermediates synthesized in cells expressing foreign and the chimeric ϕX–α3 protein were analyzed by sucrose gradient sedimentation. In extracts generated from cells expressing the ϕX–α3 chimera, which assays for defects conferred by foreign loop 6 and α-helix 7 structures, procapsids and empty capsids were present; however, virions were not, indicating a block in DNA packaging. The docking of the replication/packaging machinery is most likely prevented by sequences found in the C terminus of the α3 protein, loop 6, and α-helix 7. In addition, ϕX174 mutants resistant to the expression of ϕX–α3 chimera have been isolated (*chiD*^{R} mutants). These mutations alter viral protein A, a component of the genome biosynthesis/packaging machinery, which binds the procapsid during DNA packaging, presumably at the 2-fold axis of symmetry (Ekechukwu *et al.*, 1995). Because resistance is conferred by alterations in protein A, chimeric and wild-type scaffolding proteins probably form mixed procapsids that cannot be packaged.

In extracts prepared from cells expressing the foreign protein, procapsids (108S) and empty capsids (70S) were not detected, suggesting either a block before procapsid formation or the production of unstable particles. This suggests that all of the external scaffolding dimers were heterogeneous and some of them, keeping in mind that D proteins form asymmetric dimers, may not have been able to recognize the coat and/or spike proteins, because of the presence of the foreign α-helix 1 sequence (Burch and Fane, 2000a). Interactions between α-helix 1 of the D1 subunit and the spike protein are visible in the X-ray model. The proximity of the helix to the 3-fold axis of symmetry in the D4 subunit also suggests an interaction with the coat protein. However, this interaction cannot be observed in the atomic structure because of the above-mentioned maturation events. The *chiD*^{R} mutation does not confer resistance to the expression of the foreign α3 D protein or the α3–ϕX chimera. These data indicate that foreign scaffolding proteins may confer at least two blocks in morphogenesis, whereas chimeras inhibit only a single step.

Attempts to isolate ϕX174 mutants resistant to the α3–ϕX chimera were hindered by the low level of inhibition. Mutants resistant to the expression of foreign (nonchimeric) scaffolding proteins could not be obtained in one-step selections ($<1 \times 10^{-8}$), suggesting multiple mutations may be required. However, they have been isolated in multistep selections in which ϕX174 is propagated for 12 generations in cells harboring an "uninduced" α3 D gene (plating efficiency, 10^{-1}). The *forD*^{R} virions are also resistant to both chimeric proteins. Sequence analyses revealed multiple mutations directly upstream of gene D, the region controlling D protein translation. These mutations most likely elevate the expression of the wild-type protein, hence conferring resistance to all inhibitory proteins.

To further dissect structure–function relationships in the ϕX174 external scaffolding protein, additional chimeras have been generated in the C terminus of the protein (B. A. Fane, unpublished data). In these experiments, chimeras were built directly into the phage genome and substitute either α3 loop 6 or helix 7 sequences into the ϕX174 gene. The loop 6 chimera is viable, but confers a cold-sensitive (*cs*) phenotype. Both extragenic and intragenic revertants of the *cs* phenotype have been isolated. The intragenic mutation is in loop 6, conferring an E→D substitution in the central residue of the loop, which is the amino acid found in the ϕX174 sequence. From the atomic structure of the procapsid, it is known that this residue, but only in the D4 subunit, makes extensive contacts with Lys-118 of the coat protein. The extragenic suppressors map to surface residues of the coat protein adjacent to amino acids known to interact with the external scaffolding protein in the atomic structure. These results indicate that the coat–scaffolding interface is more extensive than revealed in the "matured" procapsid crystal structure.

In complementation experiments, neither the foreign loop-6 nor helix-7 chimera complemented this helix-1 chimera. This may indicate that α-helix 1 also plays a critical role in the D4 subunit. However, an element of uncertainty exists in intragenic complementation experiments, because of the lack of a positive control. However, genetic data (Fane *et al.*, 1993; Ekechukwu and Fane, 1995) also suggest that an unseen interaction occurs between D4 α-helix 1 and α-helix 4 of the viral coat protein. Both helices are found at the 3-fold axis of symmetry in the closed structure, which has matured during crystallization. Two point mutations in the first α helix of protein D have been extensively characterized (Fane *et al.*, 1993; Ekechukwu and Fane, 1995). These mutations confer a fragile procapsid phenotype. Although procapsid morphogenesis is not inhibited, the particles disassociate during DNA packaging. In an open structure, the coat protein helix could be shifted upward and may contact α-helix 1 of the D4 subunit of an adjacent asymmetric unit, which is the most closely associated subunit with the underlying coat protein. Both helices are amphipathic and could interact via hydrophobic interfaces. If this interaction is indeed present in the native structure, this may exclude dimers with a foreign α-helix 1 from the D3D4 position. In addition, results of experiments conducted with chimeric G4–ϕX174 external scaffolding proteins and mutant ϕX174 coat proteins capable of productively utilizing the chimeric scaffolding also suggest interactions between the first α helix of the D protein and 3-fold-related coat protein residues (B. A. Fane, unpublished data).

The relative ability of the chimeric proteins to inhibit morphogenesis suggests a temporal model for coat–external scaffolding protein recognition. The cloned α3–ϕX chimeric protein only weakly inhibits ϕX174 morphogenesis. In contrast, the ϕX–α3 chimera is a potent inhibitor. These data suggest that chimeric protein incorporation into the lattice is a function of the first α helix. The proximity of the D4 α helix to the 3-fold axis of symmetry and the results of second-site suppressor analyses (Fane *et al.*, 1993; Ekechukwu and Fane, 1995) suggest that the first substrate-specific interaction occurs between a dimer of scaffolding protein and the adjacent 5-fold-related coat protein. Figure 7, see Color Insert, illustrates the atomic structure of the 3-fold axis of symmetry in the ϕX174 procapsid. In the native structure, this axis would not be occupied by the large coat protein helix. This helix may be restrained by the D4 subunit of the adjacent asymmetric unit. After coat–scaffolding recognition, loop 6 and α-helix 7 place the D4 subunit dimer atop the capsid. D1D2 dimers would then be added to the same asymmetric and adjacent asymmetric units, mediated by 5-fold D4–D2 and D4–D1 interactions, respectively. In a chain reaction, dimers would add around the pentamer. The resulting intermediate would contain 5 copies of the spike, coat, and internal scaffolding proteins, 1 copy of the minor spike protein H, and 20 copies of protein D. A particle of this composition has been detected *in vivo* and it sediments at 18S (Fane and Hayashi, 1991; Burch *et al.*, 1999). However, a means to genetically trap this intermediate in large quantities has not yet been established.

In accord with the P22 kinetic model for capsid assembly (Prevelige *et al.*, 1993), the next reaction in procapsid formation would be rate limiting and would be a higher order reaction than the following steps. If postnucleation morphogenesis involves the rapid and successive addition of one pentameric intermediate to a growing shell, the nucleation complex formation would require at least three pentameric intermediates. Because there are no coat–coat or 3-fold-related scaffolding contacts in the procapsid atomic structure, the reaction is expected to be catalyzed by three sets of 2-fold-related interactions. The involvement of the external scaffolding protein is easily visualized, because of the specific and ordered 2-fold-related contacts in the crystal structure. The role of the internal scaffolding protein, protein B, is more obscure. The N terminus is found at 2-fold axes of symmetry but it is relatively unordered. However, the answer resides in the diffuse density. Nonspecific B–B protein may facilitate nucleation by generating local foci of pentamer critical concentration. To investigate this hypothesis a series of N terminus-deleted B proteins have been cloned and expressed *in vivo*. Some of the amino-terminal deletions can still promote procapsid morphogenesis, but it appears to be at a lower level or rate (C. Novak and B. A. Fane, unpublished data). Interestingly, the ability to support procapsid morphogenesis among these deleted proteins is not linear. Efficiency does not steadily diminish with increasingly larger deletions. Instead, there appears to be a rogue region, located between amino acids 30 and 40, that interferes with virion production, whereas proteins lacking the entire amino terminus are relatively functional. Perhaps the presence of this rogue region blocks the formation of the nucleation complex, or creates an alternate complex that promotes off-pathway reactions. This is just one of many questions to be addressed in two-scaffolding protein systems.

XVI. P4 Sid Protein

The P2/P4 bacteriophage system presents a striking example of the reprogramming of coat protein assembly by a scaffolding protein. Bacteriophage P2 encodes coat (gpN) and internal scaffolding proteins (gpO) that assemble into T=7 P4 procapsids. The satellite bacteriophage P4 encodes the external scaffolding protein Sid (*size determination*). The presence of the P4 Sid protein alters morphogenesis from a T=7 to a T=4 pathway. In essence its actions force the P2 coat and internal scaffolding proteins to form the smaller structure. The smaller T=4 procapsids cannot accommodate the larger P2 genome, hence the progeny phage carry the P4 genome.

As is the case with the ϕX174 D protein, gpSid forms an external lattice around the procapsid (Marvik *et al.*, 1995; Dokland *et al.*, 1999; Wang *et al.*, 2000). However, there are critical differences between the two systems. The ϕX174 D protein is required for capsid formation. whereas Sid is not required for capsid formation proper, but alteration. The internal scaffolding protein gpO is still required for P4 morphogenesis *in vivo* (Christie and Calendar, 1990). Although P4 procapsid-like structures can be generated *in vitro* from purified Sid and coat proteins, morphogenesis is not efficient and the reaction is strongly dependent on polyethylene glycol (PEG) (Wang *et al.*, 2000), which may supplant gpO function by creating local foci of critical concentrations.

gpSid forms a dodecahedral lattice atop the T=4 capsid (Marvik *et al.*, 1995; Wang *et al.*, 2000). The outline of the lattice traces 12 large pentagons around the 5-fold axes of symmetry. In the procapsid, the 30 hexamers are bifurcated by gpSid. At the 20 places where 3 hexamers meet, Sid forms trimeric structures. Volume measurements suggest 120 copies of Sid per procapsid. The results of biochemical and genetic studies, discussed below, suggest that the Sid assembly unit is a trimer of dimers (Wang *et al.*, 2000).

The results of genetic analyses also indicate that Sid protein interacts only with the P2 coat protein. P2 mutants resistant to the effects of the Sid protein, *sir* mutants (*si*d *r*esponsiveness) have been isolated only in gene N (Six *et al.*, 1991). The affected amino acids are clustered and may delineate a critical gpN–Sid contact site. Alternatively, they may identify a hinge region needed to form a more constrained protein fold found only in the smaller capsid. In an effort to distinguish between these two hypotheses, extragenic second-site suppressors of *sir* mutations were isolated (Kim *et al.*, 2001). These *nms* mutations (wild-type *N m*utation *s*ensitive) map to the *sid* gene and create "superSid" proteins. They cluster in the C terminus of Sid protein and are not allele specific. Therefore, it is unlikely that they restore critical gpN–Sid interactions that may have been obliterated by the original *sir* mutation.

The *nms* sites could create novel points for gpN–Sid interactions or reinforce an already existing site. However, dynamic light-scattering experiments conducted with full-length and N-truncated Sid proteins suggest another mechanism (Wang *et al.*, 2000). Full-length and N-truncated proteins produce structures with hydrodynamic radii consistent with trimers, or trimers of dimers, seen in the P4 image reconstruction. Conversely, Sid proteins lacking C-terminal amino acids form monomers in solutions. The *nms* mutations may create stronger Sid–Sid–Sid interactions. This could lead to a more rigid external lattice, capable of overcoming the less flexible hinge regions created by the *sir* mutations. As for the N terminus, structural analyses indicate that the protein forms an elongated rod. This has led to a model in which the N terminus mediates Sid–Sid and gpN–Sid interactions, which is consistent with genetic data. Sid mutations cluster in both termini of the protein (Nilssen *et al.*, 1996). The N-terminal mutations could affect Sid–Sid and/or gpN–Sid interactions. The inability to recover nonfunctional Sid missense mutations within the central region of the protein suggests that this region is relatively tolerant to mutation and not involved in specific protein–protein interactions. Although assigning gpN-binding functions to the N terminus of Sid creates an attractive model, biochemical data neither support nor refute this assumption. Truncated Sid proteins lacking the first 53 amino acids of the protein coprecipitate with coat protein *in vitro*. However, it is unclear whether coprecipitation indicates specific associations or nonspecific aggregation mechanisms.

The identification of the gpN–Sid interactions sites will most likely be made when a higher resolution structure is available. These interactions may be dispersed throughout the entire external lattice. Therefore the isolation of a single mutation that does not grossly alter protein structure but completely abrogates gpN–Sid interactions is unlikely. As seen in the crystal structure of the ϕX174 procapsid, interactions between the external scaffolding protein and the underlying coat protein shell are dispersed, throughout the entire lattice. Although local areas of more numerous interactions are found, lethal dominant mutations in these regions do not prevent scaffolding–capsid protein interactions (Burch and Fane, 2000b). Alternatively, there may be only a few localized gpN–Sid interactions needed for the nucleation of the Sid lattice. If P2 morphogenesis mechanistically resembles P22 assembly (Prevelige *et al.*, 1993), an initial nucleation event involving coat and internal scaffolding protein, gpO, would be followed by the rapid addition of coat and scaffolding subunits. Concurrently in a P4-infected cell, Sid lattices would form in association with gpN, forcing the curvature of growing capsid into the smaller structure. A coat protein hexamer may simply act as nucleation complex for the external lattice. The less flexible Sid lattice may then direct smaller capsid assembly by a default mechanism: the large capsid simply cannot form within the Sid cage.

XVII. Herpesvirus Triplex Proteins

Although the herpesvirus triplex proteins are components of the mature virion, and thus do not meet the classic criterion of a scaffolding protein, they appear to perform a scaffolding-like function, promoting the morphogenetic fidelity of the T=16 capsid. The triplexes are heterogeneous trimers composed of two molecules of Vp23 and one molecule of Vp19C (Newcomb *et al.*, 1993). The triplexes are interdigitated between and connect adjacent capsomers and are required for HSV-1 procapsid assembly (Newcomb *et al.*, 1994). The morphogenetic assembly unit appears to be the intact triplex (Spencer *et al.*, 1998). Neither purified Vp23 nor Vp19c alone can interact with the major capsid protein. The initial binding of the triplex may be mediated by Vp19. Deletions within this protein can form triplex structures with Vp23, but the resulting triplexes have lost the ability to interact with the HSV-1 coat protein. The scaffolding function of the triplexes becomes apparent when the HSV-1 procapsid image reconstruction (Trus *et al.*, 1996) is compared with image reconstruction of matured inner capsids (Zhou *et al.*, 2000). In the procapsid, there appears to be little or no contact between coat proteins in adjacent capsomers. Instead, contacts are mediated exclusively by triplex proteins at sites of local 3-fold symmetry. Hence the coat protein floor of the capsid has not yet formed. A similar phenomenon has been observed in the ϕX174 procapsid ((Dokland *et al.*, 1997, 1999), in which the external and internal scaffolding proteins bridge adjacent capsomers.

Several studies have offered insights into the underlying mechanisms of triplex function. At low efficiencies HSV-1 coat proteins and triplex protein Vp19C can assemble into aberrant T=7 spherical particles, as opposed to the wild-type T=16 structure (Saad *et al.*, 1999). In the T=7 particle hexamers appear to be held together by Vp19C–coat protein interactions. As seen in the procapsid (Trus *et al.*, 1996), the coat protein floor of the particle has not yet formed. The results of these experiments suggest that Vp19C may be both necessary and sufficient to form icosahedral structures. However, Vp23 and the internal scaffolding protein are required to control the fidelity and efficiency of this process.

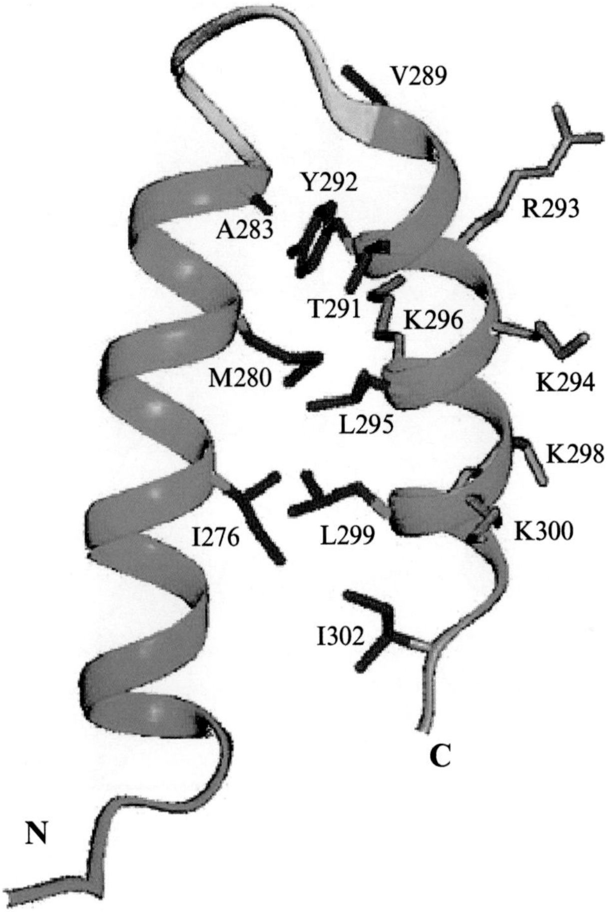

FANE AND PREVELIGE, FIG. 4. Atomic structure of the P22 scaffolding protein coat-binding domain: energy-minimized average NMR structure of the region of the P22 scaffolding protein spanning amino acids 264–303. The overall fold is a helix–turn–helix, homologous to the tetratricopeptide motif. The fold is stabilized by a hydrophobic core (purple residues) and displays a highly basic face (indicated in blue). Note the solvent-exposed hydrophobic residue, Val-289. [Adapted from Sun, Y., Parker, M. H., Weigele, P., Casjens, S., Prevelige, P. E., Jr., and Krishna, N. R. (2000). *J. Mol. Biol.* 297, 1195–1202.

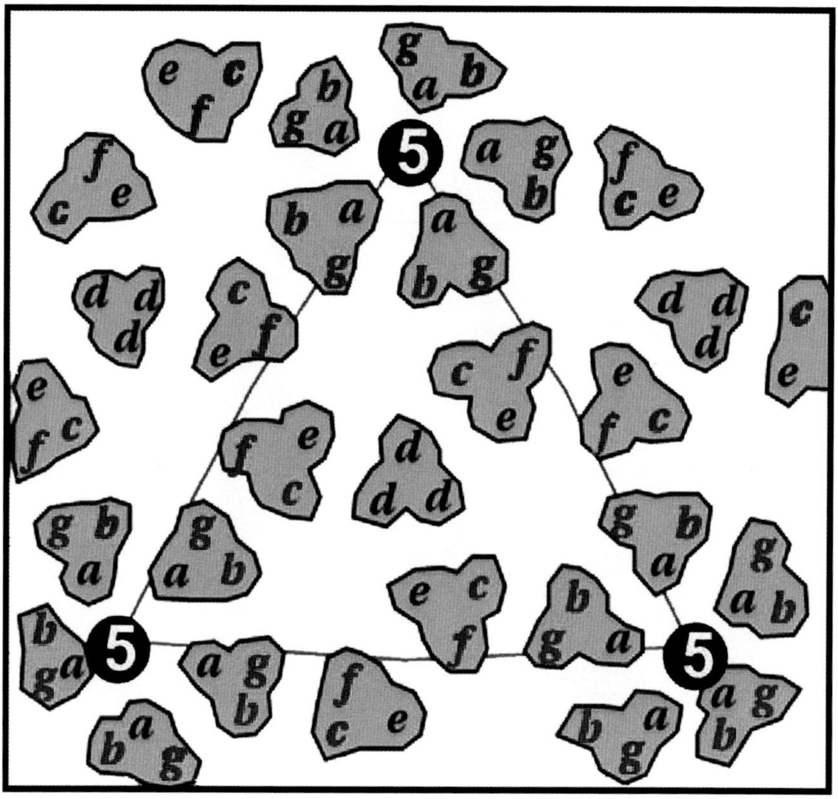

FANE AND PREVELIGE, FIG. 5. Location of the scaffolding protein bound to the P22 procapsid lattice. Schematic representation of the location of scaffolding protein within the P22 procapsid as determined by cryoelectron microscopy. The view is from the inside of the procapsid looking out. The subunits labeled *a* comprise the pentameric cluser at the vertex, while those labeled *b–g* comprise the hexameric cluster. A face is indicated by the line connecting three 5-fold vertices. The subunits present as trimeric clusters on the inside of the procapsid. Those labeled in red interact with scaffolding protein and those labeled in black do not. [Adapted from Thuman-Commike, P. A., Greene, B., Malinski, J. A., Burbea, M., McGough, A., Chiu, W., and Prevelige, P. E., Jr. (1999). *Biophys. J.* **76**, 3267–3277.]

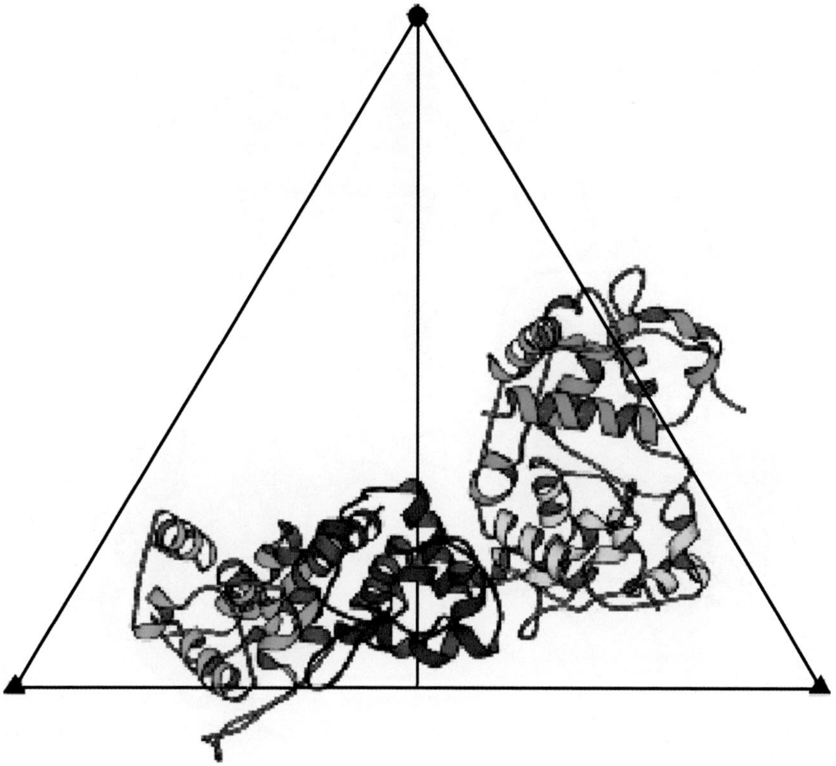

FANE AND PREVELIGE, FIG. 6. Atomic structure of the X174 external scaffolding protein. The four D subunits (D1, D2, D3, and D4) per asymmetric unit are arranged as two similar, but not identical, asymmetric dimers (D1D2 and D3D4). Each subunit makes a unique set of interactions with the underlying coat protein and neighboring D subunits both within the asymmetric unit and across 5-fold and 2-fold axes of symmetry. There are no 3-fold interactions between scaffolding subunits. Only the D1 subunit contacts the major spike protein.

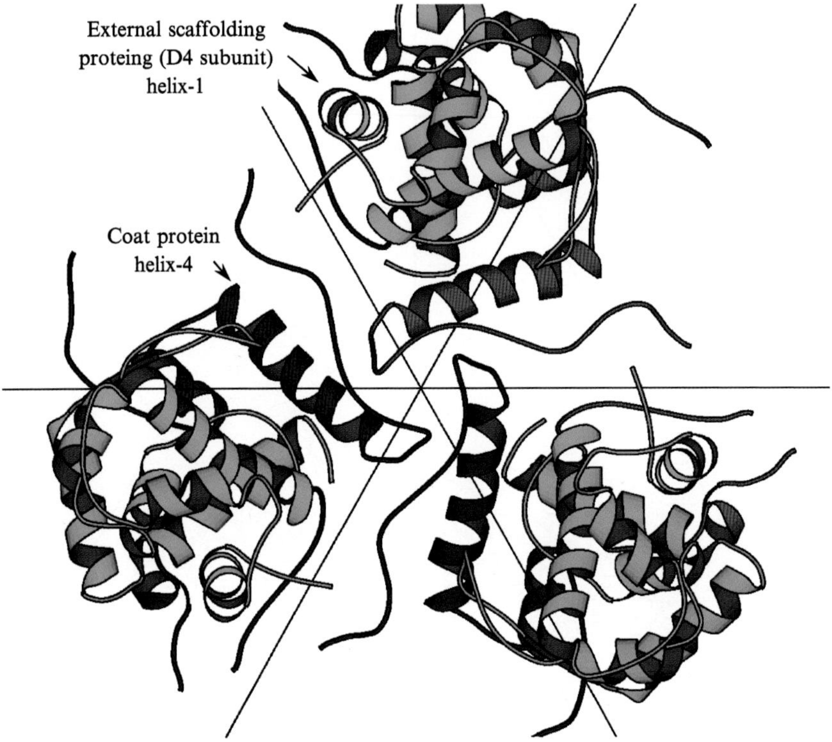

FANE AND PREVELIGE, FIG. 7. The 3-fold axes of symmetry in the matured X174 procapsid. In the native structure, the large coat helix does not occupy the 3-fold axes of symmetry, which contain pores. The result of second-site genetic analyses and the proximity of the first α helix of the D4 subunit to this axis of symmetry suggest an interaction between it and the coat protein helix.

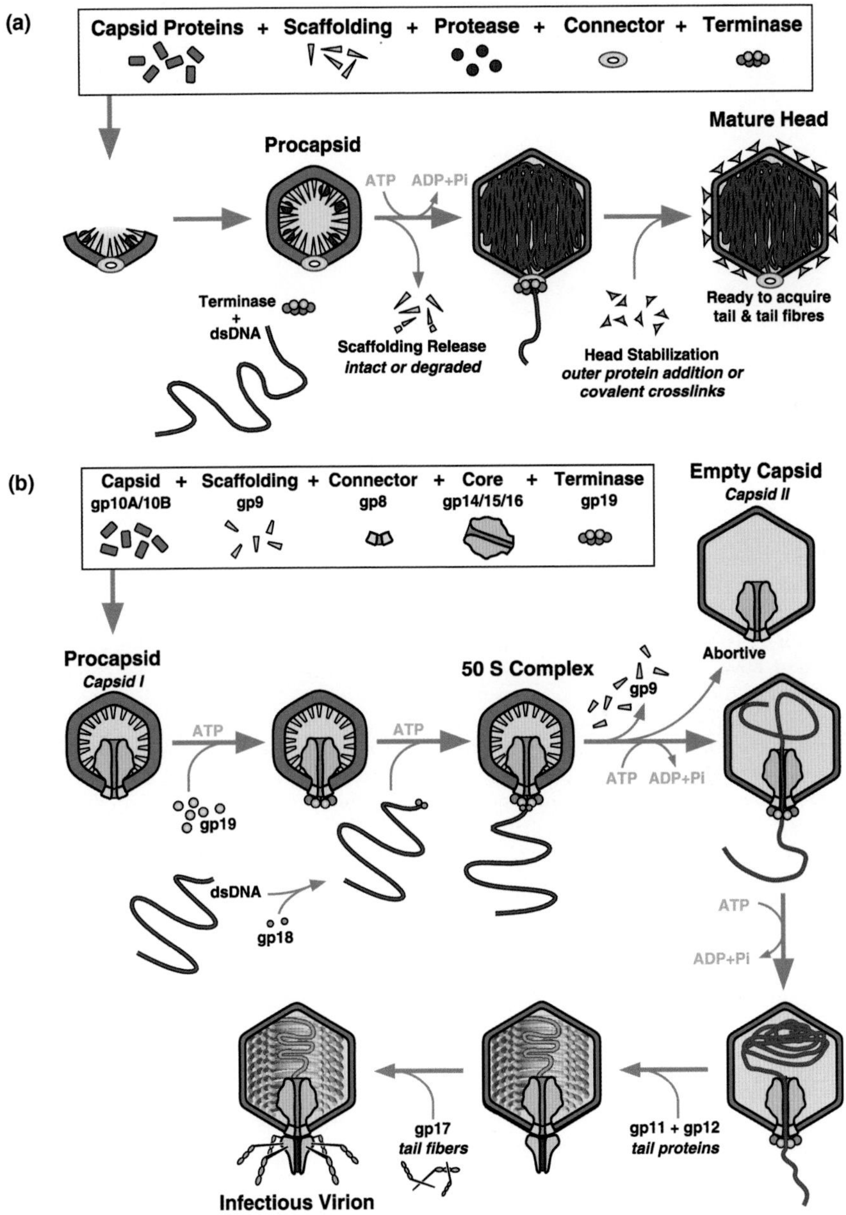
(a)
Capsid Proteins + Scaffolding + Protease + Connector + Terminase
Procapsid
ATP
ADP+Pi
Mature Head
Terminase
+
dsDNA
Scaffolding Release
intact or degraded
Head Stabilization
outer protein addition or
covalent crosslinks
Ready to acquire
tail & tail fibres
(b)
Capsid + Scaffolding + Connector + Core + Terminase
gp10A/10B
gp9
gp8
gp14/15/16
gp19
Empty Capsid
Capsid II
Abortive
Procapsid
Capsid I
ATP
gp19
ATP
50 S Complex
gp9
ATP ADP+Pi
dsDNA
gp18
ATP
ADP+Pi
gp17
tail fibers
gp11 + gp12
tail proteins
Infectious Virion

CERRITELLI *ET AL.*, FIG. 1. (a) Generic phage capsid assembly pathway. This diagram summarizes some basic features that are common to almost all pathways. For all such capsids of T=7 or higher, three components are required: connector, scaffold, and shell protein. HK97 may be an exception in lacking a scaffold protein, although the N-terminal domain of the shell protein precursor may substitute in this role (Hendrix and Duda, 1998). Many phages employ maturational proteases that are activated once the procapsid is complete. Possible elaborations on this basic plan include additional internal proteins, for example, the protease zymogen, multiple scaffolding proteins (e.g., T4; Black *et al.*, 1994), proteins that are not needed for assembly but are required to produce a viable virion (e.g., the T7 core), or proteins that are packaged in order eventually to be transferred into the host bacterium, where they play roles early in infection (e.g., T4; Black *et al.*, 1994). The connector is the assembly initiator. The scheme shown here indicates coassembly of scaffold and shell as concentric shells growing around the connector, but a variety of morphogenic mechanisms are possible (Dokland, 1999). DNA is packaged from a linear concatemer by the terminase enzyme (Black, 1989); during packaging, the expansion transformation of the procapsid shell takes place. We believe the DNA spool formation on packaging is a feature of many phages (Cerritelli *et al.*, 1997; Olson *et al.*, 2001; Zhang *et al.*, 2000) and herpesviruses (Booy *et al.*, 1991; Bhella *et al.*, 2000); however, the spool may not necessarily be oriented as in T7 with its axis along the portal vertex, and must be determined on a case-by-case basis. Here we have indicated one possible arrangement. After expansion of the procapsid, accessory proteins may bind to the outer surface (e.g., T4, λ) or covalent cross-links may form (e.g., HK97). On completion of DNA packaging, the terminase is released and the phage head acquires its tail and tail fibers—its host recognition and infection mediation system. (b) Bacteriophage T7 assembly pathway. The DNA is shown as entering through the core, although there is no evidence for this other than that the low-density central region seen in the axial projection (Fig. 3d) would be consistent with a channel. Alternatively, the core may act as a diverter so that the DNA becomes wound around the portal axis after entering along it.

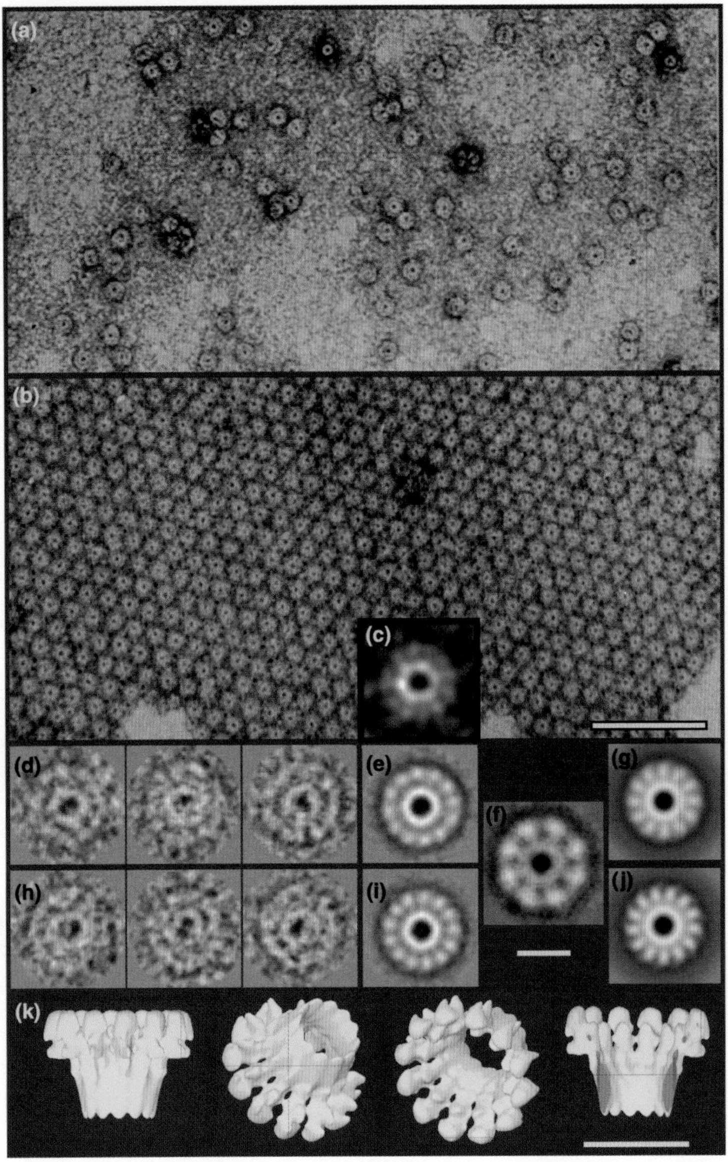

CERRITELLI *ET AL.*, FIG. 2. (a) Negatively stained field of overexpressed T7 connectors; (b) negatively stained array of connectors, shown to be of the 12-fold symmetric variant by translational averaging (c). Several individual images are shown at higher magnification in (d) and (h); those in (d) are 12-mers and those in (h) are 13-mers, as classified on the basis of their rotational power spectra. The respective averages over all the particles in each class are shown (e and i). They are compared with the corresponding T3 oligomers in (g) and (h), respectively (Valpuesta *et al.*, 2000), and with the T7 core in (f). A 3-D reconstruction of the 12-fold T3 connector (Valpuesta *et al.*, 2000) is shown in (k). Bars: 100 Å.

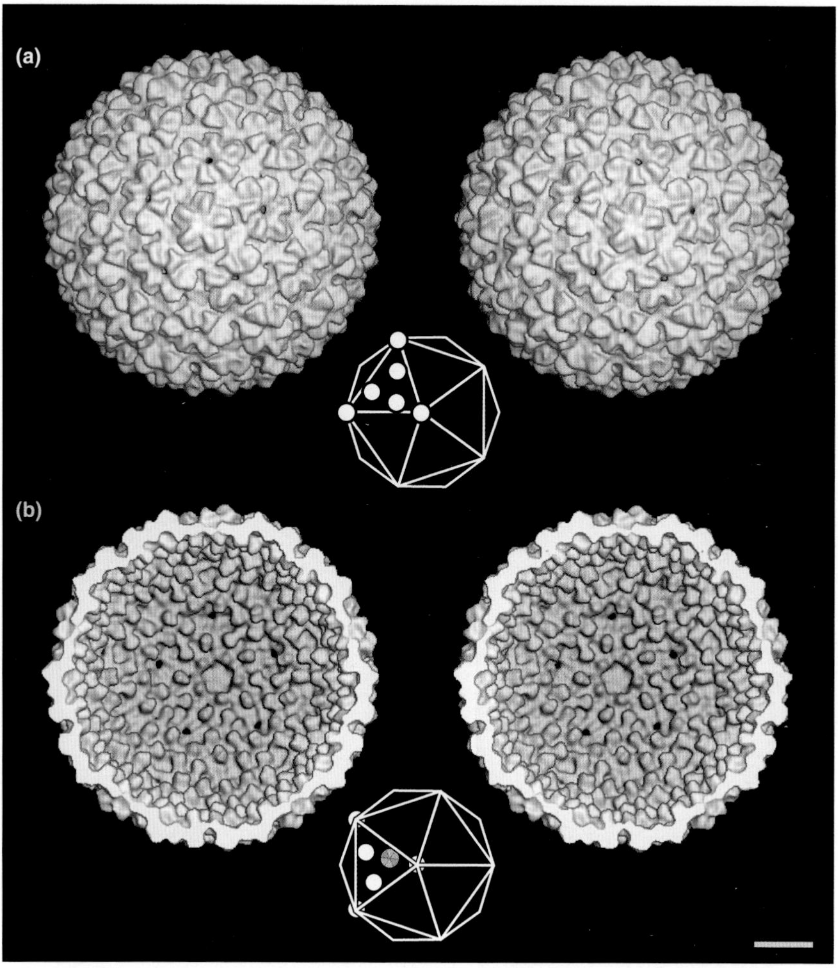

(a)
(b)

CERRITELLI *ET AL.*, FIG. 4. Structure of the 9.10 procapsid at 17-Å resolution. Stereo surface renderings of the procapsid viewed along an icosahedral 5-fold symmetry axis are shown: (a) the exterior surface, and (b) the interior surface, where the front half of the particle has been computationally removed. *Inset* in each panel is a diagram of the icosahedral surface lattice as viewed in this orientation. The procapsids were prepared according to Cerritelli and Studier (1996). Cryomicroscopy and image reconstruction were performed, respectively, according to Cheng *et al.* (1999) and Conway and Steven (1999). A total of 1659 image pairs were extracted from 6 defocus pairs: of these, 1236 were combined in the final reconstruction. The particles selected had the highest correlation coefficients in PFT (threshold value, average minus 1 standard deviation). The resolution was <17 Å according to the Fourier shell correlation coefficient. The positions of the capsomers on one facet are marked on a lattice model (beneath). The most prominent features of the interior surface are inwardly protruding "nubbins" of density underlying each hexamer subunit: putatively, these represent subunits of the gp9 scaffolding protein. Alternatively, it might be hypothesized that the pentamers also have nubbins that appear smaller only because they are more retracted into the shell, but the nubbins are, in fact, parts of gp10. However, on comparing serial spherical sections (not shown), it is clear that the hexamer-associated nubbins are much larger, implying that an additional component is present. According to SDS–PAGE (Cerritelli and Studier, 1996), gp9 is the only candidate. The pairs of apposing nubbins on one hexamer are shaded red, green, and blue, respectively. Bar: 100 Å.

KUHN AND STRAUSS, FIG. 1. Cryo-EM reconstruction of Sindbis virus, showing a surface-shaded view of the virus as viewed down a 2-fold axis. The resolution of the structure is 20 Å.

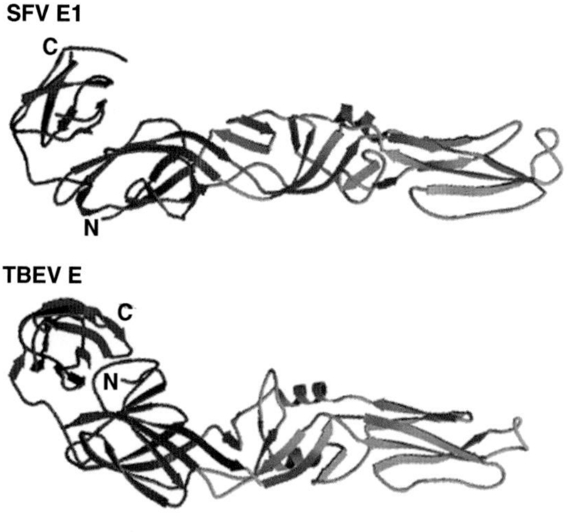

KUHN AND STRAUSS, FIG. 2. Comparison of the C_α backbone structures of SFV E1 with tick-borne encephalitis virus (TBEV) E. Ribbon diagrams have domain I colored red, domain II yellow, and domain III blue. The fusion peptides are colored green. The N and C termini are labeled N and C, respectively.

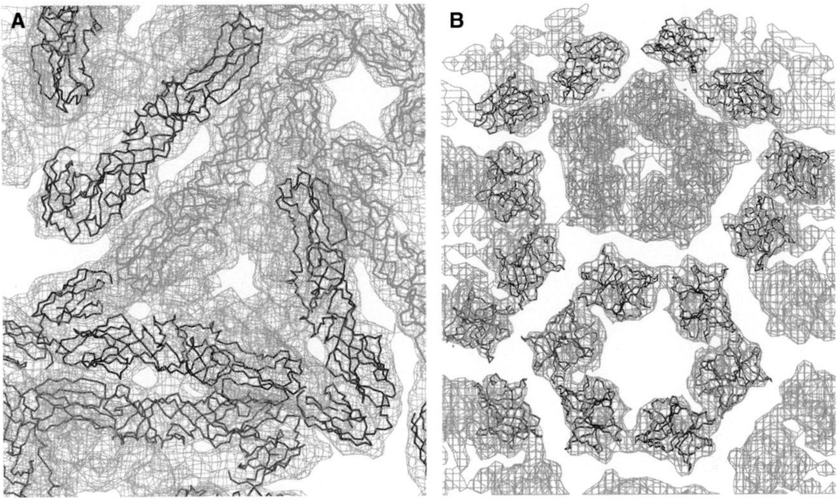

KUHN AND STRAUSS, FIG. 3. (A) Fit of the SFV E1 C_α backbone into an 11-Å SINV cryo-EM density map (gray) viewed down a quasi-3-fold axis. Each of the four independent E1 monomers is colored differently, with pink representing E1 molecules on the 3-fold axes, and red, green, and blue representing molecules of the quasi-3-fold axes. (B) Fit of the SINV capsid protein C_α backbone residues 114 to 264 into an 11-Å SINV cryo-EM density map (gray). The color convention is as described in (A). Clearly visible are the pentameric and hexameric capsomers.

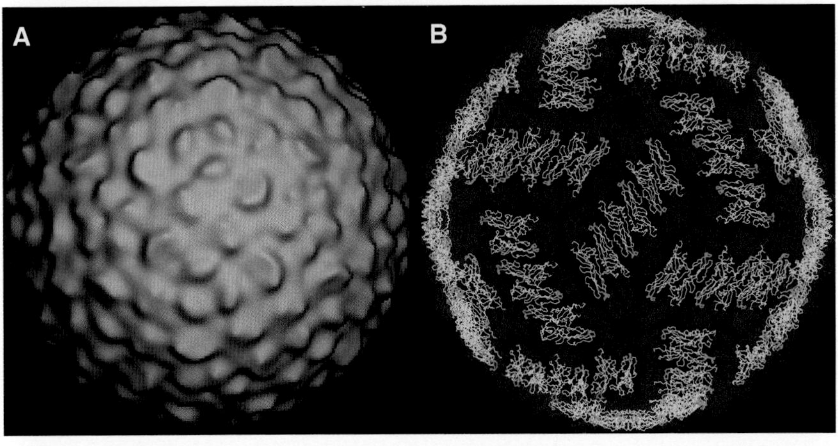

KUHN AND STRAUSS, FIG. 4. (A) Surface-shaded representation of dengue 2 cryo-EM reconstruction at 24-Å resolution. Note the smooth surface appearance of the virus particle. (B) Structure of dengue virus, showing the fit of each E monomer with domains I, II, and III in red, yellow, and blue, respectively. The fusion peptide is shown in green.

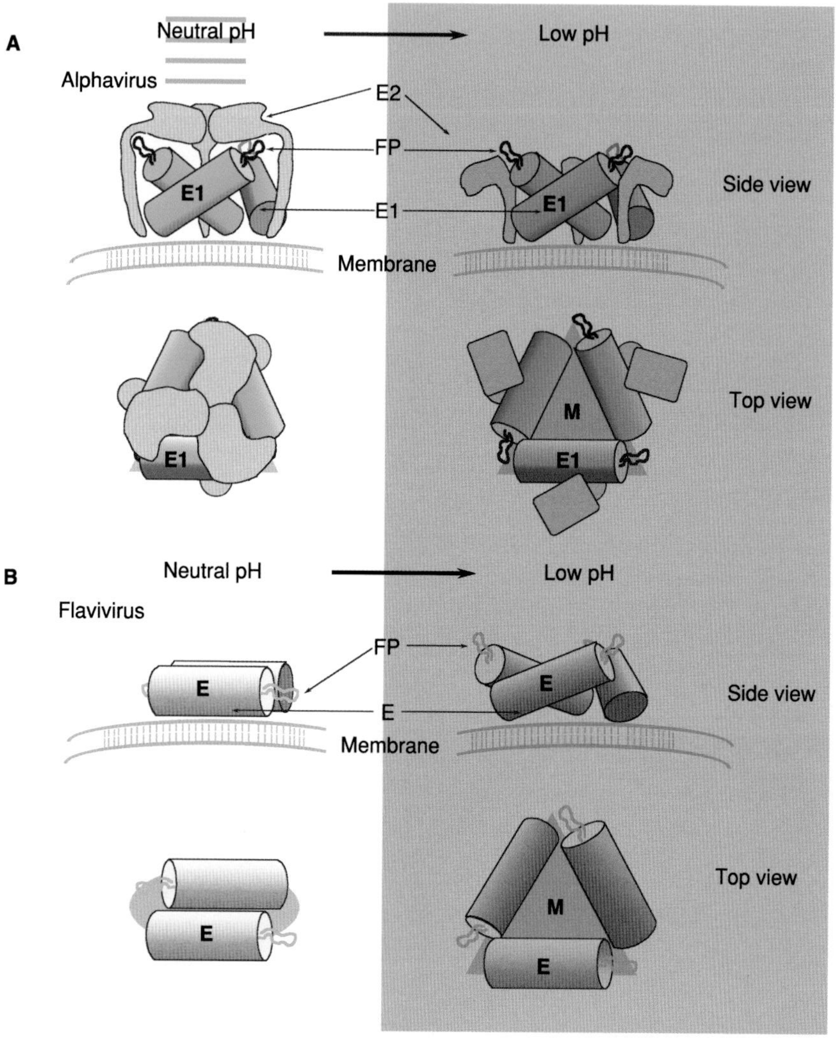

KUHN AND STRAUSS, FIG. 5. Configuration of glycoproteins of alphaviruses (A) and flaviviruses (B) on the surface of virions at neutral pH (*left*) and a hypothetical configuration at acid pH (*right*, shaded). In (A), the E1 glycoproteins are shown as green cylinders, E2 glycoproteins as tan shapes, and the fusion peptide as a black curved line. In (B), the E glycoproteins are shown as pale red cylinders with the fusion peptide as a cyan curve. In both cases, the membrane is shown in gray. M indicates the membrane.

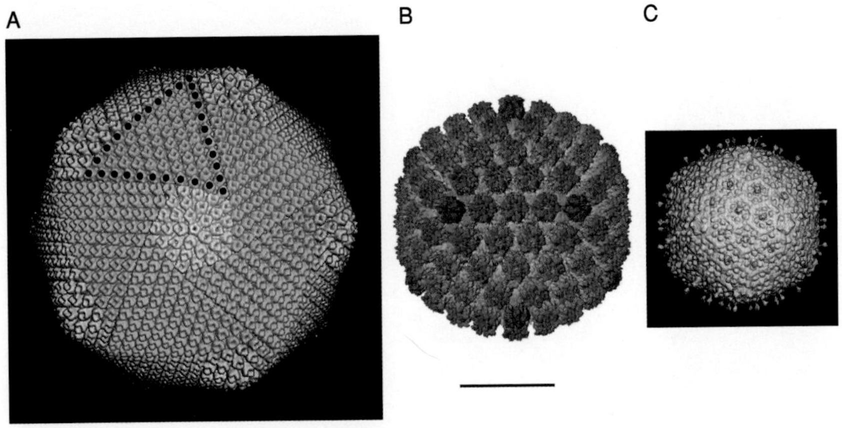

RIXON AND CHIU, FIG. 2. Three-dimensional reconstructions of virus capsids. Computer–generated reconstructions from electron cryomicroscopic images of (A) PBCV-1 virions (from Yan *et al.*, 2000), (B) HSV-1 capsids (from Zhou *et al.*, 2000), and (C) T4 isometric heads (from Olson *et al.*, 2001). PBCV-1 is viewed along a 5-fold symmetry axis. The capsomers are organized into groups to form triangular (blue) and pentagonal (yellow) facets, centered on the 3-fold and 5-fold icosahedral axes, respectively. HSV-1 and T4 are viewed along 2-fold axes. The HSV-1 map is color coded to identify the hexons (blue), pentons (red), and triplexes (green). All reconstructions are shown to the same scale. Scale bar: 500 Å.

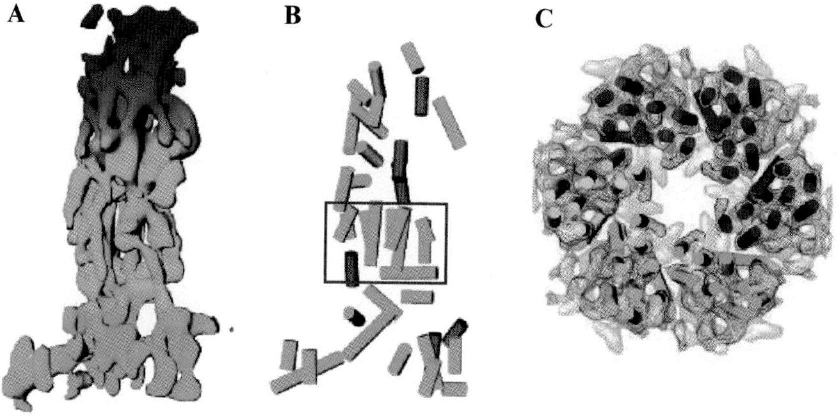

RIXON AND CHIU, FIG. 3. Secondary structural features in the HSV-1 major capsid protein. (A) A single major capsid protein (VP5) subunit isolated from an averaged hexon generated from the 8.5-Å resolution structure of the HSV-1 capsid shown in Fig. 2B (Zhou *et al.*, 2000). (B) The locations of 33 α helices, which were identified by computerized topological searching of the 8.5-Å density map, are shown superimposed on the outline of the VP5 subunit. Helices shown in green were identified with high confidence whereas the evidence for those in orange is less compelling. The red box surrounds a cluster of 10, closely associated α helices in the middle domain of VP5. These 10 α helices are reproduced in (C), arranged as they occur in the hexon. [Image supplied by Matthew Baker.]

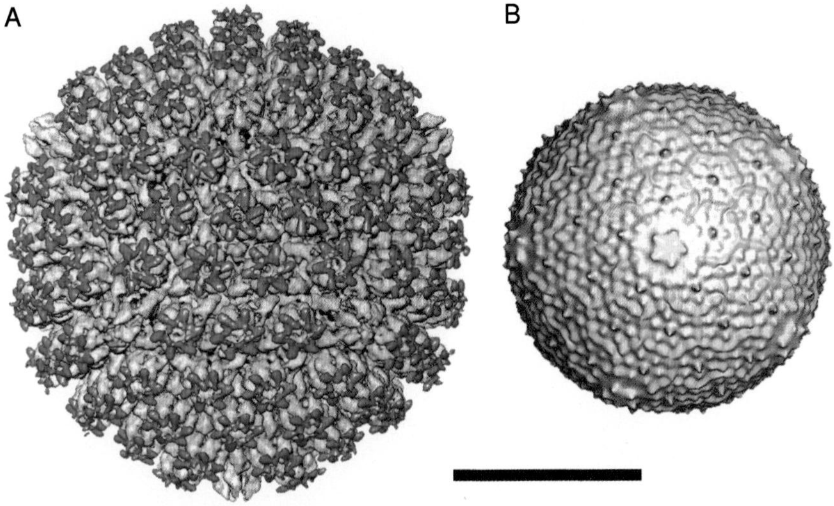

RIXON AND CHIU, FIG. 6. Distribution of accessory proteins in T4 and HSV-1 capsids. HSV-1 (A) and isometric T4 (B) capsid reconstructions are shown. The HSV-1 capsid is viewed along a 3-fold symmetry axis and the T4 head is viewed along a 5-fold axis. The distribution of the HSV-1 accessory protein, VP26, is shown in green. Six copies of VP26 form a star-shaped ring on the top of each hexon. No VP26 is present on the pentons occupying the 5-fold vertices. The locations of the two T4 accessory proteins, Hoc and Soc, are shown in pink and green, respectively. A single copy of Hoc is present in the center of each hexon whereas two Soc molecules occupy each interface between two hexons. Soc is not present at the interfaces between the hexons and pentons (shown in yellow). Scale bar: 500 Å. [Image (A) was supplied by Z. Hong Zhou and (B) was supplied by Benes Trus.]

Perhaps one of the most intriguing questions of triplex function is the role of Vp23. Before its association with Vp19C, the protein exists as a partially folded molten globule (Kirkitadze *et al.*, 1998). And its structural variability in the 8.5-Å image reconstruction of the B capsid is truly noteworthy (Zhou *et al.*, 2000). Presumably the alternate structures of the protein in the B capsid and presumed transient structures in the procapsid are influenced by its local environment, as has been observed with ϕX174 external scaffolding protein (Dokland *et al.*, 1997, 1999).

XVIII. Scaffolding-Like Functions

Although not all viruses utilize a distinct scaffolding protein, all viruses must solve the fundamental problem of associating large numbers of coat protein subunits with precise geometry. As might be expected, scaffolding-like functions are common in virus assembly even if the players involved do not meet the strict definition of a scaffolding protein.

In the case of the bacteriophage HK97, there is no distinct scaffolding protein (Duda *et al.*, 1995a,c). However, the coat protein first polymerizes via preformed hexamers and pentamers into prohead I, a form containing 415 molecules of the intact 42-kDa coat protein (Xie and Hendrix, 1995). In a subsequent step, a virally encoded protease cleaves 102 residues (the Δ-domain) off the N terminus of the coat protein to generate prohead II (Duda *et al.*, 1995b). Similar to the well-studied scaffolding proteins of P22, bacteriophage λ, and HSV, the delta domain is predicted to be largely α helical, with a propensity to form coiled coils (Conway *et al.*, 1995). Density corresponding to the Δ-domain is predicted to be largely α helical, with a propensity to form coiled coils (Conway *et al.*, 1995). Density corresponding to the Δ-domain is localized to the inside surface of prohead I, and appears to be loosely clustered beneath and tenuously tethered to the pentavalent and hexavalent capsomers (Conway *et al.*, 1995). In that regard it is similar to the structures obtained for P22 and HSV scaffolding-containing procapsids. However, it should be noted that T4 also has an N-terminal delta domain, and yet still requires a separate scaffolding protein for assembly.

Form-determining roles for viral structural proteins have been suggested for the inner core of blue tongue virus. In this case, 120 P3 subunits pack to form a T=2 icosahedral inner core (Grimes *et al.*, 1998). This core is subsequently tiled by the P7 protein, which is arranged with T=13 symmetry. Thus, the P3 inner core provides a template surface that serves to control P7 polymerization. A similar case has been made for the alphavirus fusion proteins. These viruses, which consist of nucleoprotein and fusion protein, assemble at the cell membrane and are released by budding. Lateral contacts between the fusion protein molecules dictate the formation of a T=4 icosahedral capsid and serve to order the nucleocapsid core (Lescar *et al.*, 2001; Pletnev *et al.*, 2001). In this regard, the fusion proteins resemble an external scaffold.

Retroviruses are nonicosahedral viruses that also bud through the cell membrane. Retroviruses assemble an immature capsid composed of the Gag polyprotein. During a maturation step, Gag is cleaved by a virally encoded protease to liberate the matrix (MA), capsid (CA), and nucleocapsid (NC) structural proteins, as well as a number of smaller peptides. Retroviruses display two types of assembly pathways: the C-type retroviruses assemble at the cell membrane, whereas the B and D type retroviruses assemble within the cytoplasm. The Gag polyprotein of the B and D type retroviruses contains an acidic region dubbed the ''internal scaffolding domain'' (Sakalian *et al.*, 1996; Weldon and Hunter, 1997). This domain is dispensable when the Gag polyprotein is overexpressed, but is required for assembly when Gag is expressed at levels that mimic that of normal infected cells (Sommerfelt *et al.*, 1992; Sakalian and Hunter, 1999). The simplest interpretation of these data is that this region serves to increase the effective concentration of the Gag protein. In the case of Mason–Pfizer monkey virus, sequence-based prediction suggests this region may form an amphipathic helix or coiled coil. The internal scaffolding domain does not appear to be required for C-type assembly; presumably membrane localization serves to increase the local concentration of the Gag protein (Sakalian and Hunter, 1999).

The study of scaffolding proteins has elucidated the fundamental mechanisms of macromolecular assembly. Although the viruses that these proteins assemble range from behemoths to dwarfs, and the primary sequences of these proteins bear no resemblance, these differences become superficial when one examines the conserved mechanisms by which these proteins regulate morphogenesis. Furthermore, proteins that fit the narrow definition of a classic scaffolding protein may represent only one extreme of a protein group, whose diverse members mediate morphogenetic processes. During evolution, scaffolding proteins have become incorporated domains in other proteins, as in HK97 coat, Mason–Pfizer Gag, and alphavirus glycoproteins. Nor are the molecules that perform these functions limited to peptides. The nucleic acids of many virions encode both genetic and morphogenetic information. Although this review has focused on viral morphogenesis, the discussed mechanisms most likely operate, at some level, in the assembly of other complex biological structures.

ACKNOWLEDGMENTS

The authors wish to acknowledge the support of the NIH and NSF.

REFERENCES

Bashford, D., Chothia, C., and Lesk, A. M. (1987). *J. Mol. Biol.* **196,** 199–216.
Bazinet, C., and King, J. (1985). *Annu. Rev. Microbiol.* **39,** 109–129.
Bazinet, C., Benbasat, J., King, J., Carazo, J. M., and Carrascosa, J. L. (1988). *Biochemistry* **27,** 1849–1856.
Bernhardt, T. G., Roof, W. D., and Young, R. (2000). *Proc. Natl. Acad. Sci. USA* **97,** 4297–4302.
Botstein, D., and Levine, M. (1968). *Cold Spring Harb. Symp. Quant. Biol.* **33,** 659–667.
Botstein, D., Waddell, C. H., and King, J. (1973). *J. Mol Biol.* **80,** 669–695.
Bryant, J. L., Jr., and King, J. (1984). *J. Mol. Biol.* **180,** 837–863.
Burch, A. D., and Fane, B. A. (2000a). *J. Virol.* **74,** 9347–9352.
Burch, A. D., and Fane, B. A. (2000b). *Virology* **270,** 286–290.
Burch, A. D., Ta, J., and Fane, B. A. (1999). *J. Mol. Biol.* **286,** 95–104.
Casjens, S. (1979). *J. Mol. Biol.* **131,** 1–14.
Casjens, S., and Adams, M. (1985). *J. Virol.* **53,** 185–191.
Casjens, S., and Hayden, M. (1988). *J. Mol. Biol.* **199,** 467–474.
Casjens, S., and Huang, W. M. (1982). *J. Mol. Biol.* **157,** 287–298.
Casjens, S., and King, J. (1974). *J. Supramol. Struct.* **2,** 202–224.
Casjens, S., Adams, M. B., Hall, C., and King, J. (1985). *J. Virol.* **53,** 174–179.
Caspar, D. L. (1980). *Biophys. J.* **32,** 103–138.
Chapman, M. S., and Rossmann, M. G. (1993). *Virology* **194,** 491–508.
Chen, X. S., Casini, G., Harrison, S. C., and Garcea, R. L. (2001). *J. Mol. Biol.* **307,** 173–182.
Christie, G. E., and Calendar, R. (1990). *Annu. Rev. Genet.* **24,** 465–490.
Conway, J. F., Duda, R. L., Cheng, N., Hendrix, R. W., and Steven, A. C. (1995). *J. Mol. Biol.* **253,** 86–99.
Conway, J. F., Wikoff, W. R., Cheng, N., Duda, R. L., Hendrix, R. W., Johnson, J. E., and Steven, A. C. (2001). *Science* **292,** 744–748.
Cripe, T. P., Delos, S. E., Estes, P. A., and Garcea, R. L. (1995). *J. Virol.* **69,** 7807–7813.
Dalphin, M. E., Fane, B. A., Skidmore, M. O., and Hayashi, M. (1992). *J. Bacteriol.* **174,** 2404–2406.
Desai, P., and Person, S. (1999). *Virology* **261,** 357–366.
Desai, P., Watkins, S. C., and Person, S. (1994). *J. Virol.* **68,** 5365–5374.
Dikerson, R. E., and Geis, I. (1983). "Hemoglobin: Structure, Function, Evolution and Pathology." Benjamin Cummings, Menlo Park, CA.
Ding, Y., Duda, R. L., Hendrix, R. W., and Rosenberg, J. M. (1995). *Biochemistry* **34,** 14918–14931.
Dokland, T., McKenna, R., Ilag, L. L., Bowman, B. R., Incardona, N. L., Fane, B. A., and Rossmann, M. G. (1997). *Nature* **389,** 308–313.
Dokland, T., Bernal, R. A., Burch, A., Pletnev, S., Fane, B. A., and Rossmann, M. G. (1999). *J. Mol. Biol.* **288,** 595–608.
Duda, R. L., Martincic, K., and Hendrix, R. W. (1995a). *J. Mol. Biol.* **247,** 636–647.
Duda, R. L., Martincic, K., Xie, Z., and Hendrix, R. W. (1995b). *FEMS Microbiol. Rev.* **17,** 41–46.

Duda, R. L., Hempel, J., Michel, H., Shabanowitz, J., Hunt, D., and Hendrix, R. W. (1995c). *J. Mol. Biol.* **247,** 618–635.
Earnshaw, W., and King, J. (1978). *J. Mol. Biol.* **126,** 721–747.
Earnshaw, W., Casjens, S., and Harrison, S. C. (1976). *J. Mol. Biol.* **104,** 387–410.
Ekechukwu, M. C., and Fane, B. A. (1995). *J. Bacteriol.* **177,** 829–830.
Ekechukwu, M. C., Oberste, D. J., and Fane, B. A. (1995). *Genetics* **140,** 1167–1174.
Erickson, H. P., and Pantaloni, D. (1981). *Biophys. J.* **34,** 293–309.
Fane, B. A., and Hayashi, M. (1991). *Genetics* **128,** 663–671.
Fane, B., and King, J. (1987). *Genetics* **117,** 157–171.
Fane, B. A., Shien, S., and Hayashi, M. (1993). *Genetics* **134,** 1003–1011.
Farber, M. B. (1976). *J. Virol.* **17,** 1027–1037.
Fuller, M. T., and King, J. (1981). *Virology* **112,** 529–547.
Fuller, M. T., and King, J. (1982). *J. Mol. Biol.* **156,** 633–665.
Galisteo, M. L., and King, J. (1993). *Biophys. J.* **65,** 227–235.
Georgopoulos, C. P., and Hohn, B. (1978). *Proc. Natl. Acad. Sci. USA* **75,** 131–135.
Godson, G. N., Barrell, B. G., Staden, R., and Fiddes, J. C. (1978). *Nature* **276,** 236–247.
Gordon, C. L., Sather, S. K., Casjens, S., and King, J. (1994). *J. Biol. Chem.* **269,** 27941–27951.
Greene, B., and King, J. (1994). *Virology* **205,** 188–197.
Greene, B., and King, J. (1996). *Virology* **224,** 82–96.
Greene, B., and King, J. (1999). *J. Biol. Chem.* **274,** 16141–16146.
Greene, B., and King, J. (1999). *J. Biol. Chem.* **274,** 16135–16140.
Grimes, J. M., Burroughs, J. N., Gouet, P., Diprose, J. M., Malby, R., Zientara, S., Mertens, P.P.C., and Stuart, D. I. (1998). *Nature* **395,** 470–478.
Guo, P. X., Erickson, S., Xu, W., Olson, N., Baker, T. S., and Anderson, D. (1991). *Virology* **183,** 366–373.
Haanes, E. J., Thomsen, D. R., Martin, S., Homa, F. L., and Lowery, D. E. (1995). *J. Virol.* **69,** 7375–7379.
Hafenstein, S., and Fane, B. A. (2002). *J. Virol.* **76,** 5350–5356.
Hanninen, A. L., Bamford, D. H., and Bamford, J. K. (1997). *Virology* **227,** 207–210.
Hartweig, E., Bazinet, C., and King, J. (1986). *Biophys. J.* **49,** 24–26.
Hayashi, M., Aoyama, A., Richardson, D. L., and Hayashi, M. N. (1988). Biology of the bacteriophage ϕX174. *In* "The Bacteriophages", (R. Calendar, Ed.), pp. 1–71. Plenum. Vol. 2, New York.
Hendrix, R. W., and Tsui, L. (1978). *Proc. Natl. Acad. Sci. USA* **75,** 136–139.
Hohn, B. (1983). *Proc. Natl. Acad. Sci. USA* **80,** 7456–7460.
Hong, Z., Beaudet-Miller, M., Durkin, J., Zhang, R., and Kwong, A. D. (1996). *J. Virol.* **70,** 533–540.
Ilag, L. L., Olson, N. H., Dokland, T., Music, C. L., Cheng, R. H., Brown, Z., McKenna, R., Rossmann, M. G., Baker, T. S., and Incardona, N. L. (1995). *Structure* **3,** 353–363.
Israel, V. (1977). *J. Virol.* **23,** 91–97.
Iwasaki, K., Trus, B. L., Wingfield, P. T., Cheng, N., Campusano, G., Rao, V. B., and Steven, A. C. (2000). *Virology* **271,** 321–333.
Jardine, P. J., McCormick, M. C., Lutze-Wallace, C., and Coombs, D. H. (1998). *J. Mol. Biol.* **284,** 647–659.
Jennings, B., and Fane, B. A. (1997). *Virology* **227,** 370–377.
Kellenberger, E., Eiserling, F. A., and Boy De La Tour, E. (1968). *J. Ultrastruct. Res.* **21,** 335–360.
Kim, K. J., Sunshine, M. G., Lindqvist, B. H., and Six, E. W. (2001). *Virology* **283,** 49–58.
King, J., and Casjens, S. (1974). *Nature* **251,** 112–119.

King, J., Lenk, E. V., and Botstein, D. (1973). *J. Mol. Biol.* **80,** 697–731.
King, J., Botstein, D., Casjens, S., Earnshaw, W., Harrison, S., and Lenk, E. (1976). *Philos. Trans. R. Soc. Lond B Biol. Sci.* **276,** 37–49.
King, J., Hall, C., and Casjens, S. (1978). *Cell* **15,** 551–560.
Kirkitadze, M. D., Barlow, P. N., Price, N. C., Kelly, S. M., Boutell, C. J., Rixon, F. J., and McClelland, D. A. (1998). *J. Virol.* **72,** 10066–10072.
Kodaira, K., Nakano, K., Okada, S., and Taketo, A. (1992). *Biochim. Biophys. Acta* **1130,** 277–288.
Krebs, H., Schmid, F. X., and Jaenicke, R. (1983). *J. Mol. Biol.* **169,** 619–635.
Lee, C. S., and Guo, P. (1995). *J. Virol.* **69,** 5024–5032.
Lescar, J., Roussel, A., Wien, M. W., Navaza, J., Fuller, S. D., Wengler, G., and Rey, F. A. (2001). *Cell* **105,** 137–148.
Marvik, O. J., Dokland, T., Nokling, R. H., Jacobsen, E., Larsen, T., and Lindqvist, B. H. (1995). *J. Mol. Biol.* **251,** 59–75.
Masker, W. E., and Serwer, P. (1982). *J. Virol.* **43,** 1138–1142.
Matusick-Kumar, L., Hurlburt, W., Weinheimer, S. P., Newcomb, W. W., Brown, J. C., and Gao, M. (1994). *J. Virol.* **68,** 5384–5394.
Matusick-Kumar, L., Newcomb, W. W., Brown, J. C., McCann, P. J.,III, Hurlburt, W., Weinheimer, S. P., and Gao, M. (1995). *J. Virol.* **69,** 4347–4356.
McKenna, R., Xia, D., Willingmann, P., Ilag, L. L., Krishnaswamy, S., Rossmann, M. G., Olson, N. H., Baker, T. S., and Incardona, N. L. (1992). *Nature* **355,** 137–143.
McKenna, R., Ilag, L. L., and Rossmann, M. G. (1994). *J. Mol. Biol.* **237,** 517–543.
McKenna, R., Bowman, B. R., Ilag, L. L., Rossmann, M. G., and Fane, B. A. (1996). *J. Mol. Biol.* **256,** 736–750.
McKenna, R., Olson, N. H., Chipman, P. R., Baker, T. S., Booth, T. F., Christensen, J., Aasted, B., Fox, J. M., Bloom, M. E., Wolfinbarger, J. B., and Agbandje-McKenna, M. (1999). *J. Virol.* **73,** 6882–6891.
Miller, J. H., Coulondre, C., Hofer, M., Schmeissner, U., Sommer, H., Schmitz, A., and Lu, P. (1979). *J. Mol. Biol.* **131,** 191–222.
Nakonechny, W. S., and Teschke, C. M. (1998). *J. Biol. Chem.* **273,** 27236–27244.
Nelson, R. A., Reilly, B. E., and Anderson, D. L. (1976). *J. Virol.* **19,** 518–532.
Newcomb, W. W., Trus, B. L., Booy, F. P., Steven, A. C., Wall, J. S., and Brown, J. C. (1993). *J. Mol. Biol.* **232,** 499–511.
Newcomb, W. W., Homa, F. L., Thomsen, D. R., Ye, Z., and Brown, J. C. (1994). *J. Virol.* **68,** 6059–6063.
Newcomb, W. W., Homa, F. L., Thomsen, D. R., Trus, B. L., Cheng, N., Steven, A., Booy, F., and Brown, J. C. (1999). *J. Virol.* **73,** 4239–4250.
Newcomb, W. W., Trus, B. L., Cheng, N., Steven, A. C., Sheaffer, A. K., Tenney, D. J., Weller, S. K., and Brown, J. C. (2000). *J. Virol.* **74,** 1663–1673.
Nilssen, O., Fossdal, C. G., Johansen, B. V., and Lindqvist, B. H. (1996). *Virology* **219,** 443–452.
Oien, N. L., Thomsen, D. R., Wathen, M. W., Newcomb, W. W., Brown, J. C., and Homa, F. L. (1997). *J. Virol.* **71,** 1281–1291.
Olson, N. H., Gingery, M., Eiserling, F. A., and Baker, T. S. (2001). *Virology* **279,** 385–391.
Parker, M. H., and Prevelige, P. E., Jr. (1998). *Virology* **250,** 337–349.
Parker, M. H., Stafford, W. F. III, and Prevelige, P. E., Jr. (1997). *J. Mol. Biol.* **268,** 655–665.
Parker, M. H., Jablonsky, M., Casjens, S., Sampson, L., Krishna, N. R., and Prevelige, P. E., Jr. (1997). *Protein Sci.* **6,** 1583–1586.

Parker, M. H., Casjens, S., and Prevelige, P. E., Jr. (1998). *J. Mol. Biol.* **281,** 69–79.
Parker, M. H., Brouillette, C. G., and Prevelige, P. E., Jr. (2001). *Biochemistry* **40,** 8962–8970.
Pletnev, S. V., Zhang, W., Mukhopadhyay, S., Fisher, B. R., Hernandez, R., Brown, D. T., Baker, T. S., Rossmann, M. G., and Kuhn, R. J. (2001). *Cell* **105,** 127–136.
Poteete, A. R., and Botstein, D. (1979). *Virology* **95,** 565–573.
Poteete, A. R., Jarvik, V., and Botstein, D. (1979). *Virology* **95,** 550–564.
Prasad, B. V., Prevelige, P. E., Marietta, E., Chen, R. O., Thomas, D., King, J., and Chiu, W. (1993). *J. Mol. Biol.* **231,** 65–74.
Preston, V. G., Kennard, J., Rixon, F. J., Logan, A. J., Mansfield, R. W., and McDougall, I. M. (1997). *J. Gen. Virol.* **78,** 1633–1646.
Prevelige, P. E., Jr., Thomas, D., and King, J. (1988). *J. Mol. Biol.* **202,** 743–757.
Prevelige, P. E., Jr., Thomas, D., and King, J. (1993). *Biophys. J.* **64,** 824–835.
Qiu, X., Culp, J. S., DiLella, A. G., Hellmig, B., Hoog, S. S., Janson, C. A., Smith, W. W., and Abdel-Meguid, S. S. (1996). *Nature* **383,** 275–279.
Richardson, D. L., Jr., Aoyama, A., and Hayashi, M. (1988). *J. Bacteriol.* **170,** 5564–5571.
Saad, A., Zhou, Z. H., Jakana, J., Chiu, W., and Rixon, F. J. (1999). *J. Virol.* **73,** 6821–6830.
Sakalian, M., and Hunter, E. (1999). *J. Virol.* **73,** 8073–8082.
Sakalian, M., Parker, S. D., Weldon, R. A., Jr., and Hunter, E. (1996). *J. Virol.* **70,** 3706–3715.
Salunke, D. M., Caspar, D.L.D., and Garcea, R. L. (1989). *Biophys. J.* **56,** 887–900.
Sanger, F., Coulson, A. R., Friedmann, T., Air, G. M., Barrell, B. G., Brown, N. L., Fiddes, J. C., Hutchison, C. A.,III, Slocombe, P. M., and Smith, M. (1978). *J. Mol. Biol.* **125,** 225–246.
Shibata, H., Fujisawa, H., and Minagawa, T. (1987a). *J. Mol. Biol.* **196,** 845–851.
Shibata, H., Fujisawa, H., and Minagawa, T. (1987b). *Virology* **159,** 250–258.
Shieh, H.-S., Kurumbail, R. G., Stevens, A. M., Stegeman, R. A., Sturman, E. J., Pak, J. Y., Wittwer, A. J., Palmier, M. O., Wiegand, R. C., Holwerda, B. C., and Stallings, W. C. (1996). *Nature* **383,** 279–282.
Showe, M. K., and Black, L. W. (1973). *Nat. New Biol.* **242,** 70–75.
Siden, E. J., and Hayashi, M. (1974). *J. Mol. Biol.* **89,** 1–16.
Six, E. W., Sunshine, M. G., Williams, J., Haggard-Ljungquist, E., and Lindqvist, B. H. (1991). *Virology* **182,** 34–46.
Sommerfelt, M. A., Rhee, S. S., and Hunter, E. (1992). *J. Virol.* **66,** 7005–7011.
Spencer, J. V., Newcomb, W. W., Thomsen, D. R., Homa, F. L., and Brown, J. C. (1998). *J. Virol.* **72,** 3944–3951.
Spindler, K. R., and Hayashi, M. (1979). *J. Virol.* **29,** 973–982.
Sternberg, N. (1973). *J. Mol. Biol.* **76,** 1–23.
Steven, A. C., Greenstone, H. L., Booy, F. P., Black, L. W., and Ross, P. D. (1992). *J. Mol. Biol.* **228,** 870–884.
Studier, F. W., and Maizel, J. V., Jr. (1969). *Virology* **39,** 575–586.
Sun, Y., Parker, M. H., Weigele, P., Casjens, S., Prevelige, P. E., Jr., and Krishna, N. R. (2000). *J. Mol. Biol.* **297,** 1195–1202.
Teschke, C. M., King, J., and Prevelige, P. E., Jr. (1993). *Biochemistry* **32,** 10658–10665.
Tessman, E. S., and Peterson, P. K. (1976). *J. Virol.* **20,** 400–412.
Thomas, D., and Prevelige, P. E., Jr. (1991). *Virology* **182,** 673–681.
Thomas, G. J., Jr., Li, Y., Fuller, M. T., and King, J. (1982). *Biochemistry* **21,** 3866–3878.
Thuman-Commike, P. A., Green, B., Jakana, J., Prasad, B.V.V., King, J., Prevelige, P. E., Jr., and Chiu, W. (1996). *J. Mol. Biol.* **260,** 85–98.

Thuman-Commike, P. A., Greene, B., Malinski, J. A., King, J., and Chiu, W. (1998). *Biophys. J.* **74,** 559–568.

Thuman-Commike, P. A., Greene, B., Malinski, J. A., Burbea, M., McGough, A., Chiu, W., and Prevelige, P. E., Jr. (1999). *Biophys. J.* **76,** 3267–3277.

Tonegawa, S., and Hayashi, M. (1970). *J. Mol. Biol.* **48,** 219–242.

Tong, L., Qian, C., Massariol, M.-J., Bonneau, P. R., Cordingley, M. G., and Lagace, L. (1996). *Nature* **383,** 272–275.

Trus, B. L., Booy, F. P., Newcomb, W. W., Brown, J. C., Homa, F. L., Thomsen, D. R., and Steven, A. C. (1996). *J. Mol. Biol.* **263,** 447–462.

Tuma, R., Prevelige, P. E., Jr., and Thomas, G. J., Jr. (1996). *Biochemistry* **35,** 4619–4627.

Tuma, R., Parker, M. H., Weigele, P., Sampson, L., Sun, Y., Krishna, N. R., Casjens, S., Thomas, G. J., Jr., and Prevelige, P. E., Jr. (1998). *J. Mol. Biol.* **281,** 81–94.

Tye, B. K., and Botstein, D. (1974a). *J. Supramol. Struct.* **2,** 225–238.

Tye, B. K., Chan, R. K., and Botstein, D. (1974a). *J. Mol. Biol.* **85,** 485–500.

Tye, B. K., Huberman, J. A., and Botstein, D. (1974b). *J. Mol. Biol.* **85,** 501–528.

Wang, S., Palasingam, P., Nokling, R. H., Lindqvist, B. H., and Dokland, T. (2000). *Virology* **275,** 133–144.

Warner, S. C., Desai, P., and Person, S. (2000). *Virology* **278,** 217–226.

Weldon, R. A., Jr., and Hunter, E. (1997). Molecular requirements for retrovirus assembly. *In* "Structural Biology of Viruses", (W. Chiu, R. M. Burnett and R. L. Garcea, Eds.), pp. 381–410. Oxford University Press, New York.

Wrobel, B., Yosef, Y., Oppenheim, A. B., and Oppenheim, A. (2000). *J. Biotechnol.* **84,** 285–289.

Wyckoff, E., and Casjens, S. (1985). *J. Virol.* **53,** 192–197.

Xie, Z., and Hendrix, R. W. (1995). *J. Mol. Biol.* **253,** 74–85.

Zhou, Z. H., Macnab, S. J., Jakana, J., Scott, L. R., Chiu, W., and Rixon, F. J. (1998). *Proc. Natl. Acad. Sci. USA* **95,** 2778–2783.

Zhou, Z. H., Dougherty, M., Jakana, J., He, J., Rixon, F. J., and Chiu, W. (2000). *Science* **288,** 877–880.

Ziegelhoffer, T., Yau, P., Chandrasekhar, G. N., Kochan, J., Georgopoulos, C., and Murialdo, H. (1992). *J. Biol. Chem.* **267,** 455–461.

MOLECULAR MECHANISMS IN BACTERIOPHAGE T7 PROCAPSID ASSEMBLY, MATURATION, AND DNA CONTAINMENT

By MARIO E. CERRITELLI,* JAMES F. CONWAY,† NAIQIAN CHENG,* BENES L. TRUS,*,‡ AND ALASDAIR C. STEVEN*

*Laboratory of Structural Biology, National Institute of Arthritis, Musculoskeletal, and Skin Diseases, National Institutes of Health, Bethesda, Maryland 20892, †Institut de Biologie Structurale, 38027 Grenoble, France, and ‡Computational Biology and Engineering Laboratory, Center for Information Technology, National Institutes of Health, Bethesda, Maryland 20892

I. Introduction 301
II. Overexpressed T7 and T3 Connectors have 12- and 13-Fold Symmetry 303
III. The Procapsid Core Has 8-Fold Symmetry: Another Symmetry Mismatch 303
IV. Procapsid Structure 305
Distribution of the Scaffolding Protein: A Proposal for the Morphogenetic Mechanism 306
V. Procapsid Maturation: Expansion Is Initiated in the Connector 308
VI. Packaging and Parting of DNA 309
VII. The Mature Capsid Structure: Filled and Empty Shells 310
VIII. Structure of Packaged DNA 315
IX. Summary 319
References 320

I. INTRODUCTION

The assembly pathways of double-stranded DNA (dsDNA) bacteriophages have been a particularly instructive source of information about mechanisms that control the biosynthesis of complex macromolecular particles. This role was initially facilitated by the tractable genetics and short replication cycles of phages, continued with the development of expression vectors that allowed coexpression of small sets of selected genes, and has now entered a phase of detailed analysis as structural information becomes available from cryoelectron microscopy (cryo-EM) and X-ray crystallography. The assembly pathways of some half dozen phages have been studied in depth. The emerging picture is that a number of basic molecular behaviors are widely if not universally shared, but that each phage system also has a number of individual embellishments (Hendrix and Garcea, 1994; Black *et al.*, 1994; Dokland, 1999; Wikoff and Johnson, 1999). Interestingly, close parallels have surfaced between phage assembly and corresponding events in the assembly of some dsDNA animal viruses—notably, herpesviruses (Friedmann *et al.*, 1975; Steven and Spear, 1997)—and between the membrane-containing phage PRD1 and adenovirus (Benson *et al.*, 1999).

ADVANCES IN PROTEIN CHEMISTRY, Vol. 64

The generic pathway (Fig. 1a; see Color Insert) commences with formation of a precursor particle or procapsid, whose assembly usually requires at least three components: the connector or portal protein (one oligomer), the shell or coat protein (a fixed complement, according to the capsid size or T number), and a scaffolding protein (a potentially variable copy number). On completion of the procapsid, DNA packaging is initiated and proceeds in linear fashion from the replicating concatemer into the procapsid, powered by terminase, a multifunctional protein complex with both ATPase and endonuclease activities.

During packaging, the procapsid undergoes an irreversible conformational change as it matures into the larger, thinner walled, more polyhedral, and stabler capsid. This transformation is often referred to as "expansion" because all dsDNA phage capsids characterized to date become larger on maturing. However, the procapsid of herpes simplex virus also undergoes a major structural rearrangement, but without changing size (Trus *et al.*, 1996), and the procapsid of *Nudaurelia capensis* ω virus (NωV), an RNA virus, actually shrinks (Canady *et al.*, 2000). It appears, therefore, that the fundamental mechanism underlying all these transitions is a cooperative conformational change of the surface lattice, not the size change per se. In many cases, procapsid maturation is initiated by activation of a viral protease that is incorporated into the procapsid and processes the scaffold and/or shell proteins, producing a metastable particle with heightened susceptibility to expansion. Although the expansion transformation effects a substantial stabilization, further reinforcement may follow from the binding of ancillary proteins to the outer surface or by the autocatalytic formation of covalent cross-links between neighboring subunits.

The assembly pathway of T7 is outlined in Fig. 1b (see Color Insert). The salient features of T7 capsid assembly are as follows: (1) it has a single scaffolding protein, gp9; (2) a maturational protease is absent; (3) the procapsid contains a large cylindrical proteinaceous structure called the "core"; (4) the shell protein is produced in two forms, gp10A and gp10B, of which gp10B contains a C-terminal extension that is expressed in about 10% of translation events by a read-through mechanism; and (5) there is no posttranslational binding of accessory proteins or cross-linking. The purpose of this article is to summarize more recent work on the T7 system, with emphasis on observations by cryo-EM. Noting that T7 and the closely related T3 have much in common, we also include observations pertaining to T3 on some aspects—connector structure and DNA packaging—that have advanced further in this system. Earlier work on T7 morphogenesis was reviewed by Steven and Trus (1986).

II. Overexpressed T7 and T3 Connectors Have 12- and 13-Fold Symmetry

When T7 connectors are isolated from cells overexpressing gene *8* and examined by negative staining (Fig. 2a; see Color Insert), ringlike molecules are seen. Rotational symmetry analysis showed them to be dimorphic, conforming to 12-fold (Fig. 2d and e) and 13-fold symmetry (Fig. 2h and i), respectively (Kocsis *et al.*, 1995). This property has also been observed in the closely related T3 connectors, for which the relative proportions of the two symmetries were found to vary from batch to batch (Valpuesta *et al.*, 2000) (Fig. 2g and j). Isolated connectors of both phages spontaneously form two-dimensional crystalline sheets, which consist of 12-fold connectors in alternating orientations (up/down) that partially overlap. T3 connector lattices have been described (Valpuesta *et al.*, 2000), and a lattice of T7 connectors and filtered image is shown in Fig. 2b and c. A three-dimensional reconstruction of the 12-fold T3 connector is shown in Fig. 2k (Valpuesta *et al.*, 2000).

Thirteen-fold symmetry was observed first in SPP1 connectors (Dube *et al.*, 1993). Apart from reports of 13-fold symmetry in ϕ29 connectors (e.g., Dube *et al.*, 1993), which are at odds with the 12-fold symmetry disclosed by high-resolution atomic force microscopy (AFM) analysis (Muller *et al.*, 1997) and a crystal structure (Simpson *et al.*, 2000), the T3/T7 system is the only other one in which 13-fold has so far been observed. However, T3 and T7 also produce 12-fold connectors, and, significantly, only 12-fold symmetry was observed for connectors isolated from T3 capsids (Carazo *et al.*, 1986; Donate *et al.*, 1988). Pending experimental determination of the order(s) of symmetry of connectors *in situ*, the emerging picture (Valpuesta *et al.*, 2000) is that connectors are 12-fold symmetric in phage heads, and 13-fold variants, when they occur, are probably an aberrant oligomerization product that results when overexpression overloads the normal biosynthetic assembly pathway within cells.

III. The Procapsid Core has 8-Fold Symmetry: Another Symmetry Mismatch

In an effort to determine the rotational symmetry of connectors *in situ*, we studied native T7 procapsids by cryo-EM (Fig. 3a). Our strategy was to identify images in which the particle is viewed along the portal axis and subject them to the same symmetry detection analysis as was applied to isolated connectors (see above). In the images of interest, the internal

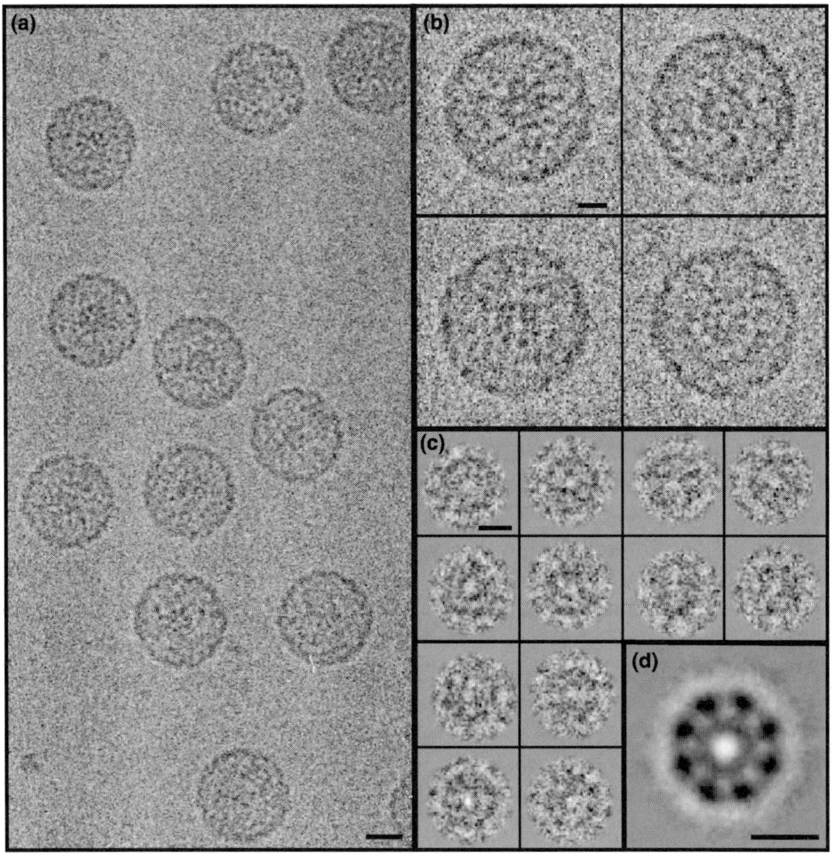

FIG. 3. (a) A field of native T7 procapsids imaged by cryoelectron microscopy. The procapsid has an eccentrically mounted internal core that is discernible in most images. The core appears centered only when the procapsid is viewed along the symmetry axis of the connector vertex. (b) Four procapsids with well-centered cores, at higher magnification. To expose the cores in procapsid images, the contribution of the surface shell was computationally subtracted, using the appropriate reprojection of the coreless 9.10 procapsid density map (see Fig. 4) for this purpose. Some core images are shown in (c). These data were analyzed by ROTASTAT (Kocsis *et al.*, 1995), and 8-fold symmetry was detected. The images were then subjected to correlation averaging (d). Bars: 100 Å. [Reproduced, with permission, from Cerritelli *et al.*, (2002).]

core should appear concentric with the procapsid shell. Because the core is mounted eccentrically on the inner surface of the connector vertex (Serwer, 1976; Steven and Trus, 1986), it should appear centered only when viewed along this axis. Moreover, the core is large enough that it should be discernible in cryomicrographs, even when coprojected with the procapsid shell. *A priori,* it was clear that the connector would be less conspicuous than the core, but we hypothesized that it might nevertheless be detectable in a quantitative symmetry analysis of axial projections (Kocsis *et al.*, 1995).

Several examples of axially viewed procapsids are shown in Fig. 3b, and some cores exposed by subtracting the shell contribution to the images are shown in Fig. 3c. The latter data set was analyzed for the presence of symmetries (Kocsis *et al.*, 1995). To our surprise, the results (Cerritelli *et al.*, 2002) showed a strongly expressed 8-fold symmetry between radii of 50 and 80 Å. After correlation averaging (Fig. 3d), these images revealed a ring of eight globular protuberances, ~40 Å in diameter, connected by narrow stalks to an annular ring. The wall of this ring is ~30 Å thick and surrounds a 35-Å axial channel. This analysis detected no significant 12- or 13-fold signal at a radius of ~60 Å, the radius at which these symmetries register most strongly in isolated negatively stained connectors (see above), thus we conclude that the signal from the connector is too weak to be detected.

In a previous estimate of the likely stoichiometry of core proteins (Steven and Trus, 1986), we interpreted earlier biochemical data (Adolph and Haselkorn, 1972; Serwer, 1976) in terms of the closest multiple of six, noting that the connector (assumed to be 12-fold) has a 6-fold symmetric tail mounted on its other (outer) side (Steven *et al.*, 1988). This calculation was based on the assumption that the core constituents also share this symmetry with the connector on which they are stacked. However, our current analysis detects a strong 8-fold symmetry, showing that this assumption was incorrect, and the copy numbers of core components are more likely to be multiples of 8 or 4.

It follows that there is a symmetry mismatch between the connector and (at least part of) the core, in addition to the one between the connector and its surrounding ring of five gp10 hexamers. Noting that symmetry mismatches between oligomeric rings have been associated with a propensity for relative rotation, both by experiment [the F1–F0 ATPase (Yasuda *et al.*, 1998)] and conjecture [phage connectors (Hendrix, 1978), ATP-dependent proteases such as ClpAP (Beuron *et al.*, 1998), and the bacterial flagellar cap (Yonekura *et al.*, 2000)], we speculate that the T7 core may also rotate around its axis, possibly serving as a spindle to facilitate the organization of incoming DNA.

IV. Procapsid Structure

To examine the procapsid shell, we reconstructed cryoelectron micrographs of 9.10 procapsids, produced by coexpressing genes *9* (scaffold) and *10* (shell) in *Escherichia coli.* Because they lack connectors, the vertex structure should be an unalloyed average of gp10 capsomers, without a contribution from the connector. A density map of the procapsid at 17-Å resolution is shown in stereo surface renderings in Fig. 4; see Color Insert. It is roundish in shape, like the procapsids of other dsDNA phages [e.g., λ (Dokland and Murialdo, 1993), P22 (Thuman-Commike *et al.*, 1996), HK97 (Conway *et al.*, 1995), P2/P4 (Marvik *et al.*, 1995), and ϕ29 (Tao *et al.*, 1998)]. Appropriately for a phage named T7, it conforms to T=7 icosahedral symmetry. Its handedness is *levo,* as we determined by the tilting method of Belnap *et al.* (1997) (data not shown). The procapsid is ~ 500 Å in diameter, and the only holes seen are small ones at the centers of the hexamers (Fig. 4; see Color Insert).

The outer surface is corrugated and the hexamers are elliptical rather than circular; again, a feature that is common to all T=7 procapsids currently on record. The gp10 subunit has an L-shaped ridge on its outer surface. The most pronounced features seen on the inner surface are nubbins of density, ~30 Å in diameter and 30 Å long, that extend inward like stalactites from points around the periphery of each hexamer. Such nubbins as are found at corresponding sites around the pentamer are vestigial in comparison (Fig. 4). Because there was no sign of proteolysis in this specimen nor of proteins other than gp10 and gp9 according to sodium dodecyl sulfate–polyacrylamide gel electro phoresis (SDS–PAGE) of the purified particles (data not shown), we infer that the hexamer-associated nubbins represent the scaffolding protein, gp9. Taking into account its stoichiometry relative to the gp10 shell (Cerritelli and Studier, 1996), each nubbin should be a gp9 monomer (34 kDa). The miniature nubbins around the pentamer (Fig. 4) may either reflect small surface protrusions of gp10 molecules, or low occupancy of these binding sites by gp9.

Distribution of the Scaffolding Protein: A Proposal for the Morphogenetic Mechanism

Assembly mechanisms governed by scaffolding proteins have been reviewed in depth (Dokland, 1999; and see Fane and Prevelige, this volume). The emerging picture is that although all scaffolding proteins serve a common purpose—to transiently interact with shell proteins in order to guide their assembly into discrete higher order structures—the mechanisms whereby they fulfill this role vary widely. Moreover, the patterns of interactions between scaffold and shell components have in most cases been difficult to discern. In this respect, T7 represents an exception in that (1) the scaffolding protein is clearly visualized, bound to specific sites on the inner surface of the shell (Fig. 4), and (2) there appear to be no lateral interactions between scaffolding protein protomers.

We note that our interpretation of the hexamer-associated nubbins as monomers of gp9 invokes a copy number of 360, whereas previous estimates are considerably smaller, that is, ~140 (Cerritelli and Studier, 1996; Serwer, 1976). In principle, this discrepancy might be attributed to partial occupancy: however, the density in the nubbins is as strong as at any other feature in the density map (Fig. 4b), indicating near-quantitative occupancy. To address this issue further, we plan to reexamine the stoichiometry of 9.10 procapsids.

This is not the first example of the binding affinity of a capsid protein varying markedly according to its site in the surface lattice, that is, depending on the quasi- or nonequivalent conformation of the site in question. Another example is given by the reactivity of the VP26 accessory capsid protein of herpes simplex virus, which binds to the major capsid protein VP5, only in its hexamer conformation and not in its pentamer conformation (e.g., Wingfield *et al.*, 1997).

Drawing on the above observations, we propose that T7 assembles by the following mechanism: namely, the binding of gp9 to gp10 hexamers locks them in a state compatible with ordered assembly into closed procapsids of the correct size. The elliptical gp10 hexamer has three pairs of quasi-equivalent gp9-binding sites on its inner surface, which are colored red, blue, and green, respectively, in Fig. 4b. If gp9 is abundantly available, all these sites are occupied, giving 360 copies per capsid. On the other hand, if gp9 is available only in limited supply [and data indicate that the complement of gp9 per procapsid is variable (Cerritelli and Studier, 1996)], binding of gp9 to fewer sites may still suffice for assembly. This scenario has much in common with the account of P22 procapsid assembly proposed by Thuman-Commike *et al.* (1999). In difference imaging between procapsids with and without scaffold, these authors visualized small patches of difference density apposed with the inner surface of the hexamers that they equated with scaffold subunits, attributing their small size to disorder.

In summary, we envisage that T7 procapsid assembly proceeds as follows: if connectors are available, they serves as initiation complexes, recruiting a surrounding ring of five complexes, each consisting of a gp10 hexamer with up to six gp9 monomers; further assembly then proceeds radially outward. Assembly may involve either binding preformed hexamers and pentamers, or building them *in situ.* Insufficient information is available to distinguish between these alternatives. However, there is good evidence that hexamers and pentamers are assembly intermediates for HK97 (Xie and Hendrix, 1995), and the latter mechanism (monomer accretion) has been advocated for P22 (Thuman-Commike *et al.*, 1999). In the absence of connectors, a similar "elongation" phase should take place, but is initiated differently–possibly, around a gp10 pentamer–and presumably requiring higher critical concentrations of gp9 and gp10.

V. Procapsid Maturation: Expansion Is Initiated in the Connector

In addition to being the likely initiator of procapsid assembly (Valpuesta and Carrascosa, 1994), the connector appears also to initiate the cooperative conformational change that transforms the procapsid into the mature capsid, according to several indirect but suggestive observations. Attempts to isolate 8.9.10 particles in their procapsid form by coexpressing these proteins in the same cell, which would have potentially allowed us to examine the rotational symmetry of the connector *in situ* without interference from the core, did not succeed. All the 8.9.10 capsids isolated (e.g., Fig. 5b) were found to be in the mature state (Cerritelli and Studier, 1996), being larger, thinner walled, and more polyhedral than 9.10 procapsids (cf. Fig. 5a) or wild-type procapsids (Fig. 5a). It follows that when the connector is present, the procapsid is much more liable to undergo this transition.

Second, we induced 9.10 procapsids to expand by storing them in low-salt buffer (10 m*M* NaCl, 50 m*M* Tris-HCl, pH 8.0) for extended periods (2 to 8 weeks) at 4°C. The resulting particles were mature as judged by their mobility in native agarose gels (Cerritelli and Studier, 1996). Examination by cryo-EM confirmed that these particles were thin-walled and morphologically typical of the expanded state, but a high proportion of them (~55%) had marked lesions or distortions (e.g., Fig. 5a). In contrast, there was a low incidence (<1%) of such defects in the starting material, and visible distortions were similarly rare in the empty 8.9.10 capsids (see Fig. 5b), with <4% visibly damaged. These observations suggest that when a procapsid contains a connector, expansion is easily induced and produces a regular icosahedral capsid in a high proportion of cases. In contrast, in the absence of a connector, the procapsid is less readily induced to expand and this traumatic event is liable to rupture the surface lattice. It is plausible that such ruptures may reflect incorrectly initiated expansion.

Because T7 does not have a maturational protease, it is most likely that DNA packaging–in which the connector is intimately involved–is the trigger for capsid expansion, and this event is initiated in or somehow facilitated by

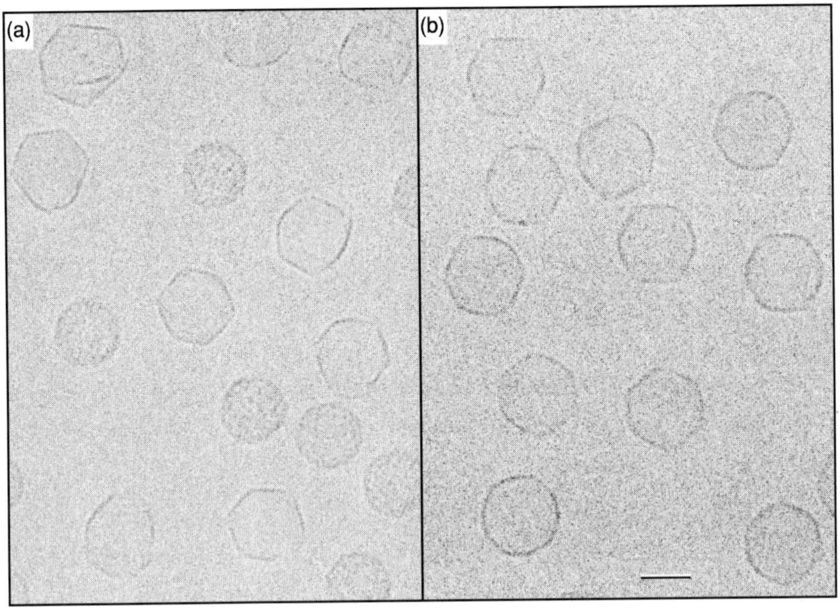

FIG. 5. Cryomicrographs of (a) 9.10 procapsids matured to capsids *in vitro* (note the high incidence of structural lesions) and (b) 8.9.10 capsids, almost all of which are intact. Bar: 500 Å.

the presence of a connector in the surface lattice. Images of giant T4 procapsids captured in the act of expanding (Steven and Carrascosa, 1979) are also consistent with expansion having started in the connector vertex.

VI. Packaging and Parting of DNA

The mechanism of DNA packaging by phage T3 has been studied by Fujisawa and co-workers (Fujisawa *et al.*, 1987; Fujisawa and Morita, 1997; Morita *et al.*, 1995) and we assume that T7 is similar in this respect. As with other phages, the terminase enzyme, which is central to this reaction, has a large subunit (gp19) and a small subunit (gp18). The large subunit binds to the T3 procapsid to form a 50S complex whereas the small subunit binds to concatemeric DNA, which then combines with the 50S complex and packaging ensues (Fig. 1b). The T3 terminase in packaging complexes is thought to form a hexameric ring. When packaging is complete, the terminase releases the filled head by cutting the DNA at the appropriate site. The consumption of ATP has been measured at 1.7 base pairs packaged per hydrolysis event, which compares with the figure of 2 measured in the ϕ29 system (Chen and Guo, 1997). T7 packaging *in vitro* has been observed by light microscopy (Sun *et al.*, 1997).

The mechanism whereby T7 DNA is conveyed from the infecting phage head into the host bacterium has been studied by Molineux and co-workers (reviewed by Molineux, 2001), who measured the transfer rate of genome entry into the cell (Garcia and Molineux, 1995). They have concluded that DNA is actively pulled out of the head, and different mechanisms are utilized at successive stages of genome transfer, the final stage being coupled to transcription of the DNA that is already inside the bacterium.

VII. The Mature Capsid Structure: Filled and Empty Shells

Mature capsids—both DNA filled and empty—are depicted by cryoelectron microscopy in Fig. 6. From such data, we reconstructed the three-dimensional structures of both kinds of capsids to ~17 Å resolution (Cerritelli *et al.*, 2002). Stereo surface renderings of their outer surfaces are shown in Fig. 7a and b, respectively; and of the inner surface of the empty capsid in Fig. 7c. The filled and empty shells are indistinguishable in size and structure. Their angular shape is well conveyed in this 3-fold view. The shell is close-knit, with no holes of appreciable size. The hexamers are 6-fold symmetric to a good approximation, so that the curing of hexamer asymmetry, already observed in the maturation of several other phage capsids [λ (Dokland and Murialdo, 1993), P22 (Thuman-Commike *et al.*, 1996), and HK97 (Conway *et al.*, 1995)], also takes places in the T7 system.

The most pronounced features on the outer surface are L-shaped ridges—one per subunit. These features tally with the capsomer morphology visualized in 2-D images calculated from negatively stained polycapsid tubes (Steven *et al.*, 1983). These ridges are qualitatively similar to corresponding features seen on the procapsid, but they are arranged somewhat differently, as is seen in a side-by-side comparison between the two particles in Fig. 8. Taking the ridges of hexamer subunits as reference points (Fig. 8a), they appear to be flatter in the mature capsid (cf. the sections in Fig. 8c). Moreover, the L-shaped ridges on nearest-neighbor

---

FIG. 6. A field of T7 heads obtained from a cryoelectron micrograph of a complete tail-deletion (genes *11* and *12*) mutant. Empty capsids appear as thin-walled particles. Full capsids exhibit the characteristic 2.5-nm spacing of densely packed DNA duplexes in motifs that vary according to viewing direction. The concentric ring motif is discernable in the views along the axis that passes through the connector–core vertex that is in the center of the particle.

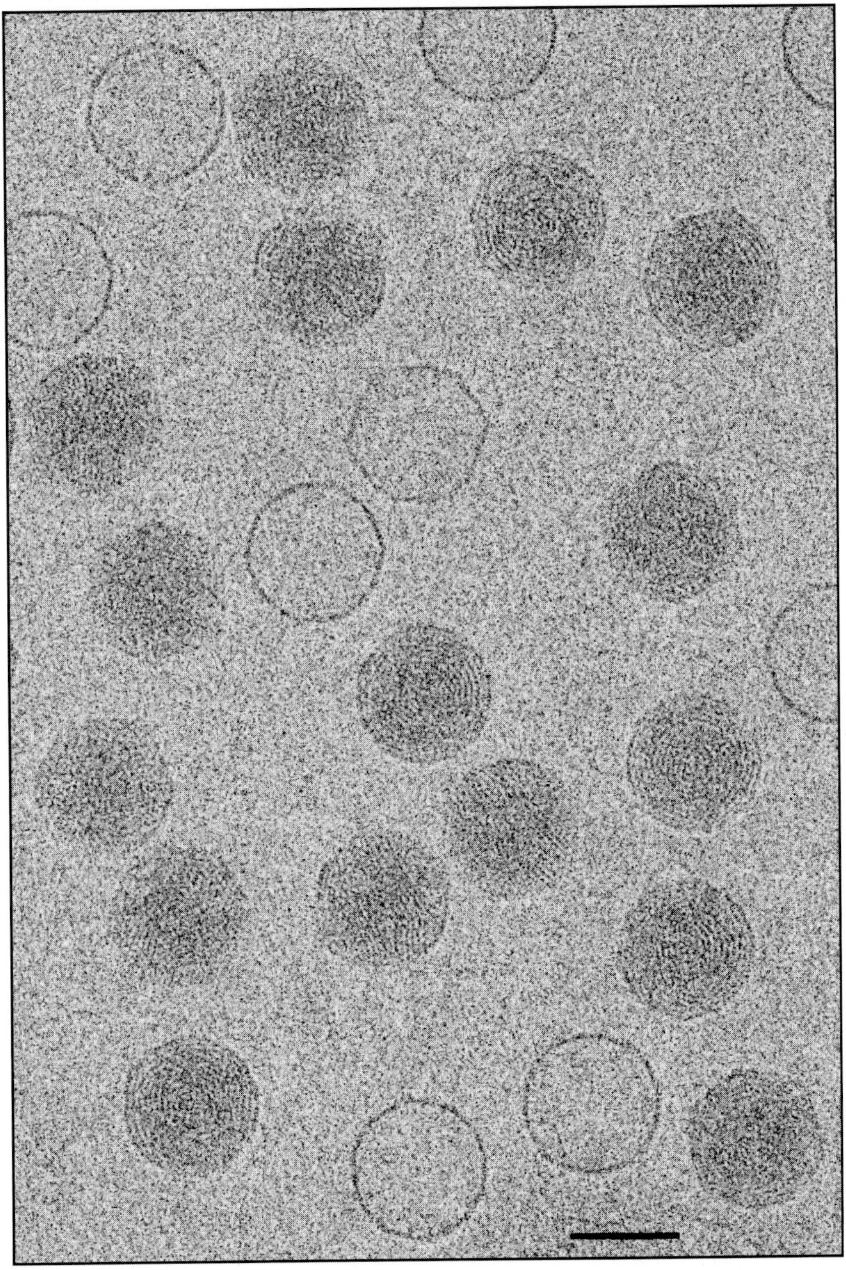

FIG. 6.

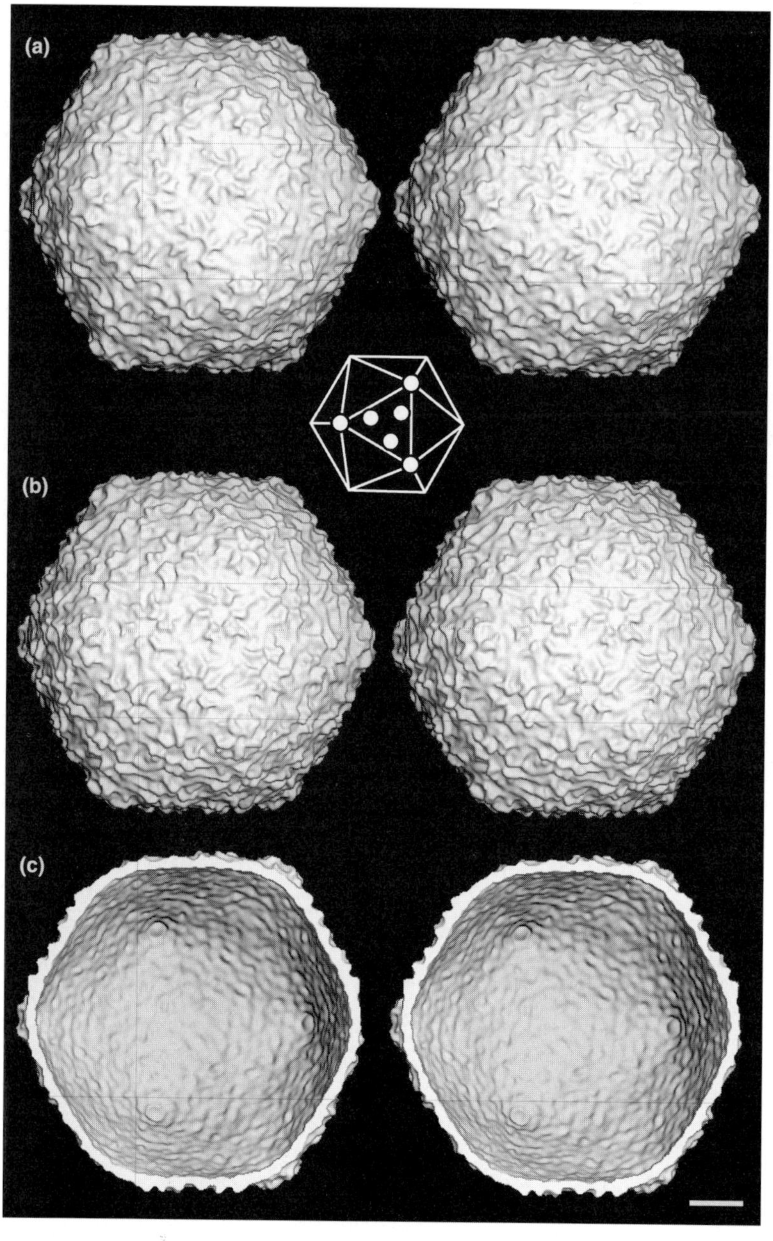

FIG. 7. Structure of mature T7 capsids at 17-Å resolution. Stereo renditions are viewed along an icosahedral 3-fold symmetry axis. They depict the exterior surfaces of (a) the DNA-filled head, (b) the empty capsid, and (c) the interior surface of the empty capsid. *Inset*: A

subunits on adjacent capsomers have moved distinctly further apart, accounting for the increased size of the mature capsid, and the inner surface has become smooth (Fig. 8c).

We suggest that the expansion transformation involves a subunit rotation mechanism, as visualized at the level of a quasi-atomic model for phage HK97 (Conway *et al.*, 2001). Specifically, rotation of gp10 subunits about axes in the plane of the capsid shell and extending radially outward from the local symmetry axis may bring densities that were formerly on the inner surface into the plane of the shell. To accommodate these densities, the centers of adjacent capsomers are pushed further apart. Concomitantly, the gp9 scaffold protein subunits are released. Both of these effects have the effect of making the mature capsid thinner walled and smoother surfaced than its precursor.

Because there are no holes in either state of the shell that are big enough to accommodate gp9 subunits with dimensions as visualized (Fig. 4b), and there is no protease to break them down into more easily exportable fragments, how is externalization accomplished? We can envisage two modes of egress, which are not mutually exclusive: either the scaffolding molecules are exported directly as the shell expands, with the binding site on gp10 moving outward and feeding the gp9 subunit into a transiently open exit channel, and then disengaging as the shell transformation proceeds; or the gp9 subunit is unfolded in a reaction that is coupled to the shell reorganization, and reptates out through a smaller exit channel. Other examples of "protein tunneling" reactions have been inferred to take place in the transfer of internal capsid proteins from bacteriophage T4 capsid into the host cell (Hong and Black, 1993; Mullaney and Black, 1998), in the export of bacterial flagellin subunits for assembly (MacNab, 2000), and in the translocation of protein substrates into the digestion chambers of ATP-dependent proteases (Ishikawa *et al.*, 2001; Ortega *et al.*, 2000).

---

diagram of the surface lattice in the same orientation as the exterior views. The empty capsid reconstruction included 155 image pairs from a total of 662 extracted from the same set of micrographs used for the 9.10 procapsid map. The density map of the DNA-filled head was calculated from 432 image pairs from a total of 1485 extracted from 5 defocus pairs. The particles chosen for reconstruction had the highest correlation coefficients, with a threshold of the mean, as required for all three coefficients calculated by PFT (Baker and Cheng, 1996). This threshold was more stringent than that applied to the procapsid data (see Fig. 4 legend) because these correlations were lower. The resolution of the filled head was 17 Å by FSC. Because the empty capsid map shows an identical shell structure, we infer that it has the same resolution although the particle number is too low for a stable FSC calculation. Bar: 100 Å.

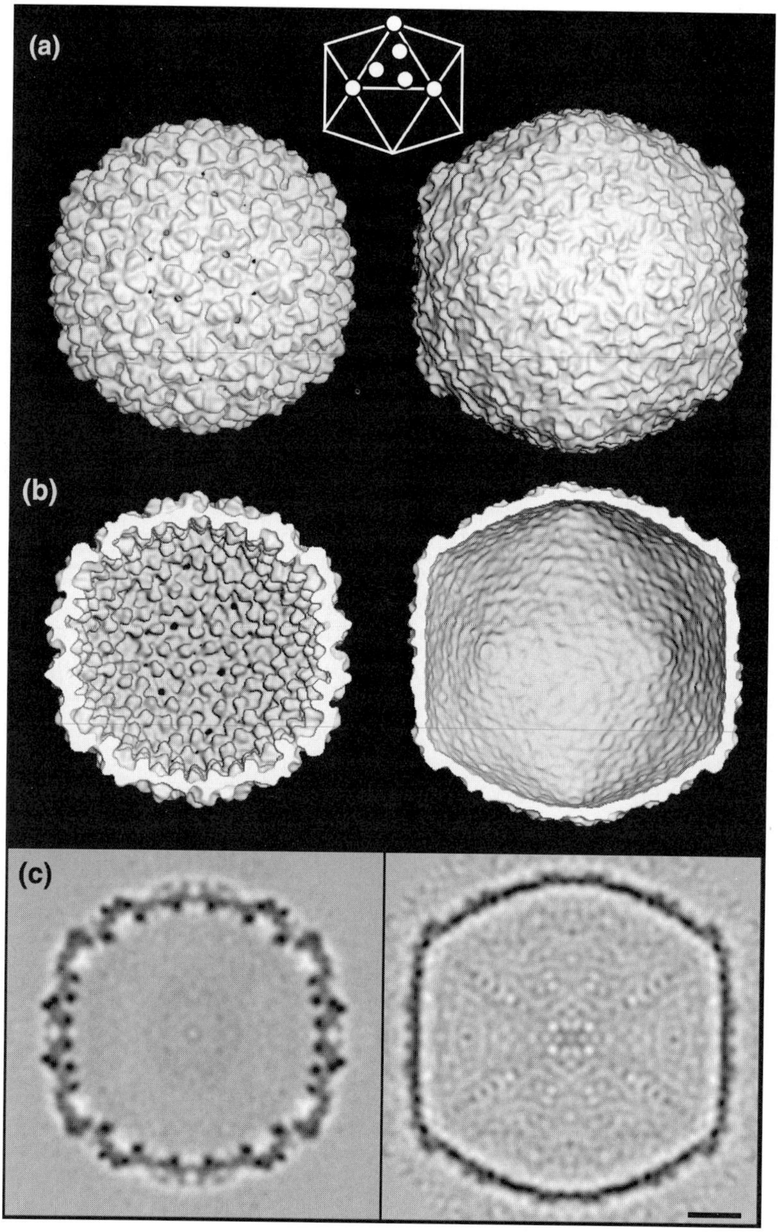

FIG. 8. Comparison of the T7 procapsid (*left column*) and empty mature capsid (*right column*) at 17-Å resolution. Views along the icosahedral 2-fold symmetry axis represent (a) exterior surfaces, (b) interior surfaces, and (c) central sections in which the densities have been coded as gray levels (protein is dark). *Inset:* A schematic view of the surface lattice, corresponding to the orientation of the exterior views. Bar: 100 Å.

As noted above, the wild-type T7 shell contains two products of gene 10—gp10A (90%) and gp10B (10%)—related by a read-through event (Condron *et al.*, 1991; Dunn and Studier, 1983) and this property has been exploited for phage display (Rosenberg *et al.*, 1996). Similar events also occur in other viruses. For instance, the L-A dsRNA virus of yeast has two copies of the Gag–Pol fusion protein, copolymerized with 118 copies of Gag (Caston *et al.*, 1997; Wickner, 1996). In this case, the functional role of the long form of capsid protein is clear, that is, expressing the polymerase as a fusion protein with the shell subunit affords a convenient mechanism for incorporating it into the viral particle. Basically the same mechanism is used to introduce the maturational protease into herpes simplex virus, although in this case the protease is fused to the scaffolding protein, which forms an inner shell lining the procapsid surface shell, with about 10% of the scaffold subunits having the protease extension (Liu and Roizman, 1991; Preston *et al.*, 1983). For T7, the gp10B form of the capsid protein is dispensable and it is not clear what functional role this protein may play when present, nor is it known where in the surface lattice the gp10B subunits reside. The latter question may potentially be addressed by difference imaging. Two variants of the mature surface lattice differing in net surface charge have been distinguished by agarose gel electrophoresis (Gabashvili *et al.*, 1997), but it is not known whether they exhibit substantive structural differences.

VIII. Structure of Packaged DNA

Inside phage heads, DNA is packed to remarkably high densities on the order of 400–500 mg/ml. This condensation phenomenon has long attracted interest and numerous proposals have been made for how their DNA is organized (e.g., Black, 1989; Earnshaw *et al.*, 1978; Harrison, 1983a; Hud, 1995; Lepault *et al.*, 1987; Richards *et al.*, 1973). In the case of T7, it has long been known from low-angle X-ray scattering that the DNA is locally ordered as parallel packings with a center-to-center spacing of 24 Å between neighboring duplexes (Ronto *et al.*, 1983; Stroud *et al.*, 1981). This spacing corresponds to hexagonal phase crystals (spacing, 23.7–31.5 Å) rather than cholesteric liquid crystal phase (32–49 Å), according to a phase diagram for condensed DNA (Livolant and Leforestier, 1996). However, apart from this spacing, low-angle X-ray diffraction data, which constitute spherically averaged powder patterns, cannot reveal the overall organization of the DNA or even whether it is arranged consistently from particle to particle.

Information of the latter kind was obtained by cryo-EM of tailless T7 heads, which were serendipitously found to present two preferential orientations when vitrified in thin films of buffer: axial views of single heads and side views of paired heads (Cerritelli *et al.*, 1997). It is likely that these orientations result from a hydrophobic patch at the connector vertex of the tail less head that sequesters from the (polar) solution either by binding to the air–water interface (thus orienting the head so as to present an axial projection) or by pairing with the corresponding patch on another head. In the latter case, the double particle remains hydrated as the film thins by orienting with its long axis in the plane of the film. The axial views produce concentric shell patterns (Fig. 9c–e), and the side views produce punctate patterns (Fig. 9h and i). These observations led to the unequivocal conclusion, confirmed by computer modeling, that the DNA is coiled around the axis passing through the portal/connector (Cerritelli *et al.*, 1997) (Fig. 9), in line with spool models previously proposed on other grounds (Furlong *et al.*, 1972; Harrison, 1983b).

In our three-dimensional reconstruction of filled T7 heads, the DNA appears as a set of nested icosahedral shells (Fig. 9a, f, and k). We infer that this shell system accurately represents the radial ordering—at least, of the outer layers—but that their icosahedral symmetry was imposed by the reconstruction procedure in which the connector axis was treated on the same basis as the other five trans vertex axes. As a consequence of this artifactual symmetrization, the DNA-associated density of the filled head reconstruction produces parallel striations when reprojected in side view (Fig. 9g and l), instead of the quasi-hexagonal punctate formations of the original projection images (cf. Fig. 9h and m). Noting that similar shell systems have also been observed in reconstructions of DNA-filled capsids of herpes simplex virus (Booy *et al.*, 1991), λ (Dokland and Murialdo, 1993), P22 (Zhang *et al.*, 2000), T4 isometric particles (Olson *et al.*, 2001), and PRD1 (San Martin *et al.*, 2001), we suspect that in these viruses also, the DNA may be wound in more or less well-ordered coaxial spools. However, a feature of the reconstruction that is probably more valid than the corresponding feature of the spherical spool model that we have used for calculational simplicity is that it shows the outer layer of DNA at a uniform distance from the inner surface of the flat-faceted icosahedral capsid. As actually wound, the DNA is likely to follow the capsid surface (and to be electrostatically repelled from it if, as seems likely, T7 resembles HK97 in having a negatively charged inner surface (Conway *et al.*, 2001); and therefore, to incur some discontinuity in curvature as it passes across edges of the capsid.

How many shells are there in the spool? The number of rings seen in the averaged axial projection is ~11 (Fig. 9e), whereas the number of shells required to accommodate the T7 genome, given geometrically perfect winding, is 6 (Cerritelli *et al.*, 1997). However, we note that the inner shells are not required in order to account for the observed image, for the same reason that the number of rings is higher than the number of nested shells, that is, because the radius of winding in a given shell shrinks toward the poles. Coprojecting a smaller number of nested spool-wound shells—say, three or four—along their common axis still generates the same number of rings (our unpublished observations). In this scenario, the remaining 25% of the genome (given three shells) could be less regularly organized. This possibility was explicitly raised by Odijk (1998), who analyzed the interplay between electrostatic repulsion and the resistance of duplex DNA to bending in energetic terms and concluded that his calculations support the coaxial spool model as a minimum energy state, while also raising the possibility that the innermost DNA may be less well ordered, in view of the penalty associated with the high curvature that spooled DNA would have toward the center of the particle. As noted above, this proposal is not contrary to any current structural observations.

The conformation of encapsidated T7 DNA has been studied by Raman difference spectroscopy and compared with free DNA at an approximately 10-fold lower density (Overman *et al.*, 1998). The results, essentially recapitulating an earlier study of P22 DNA, which is of similar size and packing density (Aubrey *et al.*, 1992), were that the DNA is in the conventional B-form and that kinking, if any, is limited by an upper bound of ~2% of bases. The only difference detected was in two spectral bands associated with phosphates along the rim of the helices (Overman *et al.*, 1998). This perturbation could be simulated in solution by increasing the concentration of Mg^{2+} ions, thus it was suggested that T7 heads may also contain magnesium in appropriate concentrations.

Although the overall arrangement of fully packaged T7 DNA is now established, many questions persist. For example, little is known about how the DNA is organized at earlier stages of packaging: in particular, the spool model in our formulation does not imply that regularly coiled shells are laid down in succession. In fact, such evidence as is available implies the contrary. As with other bacteriophages, T7 packaging (Masker and Serwer, 1982) initiates with the procapsid, which expands into its mature conformation during, and as a consequence of, DNA packaging. With λ, expansion is triggered when ~20% of the genome has entered (Hohn, 1983). If a similar threshold applies to T7, this is equivalent to almost 40% of the final density because the internal capacity of the procapsid is only half that of the mature capsid. At all stages of packaging, the same physical principles are likely to be operative, that is, minimizing the energy deficit created by bending the DNA, by the mutual electrostatic repulsion of the DNA duplexes, and by the electro-static repulsion of the DNA and the

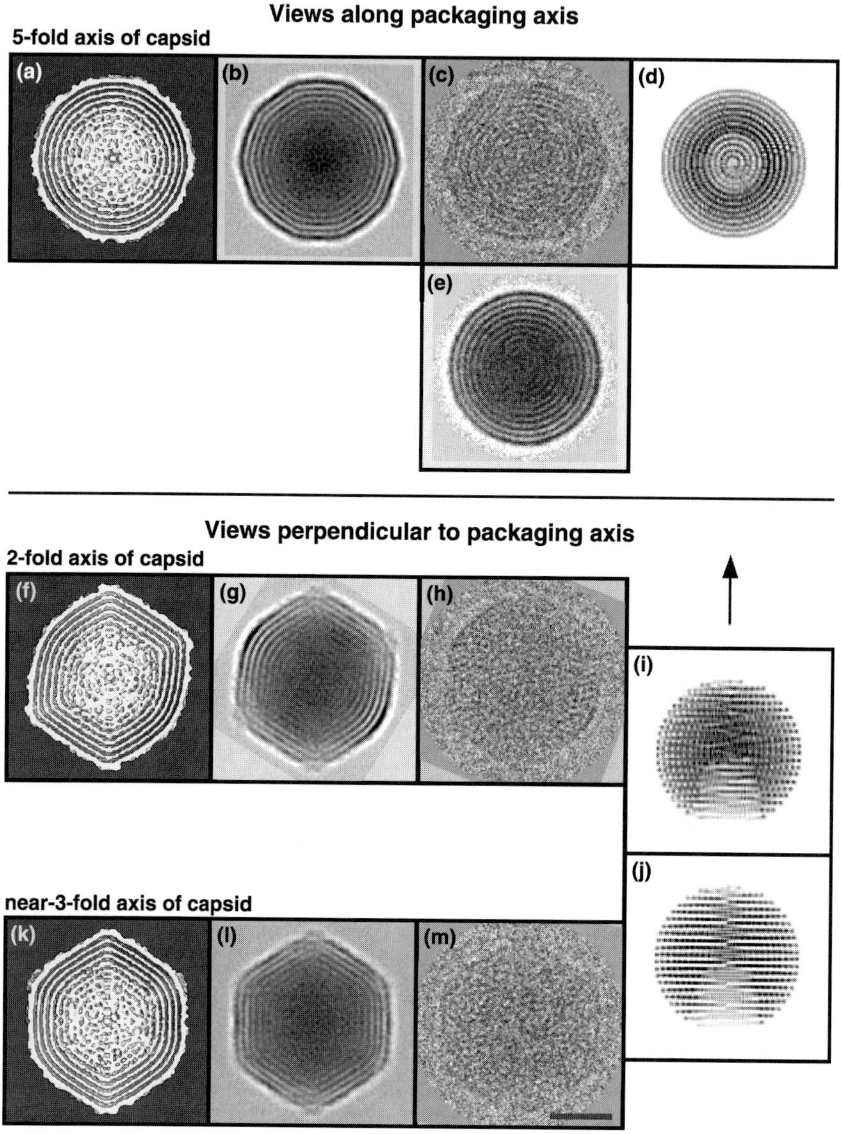

FIG. 9. Comparison of cryo-EM images of T7 heads with corresponding reprojections of the full-head reconstruction and with projections of a computer-modeled DNA spool (Cerritelli *et al.*, 1997). These tailless heads were obtained from an *11⁻.12⁻* deletion mutant. The particles selected for comparison (c, h, and m) are viewed along icosahedral symmetry axes (of the capsid) that are parallel or perpendicular to the axis of the DNA spool (packaging axis). Cutaway views of the reconstructed head viewed in 5-fold, 2-fold, and nearly 3-fold orientations are shown in (a), (f), and (k), respectively.

(putatively) negatively charged inner surface of the capsid. With only a small amount of DNA inside the capsid, the number of isoenergetic states is likely to be large, thus we assume that less ordered conformations are assumed. This scenario is consistent with the negative staining EM observations of Serwer *et al.* (1997), who reported that the "fingerprint" motif characteristic of axial views of fully packaged T7 heads was not seen on partially packaged heads containing up to 40% of the genome. However, as packaging proceeds, the conformational options are reduced and the DNA assumes a progressively more ordered conformation.

IX. Summary

Bacteriophage T7 is a double-stranded DNA bacteriophage that has attracted particular interest in studies of gene expression and regulation and of morphogenesis, as well as in biotechnological applications of expression vectors and phage display. We report here studies of T7 capsid assembly by cryoelectron microscopy and image analysis. T7 follows the canonical pathway of first forming a procapsid that converts into the mature capsid, but with some novel variations. The procapsid is a round particle with an icosahedral triangulation number of 7*levo,* composed of regular pentamers and elongated hexamers. A singular vertex in the

The corresponding reprojections are in (b), (g), and (l). Note that in (b), the outermost ring, contributed by the shell of capsid protein, is distinctly denser than the inner rings from the spooled DNA; and the characteristic two chevrons projected by the protein shell in 2-fold views (g). Head (c) is viewed in almost perfectly alignment with the packaging axis, to judge by its polyhedral profile [cf. (b)]: in contrast, at least some of the 77 nearly 5-fold heads averaged to produce image (e) exhibit significant excursions from this viewing direction because this image presents a circular profile, not a rounded decagon. Nevertheless, its DNA pattern (e) matches well with the projection of the modeled spool (d). In views perpendicular to the spool axis, the icosahedral capsid presents a variety of projections corresponding to different rotational settings around this axis, which is vertical (arrow) for each of these images (f–m). However, the modeled projection of the spool does not vary with this setting because, for reasons of computational simplicity in modeling, the DNA was confined within a sphere, not an icosahedral surface, the real situation for T7. Side views of the spool are characteristically punctate (i and j), with this feature becoming more evident when the spool axis is tilted by 5° out of plane (i) than in perfect side view (j). Punctate patterns are clearly seen in cryo-EM images of heads viewed along a 2-fold axis (h) or a 3-fold axis (m). In calculating the spool projections (Cerritelli *et al.*, 1997), DNA was excluded from the space occupied by the core, making the core evident [bottom of (i) and (j)]; in contrast, the core is not well visible in cryomicrographs of heads (c, h, and m), but is clearly seen in stain-penetrated heads (e.g., Steven and Trus, 1986; Cerritelli *et al.*, 1997; Serwer *et al.*, 1997). Bar: 250 Å.

procapsid is occupied by the connector/portal protein, which forms 12-fold and 13-fold rings when overexpressed, of which the 12-mer appears to be the assembly-competent form. This vertex is the site of two symmetry mismatches: between the connector and the surrounding five gp 10 hexamers; and between the connector and the 8-fold cylindrical core mounted on its inner surface. The scaffolding protein, gp9, which is required for assembly, forms nubbin-like protrusions underlying the hexamers but not the pentamers, with no contacts between neighboring gp9 monomers. We propose that gp9 facilitates assembly by binding to gp10 hexamers, locking them into a morphogenically correct conformation. gp9 is expelled as the procapsid matures into the larger, thinner walled, polyhedral capsid. Several lines of evidence implicate the connector vertex as the site at which the maturation transformation is initiated: *in vivo*, maturation appears to be triggered by DNA packaging whereby the signal may involve interaction of the connector with DNA. In the mature T7 head, the DNA is organized as a tightly wound coaxial spool, with the DNA coiled around the core in at least four and perhaps as many as six concentric shells.

Acknowledgments

We thank Dr. D. M. Belnap (LSBR) for help with the procapsid handedness determination experiment; Dr. E. Kocsis (OD-NIH) for help with the ROTASTAT program; Drs. J.-M. Valpuesta and J. L. Carrascosa (Centro de Biotecnologia Molecular, Madrid) for kindly providing the images for Fig. 2g, j, and k; and Dr. R. Wade (IBS) for providing computational facilities.

References

Adolph, K. W., and Haselkorn, R. (1972). *Virology* **47,** 701–710.

Aubrey, K. L., Casjens, S. R., and Thomas, G. J. (1992). *Biochemistry* **31,** 11835–11842.

Baker, T. S., and Cheng, R. H. (1996). *J. Struct. Biol.* **116,** 120–130.

Belnap, D. M., Olson, N. H., and Baker, T. S. (1997). *J. Struct. Biol.* **120,** 44–51..

Benson, S. D., Bamford, J. K. H., Bamford, D. H., and Burnett, R. M. (1999). *Cell* **98,** 825–833.

Beuron, F., Steven, A. C., Kessel, M., Booy, F. B., Wickner, S., and Maurizi, M. R. (1998). *FASEB J.* **12,** 644.

Bhella, D., Rixon, F. J., and Dargan, D. J. (2000). *J. Mol. Biol.* **295,** 155–161.

Black, L. W. (1989). *Annu. Rev. Microbiol.* **43,** 267–292.

Black, L. W., Showe, M. K., and Steven, A. C. (1994). *In* "Molecular Biology of Bacteriophage T4" (J. Karam, Ed.), pp. 218–258. American Society of Microbiology, Washington, D.C.

Booy, F. P., Newcomb, W. W., Trus, B. L., Brown, J. C., Baker, T. S., and Steven, A. C. (1991). *Cell* **64,** 1007–1015.

Canady, M. A., Tihova, M., Hanzlik, T. N., Johnson, J. E., and Yeager, M. (2000). *J. Mol. Biol.* **299,** 573–584.
Carazo, J. M., Fujisawa, H., Nakasu, S., and Carrascosa, J. L. (1986). *J. Ultrastruct. Mol. Struct. Res.* **94,** 105–113.
Caston, J. R., Trus, B. L., Booy, F. P., Wickner, R. B., Wall, J. S., and Steven, A. C. (1997). *J. Cell Biol.* **138,** 975–985.
Cerritelli, M. E., and Studier, F. W. (1996). *J. Mol. Biol.* **258,** 286–298.
Cerritelli, M. E., Cheng, N., Rosenberg, A. H., McPherson, C. E., Booy, F. P., and Steven, A. C. (1997). *Cell* **91,** 271–280.
Cerritelli, M. E., Trus, B. L., Smith, C. S., Cheng, N., Conway, J. F., and Steven, A. C. (2002). (Submitted.)
Chen, C., and Guo, P. (1997). *J. Virol.* **71,** 3864–3871.
Cheng, N., Conway, J. F., Watts, N. R., Hainfeld, J. F., Joshi, V., Powell, R. D., Stahl, S. J., Wingfield, P. E., and Steven, A. C. (1999). *J. Struct. Biol.* **127,** 169–176.
Condron, B. G., Atkins, J. F., and Gesteland, R. F. (1991). *J. Bacteriol.* **173,** 6998–7003.
Conway, J. F., and Steven, A. C. (1999). *J. Struct. Biol.* **128,** 106–118.
Conway, J. F., Duda, R. L., Cheng, N., Hendrix, R. W., and Steven, A. C. (1995). *J. Mol. Biol.* **253,** 86–99.
Conway, J. F., Wikoff, W. R., Cheng, N., Duda, R. L., Hendrix, R. W., Johnson, J. E., and Steven, A. C. (2001). *Science* **292,** 744–748.
Dokland, T. (1999). *Cell. Mol. Life Sci.* **56,** 580–603.
Dokland, T., and Murialdo, H. (1993). *J. Mol. Biol.* **233,** 682–694.
Donate, L. E., Herranz, L., Secilla, J. P., Carazo, J. M., Fujisawa, H., and Carrascosa, J. L. (1988). *J. Mol. Biol.* **201,** 91–100.
Dube, P., Tavares, P., Lurz, R., and van Heel, M. (1993). *EMBO J.* **12,** 1303–1309.
Dunn, J. J., and Studier, F. W. (1983). *J. Mol. Biol.* **166,** 477–535.
Earnshaw, W. C., King, J., Harrison, S. C., and Eiserling, F. A. (1978). *Cell* **14,** 559–568.
Friedmann, A., Coward, J. E., Rosenkranz, H. S., and Morgan, C. (1975). *J. Gen. Virol.* **26,** 171–181.
Fujisawa, H., and Morita, M. (1997). *Genes Cells* **2,** 537–545.
Fujisawa, H., Hamada, K., Shibata, H., and Minagawa, T. (1987). *Virology* **161,** 228–233.
Furlong, D., Swift, H., and Roizman, B. (1972). *J. Virol.* **10,** 1071–1074.
Gabashvili, I. S., Khan, S. A., Hayes, S. J., and Serwer, P. (1997). *J. Mol. Biol.* **273,** 658–667.
Garcia, L. R., and Molineux, I. J. (1995). *J. Bacteriol.* **177,** 4066–4076.
Harrison, S. C. (1983a). *J. Mol. Biol.* **171,** 577–580.
Harrison, S. C. (1983b). *J. Mol. Biol.* **171,** 577–580.
Hendrix, R. W. (1978). *Proc. Natl. Acad. Sci. USA* **75,** 4779–4783.
Hendrix, R. W., and Duda, R. L. (1998). *Adv. Virus Res.* **50,** 235–288.
Hendrix, R. W., and Garcea, R. L. (1994). *Semin. Virol.* **5,** 15–26.
Hohn, B. (1983). *Proc. Natl. Acad. Sci. USA* **80,** 7456–7460.
Hong, Y. R., and Black, L. W. (1993). *Virology* **194,** 481–490.
Hud, N. V. (1995). *Biophys. J.* **69,** 1355–1362.
Ishikawa, T., Beuron, F., Kessel, M., Wickner, S., Maurizi, M. R., and Steven, A. C. (2001). *Proc. Natl. Acad. Sci. USA* **98,** 4328–4333.
Kocsis, E., Cerritelli, M. E., Trus, B. L., Cheng, N., and Steven, A. C. (1995). *Ultramicroscopy* **60,** 219–228.
Lepault, J., Dubochet, J., Baschong, W., and Kellenberger, E. (1987). *EMBO J.* **6,** 1507–1512.
Liu, F. Y., and Roizman, B. (1991). *J. Virol.* **65,** 5149–5156.

Livolant, F., and Leforestier, A. (1996). *Prog. Polym. Sci.* **21,** 1115–1164.
Macnab, R. M. (2000). *Science* **290,** 2086–2087.
Marvik, O. J., Dokland, T., Nokling, R. H., Jacobsen, E., Larsen, T., and Lindqvist, B. H. (1995). *J. Mol. Biol.* **251,** 59–75.
Masker, W. E., and Serwer, P. (1982). *J. Virol.* **43,** 1138–1142.
Molineux, I. J. (2001). *Mol. Microbiol.* **40,** 1–8.
Morita, M., Tasaka, M., and Fujisawa, H. (1995). *Virology* **211,** 516–524.
Mullaney, J. M., and Black, L. W. (1998). *Biotechnology* **25,** 1008–1012.
Muller, D. J., Engel, A., Carrascosa, J. L., and Velez, M. (1997). *EMBO J.* **16,** 2547–2553.
Odijk, T. (1998). *Biophys. J.* **75,** 1223–1227.
Olson, N. H., Gingery, M., Eiserling, F. A., and Baker, T. S. (2001). *Virology* **279,** 385–391.
Ortega, J., Singh, S. K., Ishikawa, T., Maurizi, M. R., and Steven, A. C. (2000). *Mol. Cell* **6,** 1515–1521.
Overman, S. A., Aubrey, K. L., Reilly, K. E., Osman, O., Hayes, S. J., Serwer, P., and Thomas, G. J. (1998). *Biospectroscopy* **4,** S47–S56.
Preston, V. G., Coates, J. A., and Rixon, F. J. (1983). *J. Virol.* **45,** 1056–1064.
Richards, K. E., Williams, R. C., and Calendar, R. (1973). *J. Mol. Biol.* **78,** 255–259.
Ronto, G., Agamalyan, M. M., Drabkin, G. M., Feigin, L. A., and Lvov, Y. M. (1983). *Biophys. J.* **43,** 309–314.
Rosenberg, A., Griffin, K., Studier, F. W., McCormick, M., Berg, J., Novy, R., and Mierendorf, R. (1996). T7 Select phage display system: A powerful new protein display system based on bacteriophage T7. *In* "inNovations—Newsletter of Novagen," 1 6PP. Novagen, Madison, WI.
San Martin, C., Burnett, R. M., de Haas, F., Heinkel, R., Rutten, T., Fuller, S. D., Butcher, S. J., and Bamford, D. H. (2001). *Structure (Camb)* **9,** 917–930.
Serwer, P. (1976). *J. Mol. Biol.* **10,** 271–291.
Serwer, P., Khan, S. A., Hayes, S. J., Watson, R. H., and Griess, G. A. (1997). *J. Struct. Biol.* **120,** 32–43.
Simpson, A. A., Tao, Y. Z., Leiman, P. G., Badasso, M. O., He, Y. N., Jardine, P. J., Olson, N. H., Morais, M. C., Grimes, S., Anderson, D. L., Baker, T. S., and Rossmann, M. G. (2000). *Nature* **408,** 745–750.
Steven, A. C., and Carrascosa, J. L. (1979). *J. Supramol. Struct.* **10,** 1–11.
Steven, A. C., and Spear, P. G. (1997). *In* "Structural Biology of Viruses" (W. Chiu, R. M. Burnett, and R. L. Garcea, Eds.), pp. 312–351. Oxford University Press, New York.
Steven, A. C., and Trus, B. L. (1986). In "Electron Microscopy of Proteins," Vol. 5: "Viral Structure". Academic Press, London.
Steven, A. C., Serwer, P., Bisher, M. E., and Trus, B. L. (1983). *Virology* **124,** 109–120.
Steven, A. C., Trus, B. L., Maizel, J. V., Unser, M., Parry, D. A. D., Wall, J. S., Hainfeld, J. F., and Studier, F. (1988). *J. Mol. Biol.* **200,** 351–365.
Stroud, R. M., Serwer, P., and Ross, M. J. (1981). *Biophys. J.* **36,** 743–757.
Sun, M., Son, M., and Serwer, P. (1997). *Biochemistry* **36,** 13018–13026.
Tao, Y., Olson, N. H., Xu, W., Anderson, D. L., Rossmann, M. G., and Baker, T. S. (1998). *Cell* **95,** 431–437.
Thuman-Commike, P. A., Greene, B., Jakana, J., Prasad, B. V., King, J., Prevelige, P. E., Jr., and Chiu, W. (1996). *J. Mol. Biol.* **260,** 85–98.
Thuman-Commike, P. A., Greene, B., Malinski, J. A., Burbea, M., McGough, A., Chiu, W., and Prevelige, P. E. (1999). *Biophys. J.* **76,** 3267–3277.
Trus, B. L., Booy, F. P., Newcomb, W. W., Brown, J. C., Homa, F. L., Thomsen, D. R., and Steven, A. C. (1996). *J. Mol. Biol.* **263,** 447–462.

Valpuesta, J. M., and Carrascosa, J. L. (1994). *Q. Rev. Biophys.* **27,** 107–155.
Valpuesta, J. M., Sousa, N., Barthelemy, I., Fernandez, J. J., Fujisawa, H., Ibarra, B., and Carrascosa, J. L. (2000). *J. Struct. Biol.* **131,** 146–155.
Wickner, R. B. (1996). *Microbiol. Rev.* **60,** 250–265.
Wikoff, W. R., and Johnson, J. E. (1999). *Curr. Biol.* **9,** R296–R300.
Wingfield, P. T., Stahl, S. J., Thomsen, D. R., Homa, F. L., Booy, F. P., Trus, B. L., and Steven, A. C. (1997). *J. Virol.* **71,** 8955–8961.
Xie, Z., and Hendrix, R. W. (1995). *J. Mol. Biol.* **253,** 74–85.
Yasuda, R., Noji, H., Kinosita, K., and Yoshida, M. (1998). *Cell* **93,** 1117–1124.
Yonekura, K., Maki, S., Morgan, D. G., DeRosier, D. J., Vonderviszt, F., Imada, K., and Namba, K. (2000). *Science* **290,** 2148–2152.
Zhang, Z., Greene, B., Thuman-Commike, P. A., Jakana, J., Prevelige, P. E., Jr., King, J., and Chiu, W. (2000). *J. Mol. Biol.* **297,** 615–626.

CONFORMATIONAL CHANGES IN ENVELOPED VIRUS SURFACE PROTEINS DURING CELL ENTRY

By DEBORAH FASS

Department of Structural Biology, Weizmann Institute of Science, Rehovot 76100, Israel

I. Introduction: Multiple Stops on the Protein-Folding Landscape 325
II. Influenza Hemagglutinin 326
A. HA_0: First Stop on the Protein-Folding Landscape 327
B. Peptide Bond Cleavage and Rearrangement to the Native State of the HA_1–HA_2 Complex 329
C. Refolding on Exposure to Low pH 331
D. Energetic Considerations and Models for Membrane Fusion 334
E. Protein Folding and Spring Loading 337
III. Retroviruses 338
A. Membrane Fusion with Coiled Coils at the Core 340
B. Avian Leukosis Virus and Receptor-Induced pH Dependence 343
C. The Receptor-Binding Cascade of Human Immunodeficiency Virus 345
D. Mammalian Oncoretroviruses and Fusion Activation *in Trans* 346
IV. Paramyxoviruses Turn Paradigms Upside Down? 350
V. Oligomerization State Switches in Flaviviruses and Alphaviruses 353
VI. Concluding Remarks 356
References 357

I. Introduction: Multiple Stops on the Protein-Folding Landscape

Proteins derive their activities from their folds, that is, the secondary and tertiary structural scaffolds that precisely arrange various chemical groups to carry out an enzymatic reaction, binding event, and so on. In other words, proteins usually fold first, then function. In contrast, the proteins on the surfaces of viruses make multiple stops on the protein-folding landscape, and the itineraries, not simply the final destinations, determine their activities. The conformational changes that occur in virus surface proteins to promote virus entry into cells are among the most dramatic examples of how protein (re-)folding, and not simply the folded protein per se, can underlie biological activity.

The first steps in viral infection are binding of the virus to the cell surface, and, for viruses enclosed by a lipid membrane, the subsequent fusion of the virus and cell membranes to introduce the virus genetic material into the cell cytoplasm. Proteins embedded in the virus lipid bilayer envelope carry out cell surface receptor binding and membrane

ADVANCES IN PROTEIN CHEMISTRY, Vol. 64

0065-3233/03 $35.00

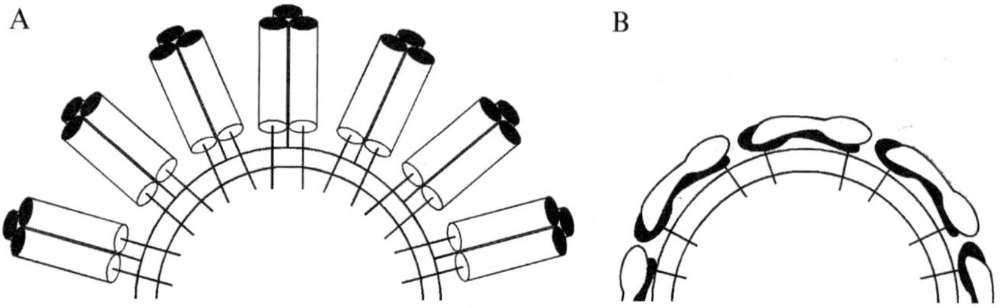

FIG. 1. Illustration of two morphologies of enveloped viruses (not drawn to scale). (A) The surface proteins of influenzavirus project like "spikes" or cylinders from the virus envelope. (B) The surface proteins of flaviviruses lie flat against the virus envelope.

fusion. These surface or "envelope" proteins from distinct virus families can be structurally different enough to complicate generalizations. For example, some virus envelope proteins protrude as spikes from the virus surface, whereas other virus surface proteins lie flat against the membrane like a coat of armor (Fig. 1). One generalization that can be made, however, is that enveloped virus surface proteins, regardless of the details of their mechanisms, undergo large-scale structural changes between their initial folding and the completion of membrane fusion. The aim of this review is to highlight common structural and energetic themes found across diverse virus families, and to detail differences in how representative viruses penetrate the cell membrane to initiate infection.

II. Influenza Hemagglutinin

Influenza (flu) virus is infamous for its antigenic variation—the superficial changes that help the virus avoid detection by the immune system while preserving its essential functions (reviewed in De Jong *et al.*, 2000). A bout of flu this year will not protect one from the illness even a year or two down the line, in contrast to a childhood chickenpox infection that most likely gives life-long immunity. Although antibodies neutralize flu virus, rapid variation in the sequence of the influenza surface proteins causes recurrent outbreaks and the inability to develop a permanent vaccine against the virus. The most striking feature of influenzavirus is the layer of spikes projecting outward from the surface (Fig. 1). These spikes are the hemagglutinin (HA) proteins of which there are estimated to be a few hundred copies per virion (Ruigrok *et al.*, 1984; Amano and Hosaka, 1992). The HA spikes carry out both receptor binding and membrane fusion during infection. HA of influenza A virus was the first enveloped virus surface protein to be studied by X-ray crystallography (Wilson *et al.*, 1981), and the first to be examined in more than one conformational state (Bullough *et al.*, 1994). HA has also been analyzed aggressively by numerous other techniques including electron microscopy (Kanaseki *et al.*, 1997; Böttcher *et al.*, 1999; Shangguan *et al.*, 1998), mutagenesis (Steinhauer *et al.*, 1996; Qiao *et al.*, 1998), circular dichroism (Korte *et al.*, 1997), and infrared spectroscopy (Gray and Tamm, 1997, 1998), to highlight only a few of the more recent studies. Consequently, influenza HA has become the paradigm against which models for the mechanisms of all other enveloped virus surface proteins are compared and contrasted. A review article dedicated to influenza HA is recommended (Skehel and Wiley, 2000).

Influenza HA is synthesized in infected cells as a single-chain precursor protein, HA_0, which self-associates into a trimer. HA_0 is posttranslationally cleaved into two subunits, HA_1 and HA_2 (Fig. 2), which remain associated with one another to constitute the mature HA spike, a trimer of heterodimers. The HA complex is brought to the cell surface via the secretory pathway and incorporated into virions, along with a section of cell membrane, as the virus buds from the cell. HA_1 is the subunit distal from the virus envelope, whereas HA_2 contains a hydrophobic region near the carboxy terminus that anchors the HA_1–HA_2 complex in the membrane (Fig. 2). At the extreme N terminus of HA_2 is a segment rich in glycine and hydrophobic amino acids, which is termed the "fusion peptide," for reasons discussed below.

When influenzavirus encounters a fresh target cell, HA_1 on the virus surface binds to its receptor sialic acid, present on the cell surface, and the virus is taken up into the cell by endocytosis. At this point, the virus is still separated by the endosomal membrane from the replication and translation machinery of the cell cytoplasm. Proton pumps in this membrane lower the lumenal pH, activating the virus membrane fusion machinery by initiating a change in the structure of the surface protein complex. Although various proposals for the structure of HA that contributes to membrane fusion have been put forth, most models agree that the fusion peptide is exposed. If target membranes are present during the HA conformational change, the fusion peptide will insert into these membranes (Durrer *et al.*, 1996) and fusion will ensue. If the conformational change is induced artificially in the absence of target membranes, the membrane fusion capability is inactivated (Stegmann *et al.*, 1987).

A. HA_0: *First Stop on the Protein-Folding Landscape*

Influenza HA, in the form of HA_0, begins folding as a single polypeptide chain. It presumably folds into the thermodynamically most stable, kinetically accessible state, although this supposition has not been proven.

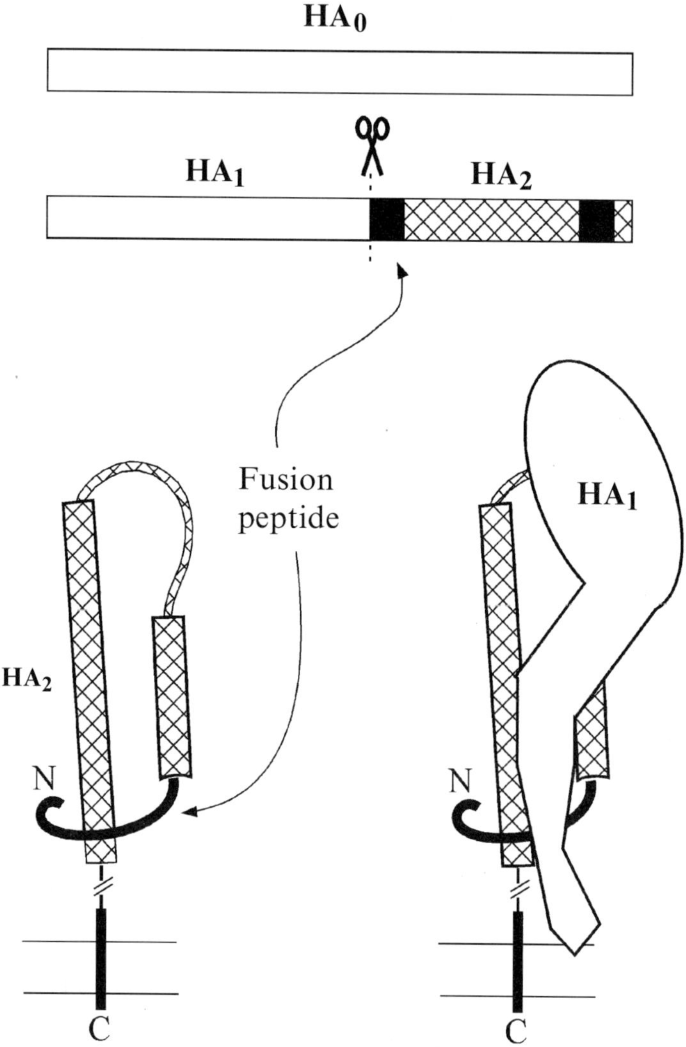

FIG. 2. Primary structure of influenza HA and spatial organization of subunits with respect to the membrane. Cleavage of the influenza HA precursor protein HA_0 yields the two subunits HA_1 and HA_2. HA_1 is white, the fusion peptide and transmembrane segments of HA_2 are black, and the remainder of HA_2 is cross-hatched. For clarity, a monomer of the HA_1–HA_2 assembly is shown. The amino and carboxy termini of HA_2 are labelled "N" and "C," respectively.

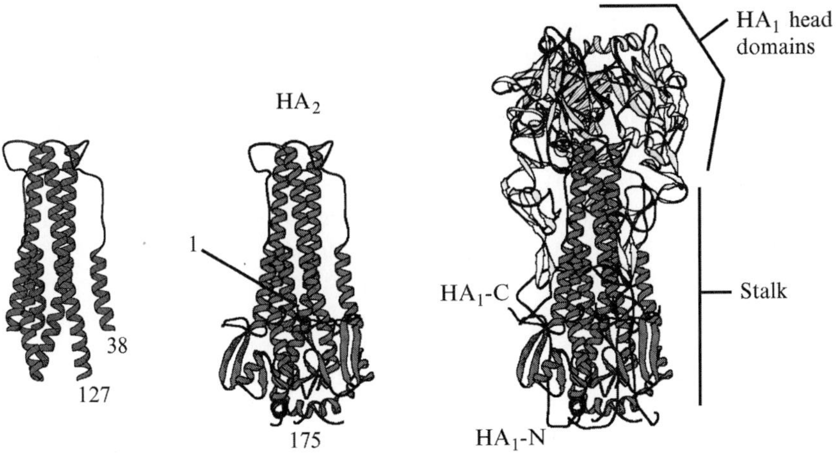

FIG. 3. Hierarchy of the influenza HA trimer structure. *Left*: The underlying helices of HA_2 that make up the core of the stalk, with the amino acid numbers at the termini of this fragment indicated. *Center*: The entire HA_2 ribbon trace. *Right*: Both HA_1 and HA_2 (Wilson *et al.*, 1981). Approximate locations of the amino (HA_1-N) and carboxy (HA_1-C) termini of HA_1 are indicated, illustrating how HA_1 extends the full length of the stalk.

The structure of trimeric HA_0 has been determined crystallographically (Chen *et al.*, 1998). Like cleaved versions of the protein (Wilson *et al.*, 1981; Watowich *et al.*, 1994), HA_0 consists of three eight-stranded "Swiss roll" receptor-binding domains atop a central three-stranded coiled-coil stalk that splays at its base (Fig. 3). The coiled coil is composed of sequences from the middle of HA_2, whereas the β sheet-rich, sialic acid-binding head domain is formed from the middle of the HA_1 sequence. The amino- and carboxy-terminal regions of HA_1, as well as the amino-terminal region of HA_2, drape down along the outside of the stalk. It was suggested, through comparison of influenza A HA with the hemagglutinin esterase from influenza C virus (Rosenthal *et al.*, 1998), that the receptor-binding domain was evolutionarily inserted as a functional unit into a primordial membrane fusion module, which consisted of HA2 plus the amino and carboxy termini of HA1.

B. *Peptide Bond Cleavage and Rearrangement to the Native State of the HA_1–HA_2 Complex*

For many viruses, a posttranslational peptide bond cleavage at a particular site in an envelope protein is required to prime the virus surface for its membrane fusion activity. Influenza HA_0 is thus cleaved proteolytically into the active HA_1–HA_2 complex. The amino acid sequence around the cleavage site determines the point in the virus infection cycle at which cleavage occurs (intracellularly versus extracellularly) and is a primary determinant of virus pathogenicity (reviewed in Steinhauer, 1999).

Comparison of the HA_0 structure (Chen *et al.*, 1998) with the structure of HA_1–HA_2 (Wilson *et al.*, 1981) reveals that, on cleavage, the new HA_1 carboxy terminus and the HA_2 amino terminus are displaced relative to one another by 22 Å. The most significant change is that the newly liberated amino terminus of HA_2 dives into a pocket near the base of the coiled-coil stalk (Fig. 4). Cleavage of HA_0 is required for this alternate packing because there is no room for a polypeptide chain to both enter

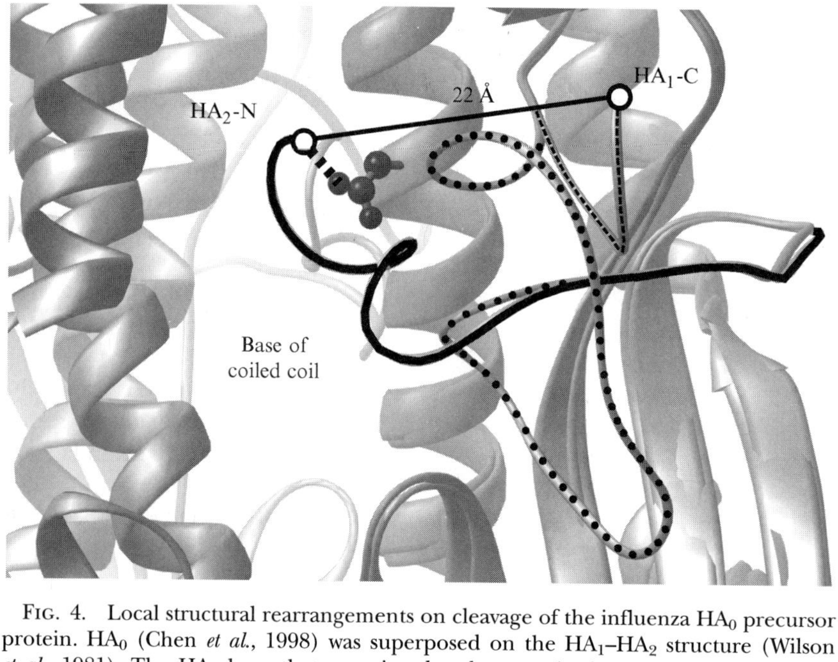

FIG. 4. Local structural rearrangements on cleavage of the influenza HA_0 precursor protein. HA_0 (Chen *et al.*, 1998) was superposed on the HA_1–HA_2 structure (Wilson *et al.*, 1981). The HA_0 loop that contains the cleavage site is marked by small black circles. After cleavage, the HA_2 amino terminus (HA_2-N; open circle) and the HA_1 carboxy terminus HA_1 (HA_1-C; open circle) are displaced. The HA_2 amino-terminal region is indicated by a thick black line, and the carboxy-terminal region of HA_1 by a fine dashed line. The thick dashed line represents a salt bridge between the amino terminus of HA_2 and an aspartate side chain, shown in ball-and-stick representation, from the long helix of HA_2. The distance between the HA_1 carboxy terminus and HA_2 amino terminus is indicated.

and exit the pocket. Furthermore, the free amino group at the amino terminus of HA_2 makes a salt bridge with an aspartic acid in the coiled coil (Fig. 4).

The conformational changes that occur on cleavage of HA_0, although critical for priming HA for membrane fusion, are localized to only 19 residues (6 from HA_1 and 13 from HA_2). All other regions of the HA trimer remain structurally similar (root mean square deviation 0.49 Å) before and after cleavage (Chen *et al.*, 1998). The receptor-binding "head" portion of the complex may act as a clamp that keeps the local structural changes that occur on chain cleavage from propagating (Carr and Kim, 1993).

C. *Refolding on Exposure to Low pH*

Influenza HA undergoes a dramatic conformational change when exposed to low pH (Ruigrok *et al.*, 1986b; White and Wilson, 1987). Structural studies have revealed many molecular details of this conformational change, if not the precise way in which it is coupled to membrane fusion. The demonstration that a peptide derived from a loop in the native HA_2 structure forms a trimeric coiled coil *in vitro* led to the development of the "spring-loaded mechanism" for the HA conformational change (Carr and Kim, 1993). In this model, peptide bond cleavage of HA_0 "expands the horizons" of the structure, opening new regions of conformational space that were inaccessible to the single polypeptide chain (i.e., conformations in which the carboxy terminus of HA_1 and the amino terminus of HA_2 are far apart). However, cleaved HA at neutral pH does not actually sample the full range of the conformational territory because the HA_1 domains trap HA_2 in a metastable, spring-loaded state. Lowering the pH dislodges the fusion peptides and dissociates the HA_1 subunits from the top of the trimer. With the removal of kinetic constraints, the HA_2 coiled coil extends to its full length, catapulting the fusion peptides toward the target cell membrane (Ruigrok *et al.*, 1986b), while relaxing to a thermodynamically more stable conformation (Fig. 5).

Whereas the spring-loaded mechanism is a simple and elegant model for regulation and execution of the influenza HA conformational change, the model leaves the virus and cell membranes 100 Å apart with a hydrophobic sequence inserted into each one but no clear route to membrane fusion. Determination of the structure of a proteolytic fragment from low pH-treated HA ectodomain provided a resolution to this problem (Bullough *et al.*, 1994). In the low pH-converted HA structure, the region of HA_2 that had been the base of the stalk in the native, metastable conformation, bends back to pack against the outside of

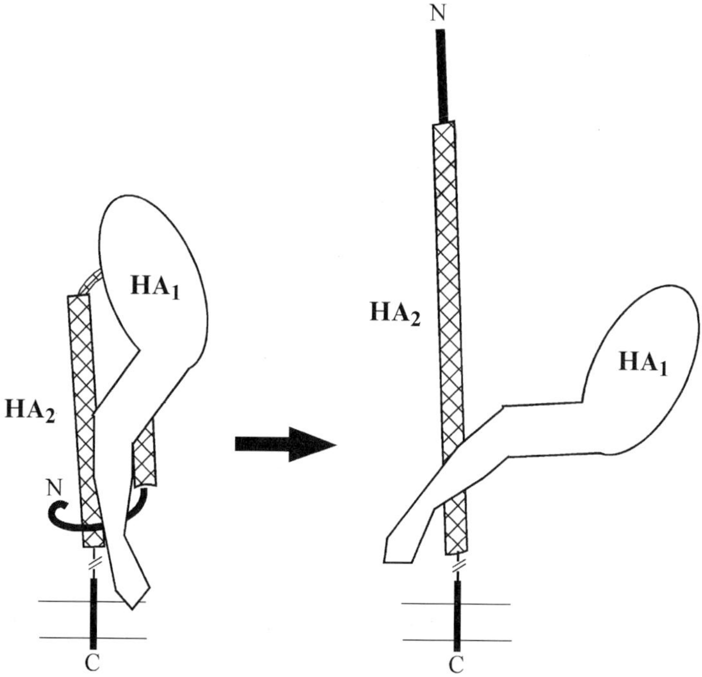

FIG. 5. Spring-loaded mechanism for the influenza HA conformational change. HA_1 domains splay, and regions that had formed a loop and a buttressing helix in the native HA structure are incorporated into the coiled coil. The fusion peptides are thereby hurled to the top of the assembly. A single HA_1–HA_2 heterodimer is shown for clarity.

the bottom of the coiled coil (Fig. 6), abolishing the pocket in which the fusion peptide had been inserted in the native HA structure. A new hydrophobic core forms by reorganizing the regions that had contributed to the pocket. The reversal of the chain direction at the base of the coiled coil brings the fusion peptide and transmembrane segment close together in space. A similar hairpin structure is seen in peptides of viruses from the paramyxovirus family (Section IV), in retroviruses (Section IIIA), and in filoviruses (Weissenhorn *et al.*, 1998; Malashkevich *et al.*, 1999).

It is possible that the structure of HA_2 in its low pH-converted form represents the results of multiple, sequential conformational changes (White and Wilson, 1987; Korte *et al.*, 1999), and that, under certain conditions, intermediate states can be trapped. The first intermediate is likely to be one in which the fusion peptides have emerged from their buried positions. An electron cryomicroscopy reconstruction of HA exposed to low pH for short time periods at 4°C revealed relatively

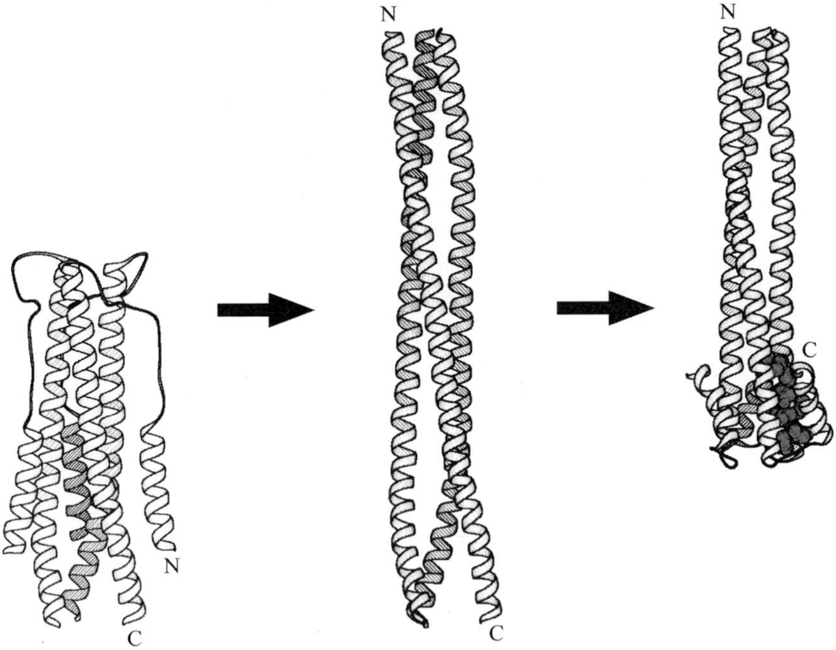

FIG. 6. Repacking of the influenza HA_2 hydrophobic core. *Left*: A ribbon trace of HA_2 residues 38 to 127, including the helices that make up the core of the stalk in the native HA structure (see Fig. 3). *Middle*: A hypothetical structure obtained by fusing the base of the coiled coil from the native HA structure with the top of the extended coiled coil from the low pH-converted HA structure. This panel helps distinguish the two major components of the HA conformational change on low pH treatment; the existence of such an intermediate structure has not been shown experimentally for influenza and may exist only transiently if at all. This extended structure, known as a "prehairpin intermediate," has been detected indirectly in other virus envelope proteins (reviewed in Chan and Kim, 1998). *Right*: Residues 38 to 127 from low pH-converted HA_2 (Bullough *et al.*, 1994). Hydrophobic residues that stabilize the jack-knifed structure are indicated in one protomer as gray space-filling atoms. The amino (N) and carboxy (C) termini of a protomer within each trimer structure are indicated.

small-scale conformational changes, including differences in the stem region near the site of the fusion peptide in the native state of HA (Böttcher *et al.*, 1999). In addition, a study of the kinetics of the HA conformational changes monitored by circular dichroism spectroscopy (Korte *et al.*, 1997) detected transient, reversible structural alterations preceding irreversible conformational changes most likely constituting extension of the HA_2 coiled coil. It is not unreasonable to suppose that extrusion of the fusion peptide might be reversible under certain conditions, as reburial of the fusion peptide would simply mimic the event that occurs on HA_0 cleavage.

After fusion peptide extrusion, the next step of the HA conformational change would be the extension of the trimeric coiled coil and insertion of the fusion peptides into the target cell membrane. This step may be slower than fusion peptide extrusion because it depends on removal of steric constraints imposed by the HA_1 head domains. In fact, preventing HA_1 dissociation altogether by engineering intermolecular disulfide bonds into the HA_1 head region abolishes fusion (Kemble *et al.*, 1992). The third part of the HA conformational change is the jack-knifing of HA_2 and reforming a hydrophobic core at the base (Fig. 6). This step may also be slow if it is coupled to membrane merging, and is likely to be independent of pH (Stegmann *et al.*, 1990).

On low-pH treatment, the HA_2 subunit undergoes the tertiary structural changes described above. HA_1 also changes, but according to quaternary, not tertiary, structure (Ruigrok *et al.*, 1986b). The HA_1 subunits dissociate from one another (White and Wilson, 1987), leaving room for extension of the HA_2 three-stranded coiled coil and harpooning of the fusion peptides. The structure of the influenza HA_1 globular receptor-binding domain has been determined both in the context of the native HA trimer (Wilson *et al.*, 1981; Watowich *et al.*, 1994; Chen *et al.*, 1998) and as a monomer dissociated by low pH, cleaved from the remainder of HA, and complexed with an antibody fragment (Bizebard *et al.*, 1995). Dissociated HA_1 essentially retains the structure it had as part of the HA_1–HA_2 assembly (Bizebard *et al.*, 1995). The rate of HA_1 dissociation is highly sensitive to temperature and pH (Korte *et al.*, 1999), and a class of mutation that allows influenza HA to initiate membrane fusion at higher pH occurs at positions in the interface between HA_1 domains (Wiley and Skehel, 1987). It thus appears that low pH weakens the interactions between HA_1 domains at the top of the trimer, and that these domains then splay by rigid body motion. The role of HA_1 in membrane fusion appears to be solely inhibitory.

D. Energetic Considerations and Models for Membrane Fusion

Considerable effort has been dedicated to identifying the "fusion-active state" of influenza HA. Because the fusion of two opposing membranes to become a single bilayer is a dynamic process, the fusogenic species is most likely to be a transition between discrete structures rather than a single conformation. If such a transition occurs in a manner uncoupled from membrane fusion, the resulting structure would not be fusion active. The structure and stability of this end state are nevertheless relevant to the fusion process, because the free energy difference between the pre- and postfusion states determines the energy available to drive the fusion process (Fig. 7).

The spring-loaded model for influenza HA implies that any means of removing the kinetic constraints on the structure of HA_2, and not just exposure to low pH, could result in a functional conformational change and activation of membrane fusion. In fact, it has been shown that overcoming the kinetic barrier with heat, or lowering the barrier by adding protein denaturants, can promote membrane fusion by HA (Carr *et al.*, 1997). The similarity in protease cleavage patterns of HA after each of these treatments suggests that at least the end state of the conformational change in each case is similar (Carr *et al.*, 1997; see, however, Ruigrok *et al.*, 1986a).

An inability to detect changes in the head region of HA_1 during the slow but efficient membrane fusion that occurs for flu at low temperatures led to the idea that splaying of HA_1, an essential feature of the spring-loaded mechanism, is not a requirement for fusion (Stegmann *et al.*, 1990). However, subsequent studies demonstrated that conversion of only a small subpopulation of HA_2 is sufficient to mediate fusion of flu (Tsurudome *et al.*, 1992) and other enveloped viruses (Bachrach *et al.*, 2000). Depending on the activation energy for the conformational change at low pH (Fig. 7) and its cooperativity, a low level of spontaneous conversion might be expected at 0°C. Care must be taken to distinguish between the bulk properties of the sample and the state of active HA subpopulations (Hughson, 1995).

Assuming that the spring-loaded mechanism applies, a number of models can be used to explain the relationship of the HA conformational change to membrane fusion. In one model, the fusion peptides insert first into the target cell membrane via the conformational change to the extended coiled coil. Either simultaneously or subsequently, jack-knifing of HA_2 brings the virus and cell membranes together (Hughson, 1997). Variations on this theme specify either that refolding of virus envelope proteins actually drives the membranes to fuse (Baker *et al.*, 1999), or that the refolding helps overcome the energy of repulsion of bringing lipid bilayers close together, and that once apposed, they fuse spontaneously (Chan and Kim, 1998). In a second model, some fusion peptides are proposed to insert first into the virus membrane. This phenomenon and its implied orientation of HA with respect to the membrane have been observed experimentally (Ruigrok *et al.*, 1986b; Tsurudome *et al.*, 1992; Weber *et al.*, 1994; Wharton *et al.*, 1995), although it may represent an inactivated state rather than a required step on the pathway to membrane

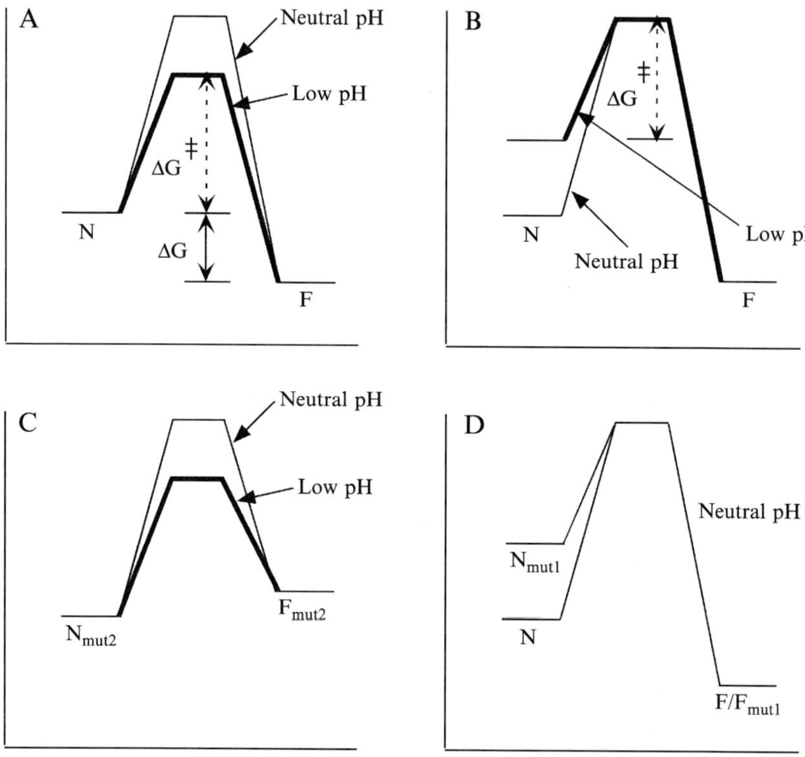

FIG. 7. Kinetic control of influenza HA (Baker and Agard, 1994): effects of low pH and mutations. Two models are presented for how low pH makes the conformational transition of HA favorable. "N" refers to the native structure of HA_1–HA_2 (Wilson *et al.*, 1981). Although "F" stands for "fusion active," this does not imply that the static F structure can induce membrane fusion. (A) Acidic conditions may lower the energy of the transition state. The difference in free energy between N and F is ΔG, the energy available from the conformational change that can be coupled to other processes. (B) Low pH may destabilize the native state of HA. In both models (A and B), the free energy of activation $\Delta G^{\ddagger}$ is smaller under low pH than under neutral pH conditions. However, the models can be distinguished by the difference in free energy between the N and F states as a function of pH. A combination of these two models is also possible. (C) A mutation that affects the stability of the extended coiled coil would destabilize the F state relative to the native state. The effect of such a mutation is shown on the basis of the model in (A), but raising the energy of F in model (B) is equally valid. In either case, destabilizing F does not imply that HA populates N thermodynamically. Shown in this diagram is a two-dimensional slice through a multidimensional energy landscape, and other partially folded or unfolded conformations may predominate if F is destabilized. (D) Mutations can destabilize N such that at neutral pH, the free energy diagram resembles the wild-type profile at low pH, seen in boldface in (B). These mutants would raise the pH at which the conformational transition occurs. Alternatively, mutations could lower the transition state at neutral pH (not shown), such that the diagram would resemble the boldface curve in (A).

fusion. According to this second model, however, the extension of the coiled coil then introduces a defect in the virus membrane, which is repaired by membrane fusion (Bentz, 2000). Alternate models for membrane fusion by flu and other viruses invoke disassembly of coiled coils and interaction of the resulting individual amphipathic helices with the membrane to bring the bilayers together and promote membrane fusion (Yu *et al.*, 1994; Ben-Efraim, *et al.*, 1999; see also Epand *et al.*, 1999).

E. Protein Folding and Spring Loading

Both synthetic peptides (Carr and Kim, 1993) and a larger HA_2 ectodomain fragment produced in *Escherichia coli* (Chen *et al.*, 1999) fold directly into structures resembling the low pH-converted state. So how does influenza HA become spring-loaded? It is noteworthy that, in the context of intact HA_0, the region corresponding to HA_2 is synthesized only after HA_1 is present and perhaps partially folded. Furthermore, HA_2 in the context of HA_0 does not fold with a free amino terminus. Finally, folding of full-length HA in the environment of the endoplasmic reticulum occurs in the presence of chaperones. Therefore, the natural local folding environment of HA_2 differs from the folding environment of soluble ectodomain fragments.

Whereas it is not known precisely how folding in a controlled environment restricts the conformational space accessible to HA_2, a comparison with alphaviruses (discussed further in Section V) is revealing. The alphavirus E1 protein on the virion surface executes membrane fusion during infection. E1 newly synthesized in an infected cell, however, forms a heterodimer with the p62 protein, which represses membrane fusion. The p62 protein is cleaved in the late Golgi apparatus, sensitizing the system to low pH. E1 and p62 are naturally synthesized as distinct polypeptides, but from a single coding unit, with p62 preceding E1. The p62 protein can fold and be translocated from the endoplasmic reticulum to the Golgi on its own. In contrast, E1 aggregates, even when coexpressed in the same cell with excess p62, unless it is produced from the same transcript immediately following p62 (Andersson *et al.*, 1997). By analogy, coexpression of influenza HA_1 and HA_2 as the polyprotein HA_0 may ensure that folding of the flu fusion-active subunit is subjugated to a fusion-inhibitory HA_1 domain.

In addition to folding as a single-chain "heterodimer," the HA_1–HA_2 complex undergoes transient associations with other proteins, namely chaperones, followed by a more permanent self-association, in its endoplasmic reticulum (ER) folding environment. It need not be supposed that chaperones actively induce folding of HA into a metastable state to appreciate the profound affect they may have on the ability of HA to achieve a spring-loaded structure. The binding of calnexin and calreticulin to N-linked glycans near the amino terminus of HA_1 affects the rate of folding of the glycoprotein and is consistent with a model in which HA folds "top-down," beginning with the globular head region and progressing toward the stem (Hebert *et al.*, 1997). Chaperone binding to the stem may help keep this region in an extended but nonaggregated conformation.

If chaperones protect monomeric folding intermediates from aggregation or premature assembly (Hebert *et al.*, 1996), then initial folding of influenza HA_0 may occur in a manner that maximizes intramolecular contacts between the HA_1 and HA_2 regions in the absence of competition from intermolecular contacts. Notably, HA_0 trimerization does not occur cotranslationally (Chen and Helenius, 2000), and HA_0 intermediates with disulfide bonds that are incompletely oxidized do not trimerize (Braakman *et al.*, 1991). Trimerization is therefore a relatively late event in the HA-folding pathway. Although the HA trimer seems to be dominated by the central coiled coil, this is a misconception. The HA2 helices in the native structure are long ($\sim$52 residues), but the carboxy-terminal ends of these helices splay from the trimer axis such that the region actually forming a coiled coil is only about half this length. This short coiled coil can be considered an "afterthought" of native HA, and a monomeric folding intermediate may be an adequately stable intermediate structure. This suggestion does not imply that the HA_0 trimer, once formed, is unstable. In fact, inspection of the HA_0 structure suggests that many interprotomer interactions in addition to the coiled coil are likely to stabilize the trimer. Instead, the argument presented is that stable quaternary structure formation (trimerization) follows the acquisition of tertiary structure within an HA_0 monomer. This situation contrasts dramatically with the fold of low pH-converted HA. The trimeric coiled coil of "sprung" HA is at least 66 residues long and is the primary stabilizing feature of the assembly. Because tertiary structure in this case relies absolutely on trimerization, the primary contribution of chaperones to spring loading may be in maintaining a monomeric state during the folding of HA such that the sprung structure cannot compete against the spring-loading folding route.

III. Retroviruses

Retroviruses are most familiar to the general reader as the family to which the lentivirus human immunodeficiency virus (HIV) belongs. Other retrovirus groups include the avian sarcoma and leukosis viral group

(ASLV), the murine leukemia-related group (MuLV), and the human T cell leukemia viruses (HTLVs). Understanding the conformational changes that retrovirus surface proteins undergo during cell entry is important in the development of drugs and vaccines (Doms and Moore, 2000; Doms and Trono, 2000; Poignard *et al.*, 2001). In addition, retroviruses, because of their ability to insert their genomes into the DNA of the infected host, have been developed as tools for gene therapy (Kordower *et al.*, 2000; reviewed in Lever, 2000). Manipulating the cell entry process to target engineered retroviruses to the cell of choice is an essential step in designing a successful gene therapy vector. For both HIV vaccine and gene therapy development, insights into the structures and functions of the retrovirus envelope proteins exposed fundamental flaws in the initial approaches to these problems (LaCasse *et al.*, 1999; Zhao *et al.*, 1999). Structural studies on retrovirus envelope proteins have also suggested good targets for the design of drugs against HIV (Eckert *et al.*, 1999).

The surface proteins of retroviruses are synthesized, like influenza, as precursors that are cleaved proteolytically into two subunits, the surface (SU) subunit and the transmembrane (TM) subunit (Fig. 8) (reviewed in Hunter and Swanstrom, 1990). The SU subunits generally contain the receptor recognition determinants in their amino-terminal regions, and their carboxy-terminal regions interact with TM. The TM subunits resemble influenza HA_2 in that they have amino-terminal fusion peptides followed by coiled coils, and transmembrane segments near the carboxy termini. The trimeric assembly of the retrovirus SU–TM complex is often referred to as "Env."

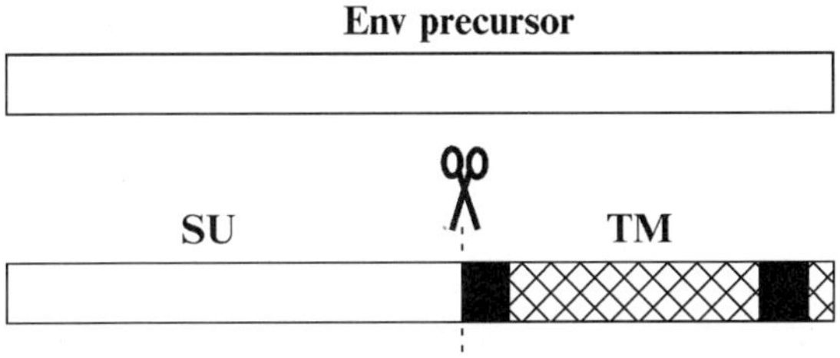

FIG. 8. Primary structure of retrovirus envelope proteins. Cleavage of retrovirus envelope proteins yields the two subunits SU and TM. SU is white, the fusion peptide and transmembrane segments of TM are black, and the remainder of TM is cross-hatched.

A. *Membrane Fusion with Coiled Coils at the Core*

Similarly to peptides from influenza HA_2, soluble retrovirus TM fragments lacking the fusion peptide and transmembrane regions fold independently into stable structures when produced in bacteria or as synthetic peptides. Structures of such fragments from simple (Fass *et al.*, 1996) and complex (Kobe *et al.*, 1999) oncoretroviruses, as well as lentiviruses (Chan *et al.*, 1997; Weissenhorn *et al.*, 1997; Caffrey *et al.*, 1998; Malashkevich *et al.*, 1998), have been determined crystallographically or by nuclear magnetic resonance (NMR). A comparison among these structures reveals a remarkable coherence (Weissenhorn *et al.*, 1999). Not only are the structures of the retrovirus TM fragments similar to one another and to the low pH-converted flu HA_2, but their energetics are similar. The midpoints of the thermal melting curves for all TM ectodomains tested are greater than 85°C (Fass and Kim, 1995; Lu *et al.*, 1995; Blacklow *et al.*, 1995).

The first retrovirus TM structure determined was of a protease-resistant, stable core from Moloney murine leukemia virus (Mo-MuLV) TM (Fass *et al.*, 1996), which is almost identical to HTLV-1 TM (Kobe *et al.*, 1999) (Fig. 9). The amino-terminal regions of these structures form three-stranded coiled coils. At the bases of the coiled coils, the chains reverse direction, form a single turn of helix perpendicular to the coiled coil, and pack back against the outside of the coiled coils with the carboxy termini pointing in the direction of the amino termini. The bases of the coiled coils are stabilized by hydrophobic cores (Fass and Kim, 1995; Fass *et al.*, 1996; Kobe *et al.*, 1999), analogous to the repacked conformation of the base of the coiled coil in low pH-converted influenza HA (Fig. 6). Three cysteine residues, forming the pattern CX_6CC with the first two cysteines disulfide bonded, lie in the region of the chain direction reversal. Although the crystallized fragment of Mo-MuLV TM was truncated 35 residues before the predicted transmembrane segment, circular dichroism studies showed that the missing region is likely to contain helical residues (Fass and Kim, 1995). The HTLV-1 TM structure contains an additional 16 residues at the carboxy terminus, including a short helix that extends the chain in the direction of the amino terminus of the domain (Kobe *et al.*, 1999). Key sequence features, such as the heptad repeat of hydrophobic residues in the coiled-coil region (Delwart *et al.*, 1990) and the CX_6CC motif (Schulz *et al.*, 1992), are present in most other retroviral groups including ALSV and the D-type mammalian leukemia viruses (e.g., Mason–Pfizer monkey virus). These viruses are therefore also likely to adopt a structure similar to Mo-MuLV and HTLV-1 TM.

The TM subunit, gp41, from the lentivirus HIV does not share sequence homology with TM proteins from the oncoretroviruses. However, HIV

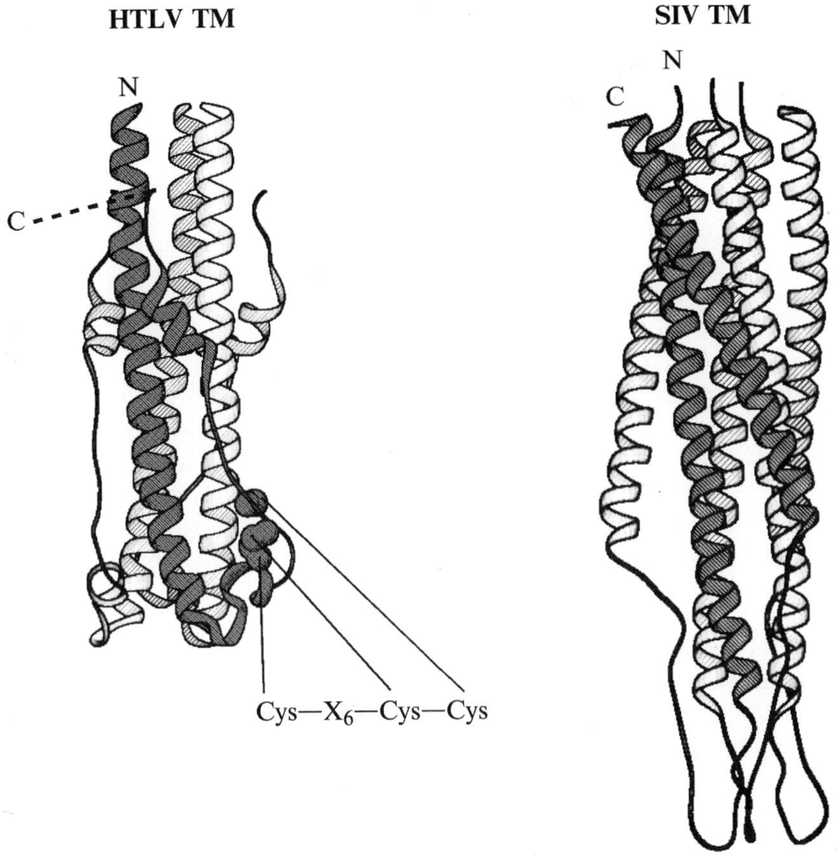

FIG. 9. Structures of soluble fragments of retrovirus TM proteins. Murine leukemia virus (MLV; Fass *et al.*, 1996) and filovirus (Weissenhorn *et al.*, 1998; Malashkevich *et al.*, 1999) TM subunits are represented on the left by the HTLV TM structure (Kobe *et al.*, 1999), with which they share remarkable similarity. Human and simian lentivirus TM subunits (Weissenhorn *et al.*, 1997; Chan *et al.*, 1997; Caffrey *et al.*, 1998) are represented by the structure of SIV TM on the right. Both structures are hairpins containing central three-stranded coiled coils surrounded by buttressing regions that pack into the grooves on the outsides of the coiled coils. The amino acid side chains in the conserved cysteine-rich motif of HTLV TM are shown as space-filling atoms and labeled according to their positions in the motif.

gp41 also has a heptad repeat of hydrophobic residues. A complex of peptides derived from a region downstream of the fusion peptide (N-helix) and a segment upstream of the transmembrane region (C-helix) form a six-helix bundle complex (Lu *et al.*, 1995; Blacklow *et al.*, 1995). The three N-helices form a coiled coil at the core, and the C-helices pack into grooves on the surface of the N-helix assembly (Chan *et al.*, 1997; Weissenhorn *et al.*, 1997). Like the oncoretroviruses, lentiviruses have a disulfide-bonded loop in the region where the chain reverses direction at the base of the coiled coil. However, this loop was omitted for crystallization studies to improve solubility (Lu *et al.*, 1995; Chan *et al.*, 1997). The NMR structure of the SIV gp41 ectodomain fragment retains a version of this loop in which the cysteine residues were mutated to alanines (Caffrey *et al.*, 1998) (Fig. 9).

No structures have yet been determined for TM subunits in complex with SU subunits. Therefore, a "spring-loaded" TM has not yet been visualized for retroviruses. Nevertheless, there are some indications that the retrovirus coiled-coil structures may represent the end points of conformational changes and that these regions pack into a different structure, or have a different significance, in the native conformation of the SU–TM complex. Mutagenesis experiments in which retroviral TM residues were substituted with amino acids predicted to destabilize coiled coils demonstrated that initial folding and transport of virus glycoproteins were not affected, but infectivity decreased (Wild *et al.*, 1994a; Ramsdale *et al.*, 1996). In addition, mutations of hydrophobic residues in HTLV-1 TM that pack against the outside of the base of the coiled coil, or of conserved glycine residues that may sterically permit the chain reversal to the helical hairpin structure, produce Env that can fold and assemble to the native state but not accomplish membrane fusion (Maerz *et al.*, 2000). Mutations of residues in regions that are recruited to extend the coiled coil of influenza HA on low-pH conversion have a similar phenotype (Qiao *et al.*, 1998). These studies are consistent with a model in which stable coiled-coil formation is required for membrane fusion activation or execution, but not for activities of the native retrovirus Env.

A second class of experiment indicates that the helical bundle of HIV TM is not formed in the native state of its Env complex. Peptides corresponding to the C-helix, and to a lesser extent the N-helix, of the HIV gp41 six-helix bundle can block infection of cultured cells or inhibit cell–cell fusion (Jiang *et al.*, 1993; Wild *et al.*, 1994b). These peptides are presumed to act by binding to their partner regions in gp41 in a dominant-negative manner, competing with the natural, intramolecular association of amino- and carboxy-terminal helices. This observation implies that the six-helix bundle is not formed before, but rather assembles during a step critical for membrane fusion and virus entry (reviewed in Chan and Kim, 1998).

One major question, along the lines of the debate on what constitutes the fusion-active state of influenza HA, is whether helical hairpin formation in retrovirus TM proteins occurs before or concomitant with membrane fusion. This question was addressed for HIV Env by trapping an intermediate structure in a cell–cell membrane fusion assay by lowering the temperature (Melikyan *et al.*, 2000). This intermediate state could be maintained for hours at 23°C, and when the temperature was subsequently raised to 37°C fusion would occur rapidly. A lag period before fusion observed on mixing the cells at 37°C was not observed on raising the temperature of cells coincubated first at 23°C, indicating that the arrested state had actually progressed functionally toward membrane fusion. The arrested fusion intermediate was sensitive to inhibition by N- and C-helix peptides, however, suggesting that a six-helix bundle had not yet formed. When lysophosphatidylcholine (LPC), a reagent that associates with membranes and inhibits their fusion, was added to the membranes and the temperature was raised to 37°C, no fusion occurred. The key observation was made on removal of the LPC: membrane fusion could still be blocked with N- and C-helix peptides. This observation indicates that preventing fusion had also prevented helical hairpin formation, which would otherwise have occurred at this temperature. Had helical hairpins already formed during the 37°C incubation in the presence of LPC, fusion after removal of the LPC would have been resistant to the peptides. Thus, blocking membrane fusion also blocked helical hairpin formation and vice versa, suggesting that these two events are tightly coupled.

B. Avian Leukosis Virus and Receptor-Induced pH Dependence

Structural studies on TM peptides and peptide competition experiments have furthered our understanding of the retrovirus membrane fusion machinery. But how is membrane fusion activity recruited during infection? Unlike influenza HA, many retrovirus envelope proteins do not have a requirement for low pH for their activities. Because conformational changes are detected in the surface protein of retroviruses on receptor binding (Sattentau and Moore, 1991; Sattentau, 1993), it is presumed that receptor binding per se induces the fusion activity of retroviruses.

Testing the biophysical effects of receptor binding on retrovirus envelope proteins is complicated by the fact that most mammalian retroviruses use multipass transmembrane proteins as their receptors (reviewed in Overbaugh *et al.*, 2001). These receptor proteins are difficult to produce in high yields and to purify, making it a formidable task to use receptor to achieve quantitative conversion of retrovirus envelope proteins from a native to a fusion-active state, as can be done for influenza HA simply by lowering the pH.

Exceptions to the general rule of deeply membrane-imbedded retrovirus receptors are the receptors of the ASLV group. These viruses recognize receptors that are single-pass transmembrane proteins. For example, the subgroup A ASLV receptor, Tva, is a protein of unknown function that is homologous to the lipoprotein binding domains of the low-density lipoprotein receptor (Bates *et al.*, 1993). The ability to manipulate ASLV Env with the soluble ectodomain of this receptor made it possible to examine the structural and functional consequences of receptor binding in this system. An overexpressed domain of Tva induces specific conformational changes in the soluble ectodomain of ASLV Env as detected by a change in protease sensitivity and promotion of an interaction between ASLV Env and liposomes (Hernandez *et al.*, 1997; Damico *et al.*, 1998). Soluble Tva can also be used to promote infection of cells lacking membrane-bound Tva (Damico and Bates, 2000). The conformational changes detected on Tva binding differ from those occurring on general destabilization of the Env structure, and mutations in the SU subunit can uncouple receptor binding from productive conformational changes (Damico *et al.*, 1999).

Surprisingly, the ASLV group proved to be an exception to the retrovirus rules in more than one sense. Receptor binding does not promote virus entry directly, but rather induces a structural change that sensitizes the envelope protein to low pH. Agents that prevent acidification of the endosome block entry of subgroups A and B ASLV into cells, but postbinding steps can proceed when these chemicals are removed (Mothes *et al.*, 2000). Furthermore, although low-pH pretreatment alone is not sufficient to inactivate ASLV Env (Gilbert *et al.*, 1990), lowering the pH in the presence of soluble receptor rapidly renders the virus envelope proteins incapable of subsequently mediating virus entry into cells (Mothes *et al.*, 2000). Because addition of soluble receptor would be expected to block virus entry in any case by competing for receptor on the cell surface, these investigators made use of a fusion protein containing both soluble Tva receptor and epidermal growth factor (EGF) (Snitkovsky and Young, 1998). Viruses prebound to this fusion protein were able to enter cells expressing the EGF receptor, even though their Tva-binding sites were already occupied. A mechanism involving receptor-induced pH sensitivity of these viruses explains the observation that ASLV, unlike many retroviruses, does not cause cell–cell fusion when expressed on the cell surface (Hernandez *et al.*, 1996). Cell–cell fusion induced by ASLV Env requires lowering the pH of the cell culture (Mothes *et al.*, 2000).

C. *The Receptor-Binding Cascade of Human Immunodeficiency Virus*

Like ASLV, HIV has a multistep mechanism for activation of membrane fusion. However, low pH is not required for HIV entry, and the virus instead uses a series of distinct interactions with components of the target membrane (reviewed in Doms and Trono, 2000). The first interaction is with the CD4 protein. This binding event enables a subsequent contact between the HIV SU subunit gp120 and a molecule of the chemokine receptor family (reviewed in Choe *et al.*, 1998). It is this second interaction with molecules termed "coreceptors" that activates the membrane fusion potential of the TM subunit gp41 (reviewed in Berger *et al.*, 1999).

Although binding of HIV to its first receptor, CD4, is not sufficient for induction of membrane fusion, CD4 binding is nevertheless a significant leg on the conformational journey of gp120/gp41 toward a fusion-active state. The energetic effects of binding of gp120 to CD4, measured by titration calorimetry, revealed an unusually large enthalpy of interaction between gp120 and CD4, as well as a quite unfavorable entropy change (Myszka *et al.*, 2000). These results were interpreted to mean that a conformational changed occurred involving the burial of 10,000 $Å^2$, the ordering of segments of the gp120 structure amounting to approximately 100 residues, or some combination of these two possibilities. An inspection of the crystallographic structure of gp120 in complex with CD4 and a fragment of a neutralizing antibody (Kwong *et al.*, 1998) supports the idea of a CD4-induced or -stabilized acquisition of structure in what is known as the "bridging sheet." Stability of this sheet appears to rely heavily on the presence of CD4, which directly contacts it (Fig. 10). Folding of the 54 residues that make up the bridging sheet, plus the 4500 $Å^2$ of surface area predicted to become buried somewhere within gp120 itself or against CD4, was the distribution proposed to account for the observed enthalpy changes (Myszka *et al.*, 2000). Verification of the structural effects of CD4 binding must await solution of the gp120 structure in the absence of CD4.

Unlike the large-scale conformational change of influenza HA on low-pH treatment, the CD4-induced conformational changes of HIV gp120 appear to be reversible (Doranz *et al.*, 1999). Mutants of gp120 that allow the virus to infect cells in a CD4-independent manner bind antibodies that recognize epitopes normally shielded in CD4-dependent virus strains (Hoffman *et al.*, 1999). Therefore, an equilibrium may exist between an open form of gp120 that can bind coreceptor and a closed form that cannot. CD4 stabilizes the open form, but mutations in gp120 itself can shift the equilibrium between the closed and open forms. If the open form is populated a significant fraction of the time, the virus can proceed

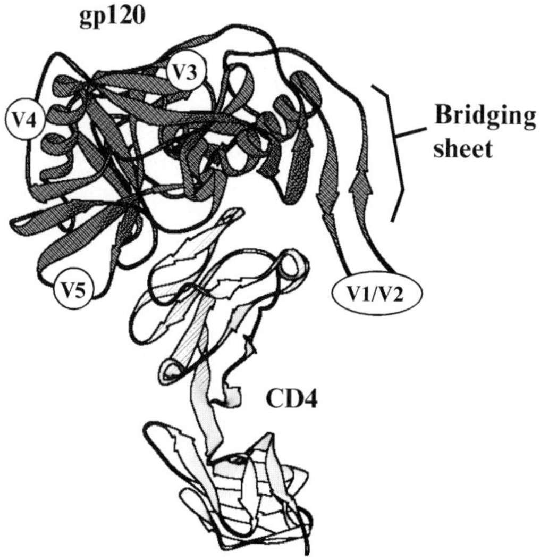

FIG. 10. Structure of trimmed HIV gp120 in complex with two-domain CD4 (Kwong *et al.*, 1998). Shown in darker shading at the top is the structure of an engineered gp120 protein in which the V1/V2, V3, and V4 loops were truncated to generate a crystallizable fragment. The positions of truncated loops are indicated on the structure. CD4, in lighter shading, contacts gp120 at the edge of the bridging sheet.

directly to coreceptor encounters. The structural effects of coreceptor binding, and how this binding event promotes membrane fusion, are more difficult to address. It is this second step that may lead to irreversible structural changes in gp41 and a commitment to membrane fusion (Gallo *et al.*, 2001).

D. *Mammalian Oncoretroviruses and Fusion Activation* in Trans

By analogy to the flu mechanism, one model for receptor-induced activation of structural changes and fusion activity in retrovirus is that receptor binding weakens the interactions between receptor-binding domain protomers at the top of the Env trimer. These receptor-binding domains are then shed, relatively intact, from the virus surface, relieving constraints on the underlying fusion machinery. Although the actual mechanism is far from understood, experiments on MuLVs suggest that the model of inhibitory receptor-binding domains, which currently appears to apply well to influenza, is either incomplete or inaccurate for retroviruses.

MuLVs, a subset of the oncoretroviruses, are viruses that are clearly related to one another evolutionarily, but enter cells by different receptors and therefore have different host ranges according to conservation of receptor sequences across species (Battini *et al.*, 1992). For example, ecotropic MuLVs, such as Moloney murine leukemia virus, infect only murine cells, whereas amphotropic MuLVs infect a wider range of species including humans. Provocatively, all the MuLV receptors identified to date (Overbaugh *et al.*, 2001) are multipass transmembrane proteins, like the HIV coreceptors. The significance of this fact is not yet known. However, some investigators have suggested that virus entry requires an additional factor (Wang *et al.*, 1991) that may be associated with this class of molecule, or that the receptors lie in distinct lipid environments that are important for membrane fusion during virus entry (Lu and Silver, 2000).

The MuLV envelope protein complex can be divided into three primary regions. A domain at the amino terminus of the SU subunit folds and is active in receptor binding when produced in isolation (Battini *et al.*, 1995; Davey *et al.*, 1997). The carboxy-terminal region of SU is characterized by a CXXC motif reminiscent of the active sites of enzymes like thioredoxin and protein disulfide isomerase. The function of this carboxy-terminal domain is not known, but the CXXC motif links SU via a disulfide bond to a CX_6CC motif in the third region of MuLV Env (Pinter *et al.*, 1997), the TM subunit, of which a representative structure is known (see Section III.A).

Structures of the receptor-binding domains from two distinct leukemia retroviruses are also known (Fass *et al.*, 1997; Barnett *et al.*, 2003) (Fig. 11). These structures, from the MuLV Friend murine leukemia virus and the subgroup B feline leukemia virus, consist of an antiparallel β sandwich with two extended interstrand loops that fold together to form a distinct lobe or subdomain (Fig. 11). This subdomain is likely to make direct contacts with the virus receptor, because viruses from this group that use distinct receptors vary primarily in the sequence (Battini *et al.*, 1992) and structure (Fass *et al.*, 1997; Barnett *et al.*, 2003) of this region. In addition, mutations that decrease receptor binding have been identified in these loops (Davey *et al.*, 1999; Panda *et al.*, 2000; Tailor *et al.*, 2000). No direct structural information is available to indicate how MuLV receptor-binding domains interact with receptors, or what effect this interaction has on the structure of these domains. However, considering the constraints on the receptor-binding domain structure imposed by disulfide bonds and the apparent solidity of the hydrophobic core throughout most of the β sandwich, no obvious schemes for intradomain structural changes are apparent.

In any case, receptor binding is clearly not sufficient to promote entry of MuLVs into cells. Extensive attempts to develop cell-specific targeting

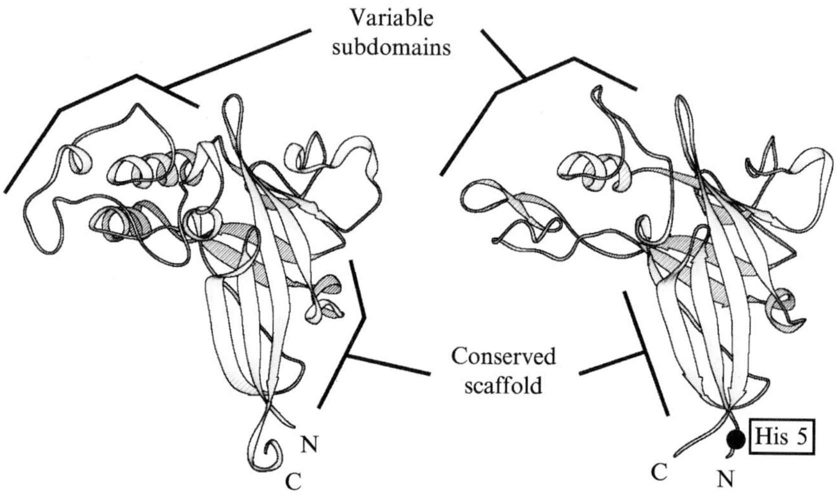

FIG. 11. Structures of oncoretrovirus receptor-binding domains. *Left:* The receptor-binding domain of Friend murine leukemia virus (Fass *et al.*, 1997). *Right:* The homologous domain from subgroup B feline leukemia virus (Barnett *et al.*, 2003). The positions of the variable subdomains and conserved β-sandwich scaffold are indicated. The histidine residue critical for receptor binding-induced activation of membrane fusion by this virus group is indicated at the extreme amino terminus in the feline leukemia virus structure, but this residue was not seen in the electron density of the Friend receptor-binding domain.

systems for gene therapy by modifying MuLVs eventually drove leaders in the field to the following conclusion: a postbinding block exists in engineered Env complexes, such that they cannot undergo the proper conformational changes to progress to membrane fusion (Zhao *et al.*, 1999). Interestingly, mutations of single amino acids in MuLVs can have a similar effect of uncoupling receptor binding from membrane fusion. For example, MuLVs have a highly conserved histidine residue near the N terminus of their SU subunits. Mutation of this histidine abolishes membrane fusion activity without affecting receptor binding (Bae *et al.*, 1997). The two structures of the MuLV receptor-binding domains indicate that this histidine lies on the flexible end of the first strand, outside the globular fold of the domain (Fass *et al.*, 1997; Barnett *et al.*, 2003) (Fig. 11). This position should play no role in the structure or stability of the receptor-binding domain itself, but may be involved in arranging the quaternary structure of the Env assembly and in coupling receptor binding to subsequent events in the cell entry pathway (Bae *et al.*, 1997).

The existence of SU mutants that are fully functional for receptor binding but fail to progress to membrane fusion does not in itself undermine a model in which the receptor-binding domains are purely fusion inhibitory. However, a striking study in which wild-type receptor-binding domain added *in trans* rescued infection by the critical histidine mutant (Lavillette *et al.*, 2000) indicates that the fusion machinery of the mutant is intact but not accessed. In a further observation along these lines, a retrovirus surface protein with a complete deletion of the receptor-binding domain was incorporated at a reduced level into virions, and was fusion competent when the receptor-binding domain was added *in trans* (Barnett *et al.*, 2001). These observations rule out models in which the receptor-binding domains simply sit as clamps on the membrane fusion machinery. Rather, the N-terminal domain of SU has a specific and active role in the recruitment of this machinery, as well as in receptor recognition. It has been proposed that the target of this domain's postbinding activity is some feature of the cell, and that binding makes the cell permissive for virus entry (Lavillette *et al.*, 2000). However, it has also been argued that the receptor-binding domain activates membrane fusion by interacting with the carboxy-terminal region of SU on the virus (Barnett *et al.*, 2001; Lavillette *et al.*, 2001).

Not only has *trans*-activation of membrane fusion with soluble receptor-binding domains been observed in experimental settings in multiple laboratories (Lavillette *et al.*, 2000; Barnett *et al.*, 2001), it also has a natural parallel. In a search for the cell surface receptor used by a T cell-tropic feline leukemia virus (FLV), a receptor-binding domain from an endogenous FLV fragment (the remnants of a virus that once integrated into the DNA of an ancestral animal, became incorporated into the germline, and is now encoded in the cat genome) was revealed as a coreceptor for virus entry (Anderson *et al.*, 2000). This soluble, secreted receptor-binding domain is expressed only in T cells and is necessary for infection by this virus. A classic, multipass transmembrane protein is also required for infection. An inspection of the SU sequence of the T cell-tropic FLV revealed that it was missing the N-terminal histidine residue critical for progression to membrane fusion in the absence of exogenous receptor-binding domain (Donahue *et al.*, 1991; Lavillette *et al.*, 2001).

The activity of soluble MuLV receptor-binding domains in rescuing membrane fusion resembles in some ways the role of CD4 in HIV infection (Anderson *et al.*, 2000). High concentrations of CD4 can inhibit lentivirus entry into cells, but lower concentrations can actually activate viruses for cell entry (Allan *et al.*, 1990). Rescue of fusion by deficient MuLVs displays a biphasic curve as a function of receptor binding-domain concentration, first increasing, and then decreasing as the receptor-binding domain concentration is raised (Barnett and Cunningham, 2001).

Trans-activation by retrovirus receptor-binding domains appears to rule out a flu-like model in which steric inhibition of underlying fusogenic structures is relieved by receptor binding. However, these experiments on retroviruses may actually enhance our understanding of the folding and function of flu HA itself. The folding, albeit to a lesser extent, and activity of the truncated MuLV Env produced without its receptor-binding domain (Barnett *et al.*, 2001) supports the idea of a primordial membrane fusion module (Rosenthal *et al.*, 1998; Weissenhorn *et al.*, 1998). The gross organization of MuLV Env is analogous to that of influenza HA. In HA, the receptor-binding domain sits atop a vestigial esterase domain fragment, which in turn rests on the HA_2 fusion machinery (Rosenthal *et al.*, 1998). In MuLV Env, the receptor-binding domain is adjacent in sequence to the SU region containing the CXXC motif, which may also be the vestige of a thiol reductase/oxidase enzyme or have some as yet unidentified enzymatic activity. This C-terminal SU domain contacts TM, at least via a disulfide bond (Opstelten *et al.*, 1998) and most likely through noncovalent interactions as well.

It is possible, then, that the true inhibitory regions of MuLV SU are the carboxy-terminal CXXC domains, and that receptor binding induces a change in the structure or orientation of the receptor-binding domain, which is propagated to the C-terminal SU domain (Ikeda *et al.*, 2000; Lavillette *et al.*, 2001). Because the CX_6CC motif in MuLV TM lies at the base of the coiled coil, the intersubunit disulfide bond between SU and TM may have a role in constraining the prefusogenic TM structure so that the helical hairpin and its stabilizing hydrophobic core cannot form (Maerz *et al.*, 2000). A thiol disulfide rearrangement can eliminate this intersubunit disulfide bond (Pinter *et al.*, 1997). It remains to be seen whether this thiol disulfide rearrangement is involved in derepression of the TM fusion activity.

IV. Paramyxoviruses Turn Paradigms Upside Down?

Paramyxoviruses cause respiratory tract diseases such as croup and pneumonia, as well as measles and mumps. The envelope proteins of these viruses share some features in common with influenza and retroviruses. These similarities include a precursor protein that is cleaved into two fragments, the second of which, called F1, bears a fusion peptide at its amino terminus. In addition, peptides from the paramyxovirus F1 proteins assemble into stable helical bundles resembling HIV gp41 and influenza HA_2 (Baker *et al.*, 1999; Lawless-Delmedico *et al.*, 2000; Zhao *et al.*, 2000). The paramyxovirus F protein differs from influenza HA and retroviral TM proteins, however, in that the central and buttressing helices in the helical bundles are not formed from regions close by in the primary structure, but instead separated by more than 250 residues.

The paramyxovirus Newcastle disease virus became the second example of an enveloped virus spike structure to be determined crystallographically (Chen *et al.*, 2001a). This study was plagued by technical problems including degradation of the protein in the crystals over time, lack of high-resolution diffraction data, low-resolution experimental phases, and a purportedly inhomogeneous distribution of conformations in the crystal. Therefore, an interpretation of this structure must proceed with caution, particularly because the Newcastle disease virus spike protein seems to turn current models for conformational changes in virus envelope proteins upside down.

One familiar feature in the Newcastle disease virus F protein (NDV-F) structure is a long coiled coil comprising residues carboxy terminal to the fusion peptide. Because of proteolytic cleavage of protein in the crystals and disorder in packing of the end of the long coiled coil, the fusion peptide and the region immediately downstream were not seen in the structure. However, approximately 55 residues after the cleavage site that generates the free amino terminus of the fusion peptide, the coiled coil could begin to be traced (Fig. 12). The coiled coil in this region overlaps with the structure of a complex of peptides from the paramyxovirus SV5 (Simian virus 5) (Baker *et al.*, 1999), but the buttressing helices are missing. Toward the carboxy terminus of the NDV-F long coiled coil, an extended strand packs in the groove between helices, as seen in the SV5 peptide structure (Fig. 12).

The authors of the NDV-F structure propose that they have crystallized a mixture of the native and "sprung" forms of the spike protein, and that these two versions are similar enough structurally to pack into the same crystal lattice (Chen *et al.*, 2001a). In light of the enormous structural changes that are observed on exposure of influenza HA to low pH, the possibility that minimal conformational changes occur in paramyxoviruses is remarkable. Furthermore, the long central helix of NDV-F is proposed to point in the opposite direction from that of influenza HA, such that extension of the central coiled coil sends the fusion peptides in the direction of the virus membrane (Fig. 12).

Although the NDV-F structure is provocative, it is not clear to which functional state the structure truly corresponds. Even assuming that the trace of the polypeptide chain faithfully represents the conformation of the majority of the molecules in the crystals, there are reasons to suspect that this structure may not be the native state of the protein. In particular, electron microscopy of another paramyxovirus F protein revealed a

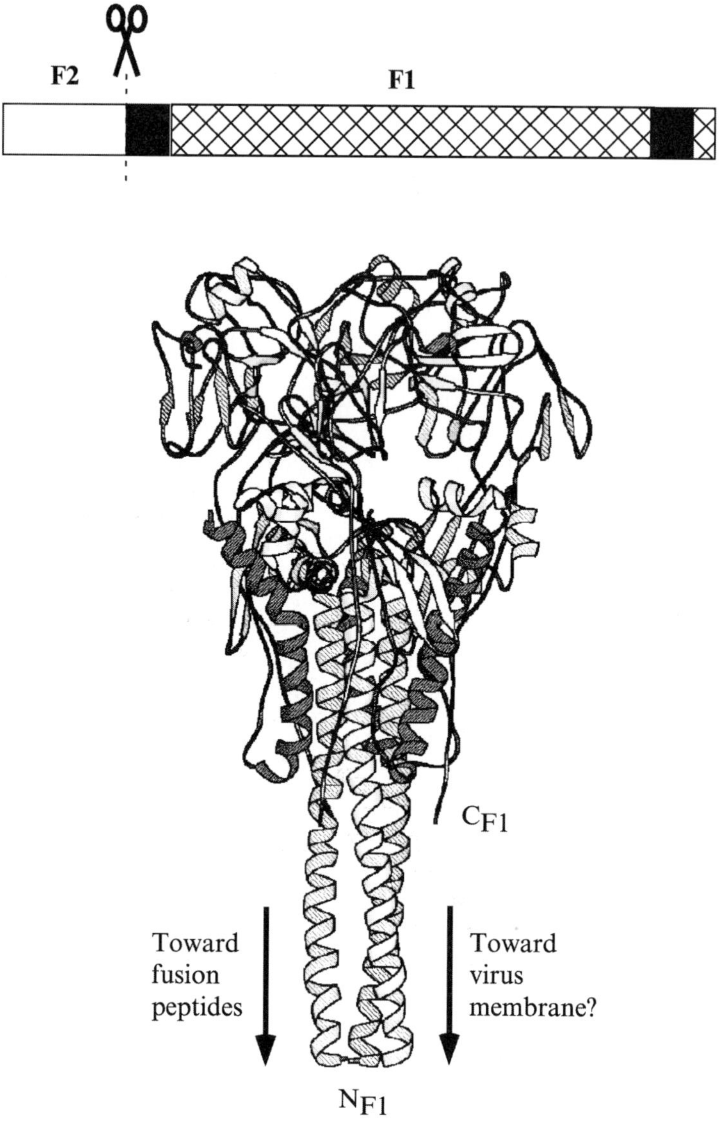

FIG. 12. Primary and tertiary structures of the Newcastle disease virus F protein. Cleavage of the precursor F protein yields the F2 (amino-terminal) and F1 (carboxy-terminal) fragments. In the top schematic, F2 is white, the fusion-peptide and transmembrane segments of F1 are black, and the remainder of F1 is cross-hatched. In the X-ray crystallographic structure of NDV-F (Chen *et al.*, 2001a), electron density for the fusion peptides was not visible, but the fusion peptide is at the amino terminus of F1 and would therefore be expected to extend from the bottom of the

mixture of morphologies including "cones" and "lollipops," with the proportion of lollipops increasing over time (Calder *et al.*, 2000). Electron microscopy results presented in a report that followed the NDV-F structure show that the morphology of the particles recovered from the crystals is similar but not identical to the morphology of the starting material (Chen *et al.*, 2001b). It is likely, therefore, that the crystallized material had largely converted to "lollipops." Furthermore, the authors propose that the native, metastable state of NDV-F has its fusion peptides buried in channels in the globular head of the structure, but no interpretable electron density corresponding to these fusion peptides can be seen. Despite the authors' claim to have obtained information about both the "lollipops" and the "cones," solving the structure seems to have been no "piece of cake," and the implications of this work are not yet clear. Initial attempts to place the structure of NDV-F in the context of a larger model for paramyxovirus-mediated membrane fusion have only recently begun (Peisajovich and Shai, 2002).

V. Oligomerization State Switches in Flaviviruses and Alphaviruses

Although there are mechanistic differences between retroviruses, paramyxoviruses, and the orthomyxovirus influenza, the viruses discussed to this point have definite structural and functional similarities including spikelike, trimeric native structures and the presence of coiled coils in their fusion-active subunits. The flaviviruses and alphaviruses, however, appear to be another class of enveloped viruses entirely. Flaviviruses include yellow fever, West Nile virus, Dengue virus, and tick-borne encephalitis virus (TBEV). Alphaviruses, of the togavirus family, include

---

structure in the region labeled "N_{F1}." The long three-stranded coiled coil of NDV-F therefore points in the opposite direction as the coiled coil in flu HA relative to the globular head domains. The transmembrane regions of NDV-F would be expected to extend from the point marked "C_{F1}" in the structure. Although electron density for the ~45 amino acids preceding the fusion peptide was not visible, the observed carboxy terminus of the NDV-F structure is in position to extend into the buttressing helices observed in the six-helix bundle structure of peptides from another paramyxovirus F1 protein (Baker *et al.*, 1999), which would also place the transmembrane segments at the bottom of the figure. Except for the retention of the head domains, the NDV-F protein therefore resembles a "sprung" virus envelope protein spike structure. Is this resemblance due to a similarity between the metastable and stable conformations of the paramyxovirus F protein as suggested (Chen *et al.*, 2001a), or because the NDV-F structure does not reveal the packing of the metastable state at all?

Sindbis and Semliki Forest virus (SFV). Unlike the surfaces of flu and many retroviruses, which are poorly constrained geometrically, flavivirus and alphavirus coats are icosahedral structures (Forsell *et al.*, 2000; Ferlenghi *et al.*, 2001; Pletnev *et al.*, 2001). This regular arrangement has made cryoelectron microscopy a particularly powerful tool in visualizing conformational changes in these virus surface proteins, for example, during proteolytic priming (Ferlenghi *et al.*, 1998) or low-pH activation (Fuller *et al.*, 1995).

Structural studies, discussed below, show flaviviruses and alphaviruses to be close cousins. However, some differences do exist in their structures and mechanisms. First, alphaviruses display their receptor-binding and membrane fusion activities in two distinct proteins, E2 and E1, respectively, whereas flaviviruses use their E protein to accomplish both tasks. Second, flaviviruses are small viruses with envelopes likely to be organized according to triangulation T=3 icosahedral symmetry (Ferlenghi *et al.*, 2001), but alphaviruses are large T=4 viruses (Mancini *et al.*, 2000). Third, flaviviruses accomplish membrane fusion quickly on lowering the pH, but alphaviruses are slower (Corver *et al.*, 2000). Finally, at least some flaviviruses have a higher pH threshold for fusion than do alphaviruses (Corver *et al.*, 2000; Bron *et al.*, 1993; Smit *et al.*, 1999).

The flavivirus and alphavirus proteins containing the membrane fusion activity do not project as spikes from the virus surface, but rather lie flat against the lipid bilayer envelope (Fig. 1) (Ferlenghi *et al.*, 2001; Lescar *et al.*, 2001; Mancini *et al.*, 2000; Pletnev *et al.*, 2001). The structure of the flavivirus TBEV E protein, determined crystallographically (Rey *et al.*, 1995), is a flat, elongated dimer that curves gently to trace out an arc consistent with the virion radius of approximately 250 Å (Fig. 13). The packing of TBEV E protein into a T=3 icosahedral structure has been modeled (Ferlenghi *et al.*, 2001). The suggestion has even been made that assembly of the envelope icosahedron can define the geometry of the entire virion in an "assembly-from-without" mechanism (Forsell *et al.*, 2000). A similar fold to the TBEV E protein was observed for its analog, E1 from the alphavirus SFV (Lescar *et al.*, 2001) (Fig. 13). Docking of the E1 crystal structure into a cryoelectron microsopic reconstruction of the intact SFV (Mancini *et al.*, 2000) indicates that E1 coats the alphavirus surface in a manner similar to TBEV E protein on the flavivirus envelope.

These structural studies on the alphavirus and flavivirus membrane fusion proteins strengthen the relationship between these virus families, but apparently distance them from the trimeric influenza and retrovirus envelope assemblies. However, low pH specifically triggers not only a conformational change in alphavirus and flavivirus surface proteins, but also an oligomerization switch to a trimeric state (Allison *et al.*, 1995). This

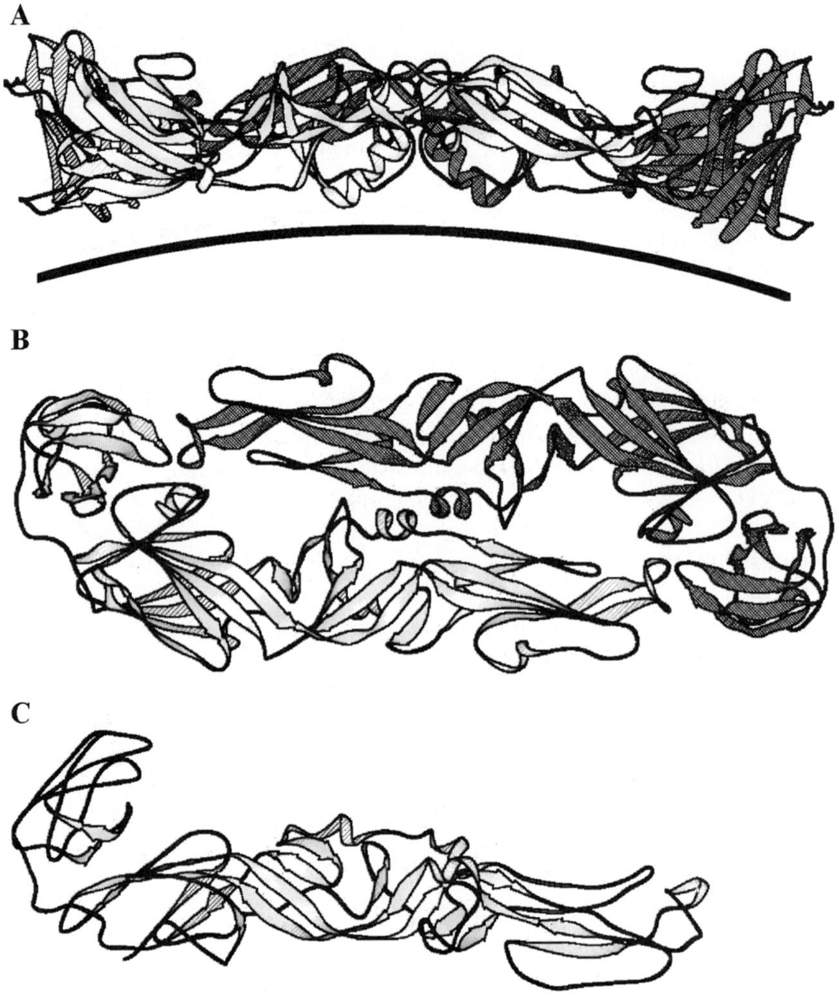

FIG. 13. Flavivirus and alphavirus envelope protein structures. (A) Side view of the TBEV E protein ectodomain (Rey *et al.*, 1995). One promoter of the dimer is shaded and the other is white. The black curve below the structure represents the approximate curvature at the membrane of a virus with radius 250 Å. (B) Top view of the TBEV E protein dimer with the same shading as in (A). (C) Ribbon trace of the SFV E1 protein. [Figure courtesy of Felix Rey.]

switch involves breakage of the dimer contacts and the subsequent formation of trimeric interactions (Stiasny *et al.*, 1996), facilitated by the icosahedral arrangement of envelope proteins. Like influenza, a region of the TBE protein predicted to be helical is required for trimerization

(Stiasny *et al.*, 1996). This region fell outside the proteolytically released ectodomain of the TBEV E protein and SFV E1 structures, and its conformation is not known in the native state. Although the trimeric form of an alphavirus E1 protein has been analyzed by proteolysis (Gibbons and Kielian, 2002), no high-resolution structures of the low-pH trimerized state are yet available for alphaviruses or flaviviruses, so a further structural comparison with flu cannot be made at this time.

A thermodynamic/mechanistic comparison of flu with alphaviruses and flaviviruses can be made, however. Although alphaviruses and flaviviruses envelope proteins are low pH activated for membrane fusion, undergo irreversible large-scale conformational changes, and, like flu HA, end up as stable trimers (Stiasny *et al.*, 1996; Gibbons *et al.*, 2000), they display one fundamental difference with flu and related viruses (Ruigrok *et al.*, 1986; Carr *et al.*, 1997; Wharton *et al.*, 2000). Heat and denaturants do not functionally mimic low pH for SFV E1 homotrimer formation and protein-mediated membrane fusion (Gibbons *et al.*, 2000). Instead, similarly to ASLV (Damico *et al.*, 1999), these treatments result in nonspecific structural changes that fail to activate membrane fusion (Gibbons *et al.*, 2000). Interestingly, once the stable, trimeric version of E1 is formed at low pH, it is highly resistant to heat, urea, and sodium dodecyl sulfate (SDS), even after being brought to neutral pH. This observation indicates that a similar metastability model applies for alphavirus envelope proteins as for flu (Fig. 7), but that at neutral pH, off-pathway local minima may compete with the deeper energy well of the active trimeric species. It has also been proposed that the low pH-induced disassembly of the dimers to monomers may be an indispensable step before trimer formation (Stiasny *et al.*, 2001).

VI. Concluding Remarks

The surface proteins of enveloped viruses perform three major tasks for the virus: shielding of the virus from the immune system, recognition of an appropriate host cell by binding to a cell surface receptor, and the subsequent fusion of the viral and cellular membranes. The first task is somewhat at odds with the last two. Escape from the immune system is aided by rapid evolution of the sequences of the surface proteins, whereas the need to maintain not one but two distinct functions requires that a core structure or structures be maintained.

Enveloped viruses have a number of tricks to balance these conflicting constraints. In keeping with the economy of virus genomes and structures, some of these tricks even serve multiple purposes. First, critical functions are buried in the cores of the virus surface protein complexes, hidden in quaternary structural contacts. This arrangement allows the sequences of essential structural segments to remain relatively conserved while the protein surfaces vary. Quaternary structural assemblies also introduce the possibility of allosteric control of the systems. A second trick, glycosylation of the protein surface, can also serve two purposes. On the one hand, carbohydrates identical to the sugars coating the host glycoproteins mask the foreign features of the virus surface to aid in immune system evasion. However, glycosylation of virus envelope proteins may have an equally important role in their folding pathways (Daniels *et al.*, 2003) and in priming spring-loaded mechanisms for virus entry. Finally, the most dramatic tricks of virus envelope proteins are, of course, the conformational changes they undergo throughout the virus life cycle and the cell entry process. These conformational changes enable the regulated exposure of buried functional regions of the proteins, as well as the coupling of energetically costly events in infection to rolling down the energy landscape of protein (re-)folding.

Acknowledgments

Work performed in the author's laboratory was supported by a Leukemia Research Foundation New Investigator grant. D.F. is incumbent of the Lilian and George Lyttle Career Development Chair. The author is grateful to James Cunningham for many fruitful discussions and to J.C., Amnon Horowitz, and Larry Varon for critical reading of the manuscript.

References

Allan, J. S., Strauss, J., and Buck, D. W. (1990). *Science* **247,** 1084–1088.

Allison, S. L., Schalich, J., Stiasny, K., Mandl, C. W., Kunz, C., and Heinz, F. X. (1995). *J. Virol.* **69,** 695–700.

Amano, H., and Hosaka, Y. (1992). *J. Electron Microsc.* **41,** 104–106.

Anderson, M. M., Lauring, A. S., Burns, C. C., and Overbaugh, J. (2000). *Science* **287,** 1828–1830.

Andersson, H., Barth, B.-U., Ekström, M., and Garoff, H. (1997). *J. Virol.* **71,** 9654–9663.

Bachrach, E., Marin, M., Pelegrin, M., Karavanas, G., and Piechaczyk, M. (2000). *J. Virol.* **74,** 8480–8486.

Bae, Y., Kingsman, S. M., and Kingsman, A. J. (1997). *J. Virol.* **71,** 2092–2099.

Baker, D., and Agard, D. A. (1994). *Structure* **2,** 907–910.

Baker, K. A., Dutch, R. E., Lamb, R. A., and Jardetzky, T. S. (1999). *Mol. Cell* **3,** 309–319.

Barnett, A. L., and Cunningham, J. M. (2001). *J. Virol.* **75,** 9096–9105.

Barnett, A. L., Davey, R. A., and Cunningham, J. M. (2001). *Proc. Natl. Acad. Sci. USA* **98,** 4113–4118.

Barnett, A. L., Wensel, D. L., Li, W., Fass, D., and Cunningham, J. M. (2003). *J. Virol.* **77,** 2717–2729.

Bates, P., Young, J. A. T., and Varmus, H. E. (1993). *Cell* **74,** 1043–1051.
Battini, J.-L., Heard, J. M., and Danos, O. (1992). *J. Virol.* **66,** 1468–1475.
Battini, J.-L., Danos, O., and Heard, J. M. (1995). *J. Virol.* **69,** 713–719.
Ben-Efraim, I., Kliger, Y., Hermesh, C., and Shai, Y. (1999). *J. Mol. Biol.* **285,** 609–625.
Bentz, J. (2000). *Biophys. J.* **78,** 886–900.
Berger, E. A., Murphy, P. M., and Farber, J. M. (1999). *Annu. Rev. Immunol.* **17,** 657–700.
Bizebard, T., Gigant, B., Rigolet, P., Rasmussen, B., Diat, O., Bösecke, P., Wharton, S. A., Skehel, J. J., and Knossow, M. (1995). *Nature* **376,** 92–94.
Blacklow, S. C., Lu, M., and Kim, P. S. (1995). *Biochemistry* **34,** 14955–14962.
Böttcher, C., Ludwig, K., Herrmann, A., van Heel, M., and Stark, H. (1999). *FEBS Lett.* **463,** 255–259.
Braakman, I., Hoover-Litty, H., Wagner, K. R., and Helenius, A. (1991). *J. Cell Biol.* **114,** 401–411.
Bron, R., Wahlberg, J. M., Garoff, H., and Wilschut, J. (1993). *EMBO J.* **12,** 693–701.
Bullough, P. A., Hughson, F. M., Skehel, J. J., and Wiley, D. C. (1994). *Nature* **371,** 37–43.
Caffrey, M., Cai, M., Kaufman, J., Stahl, S. J., Wingfield, P. T., Covell, D. G., Gronenborn, A. M., and Clore, G. M. (1998). *EMBO J.* **17,** 4572–4584.
Calder, L. J., Gonzáles-Reyes, L., Garcia-Barreno, B., Wharton, S. A., Skehel, J. J., Wiley, D. C., and Melero, J. A. (2000). *Virology* **271,** 122–131.
Carr, C. M., and Kim, P. S. (1993). *Cell* **73,** 823–832.
Carr, C. M., Chaudhry, C., and Kim, P. S. (1997). *Proc. Natl. Acad. Sci. USA* **94,** 14306–14313.
Chan, D. C., and Kim, P. S. (1998). *Cell* **93,** 681–684.
Chan, D. C., Fass, D., Berger, J. M., and Kim, P. S. (1997). *Cell* **89,** 263–273.
Chen, J., Lee, K. H, Steinhauer, D. A., Stevens, D. J., Skehel, J. J., and Wiley, D. C. (1998). *Cell* **95,** 409–417.
Chen, J., Skehel, J. J., and Wiley, D. C. (1999). *Proc. Natl. Acad. Sci. USA* **96,** 8967–8972.
Chen, L., Gorman, J. J., McKimm-Breschkin, J., Lawrence, L. J., Tulloch, P. A., Smith, B. J., Colman, P. M., and Lawrence, M. C. (2001a). *Structure* **9,** 255–266.
Chen, L., Colman, P. M., Cosgrove, L. J., Lawrence, M. C., Lawrence, L. J., Tulloch, P. A., and Gorman, J. J. (2001b). *Virology* **290,** 290–291.
Chen, W., and Helenius, A. (2000). *Mol. Biol. Cell* **11,** 765–772.
Choe, H., Martin, K. A., Farzan, M., Sodroski, J., Gerard, N. P., and Gerard, C. (1998). *Semin. Immunol.* **10,** 249–257.
Corver, J., Ortiz, A., Allison, S. L., Schalich, J., Heinz, F. X., and Wilschut, J. (2000). *Virology* **269,** 37–46.
Damico, R., and Bates, P. (2000). *J. Virol.* **74,** 6469–6475.
Damico, R. L., Crane, J., and Bates, P. (1998). *Proc. Natl. Acad. Sci. USA* **95,** 2580–2585.
Damico, R., Rong, L., and Bates, P. (1999). *J. Virol.* **73,** 3087–3094.
Daniels, R., Kurowski, B., Johnson, A. E., and Hebert, D. N. (2003). *Mol. Cell* 79–90.
Davey, R. A., Hamson, C. A., Healey, J. J., and Cunningham, J. M. (1997). *J. Virol.* **71,** 8096–8102.
Davey, R. A., Zuo, Y., and Cunningham, J. M. (1999). *J. Virol.* **73,** 3758–3763.
De Jong, J. C., Rimmelzwaan, G. F., Fouchier, R. A., and Osterhaus, A. D. (2000). *J. Infect.* **40,** 218–228.
Delwart, E. L., Mosialos, G., and Gilmore, T. (1990). *AIDS Res. Hum. Retroviruses* **6,** 703–706.
Doms, R. W., and Moore, J. P. (2000). *J. Cell Biol.* **151,** F9–F13.
Doms, R. W., and Trono, D. (2000). *Genes Dev.* **14,** 2677–2688.

Donahue, P. R., Quackenbush, S. L., Gallo, M. V., deNoronha, C. M., Overbaugh, J., Hoover, E. A., and Mullins, J. I. (1991). *J. Virol.* **65,** 4461–4469.

Doranz, B. J., Baik, S. S., and Doms, R. W. (1999). *J. Virol.* **73,** 10346–10358.

Durrer, P., Galli, C., Hoenke, S., Corti, C., Glück, R., Vorherr, T., and Brunner, J. (1996). *J. Biol. Chem.* **23,** 13417–13421.

Eckert, D. M., Malashkevich, V. N., Hong, L. H., Carr, P. A., and Kim, P. S. (1999). *Cell* **99,** 103–115.

Epand, R. F., Macosko, J. C., Russell, C. J., Shin, Y.-K., and Epand, R. M. (1999). *J. Mol. Biol.* **286,** 489–503.

Fass, D., and Kim, P. S. (1995). *Curr. Biol.* **5,** 1377–1383.

Fass, D., Harrison, S. C., and Kim, P. S. (1996). *Nat. Struct. Biol.* **3,** 465–469.

Fass, D., Davey, R. A., Hamson, C. A., Kim, P. S., Cunningham, J. M., and Berger, J. M. (1997). *Science* **277,** 1662–1666.

Ferlenghi, I., Gowen, B., de Haas, F., Mancini, E. J., Garoff, H., Sjöberg, M., and Fuller, S. D. (1998). *J. Mol. Biol.* **283,** 71–81.

Ferlenghi, I., Clarke, M., Ruttan, T., Allison, S. L., Schalich, J., Heinz, F. X., Harrison, S. C., Rey, F. A., and Fuller, S. D. (2001). *Mol. Cell* **7,** 593–602.

Forsell, K., Xing, L., Kozlovska, T., Cheng, R. H., and Garoff, H. (2000). *EMBO J.* **19,** 5081–5091.

Fuller, S. D., Berriman, J. A., Butcher, S. J., and Gowen, B. E. (1995). *Cell* **81,** 715–725.

Gallo, S. A., Puri, A., and Blumenthal, R. (2001). *Biochemistry* **40,** 12231–12236.

Gibbons, D. L., and Kielian, M. (2002). *J. Virol.* **76,** 1194–1205.

Gibbons, D. L., Ahn, A., Chatterjee, P. K., and Kielian, M. (2000). *J. Virol.* **74,** 7772–7780.

Gilbert, J. M., Mason, D., and White, J. M. (1990). *J. Virol.* **64,** 5106–5113.

Gray, C., and Tamm, L. K. (1997). *Protein Sci.* **6,** 1993–2006.

Gray, C., and Tamm, L. K. (1998). *Protein Sci.* **7,** 2359–2373.

Hebert, D. N., Foellmer, B., and Helenius, A. (1996). *EMBO J.* **15,** 2961–2968.

Hebert, D. N., Zhang, J.-X., Chen, W., Foellmer, B., and Helenius, A. (1997). *J. Cell Biol.* **139,** 613–623.

Hernandez, L. D., Hoffman, L. R., Wolfsberg, T. G., and White, J. M. (1996). *Annu. Rev. Cell Dev. Biol.* **12,** 627–661.

Hernandez, L. D., Peters, R. J., Delos, S. E., Young, J.A.T., Agard, D. A., and White, J. M. (1997). *J. Cell Biol.* **139,** 1455–1464.

Hoffman, T. L., LaBranche, C. C., Zhang, W., Canziani, G., Robinson, J., Chaiken, I., Hoxie, J. A., and Doms, R. W. (1999). *Proc. Natl. Acad. Sci. USA* **96,** 6359–6364.

Hughson, F. M. (1995). *Curr. Biol.* **5,** 265–274.

Hughson, F. M. (1997). *Curr. Biol.* **7,** R565–R569.

Hunter, E., and Swanstrom, R. (1990). *Curr. Top. Microbiol. Immunol.* **157,** 187–253.

Ikeda, H., Kato, K., Suzuki, T., Kitani, H., Matsubara, Y., Takase-Yoden, S., Watanabe, R., Kitagawa, M., and Aizawa, S. (2000). *J. Virol.* **74,** 1815–1826.

Jiang, S., Lin, K., Strick, N., and Neurath, A. R. (1993). *Nature* **365,** 113.

Kanaseki, T., Kawasaki, K., Murata, M., Ikeuchi, Y., and Ohnishi, S. (1997). *J. Cell Biol.* **137,** 1041–1056.

Kemble, G. W., Bodian, D. L., Rosé, J., Wilson, I. A., and White, J. M. (1992). *J. Virol.* **66,** 4940–4950.

Kobe, B., Center, R. J., Kemp, B. E., and Poumbourios, P. (1999). *Proc. Natl. Acad. Sci. USA* **96,** 4319–4324.

Kordower, J. H., Emborg, M. E., Bloch, J., Ma, S. Y., Chu, Y., Leventhal, L., McBride, J., Chen, E. Y., Palfi, S., Roitberg, B. Z., Brown, W. D., Holden, J. E., Pyzalski, R.,

Taylor, M. D., Carvey, P., Ling, Z., Trono, D., Hantraye, P., Deglon, N., and Aebischer, P. (2000). *Science* **290,** 767–773.

Korte, T., Ludwig, K., Krumbiegel, M., Zirwer, D., Damaschun, G., and Herrmann, A. (1997). *J. Biol. Chem.* **272,** 9764–9770.

Korte, T., Ludwig, K., Booy, F. P., Blumenthal, R. B., and Herrmann, A. (1999). *J. Virol.* **73,** 4567–4574.

Kwong, P. D., Wyatt, R., Robinson, J., Sweet, R. W., Sodroski, J., and Hendrickson, W. A. (1998). *Nature* **393,** 648–659.

LaCasse, R. A., Follis, K. E., Trahey, M., Scarborough, J. D., Littman, D. R., and Nunberg, J. H. (1999). *Science* **283,** 357–362.

Lavillette, D., Ruggieri, A., Russell, S. J., and Cosset, F.-L. (2000). *J. Virol.* **74,** 295–304.

Lavillette, D., Boson, B., Russell, S. J., and Cosset, F.-L. (2001). *J. Virol.* **75,** 3685–3695.

Lawless-Delmedico, M. K., Sista, P., Sen, R., Moore, N. C., Antczak, J. B., White, J. M., Greene, R. J., Leanza, K. C., Matthews, T. J., and Lambert, D. M. (2000). *Biochemistry* **39,** 11684–11695.

Lescar, J., Roussel, A., Wien, M. W., Navaza, J., Fuller, S. D., Wengler, G., Wengler, G., and Rey, F. A. (2001). *Cell* **105,** 137–148.

Lever, A. M. (2000). *Curr. Opin. Mol. Ther.* **2,** 488–496.

Lu, M., Blacklow, S. C., and Kim, P. S. (1995). *Nat. Struct. Biol.* **2,** 1075–1082.

Lu, X., and Silver, J. (2000). *Virology* **276,** 251–258.

Maerz, A. L., Center, R. J., Kemp, B. E., Kobe, B., and Poumbourios, P. (2000). *J. Virol.* **74,** 6614–6621.

Malashkevich, V. N., Chan, D. C., Chutkowski, C. T., and Kim, P. S. (1998). *Proc. Natl. Acad. Sci. USA* **95,** 9134–9139.

Malashkevich, V. N., Schneider, B. J., McNally, M. L., Milhollen, M. A., Pang, J. X., and Kim, P. S. (1999). *Proc. Natl. Acad. Sci. USA* **96,** 2662–2667.

Mancini, E. J., Clarke, M., Gowen, B. E., Rutten, T., and Fuller, S. D. (2000). *Mol. Cell* **5,** 255–266.

Melikyan, G. B., Markosyan, R. M., Hemmati, H., Delmedico, M. K., Lamberg, D. M., and Cohen, F. S. (2000). *J. Cell Biol.* **151,** 413–423.

Mothes, W., Boerger, A. L., Narayan, S., Cunningham, J. M., and Young, J.A.T. (2000). *Cell* **103,** 679–689.

Myszka, D. G., Sweet, R. W., Hensley, P., Brigham-Burke, M., Kwong, P. D., Hendrickson, W., Wyatt, R., Sodroski, J., and Doyle, M. L. (2000). *Proc. Natl. Acad. Sci. USA* **97,** 9026–9031.

Opstelten, D.-J.E., Wallin, M., and Garoff, H. (1998). *J. Virol.* **72,** 6537–6545.

Overbaugh, J., Miller, A. D., and Eiden, M. V. (2001). *Microbiol. Mol. Biol. Rev.* **65,** 371–389.

Panda, B. R., Kingsman, S. M., and Kingsman, A. J. (2000). *Virology* **273,** 90–100.

Peisajovich, S. G., and Shai, Y. (2002). *Trends Biochem. Sci.* **27,** 183–190.

Pinter, A., Kopelman, R., Li, Z., Kayman, S. C., and Sanders, D. A. (1997). *J. Virol.* **71,** 8073–8077.

Pletnev, S. V., Zhang, W., Mukhopadhyay, S., Fisher, B. R., Hernandez, R., Brown, D. T., Baker, T. S., Rossman, M. G., and Kuhn, R. J. (2001). *Cell* **105,** 127–136.

Poignard, P., Saphire, E. O., Parren, P. W., and Burton, D. R. (2001). *Annu. Rev. Immunol.* **19,** 253–274.

Qiao, H., Pelletier, S. L., Hoffman, L., Hacker, J., Armstrong, R. T., and White, J. M. (1998). *J. Cell Biol.* **141,** 1335–1347.

Ramsdale, E. E., Kingsman, S. M., and Kingsman, A. J. (1996). *Virology* **220,** 100–108.

Rey, F. A., Heinz, F. X., Mandl, C., Kunz, C., and Harrison, S. C. (1995). *Nature* **375,** 291–298.

Rosenthal, P. B., Zhang, X., Formanowski, F., Fitz, W., Wong, C.-H., Meier-Ewert, H., Skehel, J. J., and Wiley, D. C. (1998). *Nature* **396,** 92–96.

Ruigrok, R. W., Andree, P. J., Hooft van Huysduynen, R. A., and Mellema, J. E. (1984). *J. Gen. Virol.* **65,** 799–802.

Ruigrok, R. W. H., Martin, S. R., Wharton, S. A., Skehel, J. J., Bayley, P. M., and Wiley, D. C. (1986a). *Virology* **155,** 484–497.

Ruigrok, R. W. H., Wrigley, N. G., Calder, L. J., Cusak, S., Wharton, S. A., Brown, E. B., and Skehel, J. J. (1986b). *EMBO J.* **5,** 41–49.

Sattentau, Q. J., and Moore, J. P. (1991). *J. Exp. Med.* **174,** 407–415.

Sattentau, Q. J., Moore, J. P., Vignaux, F., Traincard, F., and Poignard, P. (1993). *J. Virol.* **67,** 7383–7393.

Schulz, T. F., Jameson, B. A., Lopalco, L., Siccardi, A. G., Weiss, R. A., and Moore, J. P. (1992). *AIDS Res. Hum. Retroviruses* **8,** 1571–1580.

Shangguan, T., Siegel, D. P., Lear, J. D., Axelson, P. H., Alford, D., and Bentz, J. (1998). *Biophys. J.* **74,** 54–62.

Skehel, J. J., and Wiley, D. C. (2000). *Annu. Rev. Biochem.* **69,** 531–569.

Smit, J. M., Bittman, R., and Wilschut, J. (1999). *J. Virol.* **73,** 8476–8484.

Snitkovsky, S., and Young, J. A. (1998). *Proc. Natl. Acad. Sci. USA* **95,** 7063–7068.

Stegmann, T., Booy, F. P., and Wilschut, J. (1987). *J. Biol. Chem.* **262,** 17744–17749.

Stegmann, T., White, J. M., and Helenius, A. (1990). *EMBO J.* **9,** 4231–4241.

Steinhauer, D. A. (1999). *Virology* **258,** 1–20.

Steinhauer, D. A., Martin, J., Lin, Y. P., Wharton, S. A., Oldstone, M.B.A., Skehel, J. J., and Wiley, D. C. (1996). *Proc. Natl. Acad. Sci. USA* **93,** 12873–12878.

Stiasny, K., Allison, S. L., Marchler-Bauer, A., Kunz, C., and Heinz, F. X. (1996). *J. Virol.* **70,** 8142–8147.

Stiasny, K., Allison, S. L., Mandl, C. W., and Heinz, F. X. (2001). *J. Virol.* **75,** 7392–7398.

Tailor, C. S., Nouri, A., and Kabat, D. (2000). *J. Virol.* **74,** 237–244.

Tsurudome, M., Glück, R., Graf, R., Falchetto, R., Schaller, U., and Brunner, J. (1992). *J. Biol. Chem.* **267,** 20225–20232.

Wang, H., Paul, R., Burgeson, R. E., Keene, D. R., and Kabat, D. (1991). *J. Virol.* **65,** 6468–6477.

Watowich, S. J., Skehel, J. J., and Wiley, D. C. (1994). *Structure* **2,** 719–731.

Weber, T., Paesold, G., Mischler, R., Semenza, G., and Brunner, J. (1994). *J. Biol. Chem.* **169,** 18353–18358.

Weissenhorn, W., Dessen, A., Harrison, S. C., Skehel, J. J., and Wiley, D. C. (1997). *Nature* **387,** 426–430.

Weissenhorn, W., Carfi, A., Lee, K.-H., Skehel, J. J., and Wiley, D. C. (1998). *Mol. Cell* **2,** 605–616.

Weissenhorn, W., Dessen, A., Calder, L. J., Harrison, S. C., Skehel, J. J., and Wiley, D. C. (1999). *Mol. Membr. Biol.* **16,** 3–9.

Wharton, S. A., Calder, L. J., Ruigrok, R.W.H., Skehel, J. J., Steinhauer, D. A., and Wiley, D. C. (1995). *EMBO J.* **14,** 240–246.

Wharton, S. A., Skehel, J. J., and Wiley, D. C. (2000). *Virology* **271,** 71–78.

White, J. M., and Wilson, I. A. (1987). *J. Cell Biol.* **105,** 2887–2896.

Wild, C., Dubay, J. W., Greenwell, T., Baird, T. Jr., Oas, T. G., McDanal, C., Hunger, E., and Matthews, T. (1994a). *Proc. Natl. Acad. Sci. USA* **91,** 12676–12680.

Wild, C. T., Shugars, D. C., Greenwell, T. K., McDanal, C. B., and Matthews, T. J. (1994b). *Proc. Natl. Acad. Sci. USA* **91,** 9770–9774.

Wiley, D. C., and Skehel, J. J. (1987). *Annu. Rev. Biochem.* **56,** 365–394.
Wilson, I. A., Skehel, J. J., and Wiley, D. C. (1981). *Nature* **289,** 366–373.
Yu, Y. G., King, D. S., and Shin, Y.-K. (1994). *Science* **266,** 274–276.
Zhao, X., Singh, M., Malashkevich, V. N., and Kim, P. S. (2000). *Proc. Natl. Acad. Sci. USA* **97,** 14172–14177.
Zhao, Y., Zhu, L., Lee, S., Li, L., Change, E., Soong, N.-W., Douer, D., and Anderson, W. F. (1999). *Proc. Natl. Acad. Sci. USA* **96,** 4005–4010.

ENVELOPED VIRUSES

By RICHARD J. KUHN* AND JAMES H. STRAUSS†

*Markey Center for Structural Biology, Department of Biological Sciences, Purdue University, West Lafayette, Indiana 47907, and †Division of Biology, California Institute of Technology, Pasadena, California 91125

I. Introduction

A distinguishing physical feature of many viruses is the presence of a host-derived lipid bilayer. This feature of the virus particle has important consequences for the assembly and release as well as the entry pathway for a particular virus. Significant advances have been made in our understanding of enveloped viruses. This progress has been due primarily to advances in structural and cell biology. Knowledge of virus structure has been aided in large part by the development of cryoelectron microscopy (cryo-EM) and imaging techniques and their linkage to independently determined atomic structures of virion components. Thus, although no X-ray crystallographic structure of an enveloped animal virus has yet been solved, several enveloped virus structures are now approaching atomic resolution, or perhaps more appropriately, pseudo-atomic resolution. Improvements in techniques as well as reagents in cell biology have also provided insights into the morphogenesis of enveloped viruses. Confocal microscopy and other imaging techniques have been used to elaborate the pathways and interactions that virion components utilize in their path toward assembly and eventually budding. In this review, some of the general features of enveloped virus structure and assembly are discussed and several icosahedral enveloped virus examples are examined.

ADVANCES IN PROTEIN CHEMISTRY, Vol. 64

0065-3233/03 $35.00

II. General Structural Features of Enveloped Viruses

The acquisition of a lipid bilayer from the infected host cell is the defining feature of enveloped viruses. However, the origin and composition of the bilayer can differ from virus to virus. Viruses are known to acquire a membrane, or bud, from various cellular compartments that include the nucleus, the endoplasmic reticulum (ER), the Golgi, and the plasma membrane. As the lipid contents of those membranes differ, so do the contents of the viral membranes. In some, if not all, cases the composition of the lipid bilayer is an important aspect of the viral assembly pathway.

In addition to the lipid bilayer, enveloped viruses generally have two or more distinct layers of protein that are organized across the membrane. Thus, most viruses have an outer layer of proteins, usually glycoproteins, which are anchored in the membrane as integral membrane proteins. These proteins function to attach the virion to target host cell receptors and facilitate the entry or fusion of the viral membrane with that of the host cell. In addition, some viruses also contain enzymatic activities associated with this outer layer of protein. For example, influenza virus carries with it a neuraminidase that is responsible for cleaving sialic acid residues on host cells.

On the interior of the lipid bilayer, a complex of protein and nucleic acid is found. This complex is usually referred to as a nucleocapsid core, and in many instances has an organized protein shell within which nucleic acid, possibly in complex with additional proteins, is found. These additional proteins may be nucleic acid-binding proteins as well as proteins necessary for genome replication. Some of the larger DNA viruses include many additional proteins within this core.

The organization of these protein layers can be either random or ordered. The inner core of the virus is usually ordered with a single protein organized into a distinct shape. An icosahedral nucleocapsid core provides the greatest volume per unit of protein and is found in many enveloped viruses. However, bullet-shaped nucleocapsids, such as those found in the rhabdoviruses, or cone-shaped nucleocapsids, such as found in some retroviruses, are also able to accommodate their respective viral genomes. Unlike the ordered array found in the nucleocapsid cores of most enveloped viruses, proteins found on the outside of the virion and anchored to the membrane are not so restricted in their morphology. Although the alphaviruses and flaviviruses have been shown to have their glycoproteins organized into icosahedral structures, most other enveloped viruses appear to have their outer proteins arranged nonsymmetrically. This review focuses on our understanding of spherical, icosahedral enveloped viruses, as advances have provided us with important new insights into their structure and function.

III. Alphavirus Structure

The alphaviruses belong to the family *Togaviridae.* This family is composed of two genera: *Alphavirus* and *Rubivirus* (Strauss and Strauss, 1994). The genus *Rubivirus,* which has a single member, rubella virus, is similar in genome organization and virion morphology to the genus *Alphavirus,* although no cryo-EM reconstructions for rubella virus have been accomplished. The alphaviruses, composed of at least 26 members, have a messenger-sense RNA genome of approximately 12 kb (Strauss and Strauss, 1994). The genome encodes the synthesis of four nonstructural proteins. A smaller subgenomic mRNA is also produced and is translated into the structural proteins of the virus, thus providing separate controls for the synthesis of replication and virion proteins. In the mature alphavirus virion, the genome RNA is surrounded by a shell of nucleocapsid proteins organized into a T=4 lattice (Cheng *et al.*, 1995; Paredes *et al.*, 1993; Vénien-Bryan and Fuller, 1994). This nucleocapsid core is surrounded by a host-derived lipid bilayer in which two envelope glycoproteins are embedded. The two proteins, called E1 and E2, organize into a trimer of heterodimers. Like the nucleocapsid core, the glycoproteins form a T=4 lattice. The E2 protein is involved in receptor attachment on the host cell whereas E1 is responsible for promoting membrane fusion in the low-pH environment of the endosome (Kielian, 1995; White *et al.*, 1980).

The plus-strand RNA alphaviruses have been extensively studied for several reasons. First, they grow to high titer in cell culture and replicate rapidly in infected cells with a shutoff of cellular translation (Wengler, 1980). Although there are numerous highly pathogenic members of the genus, Sindbis virus (SINV) and Semliki Forest virus (SFV) are for the most part innocuous in humans, producing a subclinical infection and making them ideal to study in the laboratory (Griffin, 1986). The development of reverse genetic systems for these viruses and their utilization as gene expression vectors further advanced the study of these important pathogens (Liljeström *et al.*, 1991; Rice *et al.*, 1987; Schlesinger, 2000). In addition to these important properties, early electron microscopy studies showed that alphaviruses were well-formed spherical particles having both uniform size and shape (Harrison, 1986; Murphy, 1980). Those early morphogenic studies also suggested that the flaviviruses, which are discussed in Section IV, looked similar to alphavirus and thus might be related (Murphy, 1980). This similarity, along with particle composition and virus transmission strategy, resulted in alphaviruses and flaviviruses being placed within the same family of arboviruses, or insect-transmitted viruses. The subsequent sequencing of the flavivirus and alphavirus genomes established their diversity and resulted in their placement into different virus families (Strauss and Strauss, 1988).

Knowledge of alphavirus structure has been obtained by successively higher resolution cryo-EM studies in combination with X-ray crystallographic structures (Cheng *et al.*, 1995; Mancini *et al.*, 2000; Paredes *et al.*, 1993, 2001; Vénien-Bryan and Fuller, 1994; Zhang *et al.*, 2002). The structure of the prototype alphavirus, SINV, as determined by cryo-EM, is shown in Fig. 1 (see Color Insert) (Zhang *et al.*, 2002). A striking feature apparent in the surface-shaded view shown in Fig. 1 is the projections that radiate from the surface of the particle. These projections, or spikes, are composed of the two envelope proteins E1 and E2 that wrap around one another to form a heterodimers (Rice and Strauss, 1982; Ziemiecki and Garoff, 1978). Three of these heterodimers associate to form a projecting spike as well as a covering or "skirt" that covers the lipid membrane (Vénien-Bryan and Fuller, 1994). Both proteins have transmembrane domains that traverse the lipid bilayer and the E2 glycoprotein has been shown to directly contact the capsid protein on the interior face of the membrane (Metsikkö and Garoff, 1990). Thus, there is a one-to-one interaction between each of the 240 glycoproteins subunits and the capsid proteins. How this interaction is involved in the assembly and the disassembly processes is unknown but it is likely to be important, as particles lacking the nucleocapsid core have never been observed (Strauss and Strauss, 1994).

The atomic structure of the E1 glycoprotein of SFV, determined by X-ray crystallography, has been reported by Lescar *et al.* (2001). The overall structure of the protein is strikingly similar to that of the E protein found in the flaviviruses (Rey *et al.*, 1995), although they share only 17% amino acid identity (Fig. 2; see Color Insert). The protein consists of three domains, with domain I being a central domain that links domains II and III. Domain II is composed of a series of β strands and contains the internal "fusion peptide" at its distal end. Domain III has an immunoglobulin-like topology that differs significantly from the corresponding domain in the flavivirus E protein. This knowledge of the E1 glycoprotein structure enabled Lescar *et al.* to model it into the cryo-EM density map of SFV (Lescar *et al.*, 2001). A similar fit was also accomplished with the related flavivirus E glycoprotein structure modeled into the cryo-EM density map of SINV (Pletnev *et al.*, 2001). In the latter case, the authors mapped the positions of the E1 and E2 glycosylation sites and then used these to more precisely fit the glycoprotein. In both cases, the E1 protein is organized in a triangular arrangement that is centered on 3-fold and quasi-3-fold axes (Fig. 3A; see Color Insert). This arrangement of E1 organizes the icosahedral scaffold of the virus and contributes significant density to the skirt layer of the glycoproteins. Domains I and II of the E1 molecule have a slant of roughly 35° relative to the viral surface, with the distal end of domain II containing the fusion peptide pointing away from the viral surface. These hydrophobic sequences are capped by the presence of E2, which has a molecular envelope similar to E1 but inverted relative to the viral membrane (Zhang *et al.*, 2002). Although the atomic structure of E2 is not known, the intimate association of E1 and E2 visible at 11-Å resolution suggests that there are numerous contacts that promote the heterodimer interactions (Zhang *et al.*, 2002).

The organization of the transmembrane domains of E1 and E2 were first observed as helices in a 9-Å cryo-EM structure of SFV (Mancini *et al.*, 2000). More recently, studies with SINV have confirmed the original SFV result and suggested that the helices may form a coiled-coil arrangement across the lipid bilayer (Zhang *et al.*, 2002). These high-resolution structures also revealed the direct interaction between the E2 glycoprotein and the capsid protein within the nucleocapsid core. The atomic structures for the SFV and SINV capsid proteins have been solved, although in both cases the first 100 amino acids of the ~264-residue protein are missing in the X-ray structure (Choi *et al.*, 1991, 1997). The ordered part, residues 114 to 264 in SINV, resembles a chymotrypsin-like fold, as was expected because this protein has autoproteolytic activity immediately following its synthesis (Aliperti and Schlesinger, 1978; Hahn and Strauss, 1990). As in the case of the E1 glycoprotein, a pseudo-atomic structure of the core has been determined by fitting the X-ray structure of the capsid proteins into the cryo-EM density of the nucleocapsid core (Mancini *et al.*, 2000; Zhang *et al.*, 2002). Although only the structure of the C-terminal domain is available, the fitting suggests that this domain comprises the pentamer and hexamer capsomers that project from the nucleocapsid core surface (Fig. 3B).

IV. Flavivirus Structure

Despite the original placement of flaviviruses in the same family as the alphaviruses, it became obvious that they were distinct viruses when their genomes were sequenced (Rice *et al.*, 1985). In contrast to the togaviruses, members of the family *Flaviviridae* have a singe plus-stand RNA genome segment from which all the viral proteins are translated. The order of proteins is also different, with the structural proteins located at the N terminus in the case of flaviviruses. Three genera compose the *Flaviviridae*: the genus *Flavivirus*, a large group of viruses that includes yellow fever, dengue, and West Nile; the genus *Pestivirus*, which includes bovine viral diarrhea virus; and the genus *Hepacivirus*, which has hepatitis C virus as its sole member (Rice, 1996).

Despite significant progress in understanding the structure and assembly of alphaviruses, insight into this aspect of flaviviruses has lagged far behind. Three proteins compose the virion: the core or capsid protein, the immature membrane (prM) protein, and the envelope (E) protein. The plus-strand RNA viral genome is approximately 10.7 kb and is packaged by the capsid protein into a core analogous to that found in the alphaviruses. However, assembly of the flavivirus core is dependent on the formation and budding of the immature flavivirus particle into the endoplasmic reticulum. The particles follow a secretory route and undergo a maturation cleavage of prM to M in the late Golgi by furin.

An important contribution to flavivirus structure was made by Rey *et al.* with the elucidation of the atomic structure of the E protein of the flavivirus, tick-borne encephalitis virus (TBEV; Fig. 2) (Rey *et al.*, 1995). The E protein is found as a dimer, with domain II forming the dimer interface. At the distal end of the dimer lies domain III, which has an immunoglobulin-like fold, and has been implicated in receptor binding. Joining these two domains is the central domain I, which contains the amino terminus of the protein. The fusion peptide is found at the distal end of domain II, protected from the aqueous environment by the position of the neighboring E molecule within the dimer. The paper not only provided details of the protein structure but made predictions as to the arrangement of the protein on the surface of the virus particle. The surprising suggestion was that the long axis of the E protein would be aligned parallel with the viral membrane, producing a virion with a relatively smooth surface in contrast to other viruses that have projecting spikes, such as the alphaviruses. This arrangement of the E protein on the surface of the virus has been confirmed by the cryo-EM structure of a recombinant subviral particle (RSP) of TBEV (Ferlenghi *et al.*, 2001) as well as the native particle from dengue 2 virus (Kuhn *et al.*, 2002).

For many flaviviruses, a subviral particle is released from infected cells that contains the antigenic properties of native virus but lacks the genome RNA and core protein and is thus noninfectious. These subviral particles are two-thirds the size of the native particle and appear to undergo the same type of maturation process in which the prM protein is cleaved in a late compartment by furin. Several studies have demonstrated that similar subviral particles (RSPs) can be produced by means of coexpression of prM and E in eukaryotic cell culture (Schalich *et al.*, 1996). Cryo-EM analysis of TBEV RSPs demonstrated that the particles were smooth on the outside as predicted from the earlier structural studies on the E protein

(Ferlenghi *et al.*, 2001). The atomic structure of the E protein was fitted into the cryo-EM map and it was shown that the particles obeyed T=1 icosahedral symmetry with 30 E dimer subunits. On the basis of the size of native particles having a diameter of 500Å, and the RSP structure, it was proposed that the native particles would assume a T=3 quasi-equivalent organization. This would make them essentially larger versions, having 180 subunits of E, M, and capsid, of the RSPs.

The determination of the dengue 2 virus structure, using cryo-EM methods, produced a particle shaped like RSPs but having a larger size of 500Å (Kuhn *et al.*, 2002) (Fig. 4A; see Color Insert). The atomic structure of the TBEV E protein was fitted into the dengue cryo-EM map. As suggested, there were 180 subunits of E protein in the particles but they were not organized into the predicted T=3 quasi-equivalent arrangement (Fig. 4B). Instead, the E protein is organized into sets of three parallel dimers. These sets associate to form a "herringbone" configuration on the viral surface. This arrangement differs markedly from a true T=3 structure because the dimers on the icosahedral 2-fold axes do not have a quasi-3-fold relationship to the dimers on the quasi-2-fold axes. The environment of the dimers on the icosahedral 2-fold axes is totally different from the environment of the dimers on the quasi-2-fold axes.

Although the positions of all E dimers in the outer shell of the particle were known, a precise interpretation of the density contributed by the M protein was not possible. This was due to the lack of detailed information concerning the C-terminal 101 amino acids of the E protein that were missing from the crystal structure. These residues form the stalk region, the transmembrane domain, and the NS1 signal sequence. Approximately 52 residues would compose the stalk and are found in a shell of density in which the short M protein (37 amino acids outside of the membrane) would also be predicted to be found. Together, the M and the E proteins completely cover the lipid bilayer so that there is no exposed membrane in the dengue particle.

Unlike the nucleocapsid core found in the alphaviruses, the flavivirus core is an open structure with no well-defined subunit organization. At the current resolution of flavivirus cryo-EM reconstructions, little can be said about the transmembrane domains that cross the bilayer or the possible contacts the envelope proteins might make with the underlying core (Kuhn *et al.*, 2002).

V. Virus Assembly

Although the alphaviruses and flaviviruses share similarities in overall architecture as icosahedral enveloped viruses and exhibit striking structural similarities in their fusion proteins, several aspects of their assembly process are significantly different. Alphaviruses bud virions from the plasma membrane, whereas flaviviruses bud from the ER or an intermediate compartment of the early secretory pathway. Formation of flavivirus nucleocapsid cores requires the coassembly of enveloped protein whereas in alphaviruses this does not appear to be the dominant pathway. Finally, the packaging of the flavivirus genome RNA is closely coupled to replication complexes synthesizing new RNA whereas such a strict coupling has not been described for alphaviruses.

The mechanism of the alphavirus core assembly, although studied for several decades, is poorly understood. It is known from *in vivo* studies that immediately following translation and autocatalytic proteolysis, the capsid protein is associated with the large subunit of the ribosome (Glanville and Ulmanen, 1976; Söderlund, 1973). It then interacts with genomic RNA and rapidly assembles into nucleocapsid cores. The rapid rate of core formation has made identification of assembly intermediates difficult. To date, *in vivo* studies have identified only large protein–nucleic acid aggregates that have been proposed to represent valid intermediates in the assembly process (Söderlund and Ulmanen, 1977). An *in vitro* assembly system for SINV core-like particles (CLPs) was previously established using capsid protein isolated from virus particlcs (Wengler *et al.*, 1982). CLPs produced by this system using viral genomic RNA closely resembled cytoplasmic cores purified from infected cells in size, shape, and composition. This assembly system provided the first insights into the biochemical requirements of nucleocapsid core assembly. A significant limitation of this early assembly system was the reliance on capsid protein purified from assembled virus particles, thereby eliminating the ability to assay capsid protein mutants that were defective in virus production. An *in vitro* CLP assembly system using capsid protein purified from *Escherichia coli* and a variety of nucleic acids has been developed to overcome the limitations of the previous assembly system (Tellinghuisen *et al.*, 1999). CLPs generated by this system are also identical to authentic nucleocapsid cores and intermediates in the assembly process have now been identified (Tellinghuisen *et al.*, 2001).

The synthesis and processing of alphavirus glycoproteins have been reviewed extensively (Schlesinger and Schlesinger, 1986; Strauss and Strauss, 1994). Following autocatalytic cleavage of the capsid protein, the glycoproteins are translated from the nascent polyprotein in the form of PE2-6K-E1, with the PE2 peptide being the precursor of E3 and E2. The small E3 peptide contains a signal sequence for E2 and its presence stabilizes the E1-E2 heterodimer preventing premature acid activation of E1 (Wahlberg *et al.*, 1989). During transit to the plasma membrane, PE2, 6K, and E1 are cotranslationally processed from the polyprotein by proteolytic enzymes within the lumen of the ER, and glycosylated by the addition of high-mannose oligosaccharides. The glycoproteins are then transported to the *cis* or medial Golgi cisternae, where they are covalently modified with fatty acyl chains. These modified proteins are then transported to the *trans* Golgi cisternae, where the high-mannose oligosaccharides are trimmed and then modified to form complex oligosaccharides. In a late Golgi or post-Golgi compartment, PE2 is cleaved into mature E3 and E2 proteins. Finally, the E3, E2, 6K, and E1 complex is transported to the cell surface where the cytoplasmic domain of E2 interacts with nucleocapsid cores that have assembled in the cytoplasm (Metsikkö and Garoff, 1990). This interaction has been proposed to involve the binding of the cytoplasmic domain of E2 into a hydrophobic pocket found in the capsid protein (Lee *et al.*, 1996; Skoging *et al.*, 1996; Zhao *et al.*, 1994). A productive set of E2–capsid interactions leads to budding of the mature virus particle from the plasma membrane of the cell. However, the precise molecular details that describe this budding process have yet to be elucidated.

Far less is known about the process of flavivirus assembly. Electron microscopy has shown that immature virions can be found in the lumen of the endoplasmic reticulum (Murphy, 1980). The nucleocapsid core is not assembled free in the cytoplasm; rather, its assembly appears to take place on the cytoplasmic face of membranes with which prM and E proteins are associated (Khromykh *et al.*, 2001). The carboxy-terminal signal sequence of the precursor to the capsid protein is thought to anchor that protein to the membrane (Amberg *et al.*, 1994). This should allow interactions to occur between the capsid protein and the envelope proteins, which are also anchored to the membrane but reside in the lumen of the endoplasmic reticulum or vesicles. The capsid protein also contains a conserved stretch of hydrophobic residues located roughly in the middle of the protein that has been suggested to serve as an additional or alternative membrane anchor (Markoff *et al.*, 1997).

The coupling of protein synthesis, RNA synthesis, and virion assembly on membranous structures assures that newly synthesized genome RNA can associate with capsid protein and initiate the assembly process. Encapsidation of the RNA initiates the budding of particles into the endoplasmic reticulum. Particles that have budded into the endoplasmic reticulum are then processed by carbohydrate addition and modification as they proceed through the Golgi. It is likely that transport to the Golgi and into the *trans*-Golgi network requires the presence of the glycosylated prM protein. The particles follow the exocytosis pathway to be released to the extracellular space by fusion of vesicles containing them with the plasma membrane. The cleavage of the prM protein by host-encoded furin occurs just before virion release and converts the particle to its mature form (Monath and Heinz, 1996).

VI. Virus–Cell Fusion

Elegant studies have been carried out to investigate the structural and biochemical aspects of virus–cell fusion. Although influenza virus hemagglutinin and human immunodeficiency virus (HIV) gp120 have been the best-studied models, numerous examples of this class I type fusion mechanism have been described. These experiments have shown that fusion is initiated by the formation of a trimeric coiled-coil helix adjacent to the fusion peptide on the virus exterior, the insertion of this fusion peptide into the host cell membrane, and the subsequent formation of a six-helix bundle (Skehel and Wiley, 1998).

The alphavirus and flavivirus fusion proteins, E1 and E, respectively, share a common structural fold and have been designated as class II fusion proteins (Lescar *et al.*, 2001). This designation is based on significant differences observed between these proteins and the fusion proteins of influenza virus and other class I fusion proteins. In class II fusion proteins, a second protein (pE2 in alphaviruses and prM in flaviviruses) forms a heterodimer with the fusion protein and protects it from premature activation. This second protein undergoes a maturation cleavage that activates the heterodimers for future fusion activity. Although trimers appear to be functional oligomers in both class I and II fusion proteins, the class II proteins do not appear to have sequences that form coiled coils, suggesting a different conformation for these proteins in promoting fusion (Rey *et al.*, 1995). On the basis of the arrangement of β strands in domain II and their proximity to the fusion peptide, it has been suggested that a porin-like structure having a β-barrel organization may occur during the early steps of fusion (Vashishtha *et al.*, 1998; Kuhn *et al.*, 2002).

Unfortunately, there are no structures available for either the flaviviruses or alphaviruses under conditions approximating the fusion state. For both groups of viruses, entry is believed to occur following attachment of the virus to the cellular receptor and internalization of the particle into an endosome (Kielian, 1995; Heinz and Allison, 2001). Acidification of the endosome results in rearrangement of envelope proteins and subsequent insertion of the fusion peptide into the endosomal membrane (Levy-Mintz and Kielian, 1991; Allison *et al.*, 2001). Ultimately this results in fusion of cellular and viral membranes and release of the nucleocapsid core and genome RNA into the cytoplasm of the infected cell. *In vitro* experiments using synthetic liposomes have established a sequence of events that is remarkably similar for both groups of viruses (Wilschut *et al.*, 1995; Corver *et al.*, 2000). In the presence of low pH, there is disassociation of the E1–E2 heterodimer or the E homodimer. In the presence of a membrane, insertion of the fusion peptide occurs and homotrimerization of the fusion protein follows or is concomitant with membrane insertion (Gibbons *et al.*, 2000). The role of membrane components in the insertion step appears to be critical (Kielian, 1995; Lu and Kielian, 2000). Both TBEV and SFV require cholesterol in the target membrane for efficient insertion whereas only alphaviruses require sphingolipids for the formation of fusion-competent E1 homotrimers (Corver *et al.*, 1995, 2000).

The formation of the homotrimer is an essential step for fusion although the physical arrangement of the homotrimer is not known (Allison *et al.*, 1995; Gibbons *et al.*, 2000). In alphaviruses, it is tempting to suggest that the trimer arrangement of E1 that surrounds each 3-fold or quasi-3-fold is the arrangement of E1 molecules found at low pH (Fig. 5A; see Color Insert). The low pH causes disassociation of the E1–E2 heterodimer and reveals the fusion peptides that are projecting toward the target membrane. However, in this configuration there is little evidence for E1-trimeric contacts that would stabilize such a structure. Even more difficult to reconcile is the dimer-to-trimer transition that occurs in flaviviruses at low pH. Given the organization of E dimers that exist in the dengue structure, dramatic rearrangements are required to produce E homotrimers. However, given the E1 organization seen in alphaviruses, a similar arrangement for flaviviruses at low pH is a logical extension. Thus, as shown in Fig. 5B, flaviviruses would be expected to disassociate the E homodimers and rearrange into trimers that have moved the fusion peptide in domain II closer to the target membrane, resulting in a patch of membrane exposed within the center of the trimer. Unfortunately, although these models are accurate with respect to the oligomeric forms of the fusion proteins, they remain highly speculative in terms of the fusion protein structural arrangements.

VII. Concluding Remarks

Although structural studies of enveloped viruses have lagged behind those of nonenveloped viruses, progress in combining high-resolution cryo-EM results with independently derived atomic structures of virion components has provided "pseudo-atomic resolution" structures. In terms of whole virus particles, the greatest advances have been accomplished with spherical icosahedral enveloped viruses from the alphavirus and, more recently, flavivirus groups. These inroads into the virion architecture have provided more than just a glimpse of the virus structure. They have provided insights into the process of particle assembly–disassembly, membrane fusion, and evolutionary links between virus groups.

References

Aliperti, G., and Schlesinger, M. J. (1978). Evidence for an autoprotease activity of Sindbis virus capsid protein. *Virology* **90,** 366–369.

Allison, S. L., Schalich, J., Stiasny, K., Mandl, C. W., Kunz, C., and Heinz, F. X. (1995). Oligomeric rearrangement of tick-borne encephalitis virus envelope proteins induced by an acidic pH. *J. Virol.* **69,** 695–700.

Allison, S. L., Schalich, J., Stiasny, K., Mandl, C. W., and Heinz, F. X. (2001). Mutational evidence for an internal fusion peptide in flavivirus envelope protein E. *J. Virol.* **75,** 4268–4275.

Amberg, S. M., Nestorowicz, A., McCourt, D. W., and Rice, C. M. (1994). NS2B-3 proteinase-mediated processing in the yellow fever virus structural region: In vitro and in vivo studies. *J. Virol.* **68,** 3794–3802.

Cheng, R. H., Kuhn, R. J., Olson, N. H., Rossmann, M. G., Choi, H. K., Smith, T. J., and Baker, T. S. (1995). Nucleocapsid and glycoprotein organization in an enveloped virus. *Cell* **80,** 621–630.

Choi, H.-K., Tong, L., Minor, W., Dumas, P., Boege, U., Rossmann, M. G., and Wengler, G. (1991). Structure of Sindbis virus core protein reveals a chymotrypsin-like serine proteinase and the organization of the virion. *Nature* **354,** 37–43.

Choi, H.-K., Lu, G., Lee, S., Wengler, G., and Rossmann, M. G. (1997). The structure of Semliki Forest virus core protein. *Proteins Struct. Funct. Genet.* **27,** 345–359.

Corver, J., Moesby, L., Erukulla, R. K., Reddy, K. C., Bittman, R., and Wilschut, J. (1995). Sphingolipid-dependent fusion of Semliki Forest virus with cholesterol-containing liposomes requires both the 3-hydroxyl group and the double bond of the sphingolipid backbone. *J. Virol.* **69,** 3220–3223.

Corver, J., Ortiz, A., Allison, S. L., Schalich, J., Heinz, F. X., and Wilschut, J. (2000). Membrane fusion activity of tick-borne encephalitis virus and recombinant subviral particles in a liposomal model system. *Virology* **269,** 37–46.

Ferlenghi, I., Clarke, M., Ruttan, T., Allison, S. L., Schalich, J., Heinz, F. X., Harrison, S. C., Rey, F. A., and Fuller, S. D. (2001). Molecular organization of a recombinant subviral particle from tick-borne encephalitis. *Mol. Cell* **7,** 593–602.

Gibbons, D. L., Ahn, A., Chatterjee, P. K., and Kielian, M. (2000). Formation and characterization of the trimeric form of the fusion protein of Semliki Forest virus. *J. Virol.* **74,** 7772–7719.

Glanville, N., and Ulmanen, J. (1976). Biological activity of *in vitro* synthesized protein: Binding of Semliki Forest virus capsid protein to the large ribosomal subunit. *Biochem. Biophys. Commun.* **71,** 393–399.

Griffin, D. E. (1986). Alphavirus pathogenesis and immunity. *In* "The Togaviridae and Flaviviridae" (S. Schlesinger and M. J. Schlesinger, Eds.), pp. 209–250. Plenum, New York.

Hahn, C. S., and Strauss, J. H. (1990). Site-directed mutagenesis of the proposed catalytic amino acids of the Sindbis virus capsid protein autoprotease. *J. Virol.* **64,** 3069–3073.

Harrison, S. C. (1986). Alphavirus structure. *In* "The Togaviridae and Flaviviridae" (S. Schlesinger and M. J. Schlesinger, Eds.), pp. 21–34. Plenum, New York.

Heinz, F. X., and Allison, S. L. (2001). The machinery for flavivirus fusion with host cell membranes. *Curr. Opin. Microbiol.* **4,** 450–455.

Khromykh, A. A., Varnavski, A. N., Sedlak, P. L., and Westaway, E. G. (2001). Coupling between replication and packaging of flavivirus RNA: Evidence derived from the use of DNA-based full-length cDNA clones of Kunjin virus. *J. Virol.* **75,** 4633–4640.

Kielian, M. (1995). Membrane fusion and the alphavirus life cycle. *Adv. Virus Res.* **45,** 113–151.

Kuhn, R. J., Zhang, W., Rossmann, M. G., Pletnev, S. V., Corver, J., Lenches, E., Jones, C. T., Mukhopadhyay, S., Chipman, P. R., Strauss, E. G., Baker, T. S., and Strauss, J. H. (2002). Structure of dengue virus: Implications for flavivirus organization, maturation, and fusion. *Cell* **108,** 717–725.

Lee, S., Owen, K. E., Choi, H. K., Lee, H., Lu, G., Wengler, G., Brown, D. T., Rossmann, M. G., and Kuhn, R. J. (1996). Identification of a protein binding site on the surface of the alphavirus nucleocapsid protein and its implication in virus assembly. *Structure* **4,** 531–541.

Lescar, J., Roussel, A., Wein, M. W., Navaza, J., Fuller, S. D., Wengler, G., Wengler, G., and Rey, F. A. (2001). The fusion glycoprotein shell of Semliki Forest virus: An icosahedral assembly primed for fusogenic activation at endosomal pH. *Cell* **105,** 137–148.

Levy-Mintz, P., and Kielian, M. (1991). Mutagenesis of the putative fusion domain of the Semliki Forest virus spike protein. *J. Virol.* **65,** 4292–4300.

Liljeström, P., Lusa, S., Huylebroeck, D., and Garoff, H. (1991). In vitro mutagenesis of a full-length cDNA clone of Semliki Forest virus: The samll 6000-molecular-weight membrane protein modulates virus release. *J. Virol.* **65,** 4107–4113.

Lu, Y. E., and Kielian, M. (2000). Semliki Forest virus budding: Assay, mechanisms, and cholesterol requirement. *J. Virol.* **74,** 7708–7719.

Mancini, E. J., Clarke, M., Gowen, B. E., Rutten, T., and Fuller, S. D. (2000). Cryo-electron microscopy reveals the functional organization of an enveloped virus, Semliki Forest virus. *Mol. Cell* **5,** 255–266.

Markoff, L., Falgout, B., and Chang, A. (1997). A conserved internal hydrophobic domain mediates the stable membrane integration of the dengue virus capsid protein. *Virology* **233,** 105–117.

Metsikkö, K., and Garoff, H. (1990). Oligomers of the cytoplasmic domain of the p62/E2 membrane protein of Semliki Forest virus bind to the nucleocapsid in vitro. *J. Virol.* **64,** 4678–4683.

Monath, T. P., and Heinz, F. X. (1996). Flaviviruses. *In* "Fields Virology" (B. N. Fields, D. M. Knipe, and P. M. Howley, Eds.), 3rd Ed., pp. 961–1034. Lippincott-Raven, Philadelphia.

Murphy, F. A. (1980). Togavirus morphology and morphogenesis. *In* "The Togaviruses" (R. W. Schlesinger, Ed.), pp. 241–316. Academic Press, New York.

Paredes, A., Alwell-Warda, K., Weaver, S. C., Chiu, W. I., Watowich, S. J., Shurtleff, A. C., Beasley, D. W. C., Chen, J. J. Y., Ni, H. L., Suderman, M. T., Wang, H. M., Xu, R. L., Wang, E., Watts, D. M., Russell, K. L., and Barrett, A. D. T. (2001). Venezuelan equine encephalomyelitis virus structure and its divergence from Old World alphaviruses. *J. Virol.* **75,** 9532–9537.

Paredes, A. M., Brown, D. T., Rothnagel, R., Chiu, W., Schoepp, R. J., Johnston, R. E., and Prasad, B. V. V. (1993). Three-dimensional structure of a membrane-containing virus. *Proc. Natl. Acad. Sci. USA* **90,** 9095–9099.

Pletnev, S. V., Zhang, W., Mukhopadhyay, S., Fisher, B. R., Hernandez, R., Brown, D. T., Baker, T. S., Rossmann, M. G., and Kuhn, R. J. (2001). Locations of carbohydrate sites on Sindbis virus glycoproteins show that E1 forms an icosahedral scaffold. *Cell* **105,** 127–136.

Rey, F. A., Heinz, F. X., Mandl, C., Kunz, C., and Harrison, S. C. (1995). The envelope glycoprotein from tick-borne encephalitis virus at 2 Å resolution. *Nature* **375,** 291–298.

Rice, C. M. (1996). *Flaviviridae:* The viruses and their replication. *In* "Fields Virology" (B. N. Fields, D. M. Knipe, and P. M. Howley, Eds.), 3rd Ed., pp. 931–959. Lippincott-Raven, Philadelphia.

Rice, C. M., and Strauss, J. H. (1982). Association of Sindbis virion glycoproteins and their precursors. *J. Mol. Biol.* **154,** 325–348.

Rice, C. M., Lenches, E. M., Eddy, S. R., Shin, S. J., Sheets, R. L., and Strauss, J. H. (1985). Nucleotide sequence of yellow fever virus: Implications for flavivirus gene expression and evolution. *Science* **229,** 726–733.

Rice, C. M., Levis, R., Strauss, J. H., and Huang, H. V. (1987). Production of infectious RNA transcripts from Sindbis virus cDNA clones: Mapping of lethal mutations, rescue of a temperature sensitive marker, and in vitro mutagenesis to generate defined mutants. *J. Virol.* **61,** 3809–3819.

Schalich, J., Allison, S. L., Stiasny, K., Mandl, C. W., Kunz, C., and Heinz, F. X. (1996). Recombinant subviral particles from tick-borne encephalitis virus are fusogenic and provide a model system for studying flavivirus envelope glycoprotein functions. *J. Virol.* **70,** 4549–4557.

Schlesinger, M. J., and Schlesinger, S. (1986). Formation and assembly of alphavirus glycoproteins. *In* "The Togaviridae and Flaviviridae" (S. Schlesinger and M. J. Schlesinger, Eds.), pp. 121–148. Plenum, New York.

Schlesinger, S. (2000). Alphavirus expression vectors. *Adv. Virus Res.* **55,** 565–577.

Skehel, J. J., and Wiley, D. C. (1998). Coiled coils in both intracellular vesicle and viral membrane fusion. *Cell* **95,** 871–874.

Skoging, U., Vihinen, M., Nilsson, L., and Liljeström, P. (1996). Aromatic interactions define the binding of the alphavirus spike to its nucleocapsid. *Structure* **4,** 519–529.

Söderlund, H. (1973). Kinetics of formation of Semliki Forest virus nucleocapsid. *Intervirology* **1,** 354–361.

Söderlund, H., and Ulmanen, I. (1977). Transient association of Semliki Forest virus capsid protein with ribosomes. *J. Virol.* **24,** 907–909.

Strauss, J. H., and Strauss, E. G. (1988). Replication of the RNAs of alphaviruses and flaviviruses. *In* "RNA Genetics Book" (E. Domingo, J. J. Holland, and P. Ahlquist, Eds.), Vol. 1, pp. 71–90. CRC Press, Boca Raton, FL.

Strauss, J. H., and Strauss, E. G. (1994). The alphaviruses: Gene expression, replication, and evolution. *Microbiol. Rev.* **58,** 491–562.

Tellinghuisen, T. L., Hamburger, A. E., Fisher, B. R., Ostendorp, R., and Kuhn, R. J. (1999). In vitro assembly of alphavirus cores by using nucleocapsid protein expressed in *Escherichia coli. J. Virol.* **73,** 5309–5319.

Tellinghuisen, T. L., Perera, R., and Kuhn, R. J. (2001). In vitro assembly of Sindbis virus core-like particles from cross-linked dimers of truncated and mutant capsid proteins. *J. Virol.* **75,** 2810–2817.

Vashishtha, M., Phalen, T., Marquardt, M. T., Ryu, J. S., Ng, A. C., and Kielian, M. (1998). A single point mutation controls the cholesterol dependence of Semliki Forest virus entry and exit. *J. Cell Biol.* **140,** 91–99.

Vénien-Bryan, C., and Fuller, S. D. (1994). The organization of the spike complex of Semliki Forest virus. *J. Mol. Biol.* **236,** 572–583.

Wahlberg, J. M., Boere, W. A., and Garoff, H. (1989). The heterodimeric association between the membrane proteins of Semliki Forest virus changes its sensitivity to low pH during virus maturation. *J. Virol.* **63,** 4991–4997.

Wengler, G. (1980). Effects of alphaviruses on host cell macromolecular synthesis. *In* "The Togaviruses: Biology, Structure, Replication" (R. W. Schlesinger, Ed.), pp. 459–472. Academic Press, New York.

Wengler, G., Boege, U., Wengler, G., Bischoff, H., and Wahn, K. (1982). The core protein of the alphavirus Sindbis virus assembles into core-like nucleoproteins with the viral genome RNA and with other single-stranded nucleic acids *in vitro. Virology* **118,** 401–410.

White, J., Kartenbeck, J., and Helenius, A. (1980). Fusion of Semliki Forest virus with the plasma membrane can be induced by low pH. *J. Cell Biol.* **87,** 264–272.

Wilschut, J., Corver, J., Nieva, J. L., Bron, R., Moesby, L., Reddy, K. C., and Bittman, R. (1995). Fusion of Semliki Forest virus with cholesterol-containing liposomes at low pH: A specific requirement for sphingolipids. *Mol. Membr. Biol.* **12,** 143–149.

Zhang, W., Mukhopadhyay, S., Pletnev, S., Baker, T. S., Kuhn, R. J., and Rossmann, M. G. (2002). Placement of the structural proteins in Sindbis virus. *J. Virol.* **76,** 11645–11658.

Zhao, H., Lindqvist, B., Garoff, H., von Bonsdorf, C. H., and Liljeström, P. (1994). A tyrosine-based motif in the cytoplasmic domain of the alphavirus envelope protein is essential for budding. *EMBO J.* **13,** 4204–4211.

Ziemiecki, A., and Garoff, H. (1978). Subunit composition of the membrane glycoprotein complex of Semliki Forest virus. *J. Mol. Biol.* **122,** 259–269.

STUDYING LARGE VIRUSES

By FRAZER J. RIXON* AND WAH CHIU†

*MRC Virology Unit, Institute of Virology, Glasgow G11 5JR, United Kingdom, and †National Center for Macromolecular Imaging, Verna and Marrs McLean Department of Biochemistry and Molecular Biology, Baylor College of Medicine, Houston, Texas 77030

I. What Is a Large Virus? 379
II. Why Large Viruses? 381
III. Why Study Large Viruses? 384
IV. Methods of Structural Analysis 385
V. Complexity of Organization 386
A. Nonicosahedral Particles 387
B. Particles Containing Icosahedral Capsids 390
VI. Structural Folds 393
VII. Assembly Mechanisms 394
VIII. Maturation 399
IX. Accessory Proteins 400
X. Packaging 401
XI. Future Prospects 402
XII. Summary 403
References 404

I. What Is a Large Virus?

There are two obvious criteria by which viruses may be classified as large: either through possession of a large particle or through possession of a large genome. Having a large particle might seem sufficient reason for classifying a virus as large, and many viruses do indeed fulfill this requirement. For example, if a size of greater than 1000 Å is taken as an arbitrary lower limit, the list of large viruses would include several families of double-stranded DNA (dsDNA) animal viruses such as the *Herpesviridae, Poxviridae, Iridoviridae, Baculoviridae,* some large dsDNA bacteriophages, and a number of RNA virus families including the *Filoviridae, Rhabdoviridae,* and *Paramyxoviridae.* However, there are problems with this definition. The RNA virus particles on this list would certainly be classified as large. For example, filoviruses (e.g., Ebola and Marburg virus), can form elongated particles that are up to 14,000 nm in length (Sanchez *et al.*, 2001) and, therefore, exceed in size many bacteria. However, this size is not a reflection of genetic complexity because none of the RNA viruses mentioned has a genome of >20 kb and all of these are characterized by particles that have rather simple compositions.

0065-3233/03 $35.00

ADVANCES IN PROTEIN CHEMISTRY, Vol. 64

As a basis for a discussion of virus structure, a definition based on genome size seems more useful because all viruses with large genomes also form large particles that are complex in both composition and structure. In this case, therefore, size can be equated with complexity. All large virus genomes known to date are of double-stranded DNA. Those with genomes larger than ~100 kbp, which has been set as an arbitrary lower limit for the purposes of this article, are listed in Table I (see van Regenmortel *et al.*, 2000). They form a structurally diverse group that includes brick shaped or ovoid (*Poxviridae*) and cylindrical (*Baculoviridae*) types as well as a number of icosahedrally based forms infecting both eukaryotes (*Herpesviridae* and *Iridoviridae*) and prokaryotes (*Myoviridae*).

Over time the classification shown in Table I is likely to require some revision as our increasing knowledge of their genome sequences, structures, and lifestyles provides new evidence for the relatedness of the different forms. Also, assigning relationships depends largely on which criteria are given greatest weight and this may also change over time. For example, it has been suggested that because sequence analysis has shown little or no homology between their genomes (van Hulten *et al.*, 2001),

TABLE I
Families of Large Viruses

Particle form (morphology)	Family	Genome size (kbp)	Examples mentioned in text
Cylindrical (nucleocapsid)	*Baculoviridae*	80–180	*Spodoptera litura* granulosis virus (SLGV); *Autographa californica* nuclear polyhedrosis virus (AcMNPV)
Brick-shaped or ovoid	*Poxviridae*	130–375	Vaccinia virus
Icosahedral	*Asfarviridae*	170–190	African swine fever virus (ASFV)
Icosahedral	*Iridoviridae*	140–383	Chilo iridescent virus (CIV)
Icosahedral	*Phycodnaviridae*	160–380	*Paramecium bursaria Chlorella* virus (PBCV-1)
Icosahedral (nucleocapsid)	*Herpesviridae*	125–240	Herpes simplex virus type 1 (HSV-1)
Tailed icosahedral	*Myoviridae*	39–169	Bacteriophage T4
Tailed icosahedral	*Siphoviridae*	22–121	
Ellipsoidal/cylindrical	*Polydnaviridae*[a]	150–250	
Pleomorphic	*Ascoviridae*[a]	100–180	

[a]There is little structural information available on these insect viruses, and they are not covered further in this article.

white spot virus of crustaceans should be classified separately from the baculoviruses despite having a structure similar to that of typical baculoviruses (Durand *et al.*, 1997). In contrast, the assignment of channel catfish virus to the *Herpesviridae* was based solely on structural evidence because no sequence similarity with the mammalian or avian herpesviruses has been detected (Booy *et al.*, 1996).

An interesting question is whether these large DNA viruses are evolutionarily related or have independent origins. Although they appear diverse, there is some evidence for evolutionary relationships among the different families. Thus, it is clearly reasonable to consider poxviruses and iridoviruses as separate families on the basis of their different morphologies and genome structures, with African swine fever virus (ASFV) occupying an intermediate position (Salas *et al.*, 1999) with a poxvirus-like genome organization (Delavega *et al.*, 1994) and an iridovirus-like particle (Carrascosa *et al.*, 1984). However, genomic analysis has identified several groups of genes that are common to these three families and to the phycodnaviruses. Because of their common features, it has been suggested that all these families share a common ancestor and collectively can be grouped together under the name nucleocytoplasmic large DNA viruses (NCLDVs) (Iyer *et al.*, 2001). Of particular interest from the viewpoint of this article is the inclusion of the major capsid proteins among the conserved genes, suggesting that the structural diversity of the virus particles is a relatively late development. On the basis of similarities in their mechanisms of capsid assembly and DNA packaging (see Sections VII–X), a case can also be made for a shared ancestry for the herpesviruses and the tailed bacteriophages. Some similarities in their infection and assembly mechanisms could even point toward a distant link between the phycodnaviruses and some bacteriophages (Meints *et al.*, 1984), thereby suggesting the possibility of a common progenitor for all these large icosahedral forms. However, a better case, based on structural similarities, could probably be made for linking the iridovirus grouping with adenoviruses, which have themselves been shown to share features with the RNA bacteriophage PRD1 (Benson *et al.*, 1999). In the absence of more compelling evidence the formation of any such larger groupings remains somewhat subjective.

II. Why Large Viruses?

A general perception is that viruses are very small, and, indeed, many of them approach the lowest limits at which viability appears sustainable. For example, a poliovirus particle consists of an ~7.4-kb genome inside an

~300-Å-diameter protein shell made of 60 copies each of 4 proteins (Hogle *et al.*, 1985). Nevertheless, it can survive and proliferate successfully to the extent that until vaccines were developed it was widespread in human populations. By contrast, smallpox has an ~186-kbp genome (Massung *et al.*, 1994) enclosed in a >2000-Å-long particle containing about 100 different proteins, some of them in thousands of copies. Yet, smallpox was not notably more successful than poliovirus. Indeed, large viruses might seem to be at a growth disadvantage because making the individual particle requires a much greater commitment of resources than for small viruses. Considerations regarding efficiency of replication alone suggest that viruses would show a tendency toward decreasing size and complexity. Why, therefore, are not all viruses like poliovirus? One reason is that the assumption that there is selective pressure for viruses to become smaller and simpler is almost certainly false. Maximizing replication efficiency is only one of the factors influencing the evolution of viruses and is not necessarily the most important. Because by definition all extant organisms are successful, the strategies of picornaviruses and poxviruses have clearly been equally good at ensuring their continued existence. Nevertheless, this still leaves open the question of why these particular viruses are the sizes they are. Part of the answer must lie in their origin and evolutionary history, as this will place unquantifiable but probably severe constraints on how much they can change. Information about the probable histories of virus families can be found by comparisons of genome sequence data, which are now being generated in ever-greater amounts.

In both the herpesviruses and the NCLDV grouping, it has been possible to identify sets of conserved core genes that encode functions so basic to the replication strategies of the viruses that they can be assumed to have been among those present in the primordial viruses. They include functions needed for genome replication, gene expression, and virion formation. Because the number of core genes is relatively large, ~43 in herpesviruses (McGeoch and Davison, 1999) and ~31 in the NCLDV (Iyer *et al.*, 2001), it is clear that the common ancestors of these viruses must also have been reasonably complex. Nevertheless, the existing viruses in these groups do have many species-specific genes, particularly among those involved in interactions with the host, such as receptor binding and immune evasion. These appear to be later acquisitions, which must be presumed to confer selective advantage on the virus. It is unlikely that the genes present in any one virus represent all those that are potentially beneficial. What then prevents the capture of more such functions, leading to a growth in genome size? The presence among the core conserved genes of those involved in virion formation offers a clue, as it suggests that the basic conformation of the particle is one of the fundamental characteristics that becomes fixed at an early point in the history of a virus. Indeed, particle structure appears to be one of the less malleable features of viruses, which is why it is so widely used in their classification. Once the nature of the particle has become fixed it will effectively place an upper limit on the genome size on the principle that the size of a container determines the amount that can be put into it. Interestingly, for reasons that are less obvious, the size of the capsid also appears to set a lower limit on the size of the packaged genome. In some viruses, such as T4, where DNA is cleaved from replicative concatemers by a sequence-independent headfull mechanism (Black, 1989), the reason for a link between capsid and genome sizes is obvious. In strict headfull packaging, the length of DNA packaged into each particle is determined entirely by the physical constraints on the density of DNA packing and the force exerted by the packaging complex (see Section X). In other cases, such as herpes simplex virus type 1 (HSV-1), the precise length of the packaged DNA is determined by packaging signals in the DNA sequence (Stow *et al.*, 1986). This should allow encapsidation of smaller genomes, where the packaging signals are closer together. However, analysis of defective genomes that contain multiple tandem repeats of the packaging signals has shown that only "full-length" genomes are successfully packaged into mature virus particles (Vlazny *et al.*, 1982). Encapsidation of such tandemly reiterated genomes requires that many potential cleavage sites must be ignored by the packaging mechanism. Therefore, it is evident that here, as in T4, packaging continues until the capsid is effectively full and in both cases the capsid capacity is the prime determinant of packaged genome size.

These arguments apply only to viruses in which the particle size is genetically determined. In many helical viruses, the length of the capsid is indeterminate because it is defined by the size of the packaged nucleic acid molecule and therefore the same constraints do not apply.

Although overall virus architecture does appear to be remarkably resistant to change, it is nevertheless the case, despite the arguments made here, that the genome sizes of different viruses belonging to the same family often cover quite large ranges. Clearly, this ability to change is an important property for the virus and it is interesting to examine how it is accomplished. As already stated, there is little problem for helical viruses, such as baculoviruses, which can simply vary the length of their particles. However, for icosahedral viruses changing size appears to be a much more complicated process, and the three basic types of icosahedral virus covered here achieve it in different ways.

Herpesviruses are the most conservative, with all known members of the family having capsids of approximately the same size and with the same

T=16 arrangement of subunits. What limited variation there is does not affect these features. Thus, although the genome of human cytomegalovirus (HCMV) is 60% longer than that of HSV-1, the capsids are remarkably similar. To accommodate the extra material, HCMV packages its DNA more densely into a capsid that has a slightly larger internal volume (Bhella *et al.*, 2000; Butcher *et al.*, 1998). In the large icosahedral viruses that belong to the NCLDV grouping, the capsids vary in size by modifying the arrangement of subunits within the parameters of a conserved basic architecture. For example, although similar in most other respects, the capsid of *Paramecium bursaria chlorella* virus 1 (PBCV-1) is a T=169 icosahedron, whereas that of Chilo iridescent virus (CIV) is T=147 (see Section V) and alternative T numbers are found in other members of this group. Clearly, something in the nature of the NCLDV type of capsid allows them to exhibit a degree of variability that is absent in herpesviruses. This may be related to their differing types of subunit interaction, because in the NCLDV capsids identical capsomers interact in apparently identical repeating patterns (see Section V), whereas in herpesviruses, interactions are formed between dissimilar subunits (triplexes and hexons; see Section V) in a context-dependent fashion. The variability of the NCDLV capsid structure is taken to an extreme by the poxviruses, which have abandoned the icosahedral theme entirely. The tailed bacteriophages have also modified the nature of their capsids to facilitate changes in genome size. It seems certain that the original icosahedral head was isometric as it still is in many smaller bacteriophages and in certain mutants of T4 (see Section V). However, the capacity to package larger amounts of DNA was not achieved by a straightforward expansion of the icosahedral lattice as in the NCLDV grouping. Instead, the head has undergone a nonuniform expansion through the selective insertion of extra capsomers around the midsection of the icosahedron, leading to the elongated, prolate head typical of T4.

III. Why Study Large Viruses?

Many large viruses are important pathogens and one reason for examining their structure and assembly is to obtain fundamental information, which may aid in the search for effective control and intervention strategies. However, as well as being of interest as agents of disease, viruses have traditionally served as probes for examining molecular mechanisms. Studies of viruses have contributed fundamentally to our understanding of such fundamental processes as DNA replication and gene expression. They have been particularly important in the study of macromolecular assembly mechanisms; various virus assembly models are among the best understood of such processes. Although much of what is known has come from studies on small viruses, large viruses have the potential to illuminate certain cellular processes that are poorly represented among their smaller relatives. In some respects, a large virus is like a small cell. It may have the same mixture of symmetrical and irregular components, the same indeterminate composition, and may exhibit the same plasticity of interaction among its components. Many of the larger particles are multicompartment structures. This requires the spatial and temporal coordination of a carefully orchestrated sequence of assembly steps and raises questions regarding what controls and limits the assembly processes, particularly of the asymmetrical components. At a more basic level, questions such as how protein shapes determine their functions, how protein folding is controlled, and what triggers conformational changes that cause progression of processes can all be illustrated by examples from viruses. Therefore, in many respects the relative simplicity of even the largest viruses offers a definable microcosm of the whole cell.

IV. Methods of Structural Analysis

In comparison with many smaller viruses, structural information about all the large viruses covered here is limited and, in some instances, virtually nothing is known of their organization at the molecular level. This is particularly true for the poxviruses and baculoviruses, which our understanding of virion organization is based almost entirely on examinations using the standard electron microscope (EM) techniques of negative staining and thin sectioning. These approaches produce representations of gross morphology rather than direct structural details. Therefore, they are highly dependent on interpretation by the investigator. Furthermore, because they invariably involve (often severe) modification during sample preparation the interpretation must be tempered with caution.

Nonsubjective methods of analysis suitable for examining virus structure are of two types: X-ray crystallography- and electron cryomicroscopy-based reconstruction. At present X-ray crystallography is the only technique that is able to provide the high-resolution information allowing determination of protein secondary and tertiary structures (see by Gilbert *et al.*, This volume). None of the viruses defined as large for the purpose of this article has yet been crystallized, leading to a dearth of such information. The possibility of crystallizing virions with external envelopes (*Herpesviridae*, *Baculoviridae*, and *Poxviridae*), or the complex tailed particle of T4, appears small. Success may be more achievable with some of the large icosahedral viruses, the outer surfaces of which are formed by a regular invariant protein shell. When the intact particle cannot be studied, it may be possible to crystallize particular substructures such as the herpesvirus and baculovirus nucleocapsids or the T4 phage head. However, even if these particles were successfully crystallized, solving the structures of such large entities would present technical challenges in data collection and analysis because of the large unit cell parameters in those crystals.

In the absence of crystallographic data, much can still be learned about the organization of virion structures using lower resolution information. For an increasing number of viruses this information has been provided by computer-based reconstruction of electron cryomicroscope images, which, unlike conventional EM images, are generated from rapidly frozen samples that are in nearly physiological condition (see Zhou and Chiu, This volume). At present, EM-based reconstruction of viruses can produce only low- to intermediate-resolution (~7 Å) information (Bottcher *et al.*, 1997; Zhou *et al.*, 2001). Although this approach is potentially applicable to all classes of virus, its use has largely been limited to viruses with icosahedral symmetry, where it has become a standard method for analyzing particle structure and is increasingly being applied to the study of assembly intermediates and viruses undergoing morphological transitions.

Combining these two approaches, by determining the crystal structure of individual subcomponents and fitting these into the lower resolution maps of intact particles, may offer the best current hope for obtaining high-resolution information about virion structure. An example of this is provided by adenovirus, where the crystal structure of the hexon, formed by three copies of the ~100-kDa major structural protein, has been fitted into a low-resolution capsid structure determined by electron cryomicroscopy to provide a clearer picture of capsid organization and information about the locations of minor capsid proteins (Stewart *et al.*, 1993). A similar strategy has been used to fit the crystal structure of a portion of the HSV-1 major capsid protein (VP5) into an intermediate-resolution map of the capsid (B. Bowman, personal communication).

V. Complexity of Organization

All virus particles have a range of functions, among the most fundamental of which is protecting the virus genome during the extracellular phase of the life cycle. This requires the virion to provide an impervious barrier around the nucleic acid and there are two basic ways of supplying this, either through a lipid membrane or through a continuous protein shell. Unlike smaller viruses, few large viruses rely entirely on forming an impermeable protein shell (of the viruses considered here, only bacteriophage T4 lacks a lipid envelope). The reasons why this should be so are unclear but may relate to the relative effort involved in forming the two types of structure. As a way of enclosing a defined volume, lipid membranes offer several advantages, which is presumably why they are widely used for this purpose by the host cells themselves. They form from relatively simple components by a process that is scaleable to any biologically useful size and are naturally impermeable to most water-soluble biological molecules. In addition, their properties can easily be modified by incorporating appropriate proteins, using simple and nonspecific processes. For viruses, which normally make use of preexisting cellular membranes, they have the added advantage that their use does not place major demands on the limited coding capacity available. By contrast, forming a large, impervious shell from protein alone is much more demanding because all the components are usually synthesized *de novo,* and in order to form an impervious barrier their molecular interactions must be specified precisely.

Because of their complexity, large viruses have much greater opportunities for compartmentalizing their various constituents than do smaller viruses. In this respect they again have more in common with their host cells. Most of the large viruses covered here have multilayered structures with different functions distributed among the various compartments, which in some cases are also bounded by their own lipid envelopes. For example, in herpesviruses the job of packaging and enclosing the genome is carried out by the capsid, whereas that of receptor recognition and cell entry falls to the envelope; poxviruses, on the other hand, have complex internal structures (see below) with many proteins sequestered into specific compartments. Again, T4 appears to be an exception as the phage head is a single protein shell. However, this is misleading as important functions, such as receptor binding and insertion of the DNA, are again compartmentalized by being devolved to the specialized tail structure.

A. *Nonicosahedral Particles*

In the following sections the main structural features of the virus particles are outlined. Because space limitations permit only brief descriptions, some of the fundamental properties of each family are listed in Table II.

TABLE II
Basic Properties of Virus Particles

Property	*Baculoviridae* (AcMNPV)	*Poxviridae* (vaccinia)	*Asfarviridae* (ASFV), *Iridoviridae* (CIV), *Phycodnaviridae* (PBCV-1)	*Herpesviridae* (HSV-1)	*Myoviridae* (T4)
Particle size (Å)	300–600 × 2500–3500 (capsid)	2000 × 2000 × 2500	1600–1900	~2000 (virion); 1250 (capsid)	1150 × 850 (head)
Number of structural proteins	>20	>~100	>50	>30	>30
External envelope	Yes	Yes	Sometimes	Yes	No
Internal envelope	No	Yes	Yes	No	No
T number	NA	NA	147 (CIV); 169 (PBCV-1)	16	T=13, *Q*=21
Major capsid protein copy number	?	?	8760 (CIV); 5040 (PBCV-1)	955	960
Major capsid protein (kDa)	~42	~65[a]	~50	~150	~50

Abbreviations: See Table I.
[a]Poxvirus does not form a capsid, but this structural protein (Sodeik *et al.*, 1994) shares a conserved domain with the asfarvirus, iridovirus, and phycodnavirus capsid proteins (Iyer *et al.*, 2001).

1. *Baculoviruses*

The baculovirus virion consists of one or more nucleocapsids enclosed in an envelope (Fig. 1A). Infectious virions may occur as free particles or be embedded in proteinaceous occlusion bodies (Friesen and Miller, 2001) that provide protection from environmental assaults during the frequently extended periods between hosts (Harrap, 1972; Steinhaus, 1960). The capsid structure is poorly known but examination of *Spodoptera litura* granulosis virus (SLGV) capsids by optical diffraction has shown that the outer shell is formed as a stacked ring of subunits with a 12 start helix repeating every third ring (Burley *et al.*, 1982). The capsid is capped at both ends by a regular arrangement of protein subunits that appear

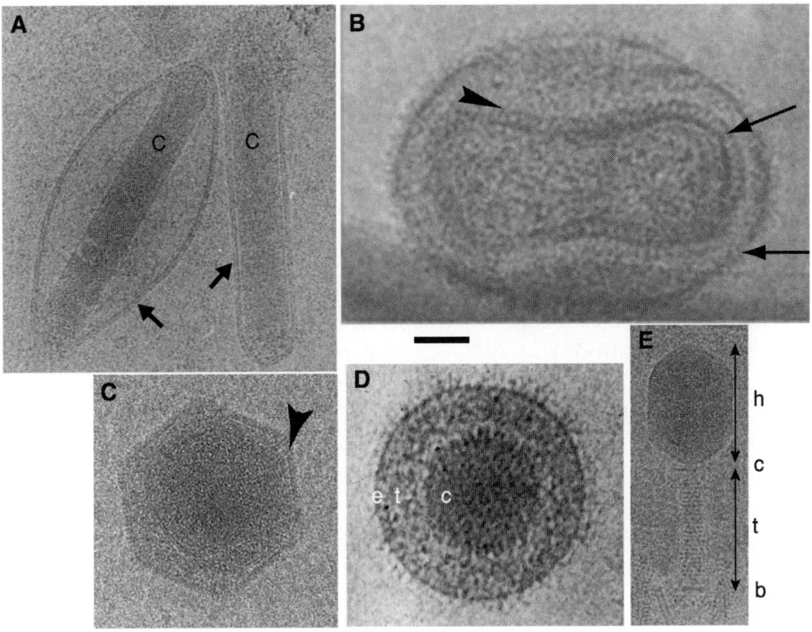

FIG. 1. Electron cryomicroscopic images of large virus particles. Images of frozen hydrated virions of (A) AcMNPV (*Baculoviridae*), (B) vaccinia (*Poxviridae*), (C) PBCV-1 (*Phycodnaviridae*), (D) HSV-1 (*Herpesviridae*), and (E) T4 (*Myoviridae*). In (A) two baculovirus particles are shown. In the right-hand particle the lipid envelope (indicated by the arrow) is closely associated with the nucleocapsid, whereas in the left-hand particle it has separated from it, making it more visible. C, nucleocapsid. In the vaccinia IMV (intracellular mature virus) particle shown in (B), two membrane layers are indicated by arrows and the layer of spicules that form a hexagonal lattice around the core (Dubochet *et al.*, 1994) is indicated by an arrowhead. The arrowhead in (C) indicates the external capsid shell in the PBCV-1 particle. In the HSV-1 image (D) the envelope (e), tegument (t), and capsid (c) are indicated. For T4 (E), the head (h), collar (c), tail (t), and baseplate (b) are indicated. The tail fibers are not visible in this image. All images are shown at equivalent magnification. Scale bar: 500 Å. [Images were supplied by David Bhella (A) and Alasdair Steven (E) or were reproduced from Griffiths *et al.* (2001) (B), Yan *et al.* (2000) (C), and Rixon (1993) (D).]

different from the cylinder subunits. The caps appear as a set of stacked rings of decreasing diameter. The ends are structurally different (Fraser, 1986), which affords polarity to the capsid. The closed circular, dsDNA is condensed into a nucleoprotein core (Tweeten *et al.*, 1980; Wilson *et al.*, 1987), which has been interpreted as a quasi-uniform cylinder of loosely packed DNA embedded in protein. Forty percent of the capsid mass is made up of a small, highly basic protein that is believed to represent the probable core protein.

2. *Poxviruses*

Our knowledge of poxvirus structure comes primarily from studies on vaccinia virus, which has a distinctive oblong, or brick-shaped particle with a complex arrangement of substructures (Moss, 2001). There are two forms of infectious particle: intracellular mature virus (IMV) and extracellular enveloped virus (EEV). EEV differs from IMV in the possession of an additional membrane (the envelope), which is acquired from cytoplasmic vacuoles. Their appearance in negative stain and thin section shows a dumbbell-shaped nucleoprotein core, bounded by a core envelope and flanked by two large lateral bodies. The outer membrane surrounding the particle frequently has a distinctive appearance consisting of randomly arranged tubular structures. It has been suggested on the basis of the appearance of IMV particles in electron cryomicroscopy that the lateral bodies and surface tubules are artifacts of sample preparation (Dubochet *et al.*, 1994), but studies have demonstrated the presence of lateral bodies in unfixed samples (Fig. 1B and Griffiths *et al.*, 2001). Almost nothing is known about the molecular organization of the poxvirus particle.

B. *Particles Containing Icosahedral Capsids*

1. *Iridoviruses, Phycodnaviruses, and Asfarviruses*

Iridoviruses, phycodnaviruses, and asfarviruses have basically similar structures (Becker *et al.*, 1993; Carrascosa *et al.*, 1984; Devauchelle *et al.*, 1985; van Etten *et al.*, 1982, 1991), in which the viral core comprises a roughly spherical DNA-containing nucleoid that is embedded in a thick protein layer designated the core shell (Fig. 1C). This is surrounded by one or two lipid envelopes, which are in turn surrounded by the icosahedral capsid. In insect iridoviruses and phycodnaviruses the capsid forms the surface of the virion but in the vertebrate iridoviruses and asfarviruses there is an outer lipid envelope, which encloses the capsid. The structures of chilo iridescent virus (CIV) and *Paramecium bursaria Chlorella* virus type 1 (PBCV-1) have been determined to 26 Å (Fig. 2A [see Color Insert] and Yan *et al.*, 2000). The capsids are made up of 1460 (CIV) and 1680 (PBCV-1) hexavalent capsomers and 12 pentavalent capsomers surrounding a lipid bilayer. Although the hexavalent capsomers are hexagonal in shape, close inspection of their density distribution reveals that they are in fact trimeric. In PBCV-1, each of the 1680 hexavalent capsomers is formed by a trimer of subunits, giving a total of 5040 copies in the capsid shell. In CIV there is an extra protein layer between the outer capsid shell and the lipid bilayer. This inner capsid shell is similar in T number, composition, and organization to the outer capsid shell, although the inner trimers are rotated through 60° with respect to the surface capsomers. Therefore, although CIV has a lower T number than PBCV, it contains more copies (8760) of the capsid protein. In both viruses, the pentons differ in appearance from the hexons. Although their composition is not known it seems likely that they are formed from a protein other than that composing the hexons, because there is no easy way in which a hexavalent, trimeric structure can be adapted to occupy a pentavalent position.

2. *Herpesviruses*

All herpesvirus virions have a characteristic appearance (Fig. 1D) with an icosahedral capsid surrounded by a thick (~500 Å) layer of protein designated the tegument (Rixon, 1993; Steven and Spear, 1997). The entire particle is enclosed by a spherical lipid envelope (Szilagyi and Berriman, 1994). The core of infectious virions is believed to consist exclusively of the double-stranded DNA genome without any associated protein (Booy *et al.*, 1991; Zhou *et al.*, 1999). The structures of the tegument and envelope are poorly understood as they are indeterminate in both size and composition and have a limited symmetrical relationship to the capsid (Chen *et al.*, 1999; Zhou *et al.*, 1999).

Although the structures of several herpesvirus capsids are now available (Baker *et al.*, 1990; Booy *et al.*, 1996; Chen *et al.*, 1999; Schrag *et al.*, 1989; Trus *et al.*, 1999, 2001; Wu *et al.*, 2000) they are basically similar and only HSV-1 is described here (Fig. 2B and Zhou *et al.*, 2000). The capsid is a T=16 icosahedron, the faces and edges of which are formed by the 150 hexons. The icosahedral vertices are usually described as being occupied by pentons, but studies have suggested that 1 of the 12 positions contains the portal complex (Newcomb *et al.*, 2001). In addition, 320 triplexes occupy the local 3-fold positions between the hexons and pentons. Unlike the other icosahedral viruses described here, the hexons and pentons are predominantly formed by the same protein. Thus, each hexon is composed of six copies of VP5, and six copies of a second protein, VP26, whereas the pentons each comprise five copies of VP5 but lack VP26. The six copies of VP26 are arranged head to tail to form a flattened star-shaped ring, which occupies the top of the hexon. The triplexes are unusual asymmetric structures that have no counterparts in the other large icosahedral viruses. In all cases studied so far they are heterotrimers, which in HSV-1 are made up from two copies of VP23 and one copy of VP19C.

3. *T4*

The large DNA bacteriophages, typified by T4, have distinctive structures in which an icosahedrally organized head is connected at one vertex through a collar to a complex tail structure (Fig. 1E). Tails are found only in certain classes of bacteriophage and they seem to have evolved as an efficient tool for transferring the virus genome across the bacterial cell wall. In T4 the tail shaft consists of two concentric cylinders with a channel down the center of the inner cylinder through which the DNA passes during infection (Coombs and Arisaka, 1994). At the bottom of the tail is a hexagonal baseplate that acts as the attachment point for six long and six short tail fibers that form part of the receptor-binding apparatus. The tail is connected to the phage head via a collar, comprising a ring of 12 subunits that forms part of the machinery of DNA packaging.

The T4 head is a prolate icosahedron (Black *et al.*, 1994) that is elongated along the direction of the 5-fold icosahedral axis occupied by the tail. The elongation is achieved by incorporating 40 additional hexons into the 10 triangular faces lying parallel to this axis. The T4 icosahedron has an underlying triangulation number of T=13 but the equivalent number for the elongated faces is 21. Therefore there are 160 hexons (960 copies of gp23*) in the prolate head compared with the 120 expected for an isometric T=13 capsid. Because the prolate head is unsuitable for analysis by reconstruction techniques that require icosahedral symmetry, most of what is known about its structure comes from conventional EM of negatively stained, shadowed, or freeze-fractured material, and from examination of variant head forms produced by the numerous virus mutants. The structure of the isometric head formed by a mutant in the hexon protein gene has been determined by electron cryomicroscopy and image reconstruction (Fig. 2C and Iwasaki *et al.*, 2000; Olson *et al.*, 2001). As expected, the subunits in the isometric head are organized in T=13 icosahedral symmetry. One of the 12 vertices is occupied by the portal complex and the 11 remaining vertices are occupied by pentamers of the penton protein, gp24*. The faces and edges of the icosahedron are formed by 120 hexavalent capsomers, each of which contains 6 copies of gp23*. Two additional proteins, Hoc and Soc, are also present on the capsid surface. A single copy of Hoc is present at the center of each of the hexameric capsomers. The distribution of Soc is more complex, because it occupies the interfaces between gp23* molecules but not those between gp23* and gp24*. Therefore, Soc completely encircles the nonperipentonal hexons but is present on only five sides of the peripentonal hexons, being absent from the hexon–penton junction. Although the capsid shell is much thinner (~30 Å) than those of the other icosahedral viruses described here, it is not penetrated by any channels or pores apart from the channel through the portal complex. In the prolate head the additional hexons in the elongated faces appear to be indistinguishable from the others and have a similar arrangement of Hoc and Soc proteins.

VI. Structural Folds

The paucity of crystallographic data for large virus structural proteins makes it difficult to talk with confidence about high-resolution features such as secondary structural elements and folds. However, the increases in resolution of electron cryomicroscopic reconstructions of viral particles mean that they are now entering the resolution range of 7–10 Å, at which more detailed pictures of the architecture of their component proteins can be obtained. In particular, it is now possible to map the distribution of long α helices (>2.5 turns) within proteins and, by using computational methods for comparing them with proteins of known structure, determine whether they fit into any previously described structural families. As the database of protein folds grows larger and the procedures for computational analysis improve, the potential for using this approach to suggest new folds or recognize previously described folds within the intermediate–resolution reconstructions produced by electron cryomicroscopy is certain to increase.

The only large virus for which this level of analysis has been achieved is HSV-1 (Zhou *et al.*, 2000), and examination of the secondary structural features provides interesting insights into the kinds of information that can be deduced regarding the molecular mechanisms underlying some of the properties of the capsid. Long α helices have been visualized in all the HSV-1 capsid shell proteins with the exception of VP26. In the major capsid protein, VP5, at least 33 α helices with more than 2.5 turns have been identified by computerized topological searching of the 8.5-Å density map (Fig. 3A and B; see Color Insert). These α helices are distributed throughout the 150-Å length of the protein, with notable concentrations in the middle of the hexon and the floor (Fig. 3B). Those in the floor cannot yet be correlated with any known structural fold, although they seem to represent a single domain (He *et al.*, 2001). However, the middle domain of VP5 assumes a conformation that does not resemble any known viral protein fold, but is similar to the α-helix bundle fold of a nonviral membrane associated protein, annexin (Baker, 2002). There is no evidence for an evolutionary relationship between annexin and VP5 and in all other respects they are structurally unrelated. Therefore, although it seems likely that these similar but independent structures represent defined protein domains, their respective functions may be unrelated. The 10 α helices in this domain are of approximately equal length and are predominantly aligned parallel to the long axis of the hexon. The relationship between them is even more notable when seen in the context of the entire hexon, where they form a continuous band of uniformly spaced α helices completely encircling the hexon channel (Fig. 3C). The close packing of these α helices suggests that they are likely to have extensive inter- and intramolecular interactions, forming a rigid ring or collar that will reinforce the hexon and lock the subunits together. In the penton the arrangement of this group of α helices is altered, making it much less likely that they would interact strongly. Because the region of VP5 containing the α helices is involved in a mass redistribution that closes the penton channel (see Section X), it is likely that the differing patterns seen in hexons and pentons are necessitated by the different roles of these capsomers. Interestingly, the distortion of the hexons, which is a feature of the procapsid, would also disrupt this ring of α helices, and this would probably contribute to the instability of the procapsid (see Section VII).

VII. Assembly Mechanisms

In most helical and some smaller icosahedral viruses, assembly and packaging occur in a single step when the nucleic acid and capsid proteins come together to form the nucleocapsid. Such coassembly appears to require an intimate and quantitative association involving all parts of the nucleic acid and the capsid proteins. This close association between contents and container is inevitably absent when a long DNA molecule is packed inside a large spherical capsid (Zhou *et al.*, 1999). Therefore, the large icosahedral viruses use an alternative approach and package their genomes into a preformed capsid, which in some cases can be isolated as a separate entity.

HSV-1 and T4 are by far the best understood of the large viruses and their assembly processes may provide pointers to some of the mechanisms likely to occur in other virus families. In both HSV-1 and T4, the first product of assembly is not the capsid described above but rather a precursor, the procapsid. The formation of procapsids and their role in the capsid assembly pathway have long been known in T4 (see Black *et al.*, 1994). However, in HSV-1, procapsids have been described only relatively recently. Although they were initially identified as the products of *in vitro* capsid assembly experiments (Newcomb *et al.*, 1994), their role in infected cells is now firmly established (Church and Wilson, 1997; Newcomb *et al.*, 2000; Preston *et al.*, 1983; Rixon and McNab, 1999). Procapsids are formed by polymerization of capsid shell proteins around an internal structure called the scaffold. The scaffold has a number of possible functions but a major purpose is to determine the size and assure the fidelity of the capsid shell (see Fane and Prevelige, This volume). This function is likely to be particularly important for large capsid shells, where the possibility of incorrect interactions is multiplied by the increased number of subunits involved.

In T4, assembly is initiated on the portal complex and probably occurs by stepwise addition of monomeric gp23 to the growing edges of the capsid shell (Fig. 4A). In contrast to many other bacteriophages, complete T4 scaffolds can form in the absence of the capsid shell proteins (Traub and Maeder, 1984; van Driel and Couture, 1978) and these preformed scaffolds appear capable of supporting subsequent head assembly if the shell proteins are supplied (Kuhn *et al.*, 1987). However, it is likely that the normal assembly process in T4 involves coassembly of these two compartments. This is certainly the case for HSV-1, where examination of intermediates formed at short intervals after mixing of *in vitro* extracts showed that assembly occurs by copolymerization of the shell and scaffolding proteins to form wedges and arcs that increase in size until the complete particle has formed (Newcomb *et al.*, 1996) (Fig. 4B). Like its T4 counterpart, the HSV-1 scaffolding protein, pre VP22a, can also form apparently intact scaffolds in the absence of the shell proteins (Kennard *et al.*, 1995; Newcomb and Brown, 1991), although it is not yet clear whether these can act as substrates for capsid assembly (Newcomb *et al.*, 1999). Assembly studies have shown that the HSV-1 portal protein is not required to initiate formation of otherwise normal capsids (Newcomb *et al.*, 1996, 1999; Tatman *et al.*, 1994; Thomsen *et al.*, 1994). However, the presence of single copies of the portal complex in mature capsids (Newcomb *et al.*, 2001) strongly suggests that, as in T4, this is the normal pathway of assembly.

In both HSV-1 and T4, the procapsids are structurally distinct from typical capsids. This is particularly evident in T4, where maturation involves head expansion, which is accompanied by angularization of the capsid shell (Black *et al.*, 1994). Although maturation of the HSV-1 procapsid does not involve a major size increase, it does alter the shape from spherical to polyhedral (Trus *et al.*, 1996) and is accompanied by changes to the antigenicity of the capsid surface (Gao *et al.*, 1994). In both T4 and HSV-1, maturation has a great effect on the stability of the particles as in both cases mature capsids are resistant to disruption whereas procapsids rapidly dissociate when exposed to fairly mild assaults (Newcomb *et al.*, 1996; Steven *et al.*, 1992; van Driel, 1977).

The structure of the T4 procapsid has not yet been analyzed in detail but that of HSV-1 has been solved to ~18-Å resolution (Newcomb *et al.*, 2000) (Fig. 5). This has revealed a number of features that differentiate the procapsid from the capsid and emphasize the similarities between its assembly pathway and that of certain tailed bacteriophages. The overall dimensions of the HSV-1 procapsid are similar to those of the mature capsid, and the capsid shell contains recognizable pentons, hexons, and triplexes organized in the same T=16 symmetry (Trus *et al.*, 1996).

A

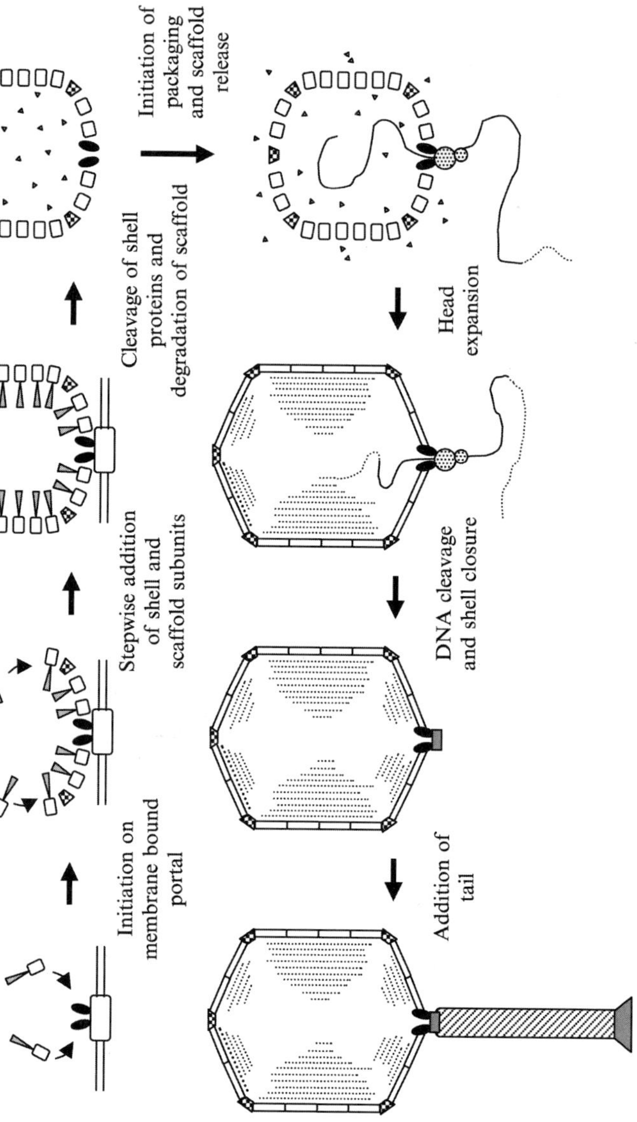
Initiation on membrane bound portal
Stepwise addition of shell and scaffold subunits
Cleavage of shell proteins and degradation of scaffold
Initiation of packaging and scaffold release
Head expansion
DNA cleavage and shell closure
Addition of tail

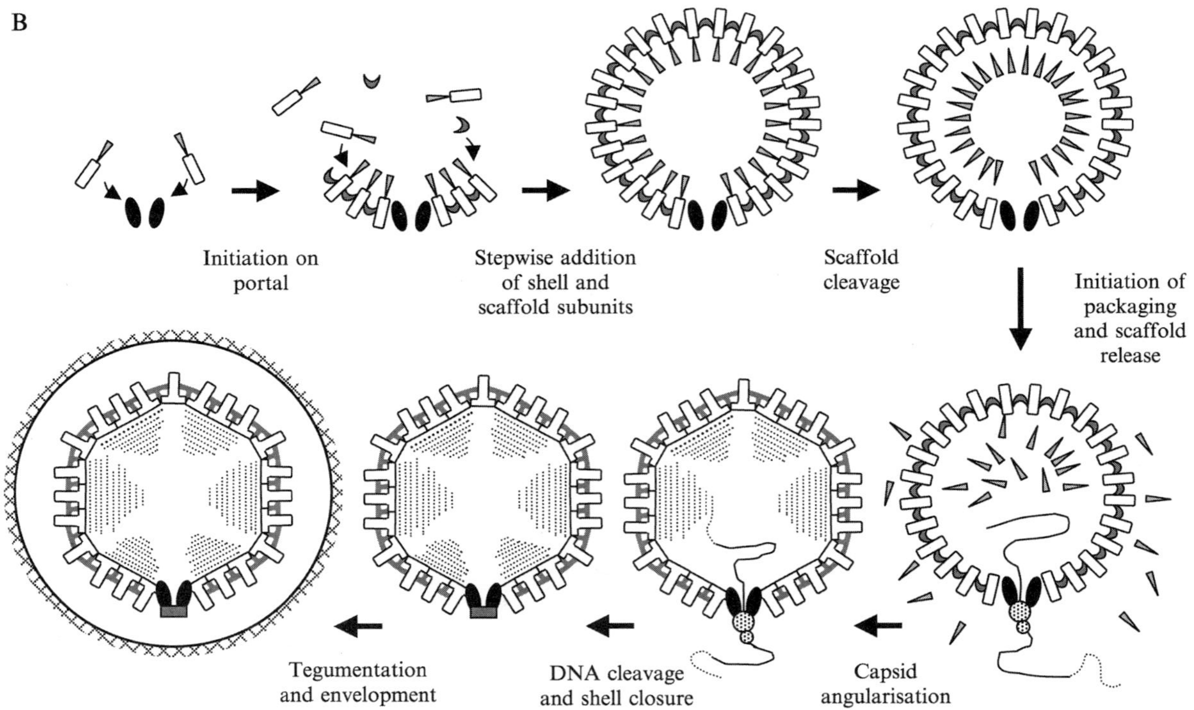

FIG. 4. Assembly pathways for T4 and HSV-1. Equivalent stages in the assembly and maturation pathways of phage T4 (A) and HSV-1 (B) are illustrated to allow direct comparison. The major capsid proteins that make up the hexons and pentons in HSV-1 and the hexons in T4 are represented as open boxes and the T4 penton proteins are indicated by hatched trapeziums. The HSV-1 triplexes are shown as crescents. In both cases the scaffolding proteins are represented by triangles. Some steps, for example, release of scaffolding proteins and closure of the portal channel, are poorly understood and may differ in detail from the mechanisms shown here.

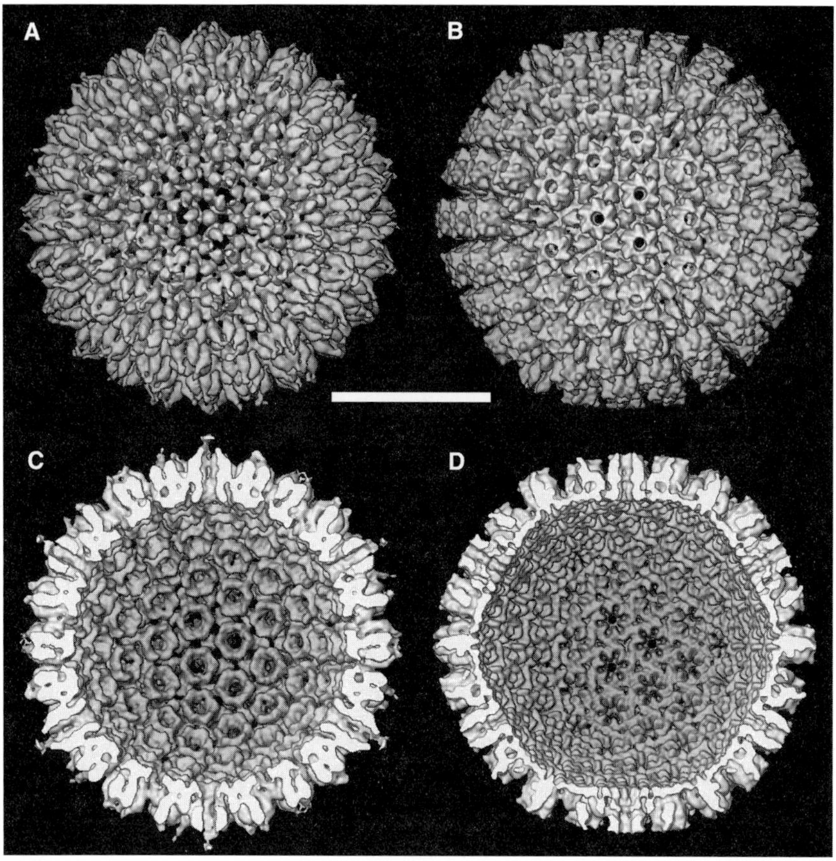

FIG. 5. Comparison of HSV-1 procapsid and capsid structures. Outside (A and B) and inside (C and D) views of reconstructions of the HSV-1 procapsid (A and C) and capsid (B and D). All views are shown looking along a 3-fold icosahedral axis. Although it retains the same icosahedral symmetry, as shown by the hexagonal arrangement of the capsomers apparent in the internal view (C), the procapsid subunits are much less clearly defined than those in the capsid. In addition, the contacts between individual subunits are much more tenuous and the continuous floor, formed by extensions from the bases of the capsomers, is not present. Note also the contrast in shape between the spherical procapsid and the polyhedral capsid. Scale bar: 500 Å. [Image supplied by Benes Trus.]

However, in addition to the overall change from spherical to polyhedral, there are also smaller changes in the detailed relationship between the subunits. In the procapsid, the hexons are oval rather than hexagonal and the subunits appear skewed. In this respect, it resembles the procapsids of certain dsDNA bacteriophages, notably P22, which also has skewed hexons arranged in a spherical shell (Thuman-Commike *et al.*, 1996). One particularly striking aspect of the HSV-1 procapsid structure is that there appears to be no direct interaction between neighboring capsomers. The capsid floor, which provides such connections in mature capsids, is not present and the only contact between adjacent capsomers is through the triplex (Trus *et al.*, 1996).

VIII. Maturation

The assembly of complex icosahedral particles requires that the component proteins interact in subtly different ways to adapt to the varying environments found at different quasi-equivalent locations. This process is likely to be more easily achieved if the contacts between the proteins are weak and generalized and if the proteins themselves retain some flexibility (Kirkitadze *et al.*, 1998). However, these weak contacts must be strengthened if the mature virus is going to have the structural integrity to resist the expected environmental challenges. The geometric constraints on large icosahedra that determine the final form of the capsid and the nature of the changes occurring during maturation are described by Moody (1999). Both T4 and HSV-1 have developed mechanisms for releasing their genomes from intact capsids; therefore the strengthening process does not need to be reversible and can be accomplished by large-scale reconfiguration of subunits (Steven *et al.*, 1990, 1991).

The maturation of both T4 and HSV-1 procapsids involves proteolytic cleavage of some of the component proteins. In T4 the proteolysis is extensive and all the capsid shell and core proteins are affected apart from the portal protein gp20 (Showe *et al.*, 1976), but in HSV-1 it is restricted to the components of the internal scaffold (Liu and Roizman, 1991). *In vitro* studies in T4 have shown that although head expansion must be preceded by proteolysis, the two processes can be separated temporally with the isolation of an intermediate form, which has undergone cleavage but not head expansion (Carrascosa, 1978). In HSV-1, procapsids containing cleaved scaffolding proteins are unknown, but cleavage of the scaffolding protein is not essential for the shell proteins to undergo the procapsid-to-capsid transition (Desai and Person, 1999; Newcomb *et al.*, 2000; Zhou *et al.*, 1998).

In T4 giant capsids, head expansion takes place in a coordinated series of changes that propagate down the long axis of the particles in a rapidly moving, narrow front (Steven and Carrascosa, 1978). It seems likely that a similar wave of change, probably initiating at the portal vertex (see Cerritelli *et al.*, This volume), occurs during the maturation of normal T4 heads and this is probably also the case for HSV-1 capsids (Rixon and McNab, 1999). The molecular changes that occur during maturation are substantial, involving large-scale subunit movements and extensive reconfiguration of individual proteins (Kistler *et al.*, 1978). In neither T4 nor HSV-1 is the precise nature of the changes known, although the availability of structures for both procapsid and capsid in HSV-1 does provide some insights (Trus *et al.*, 1996). One of the most obvious differences between the two particles is the absence of floor density in the procapsid. Formation of this floor must occur through the movement of what will become the floor domains of the major capsid protein, VP5, into their final positions, probably by rotation around a hingelike region of flexible protein at the base of the hexon (He *et al.*, 2001). This has the effect of forming new contacts between neighboring VP5 subunits from different capsomers. The triangular appearance of the triplex in the procapsid also changes and it becomes irregular with a sloping, roughly diamond-shaped body, connected to the underlying capsid floor by three projections or legs. The body and one leg are formed by VP19C, whereas the two copies of VP23 form the other two legs (Saad *et al.*, 1999; Zhou *et al.*, 2000). It seems likely that in the procapsid, all the connections between the triplex and surrounding capsomers are formed by VP19C, and that the contacts between the VP23 molecules and the newly formed capsid floor arise only as a result of capsid maturation, further strengthening the structure (Zhou *et al.*, 2000).

One consequence of the maturational changes is that the inside of the capsid becomes increasingly isolated from the outside environment. In HSV-1 this isolation is by no means complete as not all channels through the shell are closed during maturation, but in T4 and other bacteriophages the end product of head expansion is a continuous protein shell that is penetrated only at the portal vertex. Because the portal channel is presumably filled by the viral DNA during packaging, alternative channels through the procapsid shell are likely to be needed to allow the exit of the internal core components. In HSV-1 the open nature of the procapsid shell provides many possible exit routes, but as its structure has not yet been determined the nature of any channels though the T4 procapsid shell remains unknown.

IX. Accessory Proteins

The main structural differences between procapsids and mature capsids are a result of the reconfiguration of the capsid shell and loss of scaffold described above. However, both T4 and HSV-1 encode additional proteins called accessory proteins, which are not essential for capsid assembly but are incorporated into the particle at a late stage after the capsid shell has adopted its final configuration. Two accessory proteins (Hoc and Soc) are found in the T4 capsid (Iwasaki *et al.*, 2000; Olson *et al.*, 2001) (Fig. 6B; see Color Insert). The function of the single copy of Hoc present at the center of each hexon is unknown. It has little effect on capsid stability (Ishii and Yanagida, 1977; Ross *et al.*, 1985) and its location is not obviously a position of structural weakness. By contrast, the location of Soc, at the interfaces between hexons, is ideally suited to its role in stabilizing the capsid shell and particles lacking Soc are indeed more easily disrupted (Ishii *et al.*, 1978; Steven *et al.*, 1992). Soc does not bind at the hexon/penton junctions and the lack of Soc binding at this position might make it a weak point in the capsid shell susceptible to external stresses such as osmotic shock (Iwasaki *et al.*, 2000).

Like Soc, the HSV-1 accessory protein (VP26) binds to hexons but not to pentons (Fig. 6A). However, the roles of these two proteins in stabilizing the capsid appear somewhat different. In T4 the hexon subunits are tightly integrated, and form a solid monolithic structure. Therefore, the points of weakness that are most in need of reinforcement are at the junctions between the hexons. By contrast, in HSV-1 the interactions in the floor between subunits from different hexons appear to be more extensive than those in the upper domains between subunits of the same hexon. Interhexon contact is further reinforced by the triplex, a structure that has no equivalent in T4. Therefore, the point of weakness in the herpesvirus capsid is probably within, rather than between, the hexons and this may explain why VP26 binding appears designed to reinforce the intrahexon subunit interactions (Trus *et al.*, 1995; Zhou *et al.*, 1995). Both Soc (Iwasaki *et al.*, 2000) and VP26 (Wingfield *et al.*, 1997) occur predominantly as monomers in solution and although the relationship between Soc proteins on the capsid is uncertain it is clear that in HSV-1 there are strong interactions between neighboring VP26 monomers, which must further serve to tie the hexon subunits together.

X. Packaging

In both T4 and HSV-1 the maturation of the capsid shell is necessary for DNA packaging. In HSV-1 the sequence of events during packaging is not known but in T4 packaging is initiated into unexpanded heads, although head expansion is necessary to allow packaging of the complete genome (Jardine and Coombs, 1998; Jardine *et al.*, 1998). Because in both viruses the genome is a single molecule of dsDNA it can enter the capsid at only one position. In all cases in which the route of DNA packaging into a preformed capsid is known, this position is a unique vertex that is occupied by a packaging structure called the portal complex, which is inserted into a capsid vertex in place of one of the pentons. The structures of the T4 (Driedonks *et al.*, 1981) and HSV-1 (Newcomb *et al.*, 2001) portal complexes are not well known but they appear to bear a close resemblance to the better characterized $\phi29$ (Simpson *et al.*, 2000), SPP1 (Lurz *et al.*, 2001), and T3 (Valpuesta *et al.*, 2000) portals. All these portal complexes contain 12 identical subunits arranged to form a truncated cone with its narrow end facing out of the capsid and a central channel through which the DNA is translocated. The portal is a rotational motor and the mismatch of symmetries between the 12-fold portal and the 5-fold head vertex is probably important to its ability to rotate (Simpson *et al.*, 2000). Packaging is an energy-dependent process during which the portal must overcome the opposing forces resulting from DNA bending, electrostatic repulsion, and loss of entropy (Smith *et al.*, 2001). The energy contained in the physically constrained DNA as a consequence of this process is probably important in providing a driving force to assist its release at the beginning of infection. For the DNA to be retained within the capsid against the pressure exerted by these forces, it is essential that the capsid shell contains no openings large enough for it to escape through. In T4 the portal channel is sealed by the addition of proteins that in turn provide a site for tail attachment (Coombs and Eiserling, 1977). HSV-1 has the same problem of sealing the portal channel, which is possibly also achieved by binding additional proteins to the portal (McNab *et al.*, 1998). In addition, it must close the large channels through the pentons and this is accomplished by a mass translocation in the inner face of VP5 (Zhou *et al.*, 1999). Herpesviruses are the only eukaryote viruses that are known to package their DNA through a portal complex and their similarity in sharing such a sophisticated mechanism again points to the possibility of herpesviruses and tailed bacteriophages having a common origin.

XI. Future Prospects

The prospects for further advances in our understanding of the structure of large viruses look bright. The paucity of available information in some cases means that more widespread application of currently available approaches is bound to produce steep increases in knowledge. In addition, increases in our understanding of the better known structures of HSV-1 and T4 capsids should continue. It is likely that efforts will be intensified to crystallize some of the particles described here, although the prospects of obtaining information from crystals in the near future appear limited. Continued developments in the technology of electron cryomicroscopy and reconstruction should extend the range of potential applications. For example, successful development of techniques to determine asymmetric features would open up many aspects of the virus structure and life cycle to analysis.

In general, the difficulties associated with structural analysis of large viruses do not differ in nature from those of smaller particles, although their large size does pose practical problems for image acquisition and data manipulation (see Zhou and Chiu, This volume). In particular, because of problems arising from the limited depth of field of the electrons, attaining resolutions beyond 5 Å for particles such as the HSV-1 capsid will require a new computational algorithm to reconstruct the structure even if the micrographs contain suitable high-resolution information.

One potential difficulty that may arise in future is related to the quantity of data generated. At present, all electron cryomicroscopic reconstructions of viruses have been at low or intermediate resolution, where only gross mass distributions can be distinguished and it is relatively easy to discern the relationships between different features. However, once the resolution approaches ~3 Å the positions of the protein backbones and amino acid side chains are revealed, leading to a step change in the amount of information available. For example, a 3-Å structure of the HSV-1 capsid would reveal the locations of, and interactions between, the nearly 30,000 amino acids that occupy unique icosahedral positions in the capsid shell. Although there are no technical barriers preventing the manipulation or display of this amount of data, the sheer complexity of the system is likely to pose problems for investigators trying to interpret the mass of structural information available to them.

It is evident from the studies on HSV-1 and T4 that, even in the absence of crystallographic structures, it is possible to gain enough insights into the mechanics of a virus particle from intermediate-resolution information to make the effort required to obtain such information worthwhile. Therefore, it should be expected that our understanding of the interplay between the structures of large viruses and their biology will expand dramatically as studies into their molecular architecture are extended.

XII. Summary

In this article we have attempted to describe some structural aspects of large viruses. Although this may seem a straightforward task, it is complicated by the fact that large viruses do not represent a distinctive class of organisms and any grouping under this heading will include a range of unrelated viruses with different structures, replication strategies, and host types. To simplify matters we limited our definition to dsDNA viruses with genomes of 100 kbp or larger. However, even this restricted grouping includes viruses with diverse and seemingly unrelated structures. Furthermore, few if any structural features are exclusive to large viruses and most of what appears distinctive about their structure or assembly can also be found in smaller, and usually better characterized, viruses. Therefore we have not attempted to provide a comprehensive catalog of the properties of large viruses but have tried to illustrate particular structural points with examples from a few of the better known forms, notably herpes simplex virus (HSV) and phage T4.

The two techniques used to provide rigorous analyses of virus structures are X-ray crystallography and electron cryomicroscopy with computer-assisted reconstruction. To date, X-ray crystallography has been successful only with smaller viruses, and what is known about the structures of these large viruses has come primarily from electron cryomicroscopy. However, with the notable exception of the HSV capsid, such studies have been limited in extent and of relatively low resolution, and the information obtained has been confined largely to describing the spatial distributions and relationships between the subunits. Nevertheless, these studies have given us our clearest insights into the biology of these complex particles and increases in resolution promise to extend these insights by bridging the gap between gross and atomic structures, as exemplified by the identification and mapping of secondary structural elements in the HSV capsid.

Acknowledgments

This research was supported by NIH grants (AI38469 and RR02250) and the Robert Welch Foundation. Suppliers of images for figures are acknowledged in the appropriate figure legends.

References

Baker, M. L. (2002). Development and Applications of Intermediate Resolution Structural Analysis Tools: Integrating Bioinformatics and Electron Cryomicroscopy. Ph.D. thesis. Baylor College of Medicine, Houston, TX.

Baker, T. S., Newcomb, W. W., Booy, F. P., Brown, J. C., and Steven, A. C. (1990). *J. Virol.* **64,** 563–573.

Becker, B., Lesemann, D. E., and Reisser, W. (1993). *Arch. Virol.* **130,** 145–155.

Benson, S. D., Bamford, J. K. H., Bamford, D. H., and Burnett, R. M. (1999). *Cell* **98,** 825–833.

Bhella, D., Rixon, F. J., and Dargan, D. J. (2000). *J. Mol. Biol.* **295,** 155–161.
Black, L. W. (1989). *Annu. Rev. Microbiol.* **43,** 267–292.
Black, L. W., Showe, M. K., and Steven, A. C. (1994). *In* "Molecular Biology of Bacteriophage T4" (J. D. Karam, Ed.), pp. 218–258. American Society for Microbiology, Washington, D.C.
Booy, F. P., Newcomb, W. W., Trus, B. L., Brown, J. C., Baker, T. S., and Steven, A. C. (1991). *Cell* **64,** 1007–1015.
Booy, F. P., Trus, B. L., Davison, A. J., and Steven, A. C. (1996). *Virology* **215,** 134–141.
Bottcher, B., Wynne, S. A., and Crowther, R. A. (1997). *Nature* **386,** 88–91.
Burley, S. K., Miller, A., Harrap, K. A., and Kelly, D. C. (1982). *Virology* **120,** 433–440.
Butcher, S. J., Aitken, J., Mitchell, J., Gowen, B., and Dargan, D. J. (1998). *J. Struct. Biol.* **124,** 70–76.
Carrascosa, J. L. (1978). *J. Virol.* **26,** 420–428.
Carrascosa, J. L., Carazo, J. M., Carrascosa, A. L., Garcia, N., Santisteban, A., and Vinuela, E. (1984). *Virology* **132,** 160–172.
Chen, D. H., Jiang, H., Lee, M., Liu, F. Y., and Zhou, Z. H. (1999). *Virology* **260,** 10–16.
Church, G. A., and Wilson, D. W. (1997). *J. Virol.* **71,** 3603–3612.
Coombs, D. H., and Arisaka, F. (1994). *In* "Molecular Biology of Bacteriophage T4" (J. D. Karam, Ed.), pp. 259–281. American Society for Microbiology, Washington, D.C.
Coombs, D. H., and Eiserling, F. A. (1977). *J. Mol. Biol.* **116,** 375–405.
Delavega, I., Gonzalez, A., Blasco, R., Calvo, V., and Vinuela, E. (1994). *Virology* **201,** 152–156.
Desai, P., and Person, S. (1999). *Virology* **261,** 357–366.
Devauchelle, G., Stoltz, D. B., and Darcy-Tripier, F. (1985). *Curr. Top. Microbiol. Immunol.* **116,** 1–21.
Driedonks, R. A., Engel, A., Tenheggeler, B., and Vandriel, R. (1981). *J. Mol. Biol.* **152,** 641–662.
Dubochet, J., Adrian, M., Richter, K., Garces, J., and Wittek, R. (1994). *J. Virol.* **68,** 1935–1941.
Durand, S., Lightner, D. V., Redman, R. M., and Bonami, J. R. (1997). *Dis. Aquat. Org.* **29,** 205–211.
Fraser, M. J. (1986). *J. Ultrastruct. Mol. Struct. Res.* **95,** 189–195.
Friesen, P. D., and Miller, L. K. (2001). *In* "Fields Virology" (D. M. Knipe, P. M. Howley, D. E. Griffin, R. A. Lamb, M. A. Martin, B. Roizman, and S. E. Strauss, Eds.), pp. 599–628. Lippincott-Raven, Philadelphia.
Gao, M., Matusick-Kumar, L., Hurlburt, W., DiTusa, S. F., Newcomb, W. W., Brown, J. C., McCann, P. J., Deckman, I., and Colonno, R. J. (1994). *J. Virol.* **68,** 3702–3712.
Griffiths, G., Wepf, R., Wendt, T., Locker, J. K., Cyrklaff, M., and Roos, N. (2001). *J. Virol.* **75,** 11034–11055.
Harrap, K. A. (1972). *Virology* **50,** 114–123.
He, J., Schmid, V. F., Zhou, Z. H., Rixon, F., and Chiu, W. (2001). *J. Mol. Biol.* **309,** 903–914.
Hogle, J. M., Chow, M., and Filman, D. J. (1985). *Science* **229,** 1358–1365.
Ishii, T., and Yanagida, M. (1977). *J. Mol. Biol.* **109,** 487–514.
Ishii, T., Yamaguchi, Y., and Yamagida, M. (1978). *J. Mol. Biol.* **120,** 533–544.
Iwasaki, K., Trus, B. L., Wingfield, P. T., Cheng, N. Q., Campusano, G., Rao, V. B., and Steven, A. C. (2000). *Virology* **271,** 321–333.
Iyer, L. M., Aravind, L., and Koonin, E. V. (2001). *J. Virol.* **75,** 11720–11734.
Jardine, P. J., and Coombs, D. H. (1998). *J. Mol. Biol.* **284,** 661–672.

Jardine, P. J., McCormick, M. C., Lutze-Wallace, C., and Coombs, D. H. (1998). *J. Mol. Biol.* **284,** 647–659.
Kennard, J., Rixon, F. J., McDougall, I. M., Tatman, J. D., and Preston, V. G. (1995). *J. Gen. Virol.* **76,** 1611–1621.
Kirkitadze, M. D., Barlow, P. N., Price, N. C., Kelly, S. M., Boutell, C., Rixon, F. J., and McClelland, D. M. (1998). *J. Virol.* **72,** 10066–10072.
Kistler, J., Aebi, U., Onorato, L., Ten Heggeler, B., and Showe, M. K. (1978). *J. Mol. Biol.* **126,** 571–589.
Kuhn, A., Keller, B., Maeder, M., and Traub, F. (1987). *J. Virol.* **61,** 113–118.
Liu, F., and Roizman, B. (1991). *J. Virol.* **65,** 5149–5156.
Lurz, R., Orlova, E. V., Gunther, D., Dube, P., Droge, A., Weise, F., van Heel, M., and Tavares, P. (2001). *J. Mol. Biol.* **310,** 1027–1037.
Massung, R. F., Liu, L. I., Qi, J., Knight, J. C., Yuran, T. E., Kerlavage, A. R., Parsons, J. M., Venter, J. C., and Esposito, J. J. (1994). *Virology* **201,** 215–240.
McGeoch, D. J., and Davison, A. J. (1999). *In* "Origin and Evolution of Viruses" (D. E. R. Webster and J. Holland, Eds.), pp. 441–465. Academic Press, New York.
McNab, A. R., Desai, P., Person, S., Roof, L. L., Thomsen, D. R., Newcomb, W. W., Brown, J. C., and Homa, F. L. (1998). *J. Virol.* **72,** 1060–1070.
Meints, R. H., Lee, K., Burbank, D. E., and Vanetten, J. L. (1984). *Virology* **138,** 341–346.
Moody, M. F. (1999). *J. Mol. Biol.* **293,** 401–433.
Moss, B. (2001). *In* "Fields Virology" (D. M. Knipe, P. M. Howley, D. E. Griffin, R. A. Lamb, M. A. Martin, B. Roizman, and S. E. Strauss, Eds.), pp. 2849–2883. Lippincott-Raven, Philadelphia.
Newcomb, W. W., and Brown, J. C. (1991). *J. Virol.* **65,** 613–620.
Newcomb, W. W., Homa, F. L., Thomsen, D. R., Ye, Z., and Brown, J. C. (1994). *J. Virol.* **68,** 6059–6063.
Newcomb, W. W., Homa, F. L., Thomsen, D. R., Booy, F. P., Trus, B. L., Steven, A. C., Spencer, J. V., and Brown, J. C. (1996). *J. Mol. Biol.* **263,** 432–446.
Newcomb, W. W., Homa, F. L., Thomsen, D. R., Trus, B. L., Cheng, N. Q., Steven, A., Booy, F., and Brown, J. C. (1999). *J. Virol.* **73,** 4239–4250.
Newcomb, W. W., Trus, B. L., Cheng, N. Q., Steven, A. C., Sheaffer, A. K., Tenney, D. J., Weller, S. K., and Brown, J. C. (2000). *J. Virol.* **74,** 1663–1673.
Newcomb, W. W., Juhas, R. M., Thomsen, D. R., Homa, F. L., Burch, A. D., Weller, S. K., and Brown, J. C. (2001). *J. Virol.* **75,** 10923–10932.
Olson, N. H., Gingery, M., Eiserling, F. A., and Baker, T. S. (2001). *Virology* **279,** 385–391.
Preston, V. G., Coates, J. A. V., and Rixon, F. J. (1983). *J. Virol.* **45,** 1056–1064.
Rixon, F. J. (1993). *Semin. Virol.* **4,** 135–144.
Rixon, F. J., and McNab, D. (1999). *J. Virol.* **73,** 5714–5721.
Ross, P. D., Black, L. W., Bisher, M. E., and Steven, A. C. (1985). *J. Mol. Biol.* **183,** 353–364.
Saad, A., Zhou, Z. H., Jakana, J., Chiu, W., and Rixon, F. J. (1999). *J. Virol.* **73,** 6821–6830.
Salas, J., Salas, M. L., and Vinuela, E. (1999). *In* "Origin and Evolution of Viruses" (E. Domingo, R. Webster, and J. Holland, Eds.), pp. 467–480. Academic Press, New York.
Sanchez, A., Khan, A. S., Zaki, S. R., Nabel, G. J., Ksiazek, T. G., and Peters, C. J. (2001). *In* "Fields Virology" (D. M. Knipe, P. M. Howley, D. E. Griffin, R. A. Lamb, M. A. Martin, B. Roizman, and S. E. Strauss, Eds.), pp. 1279–1304. Lippincott-Raven, Philadelphia.

Schrag, J. D., Prasad, B. V. V., Rixon, F. J., and Chiu, W. (1989). *Cell* **56,** 651–660.

Showe, M. K., Isobe, E., and Onorato, L. (1976). *J. Mol. Biol.* **107,** 35–54.

Simpson, A. A., Tao, Y. Z., Leiman, P. G., Badasso, M. O., He, Y. N., Jardine, P. J., Olson, N. H., Morais, M. C., Grimes, S., Anderson, D. L., *et al.* (2000). *Nature* **408,** 745–750.

Smith, D. E., Tans, S. J., Smith, S. B., Grimes, S., Anderson, D. L., and Bustamante, C. (2001). *Nature* **413,** 748–752.

Sodeik, B., Griffiths, G., Ericsson, M., Moss, B., and Doms, R. W. (1994). *J. Virol.* **68,** 1103–1114.

Steinhaus, E. A. (1960). *J. Insect Pathol.* **2,** 225–229.

Steven, A. C., and Carrascosa, J. L. (1979). *J. Supramol. Struct.* **10,** 1–11.

Steven, A. C., and Spear, P. G. (1997). *In* "Structural Biology of Viruses" (W. Chiu, R. M. Burnett, and R. Garcea, Eds.), pp. 312–351. Oxford University Press, New York.

Steven, A. C., Greenstone, H., Bauer, A. C., and Williams, R. W. (1990). *Biochemistry* **29,** 5556–5561.

Steven, A. C., Bauer, A. C., Bisher, M. E., Robey, F. A., and Black, L. W. (1991). *J. Struct. Biol.* **106,** 221–236.

Steven, A. C., Greenstone, H. L., Booy, F. P., Black, L. W., and Ross, P. D. (1992). *J. Mol. Biol.* **228,** 870–884.

Stewart, P. L., Fuller, S. D., and Burnett, R. M. (1993). *EMBO J.* **12,** 2589–2599.

Stow, N. D., Murray, M. D., and Stow, E. C. (1986). *In* "Cancer Cells," Vol. 4: "DNA Tumour Viruses: Control of Gene Expression and Replication" (M. Botchan, T. Grodzicker, and P. Sharp, Eds.), pp. 497–507. Cold Spring Harbor Laboratory Press, Cold Spring Harbor, NY.

Szilagyi, J. F., and Berriman, J. (1994). *J. Gen. Virol.* **75,** 1749–1753.

Tatman, J. D., Preston, V. G., Nicholson, P., Elliott, R. M., and Rixon, F. J. (1994). *J. Gen. Virol.* **75,** 1101–1113.

Thomsen, D. R., Roof, L. L., and Homa, F. L. (1994). *J. Virol.* **68,** 2442–2457.

Thuman-Commike, P. A., Greene, B., Jakana, J., Prasad, B. V. V., King, J., Prevelige, P. E., Jr., and Chiu, W. (1996). *J. Mol. Biol.* **260,** 85–98.

Traub, F., and Maeder, M. (1984). *J. Virol.* **49,** 892–901.

Trus, B. L., Homa, F. L., Booy, F. P., Newcomb, W. W., Thomsen, D. R., Cheng, N. Q., Brown, J. C., and Steven, A. C. (1995). *J. Virol.* **69,** 7362–7366.

Trus, B. L., Booy, F. P., Newcomb, W. W., Brown, J. C., Homa, F. L., Thomsen, D. R., and Steven, A. C. (1996). *J. Mol. Biol.* **263,** 447–462.

Trus, B. L., Gibson, W., Cheng, N. Q., and Steven, A. C. (1999). *J. Virol.* **73,** 2181–2192.

Trus, B. L., Heymann, J. B., Nealon, K., Cheng, N. Q., Newcomb, W. W., Brown, J. C., Kedes, D. H., and Steven, A. C. (2001). *J. Virol.* **75,** 2879–2890.

Tweeten, K. A., Bulla, L. A., and Consigli, R. A. (1980). *J. Virol.* **33,** 866–876.

Valpuesta, J. M., Sousa, N., Barthelemy, I., Fernandez, J. J., Fujisawa, H., Ibarra, B., and Carrascosa, J. L. (2000). *J. Struct. Biol.* **131,** 146–155.

van Driel, R. (1977). *J. Mol. Biol.* **114,** 61–72.

van Driel, R., and Couture, E. (1978). *J. Mol. Biol.* **123,** 713–719.

van Etten, J. L., Meints, R. H., Kuczmarski, D., Burbank, D. E., and Lee, K. (1982). *Proc. Natl. Acad. Sci. USA* **79,** 3867–3871.

van Etten, J. L., Lane, L. C., and Meints, R. H. (1991). *Microbiol. Rev.* **55,** 586–620.

van Hulten, M. C. W., Witteveldt, J., Peters, S., Kloosterboer, N., Tarchini, R., Fiers, M., Sandbrink, H., Lankhorst, R. K., and Vlak, J. M. (2001). *Virology* **286,** 7–22.

van Regenmortel, M. H. V., Fauquet, C. M., Bishop, D. H. L., Carstens, E. B., Estes, M. K., Lemon, S. M., Maniloff, J., Mayo, M. A., McGeoch, D. J., Pringle, C. R., and

Wickner, R. B., Eds. (2000). "Virus Taxonomy: Seventh Report of the International Committee on Taxonomy of Viruses." Academic Press, San Diego, CA.

Vlazny, D. A., Kwong, A., and Frenkel, N. (1982). *Proc. Natl. Acad. Sci. USA* **79,** 1423–1427.

Wilson, M. E., Mainprize, T. H., Friesen, P. D., and Miller, L. K. (1987). *J. Virol.* **61,** 661–666.

Wingfield, P. T., Stahl, S. J., Thomsen, D. R., Homa, F. L., Booy, F. P., Trus, B. L., and Steven, A. C. (1997). *J. Virol.* **71,** 8955–8961.

Wu, L. J., Lo, P., Yu, X. K., Stoops, J. K., Forghani, B., and Zhou, Z. H. (2000). *J. Virol.* **74,** 9646–9654.

Yan, X. D., Olson, N. H., Van Etten, J. L., Bergoin, M., Rossmann, M. G., and Baker, T. S. (2000). *Nat. Struct. Biol.* **7,** 101–103.

Zhou, Z. H., He, J., Jakana, J., Tatman, J. D., Rixon, F. J., and Chiu, W. (1995). *Nat. Struct. Biol.* **2,** 1026–1030.

Zhou, Z. H., Macnab, S. J., Jakana, J., Scott, L. R., Chiu, W., and Rixon, F. J. (1998). *Proc. Natl. Acad. Sci. USA* **95,** 2778–2783.

Zhou, Z. H., Chen, D. H., Jakana, J., Rixon, F. J., and Chiu, W. (1999). *J. Virol.* **73,** 3210–3218.

Zhou, Z. H., Dougherty, M., Jakana, J., He, J., Rixon, F. J., and Chiu, W. (2000). *Science* **288,** 877–880.

Zhou, Z. H., Baker, M. L., Jiang, W., Dougherty, M., Jakana, J., Dong, G., Lu, G. Y., and Chiu, W. (2001). *Nat. Struct. Biol.* **8,** 868–873.

STRUCTURAL STUDIES ON ANTIBODY–VIRUS COMPLEXES

By THOMAS J. SMITH

Donald Danforth Plant Science Center, St. Louis, Missouri 63132

I. Introduction 409
II. Background 410
A. Aggregation 410
B. Induction of Conformational Changes 411
C. Virion Stabilization 411
D. Abrogation of Cellular Attachment 411
E. Other *in Situ* Effects 412
F. Other *in Vivo* Effects 412
III. Structural Studies on Virus–Antibody Complexes 412
A. Influenzavirus 413
B. Rotavirus 414
C. Cowpea Mosaic Virus 415
D. Human Rhinovirus 14 416
E. Parvovirus 420
F. Poliovirus 421
G. Alphavirus 423
H. Human Rhinovirus 2 424
I. Calicivirus 425
J. Herpes Simplex Virus 427
K. Adenovirus 428
L. Hepatitis B Virus 430
M. Foot-and-Mouth Disease Virus 431
N. Papillomavirus 433
O. Reovirus 435
P. Cucumber Mosaic Virus 437
IV. Conclusions 439
A. Neutralization Efficacy 439
B. Aggregation 440
C. Binding Affinity and Stoichiometry 440
D. Single-Hit Kinetics and p*I* Changes 440
References 443

I. Introduction

The results from a number of structural studies on several different families of viruses have begun to converge into a unified view of the mechanism of antibody-mediated neutralization of viruses. Previously, the literature was rife with conflicting results and hypotheses. This was most likely due to the inherent difficulty in trying to quantify binding constants

0065-3233/03 $35.00

and stochiometry with intact antibodies (often polyclonal), which tend to cross-link particles and viral antigens that present numerous identical epitopes. From these studies, several mechanisms have been proposed: aggregation, conformational stabilization, induced structural changes, and steric blockade. A number of laboratories have been combining structural biology with other disciplines to test each of these mechanisms. Most have come to the conclusion that, although all of these mechanisms are theoretically possible, steric blockade appears to be the most likely *in vitro.* In addition, these studies have also been able to directly address the possible role that immunity has played in the evolution of viral structure. This article briefly reviews these possible neutralization mechanisms and then discusses the structural results of most of the antibody–virus complexes determined to date.

II. Background

Over the past few decades several mechanisms of antibody-mediated neutralization have been proposed.

A. *Aggregation*

When polyclonal antibodies are added to antigen or when monoclonal antibodies are added to a multivalent antigen, it is possible to aggregate the antigen into extremely large immunocomplexes. Such precipitation occurs over a relatively narrow range of antibody:antigen ratios. When there is a molar excess of antibody over antigen, the antigen becomes coated with antibodies and few interantigen cross-links are formed. At high antigen:antibody ratios, there is insufficient antibody to form long polymers of immunocomplexes. This narrow zone of optimal immunoprecipitation gives rise to the precipitation lines observed in Oüchterlony and rocket immunoassays. Aggregation may prevent the viruses from entering the cell via normal receptor-mediated pathways. Also, there would be an increase in avidity (apparent increase in affinity due to multivalent binding of antibodies) caused by antibodies bound bivalently to neighboring particles in the large immunocomplexes. *In vivo,* such aggregation would facilitate opsonization by phagocytic leukocytes. Aggregation as a neutralization mechanism was supported by results suggesting that aggregation and neutralization occur concomitantly and that virus:antibody ratios *in vivo* favor aggregation [1–3].

B. *Induction of Conformational Changes*

The basic premise of the mechanism is that antibodies are able to induce large conformational changes in the target virus, thereby neutralizing infectivity. The evidence for this mechanism came from the observation that antibodies and Fab fragments can cause an apparent decrease in the p*I* of the viral capsid concomitant with neutralization [4, 5]. Further, this mechanism had been suggested to explain the apparent single-hit kinetics of neutralization. These single-hit models are based on the observation that infectivity starts to decrease immediately on the addition of antibody. Because there is not a lag phase in this process, it has been suggested that antibody neutralization obeys single-hit kinetics. This implies that a single antibody is sufficient for neutralization and this may be accomplished through gross changes in the virion on binding. This model is perhaps the one that can be most directly tested by structural studies.

C. *Virion Stabilization*

As the antithesis of the previous model, it is possible that antibodies might neutralize infectivity by stabilizing the virions and preventing the release of genome into the cytoplasm. Such stabilization might prevent receptor-mediated changes needed for uncoating or may prevent the changes necessary for optimization of receptor–virus interactions. In the case of human rhinovirus 14 (HRV14), this stabilization was proposed to occur by the cross-linking of icosahedral pentamers [6]. Inconsistent with the bivalent binding aspect of this model, intact antibodies and Fabs have been shown also to stabilize HRV14 [7].

D. *Abrogation of Cellular Attachment*

Antibodies might be able to neutralize viral infection by preventing the interactions between the virus and its cell receptor. Such a blockade could occur via all of the above-described mechanisms as well as by steric interference. For virus–receptor interactions, interaction with several receptor molecules might be required. This would be clearly affected by aggregating the particles in large immunocomplexes. It is also possible that if antibodies induce large conformational changes in the virion, then the structure of the cellular receptor recognition site might be deleteriously affected as well. If the viruses are stabilized by antibody binding, then postattachment optimization of the virus–receptor interactions might be blocked. Finally, antibodies might block receptor binding by direct or steric blocking of receptor–virus interactions. Direct blockade of binding would occur with the antibody binding directly to the receptor-binding region or immediately adjacent to it. Steric blockade could occur as a result of the sheer bulk of the antibodies themselves. Although antibodies might not directly bind to the receptor attachment site, they may prevent cell attachment by virture of the volume encompassed by the unbound Fab arm and highly flexible Fc region. The phenomenon of abrogation of cellular attachment, but not the root cause, has been clearly demonstrated in the case of HRV14, concerning which antibodies to all four antigenic sites have been shown to block attachment [4].

E. Other in Situ *Effects*

There is evidence that some antibodies neutralize in a manner not easily explained by the previous mechanisms. For example, it has been shown that antibodies to Sindbis virus [8, 9] and poliovirus [10, 11] can eliminate infection or progression of infection even when added to cells hours after infection. In the case of Sindbis virus, the exact mechanism of this viral clearance is unknown, but appears to be related to antibody cross-linking [9]. Similarly, the postadsorption neutralization properties of at least some of the antibodies to poliovirus appeared to be related to binding valency as well [11]. Therefore, antibodies, or antibodies interacting with viral components, may be triggering some unknown defensive mechanism within the infected cell.

F. Other in Vivo *Effects*

A number of antibodies have been shown to have little to no neutralizing activity *in situ*, but are efficacious at protecting animals from infection. For example, antibodies against Sindbis virus [12] and foot-and-mouth disease virus (FMDV) [13] that are nonneutralizing *in vitro* still protect animals from viral challenge. Also, nonneutralizing antibodies against the neuraminidase (NA) protein of influenza do affect disease progression *in vivo* [14]. These effects are clearly due to the synergistic effects between antibodies and other components of the immune system. These results also serve as a reminder that *in vitro* assay results may not always accurately prognosticate *in vivo* activity.

III. Structural Studies on Virus–Antibody Complexes

As of a decade ago, most of our structural information about antibody–virus interactions was limited to influenzavirus. Since then, there has been a flurry of structural studies aimed at elucidating the mechanism of antibody-mediated neutralization as well as using antibodies to delineate various viral features. The following is a brief review of these structural studies in approximate chronological order, with a discussion as to the information gleaned from each.

A. *Influenzavirus*

Influenzavirus studies were the first to examine interactions between viral proteins and antibodies. From this work, ideas started to develop about methods viruses have evolved to avoid immune surveillance.

Influenzaviruses are enveloped viruses that belong to the *Orthomyxoviridae* family. These viruses have a pleomorphic or spherical morphology with a diameter of 800–1200 Å. The viral genome consists of seven or eight segments of linear negative-sense single-stranded RNA. On the outer envelope, there are two types of spikes; the major protein is hemagglutinin (HA) and the less prominent is neuraminidase (NA). The ratio of HA to NA is 4–5:1. Inside the membrane envelope is a segmented nucleocapsid with helical symmetry [15].

Influenza is an infection of the respiratory tract caused by the influenzavirus. Compared with most other viral respiratory infections, such as the common cold, influenza infection often causes a more severe illness. In an average year, influenza is associated with more than 20,000 deaths nationwide and more than 100,000 hospitalizations. Influenzaviruses are divided into three types designated A, B, and C. Influenza types A and B are responsible for epidemics of respiratory illness that occur almost every winter and are often associated with increased rates for hospitalization and death. Type C infection usually causes either a mild respiratory illness or no symptoms at all [16].

Influenza type A viruses undergo two kinds of serotypic changes. One is a series of mutations that occur gradually over time, called antigenic "drift." The other kind of change, called antigenic "shift," is an abrupt change in the hemagglutinin and/or the neuraminidase proteins, causing the emergence of a new subtype. Type A viruses undergo both kinds of changes, whereas influenza type B viruses change only by the more gradual process of antigenic drift.

When the high-resolution structure of influenza virus N9 neuraminidase (NA) was determined, it was noted that the conserved residues involved in sialic acid binding were located in a crevasse [17]. Residues within this deep depression are conserved whereas the residues about the rim vary with serotype. This suggests that conserved residues are hidden from antibody recognition. The authors suggested that such architecture might have several functional implications [17]. Analogous to most enzymes, a cavity or pocket-like structural feature may have evolved to facilitate contact with receptor. It was further noted that a concave morphological feature like that occurring on the NA spike would also offer some protection against binding of host antibodies to functionally important residues. However, even if some of the conserved residues were accessible to antibodies, the quaternary structure of the NA spike required antibody contacts with some of the variable loops at the top of the sialic acid-binding site. Influenzavirus could therefore escape antibody binding without any need to alter crucial portions of its NA spike [18].

More recent studies demonstrated that about one-third of the conserved binding region in this depression is contacted by a neutralizing antibody [19]. To explain how viruses might evade antibody attack while leaving conserved residues immunologically exposed, Colman has proposed that this capability may reflect the potential for different proteins to recognize identical protein surfaces [20]. In this way, receptors and antibodies can bind to overlapping areas of the viral surface, but can exhibit differing sensitivities to mutations at these contact surfaces.

B. *Rotavirus*

The study by Prasad *et al.* [21] represented the first visualization of an intact virion complexed with neutralizing antibodies. The success of this work had tremendous impact on the studies that followed.

Rotaviruses are members of the genus *Rotavirus,* in turn a member of the family *Reoviridae.* The rotaviruses have a characteristic wheellike appearance when viewed by electron microscopy, hence the Latin name *rota,* meaning "wheel." These viruses infect only vertebrates and are transmitted by the oral–fecal route. Rotavirus infection is the most common cause of severe diarrhea among children, resulting in the hospitalization of approximately 55,000 children each year in the United States and the death of more than 600,000 children annually worldwide. Although immunity after infection is incomplete, repeat infections tend to be less severe than the original infection. The complete rotavirus particle has three shells: an outer capsid, inner capsid, and core. These viruses have an 11-segmented genome of double-stranded RNA that encodes 6 structural and 5 nonstructural proteins. The outer shell is composed of two of the structural proteins, VP7 (the glycoprotein or G protein) and VP4 (the protease cleaved or P protein). These proteins define the serotype of the virus and are the major antigens involved in virus neutralization [15].

In the study by Prasad *et al.*, the structure of rotavirus complexed with Fab fragments of a neutralizing monoclonal antibody was determined (Fig. 1; see Color Insert) [21]. The antibody used for these studies was against VP4. In addition to being one of the two outer capsid proteins of rotaviruses, VP4 has been implicated in several important functions such as cell penetration, hemagglutination, neutralization, and virulence. This work was important for at least two reasons. First, it clearly demonstrated that protein ligand–virus complexes could be analyzed by cryo-transmission electron microscopy (cryo-TEM). Second, by flagging VP4 with the antibody, they were able to show that the surface spikes on rotavirus particles are made up of VP4. This showed that cryo-TEM studies could elucidate the organization of these extremely large viral assemblies, which were not yet amenable to crystallographic techniques. They found that the antigenic sites were located near the distal ends of the spikes and that 2 Fab fragments bound to each of the 60 spikes. From mass measurements, it was determined that the spikes were probably dimers of VP4. Although the flexibility of antibodies about the elbow region had been found previously in X-ray structures, this was also the first time that this apparent flexibility was observed in solution.

C. *Cowpea Mosaic Virus*

The importance of the work on cowpea mosaic virus (CPMV) was that it represented the first time that "pseudo-atomic" models could be made to describe the exact contact areas of these cryo-TEM image reconstructions. Because the atomic structure of CPMV was known, it was possible to map out the area of contact between the virus and antibody. This technique was used in a number of subsequent cryo-TEM studies and has greatly improved the functional resolution of image reconstruction.

CPMV is the type species of the genus *Comovirus*, which belongs to the *Comoviridae* family. These viruses are transmitted by beetles and to a small degree by seed (1–5%). The current geographical distribution is limited to Cuba and parts of Africa, where it infects a relatively small number of species of the family Leguminoseae [15]. The CPMV coat contains 60 copies of 2 viral proteins: the large subunit (MW 42K) and the small subunit (MW 24K). The large subunit contains two of the canonical viral eight-stranded β barrels arranged about the icosashedral 3-fold axes, whereas the small subunit is adjacent to the 5-fold axes [22].

In the study by Wang *et al.* [23], the structure of CPMV complexed with Fab fragments was determined (Fig. 1). As expected from its size and its repetition of potential antigenic sites, CPMV is highly antigenic when injected into rabbits. From a panel of nine monoclonal antibodies, the antibody 5B2 was selected because it reacted strongly to virions in solution rather than to the denatured form that results from interaction with enzyme-linked immunosorbent assay (ELISA) plates. Unlike in subsequent studies with animal viruses, the antibody recognition site was not determinable because of the lack of natural escape mutation analysis. Therefore, this structural study was important to define the region that was most antigenic on the virion surface. Such information could be crucial in the design of plant-based vaccines using CPMV as a vector (e.g., Porta *et al.* [24]).

In this study, the three-dimensional reconstructions of native CPMV and a complex of CPMV saturated with the Fab fragment of 5B2 were determined at 23-Å resolution (Fig. 1) [23]. Interestingly, whereas the dogma was that antibodies would predominantly recognize the large protruding domains of viral capsids, this antibody actually recognized the flattened surface between the protruding pentameric towers located at the icosahedral 5-fold axes. Although it could be argued that this result is not relevant because it is a plant virus interacting with mammalian immunity, it nevertheless demonstrates that dominant epitopes need not lie on the most protruding features of a capsid. Because the authors had the atomic structure of the virus, they were able to determine the footprint of the antibody to within a few angstroms. Furthermore, this was the first structural evidence that antibodies need not cause gross conformational changes in the virions on binding.

This study was followed up with reconstructions of CPMV complexed with Fab fragments from monoclonal antibodies 5B2 and 10B7 as well as IgGs from 5B2 [25]. The IgG was observed to bind in a monodentate fashion with only one Fab arm attached to the virus surface. Fab fragments from 5B2 and 10B7 bound to nearly identical positions and both were used to identify the $\sim$30 amino acids covered by the Fab arms. In this study, the unbound Fc and Fab arms were disordered and formed islands above the bound antibodies.

D. *Human Rhinovirus 14*

The following studies were important because they represented the first time that cryo-EM studies (Fig. 1) could be interpreted with atomic structures of all of the components and this modeling was subsequently validated by the crystal structure of the Fab–virus complex. Further, these studies were able to directly test the prevailing hypotheses concerning the mechanism of antibody-mediated neutralization and viral escape of immune surveillance. Finally, the structure of the mAb17–HRV14 complex was the first visualization of an antibody bound bivalently to the virus surface.

Picornaviridae is among the largest of animal virus families and includes polio-, rhino-, foot-and-mouth disease, coxsackie, and hepatitis A viruses. The rhinoviruses, of which there are more than 100 serotypes, are major causative agents of the common cold in humans [26]. The virus is nonenveloped and has an ~300-Å-diameter protein shell that encapsidates a single-stranded, plus-sense, RNA genome of about 7200-bases. The human rhinovirus 14 (HRV14) capsid exhibits a pseudo-T=3 (P=3) icosahedral symmetry and consists of 60 copies each of 4 viral proteins, VP1, VP2, VP3, and VP4. VP1–3 each have an eight-stranded antiparallel β-barrel motif and comprise most of the capsid structure (Fig. 1). VP4 is smaller, has an extended structure, and lies at the RNA–capsid interface [27]. An ~20-Å-deep canyon lies roughly at the junction of VP1 (forming the "north" rim) with VP2 and VP3 (forming the "south" rim), and surrounds each of the 12 icosahedral 5-fold vertices. The canyon regions of HRV14 and HRV16, both major receptor group rhinoviruses, were shown to contain the binding site of the cellular receptor, intercellular adhesion molecule 1 (ICAM-1) [28–30]. Four major neutralizing immunogenic (NIm) sites, NIm-IA, NIm-IB, NIm-II, and NIm-III, were identified by studies of neutralization escape mutants with monoclonal antibodies [31, 32], and then mapped to four protruding regions on the viral surface [27].

Neutralizing monoclonal antibodies against HRV14 have been divided into three groups: strong, intermediate, and weak neutralizers [6, 33]. All strongly neutralizing antibodies bind to the NIm-IA site, which was defined by natural escape mutations at residues D1091 and E1095 of VP1 on the loop between the β-B and β-C strands of the VP1 β barrel (the letter designates the amino acid, the first digit identifies the viral protein, and the remaining three digits specify the sequence number). Because strongly neutralizing antibodies form stable, monomeric virus–antibody complexes with a maximum stoichiometry of 30 antibodies per virion, it was concluded that they bind bivalently to the virions [6, 33]. Weakly neutralizing antibodies form unstable, monomeric complexes with HRV14 and bind with a stoichiometry of ~60 antibodies per virion [33, 34]. The remaining antibodies, all of which precipitate the virions, are classified as intermediate neutralizers [6, 33].

The structures of three different Fab–HRV14 (Fab17, Fab12, and Fab1) complexes and of one mAb–HRV14 (mAb17) complex were determined (Fig. 1). Although all bind to the same antigenic site, mAb17 and mAb12 are both strongly neutralizing antibodies, whereas mAb1 is a weakly neutralizing antibody. What was immediately apparent was that these different antibodies had different binding orientations. Fab17 and Fab12 both bound to the NIm-IA site at a somewhat tangential orientation that placed the constant domains ($C_{H1} \cdot C_V$) of 2-fold-related Fabs in close proximity to one another. This suggested that the intrinsic paratope/epitope interactions of these antibodies places the Fab arms in an orientation that facilitates bidentate binding to 2-fold-related NIm-IA sites. In contrast, Fab1 binds almost vertically to the virion surface with a ''twist'' that makes it seem unlikely that these antibodies could bind bivalently. This result, therefore, suggests that the strongly neutralizing, poorly aggregating antibodies are more efficacious because bivalent binding increases the apparent affinity of the antibodies for the virion. This bivalent binding was subsequently visualized by the structure of the mAb17–HRV14 complex [35].

The structures of Fab1, Fab17, and HRV14 had been determined and used to construct pseudo-atomic models, and then these models were tested by site-directed mutagenesis [7, 34]. In the Fab17–HRV14 complex, the loop of the NIm-IA site on HRV14 sits clamped in the cleft between the heavy and light chain hypervariable regions and forms complementary electrostatic interactions with Lys58H (on the heavy chain) and Arg91L (on the light chain) of Fab17. In addition, a cluster of lysines on HRV14 (K1236, K1097, and K1085) interacts with two acidic residues, Asp45H and Asp54H, in the CDR2 region (CDR, complementarity-determining region) of the Fab heavy chain [36]. Using site-directed mutagenesis it was found that even though K1236, K1097, and K1085 were not identified as sites of naturally occurring escape mutations, they do affect antibody binding. Therefore, these results demonstrated that electrostatic interactions can dominate paratope–epitope interactions and that naturally occurring escape mutations are clearly only a small subset of residues crucial for antibody binding. Therefore, the location of epitopes cannot be used as unequivocal evidence that convolutions in the virion surface hide crucial residues.

These studies consistently demonstrated that antibodies do not induce conformational changes in the virion on binding. However, all of these cryo-TEM studies are limited to 20- to 30-Å resolution, at which relatively small conformational changes may not be observable. To address this problem, the crystal structure of the Fab17–HRV14 complex was determined [36]. The most notable property of the bound Fab17 in this structure was that the entire constant domain ($C_{H1} \cdot C_L$) was disordered, starting at the elbow region, presumably due to flexibility. The second major observation was that the pseudo-atomic model created on the basis of cryo-TEM electron density was a fairly accurate representation of the actual structure. The root of most of the error in the cryo-TEM fitting process was due to the fact that the CDR3 of Fab17 moves by ~5 Å to better accommodate the epitope, which places the Fab closer to the surface and further down into the canyon than previously modeled. Interestingly, Fab17 binds poorly to peptides representing the NIm-IA loop.

This structure also clearly showed that there were no significant changes in the virion due to Fab binding. The only observable changes were localized in a few side chains on the βB-βC loop on which the naturally occurring escape mutations lie. D1091 and E1095 rotate slightly to form salt bridges with the basic residues in the paratope cleft and K1097 and R1094 rotate to better interact with D55 and D57. There were not, however, any significant changes in the NIm-IA loop itself. If antibodies were capable of inducing conformational changes, it seems likely that such changes would start in the antigenic loop itself. This result is not entirely surprising because none of the crystallographic structures of antigen–Fab complexes determined to date have demonstrated large antibody-induced conformational changes. This structure demonstrated that antibodies do not need to deform intact virus capsids to be efficacious neutralizers.

The resolution of this Fab17–HRV14 structure permits us to more accurately analyze the antibody–virus interactions. First, it is apparent that, in contradiction to the "canyon hypothesis," Fab17 penetrates into the canyon region. It is able to do this by binding somewhat on its side with the V_H domain, making extensive interactions with the north and south walls of the canyon. The hypervariable residues contact the entire north wall, the bottom of the canyon, and part of the lower south wall. The framework residues contact the upper south wall only. Therefore, it appears that the canyon does not, in fact, adequately hide the ICAM-1-binding surface from antibody recognition. This might lead to the conclusion that Fab17 neutralizes by directly interfering with ICAM-1 binding. However, it has been shown that antibodies to all four antigenic sites can abrogate cell attachment [4] even though some sites (e.g., NIm-III) are distal to the canyon region. The simplest explanation for this is that the antibodies are large relative to the virion itself. Because the length of an antibody (~150 Å) is approximately equal to the radius of the particles, it takes only a few antibodies bound to the surface to potentially interfere with how the virus interacts with the host cell membrane. In addition, the aggregating antibodies would further exacerbate binding by clumping the virions together. Finally, from studies on the neutralization properties of these antibodies, it is clear that neutralization is not dependent on either aggregation or bivalent binding.

E. *Parvovirus*

The following studies were important because they represented another example of interpretation of cryo-EM results, using atomic models for the associated species. In addition, they confirmed and extended the results of natural escape mutation and peptide-scanning techniques used to identify the epitope region.

Members of the *Parvovirus* genus cause a number of diseases in mammals, including enteritis [37, 38] and childhood fifth disease. These nonenveloped viruses have a capsid with a diameter of ~255 Å that encases a double-stranded DNA genome and infect only cells that are in the S phase [39]. In canine parvovirus (CPV) the T=1 capsid is composed mainly of 60 copies of viral protein 2 (VP2). There is a small amount of VP3 in the capsid that is the result of proteolytic cleavage of ~17 residues from the amino terminus of VP2. Up to 20% of the 60 copies of VP2 in each capsid are replaced by VP1. VP1 is the product of an alternative transcriptional splicing event that places an additional 153 amino acids at the N terminus [15]. No evidence for VP1 was observed in the crystal structure [40].

Neutralization sites were determined by natural escape mutation analysis and peptide mapping [41–44]. The two epitopes identified by this analysis were called A and B. Site A is a spike at the icosahedral 3-fold axes and the B site is on a ridge to the "north" of the icosahedral 2-fold axes. Interestingly, peptide mapping found additional epitopes at the N terminus. Because the first visible residue is at position 37 and lies inside the virion, it was proposed that these termini might be protruding through the 5-fold axes [40].

To better understand the mechanism of antibody-mediated neutralization of CPV, the cryo-TEM structure of the CPV–Fab complex was determined (Fig. 1) [45]. For these studies, Fab fragments from the antibody A3B10, which recognizes epitope B, were used. They demonstrated that the Fab fragments were nearly as efficacious as the mAb, and therefore this complex does indeed represent a neutralized state. The bound Fab molecules were clearly visible in the image reconstruction and bound perpendicularly to the virion surface. Using the structure of HyHEL-5 as a model for the bound Fab fragments, it was clear that these antibodies were unlikely to bind bivalently to the surface of the virion. From these modeling exercises it was also clear that this epitope region was not nearly as hydrophilic as was observed in the case of the NIm-IA site of HRV14. As was the case with the HRV14 work, this antibody does not recognize the viral protein Western blots, nor does it bind to the peptide representing this epitope loop. Therefore, it was concluded that the

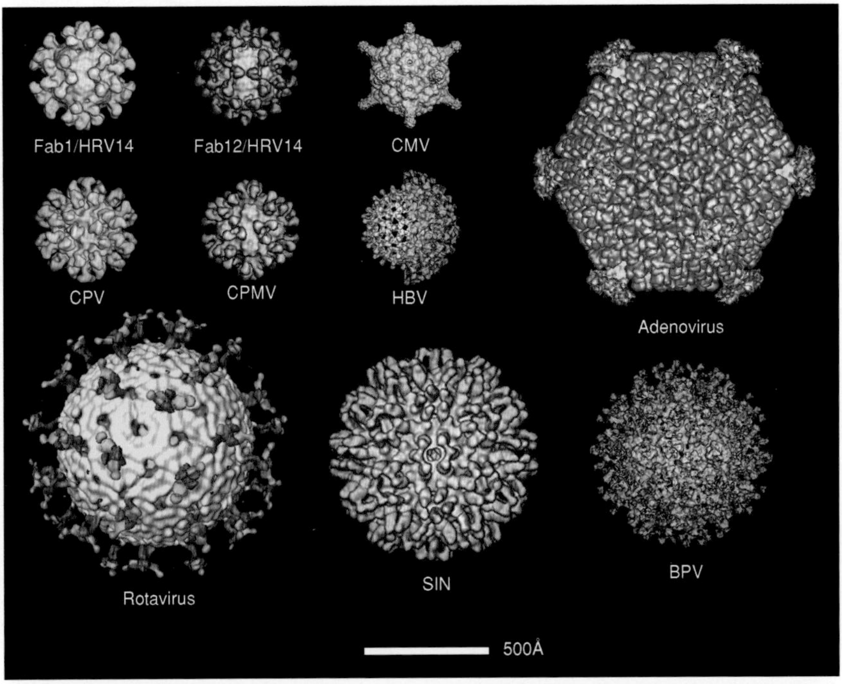

SMITH, FIG. 1. Image reconstructions of a number of the virus–antibody complexes described. In the Fab1/HRV14, Fab12/HRV14, CMV–Fab, CPV–Fab, SIN–Fab, and CPMV-Fab complexes, the bound Fab fragments have been colored whereas the virion is gray. In the HBV complex, the coat protein is blue and the bound Fab fragments are red. Here, half the outer shell of bound Fab fragments has been cut away to expose the inner core. For the rotavirus, BPV, and adenovirus, the antibodies are colored green, orange, and purple, respectively.

antibody is probably recognizing both the epitope loop and the context in which it is being presented.

From these results, several mechanisms of neutralization were eliminated. Because the Fab fragments are as efficacious as the intact IgG, neutralization must be independent of aggregation. From the orientation of the bound Fabs, and the efficacy of the Fabs in neutralizing infectivity, neutralization must also be independent of bivalent attachment. No changes were observed on antibody binding, making it unlikely that the antibody neutralizes by inducing gross conformational changes in the virion. That leaves antibody-mediated stabilization or abrogation of cell attachment as possible mechanisms. It was suggested that the proximity of the B site to icosahedral symmetry axes might allow for Fabs to stabilize the capsids. However, it is also possible that the antibodies block infectivity by virtue of their bulk.

F. Poliovirus

The following studies on poliovirus used antibodies to demonstrate that internal termini are transiently exposed during a ''breathing'' process, and structural studies on a peptide–Fab complex suggested that this antibody may neutralize by locking the virion into an inappropriate conformation. This work was the first to elucidate the kinds of structural changes that occur during uncoating and was the basis for several subsequent studies looking at capsid dynamics. More recent studies have further shown that antibody–poliovirus complexes can still be infectious if the virus is first primed by receptor and given an alternative route into the cell.

Poliovirus is a member of the *Enterovirus* genus, which belongs to the *Picornaviridae* family. There are two basic patterns of symptoms: minor illness (abortive type) and major illness (either paralytic or nonparalytic). The minor illness, accounting for 80 to 90% of clinical infections, occurs chiefly in young children, is mild, and does not involve the CNS. The major illness may progress with loss of selective tendon reflexes and assymetric weakness or paralysis of muscle groups. Humans are the only natural host for the virus and infection occurs through direct contact. There are three immunologically distinct poliovirus serotypes, with type 1 being the most paralytogenic and the most common cause of epidemics [15].

The structure of poliovirus is remarkably similar to that of human rhinovirus 14 [27, 46]. It is composed of three major viral proteins (VP1–VP3), with each forming a single canonical viral β barrel. VP4 lies at the interface between the capsid and the interior RNA. Poliovirus uses CD155 as a cellular receptor [47]. There are three dominant epitopes on the virion called sites 1, 2, and 3 [48]. There is a direct link between the receptor-binding site and the serotypic determinants in poliovirus. It has been shown that upper, exposed regions of the poliovirus canyon are crucial for receptor interactions, but residues at the bottom of the poliovirus canyon are not. In fact, changes at the top of the canyon that affect antibody-neutralizing sites also alter receptor–virus interactions [49]. Yielding similar conclusions, other studies showed that mutations at the north and south walls of the canyon overcome deleterious defects in the poliovirus receptor. These mutations, which are distal to the canyon floor, lie close to the antigenic sites and appear to represent destabilizing mutations [50].

In the first studies, antibodies were used to elucidate the dynamic nature of poliovirus [51]. Antibodies were raised against peptides representing VP4 and the N termini of VP1. In the crystal structures of all of the picornaviruses, these termini lie at the capsid–RNA interface and are therefore not exposed to external solvent [27, 46, 52–55]. These antibodies bound to the virus when the particles were heated to 37°C but did not bind when added to virus at room temperature or when the virus was heated to 37°C and then cooled to room temperature. Therefore, although difficult to visualize with the static structure of the capsid, the only explanation for these results is that these buried termini are transiently exposed. This exposure is facilitated by higher temperatures and was proposed to be part of the normal infection process. This idea of dynamic capsid structures was subsequently supported by mass spectroscopy analysis of flock house virus [56] and rhinovirus [57] and by a series of drug–poliovirus structures [58].

In subsequent studies, Fab fragments from a neutralizing antibody, C3, were used [59]. This antibody was originally raised against heat-inactivated virus particles and strongly neutralized the Mahoney strain of poliovirus type 1. The Fab was complexed with a peptide corresponding to the viral epitope and the structure was determined to a resolution of 3.0 Å. The carboxyl end of the peptide was found to extensively interact with the paratope and adopted a conformation that differed significantly from the structure of the corresponding residues in the virus. This apparent difference between the bound peptide and the authentic antigenic loop suggested that this antibody might induce structural changes important for neutralization.

Results have also shown that antibodies can actually facilitate viral infection, but only when the virions are in the 135S state. Poliovirus binding to its receptor (PVR) on the cell surface induces a conformational transition that generates an altered particle with a sedimentation value of

135S versus the 160S of the native virion [60]. These altered, 135S particles are much less infectious than native virions. In earlier studies, it was found that neutralizing antibodies to the native virion block attachment to target cells. When cells were made to express Fc receptors, the virus–antibody complex was again able to enter the cells but was still noninfectious [61]. Subsequently, it was shown that a poliovirus receptor–IgG2a (Fc portion) hybrid molecule permitted poliovirus to enter and infect via this Fc receptor [62]. This was followed by studies showing that when antibodies specific for 135S particles are added to the 135S particles, infectivity increases by two to three orders of magnitude [63]. This suggests that the lack of infectivity in 135S particles is due to the loss of cell binding. These results further imply that one function of neutralizing antibodies is to prevent these viruses from interacting with receptor and becoming "primed" for uncoating.

G. *Alphavirus*

These studies were important in that they represented the first antibody–virus complexes of an enveloped virus, helped identify the location of the receptor-binding region of this family of viruses, and were another example of using antibodies to elucidate viral topography.

The alphaviruses are a group of 26 icosahedral, positive-sense RNA viruses primarily transmitted by mosquitoes [64]. These ~700-Å-diameter viruses are some of the simplest of the membrane-enveloped viruses, and members of this group cause serious tropical diseases with characteristic symptoms such as myositis, fever, rash, encephalitis, and polyarthritis [65]. The structures of two different alphavirus–Fab complexes have been determined by cryo-TEM: Ross River virus (RR) and Sindbis virus (SIN) [66]. The amino acid sequences of the RR and SIN virus structural and nonstructural proteins are 49 and 64% identical, respectively [67]. The viral RNA genome and 240 copies of the capsid protein form the nucleocapsid core [68–73], and the E1 and E2 glycoproteins form heterodimers that associate as 80 trimeric spikes on the viral surface. Native SIN and RR lack the E3 glycoprotein because it disassociates from the spike complex after its display on the plasma membrane surface [74, 75]. E1 has a putative fusion domain that may facilitate host membrane penetration [76, 77]. E2 contains most of the neutralizing epitopes and is also probably involved in host cell recognition [78–80].

To examine the mechanism of antibody neutralization and to identify the portion of E2 involved in receptor recognition two antibodies were used for these studies: SV209 (Fig. 1) and T10C9 [66]. In the case of

SV209, antiidiotypic antibodies to this neutralizing antibody were able to compete with SIN for its cellular receptor and block viral attachment by ~50% [79]. This implies that the original SV209 antibody is recognizing at least a portion of the spike involved in cellular recognition. The naturally occurring mutation in Ross River virus that facilitates escape from the T10C9 antibody maps to residue T216 of E2 [81]. This residue is presumably near the cell receptor-binding site because residue N218 was found to vary as the virus adapted to growth in chicken cells [82]. In addition, residue T219 mutates to alanine during the course of an epidemic in humans [83]. Because small mammals act as the viral reservoir in nature, this mutation may represent changes in E2 necessary to alter host specificity.

In both virus–Fab reconstructions, the Fab fragments appear as bilobed structures that bind to the outermost tips of the timeric spikes [66]. The lobe in contact with the spikes represents the Fab variable domain and the distal lobe represent the Fab constant domain. Both domains have approximately equal mass in the RR–Fab complex, but the constant domain had lower density than the variable domain in the SIN–Fab complex. In both of these reconstructions, the binding of the antibody did not appear to cause conformational changes in the virion. To facilitate interpretation, an atomic structure of an Fab was fitted into the electron density envelope. Although the two antibodies bound to their respective viruses with markedly different orientations, their binding footprints were nearly identical. These results show that the receptor-binding region is not buried in a depression on the spike but rather is highly exposed at the tip of the spike. Therefore, it seems unlikely that the immune system had any effect on viral evolution. Finally, these results also demonstrated that homologous regions are used on these two different viruses to bind to two different receptors.

H. Human Rhinovirus 2

These studies were important in that they further defined the remarkable flexibility of antibodies, elucidated the properties of the second antigenic site on rhinoviruses, and demonstrated that bivalent binding is not necessarily correlated with efficacious neutralization.

HRV2 is a member of the minor group of human rhinoviruses. Although all of the rhinoviruses have nearly identical structures, the minor and major groups of rhinoviruses differ in several important aspects. The receptor for the major group is ICAM-1, which binds in the canyon region, whereas the receptor for the minor group is low-density lipoprotein receptor [84–88], which binds to the BC and HI loops of VP1, which lies on the surface of the virus near the 5-fold axis [89]. The atomic structure of HRV2 has been determined [55], but at the time of these cryo-TEM studies, the structure of another minor group rhinovirus, HRV1A, was used because it is 95% identical to HRV2.

The antigenic sites on HRV2 have been designated A, B, and C, and correspond to the NIm-I, NIm-II, and NIm-III sites of HRV14, respectively. Whereas all of the work on HRV14 has focused on the NIm-IA site (site A in HRV2), the structural work on HRV2 has focused on the B site (NIm-II in HRV14). Both of the antibodies used in these studies, mAb 8F5 and mAb 3B10, are weakly neutralizing antibodies [90, 91]. Interestingly, all of the antibodies to the NIm-II site in HRV14 are also weakly neutralizing as well. Because mAb 8F5, but not mAb 3B10, binds with a stoichiometry of 30 and does not cause apparent immunoprecipitation, it was proposed that only mAb 3F5 binds bivalently to the viral surface. It was also noted that neither antibody grossly impairs viral attachment to cells.

Both antibodies were used for image reconstructions as Fab–HRV2 complexes and yielded structures reminiscent of the HRV14–Fab studies [90, 91]. To help in the modeling of the Fab structures into the cryo-TEM envelopes, the structures of the Fab [92] and the Fab–antigenic loop complex [93] were determined. Even though both bind to the same epitope, the orientations of the two bound Fab fragments are different. mAb 8F5 binds nearly perpendicular to the surface. The mostly likely way this antibody binds bivalently to the virion surface is across the nearest 2-fold axis. This places the two Fab arms nearly parallel to each other and only 60 Å apart. This can be afforded only by the remarkable flexibility of the hinge region. This orientation is not what was predicted on the basis of the HRV1A structure, because this antigenic loop bends toward the 2-fold axis when the antibody binds. In contrast, mAb 3B10 binds at $\sim 45^{\circ}$ angle away from the icosahedral 2-fold axes, making it impossible to model a bivalently bound antibody. These studies again show how different antibodies binding to the same site can often have different binding orientations.

I. Calicivirus

The following study suggested that antibodies need not bind across icosahedral 2-fold axes in order to bind bivalently and again demonstrated the unique flexibility of antibodies. It is also another example of where all of the antigenic sites cannot be bound by antibodies due to steric interference. In this case, there could only be a maximum saturation of 50%.

The *Caliciviridae* family comprises only one genus, *Calicivirus* [15]. Members include the human pathogen Norwalk virus and rabbit hemorrhagic disease virus (RHDV). RHDV is highly contagious in rabbits and infected rabbits usually die within 2 to 3 days. The positive-sense ssRNA genome is encapsidated by a T=3 icosahedral capsid composed of VP60. The architecture is similar to tomato bushy stunt virus (TBSV) with dimers of VP60 forming 90 archlike capsomers on the viral surface. As with other caliciviruses, cell culture conditions have not yet been discovered and therefore structural studies have been dependent on baculovirus expression systems [94, 95]. The virus-like particles produced by this expression system have been shown to be structurally and antigenically identical to the native virion.

To better understand the mechanism of antibody-mediated neutralization, a cryo-TEM study was performed using the neutralizing mAb E3 [96]. It was suggested that this antibody might bind bivalently to the virion surface because there was relatively little aggregation on complex formation. From the image reconstruction, it was clear that the antibody was binding to the top of the dimeric bridges. This image reconstruction of the mAb–virus complex was fairly complicated. First, as with all of the other reconstructions that used intact antibodies, the Fc region was not visible, presumably due to hinge flexibility. Second, it was apparent that, because of spatial overlap, only one antibody could bind to these dimeric arches at a time. Therefore, the antibody density was relatively weak compared with the capsid, with the density of constant domains of the Fab arms being weaker than that of the variable domains. In conjunction with the relatively low propensity of these antibodies to aggregate the virions, a model of multiple bivalent binding modes was presented. It was suggested that the antibodies may be cross-linking these arch dimers about the icosahedral 3-fold axes. This would also explain the diffuse density observed above the 3-fold axes. However, because only one-half of each arch dimer can have an Fab arm bound to it, there are many possible cross-linking configurations about each 3-fold axis. This model suggests that the torsional flexibility of the mAb hinge region allows for $\sim 60^{\circ}$ rotation of one Fab arm relative to the other with a slight rotation about the elbow axes within each Fab arm. This extent of antibody flexibility has been clearly shown to be possible in other antibody studies [97, 98]. In terms of antibody neutralization, it is clear that neither aggregation nor gross conformational changes can be responsible for neutralization. This antibody blocks attachment of the virus to human group O red blood cells and therefore steric abrogation of cellular attachment is one possible mechanism of neutralization. It is also possible that the proposed bivalent attachment may stabilize the virions and prevent uncoating or attachment.

J. Herpes Simplex Virus

The following study exemplifies how antibodies can be used to find the exact location of a particular peptide region within a macro-macromolecular assembly. Although several extremely large viruses have been crystallized, structural elucidation of many others will have to depend on this kind of footprint analysis.

Herpes simplex virus type 1 (HSV-1) is a member of the *Herpesviridae* family. HSV infection is endemic in the population. The primary symptom of infection is recurrent fever blisters. On occasion, infection can lead to more serious symptoms such as encephalitis and retinitis. In particular, immunocompromised patients, such as those undergoing chemotherapy treatments or receiving organ transplants, and acquired immunodeficiency syndrome (AIDS) victims are at particular risk of developing life-threatening complications due to reactivation of latent infections.

HSV-1 is a large, complex virus with a diameter of ~1250 Å. The shell is made up of ~3000 polypeptides and the virion has a total mass of ~200 MDa [99, 100]. The shell is composed of a number of minor components with yet unknown stoichiometry and four major capsid proteins. The majority of the capsid is made up of 960 copies of VP5 (MW 149,000), which forms the 150 hexons and 12 pentons. At 320 sites of 3-fold symmetry are triplexes. These triplexes are heterotrimers and are proposed to have to have an $\alpha_2\beta$ stoichiometry [101] of VP23 and VP19c. These three proteins are essential for capsid assembly and coassemble with pre-VP22a, which acts as an internal scaffolding protein during procapsid formation [102, 103]. During maturation, pre-VP22a is cleaved and expulsed from the capsid. The fourth abundant protein in the capsid is called VP26 and has a molecular weight of 12,000 [104, 105]. VP26 is dispensable for assembly but incorporates in an equimolar ratio with VP5 [101].

Other than their differences in symmetry, the pentons and hexons have similar structures [101, 106–109]. The hexons are cylindrical projections with a diameter of 170 Å and a height of 110 Å. A channel with a diameter of 50 Å runs through the center of the hexons. Each of the VP5 subunits making up the hexons has three domains: a diamond shaped upper domain, a stemlike central domain, and a base domain. The lower domains for both hexons and pentons form the 30- to 40-Å capsid floor. In spite of these similarities cryo-TEM studies have demonstrated that VP26 associates with the tips of hexons but not the pentons.

To better augment the structural differences between the hexons and pentons that are being discerned by VP5, the virus was decorated with the antibody 6F10 [110]. The residues being recognized by this antibody were determined by a combination of limited proteolysis, immunoblotting with glutathione S-transferase (GST)–peptide fusions, and reactivity to synthetic peptides. On the basis of these results, this antibody mostly likely binds to peptide region ~862–880. In the image reconstructions, the antibody binds on the outer surface of the capsid just inside the opening of the channel that runs through the capsomers. Because these antibodies are binding near a symmetry axis, they have more of a "turret" shape than the well-defined structures observed in some of the other antibody–virus complexes. Nevertheless, it is clear that the antibodies do not induce gross conformational changes in the virion, and this antibody-tagging method can help define the orientation and location of subunit domains in these large virus particles.

K. *Adenovirus*

These studies were unique in that they used an antibody that was more efficacious as Fab fragments than as an intact antibody. In addition, this antibody was used to identify the location of a highly mobile cell receptor-binding site.

Adenoviruses are nonenveloped viruses that are a significant cause of respiratory, ocular, and gastrointestinal infections in humans [111, 112]. Adenoviruses have icosahedral shells with diameters ranging from 80 to 110 Å. The shell contains 240 hexons formed by 3 copies of viral protein II. Twelve copies of protein IX are found between 9 hexons in the center of each icosahedral facet. It has been proposed that two monomers of protein IIIa penetrate the hexon capsid at the edge of each facet. Several copies of protein VI form a ring under the peripentonal hexons. The pentons at each of the twelve 5-fold axes are composed of 5 copies of protein III and are tightly associated with 1 or 2 fibers, each composed of 3 copies of protein IV. The 22-Å fibers compose a shaft with a knob at the tip. Protein VIII has been suggested to be associated with the interior face of the hexons and other proteins are associated with the hexons to form the shell (proteins V, VII, X, and terminal).

The primary cellular attachment receptor for adenovirus has not yet been identified, but the vitronectin-binding integrins $\alpha_v\beta_3$ and $\alpha_v\beta_5$ have been shown to be coreceptors for viral internalization [113] and interact with the penton base of the capsid. It is thought that the initial interaction with the cell surface is mediated by the protruding fiber proteins. The various subgroups presumably bind to different receptors because they do not compete with one another for cellular binding. Subsequent interactions between the α_v coreceptors and the penton base trigger clathrin-mediated entry into the cell [113,114]. The shell is then dismantled in the low-pH environment of the endosomal vesicle [115]. Eventually, the partially disassembled capsid interacts with the nuclear pore, where the viral DNA is released [115].

The pentons clearly play an important role in viral entry. As expected, because the coreceptors are integrins, both antibodies to the functional domains of integrin and RGD peptides can block viral entry [113]. From cryo-TEM studies, it was proposed that the RGD loop was located on the outermost tip of these fibers [116]. Although both $\alpha_v\beta_3$ and $\alpha_v\beta_5$ integrins can serve as coreceptors, only $\alpha_v\beta_5$ enhances membrane permeablization [117]. On treatment of the capsid with pH 5.0 buffers, the pentons become highly hydrophobic and are therefore thought to interact with the endosomal membranes [118]. In image reconstructions of the virus alone, weak density is observed ~24 Å above the penton base protein, suggestive of a mobile loop decorating the penton base protein [119].

To better understand antibody neutralization of these complex virions, the structure of an Fab–adenovirus complex was determined by cryo-TEM techniques (Fig. 1) [119]. For these studies an unusual antibody, DAV-1, was chosen. Using peptide-scanning techniques, it was found that this antibody recognizes a nonapeptide with the sequence IRGDTFATR. This is consistent with the fact that this antibody recognizes several different adenovirus serotypes that have similar sequences flanking the RGD motif but poorly recognizes Epstein–Barr virus which has dissimilar flanking residues.

Contrary to all other antibodies discussed in this review, the Fab fragments of DAV-1 were better able to neutralize viral infectivity than the intact mAb. This is in spite of the fact that both the mAb and Fab were potent inhibitors of penton base–cell interactions. Indeed, the IgG had an ~4-fold higher affinity for the pentons than did the Fab, yet was ineffective at viral neutralization. A possible reason for this difference came from the fact that the Fab fragments bind with a stoichiometry of five per penton whereas the mAbs bind with a stoichiometry of 2.8 per penton. Therefore, whereas the affinities of the Fab and IgG were similar, their binding stoichiometry was not.

The image reconstruction of the Fab–virus complex clearly shows that the Fab is binding to a flexible portion of the penton base. Antibody binding to the penton base did not result in any observable structural change in the virion. Rather than appearing as a well-defined bilobed structure as observed in other Fab–virus complexes, these bound Fab fragments formed a ring of density above the penton base. This is similar to what was observed in the 4C4–FMDV reconstruction, in which the 4C4 antibody also bound to the RGD motif. From these results, it was proposed that IgGs had lower neutralization efficacy than Fabs because they tend to occlude themselves about the penton base. The minimum distance between the RGD motif on these penton proteins is ~57 Å. This is the minimum distance between the bivalently bound IgGs in the HRV2–IgG complexes. If antibodies were to bind bivalently to these penton assemblies, the bulkiness of the IgGs might occlude binding to other sites. In this way, IgGs could only partially saturate the RGD motifs, thereby leaving some available to integrin recognition. Although it was suggested that mobility in this RGD loop might allow virus to escape antibody neutralization, it should be noted that antibody binding itself is sufficient for antibody-mediated opsonization *in vivo*. Furthermore, the fact that the Fabs neutralize better than IgGs, presumably because of their higher binding stoichiometry, supports the hypothesis that antibody neutralization is primarily due to steric interference between the virus and its receptor.

L. *Hepatitis B Virus*

In hepatitis B virus, the same protein gives rise to two unique clinical antigens that are not cross-reactive. This immunocomplex was able to identify one of these unique determinants. Such studies are important in the elucidation of the assembly pathway of this virus.

Hepatitis B virus (HBV) is a member of the *Hepadnaviridae* family and of the genus *Orthohepadnavirus* [15]. HBV causes chronic, acute, and fulminate hepatitis and is still a major health issue, with hundreds of millions of individuals infected despite the development of a number of efficacious vaccines [120]. HBV first assembles the capsid around the RNA pregenome and reverse transcriptase. On assembly, the pregenome is retrotranscribed [121] and the nucleocapsid is enveloped by portions of the host cellular membrane and viral glycoprotein. There are two sizes of HBV, composed of 90 or 120 capsid protein dimers in a T=3 or T=4 icosahedral arrangement, respectively [122, 123].

HBV has three major clinical antigens. One of these antigens is the viral glycoprotein, or "surface antigen." The other two antigens, the "core antigen" (HBcAg) and the "e-antigen" (HBeAg) determinants, are both found on the capsid protein. The difference between these two antigens is that HBcAg is the capsid protein assembled into icosahedral particles and appears early in infection whereas HBeAg appears late in infection [124], correlates with disease progression, and is the capsid protein in a noncapsid form. Antibodies to these two capsid protein-derived antigens are not cross-reactive. A major goal of this work is to ascertain how the same protein can result in two different antigens (Fig. 1) [125].

This study [125] was greatly aided by previous high-resolution cryo-TEM image reconstruction studies on this virus. The Fab fragments were found to bind directly on top of the four-helical spikes formed by the interactions of the A/B and C/D subunits within the T=4 asymmetric unit. As with a number of other studies reviewed in this article, the proximity of symmetry-related epitopes prevented complete saturation of the virion with antibody. The occupancy was estimated to be 30–40%, and the relatively weak density formed a ''lyre''-shaped motif at this spike, due to the two mutually exclusive binding modes. The densities at the two different spikes were roughly equal, suggesting that all four quasi-equivalent antigenic sites are immunologically indistinguishable. As observed with most of the other antibody–virus complexes, no significant conformational changes were observed on antibody binding.

The question remaining concerns why these two forms of capsid protein are so different antigenically. One possibility is that the capsid protein is not chemically identical in these two forms. HBeAg is 29 residues shorter at the C terminus compared with HBcAg; however, the immunodominant epitopes for both forms are thought to lie on common regions of the polypeptide. It is also possible that the difference may be in the dimeric state of the two forms of capsid protein. This was also thought to be unlikely because both types of capsid protein form strong dimers in solution. Finally, it is possible that these antigenic differences are due to masking of e2 determinants in the capsid and/or conformational changes in the dimer as it is assembled in the capsid. In the case of the latter, differences in antigenic structures between quasi-equivalent subunits have been observed in herpes simplex virus [126] and cucumber mosaic virus [127].

M. Foot-and-Mouth Disease Virus

These studies were important for a number of reasons. The structure of the peptide–Fab complex was one of the first examples of direct interactions between known receptor recognition regions and a neutralizing antibody [128]. These structural studies also demonstrated that the antigen-induced fit in the paratope of the antibody probably does not require a great deal of energy for the deformation [129]. The cryo-EM studies verified the mobility in the receptor-binding region shown in the crystallographic studies, but also demonstrated that different antibodies affect this mobility in different ways [130].

Foot-and-mouth disease virus (FMDV) is a highly contagious member of the picornavirus family that infects cloven-hoofed animals. Although it is sensitive to environmental influences such as low pH, sunlight, and desiccation, it may spread over great distances with movement of infected or contaminated animals and materials. The morbidity rate can reach 100%; however, mortality can range from 5% (adults) to 75% (suckling pigs and sheep) [15].

FMDV differs from rhinovirus in several important ways [52, 131]. Unlike the convoluted surfaces of rhino- and poliovirus, FMDV has a relatively smooth surface. Protruding up from the shell is a long flexible loop connecting the βG-βH strands of VP1. At the tip of this loop is a conserved RGD sequence that is recognized by the viral receptor, integrin $\alpha\beta_3$. Although there are four or five immunogenic sites on VP1, VP2, and VP3 in the seven serotypes of FMDV, most flank the RGD motif residues. This led investigators to suggest that FMDV can use "camouflage" to hide crucial residues "in plain sight" [131]. According to this hypothesis, crucial residues might be exposed to antibodies but simply do not change in response to antibodies as doing so will be lethal to the virus. On the other hand, residues adjacent to the RGD sequence can change and thwart antibody binding.

Whereas the immunodominant RGD loop was disordered in the original structure of FMDV, it becomes ordered when the C134–C130 cystine bond is reduced. In this reduced form, βG-βH lies down on the surface but the virus is still infectious. Most escape mutations are in those residues that make direct contact with the bound antibody. However, several are distal to the antibody-binding region and are thought to indirectly abrogate binding by affecting the conformation of this loop, forcing the loop down (close to the 2-fold axis) from its normal up position (close to the 5-fold axis).

A direct test of the camouflage hypothesis came from the structures of the Fab–peptide [130, 132, 133] and FMDV–heparin complexes [134]. For these studies, two antibodies were used: SD6 and 4C4. Both antibodies strongly neutralize the virus. In the crystal structure of the SD6–peptide complex, the antibody is observed to contact 10 residues in the RGD loop whereas 4C4 interacts only with 8. This difference, however, does not appreciably affect neutralization efficacy. In the case of the SD6–peptide complex, there was a great deal of rearrangement of the CDR3 loop of the heavy chain. This loop had an unusual amino acid sequence with a great number of charged residues. On binding to the peptide a number of these residues underwent conformational changes that changed both shape and charge distribution. This rearrangement provides multiple points of interaction with the peptide, particularly with the Arg-Gly-Asp motif. On further examination of the unbound Fab electron density, this altered conformation was observed at a low occupancy. This suggests that the induced fit conformation is somewhat "natural" to the paratope and therefore antigen binding does not come at a large energy cost.

A secondary receptor binds to a different region than the RGD loop. Heparan sulfate has been suggested to be involved in a two-step attachment process in which low-affinity interactions with heparan sulfate at one site is followed by high-affinity binding to an integrin receptor via the RGD sequence [135]. The interactions between heparan sulfate and FMDV have been visualized in the structure of FMDV complexed with a heparan sulfate [134]. Like the RGD sequence [128], this oligosaccharide-binding site is not only exposed but is also is part of one of the antigenic sites [134]. Again, these results clearly demonstrate that viruses do not hide receptor-binding regions in convolutions on the virion surface.

To further examine the interactions between FMDV and neutralizing antibodies, the atomic structures of peptide–Fab complexes and the cryo-TEM structures of the Fab–virus complexes were determined [136]. In the image reconstructions, SD6 has a well-defined orientation on the virion surface whereas the density for 4C4 is extremely diffuse. One possible reason for this difference is that, while the RGD loop is in an extended conformation in the SD6 complex, a hinge rotation at the base of the loop may bring it closer to the capsid surface and stabilize the orientation.

N. Papillomavirus

Work on papillomavirus demonstrated several interesting properties of antibodies that recognize bovine papillomavirus type 1 (BPV1). Whereas one of the antibodies studied here did not block cell attachment, the other did. One of the antibodies recognized all of the possible antigenic sites, whereas the other recognized only the hexavalent capsomeres. Finally, one appears to bind bivalently whereas the other binds monovalently. These binding differences may explain the differences in neutralization and also help define the region recognized by the cell receptor.

Papillomavirus infections usually cause benign epithelial papillomas, but a subset of human papillomaviruses is associated with cervical cancer [137]. More than 90% of cervical cancers are associated with sexually transmitted genital human papillomavirus. Because of this high degree of association and the high mortality rate of cervical cancer, it is of great importance to understand the mechanism of antibody-mediated neutralization to facilitate vaccine development.

Papillomavirus has a 600-Å-diameter shell that encases a histone-bound 8-kb double-stranded, covalently closed circular genome [15]. This icosahedral shell is composed of major (L1) and minor (L2) capsid proteins at a ratio of ~30:1 [138]. Image reconstructions of papillomavirus show that the capsid is composed of pentameric, star-shaped capsomers arranged in a T=7 icosahedron [139]. Studies on BPV to 9 Å have suggested that L2 may be located at the 5-fold vertices, in the center of the pentavalent capsomers [140].

Virus-like particles (VLPs) can be generated from viral protein L1 alone, and these assemblies retain the antigenic determinants of the authentic virion [141, 142]. Passive transfer experiments have demonstrated that immune serum from VLP-challenged animals can protect naive animals from infection. Analyses of mAbs to papillomaviruses have suggested that neutralization can occur by more than one mechanism. One set of antibodies, of which mAb 9 to BPV1 L1 is a typical example, prevents virions from binding to cell surfaces presumably by abrogating interactions with the cell surface receptor (possibly α_6 integrin) [143–145]. A second group, of which 5B6 to BPV1 L1 is an example, efficiently neutralizes viral infection but does not significantly inhibit virions binding to cell surfaces.

To determine the structural basis for these apparent differences in neutralization mechanism, cryo-TEM image reconstruction of both mAb 9–BPV and mAb 5B6–BPV complexes were determined (Fig. 1) [146]. In the raw images, it was apparent that both antibodies tended to aggregate the virions, but it was more pronounced in the case of mAb 9. Whereas the unbound Fab arm and the Fc region were somewhat disordered, more domains were observable here than had been previously reported. When the masking radius was increased and the contouring level was decreased, intercapsomer cross-linking was not observed. This is consistent with the observed antibody-mediated aggregation observed in the raw images. The epitope was determined to be between the protrusions of density comprising the tip of the capsomers. Although antibody was observed to be bound to each of the pentavalent capsomers, only three of the five L1 molecules in the hexavalent capsomers were observed to bind antibody with a sufficient degree of occupancy to be clearly seen. To explain this observation, it was suggested that this antibody has a higher affinity for the pentavalent capsomers and that steric hindrance prevents binding to the two adjacent L1 molecules of the hexavalent capsomers. There was density overhanging the hexavalent capsomers that were attached to the IgG bound to the pentavalent capsomers, and is probably the Fc region.

mAb 5B6 bound in quite a different manner than mAb 9. No antibody was observed to bind to the 5-fold pentamer. In this case, the epitope is on the side of the capsomer, about 25 Å above the capsid floor. Although the distance between 2-fold-related epitopes is too close for bivalent attachment, it is possible that the other epitopes about the hexavalent pentamer are cross-linked by this antibody. In contrast to mAb 9, 5B6 adopts a more linear conformation and binds deeply in the cleft between the hexavalent capsomers, occupying all of the space between capsomers around the epitope.

These results shed some light on the mechanism of antibody-mediated neutralization of papillomavirus. No antibody-induced conformational changes in the virion were observed in these 13-Å electron density maps, therefore eliminating this as a possible mechanism. Although mAb 9 tended to aggregate the virions, the fact that this antibody can neutralize the virus postattachment makes it unlikely that aggregation is a significant neutralization mechanism. mAb 9 clearly covers much of the viral surface and therefore its ability to block cellular attachment may be steric. Because mAb 5B6 does not cover as much of the viral surface, it was proposed that there may be two receptors and that mAb 5B6 can abrogate binding only to one. Alternatively, because mAb 5B6 binds close to the putative intercapsomer linkages, this antibody may neutralize by stabilizing the virion. It was also noted that, even though mAb 5B6 occludes most of the canyon region encircling the capsomers, this antibody–virus complex can still bind to its cellular receptor. This strongly suggests that the receptor binds to the outermost, exposed regions of the capsid. Again, these results imply that receptor-binding regions are not hidden from immunological recognition.

O. *Reovirus*

These studies are the only structural studies to suggest that antibody binding may affect the capsid in ways other than just small, localized changes at the epitope–paratope interface. However, these changes are not related to bivalent binding of the antibody, nor are they related to neutralization efficacy. The conclusion from these studies is, again, that the major effect of the antibody on viral infectivity is steric abrogation of receptor–virus interactions.

Reoviruses are nonenveloped virions with icosahedral symmetry and consist of two concentric protein shells, termed the outer capsid and core [15]. These virions encapsidate a genome of 10 double-stranded RNA gene segments. Reovirus strain type 1 Lang (T1L) has a diameter of 850 Å and 600 projections composed of the $\sigma3$ protein [147]. $\sigma3$ interdigitates with a more internal layer composed of 600 copies of $\mu1$ protein to form the outer capsid. At each icosahedral 5-fold axis, pentamers of $\lambda2$ protein form turrets. At the center of each 5-fold axis is the viral attachment protein $\sigma1$. $\sigma1$ has a fibrous tail at the amino terminus that anchors the protein into the virion. The carboxy-terminal region of the protein consists of a globular head [148–151]. In the case of type 3 Dearing (T3D), the $\sigma1$ head domain binds junction adhesion molecule [152] and the $\sigma1$ tail domain binds sialic acid residues on glycosylated cell surface molecules of erythrocytes and nucleated cells [153–157]. In virions, the $\sigma1$ has a retracted conformation [147, 151], where it may interact with $\sigma3$ [158]. The crystal structure of the $\sigma3$ protein has demonstrated that it has two lobes organized around a central helix that spans the length of the protein [159] with the larger, external lobe projecting into the solvent. The smaller lobe interacts with $\mu1$ [147].

During cellular entry via the endosomes, the $\sigma3$ protein is removed from virions by acid-dependent proteolysis [160, 161]. This is the first necessary step for penetration of reovirus into the cytoplasm [162–164]. This is hypothesized to facilitate a conformational change in $\sigma1$ to a more extended form [156]. Monoclonal antibodies to each of the reovirus outer capsid proteins have been isolated and characterized [158, 165]. $\sigma1$-specific mAbs are serotype specific [158, 165] and some of these mAbs are effective at neutralizing infectivity *in vitro* [158, 165, 166].

mAb 4F2, which is specific for outer capsid protein $\sigma3$ [158], blocks the binding of $\sigma1$ protein to sialic acid and inhibits reovirus-induced hemagglutination (HA). The structure of the 4F2–T3D complex was determined to ascertain whether mAb 4F2 inhibits HA by altering $\sigma1$–$\sigma3$ interactions or by steric hindrance [167]. In this case, the intact 4F2 was >16-fold better than the corresponding Fab fragments at inhibiting T3D-induced HA. However, the affinity of the Fab fragments was only ~3-fold weaker than that of the mAbs. The difference in HA activity, therefore, is unlikely to be due solely to avidity effects alone. It is also unlikely that this difference is due to antibody-mediated immunoprecipitation because the intact antibodies did not appear to significantly aggregate the virions at concentrations necessary for HA abrogation.

To interpret these antibody–virus complexes, atomic models for both $\sigma3$ and Fabs were used [167]. Using an automated rigid body-fitting routine, the larger of the two lobes of $\sigma3$ (residues 91–286 and 337–365) were placed distal to the surface of the virion. The smaller lobe (residues 1–90 and 287–336) was placed closer to the virion surface, where it interacts with the $\mu1$ protein layer. By comparing these fitting results of the virion and antibody–virion complexes, it was proposed that the antibody binding induced a small conformational change in $\sigma3$ orientation. A small spur of density was also observed at a radius of ~385 Å and was thought to be indicative of an antibody-induced rearrangement of the $\mu1$ protein. In these reconstructions, the electron density for both the variable and constant domains of the Fab arms was clearly resolved. As was expected, this placed the hypervariable region in direct contact with residue 116, which has been shown to be important in 4F2 binding. However, the constant domains for the mAb needed to be adjusted by an elbow rotation compared with the Fab reconstruction in order to optimize the fit. This suggests that the elbow is required to flex when binding in a bivalent mode.

From these results, the most likely mode of antibody-mediated abrogation of virally induced HA appears to be steric hindrance of sialic acid binding. The mAb efficacy is >16-fold better than that of the Fab fragments, yet the measured binding constants are only ~3-fold different. Whereas the mAbs appear to have adapted to virion binding by movement about the elbow axis, both Fab and mAb cause identical apparent changes in the outer capsid. Finally, aggregation is not a plausible mechanistic difference because neither the mAbs nor Fabs appreciably aggregate the virions at neutralizing concentrations. The problem with steric interference is that the isolated $\sigma 1$ in the extended conformation has a length of ~480 Å whereas an mAb has an extended length of ~150 Å. However, previous reconstructions of intact virions have suggested that $\sigma 1$ in the turret has a more compact conformation. Therefore, it was proposed that the increased length and bulk of the intact antibody are able to block sialic acid binding to $\sigma 1$, but the shorter Fab fragments that lack the Fc portion cannot.

P. Cucumber Mosaic Virus

The importance of work on cucumber moaic virus (CMV) is that is shows the remarkably plasticity of antibodies as they recognize their target epitope. Although CMV is a T=3 virus, the antibody studied here recognizes only those subunits immediately adjacent to the icosahedral 5-fold axes. Because the atomic structure of CMV is known, it was concluded that this binding pattern is most likely due to differences in the relative position of the subunits rather than to differences in the antigenic loop itself among the quasi-equivalent subunits [127].

CMV, the type member of the genus *Cucumovirus* (family *Bromoviridae*), infects more than 800 plant species and causes economically important diseases of many crops worldwide [168]. CMV is transmitted by aphids in a nonpersistent manner. The virus does not circulate or replicate in the aphid and is quickly acquired from and transmitted to a host during feeding. Virus interacts with the anterior portion of the alimentary tract (food canal to foregut), from which it can be subsequently inoculated by egestion. Unlike some other plant viruses that are transmitted in a nonpersistent manner, CMV does not require helper proteins for transmission. Hence, the aphid recognition motifs must reside on the capsid itself. The ~24-kDa coat protein also appears to play a role in normal cell-to-cell and systemic movement independent of particle formation [169, 170].

The X-ray crystal structure of CMV revealed an exposed βH-βI loop with several unique properties [171]. In contrast with the predominant neutral or basic charge character of the capsid surface, the βH-βI loop is highly acidic. Close inspection of the electron density near this loop indicated that a metal ion might be chelated by this cluster of acidic residues. The likelihood of an important role for these residues is additionally supported by the observation that six of the eight amino acids that comprise this loop are highly conserved among strains of CMV and other cucumoviruses [172]. Furthermore, mutations in several of the loop residues (D191, D192, L194, and E195) had no significant effect on virion formation or stability, but they did reduce or eliminate aphid transmission [172]. To better understand the molecular basis for virus transmission by insects, antibodies were developed against this loop and the structure of this virus–antibody complex was determined (Fig. 1) [127].

The cryo-TEM and modeling results clearly demonstrated that antibody binds to the βH-βI loop of CMV. Unlike in other virus–antibody complexes studied to date, this antibody binds immediately adjacent to an axis of icosahedral symmetry and only one Fab binds per penton. Because the electron density of the bound Fab becomes averaged by the icosahedral 5-fold symmetry, the density of the bound Fab assumes a "turret" shape rather than the typical bilobed shape. The nature of the CMV-Fab contact is reminiscent of the lysozyme–HYHEL-10 complex, in which the antibody and antigen interact in a blunt-faced manner [173]. This contrasts with the mode of binding adopted by other antibodies such as Fab17, in which a protruding antigenic loop on HRV14 binds into the cleft between the heavy and light chain variable domains [36].

The more unusual finding was that each antibody bound to several antigenic loops at the same time. Although antibodies are known to be able to bind simultaneously to multiple viral subunits (e.g., Refs. 25 and 32), this is the first example in which an antibody bridges the same regions of two or more identical and adjacent capsid subunits. This CMV antibody is centered about the βH-βI loop, which is thought to interact with the receptor molecule in the aphid. This loop is immediately adjacent to, and facing toward, axes of icosahedral symmetry. Therefore, it may be that the aphid receptor, similar to antibody 3A8-5C10, also exhibits quasi-equivalent specificity and may interact only with pentons or only with hexons.

These results also demonstrated that the location of an antigenic loop on a viral surface is an important criterion in determining how best to engineer hybrid viruses for use as vaccines. Using the CMV structure as an example, the βH-βI loop might represent a logical position into which a foreign antigenic peptide could be introduced. However, because this loop is directed toward the 5-fold icosahedral symmetry axis, any antigen inserted in this loop would likely constrain elicited antibodies to simultaneously contact several subunits. This might therefore result in the selection of antibodies that recognize the targeted epitope in a context that does not properly mimic its native form. Hence, a more fruitful approach might be to place the antigenic determinant into a region that will help assure that it is presented in a more exposed, native form and also that would be less likely to yield antibodies with paratopes that bind more than one epitope at a time. In this sense, for example, the NIm-IA site on HRV14 is a good candidate for designing hybrid virions [7, 34].

IV. Conclusions

After reviewing most of the more recent structural studies on antibody-mediated neutralization of viruses, it is important to briefly note potential problems with the assays used to characterize the antibodies.

A. *Neutralization Efficacy*

Using the results of HRV14 as an example, antibodies have apparently different neutralization efficacies when the assays are performed in slightly different ways. To ascertain which of various antibodies were weakly or strongly neutralizing, the immunocomplexes were incubated with cell monolayers for a relatively short period of time and then the unbound virus was washed away. Under these conditions, mAb1 appeared to be much less efficacious than mAb17 in neutralizing the virus. In contrast, when antibody is kept in the plaque assay overlays, mAb1 and mAb17 both neutralize HRV14 infectivity with comparable efficacy. This difference is due to the fact that, in the latter assays, the antibody is around to inhibit secondary infections that are needed for plaque development. This demonstrates that care must be taken when assessing the efficacy of an antibody solely on the basis of *in vitro* assays. Also, in this case, it is difficult to predict how these two antibodies would protect the animal from viral challenge.

B. Aggregation

Immunoprecipitation normally occurs over a relatively narrow range of antigen:antibody ratios. In the case of HRV14, it was shown that mAb17 did not precipitate the virions over a wide range of antibody concentrations. On the basis of this result, it was suggested that this antibody bound bivalently to the surface of the virion. Although this was shown to be the case for the mAb17–HRV14 complex, it is not true that mAb17 does not precipitate the virions. Indeed, mAb17 is effective in an Oüchterlony assay and forming unaggregated mAb17–HRV14 complexes was difficult in the cryo-TEM studies. Clearly, the process of bivalent binding versus inter particle cross-linking is controlled by the kinetics of the system. Under a certain set of conditions, the antibody may not have enough time orient its other Fab arm to form an intraparticle cross-link before another virus attaches to it.

C. Binding Affinity and Stoichiometry

There are several cases reviewed above in which antibodies do not bind to all potential antigenic sites. Again, the fact that only 30 mAb17 molecules bound to HRV14 was used to suggest that this antibody binds bivalently to the virion surface. However, other studies demonstrate that this half-site saturation could have been due to steric interference. These binding measurements are clearly difficult and cannot be done in a traditional manner. Most of these techniques are sensitive to reversible binding. Even in the case of using newer techniques such as plasmon resonance, it is necessary to know whether the antibody is binding monovalently or bivalently, and this may change depending on the conditions of the experiments.

*D. Single-Hit Kinetics and p*I *Changes*

The evidence for antibody-induced conformational changes has been that the process of antibody-mediated neutralization follows single-hit kinetics and is often associated with p*I* changes. In the case of p*I* changes, it is not clear what such changes truly mean because the p*I* of the capsid alone is being compared with the p*I* of large immunocomplexes. Second, such p*I* changes have now been shown not to be associated with conformational changes [36]. Finally, p*I* measurements, using a horizontal isoelectropoint focusing device, have not shown the existence of such changes (Z. Che and T. J. Smith, unpublished data).

In the case of the single-hit kinetics model, it has been argued that a first-order neutralization reaction is not proof for neutralization by single-hit kinetics (for a review, see Burton *et al.* [174]). As antibody is added to a sample of virus, there is not an apparent lag phase in the curve. This has been used to suggest that a single antibody is sufficient for neutralization. However, because of the assay conditions, it is impossible to draw samples off fast enough to detect the presence or absence of a lag phase. Furthermore, the lack of a lag phase is mostly observed only at low temperatures, low antibody concentrations, or when using antibodies with low affinity. If antibodies are truly neutralizing according to single-hit kinetics, it is not clear why these conditions need to be met. Finally, most, if not all, antibody neutralization has been shown to require more than one antibody.

Using all of the above-described structural results, the following conclusions can be drawn.

1. Aggregation is not a significant mechanism of *in vitro* antibody-mediated neutralization. Even with the caveat stated above about the difficulty in measuring aggregation, a number of the studies described in this review clearly show that there is not a causative link between aggregation and neutralization. However, aggregation will greatly enhance *in vivo* processes such as opsonization and therefore may play a significant role in antibody protection.
2. Most, if not all, antibodies do not neutralize by inducing conformational changes. There is no *a priori* reason why antibodies cannot induce large conformational changes in the virion. Indeed, the results from poliovirus [51], flock house virus [56], and rhinovirus [57] all clearly show that these are dynamic capsids. However, for antibody-induced changes to be essential for neutralization, clonal expansion of B cells would need a mechanism to select out only those B cells that express surface antibodies that induce conformational changes. Instead, clonal expansion is driven only by the binding affinity between the virion and the B cell. The only structural evidence for an antibody-induced conformational change comes from the work on reovirus [167] and even then, there is not a casual link between induction of conformational changes and neutralization. Indeed, there is the opposite finding that the paratope region is a plastic surface that molds itself to the epitope [36, 129, 175]. If an antibody were shown to "induce" a conformational change, it seems most likely that the virus protein is undergoing reversible conformational changes (i.e., "breathing") and the antibody is

recognizing transiently exposed epitopes. In this way, the antibody would not induce a conformational change per se but rather recognize a structure that is normally occurring in the virus. However, the findings here suggest that induction of a conformational change is neither a necessary nor predominant mechanism of neutralization.

3. Most antibodies do not neutralize by stabilizing the capsid. Although it was originally suggested that antibodies that bind bivalently might neutralize by stabilizing the capsid, it is clear that strong neutralization does not require bivalent attachment. It is, however, true that bivalent attachment can help the antibody compensate for weak intrinsic affinity between the Fab and the antigen. Furthermore, if neutralization by stabilization were a dominant mechanism, then more distal site, compensatory escape mutations should have arisen. Such escape mutations have been observed in the case of the capsid stabilizing antipicornavirus drugs [176–178] but have not been observed in the case of antibody neutralization.
4. In most, if not all, of the above-described cases antibody is observed to directly contact the receptor-binding site. This strongly suggests that antibodies have not strongly influenced the evolution of the quaternary structure of these viruses but rather that the viral life cycle itself probably played a dominant role. Interestingly, it is apparent that the power of antibody recognition that comes from a large surface contact area is also the Achilles' heel of the antibody, with mutations at several locations on the viral surface abrogating binding [20].
5. Most, if not all, of the above-described studies ended with the conclusion that antibody neutralization is most likely dominated by interference of cell attachment. For a comprehensive review of this hypothesis, see Burton *et al.* [174]. However, it should also be noted that the synergism between antibodies and the total immune system may be far stronger and more important than any of these proposed *in vitro* mechanisms.

These studies have all shown the importance of "hybrid technology." Antibodies have been used to elucidate the architecture of viruses and to identify receptor-binding regions. They have directly addressed the mechanism of antibody-mediated neutralization, which has greatly impacted the development of vaccines. Finally, they have improved our understanding about the forces that have driven the evolution of viral structure. It is likely that such studies will continue to help us understand the architecture of macro-macromolecular complexes and the dynamics of these viral capsids.

References

1. Brioen, P., Dekegel, D., and Boeyé, A. (1983). Neutralization of poliovirus by antibody-mediated polymerization. *Virology* **127,** 463–468.
2. Brioen, P., Thomas, A. A. M., and Boeyé, A. (1985). Lack of quantitative correlation between the neutralization of poliovirus and the antibody-mediated pI shift of the virions. *J. Gen. Virol.* **66,** 609–613.
3. Thomas, A. A. M., Brioen, P., and Boeyé, A. (1985). A monoclonal antibody that neutralizes poliovirus by cross-linking virions. *J. Virol.* **54,** 7–13.
4. Colonno, R. J., Callahan, P. L., Leippe, D. M., and Rueckert, R. R. (1989). Inhibition of rhinovirus attachment by neutralizing monoclonal antibodies and their Fab fragments. *J. Virol.* **63,** 36–42.
5. Mandel, B. (1976). Neutralization of poliovirus: A hypothesis to explain the mechanism and the one-hit character of the neutralization reaction. *Virology* **69,** 500–510.
6. Mosser, A. G., Leippe, D. M., and Rueckert, R. R. (1989). Neutralization of picornaviruses: Support for the pentamer bridging hypothesis. *In* "Molecular Aspects of Picornavirus Infection and Detection" (B. L. Semler and E. Ehrenfeld, Eds.), pp. 155–167. American Society for Microbiology, Washington, D.C.
7. Che, Z., Olson, N. H., Leippe, D., Lee, W.-M., Mosser, A., Rueckert, R. R., Baker, T. S., and Smith, T. J. (1998). Antibody-mediated neutralization of human rhinovirus 14 explored by means of cryo-electron microscopy and X-ray crystallography of virus–Fab complexes. *J. Virol.* **72,** 4610–4622.
8. Levine, B., Hardwick, J. M., Trapp, B. D., Crawford, T. O., Bollinger, R. C., and Griffin, D. E. (1991). Antibody-mediated clearance of alphavirus infection from neurons. *Science* **254,** 856–860.
9. Ubol, S., Levine, B., Lee, S. H., Greenspan, N. S., and Griffin, D. E. (1995). Roles of immunoglobulin valency and the heavy-chain constant domain in antibody-mediated downregulation of Sindbis virus replication in persistently infected neurons. *J. Virol.* **69,** 1990–1993.
10. Tolskaya, E. A. Ivannikova, T. A., *et al.* (1992). Post-infection tratment with antiviral serum results in survival of neural cells productively infected with virulent poliovirus. *J. Virol.* **66,** 5152–5156.
11. Vrijsen, R., Mosser, A., and Boeye, A. (1993). Post-absorption neutralization of poliovirus. *J. Virol.* **67,** 3126–3133.
12. Schmaljohn, A. L., Johnson, E. D., Dalrymple, J. M., and Cole, G. A. (1982). Nonneutralizing monoclonal antibodies can prevent lethal alphavirus encephalitis. *Nature* **297,** 70–72.
13. McCullough, K. C., De Simone, F., Brocchi, E., Capucci, L., Crowther, J. R., and Kihm, U. (1992). Protective immune response against foot-and-mouth disease. *J. Virol.* **66,** 1835–1840.
14. Schulman, J. L. (1975). Immunology of influenza. *In* "The Influenza Viruses and Influenza" (E. D. Kilbourne, Ed.), pp. 373–393. Academic Press, New York.
15. Murphy, F. A., Fauquet, C. M., Bishop, D. H. L., Ghabrial, S. A., Jarvis, A. W., Martelli, G. P., and Mayo, M. A. (1995). Virus "Taxonomy: Sixth Report of the International Committee on Taxonomy of viruses," 6th Ed. Springer-Verlag, New York.
16. Kingsbury, D. W. (1990). *Orthomyxoviridae* and their replication. *In* "Virology" (B. N. Fields and D. M. Knipe, Eds.), pp. 1075–1089. Raven Press, New York.

17. Colman, P. M., Varghese, J. N., and Laver, W. G. (1983). Structure of the catalytic and antigenic sites in influenza virus neuraminidase. *Nature* **303,** 41–44.
18. Colman, P. M., Laver, W. G., Varghese, J. N., Baker, A. T., Tulloch, P. A., Air, G. M., and Webster, R. G. (1987). Three dimensional structure of a complex of antibody with influenza virus neuraminidase. *Nature* **326,** 358–363.
19. Bizebard, T., Gigant, B., Rigolet, P., Rasmussen, B., Diat, O., Bosecke, P., Wharton, S. A., Skehel, J. J., and Knossow, M. (1995). Structure of influenza virus haemagglutinin complexed with a neutralizing antibody. *Nature* **376,** 92–94.
20. Colman, P. M. (1997). Virus versus antibody. *Structure* **5,** 591–593.
21. Prasad, B. V. V., Burns, J. W., Marietta, E., Estes, M. K., and Chiu, W. (1990). Localization of VP4 neutralization sites in rotavirus by three-dimensional cryo-electron microscopy. *Nature* **343,** 476–479.
22. Chen, Z., Stauffacher, C., Li, Y., Schmidt, T., Bomu, W., Kamer, G., Shanks, M., Lomonossoff, G., and Johnson, J. (1989). Protein–RNA interactions in an icosahedral virus at 3.0 Å resolution. *Science* **245,** 154–159.
23. Wang, P., Porta, C., Chen, Z., Baker, T. S., and Johnson, J. E. (1992). Identification of a Fab interaction site (footprint) on an icosahedral virus by cryo-electron microscopy and X-ray crystallography. *Nature* **355,** 275–278.
24. Porta, C., Spall, V., Loveland, J., Johnson, J., Barker, P., and Lomonossoff, G. (1994). Development of cowpea mosaic virus as a high-yielding system for the presentation of foreign peptides. *Virology* **202,** 949–955.
25. Porta, C., Cheng, R. H., Chen, Z., Baker, T. S., and Johnson, J. E. (1994). Direct imaging of interactions between an icosahedral virus and conjugate Fab fragments by cryoelectron microscopy and X-ray crystallography. *Virology* **204,** 777–788.
26. Rueckert, R. R. (1996). *Picornaviridae* and their replication. *In* "Fundamental Virology" (B. N. Fields and D. M. Knipe, Eds.), pp. 609–654. Raven Press, New York.
27. Rossmann, M. G., Arnold, E., Erickson, J. W., Frankenberger, E. A., Griffith, J. P., Hecht, H. J., Johnson, J. E., Kamer, G., Luo, M., Mosser, A. G., Rueckert, R. R., Sherry, B., and Vriend, G. (1985). Structure of a human common cold virus and functional relationship to other picornaviruses. *Nature* **317,** 145–153.
28. Colonno, R. J., Condra, J. H., Mizutani, S., Callahan, P. L., Davies, M. E., and Murcko, M. A. (1988). Evidence for the direct involvement of the rhinovirus canyon in receptor binding. *Proc. Natl. Acad. Sci. USA* **85,** 5449–5453.
29. Olson, N. H., Kolatkar, P. R., Oliveira, M. A., Cheng, R. H., Greve, J. M., McClelland, A., Baker, T. S., and Rossmann, M. G. (1993). Structure of a human rhinovirus complexed with its receptor molecule. *Proc. Natl. Acad. Sci. USA* **90,** 507–511.
30. Kolatkar, P. R., Bella, J., Olson, N. H., Bator, C. M., Baker, T. S., and Rossmann, M. G. (1999). Structural studies of two rhinovirus serotypes complexed with fragments of their cellular receptor. *EMBO J.* **18,** 6249–6259.
31. Sherry, B., and Rueckert, R. R. (1985). Evidence for at least two dominant neutralization antigens on human rhinovirus 14. *J. Virol.* **53,** 137–143.
32. Sherry, B., Mosser, A. G., Colonno, R. J., and Rueckert, R. R. (1986). Use of monoclonal antibodies to identify four neutralization immunogens on a common cold picornavirus, human rhinovirus 14. *J. Virol.* **57,** 246–257.
33. Leippe, D. M. (1991). Stoichiometry of Picornavirus Neutralization by Murine Monoclonal Antibodies. Ph.D. thesis. University of Wisconsin, Madison, WI.
34. Smith, T. J., Olson, N. H., Cheng, R. H., Liu, H., Chase, E., Lee, W. M., Leippe, D. M., Mosser, A. G., Ruekert, R. R., and Baker, T. S. (1993). Structure of human rhinovirus complexed with Fab fragments from a neutralizing antibody. *J. Virol.* **67,** 1148–1158.

35. Smith, T. J., Olson, N. H., Cheng, R. H., Chase, E. S., and Baker, T. S. (1993). Structure of a human rhinovirus-bivalently bound antibody complex: Implications for virus neutralization and antibody flexibility. *Proc. Natl. Acad. Sci. USA* **90,** 7015–7018.
36. Smith, T. J., Chase, E. S., Schmidt, T. J., Olson, N. H., and Baker, T. S. (1996). Neutralizing antibody to human rhinovirus 14 penetrates the receptor-binding canyon. *Nature* **383,** 350–354.
37. Studdert, M. J. (1990). Tissue tropism of parvoviruses. *In* "CRC Handbook of Parvoviruses" (P. Tijssen, Ed.), pp. 3–27. CRC Press, Boca Raton, FL.
38. Parrish, C. R. (1990). Emergence, natural history, and variation of canine, mink, and feline parvoviruses. *Adv. Virus Res.* **38,** 403–450.
39. Siegl, G. (1984). Biology and pathogenicity of autonomous parvoviruses. *In* "The Parvoviruses" (K. I. Berns, Ed.), pp. 297–362. Plenum Press, New York.
40. Tsao, J., Chapman, M. S., Agbandje, M., Keller, W., Smith, K., Wu, H., Luo, M., Smith, T. J., Rossmann, M. G., Compans, R. W., and Parrish, C. R. (1991). The three-dimensional structure of canine parvovirus and its functional implications. *Science* **251,** 1456–1464.
41. Rimmelzwaan, G. J., Carlson, J., UytdeHaag, F.G.C.M., and Osterhaus, A.D.M.E. (1990). A synthetic peptide derived from the amino acid sequence of canine parvovirus structural proteins which defines a B cell epitope and elicits antiviral antibody in BALB-c-mice. *J. Gen. Virol.* **71,** 2741–2745.
42. Langevel, J. P., Casal, J. I., Vela, C., Dalsgaard, K., Smale, S. H., Puijk, W. C., and Meloen, R. H. (1993). B-cell epitopes of canine parvovirus: Distribution on the primary structure and exposure on the viral surface. *J. Virol.* **67,** 765–772.
43. López de Turiso, J. A., Cortes, E., Ranz, A., Garcia, J., Sanz, A., Vela, C., and Casal, J. I. (1991). Fine mapping of canine parvovirus B cell epitopes. *J. Gen. Virol.* **72,** 2445–2456.
44. Strassheim, M. L., Gruenberg, A., Veijalainen, P., Sgro, J. Y., and Parrish, C. R. (1994). Two dominant neutralizing antigenic determinants of canine parvovirus are found on the threefold spike of the virus capsid. *Virology* **198,** 175–184.
45. Wikoff, W. R., Wang, G., Parrish, C. R., Cheng, R. H., Strassheim, M. L., Baker, T. S., and Rossmann, M. G. (1994). The structure of a neutralized virus: Canine parvovirus complexed with neutralizing antibody fragment. *Structure* **2,** 595–607.
46. Hogle, J. M., Chow, M., and Filman, D. J. (1985). Three-dimensional structure of poliovirus at 2.9 Å resolution. *Science* **229,** 1358–1365.
47. Mendelsohn, C. L., Wimmer, E., and Racaniello, V. R. (1989). Cellular receptor for poliovirus: molecular cloning, nucleotide sequence, and expression of a new member of the immunoglobulin super-family-*Cell* **56,** 855–865.
48. Emini, E. A., Jameson, B. A., Lewis, A. J., Larsen, G. R., and Wimmer, E. (1982). Poliovirus neutralization epitopes: Analysis and localization with neutralizing monoclonal antibodies. *J. Virol.* **43,** 997–1005.
49. Harber, J., Bernhardt, G., Lu, H. H., Sgro, J. Y., and Wimmer, E. (1995). Canyon rim residues, including antigenic determinants, modulate serotype-specific binding of polioviruses to mutants of the poliovirus receptor. *Virology* **214,** 559–570.
50. Wien, M. W., Curry, S., Filman, D. J., and Hogle, J. M. (1997). Structural studies of poliovirus mutants that overcome receptor defects. *Nat. Struct. Biol.* **4,** 666–674.
51. Li, Q., Yafal, A. G., Lee, Y. M. H., Hogle, J., and Chow, M. (1994). Poliovirus neutralization by antibodies to internal epitopes of VP4 and VP1 results from reversible exposure of the sequences at physiological temperatures. *J. Virol.* **68,** 3965–3970.

52. Acharya, R., Fry, E., Stuart, D., Fox, G., Rowlands, D., and Brown, F. (1990). The structure of foot-and-mouth disease virus: Implications for its physical and biological properties [review]. *Vet. Microbiol.* **23,** 21–34.
53. Kim, S., Smith, T. J., Chapman, M. S., Rossmann, M. G., Peavear, D. C., Dutko, F. J., Felock, P. J., Diana, G. D., and McKinlay, M. A. (1989). The crystal structure of human rhinovirus serotype 1A (HRV1A). *J. Mol. Biol.* **210,** 91–111.
54. Oliveira, M. A., Zhao, R., Lee, W.-M., Kremer, M. J., Minor, I., Rueckert, R. R., Diana, G. D., Pevear, D. C., Dutko, F. J., McKinlay, M. A., and Rossmann, M. G. (1993). The structure of human rhinovirus 16. *Structure* **1,** 51–68.
55. Verdaguer, N., Blaas, D., and Fita, I. (2000). Structure of human rhinovirus serotype 2 (HRV2). *J. Mol. Biol.* **300,** 1179–1194.
56. Bothner, B., Dong, X. F., Bibbs, L., Johnson, J. E., and Siuzdak, G. (1998). Evidence of viral capsid dynamics using limited proteolysis and mass spectrometry. *J. Biol. Chem.* **273,** 673–676.
57. Lewis, J. K., Bothner, B., Smith, T. J., and Siuzdak, G. (1998). Antiviral agent blocks breathing of the common cold virus. *Proc. Natl. Acad. Sci. USA* **95,** 6774–6778.
58. Grant, R. A., Hiremath, C. N., Filman, D. J., Syed, R., Andries, K., and Hogle, J. M. (1994). Structures of poliovirus complexes with anti-viral drugs: Implications for viral stability and drug design. *Curr. Biol.* **4,** 784–797.
59. Wien, M. W., Filman, D. J., Stura, E. A., Guillot, S., Delpeyroux, F., Crainic, R., and Hogle, J. M. (1995). Structure of the complex between the Fab fragment of a neutralizing antibody for type 1 poliovirus and its viral epitope. *Nat. Struct. Biol.* **2,** 232–243.
60. Fricks, C. E., and Hogle, J. M. (1990). Cell-induced conformational change in poliovirus; Externalization of the amino terminus of VP1 is responsible for liposome binding. *J. Virol.* **64,** 1934–1945.
61. Mason, P. W., Baxt, B., Brown, F., Harber, J., Murdin, A., and Wimmer, E. (1993). Antibody-complexed foot-and-mouth disease virus, but not poliovirus, can infect normally insusceptible cells via the Fc receptor. *Virology* **192,** 568–577.
62. Arita, M., Horie, H., Arita, M., and Nomoto, A. (1999). Interaction of poliovirus with its receptor affords a high level of infectivity to the virion in poliovirus infections mediated by the Fc receptor. *J. Virol.* **73,** 1066–1074.
63. Huang, Y., Hogle, J. M., and Chow, M. (2000). Is the 135S poliovirus particle an intermediate during cell entry? *J. Virol.* **74,** 8757–8761.
64. Calisher, C. H., Karabatsos, N., Lazuick, J. S., Monath, T., and Wolff, K. L. (1988). Reevaluation of the western equine encephalitis antigenic complex of alphaviruses (family *Togaviridae*) as determined by neutralization tests. *Am. J. Trop. Med. Hyg.* **38,** 447–452.
65. Kay, B. H., and Aaskov, J. G. (1989). Ross River virus (epidemic polyarthritis). *In* "The Arboviruses: Epidemiology and Ecology" (T. P. Monath, Ed.), pp. 93–112. CRC Press, Boca Raton, FL.
66. Smith, T. J., Cheng, R. H., Olson, N. H., Peterson, P., Chase, E., Kuhn, R. J., and Baker, T. S. (1995). Putative receptor binding sites on alphaviruses as visualized by cryoelectron microscopy. *Proc. Natl. Acad. Sci. USA* **92,** 10648–10652.
67. Strauss, J. H., and Strauss, E. G. (1994). The alphaviruses: Gene expression, replication, and evolution. *Microbiol. Rev.* **58,** 491–562.
68. Coombs, K., and Brown, D. T. (1987). Topological organization of Sindbis virus capsid protein in isolated nucleocapsids. *Virus Res.* **7,** 131–149.

69. Choi, H.-K., Tong, L., Minor, W., Dumas, P., Boege, U., Rossmann, M. G., and Wengler, G. (1991). Structure of Sindbis virus core protein reveals a chymotrypsin-like serine proteinase and the organization of the virion. *Nature* **354,** 37–43.

70. Paredes, A. M., Simon, M., and Brown, D. T. (1993). The mass of the Sindbis virus nucleocapsid suggests it has T=4 icosahedral symmetry. *Virology* **187,** 324–332.

71. Paredes, A. M., Brown, D. T., Rothnagel, R., Chiu, W., Johnston, R. E., and Prasad, B. V. V. (1993). Three-dimensional structure of a membrane-containing virus. *Proc. Natl. Acad. Sci. USA* **90,** 9095–9099.

72. Tong, L., Wengler, G., and Rossmann, M. G. (1993). Refined structure of Sindbis virus core protein and comparison with other chymotrypsin-like serine proteinase structures. *J. Mol. Biol.* **230,** 228–247.

73. Cheng, R. H., Kuhn, R. J., Olson, N. H., Rossmann, M. G., Choi, H., Smith, T. J., and Baker, T. S. (1995). Nucleocapsid and glycoprotein organization in an enveloped virus. *Cell* **80,** 1–20.

74. Mayne, J. T., Rice, C. R., Strauss, E. G., Hunkapiller, M. W., and Strauss, J. H. (1984). Biochemical studies of the maturation of the small Sindbis virus glycoprotein E3. *Virology* **134,** 338–357.

75. Vrati, S., Faragher, S. G., Weir, R. C., and Dalgarno, L. (1986). Ross River virus mutant with a deletion in the E2 gene: Properties of the virion, virus-specific macromolecule synthesis, and attenuation of virulence for mice. *Virology* **151,** 222–232.

76. Garoff, H., Frischauf, A. M., Simons, K., Lehrach, H., and Delius, H. (1980). The capsid protein of Semliki Forest virus has clusters of basic amino acids and prolines in its amino-terminal region. *Proc. Natl. Acad. Sci. USA* **77,** 6376–6380.

77. Rice, C. M., and Strauss, J. H. (1981). Nucleotide sequence of the 26S mRNA of Sindbis virus and deduced sequence of the encoded virus structural proteins. *Proc. Natl. Acad. Sci. USA* **78,** 2062–2066.

78. Strauss, E. G., Stec, D. S., Schmaljohn, A. L., and Strauss, J. H. (1991). Identification of antigenically important domains in the glycoproteins of Sindbis virus by analysis of antibody escape variants. *J. Virol.* **65,** 4654–4664.

79. Ubol, S., and Griffin, D. E. (1991). Identification of a putative alphavirus receptor on mouse neural cells. *J. Virol.* **65,** 6913–6921.

80. Wang, K.-S., Schmaljohn, A. L., Kuhn, R. J., and Strauss, J. H. (1991). Antiidiotypic antibodies as probes for the Sindbis virus receptor. *Virology* **181,** 694–702.

81. Vrati, S., Fernon, C. A., Dalgarno, L., and Weir, R. C. (1988). Location of a major antigenic site involved in Ross River virus neutralization. *Virology* **162,** 346–353.

82. Kerr, P. J., Weir, R. C., and Dalgarno, L. (1993). Ross River virus variants selected during passage in chick embryo fibroblasts: Serological, genetic, and biological changes. *Virology* **193,** 446–449.

83. Burness, A. T., Pardoe, I., Faragher, S. G., Vrati, S., and Dalgarno, L. (1988). Genetic stability of Ross River virus during epidemic spread in nonimmune humans. *Virology* **167,** 639–643.

84. Ronacher, B., Marlovits, T. C., Moser, R., and Blaas, D. (2000). Expression and folding of human very-low-density lipoprotein receptor fragments: Neutralization capacity toward human rhinovirus HRV2. *Virology* **278,** 541–550.

85. Marlovits, T. C., Abrahamsberg, C., and Blaas, D. (1998). Soluble LDL minireceptors: Minimal structure requirements for recognition of minor group human rhinovirus. *J. Biol. Chem.* **273,** 33835–33840.

86. Marlovits, T. C., Zechmeister, T., Gruenberger, M., Ronacher, B., Schwihla, H., and Blaas, D. (1998). Recombinant soluble low density lipoprotein receptor fragment inhibits minor group rhinovirus infection in vitro. *FASEB J.* **12,** 695–703.
87. Mischak, H., Neubauer, C., Kuechler, E., and Blaas, D. (1988). Characteristics of the minor group receptor of human rhinoviruses. *Virology* **163,** 19–25.
88. Okun, V., Moser, R., Ronacher, B., Kenndler, E., and Blaas, D. (2001). VLDL receptor fragments of different lengths bind to human rhinovirus HRV2 with different stoichiometry: An analysis of virus–receptor complexes by capillary electrophoresis. *J. Biol. Chem.* **276,** 1057–1062.
89. Hewat, E. A., Neumann, E., Conway, J., Moser, R., Ronacher, B., Marlovits, T. C., and Blaas, D. (2000). The cellular receptor to human rhinovirus 2 binds around the 5-fold axis and not in the canyon: A structural view. *EMBO J.* **19,** 6317–6325.
90. Hewat, E. A., and Blaas, D. (1996). Structure of a neutralizing antibody bound bivalently to human rhinovirus 2. *EMBO J.* **15,** 1515–1523.
91. Hewat, E. A., Marlovits, T. C., and Blaas, D. (1998). Structure of a neutralizing antibody bound monovalently to human rhinovirus 2. *J. Virol.* **72,** 4396–4402.
92. Tormo, J., Stadler, E., Skern, H., Auer, H., Kanzler, O., Betzel, C., Blaas, D., and Fita, I. (1992). Three-dimensional structure of the Fab fragment of a neutralizing antibody to human rhinovirus serotype 2. *Protein Sci.* **1,** 1154–1161.
93. Tormo, J., Blaas, D., Parry, N. R., Rowlands, D., Stuart, D., and Fita, I. (1994). Crystal structure of a human rhinovirus neutralizing antibody complexed with a peptide derived from viral capsid protein VP2. *EMBO J.* **13,** 2247–2256.
94. Laurent, S., Vautherot, J. F., Madelaine, M. F., Le Gall, G., and Rasschaert, D. (1994). Recombinant rabbit hemorrhagic disease virus capsid protein expressed in baculovirus self assembles into virus like particles and induces protection. *J. Virol.* **68,** 6794–6798.
95. Nagesha, H. S., Wang, L. F., Hyatt, A. D., Morissy, C. J., Lenghaus, C., and Westbury, H. A. (1995). Self-assembly, antigenicity, and immunogenicity of the rabbit haemorrhagic disease virus (Czechoslovakian strain V-351) capsid protein expressed in baculovirus. *Arch. Virol.* **140,** 1095–1108.
96. Thouvenin, E., Laurent, S., Madelaine, M.-F., Rasschaert, D., Vautherot, J.-F., and Hewat, E. A. (1997). Bivalent binding of a neutralizing antibody to a calicivirus involves the torsional flexibility of the antibody hinge. *J. Mol. Biol.* **270,** 238–246.
97. Roux, K. R. (1984). Direct demonstration of multiple VH allotopes on rabbit Ig molecules: Allotope characteristics and Fab arm rotational flexibility revealed by immunoelectron microscopy. *Eur. J. Immunol.* **14,** 459–464.
98. Wade, R. H., Taveau, J. C., and Lamy, J. N. (1989). Concerning the axial rotational flexiblity of the Fab regions of immunoglobulin G. *J. Mol. Biol.* **206,** 349–356.
99. Rixon, F. J. (1993). Structure and assembly of herpesviruses. *Semin. Virol.* **4,** 135–144.
100. Steven, A. C., and Spear, P. G. (1997). Herpesvirus capsid assembly and envelopment. *In* In "Structural Biology of Viruses" (W. Chiu, R. M. Burnett, and R. L. Garcea, Eds.), pp. 312–351. Oxford University Press, New York.
101. Newcomb, W. W., Trus, B. L., Booy, F. P., Steven, A. C., Wall, J. S., and Brown, J. C. (1993). Structure of the herpes simplex virus capsid: Molecular composition of the pentons and triplexes. *J. Mol. Biol.* **232,** 499–511.
102. Newcomb, W. W., Homa, F. L., Booy, F. P., Thomsen, D. R., Trus, B. L., Steven, A. C., Spencer, J. V., and Brown, J. C. (1996). Assembly of the herpes simplex virus capsid: Characterization of intermediates observed during cell-free capsid formation. *J. Mol. Biol.* **263,** 432–446.

103. Trus, B. L., Booy, F. P., Newcomb, W. W., Brown, J. C., Homa, F. L., Thomsen, D. R., and Steven, A. C. (1996). The herpes simplex virus procapsid: Structure, conformational changes upon maturation, and roles of the triplex proteins VP19c and VP23 in assembly. *J. Mol. Biol.* **263,** 447–462.
104. Cohen, G. H., Ponce de Leon, M., Diggelmann, H., Lawrence, W. C., Vernon, S. K., and Eisenberg, R. J. (1980). Structural analysis of the capsid polypeptides of herpes simplex virus types 1 and 2. *J. Virol.* **34,** 521–531.
105. Heilman, C. J., Zweig, M., Stephenson, J. R., and Hampar, B. (1979). Isolation of a nucleocapsid polypeptide of herpes simplex virus types 1 and 2 possessing immmunologically type-specific and cross-reactive determinants. *J. Virol.* **29,** 34–42.
106. Booy, F. P., Trus, B. L., Newcomb, W. W., Brown, J. C., Conway, J. J., and Steven, A. C. (1994). Finding a needle in a haystack: Detection of a small protein (the 12-kDa VP26) in a large complex (the 200 MDa capsid of the herpes simplex virus). *Proc. Natl. Acad. Sci. USA* **91,** 5652–5656.
107. Trus, B. L., Homa, F. L., Booy, F. P., Newcomb, W. W., Thomsen, D. R., Cheng, N., Brown, J. C., and Steven, A. C. (1995). Herpes simplex virus capsids assembled in insect cells infected with recombinant baculoviruses: Structural authenticity and localization of VP26. *J. Virol.* **69,** 7362–7366.
108. Zhou, Z. H., He, J., Jakana, J., Tatman, J. D., Rixon, F. J., and Chiu, W. (1995). Assembly of VP26 in herpes simplex virus-1 inferred from structures of wild-type and recombinant capsids. *Nat. Struct. Biol.* **2,** 1026–1030.
109. Zhou, Z. H., Prasad, B. V. V., Jakana, J., Rixon, F. J., and Chiu, W. (1994). Protein subunit structures in the herpes simplex virus A-capsid determined from 400kV spot-scan electron cryomicroscopy. *J. Mol. Biol.* **242,** 456–469.
110. Spencer, J. V., Trus, B. L., Booy, F. P., Steven, A. C., Newcomb, W. W., and Brown, J. C. (1997). Structure of the herpes simplex virus capsid: Peptide A862-H880 of the major capsid protein is displayed on the rim of the capsomer protrusions. *Virology* **228,** 229–235.
111. Brandt, C. D., Kim, H. W., Vargosko, A. J., Jeffries, B. C., Arrobio, J. O., Rindge, B., Parrott, R. H., and Chanock, R. M. (1969). Infections in 18,000 infants and children in a controlled study of respiratory tract disease: Adenovirus pathogenicity in relation to serologic type and illness syndrome. *Am. J. Epidemiol.* **90,** 484–500.
112. Horwitz, M. S. (1990). *Adenoviridae* and their replication. *In* "Virology" (B. N. Fields and N. M. Knipe, Eds.), pp. 1679–1721. Raven Press, New York.
113. Wickham, T. J., Mathias, P., Cheresh, D. A., and Nemerow, G. R. (1993). Integrins $\alpha_v\beta_3$ and $\alpha_v\beta_5$ promote adenovirus internalization but not virus attachment. *Cell* **73,** 309–319.
114. Svensson, U. (1985). Role of vesicles during adenovirus 2 internalization into HeLa cells. *J. Virol.* **55,** 442–449.
115. Greber, U. F., Willetts, M., Webster, P., and Helenius, A. (1993). Stepwise dismantling of adenovirus 2 during entry into cells. *Cell* **75,** 477–486.
116. Stewart, P. L., Fuller, S. D., and Burnett, R. M. (1993). Difference imaging of adenovirus: Bridging the resolution gap between X-ray crystallography and electron microscopy. *EMBO J.* **12,** 2589–2599.
117. Wickham, T. J., Filardo, E. J., Cheresh, D. A., and Nemerow, G. R. (1994). Integrin $\alpha_v\beta_5$ selectively promotes adenovirus mediated cell membrane permeabilization. *J. Cell. Biol.* **127,** 257–264.
118. Seth, P., Wilingham, M. C., and Pastan, I. (1985). Binding of adenovirus and its external proteins to Triton X-114. *J. Biol. Chem.* **260,** 14431–14434.

119. Stewart, P. L., Chiu, C. Y., Huang, S., Muir, T., Zhao, Y., Chait, P., Mathias, P., and Nemerow, G. R. (1997). Cryo-EM visualization of an exposed RGD epitope on adenovirus that escapes antibody neutralization. *EMBO J.* **16,** 1189–1198.

120. Blumberg, B. S. (1997). Hepatitis B virus, the vaccine, and the control of primary cancer of the liver. *Proc. Natl. Acad. Sci. USA* **94,** 7121–7125.

121. Nassal, M., and Schaller, H. (1993). Hepatitis B virus replication. *Trends Microbiol.* **1,** 221–228.

122. Crowther, R. A., Kiselev, N. A., Bottcher, B., Berriman, J. A., Borisova, G. P., Ose, V., and Pumpens, P. (1994). Three-dimensional structure of hepatitis B virus core particles determined by electron cryomicroscopy. *Cell* **77,** 943–950.

123. Wingfield, P. T., Stahl, S. J., Williams, R. W., and Steven, A. C. (1995). Hepatitis core antigen produced in *Escherichia coli* subunit composition, conformational analysis, and in vitro capsid assembly. *Biochemistry* **34,** 4919–4932.

124. Hollinger, F. B. (1996). Hepatitis B virus. *In* ''Fields Virology'' (B. N. Fields, D. M. Knipe, and P. M. Powley, Eds.), pp. 2738–2808. Lippincott-Raven, Philadelphia, PA.

125. Conway, J. F., Cheng, N., Zlotnick, A., Stahl, S. J., Wingfield, P. T., Belnap, D. M., Kanngiesser, U., Noah, M., and Steven, A. C. (1998). Hepatitis B virus capsid: Localization of the putative immunodominant loop (residues 78–83) on the capsid surface, and implications for the distinction between c and e-antigens. *J. Mol. Biol.* **279,** 1111–1121.

126. Wingfield, P. T., Stahl, S. J., Thomsen, D. R., Homa, F. L., Booy, F. P., Trus, B. L., and Steven, A. C. (1997). Hexon-only binding of VP26 reflects differences between the hexon and penton conformations fo VP5, the major capsid protein of herpes simplex virus. *J. Virol.* **71,** 8955–8961.

127. Bowman, V. D., Chase, E. S., Franz, A. W. E., Chipman, P. R., Zhang, X., Perry, K. L., Baker, T. S., and Smith, T. J. (2002). An antibody to the putative aphid recognition site on cucumber mosaic virus recognizes pentons but not hexons. *J. Virol.* **76,** 12250–12258.

128. Verdaguer, N., Mateu, M. G., Andreu, E., Giralt, E., Domingo, E., and Fita, I. (1995). Structure of the major antigenic loop of foot-and-mouth disease virus complexed with a neutralizing antibody: Direct involvement of the Arg-Gly-Asp motif in the interaction. *EMBO J.* **14,** 1690–1696.

129. Verdaguer, N., Mateu, M. G., Bravo, J., Domingo, E., and Fita, I. (1996). Induced pocket to accommodate the cell attachment Arg-Gly-Asp motif in a neutralizing antibody against foot-and-mouth-disease virus. *J. Mol. Biol.* **256,** 364–376.

130. Hewat, E. A., Verdaguer, N., Fita, I., Blakemore, W., Brookes, S., King, A., Newman, J., Domingo, E., Mateau, M. G., and Stuart, D. I. (1997). Structure of the complex of an Fab fragment of a neutralizing antibody with foot-and-mouth disease virus: Positioning of a highly mobile antigenic loop. *EMBO J.* **16,** 1492–1500.

131. Acharya, R., Fry, E., Stuart, E., Fox, G., Rowlands, E., and Brown, F. (1989). The three-dimensional structure of foot-and-mouth disease virus at 2.9Å resolution. *Nature* **327,** 709–716.

132. Domingo, E., Verdaguer, N., Ochoa, W. F., Ruiz-Jarabo, C. M., Sevilla, N., Baranowski, E., Mateu, M. G., and Fita, I. (1999). Biochemical and structural studies with neutralizing antibodies raised against foot-and-mouth disease virus. *Virus Res.* **62,** 169–175.

133. Ochoa, W. F., Kalko, S. G., Mateu, M. G., Gomes, P., Andreu, D., Domingo, E., Fita, I., and Verdaguer, N. (2000). A multiply substituted G-H loop from foot-and-mouth

disease virus in complex with a neutralizing antibody: A role for water molecules. *J. Gen. Virol.* **81,** 1495–1505.

134. Fry, E. E., Lea, S. M., Jackson, T., Newman, J. W. I., Ellard, F. M., Blakemore, W. E., Abu-Ghazaleh, R., Samuel, A., King, A. M. Q., and Stuart, D. I. (1999). The structure and function of a foot-and-mouth disease virus–oligosaccharide receptor complex. *EMBO J.* **18,** 543–554.

135. Jackson, T., Ellard, F. M., Abu Ghazaleh, R., Brookes, S. M., Blakemore, W. E., Corteyn, A. H., Stuart, D. I., Newman, J. W. I., and King, A. M. Q. (1996). Efficient infection of cells in culture by type O foot-and-mouth disease virus requires binding to cell surface heparan sulfate. *J. Virol.* **70,** 5282–5287.

136. Verdaguer, N., Schoehn, G., Ochoa, W. F., Fita, I., Brookes, S., King, A., Domingo, E., Mateu, M. G., Stuart, D., and Hewat, E. A. (1999). Flexibility of the major antigenic loop of foot-and-mouth disease virus bound to a Fab fragment of a neutralising antibody: Structure and neutralisation. *Virology* **255,** 260–268.

137. Lowy, D. R., Kirnbauer, R., and Schiller, J. T. (1994). Genital human papillomavirus infection. *Proc. Natl. Acad. Sci. USA* **91,** 2436–2440.

138. Pfister, H. (1987). Papillomaviruses: General description, taxonomy, and classification. *In* "Papovaviridae," Vol. 2: "The Papillomaviruses" (N. P. Salzman and P. M. Howley, Eds.), pp. 1–38. Plenum Press, New York.

139. Baker, T. S., Newcomb, W. W., Olson, N. H., Cowsert, L. M., Olson, C., and Brown, J. C. (1991). Structures of bovine and human papilloma viruses: Analysis by cryoelectron microscopy and three-dimensional image reconstruction. *Cell* **60,** 1007–1015.

140. Trus, B. L., Roden, R. B. S., Greenstone, H. L., Vrhel, M., Schiller, J. T., and Booy, F. P. (1997). Novel structural features of bovine papillomavirus capsid revealed by a three dimensional reconstruction to 9Å resolution. *Nat. Struct. Biol.* **4,** 413–420.

141. Hagensee, M., and Galloway, D. (1993). Growing human papillomaviruses and virus-like particles in the laboratory. *Papillomavirus Rep.* **4,** 121–124.

142. Kirnbauer, R., Booy, F., Cheng, N., Lowy, D. R., and Schiller, J. T. (1992). Papillomavirus L1 major capsid protein self-assemblies into virus-like particles that are highly immunogenic. *Proc. Natl. Acad. Sci. USA* **89,** 12180–12184.

143. Roden, R. B. S., Hubbert, N. L., Kirnbauer, R., Breitburd, F., Lowy, D. R., and Schiller, J. T. (1995). Papillomavirus L1 capsids agglutinate mouse erythrocytes through a proteinaceous receptor. *J. Virol.* **69,** 5147–5151.

144. Roden, R. B. S., Weissinger, E. M., Henderson, D. W., Booy, F., Kirnbauer, R., Mushinski, J. F., Lowy, D. R., and Schiller, J. T. (1994). Neutralization of bovine papillomavirus by antibodies to L1 and L2 capsid proteins. *J. Virol.* **68,** 7570–7574.

145. Evander, M., Frazer, I. H., Payne, E., Qi, Y. M., Hengst, K., and McMillan, N. A. J. (1997). Identification of the α_6 integrin as a candidate receptor for papillomaviruses. *J. Virol.* **71,** 2449–2456.

146. Booy, F. P., Roden, R. B. S., Greenstone, H. L., Schiller, J. T., and Trus, B. L. (1998). Two antibodies that neutralize papillomavirus by different mechanisms show distinct binding patterns at 13Å resolution. *J. Mol. Biol.* **281,** 95–106.

147. Dryden, K. A., Wang, G., Yeager, M., Nibert, M. L., Coombs, K. M., Furlong, D. B., Fields, B. N., and Baker, T. S. (1993). Early steps in reovirus infection are associated with dramatic changes in supramolecular structure and protein conformation: Analysis of virions and subviral particles by cryoelectron microscopy and image reconstruction. *J. Cell Biol.* **122,** 1023–1041.

148. Banerjea, A. C., Brechling, K. A., Ray, C. A., Erikson, H., Pickup, D. J., and Joklik, W. K. (1988). High-level synthesis of biologically active reovirus protein σ1 in a mammalian expression vector system. *Virology* **167,** 601–612.
149. Centonze, V. E., Chen, Y., Severson, T. J., Borisy, G. G., and Nibert, M. L. (1995). Visualization of single reovirus particles by low-temperature, high-resolution scanning electron microscopy. *J. Struct. Biol.* **115,** 215–225.
150. Fraser, R. D., Furlong, D. B., Trus, B. L., Nibert, M. L., Fields, B. N., and Steven, A. C. (1990). Molecular structure of the cell-attachment protein of reovirus: Correlation of computer-processed electron micrographs with sequence-based predictions. *J. Virol.* **64,** 2990–3000.
151. Furlong, D. B., Nibert, M. L., and Fields, B. N. (1988). σ1 protein of mammalian reoviruses extends from the surfaces of viral particles. *J. Virol.* **62,** 246–256.
152. Barton, E. S., Forrest, J. C., Connolly, J. L., Chappell, J. D., Liu, Y., Schnell, F. J., Nusrat, A., Parkos, C. A., and Dermody, T. S. (2001). Junction adhesion molecule is a receptor for reovirus. *Cell* **104,** 441–451.
153. Chappell, J. D., Gunn, V. L., Wetzel, J. D., Baer, G. S., and Dermody, T. S. (1997). Mutations in type 3 reovirus that determine binding to sialic acid are contained in the fibrous tail domain of viral attachment σ1. *J. Virol.* **71,** 1834–1841.
154. Chappell, J. D., Duong, J., Wright, B. W., and Dermody, T. S. (2000). Identification of carbohydrate-binding domains in the attachment proteins of type 1 and type 3 reoviruses. *J. Virol.* **74,** 8472–8479.
155. Dermody, T. S., nibert, M. L., Bassel-Duby, R., and Fields, B. N. (1990). A σ1 region important for hemagglutination by serotype 3 reovirus strains. *J. Virol.* **64,** 5173–5176.
156. Nibert, M. L., Chappell, J. D., and Dermody, T. S. (1995). Infectious subvirion particles of reovirus type 3 Dearing exhibit a loss in infectivity and contain a cleaved σ1 protein. *J. Virol.* **69,** 5057–5067.
157. Ruben, D. H., Wetzel, J. D., Williams, W. V., Cohen, J. A., Dworkin, C., and Dermody, T. S. (1992). Binding of type 3 reovirus by a domain of the σ1 protein important for hemagglutination leads to infection of murine erythroleukemia cells. *J. Clin. Invest.* **90,** 2536–2542.
158. Virgin, H. W. I., Mann, M. A., Fields, B. N., and Tyler, K. L. (1991). Monoclonal antibodies to reovirus reveal structure/function relationships between capsid proteins and genetics of susceptibility to antibody action. *J. Virol.* **65,** 6772–6781.
159. Olland, A. M., Jané-Valbuena, J., Schiff, L. A., Nibert, M. L., and Harrison, S. C. (2001). Structure of the reovirus outer capsid and dsRNA-binding protein σ3 at 1.8Å resolution. *EMBO J.* **20,** 979–989.
160. Baer, G. S., and Dermody, T. S. (1997). Mutations in reovirus outer-capsid protein σ3 selected during persistent infections of L cells confer resistance to protease inhibitor E64. *J. Virol.* **71,** 4921–4928.
161. Sturzenbecker, L. J., Nibert, M., Furlong, D., and Fields, B. N. (1987). Intracellular digestion of reovirus particles requires a low pH and is an essential step in the viral infectious cycle. *J. Virol.* **61,** 2351–2361.
162. Borsa, J., Morash, B. D., Sargent, M. D., Copps, T. P., Lievaart, P. A., and Szekely, J. G. (1979). Two modes of entry of reovirus particles into L cells. *J. Gen. Virol.* **45,** 161–170.
163. Hooper, J. W., and Fields, B. N. (1996). Monoclonal antibodies to reovirus σ1 and μ1 proteins inhibit chromium release from mouse L cells. *J. Virol.* **70,** 672–677.
164. Lucia-Jandris, P., Hooper, J. W., and Fields, B. N. (1993). Reovirus M2 gene is associated with chromium release from mouse L cells. *J. Virol.* **67,** 5339–5345.

165. Burstin, S. J., Spriggs, D. R., and Fields, B. N. (1982). Evidence for functional domains on the reovirus type 3 hemagglutinin. *Virology* **117,** 146–155.
166. Spriggs, D. R., and Fields, B. N. (1982). Attenuated reovirus type 3 strains generated by selection of haemagglutinin antigenic variants. *Nature* **297,** 68–70.
167. Nason, E. L., Wetzel, D., Mukherjee, S. K., Barton, E. S., Prasad, B. V. V., and Dermody, T. S. (2001). A monoclonal antibody specific for reovirus outer-capsid protein $\sigma 3$ inhibits $\sigma 1$-mediated hemagglutination by steric hindrance. *J. Virol.* **75,** 6625–6634.
168. Palukaitis, P., Roossinck, M. J., Dietzgen, R. G., and Francki, R. I. B. (1992). Cucumber mosaic virus. *Adv. Virus Res.* **41,** 281–348.
169. Kaplan, I. B., Zhang, L., and Palukaitis, P. (1998). Characterization of cucumber mosaic virus. V. Cell-to-cell movement requires capsid protein but not virions. *Virology* **246,** 221–231.
170. Schmitz, I., and Rao, A. L. N. (1998). Deletions in the conserved amino-terminal basic arm of cucumber mosaic virus coat protein disrupt virion assembly but-do not abolish infectivity and cell-to-cell movement. *Virology* **248,** 323–331.
171. Smith, T. J., Chase, E., Schmidt, T. J., and Perry, K. (2000). The structure of cucumber mosaic virus and comparison to cowpea chlorotic mottle virus. *J. Virol.* **74,** 7578–7586.
172. Liu, S., He, X., Park, G., Josefsson, C., and Perry, K. L. (2002). A conserved capsid protein surface domain of cucumber mosaic virus is essential for aphid vector transmission. *J. Virol.* **76,** 9756–9762.
173. Padlan, E. A., Silverton, W. W., Sheriff, S., Cohen, G. H., Smith-Gill, S. J., and Davies, D. R. (1989). Stucture of an antibody–antigen complex: Crystal structure of the HyHEL-10 Fab–lysozyme complex. *Proc. Natl. Acad. Sci. USA* **86,** 5938–5942.
174. Burton, D. R., Saphire, E. O., and Parren, P. W. H. I. (2001). A model for neutralization of viruses based on antibody coating of the virion surfaces. *In* "Antibodies in Viral Infection" (D. R. Burton, Ed.), pp. 109–137. Springer-Verlag, New York.
175. Rini, J. M., Schulze-Gahmen, U., and Wilson, I. A. (1992). Structural evidence for induced fit as a mechanism for antibody–antigen recognition. *Science* **255,** 959–965.
176. Wang, W., Lee, W. M., Mosser, A. G., and Rueckert, R. R. (1998). WIN 52035-dependent human rhinovirus 16: Assembly deficiency caused by mutations near the canyon surface. *J. Virol.* **72,** 1210–1218.
177. Mosser, A. G., and Rueckert, R. R. (1993). WIN 51711-dependent mutants of poliovirus type 3: Evidence that virions decay after release from cells unless drug is present. *J. Virol.* **67,** 1246–1254.
178. Mosser, A. G., Sgro, J. Y., and Rueckert, R. R. (1994). Distribution of drug resistance mutations in type 3 poliovirus identifies three regions involved in uncoating functions. *J. Virol.* **68,** 8193–8201.

STRUCTURAL BASIS OF NONENVELOPED VIRUS CELL ENTRY

PHOEBE L. STEWART,* TERENCE S. DERMODY,† AND GLEN R. NEMEROW‡

***Department of Molecular Physiology and Biophysics; †Department of Pediatrics, Department of Microbiology and Immunology, and Elizabeth B. Lamb Center for Pediatric Research, Vanderbilt University School of Medicine, Nashville, Tennessee 37232, and ‡Department of Immunology, Scripps Research Institute, La Jolla, California 92037**

I. Introduction 455
II. Reovirus Cell Entry, Tissue Tropism, and Pathogenesis 456
III. Reovirus Structure 458
IV. Proteolysis of the $\sigma 1$ Protein Regulates Viral Growth in the Intestine and Systemic Spread 460
V. The $\sigma 1$ Tail Binds Cell Surface Sialic Acid 462
VI. The $\sigma 1$ Head Binds Junctional Adhesion Molecule 463
VII. Reovirus–Receptor Interactions Promote Cell Death by Apoptosis 464
VIII. Picornavirus–Receptor Complexes 465
IX. Poliovirus Cell Entry Mechanisms 466
X. Identification of the Poliovirus Attachment Receptor 468
XI. Poliovirus-Associated Lipid Molecules 469
XII. Receptors for Rhinoviruses 470
XIII. Receptors for Other Picornaviruses 473
A. Foot-and-Mouth Disease Virus 473
B. Echovirus Receptors 475
XIV. Human Adenoviruses 475
XV. Adenovirus Attachment Receptors 476
XVI. Cell Integrins Promote Adenovirus Internalization 478
XVII. Signaling Events Associated with Adenovirus Internalization 481
XVIII. α_v Integrins Regulate Adenovirus-Mediated Endosome Disruption 482
XIX. Conclusions 482
References 484

I. Introduction

For several virus families, significant progress has been made in understanding the molecular events associated with viral cell entry. These events include stable attachment of the virus to the cell surface, penetration of the virus into the interior of the cell, partial disassembly or conformational change of the viral capsid, release of the viral genome or viral mRNA transcripts, and activation of the viral genetic program. To effect a productive infection, a virus must traverse the extracellular environment and deliver its genome to the cellular compartment in which

0065-3233/03 $35.00

viral transcription and replication occur. Viruses have evolved various strategies for accomplishing this end. In this review, we consider the entry pathway of three divergent virus families, reoviruses, picornaviruses, and adenoviruses. We summarize what is known about the structure of these viruses, the virus–receptor complexes that are formed, and the conformational changes that occur during cell entry. Understanding the early steps in viral entry has relevance to viral pathogenesis, as these events often determine target cell selection within the host, which dictates the site of virus-induced disease. The complexes formed between viruses and host cell receptors provide insight into the general process of receptor activation for normal host receptor–ligand interactions. In addition, structural information on virus cell entry helps establish a framework for the rational design of antiviral agents that target the entry process.

II. Reovirus Cell Entry, Tissue Tropism, and Pathogenesis

Members of the *Reoviridae* family are nonenveloped viruses containing genomes of 10–12 segments of double-stranded (ds) RNA (Nibert and Schiff, 2001). This family includes mammalian orthoreoviruses (reoviruses), orbiviruses, and rotaviruses. For reoviruses, the viral proteins are designated with a Greek letter corresponding to the size of the encoding genome segment: sigma (σ) for proteins encoded by small genome segments, mu (μ) for proteins encoded by medium segments, and lambda (λ) for proteins encoded by large segments. Each of the genome segments encodes a single protein with the exception of the S1 gene, which encodes the viral attachment protein $\sigma 1$, and a small nonstructural protein, $\sigma 1s$. Like other members of the *Reoviridae*, reovirus particles are formed from concentric protein shells. Two such shells exist for reoviruses, called outer capsid and core (Nibert and Schiff, 2001).

Reoviruses infect many mammalian species, including humans; however, they are rarely associated with human disease (Tyler, 2001). Three reovirus serotypes have been recognized on the basis of neutralization and hemagglutination profiles. Each is represented by a prototype strain, type 1 Lang (T1L), type 2 Jones (T2J), and type 3 Dearing (T3D), which differ primarily in $\sigma 1$ sequence (Duncan *et al.*, 1990; Nibert *et al.*, 1990). The pathogenesis of reovirus infections has been most extensively studied by using newborn mice, in which serotype-specific patterns of disease have been identified (Tyler, 2001). The best characterized of these models is reovirus pathogenesis in the murine central nervous system (CNS).

Because reovirus contains a segmented genome, differences in disease pathogenesis exhibited by different reovirus strains can be mapped to specific viral genome segments. Coinfection of cells with different strains of reovirus results in generation of progeny viruses containing various combinations of genome segments from each parental strain (Virgin *et al.*, 1997). Progeny from such a genetic cross are termed reassortant viruses. A phenotypic difference between two parental strains can be mapped genetically by screening reassortant viruses in appropriate assays and correlating expression of the phenotype with a specific parental genome segment. This property makes reoviruses uniquely suited for studies of viral determinants of cell tropism and pathogenesis.

Reovirus infection is initiated by interactions of the $\sigma 1$ protein with one or more cell surface receptors. Following attachment, reovirus virions are internalized into cells by receptor-mediated endocytosis (Borsa *et al.*, 1979, 1981; Rubin *et al.*, 1992; Sturzenbecker *et al.*, 1987). In the endocytic compartment, the viral outer capsid is removed by acid-dependent proteases, resulting in generation of infectious subvirion particles (ISVPs) (Borsa *et al.*, 1981; Chang and Zweerink, 1971; Silverstein *et al.*, 1972; Baer and Dermody, 1997). ISVPs also can be generated extracellularly in the murine intestine (Bodkin and Fields, 1989) or by *in vitro* treatment with intestinal proteases (Borsa *et al.*, 1973; Sturzenbecker *et al.*, 1987; Baer and Dermody, 1997). ISVPs are capable of penetrating endosomal or plasma membranes, leading to delivery of viral core particles into the cytoplasm (Borsa *et al.*, 1979; Lucia-Jandris *et al.*, 1993; Tosteson *et al.*, 1993; Hooper and Fields, 1996a,b). The viral core is transcriptionally active and produces 10 species of capped mRNA, 1 for each viral genome segment. Transcription and assembly of new virus progeny takes place over a period of 18–24 h in most cell types (Tyler, 2001). Virus replication is frequently associated with programmed cell death (apoptosis) both in cultured cells (Tyler *et al.*, 1995; Rodgers *et al.*, 1997; Connolly *et al.*, 2000) and *in vivo* (Oberhaus *et al.*, 1997; Debiasi *et al.*, 2001). The role of apoptosis in reovirus replication is not entirely clear, but this cellular response may facilitate release of virus progeny or aid in virus dissemination in the host.

After oral administration to newborn mice, reovirus virions undergo proteolytic processing in the lumen of the small intestine (Bodkin *et al.*, 1989). ISVPs generated in the intestine associate with microfold (M) cells overlaying Peyer's patches. M cells transfer virions to gut-associated lymphocytes (Wolf *et al.*, 1981), where they spread systemically to various peripheral organs including brain, heart, kidney, liver, and spleen. Reovirus serotypes differ in the route of spread in the host and the CNS sites targeted for infection. Type 1 (T1) reovirus strains spread by hematogenous routes to the CNS (Tyler *et al.*, 1986), where they infect ependymal cells, leading to nonlethal hydrocephalus (Weiner *et al.*, 1977, 1980). In contrast, type 3 (T3) reoviruses spread primarily by neural routes to the CNS (Tyler *et al.*, 1986; Morrison *et al.*, 1991) and infect neurons, causing fatal encephalitis (Weiner *et al.*, 1977, 1980) associated with neuronal apoptosis (Oberhaus *et al.*, 1997). Importantly, studies using T1 × T3 reassortant viruses revealed that the route of viral spread in the host, cell tropism in the CNS, and resultant disease segregate genetically with the σ1-encoding S1 gene (Weiner *et al.*, 1977, 1980; Tyler *et al.*, 1986). Thus, these studies strongly suggest that serotype-specific patterns of reovirus pathogenesis are regulated by σ1 interactions with specific cellular receptors.

III. Reovirus Structure

Our knowledge of the three-dimensional structure of reovirus comes from a combination of cryoelectron microscopy (cryo-EM) image reconstruction and X-ray crystallography. The first published cryo-EM reconstructions were of T2J and T3D virions at ~30- to 35-Å resolution and T3D cores at ~55- Å resolution (Metcalf *et al.*, 1991). The outer surfaces of virions of both strains appeared similar, with a starfish-shaped density at the 5-fold axes and hexameric rings covering the remainder of the capsid. The core reconstruction showed hollow pentameric spikes protruding from each icosahedral vertex. This work was followed by reconstructions of virions, ISVPs, and cores of T1L at ~27- to 32-Å resolution (Dryden *et al.*, 1993) (Fig. 1; see Color Insert). Reovirus virions transition to ISVPs with the loss of σ3, cleavage of μ1/μ1C into particle-associated fragments μ1δ/δ and ϕ, and a dramatic conformational change in σ1 (Nibert *et al.*, 2001). Comparison of the T1L virion (~850 Å in diameter) and ISVP (~800 Å in diameter) image reconstructions indicated the loss of 600 finger-like subunits, which likely correspond to 600 copies of σ3. The crystal structure of T3D σ3 has been solved and placed within the cryo-EM density of the virion (Olland *et al.*, 2001), confirming that each finger-like protrusion corresponds to one subunit of σ3.

The σ1 protein forms a fibrous, lollipop-shaped structure with an overall length of ~480 Å (Fig. 2). The amino-terminal σ1 tail (~40–60 Å wide) inserts into λ2 pentamers in the virion, and the carboxy-terminal σ1 head (~95 Å in diameter) projects distally from the virion surface (Furlong *et al.*, 1988; Banerjea *et al.*, 1988; Fraser *et al.*, 1990). Four distinct and tandemly arranged morphologic regions within the σ1 tail domain have been designated T(i) to T(iv) on the basis of relative proximity to the surface of the virion (Fraser *et al.*, 1990). Correlation of σ1 primary amino acid sequence with morphologic data suggests that these domains correspond to discrete regions of predicted secondary structure, primarily α helix and

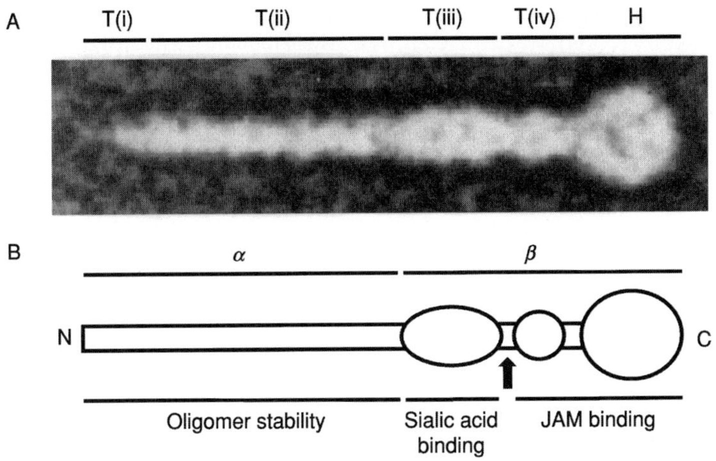

FIG. 2. Reovirus attachment protein σ1. (A) Computer-processed electron micrograph of σ1 showing morphologic regions T(i), T(ii), T(iii), T(iv), and H. The overall length of the fiber is ~480 Å. [Reproduced with permission from Fraser *et al.* (1990).] (B) Predicted secondary structures and functional domains of σ1. Morphologic regions T(i) and T(ii) are predicted to be formed by α-helical coiled coil. Regions T(iii) and T(iv) are predicted to be formed by alternating β strand and β turn. Morphologic region H is predicted to assume a more complex arrangement of secondary structures corresponding to the globular σ1 head. Sequences in type 3 σ1 required for stability of σ1 oligomers, binding sialic acid and susceptibility to protease cleavage (arrow) are contained in the σ1 tail, whereas sequences required for binding junctional adhesion molecule (JAM) and neutralization of viral infectivity reside in the σ1 head.

alternating β strand/β turn (Nibert *et al.*, 1990). The σ1 protein forms a trimer (Strong *et al.*, 1991; Leone *et al.*, 1991), and sequences in tail region T(ii) predicted to form α-helical coiled coil are required for trimer stability (Chappell *et al.*, 2000; Wilson, 1996).

EM images of negatively stained reovirus virions and ISVPs first indicated that σ1 adopts a compact form in the virion and an extended form in the ISVP (Furlong *et al.*, 1988). ISVPs, but not virions, showed filamentous projections extending up to 400 Å from the particle surface. In the cryo-EM reconstruction, discontinuous density was observed for σ1 extending ~100 Å from each icosahedral vertex. Presumably the full length of σ1 is not reconstructed because of structural flexibility. Indeed, EM images of negatively stained σ1 isolated from virions show curvature in individual fibers at specific regions within the molecule (Fraser *et al.*, 1990).

Assembly of reovirus particles *in vitro* has proved useful for studies of structure–function relationships of viral outer capsid proteins. Particles obtained by mixing baculovirus-expressed σ3 with ISVPs are similar to native virions by cryo-EM image reconstruction (Jane-Valbuena *et al.*, 1999). ISVPs recoated with $\sigma 3$ contain the cleaved form of $\mu 1/\mu 1C$ found in ISVPs. However, recoated ISVPs behave like virions in infectivity assays, suggesting that the presence of $\sigma 3$, and not the proteolytic state of $\mu 1/\mu 1C$, causes the main functional differences between virions and ISVPs. Core particles can be recoated with baculovirus-expressed $\sigma 3$ and $\mu 1$ proteins (Chandran *et al.*, 1999). Recoated cores closely resemble native virions by cryo-EM image reconstruction, despite the absence of $\sigma 1$. Cores recoated with $\sigma 3$ and $\mu 1$ are capable of infecting murine L929 (L) cells, although ~10,000-fold less efficiently than native virions, presumably due to the lack of $\sigma 1$ (Chandran *et al.*, 1999). Core particles also can be recoated with $\sigma 1$, $\sigma 3$, and $\mu 1$, giving rise to particles that faithfully reproduce each step in the reovirus entry pathway (Chandran *et al.*, 2001). These particles have been useful in linking specific $\sigma 1$ sequences with receptor-binding functions in the context of a single infectious cycle.

During the transition from the ISVP to the core (~700 Å in diameter), the $\sigma 1$ and $\mu 1$ proteins are lost, leaving five viral proteins, three of which ($\sigma 2$, $\lambda 1$, and $\lambda 2$) form the icosahedral protein shell of the core. The two remaining proteins, $\mu 2$ and $\lambda 3$, are thought to play important roles in viral transcription, with $\lambda 3$ likely serving as the catalytic subunit of the RNA polymerase (Koonin, 1992). The crystal structure of the reovirus core has been solved to 3.6 Å (Reinisch *et al.*, 2000). A smooth core shell is formed from 120 copies of $\lambda 1$, and the icosahedral lattice is stabilized by 150 copies of $\sigma 2$. Pentamers of the $\lambda 2$ protein form the mRNA-capping turrets protruding from the outer surface of the core (Fig. 3; see Color Insert). A cavity exists in the center of the turret, ~15–70 Å in diameter, through which mRNA is extruded during transcription. The crystal structure of the core also showed three or four shells of density at ~26-Å intervals inside the inner surface of $\lambda 1$, indicating that the viral dsRNA is coiled into concentric layers within the particle (Reinisch *et al.*, 2000). The $\mu 2/\lambda 3$–transcriptase complex was not visible in the crystal structure of the core. However, cryo-EM reconstructions of virion particles lacking genomic dsRNA show density projecting inward from the 5-fold axes that is presumed to correspond to $\mu 2$ and $\lambda 3$ (Dryden *et al.*, 1998).

IV. Proteolysis of the $\sigma 1$ Protein Regulates Viral Growth in the Intestine and Systemic Spread

Although many reovirus strains efficiently replicate in the intestine, not all strains do. In fact, T3D fails to grow in the intestine and does not spread to the CNS following oral inoculation. However, infection by either the intramuscular or intracranial routes results in efficient replication of T3D and lethal CNS disease. These observations led to an investigation of whether reovirus strains vary in their susceptibility to proteolytic cleavage by intestinal proteases. Treatment of T3D particles *in vitro* with either chymotrypsin or trypsin resulted in ISVPs with cleaved $\sigma 1$ proteins and a corresponding 10-fold loss in virus infectivity (Nibert *et al.*, 1995). In contrast, proteolytic treatment of T1L resulted in ISVPs with uncleaved $\sigma 1$ proteins and no change in virus infectivity. Studies using T1L × T3D reassortant viruses indicated that virus replication in the intestine segregates primarily with the S1 gene (Bodkin and Fields, 1989), suggesting that growth in the intestine is modulated by susceptibility of $\sigma 1$ to proteolytic cleavage.

Several additional clues about the nature of $\sigma 1$ protease sensitivity and its effect on virus infection were subsequently obtained from biochemical and genetic studies. A $\sigma 1$-specific neutralizing monoclonal antibody (mAb), designated G5, which binds to the T3D $\sigma 1$ head domain (Bassel-Duby *et al.*, 1986; Chappell *et al.*, 2000), does not neutralize viral infectivity following generation of ISVPs with chymotrypsin (Nibert *et al.*, 1995). This finding suggests that treatment of T3D virions with protease releases a receptor-binding domain in the carboxy terminus of the molecule that corresponds to the $\sigma 1$ head. On the basis of predictions of $\sigma 1$ secondary structure (Nibert *et al.*, 1990) and EM images of negatively stained $\sigma 1$ (Fraser *et al.*, 1990), protease cleavage sites were predicted to lie in a flexible portion of the $\sigma 1$ tail termed the neck, which is proximal to the $\sigma 1$ head. To more precisely identify sites in $\sigma 1$ that serve as targets for proteolytic attack, deduced $\sigma 1$ amino acid sequences of several field isolate strains were correlated with susceptibility of their $\sigma 1$ proteins to proteolysis (Chappell *et al.*, 1998). Protease-sensitive $\sigma 1$ proteins have a threonine at position 249, whereas protease-resistant proteins have an isoleucine at this position. The importance of this sequence polymorphism was confirmed by site-directed mutagenesis of recombinant $\sigma 1$ protein (Chappell *et al.*, 1998). On the basis of amino acid sequence analysis of tryptic fragments of $\sigma 1$, the cleavage site was localized to Arg-245 and Ile-246 (Chappell *et al.*, 1998), sequences predicted to be in the $\sigma 1$ neck (Nibert *et al.*, 1990). Therefore, regulation of protease sensitivity by Thr-249 is likely due to an indirect effect, perhaps involving the stabilization of $\sigma 1$ subunit interactions.

These studies indicate that susceptibility of the reovirus attachment protein to host proteases influences growth in the murine intestine and systemic spread. Although protease-treated T3D virions lack the capacity to infect intestinal cells, they retain the capacity to infect other cell types in culture and *in vivo*. These findings suggest that T3D $\sigma 1$ contains two distinct receptor-binding domains, one in the head that is removed by proteolysis and another in the tail that remains associated with viral particles after proteolytic treatment.

V. The σ1 Tail Binds Cell Surface Sialic Acid

Reoviruses are capable of agglutinating the erythrocytes of several mammalian species (Lerner *et al.*, 1963). Hemagglutination by T3 reovirus strains is mediated by interactions of σ1 protein with terminal α-linked sialic acid (SA) residues on several glycosylated erythrocyte proteins such as glycophorin A (Gentsch and Pacitti, 1987; Paul and Lee, 1987). SA binding is also required for reovirus attachment and infection of certain cell types including murine erythroleukemia (MEL) cells (Chappell *et al.*, 1997). Although the majority of T3 reovirus strains bind SA and produce hemagglutination, not all T3 strains have these properties. Sequence diversity within tail region T(iii) determines the capacity of field isolate reovirus strains to bind SA (Dermody *et al.*, 1990) and to infect MEL cells (Rubin *et al.*, 1992). Morphologic region T(iii) is an approximately 65-residue segment of sequence predicted to form β strand and β turn (Nibert *et al.*, 1990). Sequence polymorphism within a single predicted β strand correlates with SA-binding capacity (Chappell *et al.*, 1997). Therefore, residues in this vicinity may form part of an SA-binding site. In concordance with these results, experiments using expressed σ1 truncation mutants and chimeric molecules derived from T1L and T3D σ1 proteins demonstrated that the SA-binding domain of T3 σ1 is contained within the T(iii) region (Chappell *et al.*, 2000).

To elucidate the role of SA binding in reovirus cell attachment, genetic reassortment was used to isolate monoreassortant viruses containing the S1 gene of either non-SA-binding strain T3C44 (Dermody *et al.*, 1990) (strain T3SA$^-$) or SA-binding strain T3C44-MA (Chappell *et al.*, 1997) (strain T3SA$^+$) and all other gene segments from T1L (Barton *et al.*, 2001a). T3SA$^-$ and T3SA$^+$ vary by a single amino acid residue at position 204 (leucine for T3SA$^-$ and proline for T3SA$^+$), which correlates with the capacity to bind SA (Chappell *et al.*, 1997). The steady state avidity of these strains for L cells is nearly equivalent ($K_D \sim 3 \times 10^{-11} M$), whereas the avidity of T3SA$^+$ for HeLa cells is 5-fold higher than that of T3SA$^-$ (Barton *et al.*, 2001a). Kinetic assessments of binding indicate that the capacity to bind SA functions primarily to increase the k_{on} of virus attachment to HeLa cells. Binding of T3SA$^+$ to HeLa cells proceeds through a time-dependent adhesion-strengthening process mediated by σ1–SA interactions (Barton *et al.*, 2001a). These findings suggest that virus binding to SA adheres the virion to the cell surface, thereby enabling it to diffuse laterally until it encounters the σ1 head receptor.

VI. The σ1 Head Binds Junctional Adhesion Molecule

The idea that the σ1 head binds to cell surface receptors first came from studies of neutralization-resistant variants of T3D selected with σ1-specific mAb G5 (Spriggs and Fields, 1982; Spriggs *et al.*, 1983). These variants have mutations in the σ1 head (Bassel-Duby *et al.*, 1986) that segregate genetically with alterations in neural tropism (Kaye *et al.*, 1986). Biochemical experiments using expressed σ1 also support a role for the σ1 head in receptor binding. Truncated forms of σ1 containing only the head domain are capable of specific cell interactions (Duncan *et al.*, 1991; Duncan and Lee, 1994). These observations, along with the finding that proteolysis of T3D virions leads to release of a carboxy-terminal receptor-binding fragment of σ1 (Nibert *et al.*, 1995), indicate that the σ1 head promotes receptor interactions that are distinct from interactions with SA mediated by the σ1 tail.

To identify a receptor bound by the σ1 head, $T3SA^-$ was used as an affinity ligand in a fluorescence-activated cell sorting (FACS)-based expression-cloning approach (Barton *et al.*, 2001b). This strategy was used to avoid the potential complication of isolating heavily glycosylated molecules that do not interact specifically with σ1. A neural precursor cell (NT2) cDNA library was selectively enriched for cDNAs that confer binding of fluoresceinated $T3SA^-$ virions to transfected COS-7 cells. After four rounds of FACS enrichment and screening of subpools, four clones were identified that conferred $T3SA^-$ binding to all transfected cells. All four clones encoded human junctional adhesion molecule (hJAM), a member of the immunoglobulin superfamily (IgSF) involved in regulation of intercellular tight junction formation (Martin-Padura *et al.*, 1998; Williams *et al.*, 1999).

Several lines of evidence indicate that hJAM is a functional reovirus receptor (Barton *et al.*, 2001a). First, blockade of hJAM on the surface of Caco-2 cells, HeLa cells, or NT2 cells abolishes $T3SA^-$ binding and growth. Second, transfection of either murine or avian cells with hJAM rescues binding, entry, and infection of both T1 and T3 reovirus strains. Third, the biological effects of hJAM on reovirus infection correlate with a direct, SA-independent, high-affinity interaction between hJAM and the σ1 head domain. Together, these findings indicate that hJAM serves as a serotype-independent receptor for the σ1 head.

Because JAM binds to and confers infection by both T1 and T3 reovirus strains, it is unlikely that JAM is the sole determinant of serotype-dependent differences in reovirus tropism in the murine CNS. Instead, it is possible that carbohydrate plays the dominant role in determining reovirus CNS tropism. Although both T1 and T3 strains bind JAM, they bind different types of cell surface carbohydrate (Dermody *et al.*, 1990; Chappell *et al.*, 1997). Interactions with receptors that are carbohydrate in nature might lead to productive entry independent of JAM binding or facilitate binding to JAM by an adhesion-strengthening, coreceptor mechanism. Alternatively, JAM might serve as a serotype-independent reovirus receptor at some sites within the host, and unidentified molecules, perhaps with homology to JAM, might function as serotype-dependent reovirus receptors in the CNS.

VII. Reovirus–Receptor Interactions Promote Cell Death by Apoptosis

A common feature of many animal viruses is their capacity to induce programmed cell death (apoptosis), a process characterized by cell shrinkage, nuclear condensation, and DNA fragmentation (Shen and Shenk, 1995). Apoptosis may serve as a host defense to limit virus growth, or it may promote virus spread or enhance viral replication via activation of one or more signaling pathways involved in apoptosis induction (Teodoro and Branton, 1997).

After infection of cultured cells, reovirus strains differ in the capacity to induce apoptosis. T3D induces apoptosis to a substantially greater extent than T1L in L cells (Tyler *et al.*, 1995), Madin–Darby canine kidney cells (Rodgers *et al.*, 1997), and HeLa cells (Connolly *et al.*, 2001). Differences in the capacity of these strains to induce apoptosis are determined primarily by the σ1-encoding S1 gene (Tyler *et al.*, 1995; Rodgers *et al.*, 1997; Connolly *et al.*, 2001), suggesting that apoptosis is triggered by a signaling pathway initiated by early steps in the virus replication cycle. In support of this hypothesis, it was found that reovirus infection leads to the activation of nuclear factor κB (NF-κB) (Connolly *et al.*, 2000). Depending on cell type, NF-κB activation is first detected 2–4 h after reovirus adsorption and peaks 6–10 h after infection (Connolly *et al.*, 2000). Apoptosis induced by reovirus is significantly reduced in cells treated with a proteasome inhibitor and in cells expressing a *trans*-dominant inhibitor of NF-κB. In addition, reovirus-induced apoptosis is blocked in cells deficient in the expression of the p50 or p65 NF-κB subunits (Connolly *et al.*, 2000). These findings demonstrate that NF-κB plays a proapoptotic role during reovirus infection.

The signaling events that lead to apoptosis during reovirus infection are initiated by virus–receptor interactions. SA-binding strain T3SA$^+$ induces NF-κB activation and apoptosis to a much greater extent than does non-SA-binding strain T3SA$^-$ in both HeLa cells and L cells (Connolly *et al.*, 2001). Enzymatic removal of cell surface SA with neuraminidase, or blockade of virus binding to SA with sialyllactose, abolishes the capacity of T3SA$^+$ to activate NF-κB and induce apoptosis (Connolly *et al.*, 2001). These findings indicate that reovirus interactions with SA modulate proapoptotic signaling. However, reovirus binding to JAM also plays a critical role in this process. At high multiplicities of infection (MOIs) [$\geq$100 plaque-forming units (PFU)/cell], T3SA$^+$ can bind and enter cells via a JAM-independent pathway mediated by SA (Barton *et al.*, 2001b). Although viral replication is efficient following SA-mediated entry, T3SA$^+$ can neither activate NF-κB nor induce apoptosis in the absence of JAM binding (Barton *et al.*, 2001b). These results suggest that multivalent interactions of σ1 with SA and JAM surpass a critical threshold required for NF-κB activation and apoptosis induction.

Further studies will be needed to fully determine the consequences of the cell signaling events induced by σ1-mediated cell attachment. Moreover, because activation of NF-κB as a result of receptor–ligand interactions typically occurs more rapidly than that observed following reovirus infection (Traenckner *et al.*, 2001), it is possible that steps in viral entry following attachment, such as endocytosis or membrane penetration, are also involved in cell signaling. In either case, the central role of SA and JAM in reovirus-induced apoptosis suggests that receptor-linked signaling responses contribute to the pathogenesis of reovirus infection.

VIII. Picornavirus–Receptor Complexes

The picornavirus family of viruses is comprised of small (~300 Å), nonenveloped particles with icosahedral symmetry containing a single (plus)-stranded RNA genome. Picornaviruses are divided into five genera including rhinoviruses, enteroviruses, aphthoviruses, cardioviruses, and hepatoviruses (REACH). There are more than 100 picornavirus serotypes, which are grouped on the basis of sequence similarity, genome organization, and other biological and physical criteria. These viruses include many important human pathogens such as poliovirus, hepatitis A virus, echoviruses, coxsackieviruses, and rhinoviruses (Racaniello, 2001). Poliovirus, a major cause of paralytic disease, remains a cause of morbidity and mortality in the developing world and is the target of a worldwide eradication program. Rhinoviruses are the single most frequent cause of the common cold. Foot-and-mouth disease virus (FMDV) was the first animal virus to be discovered as a causative agent of disease in 1898 (Loeffler and Frosch, 1964), and it remains an important livestock pathogen with considerable economic impact.

IX. Poliovirus Cell Entry Mechanisms

Poliovirus replication occurs in the intestine following oral inoculation. Following primary infection, virus particles are spread via the blood to motor neurons in the central nervous system. Virus-mediated destruction of motor neurons contributes in large part to the resulting paralytic disease. The picornavirus replication cycle takes place exclusively in the cytoplasm of the host cell. Therefore, the major challenge facing these viruses is to deliver their genomes encased within a highly stable protein shell across the host cell plasma membrane into the cytoplasm, where transcription and translation of the messenger-active viral RNAs take place (Flint *et al.*, 2000). Picornavirus–receptor interactions play a major role in destabilizing the viral capsid, allowing release of the viral RNA into the cell cytoplasm. An accumulation of knowledge of picornavirus structure has not only shed light on the molecular events associated with receptor interactions and virion disassembly but also has provided valuable insights for the development of antiviral agents that interfere with cell entry.

One of the earliest observations providing a clue to the mechanisms involved in poliovirus entry was that interaction of the virus particle with receptor-expressing cells at 37°C resulted in the generation of a conformationally altered virion (termed the A particle) (Joklik and Darnell, 1961). Unmodified virus particles have a sedimentation rate (160S) on sucrose density gradients that is distinct from that of receptor-modified A particles (135S). Interestingly, 135S virus particles retain a low level of infectivity and can enter receptor-negative cells, presumably by direct plasma membrane phospholipid interactions (Curry *et al.*, 1996). 135S particles have been proposed to represent an intermediate form of the virus particle that has undergone partial disassembly during cell entry (Racaniello, 1996). Fricks and Hogle investigated the molecular changes in 135S particles, using various sequence-specific probes, including proteases and monoclonal antibodies (Fricks and Hogle, 1990). 135S particles were generated by exposing 160S particles to receptor-bearing cells and then detaching the virus particles, now transformed into 135S particles, from the cell surface. The probes demonstrate that receptor-altered virus is clearly distinguishable from native (160S) virions and in particular that the N terminus of VP1 becomes externalized. The 135S virus particles also appear to lose the internal capsid protein VP4, which is a small, myristoylated protein. Although the precise role of VP4 is unknown, a VP4 mutant particle was identified that can assemble into mature virions and undergo transition to 135S particles, but is not infectious. This finding suggests that participation of the VP4 capsid protein is required for cell entry (Moscufo *et al.*, 1993).

More recently, cryo-EM methods have been used to compare the structures of poliovirus 160S, 135S, and 80S particles (Belnap *et al.*, 2000a). 80S (or H) particles are formed after 135S particles release RNA. Reconstructions were calculated for all three particle types at ~22-Å resolution. The reconstructions were then interpreted with the atomic structures of VP1, VP2, and VP3 from the crystal structure of the virion (Hogle *et al.*, 1985). Pseudo-atomic models were generated for the 135S and 80S particles by rigid-body movements of the three capsid proteins for the best fit with the cryo-EM density. Both 135S and 80S particles are larger by ~4% than the native virion and movements of up to 9 Å were deduced for VP1, VP2, and VP3. These movements create gaps between adjacent subunits, suggesting that the gaps may help VP4 and the N terminus of VP1 become externalized during the transition between the 160S and 135S structures.

The failure of inhibitors of vacuolar proton ATPases (bafilomycin) to block poliovirus infection suggests that poliovirus entry does not require clathrin-mediated endocytosis and is pH independent (Perez and Carrasco, 1993). Consistent with this hypothesis, other studies have shown that poliovirus efficiently enters HeLa cells expressing a dominant-negative mutant dynamin, a molecule required for clathrin-mediated endocytosis (DeTulleo and Kirchhausen, 1998). These findings do not exclude the possibility that poliovirus enters host cells via a nonclathrin endoytic pathway. Current models of poliovirus entry suggest that VP1 creates a pore or channel in the plasma membrane through which the viral RNA is released directly into the cytoplasm. It has also been recognized that the 135S particle is probably not the virus intermediate from which the genome is released because these particles are still resistant to RNase digestion (Fricks and Hogle, 1990). However, the possibility exists that the 135S particle is the entry intermediate, with RNA release occuring only on lipid interaction.

On the basis of their pseudo-atomic model for the 135S virion as well as previous information, Belnap *et al.* proposed a revised model for the translocation of RNA across the cell membrane (Fig. 4; see Color Insert) (Belnap *et al.*, 2000a). In this model, the interaction of the virion with its receptor triggers the conformational change to the 135S state. This results in VP4 and the N termini of VP1 extruding from the capsid, inserting into the membrane, and forming a pore. To open a channel in the capsid at the 5-fold axis and permit the RNA to exit, it then would be necessary for the β tube of VP3 to move out of the way. During this process, further shifts in VP1, VP2, and VP3 are likely to occur, resulting in formation of 80S particles.

X. Identification of the Poliovirus Attachment Receptor

During the viremic phase of poliovirus dissemination *in vivo,* virus replication is restricted to relatively few tissues including the oropharynx, intestine, and motor neurons. Early studies indicated that replication was also restricted to cells that expressed poliovirus receptors (Holland, 1961). Subsequently, Mendelsohn *et al.* showed that poliovirus receptor expression could be conferred on nonpermissive cells by transfection with human genomic DNA (Mendelsohn *et al.*, 1986). These and other biochemical studies (Krah and Crowell, 1982) indicated that the poliovirus receptor was a cell surface protein. The cDNA encoding the poliovirus receptor, designated PVR, was ultimately cloned by Mendelsohn *et al.* (1989). PVR is a member of the IgSF, containing three immunoglobulin-like domains in its extracellular region, a single transmembrane anchor, and a short cytoplasmic tail with either 25 or 50 amino acids depending on the specific splice variant. Site-directed mutagenesis indicated that the majority of poliovirus attachment is mediated by the first (domain I) immunoglobulin-like domain (Koike *et al.*, 1991). PVR is similar to two other IgSF members known as poliovirus receptor-related proteins 1 and 2, which serve as entry receptors for several human herpesviruses (Geraghty *et al.*, 1998). Interestingly, a murine homolog of the human PVR was also identified, but this molecule failed to support poliovirus infection (Morrison and Racaniello, 1992). The normal host cell function of the murine and human PVR molecules remains unknown.

Somewhat surprisingly, PVR mRNA expression was found in tissues where poliovirus does not replicate (Mendelsohn *et al.*, 1989), suggesting that this receptor may not be the sole determinant of virus tropism *in vivo.* Several other studies suggested that expression of PVR may not be sufficient to allow poliovirus replication in certain cell types. For example, expression of PVR in multiple cell types in transgenic mice did not confer broad tissue tropism (Ren and Racaniello, 1992; Zhang and Racaniello, 1997). One possible explanation for these findings is that another cellular cofactor is needed for poliovirus infection. Previous reports suggested the possibility that CD44 might serve this function (Shepley *et al.*, 1988);

however, further analyses failed to confirm this hypothesis (Bouchard and Racaniello, 1997).

Despite some uncertainties as to the overall role of PVR *in vivo*, several studies link the importance of this receptor to the virus life cycle. Kaplan *et al.* showed that exposure of poliovirus to soluble PVR converted the 160S particle to the 135S form and that this was associated with reduced infectivity (Kaplan *et al.*, 1990). Other investigators showed that antibody-coated poliovirus was unable to enter nonpermissive CHO cells bearing Fc receptors, whereas, in contrast, foot-and-mouth disease virus (FMDV) was able to utilize this alternative entry pathway (Mason *et al.*, 1994). Thus, PVR selectively mediates conformational changes in the poliovirus particle that are associated with cell entry and confers virus infection of cultured cells. Further studies will be necessary to explain why the broad distribution of this receptor does not allow virus replication in many cell types *in vivo.*

Two cryo-EM reconstructions have been published of poliovirus complexed with soluble forms of PVR (Belnap *et al.*, 2000b; He *et al.*, 2000). Both density maps are similar and show the bound soluble PVR density extending outward from the virion surface by ~115 Å with three segmented domains (Fig. 5; see Color Insert). Poliovirus, like rhinovirus, has a narrow surface depression called the "canyon" that encircles each of the twelve 5-fold vertices. The cryo-EM reconstructions of the complex reveal that PVR penetrates into the canyon and makes contract with both the "north" wall of the canyon, which is toward the 5-fold axis, and the "south" wall, which is toward the 2- and 3-fold axes. Control cryo-EM reconstructions were also done of uncomplexed poliovirus. These studies suggest that there are no major conformational changes in the virion on binding soluble PVR; however, incubations of the virus with PVR were done at 4°C. It is presumed that the cryo-EM reconstructions of the poliovirus–PVR complexes represent the initial recognition event between the virus and its receptor.

XI. Poliovirus-Associated Lipid Molecules

A comparative study of different poliovirus capsid structures revealed a hydrophobic pocket that contained sites for cellular lipid interaction (Hogle *et al.*, 1985; Filman *et al.*, 1989). This lipid component, which is termed the pocket factor, may be sphingosine. Amino acids that modulate temperature sensitivity of poliovirus infectivity map to the interfaces between capsid protomers and are adjacent to the site of lipid binding. A similar lipid molecule appears to be present in some but not all rhinoviruses (Zhao *et al.*, 1996). A concept that has emerged from these studies is that the binding of lipid in the virus capsid provides increased stability of the particle and that receptor interactions cause destabilization of the protomers and loss of lipid. Drugs, such as WIN51711, that inhibit poliovirus infection are thought to bind the same site as the spingosine molecule and, therefore, prevent the structural transitions required for virus entry and uncoating (Dove and Racaniello, 2000). Indeed, a crystal structure of the mouse neurovirulent poliovirus type 2 Lansing (PV2L) complexed with the antiviral agent SCH48973 shows that the antiviral agent binds in approximately the same location as natural pocket factors (Fig. 6; see Color Insert) (Lentz *et al.*, 1997). Belnap *et al.* noted in their cryo-EM reconstruction of the poliovirus–PVR complex that a small tunnel opens in the floor of the canyon on binding PVR (Belnap *et al.*, 2000b). This result suggests that pocket factors are expelled on PVR binding.

XII. Receptors for Rhinoviruses

Human rhinoviruses (HRV) represent a major cause of human respiratory infections. The major group of HRVs includes more than 70 serotypes, whereas the minor group contains at least 10 additional serotypes. Investigations carried out in multiple laboratories have identified ICAM-1 as a receptor for the major group of HRVs (Tomassini *et al.*, 1989; Staunton *et al.*, 1989b; Greve *et al.*, 1989). ICAM-1 is a 90-kDa membrane protein that is the ligand for a cell integrin that is highly expressed on hematopoietic cells known as lymphocyte function-associated antigen 1 (LFA-1, CD11a/CD18). ICAM-1 is a member of the IgSF and contains five immunoglobulin-like domains in its extracellular portion. Only the amino-terminal immunoglobulin domain (domain I) contains the primary site for HRV binding, although domain II and perhaps even more membrane proximal domains may help to position the receptor for optimal ligand interaction (Staunton *et al.*, 1990). A bend is predicted to lie between domains III and IV of ICAM-1 on the basis of the presence of multiple prolines located in this region. Interestingly, although LFA-1 binding to ICAM-1 requires divalent metal cations, this is not the case for HRV association. Moreover, the binding sites for LFA-1 and HRV appear to be distinct as determined by site-directed mutagenesis (Staunton *et al.*, 1990). A single amino acid residue in ICAM-1, Gln-58, plays a major role in HVR binding. This residue is not conserved in a murine homolog of ICAM-1 or in a related adhesion molecule, ICAM-2 (Staunton *et al.*, 1989a), and neither of these molecules is capable of mediating HRV attachment (Staunton *et al.*, 1990).

Structures of two rhinovirus major group serotypes, HRV14 and HRV16, complexed with soluble fragments of ICAM-1 have been studied by cryo-EM (Olson *et al.*, 1993; Kolatkar *et al.*, 1999). Fitting of crystal structures of the component viruses HRV14 (Rossmann *et al.*, 1985) and HRV16 (Oliveira *et al.*, 1993), as well as of the two N-terminal domains (D1 and D2) of ICAM-1 (Bella *et al.*, 1998; Casasnovas *et al.*, 1998), into the cryo-EM density maps served to identify residues on the virus that interact with those on the receptor surface (Kolatkar *et al.*, 1999). The fit of the D1D2 ICAM-1 structure into the cryo-EM density was confirmed by generating cryo-EM reconstructions of HRV16 complexed with fully glycosylated and mostly unglycosylated forms of D1D2 ICAM-1. The difference map between these two cryo-EM reconstructions revealed density for three of the four glycosylation sites that aligns well with the predicted positions of glycosylation (Fig. 7; see Color Insert). The cryo-EM studies of the rhinovirus–receptor complexes show that ICAM-1 recognizes slightly shifted areas in the canyons of HRV14 and HRV16, while preserving key interactions. The D1 domain of ICAM-1 is observed to bind within the rhinovirus canyon, making contacts primarily with the south wall and floor of the canyon. Comparison of cryo-EM reconstructions of the HRV16–ICAM-1 and poliovirus–PVR complexes indicates that ICAM-1 and PVR both bind at similar sites in the viral canyons, but the orientation of the long receptor molecules relative to the viral surfaces is different (He *et al.*, 2000). In addition, the footprint of PVR on poliovirus is somewhat larger than that of ICAM-1 on rhinovirus (1300^2 versus 900 $Å^2$) and involves additional contact surfaces (Rossmann *et al.*, 2000).

A model has been proposed for a two-step binding mechanism between ICAM-1 and the major group rhinoviruses (Kolatkar *et al.*, 1999). It is hypothesized that the cryo-EM reconstruction of HRV–ICAM-1 represents the initial interaction step. A second step is proposed in which the receptor moves to create additional contacts within the canyon, causing a conformational change in the viral capsid. These events would trigger movement of VP1 away from the 5-fold axis and thus open a channel and allow externalization of the N termini of VP1, VP4, and the viral RNA (Fig. 8). Rhinovirus, like poliovirus, has a hydrophobic pocket that binds natural pocket factors as well as antiviral compounds. In the proposed two-step binding mechanism for ICAM-1 to rhinovirus, step 2 might involve ejection of weakly bound pocket molecules. In contrast, more tightly bound antiviral compounds might effectively inhibit the proposed receptor-induced conformational changes in the viral capsid.

Because the cytoplasmic domain of ICAM-1 lacks typical signal sequences that mediate endocytosis, ICAM-1 may not directly regulate virus internalization. This notion is supported by experiments in which the

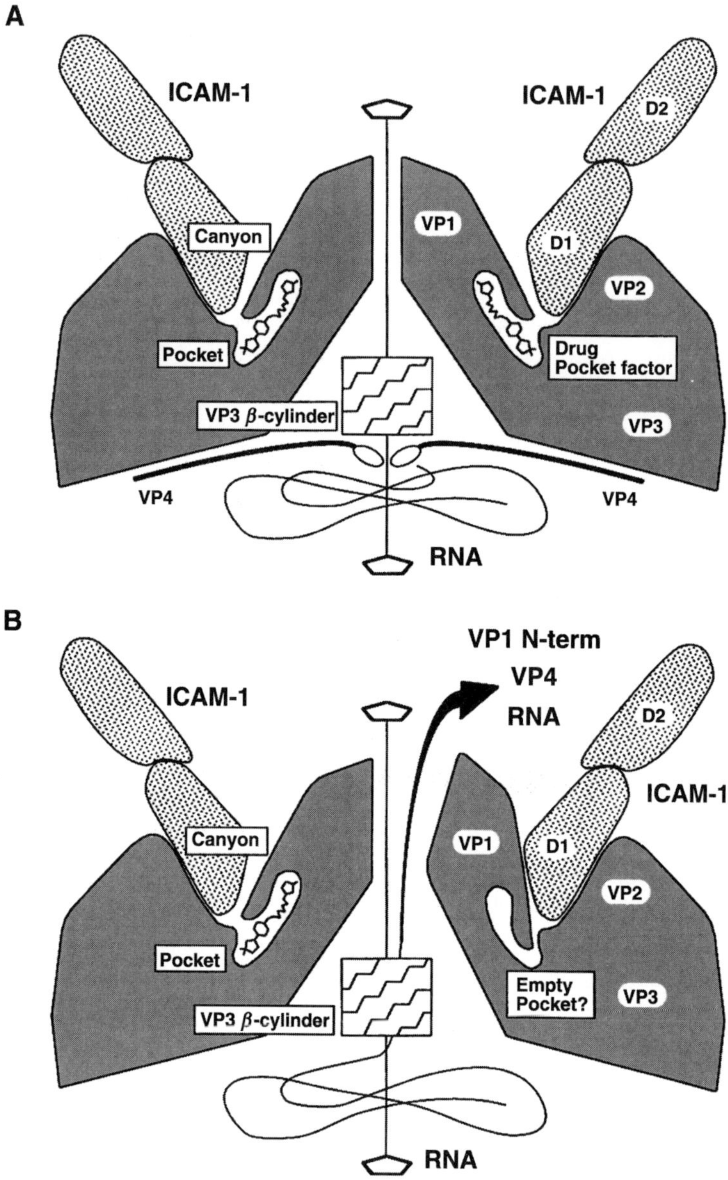

FIG. 8. The two-step binding mechanism proposed for the interaction between ICAM-1 and the major group rhinoviruses. (A) Step 1 corresponds to the structure observed in the cryo-EM reconstructions of HRV–ICAM-1 complexes. The cryo-EM structure is thought to represent the initial interaction step. (B) Step 2 is hypothesized and involves movement of the receptor and resulting conformational change of the

transmembrane anchor and cytoplasmic domain of ICAM-1 were replaced by a glycosylphophatidylinositol anchor that failed to alter HRV infectivity (Staunton *et al.*, 1992). These findings do not exclude the possibility that HRV may actually enter cells by an endocytic process, perhaps involving ligation of other as yet unidentified cell receptors. Perez and Carrasco (1993) showed that bafilomycin A1, a strong inhibitor of vacuolar ATPase, inhibited HRV14 infection, suggesting an endocytic pathway of virus infection (Fox *et al.*, 1989). HRV14 entry into HeLa cells expressing a dominant-negative mutant dynamin also is significantly reduced compared with entry into host cells expressing a normal dynamin (DeTulleo and Kirchhausen, 1998). Moreover, Schober *et al.* have reported the accumulation of partially uncoated HRV14 particles from endosomes at low temperatures (20°C) (Schober *et al.*, 1998). However, virus particles appeared to rupture endosomes at elevated (34°C) temperatures.

Whereas ICAM-1 clearly mediates attachment and infection of the major group of HRVs, the human low-density lipoprotein receptor (LDLR) has been identified as the receptor for the minor group of rhinoviruses, including HRV2 (Hofer *et al.*, 1994). The LDLR appears to mediate internalization of HRV2 via a classic endocytic pathway. Subsequently, the transfer of viral RNA occurs from the endosome/late endosome through a pore in the endosomal membrane (Prchla *et al.*, 1995).

XIII. Receptors for Other Picornaviruses

A. *Foot-and-Mouth Disease Virus*

FMDV, a member of the *Aphthovirus* genus, is an important pathogen of hooved livestock (Pickrell and Enserink, 2001). This picornavirus has an extremely high rate of transmission, with as few as 10 particles capable of causing infection in an animal. An important clue to the nature of the FMDV receptor was the observation that a conserved amino acid sequence, RGD (arginine, glycine, aspartic acid), is present in a highly variable outer loop of the VP1 capsid protein (Fox *et al.*, 1989). The RGD motif is known to be a ligand for many cell surface integrins (Pierschbacher and

---

viral capsid (shown only on the right-hand side of the diagram). This conformational change might require emptying the hydrophobic pocket and may serve to facilitate externalization of the VP1 and VP4 N-termini and the viral RNA. [Reproduced by permission of Oxford University Press from Kolatkar *et al.* (1999). Structural studies of two rhinovirus serotypes complexed with fragments of their cellular receptor. *EMBO J.* 18, 6256.]

Ruoslahti, 1984). Integrins are heterodimeric membrane glycoproteins containing noncovalently associated α and β subunits. These receptors, whose members now include more than 20 different molecules, mediate extracellular matrix or cell–cell interactions and often require divalent metal cations. Integrins are involved in a wide range of important cell functions including cell migration, cell growth and differentiation, thrombus formation, and tumor metastasis (Hynes, 1992).

Integrins also have been usurped as cell receptors by many pathogenic bacteria (Cossart *et al.*, 1996) and viruses (Nemerow and Stewart, 1999), including adenovirus (Wickham *et al.*, 1993; Roivainen *et al.*, 1994) and FMDV. Earlier studies showed that RGD-containing synthetic peptides or antibodies directed against the VP1 RGD motif inhibited FMDV infection (Fox *et al.*, 1989), and mutations of the RGD motif in FMDV VP1 resulted in diminished viral infectivity (Mason *et al.*, 1994). Subsequent studies showed that function-blocking antibodies directed against integrin $\alpha_v\beta_3$ specifically blocked FMDV infection (Berinstein *et al.*, 1995). FMDV particles also were demonstrated to bind directly to purified $\alpha_v\beta_3$ receptors (Jackson *et al.*, 1997). Neff and Baxt have shown that truncation of the cytoplasmic domains of either bovine α_v or β_3 integrin subunits did not alter FMDV infection (Neff and Baxt, 2001), suggesting that this integrin is required for virus attachment but not for virus entry into cells. However, this study did not directly assess whether virus internalization was affected by association with truncated $\alpha_v\beta_3$ integrins. It is also possible that integrin cofactors (Brown and Frazier, 2001) are involved in FMDV entry.

FMDV has been reported to also recognize integrin $\alpha_5\beta_1$ (Jackson *et al.*, 2000) as well as $\alpha_v\beta_6$ (Miller *et al.*, 2001). A leucine residue C-terminal to the RGD motif (RGDL) may influence the specificity of $\alpha_5\beta_1$ integrin interaction (Jackson *et al.*, 2000). Whereas field isolates of FMDV clearly utilize cell integrins for binding and cell entry, viruses that have been highly passaged in tissue culture accumulate mutations in the RGD motif (Martinez *et al.*, 1997); this property is associated with the acquistion of new receptor-binding functions, including those involving heparin or heparan sulfate (HS) proteoglycans (Jackson *et al.*, 1996; Sa-Carvalho *et al.*, 1997; Neff *et al.*, 1998). These findings suggest that vaccine development based on antibody production against the RGD motif may result in the selection of viral mutants with altered receptor specificity.

There are seven serotypes of FMDV (O, A, C, Asia, SAT1, SAT2, and SAT3) and although most utilize integrins for cell entry, certain strains of O_1 FMDV have been shown to use HS as the predominant cell surface ligand (Jackson *et al.*, 1996). For these strains of FMDV, attachment to HS is highly specific and is required for efficient infection. Crystal structures have been published of the FMDV strain O_1BFS complexed with various heparin and HS preparations (Fry *et al.*, 1999). The virus–oligosaccharide receptor complex structures show that subtype O_1 FMDV binds a highly abundant motif of sulfated sugars in a shallow depression on the virion surface and that this binding involves contacts with all three major capsid proteins, VP1, VP2, and VP3 (Fig. 9; see Color Insert). The observed high-avidity binding ($10^{-9}M$) of FMDV to fixed cell HS (Jackson *et al.*, 1996) is postulated to involve several of the possible 60 binding sites on the virus particle as well as different sulfated, sugar regions of the HS chain (Fry *et al.*, 1999). The crystal structures reveal that the RGD motif is ~15 Å from the closest sugar moiety of HS, and the two binding sites appear independent.

Fry *et al.* (1999) have proposed various possible mechanisms for HS-mediated cell entry of the FMDV O_1BFS strain. One idea is that HS, which is an abundant cell surface molecule, may concentrate FMDV at the cell surface and thus improve the chance of the virus particle encountering an integrin receptor, in a process analogous to the adhesion-strengthening mechanism proposed for reovirus. Another possibility is that there may be a direct interaction between HS proteoglycans and integrin receptors. The adhesion molecules vitronectin and fibronectin are also known to have dual affinities for integrin and HS (Felding-Habermann and Cheresh, 1993; Potts and Campbell, 1994) and there might be an interaction between the two receptor molecules, perhaps involving integrin activation. A third possibility is that HS proteoglycans might be sufficient for FMDV internalization without integrins. Further work will be needed to resolve this issue.

B. *Echovirus Receptors*

Echovirus 1, another member of the picornavirus family and a cause of febrile illness and meningitis, has been shown to use a human cell integrin, $\alpha_2\beta_1$ [very late antigen 2 (VLA-2)], for attachment and infection (Bergelson *et al.*, 1996). The host cellular protein recognized by this integrin is collagen, an extracellular matrix protein. A murine homolog of VLA-2 binds collagen but fails to mediate echovirus 1 cell attachment. This is consistent with the fact that the binding sites for collagen and virus are distinct on human VLA-2 (King *et al.*, 1997).

XIV. Human Adenoviruses

Adenoviruses are nonenveloped, double-stranded DNA viruses. There are more than 50 different serotypes of human adenoviruses that have been divided among 6 different subgroups (A–F) based on serologic and nucleotide sequence similarity, and other biological properties (Shenk, 2001). Adenoviruses are responsible for a significant number of respiratory, gastrointestinal, and ocular infections. Infections with adenovirus are usually self-limiting; however, they can cause serious disseminated infections in immunocompromised patients (Hierholzer, 1992) and unvaccinated military recruits. Adenovirus has been used as a model system to discover mechanisms underlying cell and molecular biological processes including cell cycle regulation and cancer (Yang *et al.*, 1996; Chinnadurai, 1983), RNA processing (Berget *et al.*, 1977; Chow *et al.*, 1977), and immunoregulation (Horwitz, 2001).

Replication-defective (Wilson, 1998; Nabel, 1999) and conditionally replicating adenovirus vectors (Kafri *et al.*, 1998) also are undergoing evaluation in human gene therapy trials for the treatment of cardiovascular disease (Chang *et al.*, 1995) and cancer (Duggan *et al.*, 1995). Although increased knowledge of adenovirus structure as well as of host cell receptor interactions (Nemerow and Stewart, 1999) has provided new opportunities to improve cell targeting of adenovirus vectors (Von Seggern *et al.*, 2000; Jakubczak *et al.*, 2001; Krasnykh *et al.*, 1996; Li *et al.*, 2000b; Ebbinghaus *et al.*, 2001), the host immune response to adenovirus or its transgene products remains a significant impediment to further advances in clinical applications (Elkon *et al.*, 1997; Wilson, 1995; Kafri *et al.*, 1998).

XV. Adenovirus Attachment Receptors

The majority of adenovirus cell entry studies have been performed with adenovirus types 2 and 5 (subgroup C), which cause respiratory infections. Ad2 entry into cells involves association with at least two different cell receptors (Wickham *et al.*, 1993). Viral attachment is mediated by the interaction of the elongated fiber protein with a 46-kDa membrane glycoprotein known as coxsackievirus–adenovirus receptor, or CAR. CAR mediates high-affinity binding of coxsackieviruses (subgroup B) as well as most adenovirus serotypes (Lonberg-Holm *et al.*, 1976; Tomko *et al.*, 1997; Bergelson *et al.*, 1997; Roelvink *et al.*, 1998). CAR is a member of the IgSF and contains two immunoglobulin-like domains. Only the membrane-distal immunoglobulin domain is required for adenovirus binding (Freimuth *et al.*, 1999). The transmembrane domain and cytoplasmic tail regions of the receptor are also not necessary for virus infection (Wang and Bergelson, 1999).

The adenovirus fiber protein is trimeric, and the monomer varies in length from 320 to 587 residues (Chroboczek *et al.*, 1995). The N-terminal region of the fiber protein associates with the penton base protein in the viral capsid, the central region forms the long thin shaft of variable length, and the C-terminal ~175 residues form the globular knob domain that interacts with CAR. A crystal structure of the Ad12 fiber knob complexed with domain 1 (D1) of CAR has been solved (Bewley *et al.*, 1999). This structure reveals that CAR D1 binds on the side of the knob at the interface between two adjacent Ad12 knob monomers (Fig. 10; see Color Insert). The AB loop of the fiber knob plays an important role in the high-affinity knob–CAR interaction and it contributes more than 50% of the interfacial protein–protein contacts. Several key residues in the AB loop are conserved among CAR-binding Ad serotypes.

A number of studies have indicated that CAR is the major host cell determinant of adenovirus infection *in vivo*. CAR has been shown to be highly expressed in the heart (Tomko *et al.*, 1997), and this observation is consistent with adenovirus-mediated gene delivery to cardiac tissue *in vivo* (Rosengart *et al.*, 1999). CAR expression has been reported to be low or absent on primary human fibroblasts (Hidaka *et al.*, 1999) and most peripheral blood cells (Leon *et al.*, 1998; Huang *et al.*, 1997), and these cell types have proved difficult to transduce with adenovirus vectors. In more recent studies, peripheral blood lymphocytes derived from transgenic mice expressing human CAR were shown to permit efficient adenovirus-mediated gene delivery (Schmidt *et al.*, 2000).

Whereas CAR is the major receptor for most adenovirus serotypes, adenoviruses belonging to subgroup B, such as Ad3 and Ad7, clearly do not recognize this receptor (Roelvink *et al.*, 1998; Stevenson *et al.*, 1995). Moreover, highly conserved sequences in the fiber knob domain that mediate CAR binding in subgroup C adenoviruses are lacking in the Ad3 and Ad7 fiber proteins (Roelvink *et al.*, 1999). Adenovirus vectors equipped with the Ad3 fiber protein allow for efficient transduction of human B lymphoblastoid cells, which express little if any CAR (Von Seggern *et al.*, 2000). Ad16, another subgroup B strain, infects vascular smooth muscle and endothelial cells more efficiently than Ad5-based vectors (Havenga *et al.*, 2001). This finding suggests that Ad16-based vectors may be particularly useful for treating cardiovascular diseases such as restenosis.

In addition to the subgroup B viruses, it is likely that members of other Ad subgroups also recognize distinct cell receptors. For example, adenovirus types belonging to subgroup D exhibit higher infectivity of neuronal (Chillon *et al.*, 1999) and ocular cells (Huang *et al.*, 1999) than do subgroup C (Ad5) viruses. Ad37 appears to recognize a cell surface sialic acid (Arnberg *et al.*, 2000) as well as a 50-kDa protein on conjunctival epithelial cells (Wu *et al.*, 2001) whose identity has yet to be determined. Huang *et al.* have noted that a single residue at position 240 of the Ad37 fiber is needed for binding and infection of conjunctival cells (Huang *et al.*, 1999). Molecular modeling, based on the crystal structure of the Ad5 fiber knob (Xia *et al.*, 1994), indicates that residue 240 is exposed on the top surface of the knob, in the CD loop. The crystal structure of the Ad12 fiber knob–CAR complex clearly shows that the CD loop is not involved in CAR binding (Bewley *et al.*, 1999). Thus, CAR is not involved in Ad37 infection of conjunctival cells despite the demonstrated binding of the Ad37 fiber knob to CAR on virus protein blot overlay assays (Wu *et al.*, 2001).

A cryo-EM reconstruction of a pseudotyped fiber-deleted Ad5 vector with the Ad37 fiber shows that this fiber is ~150 Å long, straight, and rigid (Chiu *et al.*, 2001). This is in contrast to observations, by negative-stain EM (Chroboczek *et al.*, 1995) and cryo-EM (Stewart *et al.*, 1991; Chiu *et al.*, 1999; Von Seggern *et al.*, 1999), showing that the fibers of most Ad serotypes are long (>300 Å) and flexible with a bend ~100 Å from the viral capsid surface. It has been suggested that the geometric constraints imposed by a short rigid fiber protruding from an icosahedral viral capsid may prevent the use of the side of the fiber knob for receptor binding (Wu *et al.*, 2001; Chiu *et al.*, 2001). In other words, it is possible that only a long flexible Ad fiber can effectively utilize the side of its knob for CAR binding because of the orientation of CAR on the host cell surface. This model provides a structural explanation for why Ad serotypes with fiber knobs containing the CAR-binding sequence in the AB loop do not necessarily bind CAR on cells.

There are indications that other cell surface molecules also may participate in virus attachment. Dechecchi and co-workers showed that heparan sulfate proteoglycans in combination with CAR facilitate binding of subgroup C but not subgroup B adenoviruses via interaction with the fiber protein (Dechecchi *et al.*, 2000). Chu and colleagues have suggested that vascular cell adhesion molecule 1 (VCAM-1), a receptor that is upregulated on endothelial cells of atherosclerotic vessels, may also promote Ad5 binding (Chu *et al.*, 2001).

XVI. Cell Integrins Promote Adenovirus Internalization

Early electron microscopic studies showed that adenovirus enters cells via receptor-mediated endocytosis (Patterson and Russell, 1983). Consistent with these early morphologic studies, adenovirus uptake into cells was shown to involve dynamin (Wang *et al.*, 1998), a 100-kDa GTPase that regulates endosome formation (Sever *et al.*, 1999; Marks *et al.*, 2001). Adenovirus was one of the first viruses shown to use multiple receptors for cell entry (Nemerow *et al.*, 1993; Wickham *et al.*, 1993). The adenovirus fiber protein mediates high-affinity virus binding to cells, but this binding is not sufficient for efficient virus internalization. Instead, interaction of an RGD sequence in the adenovirus penton base protein with vitronectin-binding integrins ($\alpha_v\beta_3$ and $\alpha_v\beta_5$) enhances virus uptake. More recent studies have demonstrated that integrin $\alpha_v\beta_1$, a fibronectin-binding receptor, can also promote adenovirus entry into human embryonic kidney (HEK) 293 cells, which lack $\alpha_v\beta_3$ and $\alpha_v\beta_5$ (Li *et al.*, 2001).

A number of observations indicate that penton base–integrin interactions represent an important step in adenovirus infection *in vivo*. The penton base of most adenovirus serotypes contains a conserved integrin-binding motif (RGD) (Mathias *et al.*, 1994; Cuzange *et al.*, 1994); and those serotypes that lack this sequence (e.g., Ad40/41) show delayed uptake into cells (Albinsson and Kidd, 1999). Mutation of the penton base RGD motif substantially reduces integrin association with Ad2 particles as well as the rate of virus infection (Bai *et al.*, 1993). Human lymphocytes and monocytes are generally refractory to adenovirus-mediated gene delivery; however, on upregulation of integrin expression they become susceptible to infection (Huang *et al.*, 1995, 1997). Interestingly, integrin $\alpha_m\beta_2$ can serve as an attachment receptor for Ad2 on human macrophages, which lack CAR (Huang *et al.*, 1996).

To localize the RGD residues on the Ad penton base, a cryo-EM reconstruction was performed of adenovirus type 2 (Ad2) complexed with an RGD-specific Fab fragment from an mAb directed against the penton base (Stewart *et al.*, 1997). This structural analysis revealed that the RGD regions are at the top of protrusions on the pentameric penton base protein. In addition, it was deduced from the diffuse nature of the Fab density that the RGD residues were in a structurally variable surface loop. Comparison of the known sequences of the penton base protein from various adenovirus serotypes suggested that type 12 adenovirus (Ad12) would have the least structurally variable RGD loop, as Ad12 has 45 fewer residues in the variable region flanking the conserved RGD residues than are found in Ad2 (Chiu *et al.*, 1999).

Cryo-EM reconstructions of Ad2 and Ad12 each revealed only a short portion of the long thin fiber (full length, ~300 Å) and did not show the fiber knob involved in CAR binding (Fig. 11A; see Color Insert) (Chiu *et al.*, 1999). The reason for this is that the fibers of most adenovirus serotypes are bent after a distance of just 90–100 Å from the viral surface (Chroboczek *et al.*, 1995; Stewart *et al.*, 1991). Because cryo-EM imaging relies on averaging images of many different particles, any regions of the structure that deviate from particle to particle, such as the fiber beyond the bend point, are effectively averaged away.

The cryo-EM reconstructions of both Ad2 and Ad12 complexed with soluble $\alpha_v\beta_5$ integrin revealed rings of density corresponding to bound integrin over the penton base capsid proteins (Fig. 11B) (Chiu *et al.*, 1999). As expected from the penton base sequence comparison, the integrin density was better defined for the Ad12 complex, indicating less variability for the Ad12 RGD loop. The reconstruction of the Ad12–$\alpha_v\beta_5$ integrin complex revealed that the soluble $\alpha_v\beta_5$ integrin has two structural domains: one closer to the viral surface contacting the RGD-containing protrusions, and the other farther from the viral surface and presumably closer to the host cell surface (Fig. 11C). These two domains are referred to as the proximal domain and the distal domain, respectively. For the cryo-EM study an excess of soluble integrin was used and the estimated occupancy of integrin in the Ad12–$\alpha_v\beta_5$ complex was 100%. The five proximal domains bound to one penton base appear to form a solid ring of density, as if each bound integrin has large contact areas with its neighboring integrins. This close receptor clustering, caused by the spacing of the viral RGD-binding sites, may result in activation of the integrin and perhaps initiate cell signaling events. When the proximal integrin ring is cut arbitrarily into five regions, it is easier to visualize the interaction between the integrin and the penton base (Fig. 12; see Color Insert). A cleft is observed in the proximal domain into which the RGD-containing penton base protrusion fits.

A similarity has been noted between the adenovirus and FMDV integrin-binding RGD sites (Nemerow and Stewart, 2001). Comparison of the cryo-EM structure of the adenovirus–Fab complex, which served to identify the RGD sites (Stewart *et al.*, 1997), and the crystal structure of FMDV (Acharya *et al.*, 1989) reveals that for both viruses the RGD sites are positioned around the 5-fold symmetry axes with a spacing of ~60 Å. This is in spite of the fact that otherwise these two viruses have virturally nothing in common in either their capsid proteins or overall structure. This observation suggests that the RGD spacing observed for both viruses may be optimal for α_v integrin clustering and for induction of cell signaling processes.

Knowledge of integrin-mediated virus internalization has allowed modification of adenovirus vectors to improve gene delivery to certain cell types. For example, incorporation of an RGD sequence into the virus hexon protein was shown to facilitate gene delivery to vascular smooth muscle cells in a CAR-independent/integrin-dependent manner (Vigne *et al.*, 1999). Von Seggern *et al.* have shown that a recombinant adenovirus lacking the fiber protein (fiberless) was nonetheless capable of transducing cells via α_v integrins (Von Seggern *et al.*, 2000). Interestingly, a recombinant bacteriophage displaying the adenovirus penton base or the RGD-containing domain was shown to enter cells via integrins (DiGiovine *et al.*, 2001). Despite these advances, significant gaps exist in our knowledge of how integrins regulate adenovirus tropism *in vivo* and thus further studies in this area are needed.

XVII. Signaling Events Associated with Adenovirus Internalization

Integrin clustering via interactions with the extracellular matrix frequently induces morphologic changes in the plasma membrane, causing reorganization of the actin cytoskeleton and formation of focal adhesion complexes. The cytoplasmic tail of clustered integrins can bind to one or more actin-associated proteins such as talin or α-actinin, and these receptor complexes may contain a number of cell signaling molecules and other adapter proteins (Calderwood *et al.*, 2000; Clark and Brugge, 1995). Because actin may regulate endocytic processes in mammalian cells (Fujimoto *et al.*, 2000), and disruption of the actin cytoskeleton by cytochalasin B reduces adenovirus infection (Patterson and Russell, 1983), Li and co-workers investigated the role of actin reorganization in adenovirus internalization (Li *et al.*, 1998a; Nemerow and Stewart, 1999). These investigators discovered that adenovirus–integrin interactions also induce specific signaling events that alter cell shape, enhance cortical actin polymerization, and activate phosphatidylinositol-3-OH kinase (PI3K). This lipid kinase acts as a second messenger for multiple signaling processes including those mediating cytoskeletal function (Hall, 1998) and bacterial cell invasion (Ireton *et al.*, 1996). PI3K also activates Rab5, a GTPase associated with early endosome formation. Overexpression of a dominant-negative form of Rab5 in host cells reduces adenovirus internalization and infection (Rauma *et al.*, 1999).

Multiple lines of evidence indicate that interactions of the penton base with α_v integrins rather than fiber–CAR interactions promote cell signaling and adenovirus internalization. The penton base but not the fiber protein was shown to induce PI3K activation (Li *et al.*, 1998b). Moreover, fiberless adenovirus particles trigger similar levels of activation as native adenovirus particles (Li *et al.*, 2000a). Wang and Bergelson have demonstrated that recombinant forms of CAR lacking its normal transmembrane anchor and cytoplasmic domain fully support adenovirus infection, indicating that CAR does not directly influence cell signaling events (Wang and Bergelson, 1999).

In addition to PI3K, adenovirus internalization also requires participation of several other signaling molecules including the Rho family of small GTPases (Li *et al.*, 1998a) and p130CAS (Li *et al.*, 2000a). Rho GTPases regulate changes in cell shape and promote actin reorganization (Hall, 1998) via interaction with additional downstream effector molecules such as WASP and PAK1 (Hoffman and Cerione, 2000). P130CAS is a large adapter protein that provides an important functional link between c-Src (Vuori *et al.*, 1996) and the p85 catalytic subunit of PI3K. Adenovirus-induced signaling processes may also contribute to host inflammatory responses that limit the duration of transgene expression (Zsengellér *et al.*, 2000). In support of this possibility, Bruder and Kovesdi (1997) reported that adenovirus interaction with cells triggers expression of interleukin 8. Adenovirus uptake into macrophages via PI3K also can produce inflammatory cytokines (Zsengellér *et al.*, 2000). The precise effector molecules involved in adenovirus-mediated actin polymerization leading to virus internalization have yet to be defined.

XVIII. α_v Integrins Regulate Adenovirus-Mediated Endosome Disruption

Engagement of α_v integrins by adenovirus not only promotes virus internalization but also facilitates disruption of early endosomes, allowing the virus to escape degradation in late endosomes and lysosomes. Early studies showed that adenovirus particles alter cell membrane permeability at pH 6.0 (Seth *et al.*, 1985, 1987) and that this reaction is mediated by the penton base association with α_v integrins (Seth *et al.*, 1984; Wickham *et al.*, 1994). More recently, it was shown that $\alpha_v\beta_5$ selectively plays a pivotal role in endosome disruption (Wickham *et al.*, 1994; Wang *et al.*, 2000). In particular, amino acid sequences in the cytoplasmic domain of the β_5 integrin subunit regulate virus escape from the early endosome. The mechanism by which this occurs is still obscure, but other integrin cofactors (Liu *et al.*, 2000) may work in concert with $\alpha_v\beta_5$ to promote virus penetration. Activation of the adenovirus cysteine protease also is required for endosome penetration (Hannan *et al.*, 1983; Cotten and Weber, 1995), and activation of the protease requires penton base interaction with α_v integrins (Greber *et al.*, 1996). Clearly, further research is required to determine the mechanisms involved in adenovirus-mediated endosome disruption and the precise role of integrins in this process.

XIX. Conclusions

Substantial progress has been made in the identification of specific host cell receptors for different viruses and, in many cases, the structural features of virus–receptor interactions have been defined. This review has considered the receptor interaction strategies of multiple virus families including reoviruses, picornaviruses, and adenoviruses. Although much work remains to determine why certain receptors have been selected by different viruses and to discover precisely how receptors promote infection, a number of common themes are beginning to emerge. The first is that viruses often use multiple receptors for binding and cell entry (rather than a single cell surface molecule). In general, virus attachment to primary receptors involves high-affinity binding that plays a major role in determining host cell tropism. Virus interactions with secondary receptors (e.g., adenovirus with integrins) tend to be of lower affinity but nonetheless are required for efficient virus internalization. Ligation and clustering of secondary receptors may lead to signaling events involved in virus entry, but it might also have important pathogenic consequences including inflammatory cytokine production or cell death induction (apoptosis). Another example of multiple receptor usage is the reovirus $\sigma 1$ protein, which contains two distinct receptor-binding domains. The tail domain of T3 $\sigma 1$ binds cell surface SA whereas the head domain binds junction adhesion molecule (JAM). Although only one receptor has been identified for poliovirus, PVR; its expression is not sufficient for infection in certain cell types and thus it has been suggested that another cellular cofactor is needed.

A second theme among these distinct virus families is that the interaction of the virus with one of its receptors often involves a long, extended molecule, perhaps to increase the chance of productive binding to the cell by virtue of Brownian motion. In the case of reovirus, the $\sigma 1$ protein undergoes a dramatic conformational change that results in the head domain extending ~480 Å from the surface of the ISVP. There is substantial evidence that the $\sigma 1$ head utilizes JAM as a serotype-independent receptor and perhaps this interaction is facilitated by the extended conformation of $\sigma 1$. For poliovirus and the major group of rhinoviruses in the picornavirus family, it is the receptor molecule, PVR or ICAM-1, respectively, that is elongated and flexible (He *et al.*, 2000). PVR has three extracellular domains and ICAM-1 has five extracellular domains, and both are long (115 Å or longer) slender molecules. Certain FMDV strains that utilize HS as the primary receptor may bind multiple-sulfated sugar regions along the long, flexible HS chain in order to achieve high affinity binding. The majority of adenovirus serotypes also have a long (>300 Å) and flexible fiber protein with high affinity for the attachment receptor CAR.

A third theme is that viruses are adaptable in their selection of receptors. Given the choice of a wide variety of host cell surface molecules as potential receptors, different strains or serotypes within the same virus family often acquire the ability to utilize different receptors. Reovirus T1 × T3 reassortment studies indicate that $\sigma 1$ interactions with specific cellular receptors control serotype-specific pathogenesis. The major and minor group rhinoviruses are known to utilize different receptor molecules, ICAM-1 and LDLR, respectively. Field isolates of FMDV that use integrins for binding and cell entry can acquire new receptor-binding functions after passage in tissue culture. Whereas most types of adenovirus have long, flexible fibers that utilize CAR as the attachment receptor, other types, such as Ad37, have short, rigid fibers and utilize other cell surface molecules as attachment receptors.

Important goals for future research include gaining a better understanding of the precise role of receptors in virus-induced diseases and elucidating how receptor engagement sets the stage for membrane penetration and subsequent activation of the viral genetic program. Such knowledge should lead to novel antiviral approaches and foster the development of improved gene delivery vectors.

Acknowledgments

The authors express their gratitude to members of their laboratories, past and present, who have made major contributions to work cited in the review. We also apologize for any omissions of other work, which, because of space limitations, could not be included in this review. The authors are grateful to Joan Gausepohl for excellent assistance in preparation of this review. This work was supported by NIH Grants EY11431, HL54352, AI38296, AI50080, and AI42929 and the Elizabeth B. Lamb Center for Pediatric Research. This manuscript is no. 14379 from the Scripps Research Institute.

References

Acharya, R., Fry, E., Stuart, D., Fox, G., Rowlands, D., and Brown, F. (1989). *Nature* **337,** 709–716.

Albinsson, B., and Kidd, A. H. (1999). *Virus Res.* **64,** 125–136.

Arnberg, N., Edlund, K., Kidd, A. H., and Wadell, G. (2000). *J. Virol.* **74,** 42–48.

Baer, G. S., and Dermody, T. S. (1997). *J. Virol.* **71,** 4921–4928.

Bai, M., Harfe, B., and Freimuth, P. (1993). *J. Virol.* **67,** 5198–5205.

Banerjea, A. C., Brechling, K. A., Ray, C. A., Erikson, H., Pickup, D. J., and Joklik, W. K. (1988). *Virology* **167,** 601–612.

Barton, E. S., Connolly, J. L., Forrest, J. C., Chappell, J. D., and Dermody, T. S. (2001a). *J. Biol. Chem.* **276,** 2200–2211.

Barton, E. S., Forrest, J. C., Connolly, J. L., Chappell, J. D., Liu, Y., Schnell, F. J., Nusrat, A., Parkos, C. A., and Dermody, T. S. (2001b). *Cell* **104,** 441–451.

Bassel-Duby, R., Spriggs, D. R., Tyler, K. L., and Fields, B. N. (1986). *J. Virol.* **60,** 64–67.

Bella, J., Kolatkar, P. R., Marlor, C. W., Greve, J. M., and Rossmann, M. G. (1998). *Proc. Natl. Acad. Sci. USA* **95,** 4140–4145.

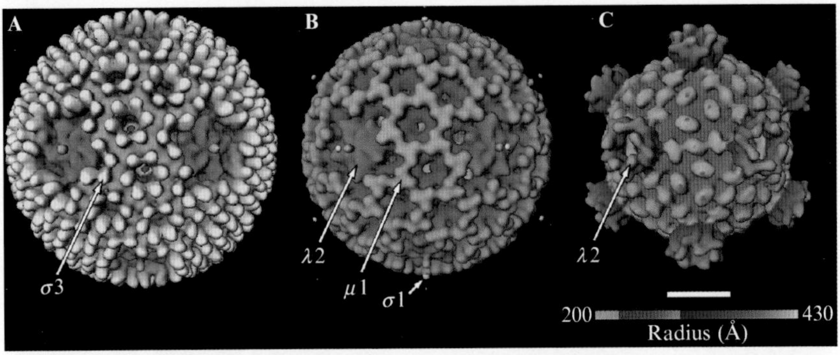

STEWART *ET AL.*, FIG. 1. Cryo-EM reconstructions of reovirus T1L (A) virion, (B) infectious subvirion particle (ISVP), and (C) core viewed along a 2-fold symmetry axis. The capsid proteins σ1, σ3, μ1, and λ2 are indicated. Note that the density from thin spikelike σ1 protein appears disconnected. The density is colored according to radial height as indicated by the color bar in order to accentuate the surface features. White scale bar: 200 Å. [Modified from the *Journal of Cell Biology*, 1993, Vol. 122, pp. 1023–1041 (Dryden *et al.*, 1993) by copyright permission of the Rockefeller University Press.]

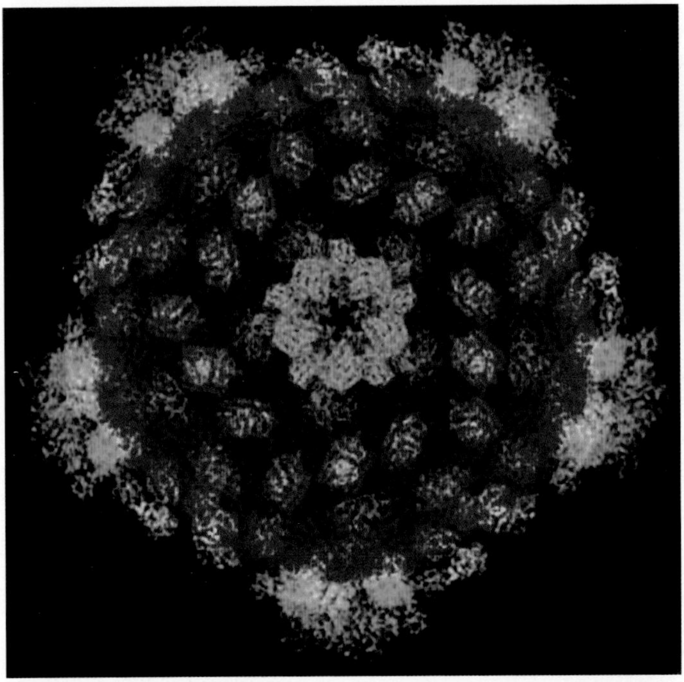

STEWART *ET AL.*, FIG. 3. The crystallographic structure of the reovirus core composed of 120 copies of λ1 (red), 150 copies of σ2 (yellow, green, and white), and 60 copies of λ2 (blue). The C_α traces are displayed and the view is along a 5-fold symmetry axis. [Reproduced with copyright permission from *Nature* (Reinisch *et al.*, 2000).]

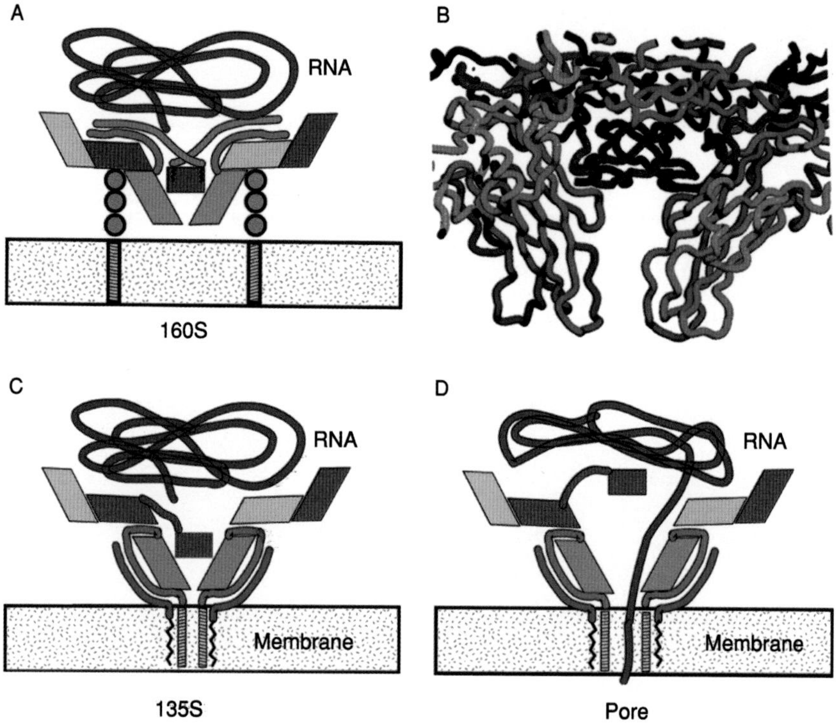

STEWART *ET AL.*, FIG. 4. A model for the translocation of poliovirus RNA across the cell membrane. (A) A schematic view of one vertex of poliovirus interacting with the cell surface. The capsid proteins VP1, VP2, VP3, and VP4 are shown in cyan, yellow, red, and green, respectively, and the poliovirus receptor is shown in gray as three circles with a transmembrane region. Note that VP3 forms a plug at the 5-fold axis. (B) The crystal structure of one vertex of the virion. (C) A schematic view of the 135S particle as it is proposed to interact with the membrane. Note that the N-terminal helices of VP1 and the N-terminal myristates of VP4 (jagged lines) are inserted into the membrane. (D) The VP3 β tube is proposed to shift out of the way to allow the viral RNA to pass through the pore into the cytoplasm. [Reproduced with permission from Belnap *et al.* (2000a).]

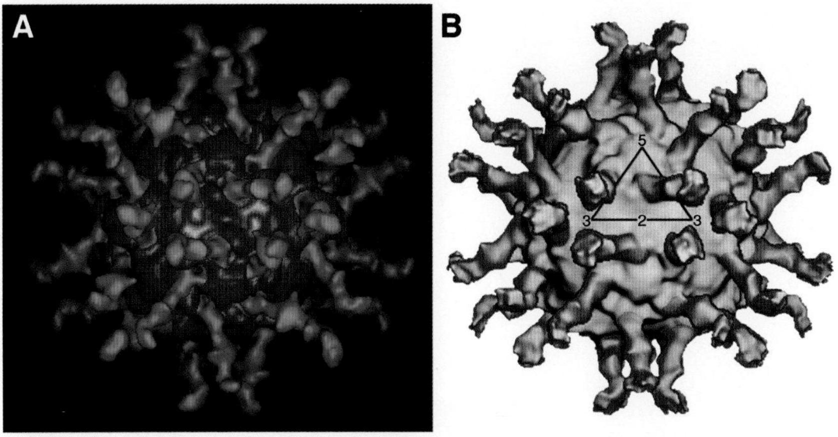

STEWART *ET AL.*, FIG. 5. Two cryo-EM reconstructions of poliovirus complexed with soluble forms of PVR. In both reconstructions the virus and the receptor are distinguished by different colors. Both reconstructions are viewed along a 2-fold symmetry axis. (A) Reproduced with permission from Belnap *et al.* (2000b); (B) reproduced with permission from He *et al.* (2000).

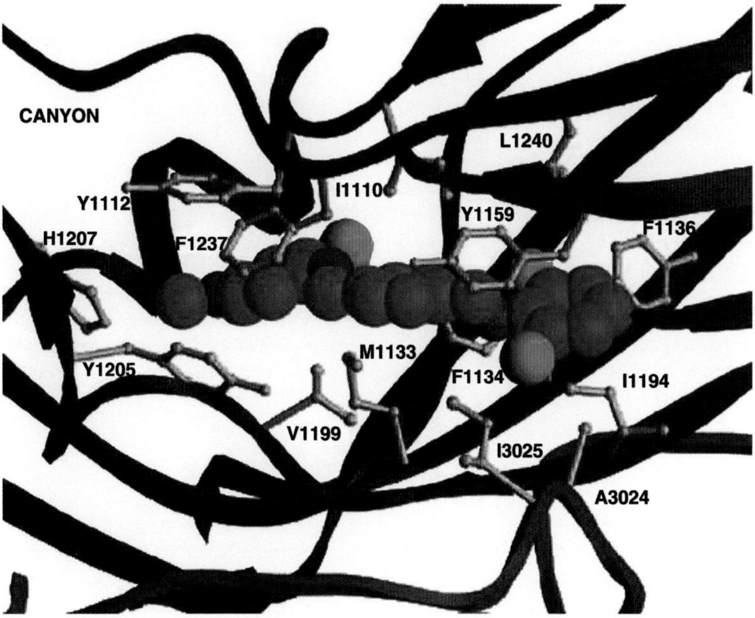

STEWART *ET AL.*, FIG. 6. The antiviral agent SCH48973 in the pocket of VP1 of poliovirus type 2 Lansing. SCH48973 is colored according to atom type: VP1 is shown as blue ribbon, VP3 is shown as red ribbon; side chains near the inhibitor are shown in yellow, and chlorine atoms are shown in cyan. [Reproduced with permission of Global Rights Department, Elsevier Science, from Lentz *et al.* (1997). Structure of poliovirus type 2 Lansing complexed with antiviral agent SCH48973: Comparison of the structural and biological properties of the three poliovirus serotypes. *Structure* 5, 961.]

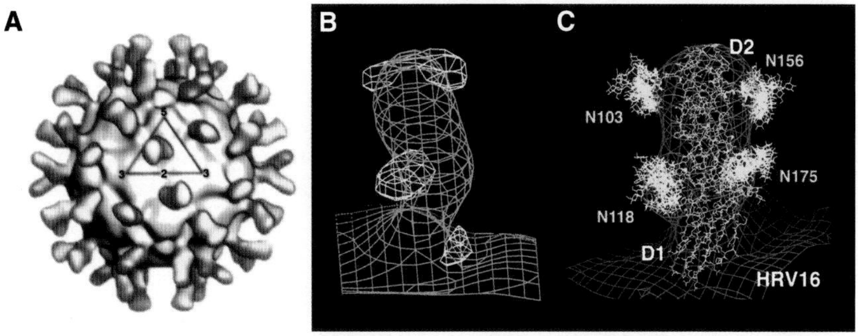

STEWART *ET AL.*, FIG. 7. Complex of rhinovirus type 16 with a soluble form of its ICAM-1 receptor. (A) Cryo-EM reconstruction of the complex viewed along an icosahedral 2-fold axis with the viral capsid shown in grey and the receptor shown in orange. [Reproduced with permission from He *et al.* (2000).] (B) A closeup view of the density (light green) corresponding to the two-domain fragment of ICAM-1 from the cryo-EM reconstruction of an HRV16–ICAM-1 complex. The difference in density between the HRV16–ICAM-1 reconstructions with fully glycosylated and mostly deglycosylated forms of ICAM-1 is shown in yellow. (C) Fitting of the ICAM-1 atomic model for domains D1 and D2 (white) within the cryo-EM density of the complex (blue). The disordered carbohydrates are represented by an ensemble of conformations (yellow). [(B) and (C) reproduced by permission of Oxford University Press, from Kolatkar *et al.* (1999). Structural studies of two rhinovirus serotypes complexed with fragments of their cellular receptor. *EMBO J.* 18, 6256.]

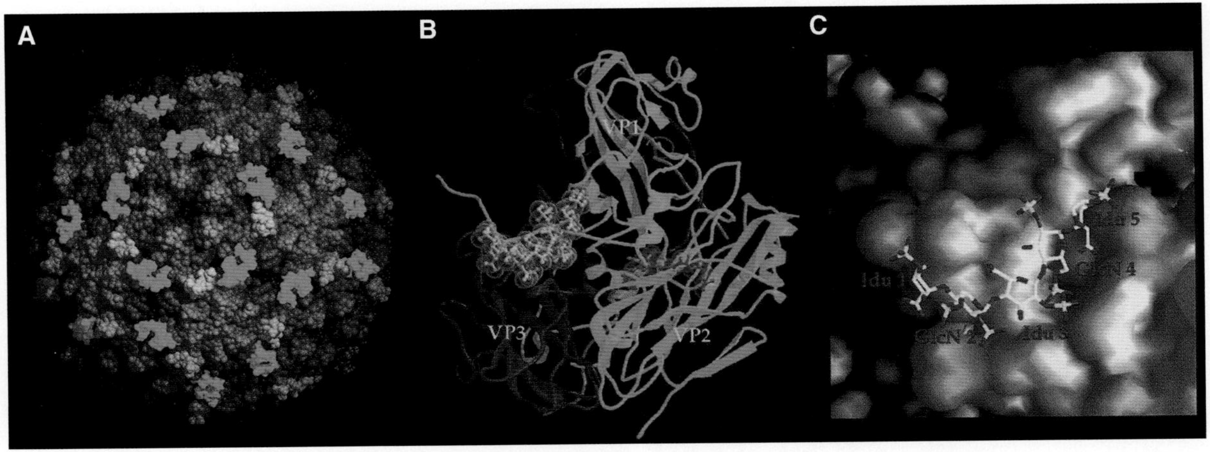

STEWART *ET AL.*, FIG. 9. The crystal structure of a foot-and-mouth disease virus–heparan sulfate proteoglycan complex. (A) A space-filling representation of the reduced O_1BFS strain of FMDV shown predominantly in yellow with the VP1 GH loop in cyan, the RGD integrin-binding tripeptide in brighter blue, and the bound heparin motif in white. (B) A ribbon drawing of the viral capsid proteins VP1 (cyan), VP2 (green), and VP3 (red) shown together with five sugars from heparin in both white ball-and-stick and transparent CPK representations. The RGD integrin-binding motif is shown in blue ball-and-stick and transparent CPK representations. (C) An accessible surface representation (gray) showing the shallow depression occupied by five sugar rings with bonds drawn as rods and colored according to the standard convention. [Reproduced by permission of Oxford University Press from Fry *et al.* (1999). The structure and function of a foot-and mouth disease virus–oligosaccharide receptor complex. *EMBO J.* 18, 548.]

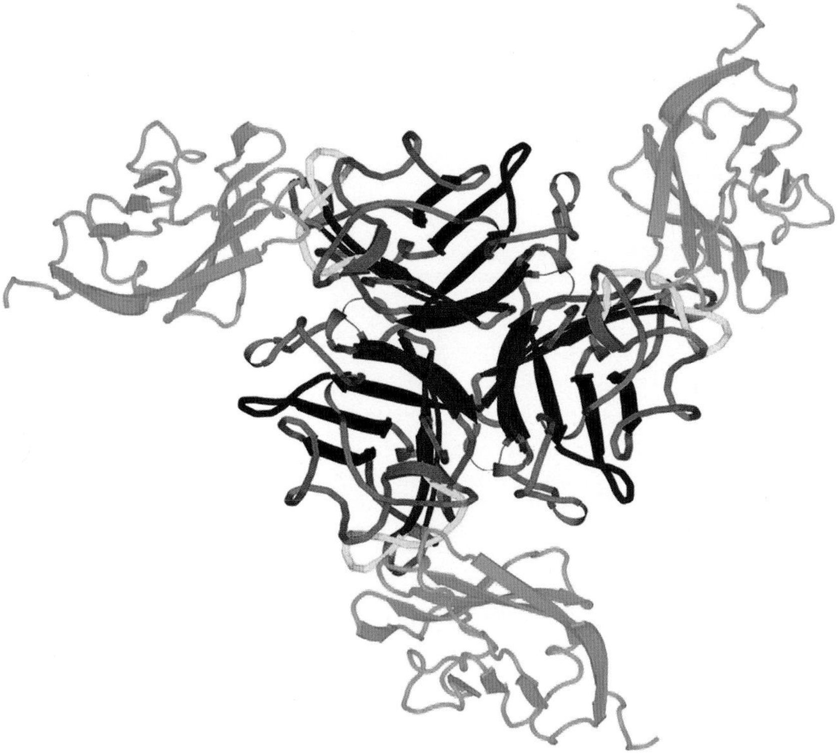

STEWART *ET AL.*, FIG. 10. A ribbon diagram of the crystal structure of the Ad12 knob complexed with the D1 domain of CAR viewed along the 3-fold axis of the fiber. The two β sheets in the core of each knob monomer are shown in red and purple; the AB loop involved in CAR binding is highlighted in yellow; the HI loop is in purple; and the rest of the knob is in gray. The three bound CAR D1 domains are in cyan. [Reproduced from Bewley *et al.* (1999).]

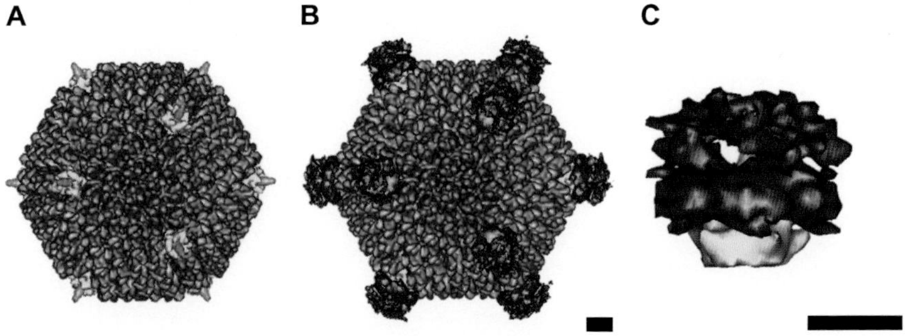

STEWART *ET AL.*, FIG. 11. Cryo-EM reconstructions of Ad12 and Ad12 complexed with soluble $\alpha_v\beta_5$ integrin. (A) A reconstruction of Ad12 at 21-Å resolution with the penton base in yellow, the short reconstructed portion of the fiber in green, and the rest of the capsid in blue. (B) A reconstruction of the Ad12–$\alpha_v\beta_5$ complex at 24-Å resolution with the integrin in red. The reconstructions in (A) and (B) are both viewed along 3-fold symmetry axes. (C) A side view of the penton base, fiber, and integrin density at one vertex. Scale bars: 100 Å. [Reproduced with permission from Chiu *et al.* (1999).]

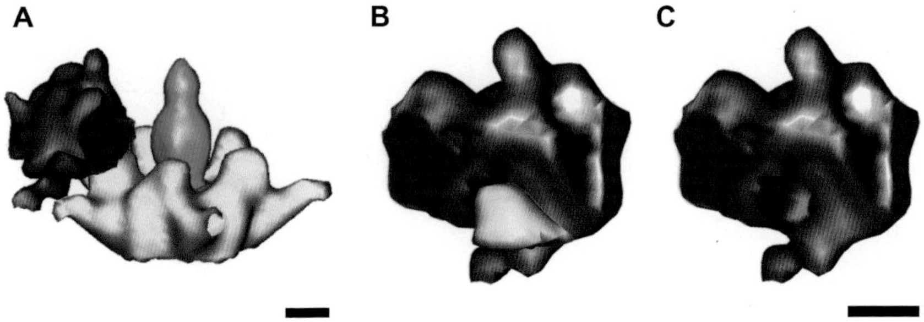

STEWART *ET AL.*, FIG. 12. The interaction between the integrin proximal domain and the Ad12 penton base protein. (A) The integrin density is shown extracted along estimated boundaries to model the proximal domain of one $\alpha_v\beta_5$ integrin heterodimer and shown with the penton base and fiber. (B) The modeled proximal domain rotated to show the interaction with a single penton base protrusion. (C) The same view as in (B), but with the protrusion removed to reveal the RGD-binding cleft. The color scheme is the same as in Fig. 11. Scale bars: 25 Å. [Reproduced with permission from Chiu *et al.* (1999).]

Belnap, D. M., Filman, D. J., Trus, B. L., Cheng, N., Booy, F. P., Conway, J. F., Curry, S., Hiremath, C. N., Tsang, S. K., Steven, A. C., and Hogle, J. M. (2000a). *J. Virol.* **74,** 1342–1354.

Belnap, D. M., McDermott, B. M., Jr., Filman, D. J., Cheng, N., Trus, B. L., Zuccola, H. J., Racaniello, V. R., Hogle, J. M., and Steven, A. C. (2000b). *Proc. Natl. Acad. Sci. USA* **97,** 73–78.

Bergelson, J. M., Shepley, M. P., Chan, B. M. C., Hemler, M. E., and Finberg, R. W. (1996). *Science* **255,** 1718–1720.

Bergelson, J. M., Cunningham, J. A., Droguett, G., Kurt-Jones, E. A., Krithivas, A., Hong, J. S., Horwitz, M. S., Crowell, R. L., and Finberg, R. W. (1997). *Science* **275,** 1320–1323.

Berget, S. M., Moore, C., and Sharp, P. A. (1977). *Proc. Natl. Acad. Sci. USA* **74,** 3171–3175.

Berinstein, A., Roivainen, M., Hovi, T., Mason, P. W., and Baxt, B. (1995). *J. Virol.* **69,** 2664–2666.

Bewley, M. C., Springer, K., and Zhang, Y.-B. (1999). *Science* **286,** 1579–1583.

Bodkin, D. K., and Fields, B. N. (1989). *J. Virol.* **63,** 1188–1193.

Bodkin, D. K., Nibert, M. L., and Fields, B. N. (1989). *J. Virol.* **63,** 4676–4681.

Borsa, J., Copps, T. P., Sargent, M. D., Long, D. G., and Chapman, J. D. (1973). *J. Virol.* **11,** 552–564.

Borsa, J., Morash, B. D., Sargent, M. D., Copps, T. P., Lievaart, P. A., and Szekely, J. G. (1979). *J. Gen. Virol.* **45,** 161–170.

Borsa, J., Sargent, M. D., Lievaart, P. A., and Copps, T. P. (1981). *Virology* **111,** 191–200.

Bouchard, M. J., and Racaniello, V. R. (1997). *J. Virol.* **71,** 2793–2798.

Brown, E. J., and Frazier, W. A. (2001). *Trends Cell Biol.* **11,** 130–135.

Bruder, J. T., and Kovesdi, I. (1997). *J. Virol.* **71,** 398–404.

Calderwood, D. A., Shattil, S. J., and Ginsberg, M. H. (2000). *J. Biol. Chem.* **275,** 22607–22610.

Casasnovas, J. M., Bickford, J. K., and Springer, T. A. (1998). *J. Virol.* **72,** 6244–6246.

Chandran, K., Walker, S. B., Chen, Y., Contreras, C. M., Schiff, L. A., Baker, T. S., and Nibert, M. L. (1999). *J. Virol.* **73,** 3941–3950.

Chandran, K., Zhang, X., Olson, N. H., Walker, S. B., Chappell, J. D., Dermody, T. S., Baker, T. S., and Nibert, M. L. (2001). *J. Virol.* **75,** 5335–5342.

Chang, C. T., and Zweerink, H. J. (1971). *Virology* **46,** 544–555.

Chang, M. W., Barr, E., Lu, M. M., Barton, K., and Leiden, J. M. (1995). *J. Clin. Invest.* **96,** 2260–2268.

Chappell, J. D., Gunn, V. L., Wetzel, J. D., Baer, G. S., and Dermody, T. S. (1997). *J. Virol.* **71,** 1834–1841.

Chappell, J. D., Barton, E. S., Smith, T. H., Baer, G. S., Duong, D. T., Nibert, M. L., and Dermody, T. S. (1998). *J. Virol.* **72,** 8205–8213.

Chappell, J. D., Duong, J. L., Wright, B. W., and Dermody, T. S. (2000). *J. Virol.* **74,** 8472–8479.

Chillon, M., Bosch, A., Zabner, J., Law, L., Armentano, D., Welsh, M. J., and Davidson, B. L. (1999). *J. Virol.* **73,** 2537–2540.

Chinnadurai, G. (1983). *Cell* **33,** 759–766.

Chiu, C. Y., Mathias, P., Nemerow, G. R., and Stewart, P. L. (1999). *J. Virol.* **73,** 6759–6768.

Chiu, C. Y., Wu, E., Brown, S. L., Von Seggern, D. J., Nemerow, G. R., and Stewart, P. L. (2001). *J. Virol.* **75,** 5375–5380.

Chow, L. T., Gelinas, R. E., Broker, T. R., and Roberts, R. J. (1977). *Cell* **12,** 1–8.

Chroboczek, J., Ruigrok, R. W. H., and Cusack, S. (1995). *Curr. Top. Microbiol. Immunol.* **199,** 163–200.

Chu, Y., Heistad, D. D., Cybulsky, M. I., and Davidson, B. L. (2001). *Arterioscler. Thromb. Vasc. Biol.* **21,** 238–242.

Clark, E. A., and Brugge, J. S. (1995). *Science* **268,** 233–239.

Connolly, J. L., Rodgers, S. E., Clarke, P., Ballard, D. W., Kerr, L. D., Tyler, K. L., and Dermody, T. S. (2000). *J. Virol.* **74,** 2981–2989.

Connolly, J. L., Barton, E. S., and Dermody, T. S. (2001). *J. Virol.* **75,** 4029–4039.

Cossart, P., Boquet, P., Normark, S., and Rappuoli, R. (1996). *Science* **271,** 315–316.

Cotten, M., and Weber, J. M. (1995). *Virology* **213,** 494–502.

Curry, S., Chow, M., and Hogle, J. M. (1996). *J. Virol.* **70,** 7125–7131.

Cuzange, A., Chroboczek, J., and Jacrot, B. (1994). *Gene* **146,** 257–259.

Debiasi, R. L., Edelstein, C. L., Sherry, B., and Tyler, K. L. (2001). *J. Virol.* **75,** 351–361.

Dechecchi, M. C., Tamanini, A., Bonizzato, A., and Cabrini, G. (2000). *Virology* **268,** 382–390.

Dermody, T. S., Nibert, M. L., Bassel-Duby, R., and Fields, B. N. (1990). *J. Virol.* **64,** 5173–5176.

DeTulleo, L., and Kirchhausen, T. (1998). *EMBO J.* **17,** 4585–4593.

DiGiovine, M., Salone, B., Martina, Y., Amati, V., Zambruno, G., and Cundari, E. (2001). *Virology* **282,** 102–112.

Dove, A. W., and Racaniello, V. R. (2000). *J. Virol.* **74,** 3929–3931.

Dryden, K. A., Wang, G., Yeager, M., Nibert, M. L., Coombs, K. M., Furlong, D. B., Fields, B. N., and Baker, T. S. (1993). *J. Cell Biol.* **122,** 1023–1041.

Dryden, K. A., Farsetta, D. L., Wang, G., Keegan, J. M., Fields, B. N., Baker, T. S., and Nibert, M. L. (1998). *Virology* **245,** 33–46.

Duggan, C., Maguire, T., McDermott, E., O'Higgins, N., Fennelly, J. J., and Duffy, M. J. (1995). *Int. J. Cancer* **61,** 597–600.

Duncan, R., and Lee, P. W. (1994). *Virology* **203,** 149–152.

Duncan, R., Horne, D., Cashdollar, L. W., Joklik, W. K., and Lee, P. W. (1990). *Virology* **174,** 399–409.

Duncan, R., Horne, D., Strong, J. E., Leone, G., Pon, R. T., Yeung, M. C., and Lee, P. W. (1991). *Virology* **182,** 810–819.

Ebbinghaus, C., Al-Jaibaji, A., Operschall, E., Schoffel, A., Peter, I., Greber, U. F., and Hemmi, S. (2001). *J. Virol.* **75,** 480–489.

Elkon, K. B., Liu, C.-C., Gall, J. G., Trevejo, J., Marino, M. W., Abrahamsen, K. A., Song, X., Zhou, J.-L., Old, L. J., Crystal, R. G., and Falck-Pedersen, E. (1997). *Proc. Natl. Acad. Sci. USA* **94,** 9814–9819.

Felding-Habermann, B., and Cheresh, D. A. (1993). *Curr. Opin, Cell Biol.* **5,** 864–868.

Filman, D. J., Syed, R., Chow, M., Macadam, A. J., Minor, P. D., and Hogle, J. M. (1989). *EMBO J.* **8,** 1567–1579.

Flint, S. J., Enquist, L. W., Krug, R. M., Racaniello, V. R., and Skalka, A. M. (2000). *In* "Principles of Virology" (S. J. Flint, L. W. Enquist, R. M. Krug, V. Racaniello, and A. M. Skalka, Eds.), pp. 101–131. ASM, Washington, D.C.

Fox, G., Parry, N. R., Barnett, P. V., McGinn, B., Rowlands, D. J., and Brown, F. (1989). *J. Gen. Virol.* **70,** 625–637.

Fraser, R. D., Furlong, D. B., Trus, B. L., Nibert, M. L., Fields, B. N., and Steven, A. C. (1990). *J. Virol.* **64,** 2990–3000.

Freimuth, P., Springer, K., Berard, C., Hainfeld, J., Bewley, M., and Flanagan, J. (1999). *J. Virol.* **73,** 1392–1398.

Fricks, C. E., and Hogle, J. M. (1990). *J. Virol.* **64,** 1934–1945.

Fry, E. E., Lea, S. M., Jackson, T., Newman, J. W. I., Ellard, F. M., Blakemore, W. E., Abu-Ghazeleh, R., Samuel, A., King, A. M. Q., and Stuart, D. I. (1999). *EMBO J.* **18,** 543–554.

Fujimoto, L. M., Roth, R., Heuser, J. E., and Schmid, S. L. (2000). *Traffic* **1,** 161–171.

Furlong, D. B., Nibert, M. L., and Fields, B. N. (1988). *J. Virol.* **73,** 246–256.

Gentsch, J. R., and Pacitti, A. F. (1987). *Virology* **161,** 245–248.

Geraghty, R. J., Krummenacher, C., Cohen, G. H., Eisenberg, R. J., and Spear, P. G. (1998). *Science* **280,** 1618–1620.

Greber, U. F., Webster, P., Helenius, A., and Weber, J. (1996). *EMBO J.* **15,** 1766–1777.

Greve, J. M., Davis, G., Meyer, A. M., Forte, C. P., Yost, S. C., Marlor, C. W., Kamarck, M. E., and McClelland, A. (1989). *Cell* **56,** 839–847.

Hall, A. (1998). *Science* **279,** 509–514.

Hannan, C., Raptis, L. H., Dery, C. V., and Weber, J. (1983). *InterVirology* **19,** 213–223.

Havenga, M. J., Lemckert, A. A., Grimbergen, J. M., Vogels, R., Huisman, L. G., Valerio, D., Bout, A., and Quax, P. H. (2001). *J. Virol.* **75,** 3335–3342.

He, Y., Bowman, V. D., Mueller, S., Bator, C. M., Bella, J., Peng, X., Baker, T. S., Wimmer, E., Kuhn, R. J., and Rossmann, M. G. (2000). *Proc. Natl. Acad. Sci. USA* **97,** 79–84.

Hidaka, C., Milano, E., Leopold, P. L., Bergelson, J. M., Hackett, N. R., Finberg, R. W., Wickham, T. J., Kovesdi, I., and Roelvink, P. (1999). *J. Clin. Invest.* **103,** 579–587.

Hierholzer, J. C. (1992). *Clin. Microbiol. Rev.* **5,** 262–274.

Hofer, F., Gruenberger, M., Kowalski, H., Machat, H., Huettinger, M., Kuechler, E., and Blass, D. (1994). *Proc. Natl. Acad. Sci. USA* **91,** 1839–1842.

Hoffman, G. R., and Cerione, R. A. (2000). *Cell* **102,** 403–406.

Hogle, J. M., Chow, M., and Filman, D. J. (1985). *Science* **229,** 1358–1365.

Holland, J. J. (1961). *Virology* **15,** 312–326.

Hooper, J. W., and Fields, B. N. (1996a). *J. Virol.* **70,** 672–677.

Hooper, J. W., and Fields, B. N. (1996b). *J. Virol.* **70,** 459–467.

Horwitz, M. S. (2001). *Virology* **279,** 1–8.

Huang, S., Endo, R. I., and Nemerow, G. R. (1995). *J. Virol.* **69,** 2257–2263.

Huang, S., Kamata, T., Takada, Y., Ruggeri, Z. M., and Nemerow, G. R. (1996). *J. Virol.* **70,** 4502–4508.

Huang, S., Stupack, D. G., Mathias, P., Wang, Y., and Nemerow, G. (1997). *Proc. Natl. Acad. Sci. USA* **94,** 8156–8161.

Huang, S., Reddy, V., Dasgupta, N., and Nemerow, G. R. (1999). *J. Virol.* **73,** 2798–2802.

Hynes, R. O. (1992). *Cell* **69,** 11–25.

Ireton, K., Payrastre, B., Chap, H., Ogawa, W., Sakaue, H., Kasuga, M., and Cossart, P. (1996). *Science* **274,** 780–782.

Jackson, T., Ellard, F. M., Ghazaleh, R. A., Brookes, S. M., Blakemore, W. E., Corteyn, A. H., Stuart, D. I., Newman, J. W., and King, A. M. (1996). *J. Virol.* **70,** 5282–5287.

Jackson, T., Sharma, A., Ghazaleh, R. A., Blakemore, W. E., Ellard, F. M., Simmons, D. L., Newman, J. W., Stuart, D. I., and King, A. M. (1997). *J. Virol.* **71,** 8357–8361.

Jackson, T., Blakemore, W., Newman, J. W., Knowles, N. J., Mould, A. P., Humphries, M. J., and King, A. M. (2000). *J. Gen. Virol.* **81,** 1383–1391.

Jakubczak, J. L., Rollence, M. L., Stewart, D. A., Jafari, J. D., Von Seggern, D. J., Nemerow, G. R., Stevenson, S. C., and Hallenbeck, P. L. (2001). *J. Virol.* **75,** 2972–2981.

Jane-Valbuena, J., Nibert, M. L., Spencer, S. M., Walker, S. B., Baker, T. S., Chen, Y., Centonze, V. E., and Schiff, L. A. (1999). *J. Virol.* **73,** 2963–2973.

Joklik, W. K., and Darnell, J. E. (1961). *Virology* **13,** 439–447.
Kafri, T., Morgan, D., Krahl, T., Sarvetnick, N., Sherman, L., and Verma, I. (1998). *Proc. Natl. Acad. Sci. USA* **95,** 11372–11382.
Kaplan, G., Freistadt, M. S., and Racaniello, V. R. (1990). *J. Virol.* **64,** 4697–4702.
Kaye, K. M., Spriggs, D. R., Bassel-Duby, R., Fields, B. N., and Tyler, K. L. (1986). *J. Virol.* **59,** 90–97.
King, S. L., Kamata, T., Cunningham, J. A., Emsley, J., Liddington, R. C., Takada, Y., and Bergelson, J. M. (1997). *J. Biol. Chem.* **272,** 28518–28522.
Koike, S., Ise, I., and Nomoto, A. (1991). *Proc. Natl. Acad. Sci. USA* **88,** 4104–4108.
Kolatkar, P. R., Bella, J., Olson, N. H., and Bator, C. M. (1999). *EMBO J.* **18,** 6249–6259.
Koonin, E. V. (1992). *Semin. Virol.* **3,** 327–340.
Krah, D. L., and Crowell, R. L. (1982). *Virology* **118,** 148–156.
Krasnykh, V. N., Mikheeva, G. V., Douglas, J. T., and Curiel, D. T. (1996). *J. Virol.* **70,** 6839–6846.
Lentz, K. N., Smith, A. D., Geisler, S. C., Cox, S., Buontempo, P., Skelton, A., DeMartino, J., Rozhon, E., Schwartz, J., Girijavallabhan, V., O'Connell, J., and Arnold, E. (1997). *Structure* **5,** 961–978.
Leon, R. P., Hedlund, T., Meech, S. J., Li, S., Schaack, J., Hunger, S. P., Duke, R. C., and DeGregori, J. (1998). *Proc. Natl. Acad. Sci. USA* **95,** 13159–13164.
Leone, G., Mah, D. C., and Lee, P. W. (1991). *Virology* **182,** 346–350.
Lerner, A. M., Cherry, J. D., and Finland, M. (1963). *Virology* **19,** 58–65.
Li, E., Stupack, D., Bokoch, G., and Nemerow, G. R. (1998a). *J. Virol.* **72,** 8806–8812.
Li, E., Stupack, D., Cheresh, D., Klemke, R., and Nemerow, G. (1998b). *J. Virol.* **72,** 2055–2061.
Li, E., Stupack, D. G., Brown, S. L., Klemke, R., Schlaepfer, D. D., and Nemerow, G. R. (2000a). *J. Biol. Chem.* **275,** 14729–14735.
Li, E., Stupack, D. G., Brown, S. L., Klemke, R., Schlaepfer, D. D., and Nemerow, G. R. (2000b). *Gene Ther.* **7,** 1593–1599.
Li, E., Brown, S. L., Stupack, D. G., Puente, X. S., Cheresh, D. A., and Nemerow, G. R. (2001). *J. Virol.* **75,** 5405–5409.
Liu, S., Calderwood, D. A., and Ginsberg, M. H. (2000). *J. Cell Sci.* **113,** 3563–3571.
Loeffler, F., and Frosch, P. (1964). Report of the commission for research on foot-and-mouth disease. *In* "Selected Papers on Virology" (N. Hahon, Ed.), pp. 64–68. Prentice-Hall, Englewood Cliffs, NJ.
Lonberg-Holm, K., Crowell, R. L., and Philipson, L. (1976). *Nature* **259,** 679–681.
Lucia-Jandris, P., Hooper, J. W., and Fields, B. N. (1993). *J. Virol.* **67,** 5339–5345.
Marks, B., Stowell, M. H., Vallis, Y., Mills, I. G., Gibson, A., Hopkins, C. R., and McMahon, H. T. (2001). *Nature* **8,** 231–235.
Martinez, M. A., Verdaguer, N., Mateu, M. G., and Domingo, E. (1997). *Proc. Natl. Acad. Sci. USA* **94,** 6798–6802.
Martin-Padura, I., Lostaglio, S., Schneemann, M., Williams, L., Romano, M., Fruscella, P., Panzeri, C., Stoppacciaro, A., Ruco, L., Villa, A., Simmons, D., and Dejana, E. (1998). *J. Cell Biol.* **142,** 117–127.
Mason, P. W., Rieder, E., and Baxt, B. (1994). *Proc. Natl. Acad. Sci. USA* **91,** 1932–1936.
Mathias, P., Wickham, T. J., Moore, M., and Nemerow, G. (1994). *J. Virol.* **68,** 6811–6814.
Mendelsohn, C., Johnson, B., Lionetti, K. A., Nobis, P., Wimmer, E., and Racaniello, V. R. (1986). *Proc. Natl. Acad. Sci. USA* **83,** 7845–7849.
Mendelsohn, C. L., Wimmer, E., and Racaniello, V. R. (1989). *Cell* **56,** 855–865.
Metcalf, P., Cyrklaff, M., and Adrian, M. (1991). *EMBO J.* **10,** 3129–3136.

Miller, L. C., Blakemore, W., Sheppard, D., Atakilit, A., King, A. M., and Jackson, T. (2001). *J. Virol.* **75,** 4158–4164.

Morrison, L. A., Sidman, R. L., and Fields, B. N. (1991). *Proc. Natl. Acad. Sci. USA* **88,** 3852–3856.

Morrison, M. E., and Racaniello, V. R. (1992). *J. Virol.* **66,** 2807–2813.

Moscufo, N., Yafal, A. G., Rogove, A., Hogle, J., and Chow, M. (1993). *J. Virol.* **67,** 5075–5078.

Nabel, G. J. (1999). *Proc. Natl. Acad. Sci. USA* **96,** 324–326.

Neff, S., and Baxt, B. (2001). *J. Virol.* **75,** 527–532.

Neff, S., Sa-Carvalho, D., Rieder, E., Mason, P. W., Blystone, S. D., Brown, E. J., and Baxt, B. (1998). *J. Virol.* **72,** 3587–3594.

Nemerow, G. R., and Stewart, P. L. (1999). *Microbiol. Mol. Biol. Rev.* **63,** 725–734.

Nemerow, G. R., and Stewart, P. L. (2001). *Virology* **288,** 189–191.

Nemerow, G. R., Wickham, T. J., and Cheresh, D. A. (1993). *In* "Biology of Vitronectins and Their Receptors" (K. T. Preissner, S. Rosenblatt, C. Kost, J. Wegerhoff, and D. F. Mosher, Eds.), pp. 177–184. Elsevier Science, Amsterdam.

Nibert, M. L., and Schiff, L. A. (2001). *In* "Fields Virology" (B. N. Fields, D. M. Knipe, and P. M. Howley, Eds.), pp. 1679–1728. Raven, New York.

Nibert, M. L., Dermody, T. S., and Fields, B. N. (1990). *J. Virol.* **64,** 2976–2989.

Nibert, M. L., Chappell, J. D., and Dermody, T. S. (1995). *J. Virol.* **69,** 5057–5067.

Oberhaus, S. M., Smith, R. L., Clayton, G. H., Dermody, T. S., and Tyler, K. L. (1997). *J. Virol.* **71,** 2100–2106.

Oliveira, M. A., Zhao, R., Lee, V. M., Kremer, M. J., Minor, I., Rueckert, R. R., Diana, G. D., Pevear, D. C., Dutko, F. J., and McKinlay, M. A. (1993). *Structure* **1,** 51–68.

Olland, A. M., Jane-Valbuena, J., Schiff, L. A., Nibert, M. L., and Harrison, S. C. (2001). *EMBO J.* **20,** 979–989.

Olson, N. H., Kolatkar, P. R., Oliveira, M. A., Cheng, R. H., Greve, J. M., McClelland, A., Baker, T. S., and Rossmann, M. G. (1993). *Proc. Natl. Acad. Sci. USA* **90,** 507–511.

Patterson, S., and Russell, W. C. (1983). *J. Gen. Virol.* **64,** 1091–1099.

Paul, R. W., and Lee, P. W. (1987). *Virology* **159,** 94–101.

Perez, L., and Carrasco, L. (1993). *J. Virol.* **67,** 4543–4548.

Pickrell, J., and Enserink, M. (2001). *Science* **291,** 1677.

Pierschbacher, M. D., and Ruoslahti, E. (1984). *Nature* **309,** 30–33.

Potts, J. R., and Campbell, I. D. (1994). *Curr. Opin. Cell Biol.* **6,** 648–655.

Prchla, E., Plank, C., Wagner, E., Blaas, D., and Fuchs, R. (1995). *J. Cell Biol.* **131,** 111–123.

Racaniello, V. R. (1996). *Proc. Natl. Acad. Sci. USA* **93,** 11378–11381.

Racaniello, V. R. (2001). *In* "Fields Virology" (D. M. Knipe and P. M. Howley, Eds.), pp. 685–722. Lippincott Williams & Wilkins, Philadelphia.

Rauma, T., Tuukkanen, J., Bergelson, J. M., Denning, G., and Hautala, T. (1999). *J. Virol.* **73,** 9664–9668.

Reinisch, K. M., Nibert, M. L., and Harrison, S. C. (2000). *Nature* **404,** 960–967.

Ren, R., and Racaniello, V. R. (1992). *J. Virol.* **66,** 296–304.

Rodgers, S. E., Barton, E. S., Oberhaus, S. M., Pike, B., Gibson, C. A., Tyler, K. L., and Dermody, T. S. (1997). *J. Virol.* **71,** 2540–2546.

Roelvink, P. W., Lizonova, A., Lee, J. G. M., Li, Y., Bergelson, J. M., Finberg, R. W., Brough, D. E., Kovesdi, I., and Wickham, T. J. (1998). *J. Virol.* **72,** 7909–7915.

Roelvink, P. W., Lee, G. M., Einfeld, D. A., Kovesdi, I., and Wickham, T. J. (1999). *Science* **286,** 1568–1571.

Roivainen, M., Piirainen, L., Hovi, T., Vitanen, I., Riikonen, T., Heino, T., and Hyypia, T. (1994). *Virology* **203,** 357–365.
Rosengart, T. K., Lee, L. Y., Patel, S. R., Kligfield, P. D., Okin, P. M., Hackett, N. R., Isom, O. W., and Crystal, R. G. (1999). *Ann. Surg.* **230,** 466–472.
Rossmann, M. G., Arnold, E., Erickson, J. W., Frankenberger, E. A., Griffith, J. P., Hecht, H.-J., Johnson, J. E., Kamer, G., Luo, M., Mosser, A. G., Rueckert, R. R., Sherry, B., and Vriend, G. (1985). *Nature* **317,** 145–153.
Rossmann, M. G., Bella, J., Kolatkar, P. R., He, Y., Wimmer, E., Kuhn, R. J., and Baker, T. S. (2000). *Virology* **269,** 239–247.
Rubin, D. H., Wetzel, J. D., Williams, W. V., Cohen, J. A., Dworkin, C., and Dermody, T. S. (1992). *J. Clin. Invest.* **90,** 2536–2542.
Sa-Carvalho, D., Rieder, E., Baxt, B., Rodarte, R., Tanuri, A., and Mason, P. W. (1997). *J. Virol.* **71,** 5115–5123.
Schmidt, M. R., Pickos, B., Cabatingan, M. S., and Woodland, R. T. (2000). *J. Immunol.* **165,** 4112–4119.
Schober, D., Kronenberger, P., Prchla, E., Blass, D., and Fuchs, R. (1998). *J. Virol.* **72,** 1354–1364.
Seth, P., Fitzgerald, D., Ginsberg, H., Willingham, M., and Pastan, I. (1984). *Mol. Cell Biol.* **4,** 1528–1533.
Seth, P., Pastan, I., and Willingham, M. C. (1985). *J. Biol. Chem.* **260,** 9598–9602.
Seth, P., Pastan, I., and Willingham, M. C. (1987). *J. Virol.* **61,** 883–888.
Sever, S., Muhlberg, A. B., and Schmid, S. L. (1999). *Nature* **398,** 481–486.
Shen, Y., and Shenk, T. E. (1995). *Curr. Opin. Genet. Dev.* **5,** 105–111.
Shenk, T. E. (2001). *In* "Fields Virology" (D. M. Knipe and P. M. Howley, Eds.), pp. 2265–2300. Lippincott Williams & Wilkins, Philadelphia.
Shepley, M. P., Sherry, B., and Weiner, H. L. (1988). *Proc. Natl. Acad. Sci. USA* **85,** 7743–7747.
Silverstein, S. C., Astell, C., Levin, D. H., Schonberg, M., and Acs, G. (1972). *Virology* **47,** 797–806.
Spriggs, D. R., and Fields, B. N. (1982). *Nature* **297,** 68–70.
Spriggs, D. R., Kaye, K., and Fields, B. N. (1983). *Virology* **127,** 220–224.
Staunton, D. E., Dustin, M. L., and Springer, T. A. (1989a). *Nature* **339,** 61–64.
Staunton, D. E., Merluzzi, V. J., Rothlein, R., Barton, R., Marlin, S. D., and Springer, T. A. (1989b). *Cell* **56,** 849–853.
Staunton, D. E., Dustin, M. L., Erickson, H. P., and Springer, T. A. (1990). *Cell* **61,** 243–254.
Staunton, D. E., Gaur, A., Chan, P. Y., and Springer, T. A. (1992). *J. Immunol.* **148,** 3271–3274.
Stevenson, S. C., Rollence, M., White, B., Weaver, L., and McClelland, A. (1995). *J. Virol.* **69,** 2850–2857.
Stewart, P. L., Burnett, R. M., Cyrklaff, M., and Fuller, S. D. (1991). *Cell* **67,** 145–154.
Stewart, P. L., Chiu, C., Huang, S., Muir, T., Zhao, Y., Chait, B., Mathias, P., and Nemerow, G. (1997). *EMBO J.* **16,** 1189–1198.
Strong, J. E., Leone, G., Duncan, R., Sharma, R. K., and Lee, P. W. (1991). *Virology* **184,** 23–32.
Sturzenbecker, L. J., Nibert, M., Furlong, D., and Fields, B. N. (1987). *J. Virol.* **61,** 2351–2361.
Teodoro, J. G., and Branton, P. E. (1997). *J. Virol.* **71,** 1739–1746.
Tomassini, J. E., Graham, D., DeWitt, C. M., Lineberger, D. W., Rodkey, J. A., and Colonno, R. J. (1989). *Proc. Natl. Acad. Sci. USA* **86,** 4907–4911.

Tomko, R. P., Xu, R., and Philipson, L. (1997). *Proc. Natl. Acad. Sci. USA* **94,** 3352–3356.

Tosteson, M. T., Nibert, M. L., and Fields, B. N. (1993). *Proc. Natl. Acad. Sci. USA* **90,** 10549–10552.

Traenckner, E. B., Pahl, H. L., Henkel, T., Schmidt, K. N., Wilk, S., and Baeuerle, P. A. (2001). *EMBO J.* **14,** 2876–2883.

Tyler, K. L. (2001). *In* "Fields Virology" (B. N. Fields, D. M. Knipe, and P. M. Howley, Eds.), pp. 1729–1745. Raven, New York.

Tyler, K. L., McPhee, D. A., and Fields, B. N. (1986). *Science* **233,** 770–774.

Tyler, K. L., Squier, M. K. T., Rodgers, S. E., Schneider, B. E., Oberhaus, S. M., Grdina, T. A., Cohen, J. J., and Dermody, T. S. (1995). *J. Virol.* **69,** 6972–6979.

Vigne, E., Mahfouz, I., Dedieu, J.-F., Brie, A., Perricaudet, M., and Yeh, P. (1999). *J. Virol.* **73,** 5156–5161.

Virgin, H. W., Tyler, K. L., and Dermody, T. S. (1997). *In* "Viral Pathogenesis" (N. Nathanson, Ed.), pp. 669–699. Lippincott-Raven, New York.

Von Seggern, D. J., Chiu, C. Y., Fleck, S. K., Stewart, P. L., and Nemerow, G. R. (1999). *J. Virol.* **73,** 1601–1608.

Von Seggern, D. J., Huang, S., Fleck, S. K., Stevenson, S. C., and Nemerow, G. R. (2000). *J. Virol.* **74,** 354–362.

Vuori, K., Hirai, H., Aizawa, S., and Ruoslahti, E. (1996). *Mol. Cell Biol.* **16,** 2606–2613.

Wang, K., Huang, S., Kapoor-Munshi, A., and Nemerow, G. R. (1998). *J. Virol.* **72,** 3455–3458.

Wang, K., Guan, T., Cheresh, D. A., and Nemerow, G. R. (2000). *J. Virol.* **74,** 2731–2739.

Wang, X., and Bergelson, J. M. (1999). *J. Virol.* **73,** 2559–2562.

Weiner, H. L., Drayna, D., Averill, D. R., and Fields, B. N. (1977). *Proc. Natl. Acad. Sci. USA* **74,** 5744–5748.

Weiner, H. L., Ault, K. A., and Fields, B. N. (1980). *J. Immunol.* **124,** 2143–2148.

Wickham, T. J., Mathias, P., Cheresh, D. A., and Nemerow, G. R. (1993). *Cell* **73,** 309–319.

Wickham, T. J., Filardo, E. J., Cheresh, D. A., and Nemerow, G. R. (1994). *J. Cell Biol.* **127,** 257–264.

Williams, L. A., Martin-Padura, I., Dejana, E., Hogg, N., and Simmons, D. L. (1999). *Mol. Immunol.* **36,** 1175–1188.

Wilson, J. M. (1995). *J. Clin. Invest.* **96,** 2547–2554.

Wilson, J. M. (1996). *N. Engl. J. Med.* **334,** 1185–1187.

Wilson, J. M. (1998). *Mol. Med.* **334,** 1185–1187.

Wolf, J. L., Rubin, D. H., Finberg, R., Kauffman, R. S., Sharps, A. H., and Trier, J. S. (1981). *Science* **212,** 471–472.

Wu, E., Fernandez, J., Fleck, S. K., Von Seggern, D. J., Huang, S., and Nemerow, G. R. (2001). *Virology* **279,** 78–89.

Xia, D., Henry, L. J., Gerard, R. D., and Deisenhofer, J. (1994). *Structure* **2,** 1259–1270.

Yang, X. J., Ogryzko, V. V., Nishikawa, J., Howard, B. H., and Nakatani, Y. (1996). *Nature* **382,** 319–324.

Zhang, S., and Racaniello, V. R. (1997). *J. Virol.* **71,** 4915–4920.

Zhao, R., Pevear, D. C., Kremer, M. J., Giranda, V. L., Kofron, J. A., Kuhn, R. J., and Rossmann, M. G. (1996). *Structure* **4,** 1205–1220.

Zsengellér, Z., Otake, K., Hossain, S.-A., Berclaz, P.-Y., and Trapnell, B. C. (2000). *J. Virol.* **74,** 9655–9667.

AUTHOR INDEX

A

Aaskov, J. G., 423
Aasted, B., 263
Abad-Zapatero, C., 131, 136, 205, 220
Abdel-Meguid, S. S., 131, 136, 147, 205, 220, 281
Abhlquist, P., 186
Abrahamsberg, C., 424
Abrahamsen, K. A., 480
Abrams, C. C., 3, 225
Abu-Ghazaleh, R., 54, 79, 81, 134, 155, 432, 433, 477, 478, 479
Acharya, R., 38, 39, 54, 79, 81, 134, 154, 155, 223, 422, 432, 485
Acosta-Rivero, N., 3
Acs, G., 457
Adachi, T., 137
Adams, M. B., 271, 275
Adams, P. D., 162
Adams, S. E., 12, 48
Addison, C., 28
Adler, C. J., 221
Adolph, K. W., 305
Adrian, M., 48, 49, 221, 389, 390, 458
Aebi, U., 400
Aebischer, P., 339
Agadjanyan, M., 15
Agamalyan, M. M., 315
Agard, D. A., 157, 336
Agbandje, M., 130, 144, 168, 186, 239
Agbandje-McKenna, M., 38, 56, 80, 81, 144, 168, 236, 239, 240, 263
Agrawal, R. K., 43, 46, 47, 209
Agresta, B. E., 3
Ahlquist, P., 198, 226
Ahn, A., 356, 373
Ailey, B., 120
Air, G. M., 236, 265, 284, 285, 414
Aitken, J., 384
Aizawa, S., 350, 487
Albert, F. G., 205
Albinsson, B., 483
Aldrich, R., 48
Alexander, R. S., 134
Alexandratos, J., 151
Alford, D., 327
Al-Jaibaji, A., 480
Allan, J. S., 349
Allison, S. L., 76, 354, 355, 356, 368, 369, 373
Almond, J. W., 154
Alsonso, A., 23
Alwell-warda, K., 366
Amano, H., 326
Amarasinghe, G. K., 76
Amati, V., 486
Amberg, S. M., 371
Amos, L. A., 101, 173, 221
Anderson, D. L., 38, 39, 53, 58, 59, 60, 146, 168, 176, 225, 226, 243, 244, 246, 271, 274, 304, 306, 403
Anderson, M. M., 349
Anderson, W. F., 339, 348
Andersson, H., 337
Andree, P. J., 326
Andreu, E., 431, 432, 433
Andrew, P. W., 43, 51
Andries, K., 422
Angenon, G., 13
Antczak, J. B., 350
Aoyama, A., 236, 238, 261, 269, 270
Aponte, C., 235
Appella, E., 182
Apweiler, R., 139, 141, 142, 144, 148, 151
Aravind, L., 381, 382, 388
Argos, P., 158, 226
Arisaka, F., 60, 65, 145, 176, 392
Arita, M., 422
Armentano, D., 482
Armstrong, R. T., 71, 327, 342

Arnberg, N., 482
Arnold, E., 38, 79, 127, 130, 131, 134, 154, 155, 186, 418, 421, 422, 471, 491, 473
Arntzen, C. J., 14, 26
Arrobio, J. O., 428
Arrowsmith, C. H., 146
Arthur, L. O., 151, 182
Ashrafian, H., 5
Ashton, A. C., 43, 47
Asso, J., 223, 225
Astell, C., 457
Atakilit, A., 477
Athappilly, F. K., 65, 147, 173, 185
Atkins, J. F., 315
Aubert, H., 55
Aubrey, K. L., 319, 320
Auer, H., 425
Ault, K. A., 458
Averill, D. R., 457, 458
Axelson, P. H., 327
Azzi, A., 144, 168, 169

B

Bacher, A., 55
Bachmann, L., 55
Bachrach, E., 335
Badasso, M. O., 38, 39, 58, 59, 60, 146, 168, 176, 243, 304, 403
Bae, Y., 348
Baer, G. S., 436, 457, 461, 462, 463, 464
Baeuerle, P. A., 465
Bai, M., 483
Baird, T., Jr., 342
Bairoch, A., 142, 144, 148, 151
Baker, A. T., 414
Baker, D., 336
Baker, K. A., 140, 335, 350, 351, 353
Baker, M. L., 55, 58, 72, 94, 98, 102, 105, 115, 117, 118, 119, 120, 121, 230, 385, 391, 394, 395
Baker, T. S., 5, 15, 27, 38, 39, 43, 54, 56, 58, 59, 60, 61, 63, 67, 68, 78, 79, 80, 81, 82, 94, 97, 105, 111, 113, 118, 128, 130, 137, 138, 142, 144, 145, 146, 155, 162, 167, 168, 169, 171, 176, 198, 205, 206, 236, 237, 238, 239, 244, 246, 261, 263, 265, 274, 285, 294, 302, 304, 306, 307, 313, 319, 354, 365, 366, 368, 369, 370, 372, 389, 390, 391, 392, 393, 401, 411, 416, 417, 418, 419, 421, 423, 424, 431, 433, 435, 436, 437, 438, 439, 458, 459, 460, 461, 470, 473, 474, 489
Baldwin, J. M., 121
Baldwin, P. R., 102, 109, 117
Ball, J. M., 14, 25, 49, 223
Ball, L. A., 135, 203
Ballard, D. W., 457, 465
Baltimore, D., 199, 223, 225
Bamford, D. H., 39, 45, 53, 64, 65, 67, 146, 165, 175, 185, 230, 234, 235, 246, 261, 302, 319, 381
Bamford, J. H., 53, 67
Bamford, J. K. H., 64, 65, 146, 175, 185, 246, 261, 302, 381
Ban, N., 47, 76, 153, 154, 222
Bancroft, D. P., 74, 75, 77, 151, 181
Bancroft, J. B., 204
Banerjea, A. C., 435, 458
Banerjee, A. K., 232
Baranowski, E., 80, 432
Barba, A. P., 55
Barioch, A., 139, 141
Barker, P., 416
Barlow, P. N., 181, 293, 399
Barnett, A. L., 347, 348, 349, 350
Barnett, P. V., 155, 474, 476, 477
Barr, E., 480
Barrell, B. G., 236, 265, 284, 285
Barrett, A. D. T., 366
Barry, G. F., 9
Barth, B. U., 337
Barthelemy, I., 303, 304, 305, 403
Bartlett, N. M., 232
Bartman, G., 15, 31
Barton, E. S., 435, 436, 457, 460, 461, 462–82, 464, 465
Barton, G., 181
Barton, K., 480
Barton, R., 491
Basak, A. K., 45, 54, 68, 71, 72, 142, 163, 164, 165, 230
Basavappa, R., 225
Baschong, W., 241, 315
Bashford, D., 265
Bassel-Duby, R., 436, 461, 462, 463, 464
Bates, P., 344, 356
Bator, C. M., 79, 80, 155, 418, 470, 473, 474, 476, 489

Battini, J. L., 347
Battiste, J. L., 151
Baudoux, P., 230
Bauer, A. C., 217, 399
Baumann, S., 5
Baumert, T. F., 3
Bax, A., 151
Baxa, U., 176, 177
Baxt, B., 422, 470, 477
Bayley, P. M., 160, 335, 356
Bazan, J. F., 132, 186
Bazinet, C., 243, 277
Beachy, R. N., 15, 136, 206
Bean, W. F., 55, 238
Beasley, D. W. C., 366
Beaudet-Miller, M., 269
Becker, B., 390
Beckman, E., 43, 46, 121
Beckman, R., 48
Belanger, H., 15, 31
Bell, J. A., 23
Bella, J., 3, 54, 79, 80, 135, 154, 155, 226, 418, 470, 473, 474, 476, 489
Bellamy, A. R., 68, 74, 221, 230, 232, 234
Bellamy, H., 143
Belnap, D. M., 56, 65, 79, 155, 307, 430, 467, 468, 470, 471
Belsham, G. J., 3, 225
Belyaev, A. S., 9, 74, 77
Benbasat, J., 277
Ben-Efraim, I., 337
Benjamin, T. L., 39, 148, 171, 215
Benner, W. H., 52
Bennett, M. J., 139
Benson, S. D., 64, 65, 146, 175, 185, 246, 302, 381
Bentley, G. A., 50, 221
Bentz, J., 327, 337
Berard, C., 480
Berclaz, P. Y., 487
Berg, J. M., 182, 315, 342
Bergelson, J. M., 134, 479, 480, 481, 482, 487
Berger, E. Al., 345
Berger, J. M., 74, 76, 151, 182, 184, 340, 341, 347, 348
Berget, S. M., 480
Bergmann, E. M., 134
Bergoin, M., 67, 389, 390, 391
Berinstein, A., 477
Berman, H. M., 138, 139, 141, 142, 144, 148, 151
Bernal, R. A., 143, 170, 261, 266, 281, 285, 289, 293
Bernhardt, T. G., 270, 421
Berns, K. I., 167
Berriman, J. A., 61, 68, 354
Berthet-Colominas, C., 48, 65, 74, 75, 77
Bess, J. W., 76, 151, 182
Bess, J. W., Jr., 76
Betenbaugh, M., 3
Betzel, C., 425
Beuron, F., 305
Bewley, M. C., 65, 147, 480, 481, 482
Bhat, T. N., 138, 139, 141, 142, 144, 148, 151
Bhatia, S., 144, 168, 169
Bhella, D., 302, 384, 389
Bhuvaneswari, M., 136
Bhyravbhatla, B., 137, 158
Bibbs, L., 200, 201, 422, 460
Bickford, J. K., 473
Bigot, D., 71
Bina, M., 111, 113
Binley, J. M., 183
Bisaillon, M., 230
Bischoff, H., 20, 365, 370
Bisher, M. E., 217, 220, 310, 399, 402
Bishop, D. H. L., 29, 126, 138, 139, 141, 142, 144, 148, 151, 171, 225, 229, 235, 239, 380, 413, 414, 416, 421, 425, 430, 431, 433, 435
Bishop, N. E., 225, 226
Bitter, G. A., 12
Bittle, J., 199
Bittman, R., 354, 373
Bizebard, T., 334
Blaas, D., 80, 81, 82, 134, 422, 424, 425
Black, L. W., 217, 220, 241, 242, 243, 260, 285, 301, 302, 313, 315, 383, 392, 395, 396, 398, 399, 402
Blacklow, S. C., 340, 341, 342
Blake, P. R., 76, 151, 182
Blakemore, W. E., 54, 79, 81, 134, 155, 431, 432, 433, 477, 478, 479
Blanc, E., 127, 134
Blasco, R., 381
Blass, D., 476
Blobel, G., 48
Bloch, J., 339
Bloom, M. E., 263

Bloomer, A. C., 52, 132, 158
Blow, D. M., 132, 157
Blumberg, B. S., 430
Blumenthal, R., 53, 332, 346
Blystone, S. D., 477
Bodian, D. L., 334
Bodkin, D. K., 457, 461
Boege, U., 20, 131, 134, 157, 365, 369, 370, 423
Boere, W., 387
Boerger, A. L., 344
Boeyé, A., 410, 412
Bokoch, G., 486, 487
Bollinger, R. C., 412
Bomu, W., 131, 134, 135, 153, 154, 209, 223, 224, 226, 416
Bonami, J. R., 55, 381
Bone, R., 157
Bonev, B. B., 51
Bong, D. T., 227
Bonizzato, A., 483
Bonneau, C., 206
Bonneau, P. R., 281
Bonnez, W., 22
Booth, C. R., 117, 263
Booth, T. F., 30, 65, 68, 72, 73, 230
Booth, V., 146
Booy, F. P., 1, 22, 23, 27, 28, 94, 99, 103, 105, 171, 230, 242, 244, 263, 266, 269, 285, 291, 302, 303, 305, 309, 315, 316, 317, 319, 327, 332, 381, 391, 392, 396, 398, 399, 400, 402, 427, 431, 433, 434, 467, 468
Boquet, P., 476
Borer, P. N., 50, 75, 76, 181, 182
Borisov, A. V., 230
Borisy, G. G., 435
Borsa, J., 436, 457
Bosch, A., 482
Bösecke, P., 334
Boshuizen, R. S., 15
Boson, B., 349, 350
Bothner, B., 52, 126, 200, 201, 202, 422, 460
Botstein, D., 270, 271, 280
Böttcher, B., 46, 55, 57, 94, 97, 99, 104, 105, 114, 120, 128, 221, 230, 385
Böttcher, C., 327, 333
Bouchard, M. J., 470
Boulanger, P., 77
Boulton, M. I., 38, 56, 236
Bourne, P. E., 138, 139, 141, 142, 144, 148, 151
Bout, A., 482
Boutell, C. J., 293, 399
Bowen, Z., 237
Bowman, B. R., 55, 79, 80, 143, 169, 237, 238, 261, 265, 266, 281, 285, 293
Bowman, V. D., 155, 431, 437, 438, 470, 473, 474
Boy De La Tour, E., 260
Boyd, J., 181
Braakman, I., 338
Bracker, C. E., 204
Brändén, C. I., 159, 182
Brandt, C. D., 428
Branton, P. E., 464
Bravo, J., 431, 460
Brayer, G. D., 157
Brechling, K. A., 435, 458
Breeze, A. L., 50
Breitburd, F., 23, 434
Brennan, F. R., 15
Brenner, S. E., 120
Bressanelli, S., 137
Bricogne, G., 39, 49, 50, 52, 132, 136, 153, 158, 220
Brie, A., 486
Brigham-Burke, M., 345
Brioen, P., 410
Brocchi, E., 134, 155, 412
Broderson, D. E., 47, 76
Broering, T. J., 230
Broker, T. R., 480
Bron, P., 77, 354
Bronnert, C., 3
Bron, R., 373
Brookes, S. M., 81, 431, 432, 433, 477
Brough, D. E., 384, 482
Brouillette, C. G., 278, 279, 280
Brown, C. S., 3
Brown, D. R., 12, 22
Brown, D. T., 38, 54, 61, 63, 158, 285, 294, 354, 365, 366, 368, 387, 423
Brown, E. B., 160, 331, 334, 335
Brown, E. J., 477
Brown, F., 38, 39, 54, 79, 81, 134, 154, 155, 223, 422, 432, 474, 476, 477, 485
Brown, J. C., 15, 27, 28, 171, 244, 245, 263, 266, 268, 269, 271, 285, 291, 302, 303,

319, 391, 392, 396, 398, 399, 400, 402, 403, 427, 433
Brown, J. H., 159
Brown, K. E., 80, 81
Brown, N. L., 236, 265, 284, 285
Brown, S. L., 480, 482, 483, 487
Brown, T. D., 134, 147
Brown, W. D., 339
Brown, Z., 261, 265
Brugge, J. S., 486
Brugidou, C., 52, 136, 206
Brunger, A. T., 162
Brunner, J., 327, 335
Bryant, J. L., Jr., 277
Bryon, O., 51
Bu, W., 144, 168, 169
Buck, D. W., 349
Budahazi, G., 132, 134
Budisa, N., 176, 177
Budkowska, A, 3
Bujacz, G., 151
Bulla, L. A., 389
Bullivant, S., 232
Bullough, P. A., 160, 327, 331, 333
Bundule, M., 133, 152
Buontempo, P., 471, 491
Burbank, D. E., 381, 390
Burbea, M., 266, 278, 283, 309, 310
Burch, A. D., 143, 170, 244, 245, 261, 264, 265, 266, 267, 268, 269, 270, 281, 284, 285, 287, 289, 291, 293, 392, 396, 403
Burgeson, R. E., 347
Burley, S. K., 388
Burmeister, W. P., 147
Burness, A. T., 423
Burnett, R. M., 1, 45, 64, 65, 67, 130, 146, 147, 172, 173, 175, 185, 246, 302, 319, 381, 385, 429, 482, 484
Burnette, W. N., 12
Burns, J. W., 414, 415
Burns, N. R., 48, 181
Burns, C. C., 349
Burroughs, J. N., 1, 39, 41, 48, 54, 57, 61, 65, 68, 69, 70, 71, 72, 73, 119, 142, 164, 221, 230, 231, 232, 234, 294
Burstin, S. J., 436
Burton, D. R., 459, 461
Bushweller, J. H., 141, 160
Bustamante, C., 53, 60, 243, 244, 403
Butcher, S. J., 39, 43, 45, 61, 64, 65, 67, 94, 230, 234, 246, 319, 354, 384
Butler, P. G., 52, 228
Buzaite, O., 12

C

Cabatingan, M. S., 482
Cable, M. B., 137
Cabrini, G., 483
Caffrey, M., 74, 76, 151, 182, 340, 341, 342
Cafiso, D. S., 141, 160
Cai, M, 74, 76, 151, 182, 340, 341, 342
Cai, Z., 65, 147, 173, 185
Caldentey, J., 39, 65
Calder, L. J., 331, 334, 335, 340, 353
Calderwood, D. A., 486, 488
Calendar, R., 241, 315
Calisher, C. H., 423
Callahan, P. L., 411, 412, 418, 420
Calvo, V., 381
Campbell, I. D., 181, 479
Campbell, S., 5
Campos-Olivas, R., 151, 180
Campusano, G., 285, 393, 401, 402
Canady, M. A., 25, 138, 198, 207, 209, 210, 303
Canziani, G., 345
Capucci, L., 412
Carazo, J. M., 59, 277, 304, 381, 390
Carfi, A., 140, 147, 162, 163, 186, 332, 341, 350
Carlson, J., 421
Carmichael, J., 5
Carr, C. M., 331, 335, 337, 356
Carr, P. A., 339
Carrasco, L., 467, 474
Carrascosa, A. L., 381, 390
Carrascosa, J. L., 59, 146, 243, 277, 303, 304, 305, 310, 311, 381, 390, 400, 403
Carringtington, J. C., 15
Carroll, M. W., 11
Carstens, E. B., 126, 138, 139, 141, 142, 144, 148, 151, 171, 229, 235, 239, 380
Carter, A. P., 47, 76
Carter, C. A., 3, 76, 151, 180, 181
Carvey, P., 339
Casal, J. I., 15, 421
Casasnovas, J. M., 473

Cashdollar, L. W., 456
Casini, G., 3, 5, 23, 148, 172, 263
Casjens, S., 220, 221, 237, 240, 241, 243, 259, 260, 261, 262, 266, 269, 271, 272, 274, 275, 276, 277, 278, 281, 282, 283, 320
Caspar, D. L., 22, 38, 51, 137, 158, 171, 173, 228, 260
Caston, J. R., 1, 230, 315
Cavarelli, J., 135, 153, 154
Cavarelli, W., 209
Center, R. J., 340, 342, 350
Centonze, V. E., 435, 459
Cereghino, J. L., 12, 13
Ceres, P., 48
Cerione, R. A., 487
Cernter, R. J., 150
Cerritelli, M. E., 146, 242, 302, 303, 304, 306, 307, 309, 310, 316, 317, 319, 400
Ceska, T. A., 121
Chaiken, I., 345
Chait, B. T., 151, 484, 485
Chait, P., 429
Champness, J. N., 52, 132, 158
Chan, B. M., 479
Chan, D. C., 74, 76, 151, 182, 333, 335, 340, 341, 342
Chan, P. Y., 474
Chan, R. K., 271
Chance, M. R., 76, 151, 182
Chancellor, K. J., 76
Chandran, K., 15, 142, 145, 167, 460
Chandrasekar, V., 131, 135
Chandrasekhar, G. N., 275
Chang, A., 387
Chang, C. M., 52
Chang, C. T., 457
Chang, M. W., 480
Change, E., 339, 348
Channock, R. M., 229, 428
Chap, H., 486
Chapman, J. D., 457
Chapman, M. S., 127, 128, 130, 134, 144, 155, 167, 168, 169, 186, 239, 422
Chappell, J. D., 142, 435, 436, 459, 460, 461, 462–82, 464, 465
Charpilienne, A., 230
Chase, E. S., 78, 79, 81, 82, 138, 419, 423, 424, 431, 437, 438, 439, 459, 460
Chatterjee, P. K., 356, 373
Chaudhry, C., 335, 356
Chauvin, C., 48, 49, 50, 51
Che, Z., 411, 419, 439, 459
Chen, C., 58, 59, 310
Chen, D. H., 68, 73, 74, 102, 113, 230, 235, 392, 395, 403
Chen, E. Y., 339
Chen, J., 329, 330, 331, 334, 337
Chen, J. J. Y., 366
Chen, L., 140, 162, 351, 352, 353
Chen, R. O., 265, 271, 280
Chen, S., 43
Chen, W., 338
Chen, X. S., 3, 5, 23, 148, 172, 235, 263
Chen, Y., 15, 435, 459, 460
Chen, Z., 131, 134, 135, 204, 223, 224, 226, 416, 417
Cheng, N. Q., 2, 22, 23, 28, 56, 57, 79, 94, 99, 103, 105, 120, 128, 146, 155, 198, 211, 212, 213, 217, 218, 221, 242, 244, 263, 266, 269, 271, 280, 285, 293, 294, 302, 303, 304, 306, 308, 310, 313, 316, 317, 319, 392, 393, 396, 398, 400, 401, 402, 427, 430, 434, 467, 468, 470, 471
Cheng, R. H., 38, 39, 43, 61, 81, 94, 155, 198, 237, 261, 265, 313, 354, 365, 366, 417, 419, 421, 423, 424, 439, 473
Cheng, X., 147
Cheresh, D. A., 428, 429, 477, 479, 480, 483, 487
Chernaia, M. M., 134
Cherry, J. D., 462
Chillon, M., 482
Chinnadurai, G., 480
Chipman, P. R., 60, 65, 80, 81, 137, 144, 145, 168, 169, 176, 239, 263, 369, 370, 372, 431, 437, 438
Chiu, C. Y., 64, 429, 482, 483, 484, 485
Chiu, W. I., 1, 28, 43, 44, 45, 47, 54, 55, 68, 72, 94, 95, 96, 97, 98, 101, 102, 104, 105, 106, 109, 111, 112, 113, 114, 115, 116, 117, 118, 119, 121, 221, 230, 232, 234, 242, 244, 265, 266, 271, 273, 278, 280, 283, 285, 291, 293, 302, 306, 309, 310, 319, 365, 366, 385, 391, 392, 394, 395, 399, 400, 402, 403, 421, 414, 415, 423, 427
Choe, H., 345
Choi, H., 39, 61, 132, 137, 157, 158, 186, 365, 366, 369, 387, 423
Chothia, C., 120, 265

Chow, L. T., 480
Chow, M., 130, 131, 133, 153, 154, 156, 186, 199, 223, 382, 421, 422, 460, 466, 467, 471
Chow, T. P., 15
Christensen, A. M., 74, 75, 77, 150, 151, 181, 240, 263
Christian, P., 134, 155
Chroboczek, J., 480, 482, 483, 484
Chu, Y., 339, 483
Chuang, D., 95
Chugh, D. A., 13
Chui, W., 260
Chuma, T., 71
Church, G. A., 396
Chutkowski, C. T., 340
Cielens, I., 152
Clark, A. J., 135
Clark, E. A., 486
Clarke, M., 43, 47, 56, 61, 76, 98, 354, 366, 368, 369, 369
Clarke, P., 457, 465
Clayton, G. H., 457, 458
Clemons, W. M., Jr., 47, 76
Clerx, J. P. M., 111, 113
Cliff, M., 43
Clore, G. M., 74, 76, 151, 182, 340, 341, 342
Coates, J. A., 315, 396
Cockburn, J., 64
Coen, D., 147
Cohen, D., 43, 46, 47, 48
Cohen, F. S., 343
Cohen, G. H., 147, 438, 469
Cohen, J. A., 70, 71, 82, 142, 165, 230, 235, 436, 457, 462
Cohen, J. J., 457, 464
Cole, G. A., 412
Coll, M., 146, 243
Colman, P. M., 140, 141, 158, 161, 162, 351, 352, 353, 413, 414, 461
Colonno, R. J., 154, 411, 412, 418, 420, 491
Compans, R. W., 51, 130, 144, 168, 186, 239
Compton, S. R., 226
Condit, R. C., 186
Condon, B., 132, 134
Condra, J. H., 418
Condreay, J. P., 8
Condron, B. G., 315
Connolly, J. L., 435, 457, 462–82, 464, 465
Consigli, R. A., 389
Contreras, C. M., 15, 460
Conway, J. F., 2, 27, 38, 56, 57, 80, 81, 94, 99, 103, 105, 120, 128, 146, 198, 211, 212, 213, 217, 218, 221, 280, 293, 294, 304, 306, 308, 310, 313, 319, 430, 467, 468
Conway, J. J., 424, 427
Cook, J. C., 12, 22
Coombs, D. H., 280, 392, 402, 403, 423
Coombs, K. M., 158, 232, 435, 436, 458, 459
Copps, T. P., 436, 457
Cordingley, M. G., 281
Cortes, E., 421
Corteyn, A. H., 433, 477
Corti, C., 327
Corver, J., 354, 369, 370, 372, 373
Cosgrove, L. J., 353
Cossart, P., 476, 486
Cosset, F. L., 349, 350
Cotmore, S. F., 240
Cotten, M., 488
Coulondre, C., 265
Coulson, A. R., 236, 265, 284, 285
Couture, E., 396
Covell, D. G., 74, 76, 340, 341, 342
Coward, J. E., 301
Cowsert, L. M., 171, 433
Cox, S., 15, 31, 471, 491
Craigie, R., 151
Crainic, R., 3, 422
Crane, J., 344
Crawford, L., 12, 22
Crawford, T. O., 412
Crayton, M. A., 243
Cregg, J. N., 12, 13
Cremers, A. F., 51
Crennell, S., 140, 162
Crick, F., 38
Cripe, T. P., 5, 23, 263
Cristiano, R. J., 23
Cross, T. A., 51, 141, 162
Crowell, R. L., 469, 480
Crowther, J. R., 412
Crowther, R. A., 3, 46, 54, 55, 57, 94, 97, 99, 101, 104, 105, 111, 113, 114, 120, 126, 128, 149, 173, 179, 221, 225, 230, 385
Crystal, R. G., 480, 481
Cullis, P. R., 51
Culp, J. S., 281
Culver, J. N., 52, 137, 228

Cundari, E., 486
Cunningham, J. A., 134, 479, 480
Cunningham, J. M., 181, 184, 344, 346, 348, 349, 350
Curiel, D. T., 480
Curry, S., 54, 81, 134, 155, 225, 422, 466, 467, 468
Cusack, S., 49, 65, 74, 75, 77, 147, 159, 174, 331, 334, 335, 480, 482, 484
Cuzange, A., 483
Cybulsky, M. I., 483
Cyrklaff, M., 64, 173, 246, 389, 390, 458, 482, 484

D

Dai, J. B., 131, 134, 135, 204
Dalgarno, L., 423
Dalphin, M. E., 269
Dalrymple, J. M., 412
Dalsgaard, K., 15, 421
Damaschun, G., 327, 333, 334
Dameyama, K., 49
Damico, R. L., 344, 356
Daniels, R., 357
Danos, O., 347
Danthi, P., 131
Darcy-Tripier, F., 390
Dargan, D. J., 302, 384
Darlix, J. L., 76, 151, 182
Darnell, J. E., 466
Dasgupta, N., 482
Dasgupta, R., 24, 186, 202, 226
Dattagupta, N., 242
Dauter, Z., 136, 146
Davey, R. A., 184, 347, 348
Davidson, A. R., 146
Davidson, B. L., 482, 483
Davies, D. R., 151, 438
Davies, J. W., 38, 56, 236
Davies, M. E., 418
Davis, G., 491
Davison, A. J., 381, 382, 392
Dawson, W. O., 15, 228
Day, J., 55, 138, 153, 222, 224, 225
Day, L. A., 51, 53
de Foresta, F., 3
De Francesco, R., 137
De Guzman, R. N., 50, 75, 76, 181, 182
de Haas, F., 45, 64, 65, 67, 77, 235, 246, 319, 354
De Jaeger, G., 13
De Jong, J. C., 326
de Kruijiff, B., 51
de la Rosa, M. C., 3
De Rocquigny, H., 53, 76, 151, 182
De Rosier, D. J., 221
De Simone, F., 412
De Wilde, C., 13
De Wilde, M., 3
Debiasi, R. L., 457
Debouck, C., 147
Dechecchi, M. C., 483
Dedieu, J. F., 486
Deglon, N., 339
DeGregori, J., 481
Deisenhofer, J., 129, 147, 174, 482
Dejana, E., 463
Deka, D., 15, 31
Dekegel, D., 410
Delavega, I., 381
Delbaere, L. T., 157
Delius, H., 423
della-Cioppa, G., 15
Delmedico, M. K., 343
Delos, S. E., 263
Delpeyroux, F., 422
Delwart, E. L., 340
DeMartino, J., 471, 491
Denning, G., 487
Dennis, C. A., 50
Denny, J. K., 141, 162
deNoronha, C. M., 349
Denyer, M., 81
Depicker, A., 13
Derbyshire, D. J., 50
Dermody, T. S., 142, 435, 436, 460, 456, 457, 458, 459, 460, 461, 462–82, 464, 465
DeRosier, D. J., 101, 105, 121, 305
Dery, C. V., 488
Desai, P., 263, 264, 400, 403
Desjardins, E., 76, 183
Dessen, A., 74, 76, 140, 151, 162, 163, 182, 186, 340, 341, 342
DeTulleo, L., 467, 476
Devauchelle, G., 390
Devaux, C., 48, 65
DeWitt, C. M., 491
DeYoreo, J. J., 55

Diana, G. D., 130, 131, 134, 155, 422, 473
Diat, O., 334
Diaz-Laviada, M., 71
Dietrick, I., 101
Dietzgen, R. G., 437
Dietzschold, B., 15
DiGiovine, M., 486
Dikerson, R. E., 265
DiLella, A. G., 147, 281
Dimmock, N. J., 15
Ding, Y., 261
Diprose, J. M., 1, 39, 41, 48, 54, 57, 61, 64, 65, 68, 69, 70, 71, 72, 73, 119, 142, 164, 221, 230, 231, 232, 234, 294
Dishlers, A., 8
Ditusa, S. F., 396
Dobson, C. M., 51
Dodds, J. A., 153, 222
Dokland, T., 3, 54, 55, 135, 143, 154, 170, 226, 234, 237, 238, 261, 265, 281, 285, 289, 290, 293, 301, 302, 306, 307, 310, 319
Dolja, V. V., 15
Dolnik, O., 140, 162, 163, 186
Domingo, E., 80, 81, 134, 155, 431, 432, 433, 460, 477, 483
Doms, R. W., 339, 345, 388
Dona, M., 71
Donahue, P. R., 349
Dong, C. Z., 53, 150
Dong, G., 43, 55, 72, 230, 385, 391
Dong, X. F., 25, 198, 200, 201, 422, 460
Donson, J., 15
Doranz, B. J., 345
Dormitzer, P. R., 142
Dorner, A. J., 221
Douer, D., 339, 348
Dougherty, M., 43, 47, 55, 72, 94, 97, 98, 102, 105, 115, 118, 221, 230, 244, 285, 291, 293, 385, 391, 400
Douglas, J. T., 480
Dove, A. W., 471
Downing, K. H., 121
Downing, M. R., 12
Doyle, M. L., 345
Drabkin, G. M., 315
Drak, J., 111, 113
Drayna, D., 457, 458
Dressler, D., 237
Driedonks, R. A., 403
Droge, A., 403
Droguett, G., 480
Drouet, E., 147
Dryden, K., 68, 118, 230, 435, 436, 458, 459, 461
Dubay, J. W., 342
Dube, P., 43, 46, 304, 403
Dubochet, J., 221, 241, 315, 389, 390
Dubuisson, J., 3
Duda, R. L., 2, 38, 39, 54, 56, 57, 69, 146, 175, 198, 211, 212, 213, 214, 217, 218, 221, 241, 261, 280, 293, 294, 306, 310, 313, 319
Duenas-Carrera, S., 3
Duffy, M. J., 480
Duggan, C., 480
Duke, G. M., 131, 134
Duke, R. C., 481
Dumas, P., 132, 157, 158, 186, 369, 423
Duncan, R., 456, 459, 463
Dunn, J. J., 315
Duong, D. T., 459, 461, 462
Duong, J. L., 436, 459, 461, 462
Durand, S., 381
Durham, A. C., 204, 205
Durkin, J., 269
Durmort, C., 147
Durrani, Z., 15
Durrer, P., 327
Durst, M., 22
Dustin, M. L., 491, 473
Dutch, R. E., 140, 162, 335, 350, 351, 353
Dutko, F. J., 130, 134, 155, 203, 422, 473
Dworkin, C., 436, 457, 462
Dyda, F., 151

E

Earnshaw, W. C., 74, 221, 234, 242, 260, 263, 271, 273, 277, 315
Ebbinghaus, C., 480
Eckert, D. M., 339
Eddy, S. R., 369
Edelstein, C. L., 457
Edlund, K., 482
Egan, K. M., 12
Egner, C., 243
Ehrlich, L. S., 3, 76, 151, 180, 181
Eiden, M. V., 343, 347

Einfeld, D. A., 482
Eisenberg, D., 139
Eisenberg, R. J., 147, 469
Eiserling, F. A., 241, 260, 285, 302, 315, 319, 391, 393, 401, 403
Ekechukwu, M. C., 261, 263, 264, 285, 287, 288, 289
Ekström, M., 337
Elkon, K. B., 480
Ellard, F. M., 79, 432, 433, 477, 478, 479
Elliott, R. M., 27, 396
Ely, K. R., 133
Emborg, M. E., 339
Emini, E. A., 221, 421
Emsley, J., 479
Endo, R. I., 483
Endres, D., 48
Engel, A., 304, 403
Engelman, A., 151
Enquist, L. W., 221, 466
Enserink, M., 476
Enzmann, P. J., 158
Epand, R. F., 337
Epand, R. M., 337
Erickson, H. P., 95, 260, 491, 473
Erickson, J. W., 38, 79, 130, 134, 154, 155, 186, 220, 223, 225, 226, 418, 421, 422, 473
Erickson, S., 274
Ericsson, M., 388
Erikson, H., 435, 458
Erk, I., 71
Erukulla, R. K., 373
Esposito, J. J., 382
Estes, M. K., 2, 3, 14, 23, 26, 54, 126, 135, 138, 139, 141, 142, 144, 148, 151, 154, 171, 229, 230, 232, 233, 234, 235, 239, 380, 414, 415
Estes, P. A., 5, 263
Eswar, N., 48
Evander, M., 434
Evans, D. M. A., 154

F

Falchetto, R., 335
Falck-Pedersen, E., 480
Falcon, V., 3
Falgout, B., 387
Fancher, M., 203
Fane, B. A., 143, 169, 170, 237, 238, 261, 263, 264, 265, 266, 267, 268, 269, 270, 281, 285, 287, 285, 287, 288, 289, 291, 293, 307
Faragher, S. G., 423
Farber, J. M., 345
Farber, M. B., 261
Farsetta, D. L., 68, 230, 461
Farzan, M., 345
Fass, D., 74, 76, 150, 151, 182, 184, 340, 341, 342, 347, 348
Faulkner, L., 38, 56, 236
Fauquet, C. M., 52, 126, 136, 138, 139, 141, 142, 144, 148, 151, 171, 206, 229, 235, 239, 380, 413, 414, 416, 421, 421, 425, 430, 431, 433, 435
Faure, G., 3
Fayard, B., 30
Feigin, L. A., 315
Feiss, M., 242, 243
Felding-Habermann, B., 479
Felock, P. J., 130, 134, 422
Feng, Z., 138, 139, 141, 142, 144, 148, 151
Fennelly, J. J., 480
Ferguson, J., 5
Ferguson, M., 154
Ferlenghi, I., 76, 354, 368, 369
Fernandez, J. J., 59, 303, 304, 305, 403, 482–2
Fernon, C. A., 423
Ferrari, E., 137
Ferre, R. A., 132, 134
Fiddes, C. A., 236, 265
Fiddes, J. C., 284
Fields, B. N., 68, 229, 435, 436, 456, 457, 458, 459, 460, 461, 462, 463, 464
Fiers, M., 380
Fieschko, J. C., 12
Fife, K. H., 12, 22
Figlerowicz, M., 14
Filali Maltouf, A., 244
Filardo, E. J., 429, 487
Fillmore, G. C., 235
Filman, D. J., 56, 79, 130, 133, 134, 147, 153, 154, 155, 156, 186, 199, 223, 225, 382, 421, 422, 467, 468, 470, 471
Finberg, R. W., 457, 479, 480, 481, 482
Finch, J. T., 3, 12, 22, 78, 101, 221
Finland, M., 462
Finn, R., 43, 46, 47, 48, 99, 103, 105

Fisher, A. J., 135, 198, 200, 223, 225
Fisher, B. R., 5, 20, 38, 54, 61, 63, 285, 294, 354, 368, 370
Fisher, R. A., 242, 243
Fita, I., 81, 134, 137, 226, 422, 424, 425, 431, 432, 433, 460
Fitchen, J., 15
Fitz, W., 329, 350
Fitzgerald, D., 487
Flanagan, J. M., 65, 147, 480
Fleck, S. K., 480, 482–2, 486
Fletterick, R. J., 132, 186
Fleysh, N., 15, 31
Flint, S. J., 221, 466
Flore, O., 225
Foellmer, B., 338
Folkers, G., 147
Follis, K. E., 339
Forghani, B., 392
Formanowski, F., 329, 350
Forrer, P., 146
Forrest, J. C., 435, 462–82, 464, 465
Forsell, K, 354
Forte, C. P., 491
Fossdal, C. G., 290
Fouchier, R. A., 326
Fox, G., 38, 39, 54, 79, 81, 134, 154, 155, 223, 422, 432, 474, 476, 477, 485
Fox, J. M., 5, 205, 263
Francki, R. I. B., 437
Francotte, M., 3
Frank, J., 43, 46, 47, 48, 110
Frankel, A. D., 151
Frankenberger, E. A., 38, 79, 130, 134, 154, 155, 186, 418, 421, 422, 473
Franz, A. W. E., 431, 437, 438
Fraser, M. J., 389
Fraser, R. D., 435, 458, 459, 460, 461
Frazer, I. H., 22, 23, 434
Frazier, W. A., 477
Freimuth, P., 65, 147, 480, 483
Freistadt, M. S., 470
French, T. J., 29, 30
Frenkel, N., 383
Fricks, C. E., 199, 422, 467, 468
Fridborg, K., 133, 139, 152, 153, 186, 220, 222
Friedman, A., 301
Friedman, J. M., 151, 265
Friedman, T., 285
Friesen, P. D., 388, 389
Frischauf, A. M., 423
Frosch, P., 466
Fruscella, P., 463
Fry, E., 38, 39, 40, 42, 54, 74, 77, 79, 81, 133, 134, 136, 151, 154, 155, 223, 225, 422, 432, 433, 477, 478, 479, 485
Fu, Z. F., 15
Fuchs, R., 476
Fuerstenau, S. D., 52
Fujimoto, L. M., 486
Fujinaga, M., 157
Fujinami, R. S., 134
Fujisawa, H., 280, 303, 304, 305, 311, 403
Fukuyama, K., 136, 137
Fuller, M. T., 281, 282
Fuller, S. D., 3, 5, 38, 39, 43, 44, 45, 46, 47, 56, 57, 61, 63, 64, 65, 67, 74, 76, 77, 94, 98, 111, 113, 118, 158, 173, 230, 234, 235, 246, 294, 319, 354, 365, 366, 367, 368, 369, 372, 373, 385, 429, 482, 484
Furlong, D., 317, 435, 436, 457, 458, 459, 460, 461

G

Gabashvili, I. S., 43, 46, 47, 315
Galisteo, M. L., 282
Gall, J. G., 480
Gallagher, T. M., 24, 153, 226
Galli, C., 327
Gallo, M. V., 349
Gallo, S. A., 346
Galloway, D. A., 22, 23, 434
Gamble, T. R., 74, 76, 77, 151, 180, 181
Ganser, B. K., 3, 78
Gao, M., 245, 263, 396
Garcea, R. L., 1, 3, 5, 22, 23, 148, 172, 260, 263, 301
Garces, J., 389, 390
Garcia, C., 3
Garcia, J., 421
Garcia, L. R., 244, 310
Garcia, N., 381, 390
Garcia-Barreno, B., 353
Garoff, H., 157, 337, 350, 354, 365, 366, 367, 387, 423
Garver, K., 58, 59
Gaur, A., 474
Geis, I., 265

Geisler, S. C., 471, 491
Gelinas, R. E., 480
Gellissen, G., 12
Gentsch, J. R., 462
George, H. A., 12, 22
Georgopoulos, C. P., 261, 275
Geraghty, R. J., 469
Gerald, C., 345
Gerald, D., 53
Gerald, N. P., 345
Gerard, R. D., 129, 147, 174, 482
Gesteland, R. F., 315
Ghabrial, S. A., 413, 414, 416, 421, 425, 430, 431, 433, 435
Ghazaleh, R. A. *See* Abu-Ghazaleh, R.
Gheysen, D., 3
Ghim, S. J., 15, 23
Ghosh, A. K., 55, 230
Ghosh, M., 45, 54, 68
Gibbons, D. L., 356, 373
Gibson, C. A., 457, 464, 483
Gibson, W., 392
Giga-Hama, Y., 12
Gigant, B., 334
Gilbert, J. M., 344
Gilbert, R. J., 43, 51, 165
Gillian, A. L., 235
Gillies, S. C., 232
Gilliland, G., 138, 139, 141, 142, 144, 148, 151
Gilmore, T., 340
Gingery, M., 285, 302, 319, 391, 393, 401
Ginsberg, M. H., 486, 487, 488
Giralt, E., 431, 433
Giranda, V. L., 134, 155, 471
Girijavallabhan, V., 471, 491
Gissmann, L., 22, 23
Gitti, R. K., 76, 151, 180, 181
Glanville, N., 370
Glück, R., 327, 335
Godson, G. N., 284
Goelet, P., 186
Gold, M., 146
Goldberg, I., 3, 5, 23, 148, 172
Goldsmith, A., 70, 232
Golmohammadi, R., 133, 139, 152
Gombold, J. L., 235
Gomes, P., 432
Gomis-Ruth, F. X., 146, 243
Gonda, M. A., 3
Gong, W., 147
Gonzáles-Reyes, L., 353
Gonzalez, A., 381
Gonzalez, E., 3
Gopinath, K., 15, 136
Gorbalenya, A. E., 76, 151, 180, 181
Gordon, C. L., 261, 262
Gormann, J. J., 140, 162, 351, 352, 353
Gouet, P., 1, 39, 41, 48, 54, 57, 61, 65, 68, 69, 70, 71, 72, 73, 119, 142, 164, 221, 230, 231, 232, 234, 294
Gounon, P., 3
Govindan, V., 116
Gowen, B. E., 43, 46, 47, 48, 56, 61, 76, 77, 98, 103, 105, 354, 366, 369, 384
Graef, R., 335
Graham, D. Y., 3, 26, 235, 491
Grant, R. A., 133, 134, 422
Grantham, G. L., 15
Grassucci, R. A., 43, 46, 47
Grattinger, M., 3, 5
Gray, C. W., 143, 327
Grdina, T. A., 457, 464
Greber, U. F., 428, 480, 488
Green, B., 273
Greenberg, H. B., 3, 230, 235
Greene, B., 242, 260, 266, 274, 277, 278, 280, 281, 283, 302, 306, 309, 310, 319, 399
Greene, R. J., 350
Greenspan, N. S., 412
Greenstone, H. L., 22, 94, 99, 103, 105, 171, 285, 398, 399, 402, 433, 434
Greenwell, T., 342
Greenwood, A., 55, 138, 153, 222, 224, 225
Greve, J. M., 155, 418, 491, 473
Gribskov, C. L., 132, 134
Griess, G. A., 317, 320
Griffin, D. E., 365, 412, 423
Griffin, K., 315
Griffin, R. G., 51
Griffith, J. P., 38, 79, 130, 134, 154, 155, 186, 418, 421, 422, 473
Griffiths, G., 388, 389, 390
Grillo, J. M., 3
Grimbergen, J. M., 482
Grimes, J., 1, 40, 41, 44, 45, 48, 54, 57, 61, 64, 65, 68, 69, 70, 71, 72, 73, 119, 142, 163, 164, 165, 221, 230, 231, 232, 234, 294

Grimes, S., 38, 39, 53, 58, 59, 60, 146, 168, 176, 243, 244, 246, 304, 403
Groarke, J., 155
Gronenborn, A. M., 74, 76, 151, 182, 340, 341, 342
Grorke, J., 134
Gross, I., 3, 5, 77
Gruenberg, A., 421
Gruenberger, M., 424, 476
Grutter, M. G., 64, 130
Grzesiek, S., 151
Gschmeissner, S. E., 12, 22
Gttis, J., 134
Guan, T., 487
Guan, Y., 143
Guarne, A., 134
Guasch, A., 146, 243
Guillot, S., 422
Gunn, V. L., 436, 462, 463, 464
Gunther, D., 403
Guo, P., 58, 59, 243, 269, 274, 310

H

Ha, Y., 141
Haan, K.M., 147
Haanes, E. J., 265, 268
Haasnoot, C. A., 51
Habuka, N., 137
Hacker, J., 327, 342
Hackett, N. R., 481
Hadfield, A. T., 130, 156
Hafenstein, S., 264
Hagensee, M. E., 22, 23, 434
Haggard-Ljungquist, E., 290
Hahn, C. S., 157, 369
Hainfeld, J. F., 305, 308, 480
Hajdu, J., 52, 53
Hajibagheri, M. A., 12, 22
Hale, R. D., 143
Hall, A., 486, 487
Hall, C., 271, 274, 275
Hall, J., 133
Hallenbeck, P. L., 480
Hamada, K., 311
Hamamoto, H, 15
Hamatake, R. K., 236, 238
Hamblin, C., 71
Hamburger, A. E., 5, 20, 370
Hamilton, W. D., 15
Hamson, C. A., 184, 347, 348
Han, X., 141, 160
Hanecak, R., 221
Hannan, C., 488
Hanninen, A. L., 261
Hansen, J., 47, 76
Hantraye, P., 339
Hanzlik, T. N., 25, 207, 210, 303
Harber, J., 421, 422
Hardt, S., 102, 104, 109
Hardwick, J. M., 412
Hardy, M. E., 3, 54, 135, 154, 226
Hare, D. R., 76, 151, 182
Hare, J., 144, 168, 169
Harfe, B., 483
Harley, C., 134
Harrap, K. A., 388
Harris, J. R., 43, 46
Harrison, S. C., 1, 3, 5, 23, 39, 49, 50, 51, 54, 65, 68, 72, 73, 74, 76, 126, 130, 131, 136, 137, 141, 142, 145, 148, 150, 151, 153, 156, 159, 163, 165, 167, 171, 172, 182, 185, 205, 215, 220, 221, 225, 230, 234, 241, 242, 263, 271, 273, 315, 340, 341, 342, 354, 355, 365, 367, 368, 372, 436, 458, 460, 461
Hartsch, T., 47, 76
Hartweig, E., 277
Harvey, J. D., 74, 221, 234
Hasegawa, T., 49
Haselkorn, R., 305
Haseloff, J., 186
Hashida, E., 15
Haskell, K., 94, 105, 106, 111, 112, 116
Hata, Y., 137
Hatanaka, H., 133, 151
Hautala, T., 487
Havenga, M. J., 482
Hay, A. J., 162
Hayashi, M. N., 236, 237, 238, 261, 262, 263, 264, 269, 270, 284, 288, 289
Hayden, M., 243
Hayes, M., 43
Hayes, S. J., 315, 317, 319, 320
He, J., 98, 106, 115, 118, 134, 221, 244, 285, 291, 293, 391, 394, 400, 402, 427
He, X., 438

He, Y., 38, 39, 47, 58, 59, 60, 79, 80, 94, 97, 102, 105, 146, 155, 168, 176, 243, 304, 403, 470, 473, 474, 489
Healey, J. J., 347
Heard, J. M., 347
Heber, D. N., 338
Hebert, B., 79, 80, 168
Hebert, D.N., 357
Hecht, H. J., 38, 79, 130, 134, 154, 155, 186, 418, 421, 422, 473
Hedlund, T., 481
Hein, M. B., 15
Heinkel, R., 45, 64, 65, 67, 246, 319
Heino, T., 477
Heinz, B.A., 203
Heinz, F. X., 76, 131, 137, 156, 354, 355, 356, 367, 368, 369, 372, 373
Heistad, D. D., 483
Helenius, A., 334, 335, 338, 365, 428, 488
Hellmig, B., 147, 281
Helmer-Citterich, M., 143
Hemler, M. E., 479
Hemmati, H., 343
Hemmi, S., 480
Hempel, J., 211, 212, 293
Henderson, D. W., 434
Henderson, L. E., 76, 151, 182
Henderson, R., 53, 121, 132, 157
Hendrickson, W. A., 74, 75, 76, 151, 183, 184, 345, 346
Hendrix, R. W., 2, 38, 39, 54, 56, 57, 69, 146, 175, 198, 211, 212, 213, 214, 217, 218, 221, 241, 243, 261, 280, 293, 294, 301, 305, 306, 310, 310, 313, 319
Hendry, D. A., 204, 205
Hendry, E., 133
Hengst, K., 434
Henkel, T., 465
Henry, L. J., 129, 147, 174, 482
Hensley, P., 345
Herbert, B., 144
Herion, D., 3
Hermesh, C., 337
Hernandez, J., 134, 155
Hernandez, L. D., 344
Hernandez, R., 38, 54, 61, 63, 285, 294, 354, 368
Herr, W., 147
Herrmann, A., 327, 332, 333, 334
Heuser, J. E., 486
Hewat, E. A., 30, 68, 80, 81, 82, 424, 425, 426, 431, 432, 433
Hickman, A. B., 151
Hidaka, C., 481
Hierholzer, J. C., 479
Higgins, D. R., 12, 13
Hilbers, C. W., 51
Hilf, M. E., 15
Hill, C. L., 65, 68, 72, 73, 151, 230
Hill, C. P., 3, 74, 75, 76, 77, 78, 151, 180, 181
Hirai, H., 487
Hiremath, C. N., 133, 138, 422, 467, 468
Hirsch, M. S., 229
Hockley, D. J., 77
Hodgson, K. O., 53
Hoenke, S., 327
Hoey, K., 52
Hofer, F., 476
Hofer, M., 265
Hoffman, G. R., 487
Hoffman, L. R., 327, 342, 344
Hoffman, T. L., 345
Hofmann, K. J., 12, 22
Hogan, M., 242
Hogg, N., 463
Hogle, J. M., 56, 79, 130, 131, 133, 134, 136, 147, 153, 154, 155, 156, 186, 199, 223, 225, 382, 421, 422, 460, 466, 467, 468, 470, 471
Hohenberg, H., 3, 5, 78
Hohn, B., 243, 261, 280
Holden, J. E., 339
Holland, J. J., 469
Hollenberg, C. P., 12
Hollinger, F. B., 430
Holmes, K., 52
Holsinger, L. J., 162
Holwerda, B. C., 281
Homa, F. L., 15, 27, 28, 244, 245, 263, 265, 266, 268, 269, 285, 291, 303, 309, 392, 396, 398, 399, 400, 402, 403, 427, 431
Hong, J. S., 480
Hong, L. H., 339
Hong, Z., 137, 269
Hooft van Huysduynen, R. A., 326
Hoog, S. S., 147, 281
Hooper, D. C., 15
Hooper, J. W., 436, 457
Hoover, E. A., 349
Hoover, Litty, H., 338

Hopkins, C. R., 483
Horie, H., 422
Horne, D., 456, 463
Horwitz, M. S., 428, 480
Horzinek, M., 158
Hosaka, Y., 326
Hossain, S. A., 487
Hostomska, Z., 137
Hosur, M. V., 135, 153, 200
Hourcade, D., 237
Houseweart, M., 74, 180, 181
Hovi, T., 199, 477
Howard, B. H., 479–99
Hoxie, J. A., 345
Hruby, D. E., 11
Hu, J. S., 151
Huang, C. C., 147
Huang, H, V., 365
Huang, S., 429, 480, 481, 482–2, 483, 484, 485, 486
Huang, W. M., 243, 271
Huang, Y., 422
Hubbard, T. J., 120
Hubbert, N. L., 23, 434
Huber, R., 55, 146, 176, 177
Huberman, J. A., 271
Huckhagel, C., 78
Hud, N. V., 241, 315
Huettinger, M., 476
Hughson, F. M., 160, 327, 331, 333, 335
Huisman, L. G., 482
Humphries, M. J., 477
Hunger, E., 342
Hunger, S. P., 481
Hunkapiller, M. W., 423
Hunt, D., 211, 212, 293
Hunter, E., 180, 294, 339
Hurlburt, W., 263, 396
Hutchinson, C. A., 236, 265, 284
Hutchinson, C. A., III, 285
Huylebroeck, D., 365
Hyatt, A. D., 426
Hynes, R. O., 476
Hyypia, T., 133, 477

I

Ibarra, B., 146, 243, 303, 304, 305, 403
Icenogle, J. P., 134, 154, 225
Ikeda, H., 350
Ikeuchi, Y., 327
Ilag, L. L., 54, 130, 143, 169, 237, 238, 261, 265, 266, 281, 285, 293
Imada, K., 305
Imai, H., 63
Imai, M., 49, 51, 63
Incardona, N., 130, 143, 204, 237, 238, 261, 265, 266, 281, 285, 293
Incitti, I., 137
Inglis, S., 5
Inoue, K., 51, 63
Iourin, O., 151
Ireton, K., 486
Ise, I., 469
Ishii, T., 402
Ishikawa, N., 49, 63
Ishikawa, T., 315
Isobe, E., 399
Isom, O. W., 481
Israel, V., 277
Ito, S., 3
Ito, Y., 49, 63
Ivannikova, T. A., 412
Iwasaki, K., 393, 401, 402
Iwaski, K., 285
Iyengar, S., 15
Iyer, L. M., 381, 382, 388

J

Jablonsky, M. J., 266, 269, 275, 276
Jack, A, 54, 221
Jackson, A. O., 14
Jackson, T., 54, 79, 81, 432, 433, 477, 478, 479
Jacobs, E., 3
Jacobsen, E., 289, 306
Jacobson, M. F., 223, 225
Jacrot, B., 48, 49, 50, 51, 483
Jaenicke, R., 265
Jafari, J. D., 480
Jakana, J., 28, 43, 45, 47, 54, 55, 68, 72, 73, 74, 94, 95, 96, 97, 98, 101, 102, 104, 105, 106, 109, 111, 112, 113, 114, 115, 116, 118, 121, 221, 230, 232, 234, 242, 244, 266, 273, 285, 291, 293, 302, 306, 310, 319, 385, 391, 395, 399, 400, 402, 403, 427
Jakubczak, J. L., 480

James, M. N., 134, 157
Jameson, B. A., 340, 421
Jandrig, B., 12
Jane-Valbuena, J., 73, 142, 167, 436, 458, 459
Jansen, K. U., 12, 22
Janshoff, A., 226, 227
Janson, C. A., 134, 147, 281
Jardetzky, T. S., 140, 147, 162, 335, 350, 351, 353
Jardine, P. J., 38, 39, 58, 59, 60, 146, 168, 176, 243, 246, 280, 304, 402, 403
Jarvis, A. W., 413, 414, 416, 421, 425, 430, 431, 433, 435
Jaskolski, M., 151
Jayaram, H., 235
Jeffries, B. C., 428
Jeng, T. W., 101
Jenkins, T. M., 151
Jennings, B., 264, 265
Jenson, A. B., 23
Jenson, G. J., 121
Ji, W. G., 41
Jiang, H., 392
Jiang, J., 98, 118
Jiang, S., 342
Jiang, W., 43, 55, 72, 94, 95, 97, 98, 105, 115, 117, 119, 120, 121, 230, 385, 391
Jiang, X., 3, 14, 26
Jimenez, J. L., 43
Jin, B. R., 55
Jin, L., 151, 181
Jin, Z., 151, 181
Johansen, B. V., 290
Johnson, A.E., 357
Johnson, B., 469
Johnson, E. D., 412
Johnson, J., 2, 5, 10, 24, 25, 38, 39, 49, 52, 54, 56, 57, 69, 79, 126, 127, 128, 130, 131, 133, 134, 135, 136, 137, 138, 146, 153, 154, 155, 164, 175, 186, 198, 200, 201, 202, 203, 204, 205, 207, 209, 210, 211, 212, 214, 217, 218, 220, 221, 223, 224, 226, 227, 241, 280, 301, 303, 313, 319, 416, 417, 418, 421, 422, 460, 473
Johnson, J. M., 48
Johnson, K. N., 49, 135, 203
Johnson, O., 116
Johnston, R. E., 365, 366, 423
Joklik, W. K., 232, 435, 456, 458, 466
Jones, C. T., 369, 370, 372
Jones, D. T., 120
Jones, I. M., 74, 77
Jones, S., 120
Jones, T. A., 120, 153, 220, 222
Jones, T. D., 15
Josefsson, C., 438
Joshi, S. M., 5
Joshi, V., 308
Joyce, J. G., 12, 22
Juhas, R. M., 244, 245, 392, 396, 403
Jullian, N., 53, 76, 151, 182
Juuti, J. T., 235

K

Kabat, D., 347
Kaesberg, P., 186, 204, 226
Kafri, T., 480
Kajigaya, S., 80, 81
Kalinski, A., 242, 243
Kalko, S. G., 432
Kamarck, M. E., 491
Kamata, T., 479, 484
Kamer, G., 38, 79, 130, 131, 134, 154, 155, 186, 223, 224, 226, 416, 418, 421, 422, 473
Kameyama, K., 63
Kamstrup, S., 15
Kan, J. H., 51
Kanamaru, S., 60, 65, 145, 176
Kanaseki, T., 327
Kanellopoulos, P. N., 147
Kang, S. K., 55
Kanngiesser, U., 430
Kanzler, O., 425
Kaplan, G., 470
Kaplan, I. B., 437
Kapoor-Munshi, A., 483
Kapusta, J., 14
Karabatsos, N., 423
Karacostas, V., 3
Karande, A., 136
Karavanas, G., 335
Kartenbeck, J., 365
Kasuga, M., 486
Kato, K., 350
Kattoura, M., 235
Katz, R. A., 151
Kauffman, R. S., 457

Kaufman, J., 74, 76, 151, 182, 340, 341, 342
Kawabe, T., 238
Kawasaki, K., 327
Kay, B. H., 423
Kay, L. E., 151
Kaye, K., 463
Kayman, S. C., 347, 350
Kazaks, A, 8
Ke, E. Y., 98, 105, 106, 107, 109, 116
Kearney, C. M., 15
Keegan, J. M., 68, 461
Keene, D. R., 347
Kellenberger, E., 241, 260, 315
Keller, B., 396
Keller, W., 76, 130, 144, 151, 168, 180, 181, 186, 239
Kelly, D. C., 388
Kelly, S. M., 293, 399
Kemble, G. W., 334
Kemp, B. E., 150, 340, 342, 350
Kemp, P., 244
Kennard, J., 268, 396
Kenndler, E., 424
Kerlavage, A. R., 382
Kerr, L. D., 457, 465
Kerr, P. J., 423
Kessel, M., 305
Kettner, C. A., 157
Khan, A. S., 379
Khan, S. A., 315, 317, 320
Khanna, N., 13
Khorasanizadeh, S., 180
Khromykh, A. A., 387
Kidd, A. H., 482, 483
Kielian, M., 356, 365, 372, 373
Kihm, U., 412
Kim, B., 182
Kim, H. S., 55, 151
Kim, H. W., 428
Kim, J., 230
Kim, K. J., 290
Kim, P. S., 74, 76, 140, 150, 151, 162, 163, 182, 184, 186, 331, 332, 333, 335, 337, 339, 340, 341, 342, 347, 348, 350, 356
Kim, S., 130, 134, 422
Kim, W. J., 55
Kim, Y., 143, 171
King, A., 3, 54, 79, 81, 134, 155, 225, 431, 432, 433
King, A. M. Q., 432, 433, 477, 478, 479
King, D. S., 337
King, J., 241, 242, 243, 259, 260, 261, 262, 263, 265, 270, 271, 272, 273, 274, 275, 276, 277, 278, 280, 281, 282, 283, 289, 291, 302, 306, 310, 315, 319, 399
King, S. L., 479
Kingsbury, D. W., 126, 139, 141, 142, 144, 148, 151, 168, 185
Kingsman, A. J., 12, 48, 151, 181, 342, 347, 348
Kingsman, S. M., 12, 48, 181, 342, 347, 348
Kingston, R. L., 5
Kinosita, K., 305
Kirchhausen, T., 467, 476
Kirchweger, R., 134
Kirkitadze, M. D., 293, 399
Kirnbauer, R., 22, 23, 433, 434
Kirz, J., 53
Kiselev, N. A., 230
Kistler, J., 400
Kitagawa, M., 350
Kitamura, N., 221
Kitani, H., 350
Klemke, R., 480, 487
Klenk, H. D., 140, 162, 163, 163volch, 186
Kleywegt, G. J., 120
Kliger, Y., 53, 337
Kligfield, P. D., 481
Klishko, V. Y., 3, 78
Kloosterboer, N., 380
Klug, A., 38, 52, 95, 101, 105, 132, 158, 171, 173, 221, 228
Knapek, E., 101
Knight, J. C., 382
Knipe, D. M., 229
Knossow, M., 334
Knowles, N. J., 477
Kobe, B., 150, 340, 342, 350
Kochan, J., 275
Kocsis, E., 303, 304, 306
Kodaira, K., 284, 285
Kodandapani, R., 133
Kofron, J. A., 134, 471
Kohli, E., 70, 71, 82, 142, 165
Kohlstaedt, L. A., 151
Koike, S., 469
Kolatkar, P. R., 79, 80, 155, 418, 473, 474, 476
Komano, T., 238
Konings, R. N., 143
Konnert, J., 55

Koonin, E. V., 381, 382, 388, 460
Kopelman, R., 347, 350
Koprowski, H., 13, 14, 15, 31
Kordower, J. H., 339
Korte, T., 327, 332, 333, 334
Kost, T. A., 8
Kosturko, L. D., 242
Kostyuchenko, V. A., 60, 65, 145, 176, 178
Koszelk, S., 153, 222
Koths, K., 237
Kotloff, K. L., 15
Kovacs, F. A., 141, 162
Kovari, L. C., 76, 151, 180, 181
Kovesdi, I., 384, 481, 482
Kowalski, H., 476
Kozlovska, T., 152, 354
Krah, D. L., 469
Krahl, T., 480
Krasnykh, V. N., 480
Krausslich, H. G., 3, 5, 77, 78
Krawczynski, K., 3
Krebs, H., 265
Kremer, M. J., 131, 133, 134, 155, 422, 471, 473
Krijgsman, P. C., 49
Krishna, N. R., 209, 266, 269, 275, 276, 277, 278, 281, 282
Krishna, S. S., 138
Krishnaswarmy, S., 130, 237, 238, 261, 265
Krithivas, A., 480
Krol, M. A., 198
Kronenberger, P., 476
Krug, R. M., 221, 466
Kruger, D. H., 8, 12
Krumbiegel, M., 327, 333, 334
Krummenacher, C., 147, 469
Kruse, J., 48, 49, 50, 51
Kruse, K. M., 48, 49, 50, 51
Krygsman, P. C., 49
Ksiazek, T. G., 379
Kuczmarski, D., 390
Kuechler, E., 424, 476
Kuehl, K., 3
Kuhn, A., 396
Kuhn, R. J., 5, 20, 38, 39, 54, 61, 63, 79, 80, 134, 137, 155, 285, 294, 354, 365, 366, 368, 369, 370, 387, 372, 423, 424, 470, 473, 474, 489
Kukol, A., 162
Kumagai, H., 12, 15
Kumar, A., 5, 137
Kundu, S. C., 55
Kunz, C., 131, 137, 156, 354, 355, 356, 367, 369, 372, 373
Kuriyan, J., 151
Kurochkina, L. P., 60, 145, 178
Kurowski, B., 357
Kurt-Jones, E. A., 480
Kurumbail, R. G., 281
Kuznetsov, Y. G., 53, 55
Kwong, A. D., 269, 383
Kwong, P. D., 74, 75, 76, 151, 183, 184, 345, 346

L

Labbe, M., 230
Labedan, B., 244
LaBranche, C. C., 345
LaCasse, R. A., 339
Lagace, L., 281
Lam, D. M., 14
Lamb, R. A., 140, 162, 335, 350, 351, 353
Lamberg, D. M., 343
Lambert, D. M., 350
Lamy, J. N., 426
Land, T. A., 55
Landsberger, F. R., 51
Lane, L. C., 53, 390
Langevel, J. P., 421
Langeveld, J. P., 15
Langley, K. E., 12
Lankhorst, R. K., 380
Larsen, G. R., 421
Larsen, T., 289, 306
Larson, G. R., 221, 224
Larson, S. B., 53, 55, 138, 153, 154, 221, 222, 225
Lata, K., 38, 56
Lata, R., 56, 57
Laurent, S., 82, 426
Lauring, A. S., 349
Laver, W. G., 141, 158, 161, 413–32
Lavigne, G., 65, 147, 174
Lavillette, D., 349, 350
Law, L., 482
Lawless-Delmedico, M. K., 350
Lawrence, L. J., 140, 162, 351, 353
Lawrence, M. C., 140, 162, 351, 352, 353

Lawson, C. L., 151, 181
Lawton, J. A., 2, 229, 230, 232, 233, 234
Lazuick, J. S., 423
Le Blois, H., 30, 71
Le Gall, G., 426
Lea, S. M., 54, 79, 81, 134, 155, 432, 433, 477, 478, 479
Leanza, K. C., 350
Lear, J. D., 327
Lechmann, M., 3
Lee, B. M., 76, 151, 180, 181
Lee, C. S., 269
Lee, H., 387
Lee, J. G. M., 480, 482
Lee, J. J., 221
Lee, K. H., 140, 162, 163, 186, 329, 330, 331, 332, 334, 341, 350, 381, 390
Lee, L. Y., 481
Lee, M., 392
Lee, P. W., 456, 459, 462, 463
Lee, S. C., 9, 61, 137, 158, 339, 348, 369, 387
Lee, S. H., 412
Lee, V. M., 473
Lee, W. M., 81, 130, 134, 156, 226, 411, 419, 422, 439, 461
Lee, Y. H., 55
Lee, Y. M. H., 422, 460
Leforestier, A., 233, 315
Legocki, A. B., 14
Lehrach, H., 423
Leiden, J. M., 480
Leiman, P. G., 38, 39, 58, 59, 60, 65, 146, 168, 176, 178, 243, 304, 403
Leippe, D. M., 81, 411, 412, 418, 419, 420, 439
Leis, J., 151
Leith, A., 48
Lemay, G., 230
Lemckert, A. A., 482
Lemieux, S., 58
Lemon, S. M., 126, 138, 139, 141, 142, 144, 148, 151, 171, 229, 235, 239, 380
Lenard, J., 51
Lenches, E. M., 369, 370, 372
Lenghaus, C., 426
Lenk, E. V., 270, 271
Lentz, K. N., 471, 491
Leon, R. P., 481
Leone, G., 459, 463
Leopold, P. L., 481
Lepault, J., 70, 71, 142, 165, 221, 241, 315
Lerner, A. M., 462
Lesburg, C. A., 137
Lescar, J., 61, 63, 76, 294, 354, 367, 372
Leschnitzer, D. H., 143
Lesemann, D. E., 390
Lesk, A. M., 265
Leslie, A. G. W., 3, 54, 120, 128, 131, 136, 149, 179, 205, 220
Letellier, M., 14
Leventhal, L., 339
Lever, A. M., 339
Levin, D. H., 457
Levine, B., 412
Levine, M., 271
Levis, R., 365
Levy-Mintz, P., 373
Lewis, A. J., 421
Lewis, J. K., 200, 422, 460
Lewis, R., 54, 81
Lewit-Bentley, A., 50, 221
Li, E., 480, 483, 486, 487
Li, J. K., 235
Li, L., 339, 348
Li, M, 5
Li, Q., 422, 460
Li, S., 3, 78, 481
Li, Y., 131, 134, 135, 223, 224, 226, 281, 416, 480, 482
Li, Z., 98, 105, 117, 347, 350
Liang, T. J., 3, 106, 107, 109
Liang, Y., 98, 105, 116
Liddington, R. C., 39, 148, 171, 215, 479
Liemann, S., 142, 145, 167
Lievaart, P. A., 436, 457
Lightner, D. V., 55, 381
Liljas, L., 1, 39, 50, 54, 57, 69, 133, 134, 135, 136, 139, 146, 152, 153, 154, 155, 175, 186, 209, 212, 214, 220, 221, 222, 241
Liljeström, P., 365, 387
Lin, H., 235
Lin, K., 342
Lin, T., 49, 131, 134, 135, 136, 155, 203, 204, 223
Lin, Y. P., 327, 330
Lindqvist, B. H., 289, 290, 306, 387
Lineberger, D. W., 491
Ling, Z., 339
Lionetti, K. A., 469
Lipton, H. L., 134

Lisowa, O., 14
Littman, D. R., 339
Liu, C. C., 480
Liu, D. J., 51
Liu, F. Y., 315, 399
Liu, H., 81, 419, 439
Liu, J., 74, 76, 151, 182
Liu, L. I., 382
Liu, S., 438, 488
Liu, Y., 147, 435, 465
Livolant, F., 233, 315
Lizonova, A., 480, 482
Llamas-Saiz, A. L., 144, 168, 239, 240
Lo Conte, L., 120
Lo, P., 392
Locker, J. K., 389, 390
Loeffler, F., 466
Loesch-Fries, L. S., 5, 137
Logan, A. J., 268
Logan, D., 54, 81, 155
Lomonossoff, G., 14, 15, 131, 134, 135, 204, 223, 224, 226, 416
Lonberg-Holm, K., 480
Long, D. G., 457
Longnecker, R., 147
Lopalco, L., 340
López de Turiso, J. A., 421
Lorenzo, L. J., 3
Lostaglio, S., 463
Loudon, P. T., 30, 68, 230
Love, R. A., 137
Loveland, J., 416
Lovgren, S., 153, 220, 222
Lowery, D. E., 265, 268
Lowy, D. R., 22, 23, 433, 434
Lu, G., 43, 55, 61, 72, 137, 157, 230, 369, 387, 385, 391
Lu, H. H., 421
Lu, M. M., 74, 76, 151, 182, 340, 341, 342, 347, 480
Lu, P., 265
Lu, X. Y., 68, 73, 74, 98, 230
Lu, Y. E., 373
Lucas, R. W., 53, 55, 138
Lucia-Jandris, P., 436, 457
Luckow, V. A., 9
Ludtke, S. J., 47, 95, 96, 98, 101, 102, 109, 117, 119, 120, 121
Ludwig, K., 327, 332, 333, 334
Luger, K., 42
Lui, F. Y., 392
Luo, M., 38, 79, 130, 131, 134, 141, 144, 154, 155, 161, 168, 186, 239, 418, 421, 422, 473
Lurz, R., 304, 403
Lusa, S., 365
Lutz-Wallace, C., 280, 402
Lvov, Y. M., 315
Lyles, D. S., 51
Lyon, M. K., 5, 23

M

Ma, S. Y., 339
Ma, Y. M., 5
Maaronen, M., 133
Macadam, A., 130, 156, 471
Machat, H., 476
Macnab, R. M., 313
Macosko, J. C., 337
Madelaine, M. F., 82, 426
Mader, A. W., 42
Maeda, A., 136, 153
Maeder, M., 396
Maerz, A. L., 342, 350
Maguire, T., 480
Mah, D. C., 459
Mahfouz, I., 486
Maigret, B., 76, 151, 182
Maillard, P., 3
Mainprize, T. H., 389
Maizel, J. V., 305
Maizel, J. V., Jr., 260
Major, F., 58
Maki, S., 305
Makowski, L., 128, 143, 170
Malashkevich, V. N., 140, 151, 162, 163, 186, 332, 339, 340, 341, 350
Malby, R., 1, 39, 41, 48, 54, 57, 61, 65, 68, 69, 70, 71, 72, 73, 119, 142, 164, 221, 230, 231, 232, 234, 294
Malcolm, B. A., 134
Malinski, J. A., 260, 266, 278, 283, 309, 310
Malkin, A. J., 53, 55
Mallet, F., 74, 75, 77
Mamon, H., 22
Mancini, E. J., 38, 39, 43, 44, 45, 46, 47, 56, 57, 61, 98, 354, 366, 369

Mandel, B., 411
Mandl, C., 131, 137, 156, 354, 355, 356, 367, 369, 372, 373
Maniloff, J., 126, 138, 139, 141, 142, 144, 148, 151, 171, 229, 235, 239, 380
Mann, M. A., 436
Mannarino, A. F., 137
Mano, Y., 238
Manser, C., 43, 47
Mansfield, R. W., 268
Mao, H., 151
Marchler-Bauer, A., 355, 356
Mari, J., 55
Marietta, E., 265, 271, 280, 414, 415
Marin, M., 335
Marino, M. W., 480
Markl, J., 43, 46
Markoff, L., 387
Markosyan, R. M., 343
Marks, B., 483
Marlin, S. D., 491
Marlor, C. W., 491, 473
Marlovits, T. C., 80, 81, 424, 425
Marquardt, M. T., 372
Marshall, D., 202
Marshall, J. J., 25, 30, 52
Martelli, G. P., 413, 414, 416, 421, 425, 430, 431, 433, 435
Martin, C. S., 45, 64, 65, 67, 246
Martin, J., 327, 330
Martin, K. A., 345
Martin, S. R., 265, 268, 335, 356
Martina, Y., 486
Martincic, K, 211, 293
Martinez, M. A., 477, 483
Martin-Padura, I., 463
Marvik, O. J., 289, 306
Marvin, D. A., 51, 143
Masker, W. E., 280
Mason, D., 344
Mason, H. S., 14, 26
Mason, P. W., 422, 470, 477
Massariol, M. J., 281
Massiah, M. A., 150, 151, 181
Massung, R. F., 382
Matadeen, R., 43, 46, 47, 48, 99, 103, 105
Mat-Arip, Y., 59
Mateu, M. G., 81, 431, 432, 433, 460, 477, 483
Mathias, P., 64, 428, 429, 477, 480, 481, 482, 483, 484, 485
Mathieu, M., 70, 71, 82, 137, 142, 165
Matsubara, Y., 350
Matsudaira, P., 44, 47
Matsunaga, Y., 15
Mattes, R., 5
Matthews, B. W., 132, 157
Matthews, D. A., 134
Matthews, S., 181
Matthews, T., 342, 350
Matusickkumar, L., 396
Matusick-Kumar, L., 263
Maurizi, M. R., 305, 315
Maxwell, K. L., 146
Mayer, A., 76
Mayne, J. T., 423
Mayo, M. A., 126, 138, 139, 141, 142, 144, 148, 151, 171, 221, 222, 229, 235, 239, 380, 413, 414, 416, 421, 425, 430, 431, 433, 435
McAuley-Hecht, K., 147
McBride, H. J., 15, 339
McCaan, P. J., III, 263
McClelland, D. A., 155, 293, 418, 491, 473, 482
McClelland, D. M., 399
McClure, J., 76, 151, 180, 181
McCormick, M. C., 280, 315, 402
McCourt, D. W., 371
McCullough, K. C., 412
McCutcheon, J. P., 76, 77, 151, 180, 181
McDanal, C., 342
McDermott, B. M., Jr., 56, 79, 155, 470, 471
McDermott, E., 480
McDonnell, P. A., 143, 171
McDougall, I. M., 268, 396
McDowall, A. W., 221
McElroy, H. E., 132, 134
McGeoch, D. J., 126, 138, 139, 141, 142, 144, 148, 151, 171, 229, 235, 239, 278, 380, 382
McGinn, B., 155, 474, 476, 477
McGough, A., 221, 266, 283, 283, 309, 310
McGregor, A., 28
McKee, T. A., 134
McKenna, R., 38, 54, 55, 56, 127, 130, 143, 144, 168, 169, 236, 237, 238, 261, 263, 264, 265, 266, 281, 285, 284, 293
McKimm-Breschkin, J., 140, 162, 351
McKinlay, M. A., 130, 131, 134, 202, 203, 422, 473

McKinney, B. R., 135, 137, 153, 154, 158, 209
McLain, L., 15
McMillan, N. A. J., 434
McMurray, C. T., 58
McNab, D., 396, 400, 403
Mcnab, S. J., 28, 266
McNally, M. L., 140, 162, 163, 186, 332, 341
McPhee, D. A., 457, 458
McPherson, A., 53, 55, 138, 153, 154, 222, 224, 225, 242
McPherson, C. E., 302, 316, 317, 319
McPhie, P., 235
McQueney, M. S., 147
Meech, S. J., 481
Meier-Ewert, H., 329, 350
Meining, W., 55
Meints, R. H., 381, 390
Meisel, H., 8
Melancon, P., 157
Melikyan, G. B., 343
Mellema,J. E., 49, 326
Melnick, J. L., 229
Meloen, R. H., 15, 421
Mely, Y., 53
Mendelsohn, C. L., 421, 469
Menendez, I., 3
Menon, S., 147
Merkel, G., 151
Merluzzi, V. J., 491
Merryweather, A., 15
Mertens, P. P., 1, 39, 41, 48, 54, 57, 61, 65, 68, 69, 70, 71, 72, 73, 74, 142, 164, 221, 230, 231, 232, 234, 294
Mesyanzhinov, V.V., 38, 56, 58, 60, 65, 145, 176, 178
Metcalf, P., 458
Metsikkö, K., 367, 387
Meyer, A. M., 491
Meyers, N., 181
Miao, J., 53
Michel, H., 211, 212, 293
Michie, A. D., 120
Mierendorf, R., 315
Mikhailov, A. M., 136
Mikhailov, M., 29
Mikheeva, G. V., 480
Milano, E., 481
Milhollen, M. A., 140, 162, 163, 186, 332, 341
Miller, A., 49, 343, 347, 388
Miller, J. H., 265
Miller, L. C., 477
Miller, L. K., 388, 389
Miller, M., 151
Miller, S., 146, 176, 177
Mills, H., 181
Mills, I. G., 483
Minagawa, T., 280, 311
Mindich, L., 235
Miner, J. N., 11
Minor, I., 130, 131, 133, 134, 155, 156, 422, 473
Minor, P. D., 130, 154, 156, 471
Minor, W., 41, 132, 157, 158, 186, 369, 423
Mirza, U. a., 151
Mischak, H., 424
Mischler, R., 335
Mitchell, J., 384
Mitchell, R. S., 133
Mitraki, A., 65, 147, 174
Miyajima, S., 49
Miyamura, T., 51, 63
Mizuno, M., 3
Mizutani, S., 418
Modelska, A., 14, 15
Moesby, L., 373
Molineux, I. J., 244, 310
Momany, C., 76, 151, 180, 181
Monaco, S., 74, 75, 77
Monath, T., 229, 372, 423
Monroe, S. S., 226
Montross, L., 22
Moody, M. F., 399
Moomaw, E. W., 137
Moore, C., 480
Moore, J. P., 339, 340, 343
Moore, K. W., 147, 183
Moore, M., 483
Moore, N. C., 350
Moore, P. B., 47, 76
Morais, M. C., 38, 39, 58, 59, 60, 146, 168, 176, 243, 248, 304, 403
Morales, J., 3
Morash, B. D., 436, 457
Moreland, R. B., 22
Morellet, N., 53, 76, 151, 182
Morgan, C., 301
Morgan, D., 305, 480
Morgan Warren, R. J., 47, 76
Morgunova, E., 136
Morissy, C. J., 426

Morita, M., 311
Morris, T. J., 225
Morrison, L. A., 458
Morrison, M. E., 469
Moscufo, N., 467
Moser, R., 80, 81, 424
Mosialos,g., 340
Mosimann, S. C., 134
Moss, B., 3, 11, 388, 390
Mosser, A. G., 38, 79, 81, 130, 134, 154, 155, 186, 203, 411, 412, 418, 419, 421, 422, 439, 461, 473
Mothes, W., 344
Moulai, J., 39, 148, 171, 215
Mould, A. P., 477
Muckelbauer, J. K., 133, 155
Mudhandiram, D. R., 151
Mueller, S., 79, 155, 470, 473, 474, 489
Muhlberg, A. B., 483
Muir, T., 429, 484, 485
Mukai, R., 236
Mukherjee, S. K., 436, 460
Mukhopadyay, S., 38, 54, 61, 63, 285, 294, 354, 366, 368, 369, 370, 372
Mullaney, J. M., 313
Mullen, M. M., 147
Muller, B., 3, 5
Muller, D. J., 304
Muller, M., 23
Mullins, J. I., 349
Munowitz, M. G., 51
Munshi, S., 54, 127, 135, 138, 153, 154, 198, 205, 206, 209
Murali, R., 65, 147, 173, 185
Murata, M., 327
Murcko, M. A., 418
Murdin, A., 422
Murialdo, H., 275, 306, 310, 319
Murphy, F. A., 126, 139, 141, 142, 144, 148, 151, 168, 185, 365, 413, 414, 416, 421, 421, 425, 430, 431, 433, 435
Murphy, P. M., 345
Murray, M. D., 383
Murshudov, G. N., 50
Murthy, M. R., 136, 138, 226
Murzin, A. G., 120
Mushinski, J. F., 434
Music, C. L, 237, 261, 265
Mussgay, M., 158
Myszka, D. G., 345

N

Nabel, G. J., 379, 480
Nagao, M., 49
Nagashima, K., 3
Nagesha, H. S., 426
Naitow, H., 136
Nakagawa, N., 15
Nakano, K., 284, 285
Nakasu, S., 304
Nakatani, Y., 479–99
Nakonechny, W. S., 261
Namba, K., 52, 128, 137, 227, 228, 229, 305
Nambudripad, R., 143, 170
Nar, H., 55
Narayan, S., 344
Narvanen, A., 199
Nason, E. L., 230, 436, 460
Nassal, M., 430
Natarajan, P., 25, 135, 198
Navaza, J., 61, 63, 70, 71, 76, 142, 165, 294, 354, 367, 372
Nave, C., 143
Navruzbekov, G. A., 60, 145, 178
Nayudu, M. V., 136
Neff, S., 477
Nelson, B., 226
Nelson, R. A., 271
Nemerow, G. R., 64, 428, 429, 476, 477, 480, 481, 482–2, 483, 484, 485, 486, 487
Nermut, M. V., 77
Nestorwicz A., 371
Neubauer, C., 424
Neumann, E., 80, 81, 424
Neurath, A. R., 342
Newcomb, W. W., 15, 27, 28, 171, 244, 245, 263, 266, 268, 269, 271, 285, 291, 302, 303, 319, 391, 392, 396, 398, 399, 400, 402, 403, 427, 433
Newman, J., 431, 432
Newman, J. L., 151
Newman, J. W., 54, 79, 81, 134, 155, 432, 433, 477, 478, 479
Newsome, J. A., 23
Ng, A. C., 372
Ni, C., 133
Ni, H. L., 366

Nibert, M., 1, 15, 39, 54, 65, 68, 72, 73, 118, 142, 145, 165, 167, 221, 230, 234, 235, 435, 436, 456, 457, 458, 459, 460, 461, 461, 462, 463, 464
Nicholson, P., 27, 28, 396
Nicole, M., 206
Nieva, J. L., 373
Nilssen, O., 290
Nilsson, L., 387
Nishikawa, J., 479–499
Nissen, P., 47, 76
Nitkiewicz, J., 3
Noah, M., 430
Nobis, P., 469
Noble, S., 230
Nogales, E., 121
Noji, H., 305
Nokling, R. H., 289, 290, 306
Nomoto, A., 422, 469
Nonnenmacher, B., 23
Normark, S., 476
North, A., 5
Nouri, A., 347
Novak, C., 268
Novelli, A., 74, 75, 77
Novy, R., 315
Nuesch, J., 240
Nunberg, J. H., 339
Nusrat, A., 435, 465

O

Oas, T. G., 342
Oberhaus, S. M., 457, 458, 464
Oberste, D. J., 285, 287
Ochoa, W. F., 81, 432, 433
O'Connell, J., 471, 491
Oda, Y., 136
Odijk, T., 319
O'Donnell, K., 147
Ogawa, W., 486
Ogryzko, V. V., 479–499
O'Hara, R. S., 71
O'Higgins, N., 480
Ohnishi, S., 327
Oien, N. H., 268
Ojala, P. M., 234
Okada, S., 284, 285
Okaka, Y., 15
Okin, P. M., 481
Okun, V., 424
Old, L. J., 480
Oldstone, M. B. A., 327, 330
Olins, P. O., 9, 163volch
Oliveira, M. A., 130, 134, 155, 156, 418, 422, 473
Olland, A. M., 73, 142, 167, 436, 458
Olson, A. J., 50, 136, 153, 220
Olson, C., 171, 433
Olson, N. H., 5, 38, 39, 49, 56, 58, 59, 60, 61, 67, 78, 79, 80, 81, 82, 94, 111, 113, 130, 146, 155, 168, 171, 176, 198, 236, 237, 238, 243, 246, 261, 263, 265, 274, 285, 302, 304, 306, 307, 319, 365, 366, 389, 391, 393, 401, 403, 411, 418, 419, 423, 424, 433, 438, 439, 459, 460, 460, 473, 474, 476
Omichinski, J. G., 182
Onorato, L., 399, 400
Oostergetel, G. T., 49
Opalka, N., 136, 206
Opella, S. J., 51, 143, 170, 171
Operschall, E., 480
Oppenheim, A. B., 263
Opstelten, D. -J. E., 350
Orengo, C. A., 120
Orlova, E. V., 43, 46, 47, 48, 99, 103, 105, 403
Ortega, J., 315
Orth, G., 23
Ortiz, A., 354
Ose, V., 152
O'Shannessy, D., 147
Osman, O., 319, 320
Ostendorp, R., 5, 20, 370
Osterhaus, A. D., 326, 421
Otake, K., 487
Otto, M. J., 131
Otwinowski, Z., 41
Overbaugh, J., 343, 347, 349
Overman, S. A., 319, 320
Overton, I., 70, 232
Owen, K. E., 387
Oxelfelt, P., 135

P

Paatero, A. O., 235
Pacitti, A. F., 462

Padlan, E. A., 438
Paesold, g., 335
Pahl, H. L., 465
Pak, J. Y., 281
Palasingam, P., 289, 290
Palfi, S., 339
Palmenberg, A. C., 131, 134, 155
Palmer, I., 151
Palmer-Hill, F. J., 23
Palmier, M. O., 281
Palukaitis, P., 437
Panda, B. R., 347
Pang, J. X., 140, 162, 163, 186, 332, 341
Pantaloni, D., 260
Panzeri, C., 463
Pape, T., 43, 46, 47, 48
Pappalardo, L., 50, 75, 76, 181, 182
Pardes, A. M., 366
Pardoe, I., 423
Pardon, J. F., 48
Paredes, A. M., 365, 423
Parge, H. E., 137
Park, G., 438
Park, Y., 116
Parker, M. H., 43, 266, 269, 275, 276, 277, 278, 279, 280, 281, 282, 283
Parkos, C. A., 435, 465
Parr, R., 243
Parren, P. W. H. I., 459, 461
Parrish, C., 130, 144, 168, 186, 420, 421
Parrott, R. H., 428
Parry, D. A. D., 305
Parry, N., 54, 81, 425, 474, 476, 477
Parsons, J. M., 382
Paschall, C., 181
Pastan, I., 429, 487
Patel, S. R., 481
Pattanayek, R., 51, 52, 128, 227, 228, 229
Patterson, S., 483, 486
Patton, J. T., 235
Patwardan, A., 43, 46, 47, 48
Paul, R., 347, 462
Paulson, J. C., 159
Pauptit, R. A., 50
Pautsch, A., 147
Paxton, J., 51
Payne, C. C., 74
Payne, E., 434
Payrastre, B., 486
Pearl, L., 147
Pederson, D. M., 51
Peeters, K., 13
Peisajovich, S. G., 53, 353
Pelegrin, M., 335
Pelletier, S. L., 327, 342
Penczek, P., 43, 46, 47, 48
Peng, X., 79, 155, 470, 473, 474, 489
Pennock, D., 3
Perera, R., 370
Perevozchikova, N. A., 230
Perez, L., 467, 474
Perez-Alvarado, G., 76, 151, 182
Perham, R. N., 51
Perozzo, R., 147
Perricaudet, M., 486
Perry, K., 138, 431, 437, 438
Perry, N., 155
Person, S., 263, 264, 400, 403
Pesavento, J. B., 233
Peter, I., 480
Peters, C. J., 379
Peters, S., 380
Peterson, C., 243
Peterson, D. L., 12, 151, 181
Peterson, P., 262, 423, 424
Petitpas, I., 70, 71, 82, 142, 165
Petrie, B. L., 235
Petrovskis, I., 152
Pevear, D. C., 130, 134, 155, 422, 471, 473
Pfeiffer, P., 204
Pfister, H., 433
Pfistermueller, D., 134
Phalen, T., 372
Phelps, D. K., 202
Philipson, L., 480, 481
Pickkup, D. J., 435
Pickos, B., 482
Pickrell, J., 476
Pickup, D. J., 458
Piechaczyk, M., 335
Piemont, E., 53
Pierschbacher, M. D., 476
Piirainen, L., 199, 477
Pike, B., 457, 464
Pilger, B., 147
Pinter, A., 347, 350
Pique, M. E., 118
Plank, C., 476

Pletnev, S. V., 38, 54, 61, 63, 143, 170, 261, 266, 281, 285, 289, 293, 294, 354, 366, 368, 369, 370, 372
Plomp, M., 53, 55
Plonk, K., 228
Plucienniczak, A., 14
Pluckthun, A., 146
Pniewski, T., 14
Poignard, P., 339, 343
Pon, R. T., 463
Poncet, D., 230, 235
Porta, C., 14, 15, 416, 417
Portner, A., 140, 162
Possee, R., 8
Post, C. B., 202
Poteete, A. R., 271, 280
Pothier, P., 70, 71, 82, 142, 165
Potts, J. R., 479
Poulos, B. T., 55
Poumbourious, P., 150, 340, 342, 350
Pous, J., 146, 243
Powell, R. D., 308
Prahadeeswaran, D., 138
Prasad, B. V., 2, 3, 45, 54, 65, 68, 70, 71, 72, 73, 102, 109, 111, 113, 114, 115, 135, 142, 154, 165, 226, 229, 230, 232, 233, 234, 235, 265, 271, 273, 280, 306, 310, 365, 366, 392, 399, 414, 415, 423, 427, 436, 460
Prchla, E., 476
Preikschat, P., 8
Preston, V. G., 27, 28, 268, 315, 396
Prevelige, P. E., Jr., 242, 265, 266, 269, 271, 273, 275, 276, 277, 278, 279, 280, 281, 282, 283, 289, 291, 302, 306, 307, 399
Prevelige, P.E., 98, 309, 310, 310, 319
Price, N. C., 293, 399
Pringle, C. H., 235, 239
Pringle, C. R., 126, 138, 139, 141, 142, 144, 148, 151, 171, 229, 380
Pritsch, C., 221, 222
Prongay, A. J., 76, 151, 180, 181
Prota, A. E., 142, 147
Puente, X. S., 483
Puijk, W. C., 421
Pumpens, P., 8, 133, 152
Puri, A., 346
Pushko, P., 12, 22
Pyzalski, R., 339

Q

Qanungo, K. R., 55
Qi J., 382
Qi, Y. M., 434
Qian, C., 281
Qiao, H., 327, 342
Qiu, X., 147
Qu, C., 136
Quackenbush, S. L., 349
Quax, P. H., 482
Qui, X., 281
Quine, J. R., 141, 162

R

Racaniello, V. R., 56, 79, 155, 221, 421, 466, 467, 469, 470, 471
Rahman, M. A., 43, 47
Ramakrishnan, V., 47, 76
Ramig, R. F., 235
Ramsdale, E. F., 342
Ranz, A., 421
Rao, A. L. N., 437
Rao, J. K., 151
Rao, N. S., 151
Rao, V. B., 285, 393, 401, 402
Rao, Z., 74, 77, 151
Rappuoli, R., 476
Raptis, L. H., 488
Rasmussen, B., 334
Rasschaert, D., 82, 426
Ratner, L., 3, 15
Rauma, T., 487
Ravantti, J. J., 39
Ray, C. A., 435, 458
Rayment, I., 39, 131, 136, 205, 220
Razanskas, R., 12
Readody, D. S., 133
Reddy, K. C., 373
Reddy, V., 10, 24, 52, 135, 137, 138, 139, 141, 142, 144, 148, 153, 154, 164, 198, 200, 202, 209, 227, 482
Redman, R. M., 381
Reichman, R. C., 22
Reilly, B. E., 271
Reilly, K. E., 319, 320
Rein, A., 5
Reinemer, P., 146, 176, 177

Reinisch, K. M., 1, 39, 54, 65, 68, 72, 142, 165, 167, 221, 230, 234, 460, 461
Reisser, W., 390
Ren, R., 469
Renhofa, R., 152
Rey, F. A., 70, 71, 76, 82, 131, 137, 142, 156, 165, 294, 354, 355, 367, 368, 372
Rey, R. A., 61, 63, 76
Reza Ghadiri, M., 227
Rhee, S. S., 294
Rice, C. M., 365, 366, 368, 371, 423
Rice, C. R., 423
Rice, L.M., 162
Rice, P. A., 151
Richards, B. M., 48
Richards, K. E., 241, 315
Richardson, D. L., 261, 269, 270
Richardson, D. L., Jr., 236, 269
Richardson, S. M., 12, 48
Richmond, R. K., 42
Richmond, T. J., 42
Richter, K., 389, 390
Rieder, E., 79, 80, 470, 477
Rigolet, P., 334
Riikonen, T., 477
Rimmelzwaan, G. F., 326
Rimmelzwaan, G. J., 421
Rindge, B., 428
Rini, J. M., 460
Rixon, F. J., 27, 28, 47, 94, 96, 97, 98, 101, 102, 105, 106, 109, 113, 114, 115, 118, 121, 221, 244, 266, 268, 285, 291, 293, 302, 315, 384, 389, 391, 392, 394, 395, 396, 399, 400, 402, 403, 427
Roberts, M. M., 64, 130
Roberts, R. J., 480
Robey, F. A., 217, 399
Robins, W. P., 244
Robinson, C. V., 52, 53
Robinson, I. K., 49, 50, 205
Robinson, J., 74, 75, 76, 151, 183, 184, 345, 346
Rodarte, R., 477
Roden, R. B., 22, 23, 94, 99, 103, 105, 171, 433, 434
Rodgers, P. B., 15
Rodgers, S. E., 457, 464, 465
Rodkey, J. A., 491
Rodnina, M. V., 43
Roelvink, P. W., 480, 481, 482
Rogove, A., 467
Roitberg, B. Z., 339
Roivainen, M., 199, 477
Roizman, B., 229, 315, 317, 399
Rollence, M. L., 480, 482
Romano, M., 463
Ronacher, B., 80, 81, 424
Rong, L., 344, 356
Ronto, G., 315
Roof, L. L., 27, 28, 403
Roof, W. D., 270
Roos, N., 389, 390
Roossinck, M. J., 437
Roques, B. P., 53, 76, 150, 151, 182
Rosé, J., 334
Rose, R. C., 5, 22, 23
Rosenberg, A. H., 242, 261, 302, 315, 316, 317, 319
Rosengart, T. K., 481
Rosenkranz, H. S., 301
Rosenthal, P. B., 329, 350
Rosolowsky, M., 12, 22
Ross, M. J., 315
Ross, P., 220
Ross, P. D., 285, 398, 402
Rossel, A., 61, 63, 76
Rossjohn, J., 43
Rossmann, M. G., 3, 38, 39, 45, 54, 55, 56, 58, 59, 60, 61, 63, 65, 67, 76, 79, 80, 81, 126, 127, 128, 130, 131, 132, 133, 134, 135, 136, 137, 143, 144, 145, 146, 151, 154, 155, 156, 157, 158, 167, 168, 169, 170, 176, 178, 180, 181, 186, 205, 220, 223, 225, 226, 237, 238, 239, 240, 246, 261, 265, 266, 281, 285, 289, 293, 294, 304, 306, 354, 365, 366, 368, 369, 370, 387, 372, 389, 390, 391, 418, 421, 422, 423, 470, 471, 473, 474, 489
Roth, M., 50, 221
Roth, R., 486
Rothberg, P. G., 221
Rothlein, R., 491
Rothnagel, R., 226, 230, 232, 234, 365, 366, 423
Roussel, A., 137, 294, 354, 367, 372
Roux, K. R., 426
Rowlands, D., 38, 39, 54, 79, 81, 134, 154, 155, 223, 422, 425, 432, 474, 476, 477, 485
Rowlands, J. A., 41

Rowsell, S., 50
Roy, P., 9, 29, 30, 31, 45, 54, 68, 71, 72, 74, 77, 142, 163, 164, 165, 230
Rozhon, E., 471, 491
Ruben, D. H., 436
Rubin, D. H., 457, 462
Ruco, L., 463
Rueckert, R. R., 24, 81, 130, 134, 153, 154, 155, 156, 186, 202, 203, 220, 224, 226, 227, 411, 412, 417, 418, 419, 420, 421, 422, 439, 461, 473
Ruggeri, Z. M., 484
Ruggieri, A., 349
Ruigrok, R. W., 49, 326, 331, 334, 335, 356, 480, 482, 484
Ruiz-Jarabo, C. M., 80, 432
Ruoslahti, E., 476, 487
Russel. M., 170
Russell, C. J., 337
Russell, K. L., 366
Russell, P. M., 53, 67
Russell, R., 181
Russell, S. J., 349, 350
Russell, W. C., 483, 486
Ruttan, T., 39, 43, 45, 47, 56, 61, 64, 65, 67, 76, 77, 98, 246, 319, 354, 366, 368, 369
Rux, J. J., 65, 147, 173, 185
Rydman, P. S., 39, 65
Rysa, T., 199
Ryu, J. S., 372

S

Saad, A., 28, 47, 96, 102, 121, 293, 400
Sa-Carvalho, D., 477
Sacher, R., 226
Saeki, K., 136
Sagi, I., 76, 151, 182
Sahli, R., 39, 148, 171, 215
Saibil, H. R., 43, 48
Sakaguchi, K., 182
Sakalian, M., 294
Sakaue, H., 486
Saksela, K., 151
Salas, J., 381
Salas, M. L., 381
Sali, A., 48
Salone, B., 486
Salunke, D. M., 22, 260
Samal, B., 12
Samal, S. K., 230
Sampson, L., 266, 269, 275, 276, 277, 282
Sampson, M., 51
Samuel, A., 79, 432, 433, 477, 478, 479
San Martin, C., 319
Sanchez, A., 379
Sanchez-Vizcaino, J. M., 71
Sandbrink, H., 380
Sanders, D. A., 347, 350
Sanger, F., 236, 265, 284, 285
Santisteban, A., 381, 390
Santti, J., 133
Sanz, A., 421
Saphire, E. O., 339, 459, 461
Sapp, M., 22, 23
Sargent, D. F., 42
Sargent, M. D., 436, 457
Sarvetnick, N., 480
Sasagawa, T., 12, 22
Sasnauskas, K., 12
Sastri, M., 138
Sather, S. K., 261, 262
Sato, M., 49, 63
Sato Miyamoto, Y., 49, 63
Satoh, O., 51, 63
Satoi, J., 3
Sattentau, Q. J., 76, 343
Savithri, H.S., 136, 138, 220
Savv, R., 147
Sayre, D., 53
Scapozza, L., 147
Scarborough, J. D., 339
Schaack, J., 481
Schalich, J., 76, 354, 369, 373
Schalk-Hihi, C., 147
Schaller, H., 430
Schaller, U., 335
Schatz, M., 43, 46, 47, 48
Scheaffer, A. K., 396
Scheemann, A., 198
Schelling, P., 147
Schenk, P., 5
Scherneck, S., 12
Schiff, L. A., 15, 73, 142, 167, 436, 456, 458, 459, 460
Schiller, J. T., 22, 23, 94, 99, 103, 105, 171, 433, 434
Schirmacher, P., 22, 23

Schlaepfer, D. D., 480, 487
Schlegel, R., 23
Schlesinger, S., 365
Schlunegger, M. P., 139
Schmaljohn, A. L., 412, 423
Schmaljohn, C., 3
Schmeissner, U., 265
Schmid, C., 55
Schmid, F. X., 265
Schmid, M. F., 44, 47, 101, 221
Schmid, S. L., 483, 486
Schmid, V. F., 394, 400
Schmidt, K. N., 465
Schmidt, M. R., 482
Schmidt, R., 43, 46, 47, 48, 226
Schmidt, T., 78, 79, 81, 82, 131, 134, 135, 138, 153, 204, 223, 224, 226, 416, 419, 438, 459, 460
Schmitz, A., 265
Schmitz, I., 437
Schneemann, A., 24, 25, 52, 202, 209, 227
Schneemann, M., 463
Schneider, B. E., 457, 464
Schneider, B. J., 140, 162, 163, 186, 332, 341
Schnell, F. J., 435, 465
Schober, D., 476
Schoehn, G., 43, 81, 82, 147, 433
Schoepp, R. J., 365, 366
Schoffel, A., 480
Scholthof, H. B., 14
Scholthof, K. B., 14
Schonberg, M., 457
Schrag, J. D., 392
Schuck, P., 235
Schuler, W., 150
Schulman, J. L., 412
Schultz, L. D., 12, 22
Schulz, G. E., 147
Schulz, T. F., 340
Schulze-Gahmen, U., 460
Schutt, C. E., 39, 74, 136, 153, 220, 221, 234
Schwartz, J., 471, 491
Schwihla, H., 424
Scott, L. R., 266
Scotti, P., 134, 155
Scraba, D. G., 131, 134
Seckler, R., 146, 176, 177
Sedlak, P. L., 387
Selling, B. H., 153
Semenza, G., 335
Semler, B. L., 221
Sen, R., 350
Seoane, J., 3
Serwer, P., 280, 304, 305, 307, 310, 315, 317, 319, 320
Seth, P., 429, 487
Sever, S., 483
Severson, T. J., 435
Sevilla, N., 432
Sgro, J., 118, 421, 461
Sha, B., 141, 161
Shabanowitz, J., 211, 212, 293
Shah, K. V., 15
Shai, Y., 53, 337, 353
Shangguan, T., 327
Shanks, M., 131, 134, 135, 223, 224, 226, 416
Shao, Z., 59
Sharma, A., 477
Sharma, R. K., 459
Sharp, P. A., 480
Sharps, A. H., 457
Shatkin, A. J., 232
Shattil, S. J., 486
Sheaffer, A. K., 28, 245, 266, 271, 398, 400
Sheehan, B., 118
Sheets, R. L., 369
Shen, S., 74, 76, 134, 151, 182
Shen, Y., 464
Sheng, S., 59
Shenk, T. E., 464, 479
Shenvi, A. B., 157
Shepard, D. A., 203
Shepley, M. P., 470, 479
Sheppard, D., 81, 477
Sheriff, S., 438
Sherman, D. M., 55, 238
Sherman, L, 480
Sherman, M. B., 44, 47, 55, 102, 104, 109, 221
Sherry, B., 130, 134, 154, 155, 186, 418, 421, 422, 457, 470, 473
Shi, J., 12, 13, 14, 26
Shibata, H., 280, 311
Shieh, H. S., 281
Shien, S., 264, 288, 288
Shimotohno, K., 3
Shin, S. J., 369
Shin, Y. K., 337
Shindyalov, I. N., 138, 139, 141, 142, 144, 148, 151

Shneider, M. M., 60, 178
Shon, K., 143, 171
Showe, M. K., 260, 301, 302, 392, 395, 396, 399, 400
Shugars, D. C., 342
Shurtleff, A. C., 366
Sibai, G., 74, 75, 77
Siccardi, A. G., 340
Siden, E. J., 263, 269
Sidman, R. L., 458
Sidorkiewicz, M., 3
Siegel, D. P., 327
Siegl, G., 421
Sigler, P. B., 132, 157
Silva, A. M., 226
Silver, J., 347
Silverstein, S. C., 457
Silverton, W. W., 438
Simmons, D. L., 463, 477
Simon, M., 423
Simons, K., 423
Simpson, A. A., 38, 39, 58, 59, 60, 144, 146, 168, 169, 176, 239, 243, 304, 403
Sims, J., 237
Singh, M., 151, 350
Singh, S. K., 315
Sista, P., 350
Siuzdak, G., 52, 53, 200, 201, 422, 460
Six, E. W., 290
Sjöberg, M., 354
Skalka, A. M., 151, 221, 466
Skehel, J. J., 74, 76, 126, 130, 140, 141, 151, 158, 159, 160, 162, 163, 182, 186, 232, 326, 327, 329, 330, 331, 332, 333, 334, 335, 336, 337, 340, 341, 342, 350, 353, 356, 372
Skelton, A., 471, 491
Skern, H., 425
Skern, T., 134
Skidmore, M. O., 269
Skinner, M. M., 143, 170
Skoging, U., 387
Skoglund, U., 50, 153, 220, 221, 222
Slasson, W., 131, 132, 134
Slater, M. R., 51
Sleysh, N., 15
Slocombe, P. M., 236, 265, 284, 285
Smale, C. J., 71
Smale, S. H., 421
Smit, J. M., 354
Smith, A. D., 471, 491
Smith, B. J., 140, 162, 351, 352
Smith, C. S., 304, 306, 310
Smith, D. E., 53, 60, 243, 244, 403
Smith, D. L., 53
Smith, K., 130, 144, 168, 186, 239
Smith, M., 236, 265, 284, 285
Smith, R. L., 457, 458
Smith, S. B., 53, 60, 243, 244, 403
Smith, T. H., 461, 462
Smith, T. J., 39, 61, 78, 79, 81, 82, 130, 131, 134, 138, 144, 168, 186, 200, 239, 365, 366, 411, 419, 422, 423, 424, 431, 437, 438, 439, 459, 460
Smith, W. W., 132, 134, 147, 281
Smith-Gill, S. J., 438
Smyth, M., 133
Snitkovsky, S., 344
Sodeik, B., 388
Söderlund, H., 370
Sodroski, J., 74, 75, 76, 151, 183, 184, 345, 346
Somasundaram, T., 144, 168, 169
Sommer, H., 265
Sommerfelt, M. A., 294
Son, M., 310
Song, J. L., 95
Song, X., 480
Song, Z., 141, 162
Soong, N. W., 339, 348
Sorger, P. K., 226
Sousa, N., 146, 243, 303, 304, 305, 403
South, T. L., 76, 151, 182
Sowder, R. C. II., 76, 151, 182
Spaan, W. J., 3
Spahn, C. M., 43, 46, 47, 48
Spall, V., 416
Spear, P. G., 302, 427, 469
Spearman, P., 3, 15
Spears, H. Jr., 94, 105, 106, 111, 112, 116
Speir, J. A., 5, 54, 138, 198, 205, 206
Spencer, E., 235
Spencer, J. V., 27, 28, 396, 398, 427
Spencer, S., 118, 459
Spik, K., 3
Spindler, K. R., 269
Spriggs, D. R., 436, 461, 463
Springer, K., 65, 147, 480, 481, 482
Springer, T. A., 491, 473, 474
Squier, M. K. T., 457, 464

St. Arnoud, D., 58
Staden, R., 52, 132, 158, 284
Stadler, E., 425
Stafford, W. F., III, 276, 282
Stahl, S. J., 48, 74, 76, 94, 99, 103, 105, 120, 128, 151, 182, 221, 308, 309, 340, 341, 342, 402, 430, 431
Stalling, C.C., 50, 75, 76, 181, 182
Stallings, W. C., 281
Staniulis, J, 12
Stanley, M., 5
Stanway, G., 133
Starich, M. R., 181
Stark, H., 43, 46, 47, 48, 327, 333
Stark, W., 143, 170
Stauffacher, C., 131, 134, 204, 223, 224, 226, 416
Staunton, D. E., 491, 473, 474
Stec, D. S., 423
Steers, G., 12, 22
Stegeman, R. A., 281
Stegmann, T., 327, 334, 335
Stehle, T., 142, 148, 171
Stehlin, C., 147
Steinbacher, S., 146, 176, 177
Steinbiss, H. H., 5
Steinem, C., 226, 227
Steinhauer, D. A., 327, 329, 330, 331, 334, 335
Steinhaus, E. A., 388
Steipe, B., 146, 176, 177
Steitz, T. A., 47, 76, 151
Stel'mashchuk, V., 136, 230
Stenzel, D. J., 22, 23
Steplewski, K., 15
Sternberg, N., 261, 262
Steven, A. C., 2, 27, 28, 38, 56, 57, 65, 79, 94, 99, 103, 105, 120, 128, 146, 155, 198, 211, 212, 213, 217, 218, 220, 221, 230, 242, 244, 263, 266, 269, 271, 280, 283, 285, 291, 293, 294, 301, 302, 303, 304, 305, 306, 308, 309, 311, 310, 313, 315, 316, 317, 319, 381, 391, 392, 393, 395, 396, 398, 399, 400, 401, 402, 427, 430, 431, 458, 459, 460, 461, 467, 468, 470, 471
Steven, Alasdair, 389
Stevens, A. M., 281
Stevens, D. J., 141, 329, 330, 331, 334
Stevenson, S. C., 480, 482, 486
Stewart, D. A., 480
Stewart, P. L., 64, 65, 67, 173, 246, 385, 429, 476, 480, 482, 483, 484, 485, 486
Stiasny, K., 354, 355, 356, 368, 369, 373
Stockley, P.G., 228
Stoltz, D. B., 390
Stoops, J. K., 392
Stoppacciaro, A., 463
Stow, E. C., 383
Stow, N. D., 383
Stowell, M. H., 483
Strandberg, B., 153, 220, 222
Strassheim, M. L., 144, 168, 421
Strauss, E. G., 157, 365, 366, 367, 369, 370, 372, 423
Strauss, J. H., 157, 349, 365, 366, 367, 369, 370, 372, 423
Streeck, R. E., 22, 23
Strelkov, S. V., 38, 56, 58, 60, 145, 178
Strelnikova, A., 152
Strick, N., 342
Strong, J. E., 459, 463
Stroud, R. M., 315
Struthers-Schlinke, J. S., 244
Stuart, A. D., 134, 142
Stuart, D., 1, 38, 39, 40, 41, 42, 45, 48, 54, 57, 61, 64, 65, 68, 69, 70, 71, 72, 73, 74, 77, 79, 81, 133, 134, 136, 142, 151, 154, 155, 163, 164, 165, 221, 223, 225, 230, 231, 232, 234, 294, 422, 425, 431, 432, 433, 477, 478.479, 485
Stuart, E., 432
Stubbs, G., 52, 128, 137, 227, 228, 229
Studdert, M. J., 420
Studier, F. W., 260, 305, 307, 309, 310, 315
Stupack, D. G., 480, 481, 483, 486, 487
Stura, E. A., 422
Sturman, E. J., 281
Sturtevant, J. M., 51
Sturzenbecker, L. J., 436, 457
Suck, D., 39, 205, 220
Suderman, M. T., 366
Sugiyama, Y., 15
Sugrue, R. J., 162
Sullivan, G. M., 144, 168
Summers, M. F., 50, 75, 76, 150, 151, 180, 181, 182, 185
Sun, M., 310
Sun, X. Y., 22, 23
Sun, Y., 266, 277, 278, 281, 282
Sun, Z. Y., 142

Sundareshan, S., 136
Sundquist, W. I., 3, 74, 75, 76, 77, 78, 150, 151, 180, 181
Sunshine, M. G., 290
Sutton, G. C., 41, 64, 65, 68, 70, 72, 73, 230, 232
Suzich, J. A., 23
Suzuki, T., 350
Svensson, U., 428
Svergun, D. I., 43, 46, 47
Swaminathan, S., 13
Swanstrom, R., 339
Sweet, R. W., 74, 75, 76, 143, 151, 183, 184, 345, 346
Swift, H., 317
Swindells, M. B., 120
Syed, R., 130, 156, 225, 422, 471
Symmons, M. F., 51
Szekely, J. G., 436, 457

T

Ta, J., 264, 265, 269, 270, 288
Tailor, C. S., 347
Takada, Y., 479
Takagi, T., 49, 63
Takahashi, Y., 136
Takase-Yoden, S., 350
Takatani, M., 3
Takemoto, S., 15
Taketo, A., 284, 285
Takimoto, T., 140, 162
Tamanini, A., 483
Tamm, L. K., 141, 160, 327
Tamura, J. K., 23
Tan, K., 74, 76, 151, 182
Tan, R, 151
Tanaka, T., 3
Tang, L., 49, 135, 203, 223
Tans, S. J., 53, 60, 243, 244, 403
Tanuri, A., 477
Tao, Y., 38, 39, 56, 58, 59, 60, 145, 146, 168, 176, 178, 243, 248, 304, 306, 403
Taraporewla, Z., 235
Tarchini, R., 380
Tardieu, A., 48, 49, 50, 51
Tars, K., 52, 53, 133, 152
Tasaka, M., 311
Tate, J., 133, 134, 155, 198
Tatman, J. D., 27, 28, 106, 396, 402, 427
Tattersall, P., 144, 168, 239, 240
Taub, J., 22
Tavares, P., 304, 403
Taveau, J. C., 426
Taylor, D., 209
Taylor, G., 140, 162
Taylor, K. M., 15
Taylor, M. D., 339
Tellinghuisen, T. L., 5, 20, 370
ten Heggeler, B., 400, 403
Tenney, D. J., 28, 245, 266, 271, 396, 398, 400
Teodoro, J. G., 464
Terwilliger, T. C., 143
Teschke, C. M., 261, 281, 282
Tessman, E. S., 262
Tessmer, U., 78
Thines, D., 3
Thiriart, C., 3
Thomas, A. A. M., 410
Thomas, D., 77, 265, 271, 276, 277, 280, 282, 288, 291
Thomas, G. J., 319, 320
Thomas, G. J., Jr., 53, 67, 235, 266, 277, 280, 281, 282
Thomas, J. J., 52
Thompson, M, 3
Thomsen, D. R., 15, 27, 28, 244, 245, 263, 265, 266, 268, 269, 285, 291, 303, 309, 392, 396, 399, 400, 402, 403, 427, 431
Thornton, J. M., 120
Thouvenin, E., 82, 426
Thuman-Commike, P. A., 94, 109, 242, 260, 266, 273, 278, 283, 283, 302, 306, 309, 310, 319, 399
Tickle, I. J., 43
Tihova, M., 25, 198, 207, 210, 230, 303
Tijssen, P., 144, 168, 169, 239
Timmins, P. A., 48, 49, 50, 65
Tito, M. A., 52, 53
Tolskaya, E. A., 412
Tomassini, J. E., 491
Tomei, L., 137
Tomita, M., 49
Tomko, R. P., 480, 481
Tommasino, M., 12, 22
Tonegawa, S., 261, 263, 269, 284
Tong, L., 76, 132, 133, 151, 157, 158, 180, 181, 186, 281, 369, 423
Tooze, J., 159, 182

Tormo, J., 134, 425
Tosteson, M. T., 457
Toth, K. S., 134
Traenckner, E. B., 465
Trahey, M., 339
Traincard, F., 343
Trapnell, B. C., 487
Trapp, B. D., 412
Traub, F., 396
Trevejo, J., 480
Trier, J. S., 457
Trin-Dinh-Desmarquet, C., 23
Trono, D., 339, 345
Trottier, M., 58
Trudel, M., 15, 31
Trueman, L., 12
Trus, B. L., 1, 27, 28, 56, 79, 94, 99, 103, 105, 155, 171, 230, 244, 263, 266, 269, 271, 285, 291, 302, 303, 304, 305, 306, 309, 310, 315, 317, 319, 381, 391, 392, 393, 396, 398, 399, 400, 401, 402, 427, 431, 433, 434, 458, 459, 460, 461, 467, 468, 470, 471
Tsang, S. K., 131, 467, 468
Tsao, J., 127, 130, 144, 186, 239
Tsernoglou, D., 147
Tsui, L., 261
Tsuji, T., 3
Tsukihara, T., 49, 136, 205, 220
Tsurudome, M., 335
Tsuruta, H., 1, 38, 39, 44, 47, 54, 56, 57, 69, 96, 101, 102, 121, 146, 175, 198, 207, 209, 212, 214, 221, 241, 282, 283
Tucker, P. A., 147
Tucker, R. C., 153
Tulloch, P. A., 140, 162, 351, 352, 353, 414
Tuma, R., 53, 67, 235, 266, 277, 280, 282, 283
Turner, B. G., 50, 150, 180, 181, 185
Turner, R. B., 76
Turpen, A. M., 15
Turpen, T. H., 15
Tuukkanen, J., 487
Tweeten, K. A., 389
Tye, B. K., 271
Tyler, K. L., 436, 456, 457, 458, 461, 463, 464, 465

U

Ubol, S., 412, 423
Ulmanen, J., 370
Ulrich, R., 12
Umeda, M., 51, 63
Unckell, F., 23
Unge, T., 50, 139, 153, 186, 220, 221, 222
Unser, M., 305
Urakawa, T., 30
Usha, R., 131, 134, 204
Ushkaryoy, 43, 47
Utsumi, H., 51, 63
Uttenthal, A., 15
UytdeHaag, F. G. C. M., 421

V

Vachette, P., 71
Vainshtein, B. K., 136
Vajdos, F. F., 74, 76, 77, 151, 180, 181
Valegård, K., 52, 53, 133, 139, 152, 186, 228
Valerio, D., 482
Vallis, Y., 483
Valpuesta, J. M., 59, 146, 243, 303, 304, 305, 310, 403
van der Vliet, P. C., 147
van der Werf, S., 221
van Driel, R., 396, 398, 403
van Etten, J. L., 67, 389, 390, 391
van Heel, M., 43, 46, 47, 48, 99, 103, 105, 114, 304, 327, 333, 403
van Hulten, M. C. W., 380
van Kammen, A., 15
Van Lent, J. W., 3
van Raaij, M. J., 65, 147, 174
van Regenmortel, M. H. V., 126, 138, 139, 141, 142, 144, 148, 151, 171, 229, 235, 239, 380
Vanetten, J. L., 381
Vaney, M. C., 82
Varghese, J. N., 141, 158, 161, 413–32
Vargosko, A. J., 428
Varmus, H. E., 344

Varnavski, A. N., 387
Vashishtha, M., 372
Vassileva, A., 13
Vautherot, J. F., 82, 426
Veijalainen, P., 421
Vela, C., 15, 421
Velez, M., 304
Vénien-Bryan, C., 61, 365, 366, 373
Venkataram Prasad, B. V., 230, 233, 234
Venter, J. C., 382
Verdaguer, N., 81, 134, 422, 424, 431, 432, 433, 460, 477, 483
Vergalla, J., 3
Verma, I., 480
Vignaux, F., 343
Vigne, E., 486
Vihinen, M., 387
Villa, A., 463
Villafranca, J. E., 132, 134
Vinuela, E., 381, 390
Virgin, H. W., 436, 457
Viscidi, R. P., 15
Vitale, R. L., 137
Vitanen, I., 477
Vlak, J. M., 3, 380
Vlazny, D. A., 383
Vogel, F., 12
Vogels, R., 482
Vogt, J., 147
Vogt, V. M., 5
Volchkov, V., 140, 162, 163, 186
Volpers, C., 22, 23
Volynski, K. E., 43, 47
von Bonsdorf, C. H., 387
von Schwedler, U. K., 76, 77, 151, 180, 181
Von Seggern, D. J., 480, 482–2, 483, 486
Von Wechmar, M. B., 204, 205
Vonderviszt, F., 305
Vonrhein, C., 47, 76
Vorherr, T., 327
Vrati, S., 423
Vrhel, M., 94, 99, 103, 105, 171, 433
Vriend, G., 130, 131, 134, 154, 155, 186, 418, 421, 473
Vrijsen, R., 412
Vuori, K., 487

W

Waddell, C. H., 270, 271
Wade, R. H., 426
Wadell, G., 482
Wag, H. M., 366
Wagner, E., 476
Wagner, G., 142, 204
Wagner, K. R., 338
Wahlberg, J. M., 354, 387
Wahn, K., 20, 365, 370
Walker, J., 151, 180
Walker, S. B., 15, 459, 460
Wall, J. S., 27, 230, 291, 305, 315, 427
Wallin, M., 350
Walton, A., 71
Wang, A. H., 143
Wang, B., 102, 104, 109
Wang, E., 366
Wang, F., 144, 168, 239, 240
Wang, G., 5, 54, 68, 111, 113, 138, 198, 205, 206, 421, 435, 436, 458, 459, 461
Wang, H., 52, 76, 77, 137, 151, 180, 181, 347
Wang, J., 74, 76, 151, 182
Wang, K., 423, 483, 487
Wang, L., 53
Wang, L. F., 426
Wang, M., 3, 26
Wang, P., 416, 417
Wang, S., 289, 290
Wang, W., 461
Wang, X., 480, 487
Wang, Y., 481, 483
Warner, S. C., 264
Warren, S., 52
Watanabe, R., 350
Watanabe, Y., 15
Wathen, M. W., 268
Watkins, S., 22, 263
Watowich, , S. J., 137, 158, 329, 334, 366
Watson, J., 38
Watson, R. H., 317, 320
Watts, A., 51
Watts, D. M., 366
Watts, N, R., 308
Weaver, L., 482
Weaver, S. C., 366
Weber, J. M., 488
Weber, P. C., 137
Weber, T., 335

Webster, P., 428, 488
Webster, R. G., 414
Wecker, K., 150
Wei, N., 225
Weigele, P., 277, 278, 282
Weiland, F., 158
Wein, M. W., 367, 372
Weiner, D., 15
Weiner, H. L., 457, 458, 470
Weiner, S., 232
Weinheimer, S. P., 263
Weinkauf, S., 55
Weintraub, A., 176, 177
Weinzett, L. N., 15
Weir, R. C., 423
Weis, W., 159
Weise, F., 403
Weiss, R. A., 340
Weissenhorn, W., 74, 76, 140, 151, 162, 163, 182, 186, 340, 341, 342, 350
Weissig, H., 138, 139, 141, 142, 144, 148, 151
Weissinger, E. M., 434
Weldon, R. A., 180
Weldon, R. A., Jr., 294
Welker, R., 77, 78
Weller, S. K., 28, 244, 245, 266, 271, 392, 396, 398, 400, 403
Wellink, J., 15
Welsh, L. C., 51
Welsh, M. J., 482
Wendt, T., 389, 390
Wengler, G., 20, 61, 63, 76, 132, 137, 157, 158, 186, 294, 354, 365, 367, 369, 370, 387, 372, 423
Wepf, R., 389, 390
Wery, J. P., 135, 200
Westaway, E. G., 387
Westbrook, J., 138, 139, 141, 142, 144, 148, 151
Westbury, H. A., 426
Weston, S. A., 50
Wetzel, J. D., 436, 460, 457, 462, 463, 464
Wharton, S. A., 327, 330, 331, 334, 335, 353, 356
Whitbeck, J. C., 147
White, B., 482
White, C. A., 133
White, J., 3, 64, 130, 365
White, J. M., 327, 332, 334, 335, 342, 344, 350
White, N. S., 48
White, W. I., 23
Wickersham, J., 133
Wickersham, J. A., 137
Wickham, T. J., 384, 428, 477, 480, 481, 482, 483, 487
Wickner, R. B., 126, 138, 139, 141, 142, 144, 148, 151, 171, 229, 230, 235, 239, 315, 380
Wickner, S., 305
Wieden, H. J., 43
Wiegand, R. C., 281
Wiegele, P., 266, 281
Wiegers, K., 3, 5
Wien, M. W., 61, 63, 76, 134, 294, 354, 422
Wikoff, W. R., 2, 38, 39, 54, 56, 57, 69, 146, 175, 211, 212, 214, 217, 218, 221, 241, 301, 313, 319, 421
Wild, C., 342
Wild, C. T., 342
Wild, D., 50
Wiley, D. C., 74, 76, 126, 130, 140, 141, 147, 151, 158, 159, 160, 162, 163, 182, 186, 326, 327, 329, 330, 331, 332, 333, 334, 335, 336, 337, 340, 341, 342, 350, 353, 356, 372
Wilingham, M. C., 429
Wilk, S., 465
Wilk, T., 3, 5, 74, 76, 77
Willetts, M., 428
Williams, J., 290
Williams, L. A., 463
Williams, P. A., 134
Williams, R. C., 241, 315
Williams, R. W., 399
Williams, W. V., 436, 457, 462
Williamson, J. R., 151
Willingham, M. C., 487
Willingmann, P., 54, 130, 169, 237, 238, 261, 265
Willis, S. H., 147
Wilschut, J., 327, 354, 373
Wilson, D. W., 396
Wilson, I. A., 141, 158, 326, 329, 330, 332, 334, 336, 460
Wilson, J. M., 459, 480
Wilson, K. S., 136
Wilson, M. E., 389
Wimberly, B. T., 47, 76
Wimmer, E., 79, 80, 155, 421, 422, 469, 470, 473, 474, 489

Wingfield, P. E., 308, 342
Wingfield, P. T., 74, 76, 94, 99, 103, 105, 120, 128, 151, 182, 221, 285, 309, 340, 341, 393, 401, 402, 430, 431
Wingfield, P. W., 48
Winkler, F. K., 38, 136, 153, 220
Wintermeyer, W., 43
Wissenhorn, W., 332
Wittek, R., 389, 390
Witteveldt, J., 380
Wittmann-Liebold, B., 157
Wittwer, A. J., 281
Witz, J., 48, 49, 50, 51
Wlodawer, A., 146, 147, 151
Wolf, J., 457
Wolf, S. G., 121
Wolff, K. L., 423
Wolfinbarger, J. B., 263
Wolfsberg, T. G., 344
Wong, C. H., 329, 350
Woo, S. D., 55
Woodland, R. T., 482
Worland, S., 132, 134
Worthylake, D. K., 74, 75, 76, 77, 151, 180, 181
Wright, B. W., 436, 459, 461, 462
Wrigley, N. G., 334, 335
Wrigley, N. G., 331
Wrobel, B., 263
Wu, E., 482–2
Wu, H., 130, 144, 167, 168, 186, 239
Wu, L. J., 392
Wu, X., 58
Wu, Z. R., 50, 75, 76, 181, 182
Wyatt, R., 74, 75, 76, 151, 183, 184, 345, 346
Wyckoff, E., 275
Wynne, S. A., 3, 26, 46, 54, 55, 57, 94, 97, 99, 104, 105, 114, 120, 128, 149, 179, 221, 385
Wypych, J., 12

X

Xia, D., 54, 126, 129, 130, 147, 169, 174, 237, 238, 261, 265, 482
Xiao, C., 79, 80
Xie, Q., 128, 130, 144, 167, 168, 169
Xie, Z., 212, 293, 310
Xing, L., 354
Xu, R., 366, 480, 481
Xu, W., 38, 246, 274, 306

Y

Yabrov, R., 199
Yaegashi, N., 22, 23
Yafal, A. G., 422, 460, 467
Yamada, G., 3
Yamaguchi, Y., 402
Yan, X. D., 389, 390, 391
Yan, Y., 39, 67, 148, 171, 215
Yanagida, M., 402
Yang, F., 146
Yang, X. J., 479–99
Yasuda, R., 305
Yau, P., 275
Yazaki, K., 238
Ye, Z., 15, 27, 291, 396
Yeager, M., 25, 49, 52, 68, 118, 135, 136, 203, 206, 207, 210, 223, 230, 232, 303, 435, 436, 458, 459
Yeagle, P. L., 51
Yee, A. A., 146
Yeh, P., 486
Yeung, M. C., 463
Yonekura, K., 305
Yoneyama, T., 51, 63
Yoo, S., 74, 76, 77, 151, 180, 181
Yosef, Y., 263
Yoshida, M., 305
Yost, S. C., 491
Young, J. A. T., 344
Young, M. J., 5, 48, 205
Young, N. S., 80, 81
Young, R., 270
Yu, D., 245
Yu, F., 5, 51
Yu, M., 3, 51
Yu, X., 68, 73, 74, 230, 392
Yu, Y. G., 337
Yuran, T. E., 382
Yusibov, V., 5, 13, 14, 15, 31, 137

Z

Zabner, J., 482
Zadori, Z., 144, 168

Zaki, S. R., 379
Zambruno, G., 486
Zdanov, A., 147
Zechmeister, T., 424
Zemlin, F., 43, 46, 121
Zeng, C. Q., 230, 232, 234
Zentgraf, H., 23
Zhang, C., 58, 134
Zhang, F., 58
Zhang, H., 68, 73, 74, 98, 143, 230
Zhang, J., 68, 73, 74, 230, 338
Zhang, J. Q., 98
Zhang, L., 437
Zhang, Q., 68, 73, 74, 230
Zhang, R., 269
Zhang, S., 469
Zhang, W., 5, 38, 54, 56, 61, 63, 77, 236, 285, 294, 345, 354, 366, 368, 369, 369, 370, 372
Zhang, X., 68, 73, 74, 230, 350, 431, 437, 438, 460
Zhang, Y. B., 65, 147, 480, 481, 482
Zhang, Y. P., 137, 158
Zhang, Z., 98, 105, 117, 242, 302, 319, 329
Zhao, H., 387
Zhao, R., 130, 134, 156, 422, 471, 473
Zhao. W., 41
Zhao, X., 5, 350
Zhao, Y., 339, 348, 429, 484, 485
Zheng, R., 151
Zhou, J., 22, 23, 480
Zhou, Z. H., 28, 43, 47, 55, 68, 72, 73, 74, 94, 97, 98, 102, 104, 105, 106, 107, 109, 111, 112, 113, 114, 115, 116, 118, 221, 230, 244, 266, 285, 291, 293, 385, 391, 392, 394, 395, 400, 402, 403, 421, 427
Zhu, L., 339, 348
Ziegelhoffer, T., 275
Ziemiecki, A., 366
Zientara, S., 1, 39, 41, 48, 54, 57, 61, 65, 68, 69, 70, 71, 72, 73, 142, 164, 221, 230, 231, 232, 234, 294
Zimmern, D., 186, 228
Zirwer, D., 327, 333, 334
Zlotnick, A., 1, 48, 94, 99, 103, 105, 120, 128, 133, 135, 221, 226, 430
Zsengellér, Z., 487
Zuccola, H. J., 56, 79, 147, 155, 470, 471
Zuo, Y., 347
Zweerink, H. J., 457

SUBJECT INDEX

A

B

C

D

E

Ebola virus, 379
  matrix protein/glycoprotein, 162–63, 163
Echovirus, receptors of, 475
EEV. *See* Extracellular enveloped virus (EEV), 390
Electron cryomicroscopy, 2, 94–101, 401–2
  field depth/resolution graph for, 100
  instrument choices in, 97–101
  theoretical consideration of, 94–101
  three dimensional reconstruction and, 101
Electron density maps, 42
Electron microscopy resolution (EM resolution), 45–46
Electron spin resonance (ESR), nuclear magnetic resonance (NMR) and, 50–51
eliminateOrt. *See* Orientation Elimination program *(eliminateOrt)*
EM resolution. *See* Electron microscopy resolution (EM resolution)
Endogenous transcription/exit pathways, of double-stranded RNA viruses, 231–232
Envelope protein structures
  in alphavirus, 369
  in flavivirus, 369, 383, 384
  of murine leukemia-related viral group (MuLV), 361
  retroviridae and, 182–184
  in retroviruses, 353, 353
Enveloped single positive-sense strand RNA viruses, 156–58
  flavivirus, 156–157
  helical tobamaviruses and, 158
  togaviruses and, 157–158
Enveloped viruses, 363–374
  alphavirus structure as, 365–367
  cell entry conformational changes in, 325–357
  flavivirus structure as, 367–369
  HA_0 as, 327–337
  influenza hemagglutinin as, 326–337
  lipid bilayer of, 364
  protein layers in, 364–365
  structural features of, 365–369
  two morphologies of, 326
Epithelial tumors, human papillomaviruses (HPVs) and, 22
Escherichia coli, 14, 15, 22, 241, 305–306
  expression problems with, 8
  heterologous expression investigation of, 3
  prokaryotic expression systems and, 4–8
  protein production and, 5, 8
  viral capsid proteins in, 5
Escherichia coli phage ϕX174, scaffolding-assisted viral assembly and, 261
ESR. *See* Electron spin resonance (ESR)
ESRF. *See* European Synchrotron Radiation Facility (ESRF)
Eukaryotic expression systems
  baculovirus as, 8–10
  plant-based, 13–15
  vaccinia virus and, 10–12
  yeast expression system and, 12–13
European Synchrotron Radiation Facility (ESRF), 40
Extracellular enveloped virus (EEV), 390

F

Fab 17-A, 82
Fab fragment, 81–82
FHV. *See* Flock House virus (FHV)
Fiber diffraction
  neutron, 51–52
  of tobacco mosaic virus (TMV), 51–52
  X-ray, 51–52
Filoviridae, 379
Flavivirus(es)
  envelope protein structures in, 369
  fusion proteins, 388
  oligomerization state switches in, 367–370
  structure of, 383–386
Flock House virus (FHV), 10, 24, 200
  MALDI-MS data of, 201
  RNAs of, 25
Flu virus. *See* influenza virus
Fluorescence resonance energy transfer (FRET), 53
FMDV. *See* Foot-and-mouth disease virus (FMDV)
Focal pair method, 110
Foot-and-mouth disease virus (FMDV), 79, 155, 431–433
  crystal structure of, 475
  heterologous expression investigation of, 3
  receptors for, 473–475
Fourier-Bessel synthesis, subnanometer reconstructions with, 113–14
Fourier intensity equations, 95

Fourier methods, of cryoelectron microscopy, 43
Fourier ring correlation (FRC), 104
FRC. *See* Fourier ring correlation (FRC)
FRET. *See* Fluorescence resonance energy transfer (FRET)
Fusion, 60–63
Fusion activation, in trans, 360–364
Fusion proteins
  alphaviruses and, 388
  flaviviruses and, 388

G

Gag particles, 74, 76, 77
β-galactosidase gene, 9, 11
GAP. *See* Glyceraldehyde-3-phosphate dehydrogenase (GAP)
Gastroenteritis, Norwalk virus (NV) and, 25
Generic phage capsid assembly pathway, 302, 303
Genome capsid layer organization, in double-stranded RNA viruses, 230–232
Genome entry model, of single-stranded DNA (ssDNA) viruses, 237
Genome packaging
  in double-stranded DNA viruses, 242
  of single-stranded RNA (ssRNA) viruses, 225
Genome release
  in single-stranded DNA (ssDNA) viruses, 241
  in single-stranded RNA (ssRNA) viruses, 226
Genome replication/packaging, in double-stranded RNA viruses, 234
Genome size, 379–380
Genome structural organization
  in double-stranded DNA viruses, 241
  in double-stranded RNA viruses, 230
  of single-stranded RNA (ssRNA) viruses, 224
Genome structure in helical ssRNA viruses, 228–229
Genomic/capsid features, of double-stranded RNA viruses, 230–231
Glyceraldehyde-3-phosphate dehydrogenase (GAP), 13
Glycoprotein configuration, of alphaviruses, 390
gp65 protein promoter, 9
gp5 subunit, structure of, 212

H

HA_0, 341–342
HA. *See* Hemagglutinin (HA)
HA_1-HA_2 complex, peptide bond cleavage/rearrangement to, 343–345
Hantaan virus, heterologous expression investigation of, 3
HBV. *See* Hepatitis B virus (HBV)
HCMV. *See* Human cytomegalovirus (HCMV)
HCV. *See* Hepatitis C virus (HCV)
Helical tobamaviruses, enveloped single positive-sense strand RNA viruses and, 158
Helical viruses, 383
helixHunter, 119
Hemagglutinin (HA)
  conformational changes of, 327–331
  proteins, 327-331
Hemagglutinin-neuraminidase, negative-strand RNA viruses and, 162
Hepadnaviridae, 178–179
Hepatitis B virus (HBV), 12, 63
  antibody-virus complexes and, 430–431
  heterologous expression investigation of, 3
  mosaic core assembly of, 8
  surface antigen of, 14
Hepatitis C virus (HCV), 3
Herpes simplex virus-1 (HSV-1), 15, 268–269, 383–384, 426–428
  heterologous expression investigation of, 4
  heterologous expression systems and, 27–29
  procapsid/capsid structure comparison of, 398
  secondary structural features of, 393–394
Herpesviridae, 209, 379, 381
Herpesviridae scaffolding protein, C-terminal interactions and, 268–269
Herpesviruses, 381, 382, 383, 391
  as triplex proteins, 292–293

Heterologous expression investigation
assembly process separation through, 3
of bluetongue virus (BTV), 4
of *Escherichia coli,* 3
of foot-and-mouth disease virus, 3
hantaan virus of, 3
of hepatitis B virus, 3
of herpes simplex virus 1 (HSV-1), 4
of human immunodeficiency virus (HIV), 3
of human papillomavirus, 3
of human parvovirus, 3
of Norwalk virus, 3
of structural proteins, 3
of viruslike particles (VLPs), 3
Heterologous expression systems, 2
advantages of, 2
alphaviruses in, 20–21
for assembly of viruslike particles, 6t–7t
in biological activity, 17–18
bluetongue virus (BTV) in, 29–31
caliciviruses in, 26
complexity of, 19
diversity of, 4–16
end use of, 16
expandability of, 18
guidelines for, 16–19
herpes simplex virus 1 (HSV-1), 27–29
multiple protein synthesis in, 17
nodaviruses, 24–26
papillomaviruses in, 21–23
polyomaviruses in, 21–23
product quantities from, 3
respiratory syncytial virus (RSV), 31
single protein in, 17
tetraviruses in, 24–26
turnaround time of, 18–19
viral assembly in, 20–30
High Five cells. *See Trichoplusia ni* cells
HIV. *See* Human immunodeficiency virus (HIV)
HK97, 73
assembly/maturation of, 211
buried surface area between prohead &head in, 215
change in subunit interactions/locations of, 214
double-stranded DNA bacteriophage and, 209–215
head, double-stranded DNA viruses and, 175
prohead II, 211
prohead-to-head transition of, 213–215
quaternary structure of, 213
structural analysis of, 56–57
subunit structure, 212–213
HPVs. *See* Human papillomaviruses (HPVs)
HRV. *See* Human rhinovirus (HRV)
HRV-16, 80, 156
HSV-1. *See* Herpes simplex virus 1 (HSV-1)
Human cytomegalovirus (HCMV), 384
Human immunodeficiency virus (HIV), 3, 388
E. coli and, 3
heterologous expression investigation of, 3
protein arrangement in, 75
receptor-binding cascade of, 345–346
as retroviruses, 338–339
structural analysis of, 74–78
trimmed structure of, 346
Human papillomaviruses (HPVs)
cervical cancer and, 22
epithelial tumors and, 22
heterologous expression investigation of, 3
Human parvovirus, heterologous expression investigation of, 3
Human rhinovirus-2, 80, 424–425
Human rhinovirus-14, 78, 82, 200, 416–419
Human rhinovirus-16, 16, 80, 86
ICAM interaction and, 80
Human rhinovirus (HRV), 39

I

ICAM1. *See* Intracellular adhesion molecule 1 (ICAM1)
Icosahedral capsid particles, 390
Icosahedral single-stranded RNA (ssRNA) viruses
pseudo T = 3 single-stranded RNA (ssRNA) viruses as, 223
RNA conformation in, 222–224
satellite ssRNA viruses as, 222–223
T = 3 single-stranded RNA (ssRNA) virus as, 223
Icosahedral viruses, 383
IMV. *See* Intracellular mature virus (IMV), 390
In vitro systems, 15–16

Influenza A, 158–62
negative-strand RNA viruses and, 158–163
Influenza HA_0, cleavage structural rearrangements on, 329–331
Influenza hemagglutinin (HA)
energetics of, 334–337
as enveloped viruses, 326–337
fusion peptide extrusion of, 332–334
hydrophobic core repacking of, 331–334, 347
jack-knifing in, 334
kinetic control of, 336
low pH exposure refolding of, 331–334
membrane fusion models of, 334–337
primary structure of, 327–329
protein folding of, 337–338
spring-loaded mechanism of, 331–332, 337–338
trimer structure hierarchy of, 329
Influenzavirus, 413–414
features of, 326–327
Inoviridae bacteriophages, 170–171
Integrin proximal domain, Ad12 penton base protein and, 480
α_v integrins, adenovirus-mediated endosome disruption and, 482–482
Internal scaffolding protein functions, ϕX174 v. P22/herpesviruses and, 267–268
Intracellular adhesion molecule 1 (ICAM1), 78
interaction, human rhinovirus-16 (HRV-16) and, 80
interaction, with rhinovirus group, 473
Intracellular mature virus (IMV), 390
Invitrogen Life Technologies, 8, 9
Iridoviridae, 379
Iridoviruses, 381
Isohedral enveloped viruses, analysis of, 38
Isohedral nonenveloped plant/insect viruses, 152–53

J

Jack-knifing
in influenza hemagglutinin (HA), 334
Junction adhesion molecule, σ 1head and, 463–464

L

Large-scale reversible quaternary structure, of viruses, 203–209
Large viruses, 379–408
accessory proteins of, 401–402
assembly mechanisms of, 394–399
conserved core genes and, 382
definition of, 379
DNA packaging of, 401–402
families of, 380
genome size and, 379–380
icosahedral capsid particles and, 390
images of, 389
maturation of, 399–401
nonicosahedral particles and, 387–390
organization complexity of, 386–392
structural analysis methods for, 385–386
structural folds in, 393–394
study of, 384–385
LDPR. *See* Low-density lipoprotein receptor (LDPR)
Lipid bilayer
of enveloped viruses, 364
host-derived, 363
Lipid distribution, in viruses, 49
Lipid molecules, poliovirus-associated, 470
Low-density lipoprotein receptor (LDPR), 80
Luteoviridae, 152, 153

M

M_1, negative-strand RNA viruses and, 161
M_2, negative-strand RNA viruses and, 161–162
Mammalian on cretroviruses, 346–350
Marburg virus, 379
Mass spectrometry
of FHV, 52–53
of VLPs, 52–53
Matrix proteins, retroviridae and, 180–182, 181
Mature capsid structure, of bacteriophage T7 procapsid assembly, 310–315
Membrane-containing isometric viruses, structures of, 60–63
Membrane fusion models, of influenza hemagglutinin (HA), 340–343

Methanol-inducible alcohol oxidase (AOX1) promoter, 13
Microviridae bacteriophages, 169–170
Morphogenetic mechanism, of bacteriophage T7 procapsid assembly, 306–308
MOSFLM, data processing software, 41
Mouse polyomavirus capsid protein. *See* VP1
Multiple protein synthesis, in heterologous expression systems, 17
MuLV. *See* Murine leukemia-related viral group (MuLV)
Murine leukemia-related viral group (MuLV)
  envelope protein complex of, 347
  receptor-binding domains of, 347–350
Mutagenesis, 54
Mutant coat proteins, heterologous expression systems and, 4

N

NωV. *See Nudaurelia ω capensis* virus (NωV), 25
National Institutes of Health (NIH), 11
NCLDVs. *See* Nucleocytoplasmic large DNA viruses (NCLDVs)
NDV. *See* Newcastle disease virus (NDV)
NDV-F. *See* Newcastle disease virus F protein (NDV-F)
Negative-strand RNA viruses, 158–163
  ebola virus matrix protein/glycoprotein, 162–163
  hemagglutinin-neuraminidase and, 162
  influenza A and, 158–163
  M_1 and, 161
  M_2 and, 161–162
  neuraminidase and, 161
  paramyxovirus fusion protein and, 162
Neuraminidase, negative-strand RNA viruses and, 161
Neutron scattering
  RNA structure and, 49–50
  satellite tobacco necrosis virus (STNV) and, 49–50
  small-angle neutron scattering (SANS) and, 48–49
Newcastle disease virus (NDV), 351
Newcastle disease virus F (NDV-F) protein, 351
  primary/tertiary structures of, 352–353
NIH. *See* National Institutes of Health (NIH)
NMR. *See* Nuclear magnetic resonance (NMR)
Nodaviridae, 152, 153, 223
Nodaviruses, 10, 24–26
Nonenveloped virus cell entry, 455–482
Nonicosahedral particles, baculoviruses as, 387–390
Norwalk virus (NV)
  capsid protein, 14
  gastroenteritis and, 25
  heterologous expression investigation of, 3
Nuclear magnetic resonance (NMR)
  electron spin resonance (ESR) and, 50–51
  spectroscopy, 126–27
  tobacco mosaic virus and, 51
Nucleocapsid proteins, retroviridae and, 180–182
Nucleocytoplasmic large DNA viruses (NCLDVs), 381–382
Nudaurelia capensis virus (NCV), 25, 207–209
  3-D sectional view of, 206–207
  capsid/procapsid views, 206–207
  X-ray scattering of, 207
NV. See Norwalk virus (NV)

O

Oligomerization state switches
  in alphaviruses, 367–370
  in flaviviruses, 367–370
  σ1 head, junction adhesion molecule and, 482–483
  σ1 protein, proteolysis of, 480–481
  σ1 tail, cell surface sialic acid and, 481–482
Optical tweezers, 53
Orientation Elimination program (eliminate Ort), 111
Orthoreovirus, 73, 232
  core structure of, 39
Overexpressed T7/T3 connectors
  12- and 13-fold symmetry in, 303
  rotational symmetry images of, 314

P

P 4 SID protein, 290–292
P 10 promoter, 9
P22 scaffolding protein
  chemistry of, 280–281
  coat protein binding in, 277–279
  domain structure of, 274
  functional domains of, 274–280
  minor proteins, recruitment of, 276–277
  oligomerization of, 275–276
  procapsid exit in, 279–280
  role of, 272–274
  small-angle X-ray scattering studies of, 281
  synthesis, autoregulation of, 274–275
P22 tailspin protein, Double-stranded DNA viruses and, 176, 177
P22 morphogenetic pathway, 271
P22 procapsid lattice, scaffolding protein location on, 292
Pac site, 246
Packaged DNA structure, of bacteriophage T7 procapsid assembly, 329–333, 330
Papillomavirus, 5, 12, 172
  warts and, 22
Papillomaviruses, in heterologous expression systems, 21–23
Papillovirus, 433–435
Paramecium bursaria Chlorella virus 1 (PBCV-1), 384, 390
Paramyxoviridae, 379
Paramyxovirus, Newcastle disease virus (NDV) as, 351
Paramyxovirus fusion protein, negative-strand RNA viruses and, 162
Pariacoto virus, 223
Particle reconstruction images, 198
Parvoviridae family, 238–240
  Desovirinae subfamily of, 239
  Parvovirinae subfamily of, 239
  of single-stranded DNA (ssDNA) viruses, 242
Parvovirus (PaV), 167–69, 419–421
PaV. *See* Parvovirus (PaV)
PBCV-1. *See Paramecium bursaria Chlorella* virus 1 (PBCV-1)
Peptide bond cleavage/rearrangement, to HA_1-HA_2 complex, 343–345
pEt vectors, 5
Phase information
  crystal structure and, 40
  reconstitution of, 40
Pichia pastoris, 12, 13
Picornaviruses, 78–79, 154–156
  receptor complexes and, 465–466
Plant-based expression system, 13–15
  advantages of, 14–15
  strategies, 13–14
Plant satellite viruses, 153
Plant virus-based vectors, 14
Plant viruses, large-scale morphological changes in, 205
Poliovirus, 421–423
  attachment receptor of, 381–382, 468–470
cell entry mechanisms and, 466–468
  cryo-EM reconstructions of, 469
Polyhedrin gene, 8
Polyomavirus, 171–172
  in heterologous expression systems, 21–23
  particles, 5
Poxviridae, 379
Poxviruses, 390
PRD1 capsid protein, Double-stranded DNA viruses and, 175
Procapsid/capsid structure, with herpes simplex virus 1 (HSV-1), 398
project3fFile, 111, 112
Prokaryotic expression systems, 4–8
  Escherichia coli and, 4–8
Protein folding, of influenza hemagglutinin (HA), 351–352
Protein-folding landscape, multiple stops on, 339–340
Protein layers, in enveloped viruses, 364
Protein production, *Escherichia coli* and, 5, 8
Protein v. virus structure analysis, 40
Proteolysis, of σ 1 protein, 461–462
pSBetB vector, 5
Pseudo $T = 3$ single-stranded RNA (ssRNA) viruses, as icosahedral single-stranded RNA (ssRNA) viruses, 224

R

Rabbit hemorrhagic disease virus (RHDV), 82
Raman spectroscopy, 67
Rauscher murine leukemia virus, electron spin resonance (ESR) and, 51

RDV. *See* Rice dwarf virus (RDV)
Recombinant plasmids, 12
Reoviridae family, 232, 236, 237
Reoviridae virus, 68
  double stranded RNA viruses and, 163–167
  genus *orbivirus* of, 163–164
  genus *orthoreovirus* of, 165–167
  genus *rotavirus* of, 165
  structure of, 69
Reovirus, 435–437
  apoptosis and, 464–465
  attachment protein of, 459
  cell entry/tissue tropism/pathogenesis in, 456–458
  cryo-EM reconstructions of, 459
  crystallographic core of, 460
  receptor interactions of, 464–465
  structure of, 458–461
Respiratory syncytial virus (RSV), in heterologous expression systems, 31
Reticulocyte/wheat germ lysate, 16
Retroviridae
  capsid proteins and, 180–182, 181
  envelope proteins and, 182–184
  gag structural proteins and, 180–182, 181
  matrix proteins and, 180–182, 181
  nucleocapsid proteins and, 180–182, 181
  transmembrane/surface glycoprotein gp120 and, 182–184
Retrovirus(es)
  avian leukkosis virus (ALV), 357–360
  avian sarcoma/leukosis viral group (ASLV) as, 352–353
  envelope protein structure of, 353
  human immunodeficiency virus (HIV) as, 352
  mammalian oncretroviruses, 360–364
  membrane fusion in, 354–357
  receptor-induced pH dependence in, 357–360
  TM protein structure, 355, 356–357
Reverse transcribing viruses, 178–184
  retroviridae and, 178–184
Rhabdoviridae, 379
RHDV. *See* Rabbit hemorrhagic disease virus (RHDV)
Rhinovirus type 16 complex, 471
Rhinoviruses
  ICAM-1 interaction with, 473
  receptors for, 470–473
Rice dwarf virus (RDV), 72, 108
Rice yellow mottle virus (RYMV), 205
RNA genome
  double stranded, viral structure taxonomy of, 142
  single negative-sense strand, viral structure taxonomy of, 140–141
  single positive-sense strand, viral structure taxonomy of, 133–138, 139, 152
  single stranded, viral structure taxonomy of, 143–144
RNA structure, neutron scattering and, 49–50
RNA translocation, 467
RNA viruses, 4
Ross River virus, 5
Rotavirus, 414–415
  architectural features of, 230
Rous sarcoma virus, 5
RSV. *See* Respiratory syncytial virus (RSV)
RYMV. *See* Rice yellow mottle virus (RYMV)

S

S-adenosylhomocysteine (SAH), 73
S-adenosylmethionine (SAM), 72–73
Sacharomyces cerevisiae, 12
SAH. *See* *S*-adenosylhomocysteine (SAH)
Salmonella bacteriophage P22, Scaffolding-assisted viral assembly and, 259
Salmonella bacteriophage PRD1, 64, 65–67, 66
SAM. *See* *S*-adenosylmethionine (SAM)
Sandwich strategy, 54
SANS. *See* Small-angle neutron scattering (SANS)
Satellite panicum mosaic virus (SPMV), 153
Satellite tobacco mosaic virus (STMV), 153, 222–225
  encapsidated genome of, 223
Satellite tobacco necrosis virus (STNV), neutron scattering and, 49–50, 222
Satellite viruses, viral structure taxonomy and, 139–141, 139
SAXS. *See* Small-angle X-ray scattering (SAXS)
SBMV. *See* Southern bean mosaic virus (SBMV)

Scaffolding-assisted viral assembly, 259–294
coat proteins/chaperones and, 261–263
Escherichia coli phage ϕX174 and, 259
ϕX174/*herpesviridae* and, 264–267
ϕX174 internal scaffolding protein and, 263–264, 267–268
ϕX174 morphogenesis and, 259, 262
ϕX174/P22/*herpesviridae* and, 266–267
genetic data for scaffolding protein flexibility and, 264–265
mechanism of, 281–283
prescaffolding stages in, 261–263
Salmonella bacteriophage P22 and, 259
structural data for, 266–267
Scaffolding-like functions, 293–294
Scaffolding protein(s), 259–270
external, 283
functions of, 260–261
ϕX174 external, 283–289
Scaffolding protein distribution, in bacteriophage T7 procapsid assembly, 306–308
Scanning electron microscopy (SEM) methods, 55
SDS-PAGE. *See* Sulfate–polyacrylamide gel electro phoresis (SDA-PAGE)
SEM methods. *See* Scanning electron microscopy (SEM) methods
Semliki Forest virus (SFV), 61, 353–354
E1 C_α backbone, 382
E1 glycoprotein atomic structure of, 381
resolution reconstruction of, 62
Semliki Forest virus (SFV)/tick-borne encephalitis virus (TBEV), backbone structure comparison of, 381
Sendai virus, electron spin resonance (ESR) and, 51
Sf. *See Spodoptera frugiperda* (Sf)
SFV. *See* Semliki Forest virus (SFV)
Signaling events, adenovirus internalization and, 481
Simian vitrus 40 (SV40), 12
Sindbis virus, 5, 353–354
cryo-EM reconstruction of, 380
Single-crystal structures, of viruses, 38
Single positive-sense strand plant virus capsids, 152
Single protein, required amount of, in heterologous expression systems, 17
Single-stranded DNA (ssDNA) viruses, 167–71, 235–240
capsid maturation in, 236–237
capsid structure ordered DNA in, 237-238
genome entry model of, 237–238
genome release, 238
inoviridae bacteriophages, 170–71
microviruses of, 236–237
parvoviridae family of, 238-240
parvoviruses, 167–169, 239–240
Single-stranded RNA (ssRNA) viruses, 221–229
animal, 207–211
genome packaging of, 225–226
genome release in, 226–227
genome structural organization of, 224–225
genome structure in helical ssRNA viruses, 228–229
large-scale quaternary structure changes in, 209–215
positive v. negative ssRNA viruses, 221–222
specific genome recognition in, 227–228
Small-angle neutron scattering (SANS), 74
as neutron scattering, 48–49
Small-angle X-ray scattering (SAXS), 47–48
Soluble $\alpha_\nu\beta_5$, Ad12/Ad12 complexed with, 503
Southern bean mosaic virus (SBMV), 204, 226
x-ray crystallography of, 39
Spin-labeled influenza virus, electron spin resonance (ESR) and, 51
SPMV. *See* Satellite panicum mosaic virus (SPMV)
Spodoptera frugiperda (Sf), 8
ssDNA viruses. *See* Single-stranded DNA (ssRNA) viruses
ssRNA viruses. *See* Single-stranded RNA (ssRNA) viruses
Standard expression vectors, 5
STMV. *See* Satellite tobacco mosaic virus (STMV)
STNV. *See* Satellite tobacco necrosis virus (STNV)
Structural proteins, heterologous expression investigation of, 3
Subnanometer reconstructions
3-D visualization methods in, 117–118
for close-to-focus images, 112–113
CTF/B factor corrections in, 101–4
data processing flow chart of, 106

Subnanometer reconstructions (*cont.*)
digitization/particle selection of, 107–109
fold's derivations in, 120–121
with Fourier-Bessel synthesis of 3-D reconstruction, 113–114
illustrated helices and β sheets in, 119
image data in, 116
image screening of, 109–110
imaging/image quality assessment of, 108
imaging of, 105–107
method overview of, 101–105
near-atomic resolution in, 121
orientation/center determination in, 105, 110–113
orientation elimination/selection and, 110–111
orientation estimation/center parameters estimation and, 110–111
resolution assessment of, 114–115, 115
secondary structure elements and, 118–119
software use in, 115–117
strategy comparisons of, 98t–99
visualization/structure interpretation, 117–21
X-ray solution scattering intensity of HSV-1, 103
Subnanometer resolution reconstruction, contrast transfer function (CTF) in, 104
Sulfate–polyacrylamide gel electro phoresis (SDS-PAGE), 306
SV40. *See* Simian vitrus 40 (SV40)
Synchrotron radiation
advances in, 40
research with, 39
sources of, 40
Synchrotron radiation research, 39

T

T4 DNA bacteriophages, 391–392
T4 fibritin/baseplate proteins, double-stranded DNA viruses and, 176–178
T4/HSV-1
accessory protein distribution in, 400
assembly pathways of, 397–398
T = 3 plant viruses, reversible swelling in, 204–205
T = 3 single-stranded RNA virus, as icosahedral single-stranded RNA viruses, 224
TBEV. *See* Tick-bourne encephalitis virus (TBEV)
TBSV. *See* Tomato bushy stunt virus (TBSV)
TCV. *See* Turnip crinkle virus (TCV)
TEM. *See* Transmission electron microscopy (TEM)
Tetraviridae, 152, 153
Tetraviruses
in heterologous expression systems, 24–26
Nudaurelia T capensis virus (N *T*V), 25
Three dimensional Coulomb potential function, 94
Tick-borne encephalitis virus (TBEV), 61, 353–356, 368–369
TMV. *See* Tobacco mosaic virus (TMV)
Tobacco mosaic virus (TMV), 228–229
nuclear magnetic resonance (NMR) and, 51
Togaviridae family, 379
Togaviruses, enveloped single positive-sense strand RNA viruses and, 157–158
Tomato bushy stunt virus (TBSV), 153, 204, 226
neutron scattering and, 50
x-ray crystallography of, 39
Totiviridae family, 230
Trans, fusion activation in, 346–350
Transmembrane/surface glycoprotein gp120, retroviridae and, 182–184
Transmission electron microscopy (TEM), 55
Trichoplusia ni cells, 8
Trimer structure hierarchy, of influenza HA, 343
Triplex proteins, herpesvirus as, 292–294
Turnip crinkle virus (TCV), 153, 226
12/13 fold symmetry, in overexpressed T7/T3 connectors, 303
ϕ29 motor protein, double-stranded DNA viruses and, 175–76

V

Vaccinia virus
eukaryotic expression systems and, 10–12
foreign protein expression strategies for, 11–12
study advantages of, 11
T7 system, 2, 19

Viral capsid proteins, in *E. coli,* 5
Viral genome organization, 219
Viral phylogeny, 185–186
Viral protein structural folds, 125–186
determination of, 125–126
four-helix bundle, 132
immunoglobulin fold, 131
jelly-roll β barrel, 128–131, 129
serine protease fold, 132
terminology, 125
virus families and, 125
Viral replication cycle, 8
assembly process separation in, 3
Viral structure taxonomy
DNA genome double stranded, 145–148
DNA/RNA reverse-transcribing viruses, 149–151
RNA genome double stranded, 142
RNA genome single negative-sense strand, 140–141
RNA genome single positive-sense strand, 133–138, 139, 152
RNA genome single stranded, 143–144
satellite viruses and, 139–141, 139
Virion stabilization, 411
Virus(es), 37
asymmetric units of, 38
cyroelectron microscopy (cryo-EM) of, 38
description of, 219–224
icosahedral, 38
large, structure of, 67–74
large-scale reversible quaternary structure of, 204–205
membrane-containing, 39
noninfectious form of, 3
nonicosahedral, 39
v. protein, structural analysis of, 40
single-crystal structures of, 38
structural analysis techniques, 220
structure of, 2, 220
Virus assembly, 369–372
of alphaviruses, 369–372
of flaviviruses, 369–372
investigation of, 1, 2
Virus-cell fusion, 372–373
Virus-like particles (VLPs), 3, 22–23
heterologous expression investigation of, 3
Virus particle dynamics, 197–221
fluctuations/ineffectivity in, 199–200
mass spectrometry study of, 200
plant virus capsids and, 203–204
proteolysis study of, 200
viral genome effects on, 202–203
Virus source, American Type Culture Collection (ATTC), 19
VLPs. *See* Virus-like particles (VLPs)
VP1, 21–22, 154, 155, 220, 227
VP2, 22, 154, 155
VP3, 68, 70, 154
VP4, 220, 227
of polio virus fluctuations, 199–200
VP7, 68–71, 71, 166

W

Warts, papillomavirus and, 22
West Nile virus, 353
White spot virus, 381
WIN 52084, 200

X

ϕX174 assembly pathway, schematic of, 239
ϕX174 external scaffolding protein, atomic structure of, 285
ϕX174 external scaffolding proteins, procapsid 3-fold symmetry axes of, 290
ϕX174 internal scaffolding protein, C-terminal interactions and, 267–268
X-ray analysis, of virus structure, 1, 3, 221
X-ray crystallography, 93–94, 221, 385–386
cryoelectron microscopy with, 54–56
of southern bean mosaic virus (SBMV), 39
of tomato bushy stunt virus (TBSV), 39
X-ray detectors, 40–41
solid state amorphous selenium, 41
X-ray free electron lasers (XFELs), 53
X-ray scattering
light scattering and, 47–48
small-angle X-ray scattering (SAXS) as, 47–48
X-ray structures, cryoelectron microscopy maps and, 44–45

Y

Yeast expression system
eukaryotic expression systems and, 12–13
study advantages of, 12–13